ADVANCES IN CHEMICAL PHYSICS

VOLUME XLIV

Advances in CHEMICAL PHYSICS

EDITED BY

I. PRIGOGINE

University of Brussels
Brussels, Belgium
and
University of Texas
Austin, Texas

AND

STUART A. RICE

Department of Chemistry
and
The James Franck Institute
The University of Chicago
Chicago, Illinois

VOLUME XLIV

AN INTERSCIENCE® PUBLICATION

JOHN WILEY & SONS

NEW YORK • CHICHESTER • BRISBANE • TORONTO

AN INTERSCIENCE® PUBLICATION

Library of Congress Catalog Card Number: 58-9935
ISBN 0-471-06025-9

Printed in the United States of America

10 9 8 7 6 5 4 3 2 1

CONTRIBUTORS TO VOLUME XLIV

S. A. ADELMAN, Department of Chemistry, Purdue University, West Lafayette, Indiana

A. M. BRODSKY, Institute of Electrochemistry Academy of Sciences of the USSR, Moscow, USSR

THOMAS D. BOUMAN, Department of Chemistry, Southern Illinois University, Edwardsville, Illinois

RUSSELL DAVIES, Department of Applied Mathematics, University College of Wales, Aberystwyth, Wales

L. ENGELBRECHT, Fakultät für Chemie, Universität Bielefeld, Bielefeld, Germany

GARETH EVANS, Edward Davies Chemical Laboratories, Aberystwyth, Wales

MYRON EVANS, Edward Davies Chemical Laboratories, Aberystwyth, Wales

AAGE E. HANSEN, Department of Physical Chemistry, H. C. Ørsted Institute, Copenhagen, Denmark

JUERGEN HINZE, Fakultät für Chemie, Universität Bielefeld, Bielefeld, Germany

A. V. TSAREVSKY, Institute of Electrochemistry Academy of Sciences of the USSR, Moscow, USSR

INTRODUCTION

Few of us can any longer keep up with the flood of scientific literature, even in specialized subfields. Any attempt to do more, and be broadly educated with respect to a large domain of science, has the appearance of tilting at windmills. Yet the synthesis of ideas drawn from different subjects into new, powerful, general concepts is as valuable as ever, and the desire to remain educated persists in all scientists. This series, *Advances in Chemical Physics*, is devoted to helping the reader obtain general information about a wide variety of topics in chemical physics, which field we interpret very broadly. Our intent is to have experts present comprehensive analyses of subjects of interest and to encourage the expression of individual points of view. We hope that this approach to the presentation of an overview of a subject will both stimulate new research and serve as a personalized learning text for beginners in a field.

ILYA PRIGOGINE
STUART A. RICE

CONTENTS

ADVANCES IN CHEMICAL PHYSICS

VOLUME XLIV

MOLECULAR PROPERTIES OBSERVED AND COMPUTED

L. ENGELBRECHT AND JUERGEN HINZE

Fakultät für Chemie
Universität Bielefeld
Bielefeld, Germany

CONTENTS

I. INTRODUCTION

Atoms and molecules are the building blocks of all matter: solid, liquid, or gaseous. If we are to understand bulk matter, its property, and its reactions in detail, we need to know explicitly the static and dynamic structure of molecules. To be sure, explicit knowledge of the properties and interactions of molecules, although essential, is not enough if we are to understand bulk matter in detail. We consider here the electronic, vibrational, and rotational energy structure of isolated molecules, including their fine and hyperfine structure, as well as the interactions of molecules with external electric and magnetic fields. We base our treatment on the extensive theory that has been developed for the interpretation of molecular spectra,[1] and hope to develop analogous expressions amenable to computation from first principles.

Much information about the rovibronic and fine structure of molecules has been obtained through extensive spectroscopic studies. Qualitative theoretical understanding of the observed effects has advanced enormously and has contributed significantly to the development and refinement of quantum mechanics, the fundamental theory for our understanding of atomic and molecular phenomena. Equally great strides have been made recently, with the advent of electronic computers, in obtaining detailed molecular structure information through *ab initio* quantum-mechanical calculations.[2] Nevertheless, there is still a great gap between the detail observed and the information obtained directly from theory, where heretofore the major effort had been focused on the calculation of electronic wave functions. The complexity of the problem to compute accurate, correlated electronic wave functions and energies has tended to obscure the importance of using such wave functions and energies in the evaluation of various molecular properties. Even the development of the theoretical basis for such calculations, which is fraught with many intricate complexities, seems rather incomplete, in spite of some notable, isolated efforts.[3] We hope to give an extensive and consistent treatment in the direction of such a development.

To be sure, there is little need for computing molecular properties if they have been determined spectroscopically or are easily accessible to observation, except for gauging a theory, a computational method, or the accuracy of a computed wave function. However, there is a host of information about the structure of molecules, important to the understanding of their interactions and reactions, which is difficult or impossible to obtain from experiment. Spectroscopy is and will remain superior in yielding quantitative information about the structure, the potential energy surface, and the coupling terms of stable isolated molecules in the vicinity of their nuclear equilibrium conformation in the electronic ground state. Highly reactive molecules, free radicals, or electronically excited species, many of which are important in molecular interactions and reactions, are much more difficult to investigate experimentally. It is almost impossible experimentally to obtain information about the potential energy surface or coupling terms for repulsive electronic states (i.e., energy surfaces which have no strong minima); equally inaccessible is information for structures far removed from their equilibrium conformation. However, these quantities play a significant role in molecular interactions and reactions. It is here where a quantitative theoretical study is of extreme value and yields information complementary to that obtainable experimentally. To this end we hope to contribute by detailing formulas for the computation of various molecular properties analogous to the different spectroscopic constants. Even in those cases where spectra can be observed readily, quantitative studies of the coupling effects are of value in aiding interpretation of the observed energy differences.

To accomplish and structure the task ahead, we present in Section II our understanding of the adiabatic approximation. This puts the electronic Schrödinger equation and potential surface into context; Section III gives a brief review of the state of the art of obtaining solutions to this part of the problem. In Section IV we discuss symmetry aspects of the potential energy surface, Jahn–Teller and Renner–Teller splittings, classification of electronic wave functions, and eigenfunctions of the angular momentum operator. In addition, we introduce, with the Eckart conditions, internal, normal coordinates. With this the stage is set to deal explicitly in Section V with the separation of translation, rotation, vibration, and electronic motion on the basis of the adiabatic formulation, which permits us to trace in detail the origin and form of the various coupling terms. Section VI gives a brief overview of general aspects used in the interpretation and assignment of high-resolution molecular spectral data.

General aspects needed for the theoretical derivation of molecular properties are presented in Section VII, with the details given in Appendixes 2 and 3. Here we consider the symmetry classifications of the coupling terms and perturbations, which leads naturally to the introduction of symmetry-group coupling coefficients and reduced matrix elements. These are analogous to Clebsch–Gordan coefficients and reduced matrix elements of tensor operators for the infinite rotation group. With this the theoretical background is prepared to present in Section VIII explicit formulas for the calculation of molecular properties as they are derived from the relativistic Hamiltonian of the whole molecule. This Hamiltonian and its formal separation is detailed in Appendix 1. The formulas presented give the properties in analogy to the experimentally determined spectroscopic constants. The latter, however, are state-dependent, while the formulas for the properties derived here are obtained as a function of the internal coordinates. This is theoretically more sound and should be advantageous for detailed computations. The fundamental differences that arise between the linear and nonlinear cases are pointed out separately in the places where such differences occur. Finally, in Section IX we review briefly the possibilities of calculating the properties for which we have detailed the formulas and reference some of the computations already carried out successfully.

II. ADIABATIC SEPARATION

The stationary state I of a molecule in its quantum-mechanical description is completely characterized by the wave function $\Psi_I(\mathfrak{Q}, \mathfrak{q})$, which is a solution of the time-independent Schrödinger equation

$$[\mathfrak{H}(\mathfrak{Q},\mathfrak{q}) - E_I]\Psi_I(\mathfrak{Q},\mathfrak{q}) = 0 \tag{II.1}$$

Here the Hamiltonian and the wave function depend on all nuclear and electronic coordinates, $\mathfrak{Q}$ and $\mathfrak{q}$, respectively, including spin, in a space fixed frame. The Hamiltonian may contain static external electric and magnetic fields as well as relativistic couplings between electrons and nuclei, at least in a Pauli-type approximation. To achieve a separation of nuclear and electronic motion, we reduce the problem of solving (II.1) using, following Born,[4] the ansatz

$$\Psi_I(\mathfrak{Q},\mathfrak{q}) = \sum_n \psi_n(\mathfrak{R},\mathfrak{r})\theta_{nI}(\mathfrak{R},\mathfrak{Q}) \tag{II.2}$$

which requires some explanation. The sum is over electronic states represented by n, the collection of electronic quantum numbers and characterized through the electronic wave function $\psi_n(\mathfrak{R},\mathfrak{r})$, which depends explicitly on the collection of electron coordinates $\mathfrak{r}$ and parametrically on the internal nuclear coordinates $\mathfrak{R}$. This function is a solution of the electronic Schrödinger equation

$$[\mathfrak{H}_e(\mathfrak{R},\mathfrak{r}) - V_n(\mathfrak{R})]\psi_n(\mathfrak{R},\mathfrak{r}) = 0 \tag{II.3}$$

which may include relativistic as well as external electric and magnetic field effects. However, it depends on coordinates (and spin) in a molecular fixed frame, and the $\mathfrak{R}$ dependence is parametric (i.e., there are at most multiplicative operators of $\mathfrak{R}$ in $\mathfrak{H}_e$). The function $\theta_{nI}(\mathfrak{R},\mathfrak{Q})$ is less straightforward. It is the function component in the electronic state n of the nuclear motion function corresponding to state I of the molecule. It is independent of the electronic coordinates. Its dependence on the molecular fixed frame internal coordinates $\mathfrak{R}$ as well as on the space fixed coordinates of the nuclei $\mathfrak{Q}$ is to signify that it contains the translation of the center of mass as well as the rotation of the coordinate frame, which affords transformation from the space fixed to the molecular fixed frame.

We postpone to Section V a more detailed and specific discussion and substitute the right-hand side of (II.2) for Ψ_I in (II.1), left-multiply with $\psi_m^*(\mathfrak{R},\mathfrak{r})$ and integrate over the electronic coordinates. We obtain, considering (II.3),

$$(X_{mm} + V_m - E_I)\theta_{mI} = -\sum_n{}' X_{mn}\theta_{nI} \tag{II.4}$$

for all m. This separation is exact, inasmuch as the sum in (II.2) is complete; however, it is rather formal. All the coupling between electronic and nuclear motion is hidden in the matrix elements, which may be written formally as

$$X_{mn} = \langle \psi_m(\mathfrak{R}, \mathfrak{r}) | \mathfrak{H}(\mathfrak{Q}, \mathfrak{q}) - \mathfrak{H}_e(\mathfrak{R}, \mathfrak{r}) | \psi_n(\mathfrak{R}, \mathfrak{r}) \rangle_r$$

with the integration implied over the electronic coordinate, $\mathfrak{r}$, in the molecule fixed frame. Obviously, neither the integration nor the operator difference can be carried out without first transforming $\mathfrak{H}(\mathfrak{Q}, \mathfrak{q})$ to the molecular fixed frame; that is, $\mathfrak{H}(\mathfrak{Q}, \mathfrak{q})$ needs to be transformed first with those parts of $\theta_{nI}(\mathfrak{R}, \mathfrak{Q})$ that effect this transformation and which should be independent of n, depending only on the instantaneous nuclear coordinate $\mathfrak{R}$ and the state quantum numbers collected in I, especially translational energy, the total angular momentum, and its projection on a space fixed axis. This transformation will depend on the particular basis chosen and will be discussed in detail in Section V.

The conventional adiabatic approximation is obtained from (II.4) by neglecting the right-hand side and by using in X_{mn} only the operator for the nuclear kinetic energy in a molecule fixed frame. Then θ_{mI} will be a function of $\mathfrak{R}$, the internal nuclear coordinates only; however, the potential may then be amended by a centrifugal term. We will not elaborate on this. Since we will desire the coupling terms, we need (II.4) fully or at least with a truncated sum of type (II.2).

III. THE ELECTRONIC EQUATION

The solutions to the electronic Schrödinger equation, (II.3), are the basis for the adiabatic approximation; they define the potential in which the nuclei move. The potentials, as well as the electronic wave functions, are required to solve the problem for the nuclear motion in general. In order to define the terms needed later we detail here briefly the electronic Schrödinger equation and sketch structure and methods for solving it approximately.

For a molecule with N nuclei and n electrons, the electronic Hamiltonian of (II.3) may be expressed in general as a constant term plus a sum of one- and two-electron operators:

$$\mathfrak{H}_e(\mathfrak{R}, \boldsymbol{r}) = K(\mathfrak{R}) + \sum_{i}^{n} h(\mathfrak{R}, \hat{\boldsymbol{r}}_i) + \sum_{i<j}^{n} g(\hat{\boldsymbol{r}}_i, \hat{\boldsymbol{r}}_j) \qquad \text{(III.1)}$$

where the constant term, which represents the nuclear–nuclear repulsion, and the one-electron terms, which contain the electron–nuclear attraction, depend parametrically on the internal coordinates represented by $\mathfrak{R}$.

The individual terms in (III.1) may still include contributions due to static external electric or magnetic fields as well as relativistic effects. The nuclear and electronic position vectors $\mathfrak{R}$ and $\mathfrak{r}$, respectively, should then be thought of as containing the spin coordinate in addition to the spatial coordinate in a molecular fixed frame. In general, however, the field-free, nonrelativistic equation is solved explicitly (the other effects to be added later perturbatively), so we will focus for the moment on it and have in atomic units

$$h(\mathfrak{R},\boldsymbol{r}_i)=-\tfrac{1}{2}\nabla_i^2-\sum_{\alpha}^{N} Z_\alpha r_{i\alpha}^{-1} \tag{III.2}$$

and

$$g(\boldsymbol{r}_i,\boldsymbol{r}_j)=r_{ij}^{-1} \tag{III.3}$$

The nuclear–nuclear repulsion term

$$K(\mathfrak{R})=\sum_{\alpha<\beta}^{N} Z_\alpha Z_\beta r_{\alpha\beta}^{-1} \tag{III.4}$$

is independent of the electronic coordinates. It remains constant and can be disregarded here, to be added whenever appropriate to the electronic energy, $\varepsilon(\mathfrak{R})$, to give the potential

$$V(\mathfrak{R})=K(\mathfrak{R})+\varepsilon(\mathfrak{R}) \tag{III.5}$$

in which the nuclei move.

The electronic energy is obtained by solving

$$[\mathfrak{H}_e(\mathfrak{R},\mathfrak{r})-\varepsilon_n(\mathfrak{R})]\psi_n(\mathfrak{R},r)=0 \tag{III.6}$$

for various nuclear conformations. Here $\mathfrak{H}_e$ is the electronic Hamiltonian, (III.1) without the nuclear–nuclear repulsion term K.

With the advent of electronic computers a major effort has been focused upon solving (III.6) approximately, and highly efficient and computationally sophisticated techniques have been developed to obtain correlated wave functions and energies of desired accuracy. These methods, which have become standard by now, have been reviewed extensively and expertly,[2,5] so that we can confine ourselves here to a brief outline.

Variational as well as perturbation techniques may and are being used to solve (III.6) approximately. The computational effort required to obtain solutions of comparable accuracy determines by and large the specific method chosen, and variational methods, which seem to be more standard, appear to have a slight edge. To be sure, highly accurate energies can be obtained using perturbation theory. However, to obtain compact and accurate wave functions in this way is much more difficult. Since for the computation of molecular properties the evaluation of expectation values and transition matrix elements is required and for this electronic wave functions are needed, we restrict our discussion to the form of variational procedures. Except for two-electron systems, where it is more effective to include the interelectronic distance explicitly, the electronic wave function for a state n, which can be characterized by a symmetry species ν and an identification index I, is in general written as a superposition of configurations

$$\psi_n(\mathfrak{R},\mathrm{r}) \equiv \psi_{\nu I}(\mathfrak{R},\mathrm{r}) = \sum_J \Phi_{\nu J}(\mathfrak{R},\mathrm{r}) C_{JI} \tag{III.7}$$

where the configuration state functions (CSF's) $\Phi_{\nu J}(\mathfrak{R},\mathrm{r})$ are minimal linear combinations of Slater determinants (SD's)

$$\Phi_{\nu J}(\mathfrak{R},\mathrm{r}) = \sum_K \phi_K(\mathfrak{R},\mathrm{r}) B_{K,\nu J} \tag{III.8}$$

such that they transform as symmetry species ν. The SD's are antisymmetric determinental product functions

$$\phi_K(\mathfrak{R},\mathrm{r}) = n!^{-1/2} \det\{\varphi_{i_1}(r_1) \cdots \varphi_{i_n}(r_n)\} \tag{III.9}$$

of the orbitals $\varphi_i(\mathrm{r})$, which are generally chosen to be orthonormal; that is,

$$\langle \varphi_i | \varphi_j \rangle = \delta_{ij} \tag{III.10}$$

and such that they transform according to the symmetry species of the system. The dependence of the orbitals on the nuclear conformation has been suppressed here. The set

$$K = \{i_1, i_2, \ldots, i_n\} \qquad \text{with } i_1 < i_2 \cdots < i_n \tag{III.11}$$

together with the orbitals specifies the SD, ϕ_K, uniquely, while the $B_{K,\nu J}$ in (III.8) are determined by the coupling chosen and the demand that the CSF, $\Phi_{\nu J}$, transform according to symmetry species ν.

Substituting for ψ_n in (III.6) the right-hand side of (III.7) for a given symmetry and varying with respect to the expansion coefficients yields the standard configuration mixing eigenvalue equation

$$(\mathbf{H}-\varepsilon)\mathbf{C}=\mathbf{0} \tag{III.12}$$

which may be solved by direct diagonalization[6] of **H** or by obtaining with iterative methods[7] the lowest few roots ε_I and the corresponding eigenvectors C_I.

The energy matrix **H** is defined here with the elements

$$H_{IJ}=\langle\Phi_I|\mathfrak{H}_e|\Phi_J\rangle \tag{III.13}$$

which are obtained as specific combinations of one- and two-electron integrals of the operators of (III.2) and (III.3) with the orbitals. The specific combinations are determined through the CSF's and SD's used. It should be mentioned here that direct methods,[8] which solve (III.12) iteratively, using the orbital integrals directly, without explicitly constructing **H** appear rather promising, permitting very large configuration mixing expansions of type (III.7).

It is well known that the configuration mixing expansion is slowly convergent at best and major emphasis has been placed on the selection of important CSF's in constructing **H**. Clearly, the choice of the single-particle functions, the orbitals, is of central importance for the convergence of (III.7). The orbitals are therefore determined self-consistently, either by solving iteratively the standard Hartree–Fock equations[9] or the Fock-like MCSCF equations.[10] Since it would lead us too far here to review these methods, it may suffice to state that although in principle the MCSCF method would be able to yield all the orbitals needed to construct the CSF's in (III.7) and thus give an adequate wave function directly, this is in general computationally inefficient. In the SCF method, by definition, and in the MCSCF method by design, only the most important configuration and orbitals are included, while the remaining orbitals and configurations in (III.7), necessary for an adequate representation of the electronic wave function, have to be chosen differently. The virtual orbitals of SCF-type calculations are in general inappropriate unless some configuration selections are performed.[11] In general, such a selection is based on a second-order perturbation theory estimate of the energy contribution of the CSF's, and extrapolation techniques have been developed to estimate the total error introduced by such selections.[12] However, in the direct methods,[8] which permit the calculation with a much larger number of SCF's than the conventional methods, such a selection does not seem to be possible. Here the procedures to obtain better orbitals than the virtual SCF orbitals for the configuration mixing (CM) calculation become of utmost importance. Such procedures involve the approximate determination of natural orbitals[13] either using CM calculations iteratively,[14] yielding the iterative natural orbitals (INO's), or using second-order perturbation theory,[15] yielding the perturbation natural orbitals (PNO's). Naturally, these techniques, using a conventional CM, can be used in conjunction with CSF selection and extrapolation mentioned above. Much of this has been reviewed recently[2] and need not be elaborated further here. However, before we turn to the use of electronic wave functions for the calculation of various molecular properties, we should mention that the orbitals themselves are generally expressed as linear combinations of either Slater-type functions (STF's)

$$\varphi_i = \sum_p x_p c_{pi} \tag{III.14}$$

with

$$x_p = r^n \exp(-\zeta r) Y_m^l(\vartheta, l) \tag{III.15}$$

where p characterizes the origin, l, m, n, and ζ, or Gaussian-type functions (GTF's)

$$x_p = x^l y^m z^n \exp(-\zeta r^2) \tag{III.16}$$

where p characterizes the origin, l, m, n, and ζ. Of these, STF's can represent the proper long-rage behavior as well as the cusps at the nuclei, both of which are impossible with a finite number of GTF's. However, the calculation of the necessary integrals with STF's is much more cumbersome than with GTF's, except in atoms or linear molecules; therefore, GTF's are used frequently for polyatomic molecules. Compendia of results of such electronic structure calculations have been compiled.[2b]

The first few steps in using such wave functions for the calculation of expectation values or transition matrix elements of operators other than the energy would be quite analogous to those used in an energy-wave-function calculation. First it is necessary to evaluate the one- and two-electron integrals over the appropriate operators and the basis functions defined by (III.15) and (III.16). These integrals are transformed with the orbital expansion coefficients of (III.14) from the basis space into the orbital space. The matrix elements over the n-electron functions are then obtained, akin to those of (III.13), as specified sums of these one- or two-electron integrals, with the terms in the sum determined by the configuration mixing coefficients in (III.7), the type of CSF's and SD's used, and the symmetry of the operator involved.

IV. SYMMETRY—INTERNAL COORDINATES

In our attempt to understand the stationary states of molecules, symmetry is an extremely useful concept. Symmetry considerations allow a classification of states of the wave function as a whole and of individual parts of the wave function according to the irreducible representations of the group spanned by the symmetry operators. Group theory and the symmetry classification of states tell us the degeneracy of a particular state and whether this degeneracy will be split, and between which type of states interactions are possible, depending on the transformation properties of a perturbation.

The usefulness of the symmetry concept is based on the theorem that commuting operators in quantum mechanics have common sets of eigenfunctions, or more specifically, if

$$\Lambda\mathfrak{H}=\mathfrak{H}\Lambda \tag{IV.1}$$

and

$$\Lambda\Psi=\lambda\Psi \tag{IV.2}$$

where the operator Λ commutes with the Hamiltonian and where Ψ is an eigenfunction in the space of $\mathfrak{H}$, then Ψ can also be an eigenfunction of Λ with eigenvalue λ, a constant of motion. Relations (IV.1) and (IV.2) can be used with the total Hamiltonian for the classification of the total wave function or with parts of the Hamiltonian to classify the corresponding parts of the wave function, with the symmetry classification of the total wave function obtainable by appropriate coupling of the wave-function parts into a whole.

Before we proceed to further details, let us list the types of operations that commute with the total Hamiltonian and leave it invariant. For this we restrict ourselves to the field-free case; external electric or magnetic fields can always be considered as perturbations. The invariants are:

1. Translation of the origin of the coordinate system.
2. Rotation of the coordinate frame about a space fixed axis.
3. Time reversal.
4. Inversion of all spatial coordinates.
5. Permutation of the coordinates, including spin, of identical nuclei.
6. Permutation of the coordinates, including spin, of the electrons.

The first three follow from the isotropy of space and time, where (1) leads to the infinite-order translation group, permitting a classification of the overall translation along some space fixed axis, not affecting internal molecular states. Invariant (2) leads to the infinite-order external rotation group. It effects a classification of all molecular states according to the quantum numbers $J(F)$ and $M_J(M_F)$, which designate the value of the total angular momentum (including nuclear spin for F) and its projection on some space fixed axis. In the field-free case we obtain a $2J+1$ $(2F+1)$-fold degeneracy. An external electric field will split such a state into $J+1$ components according to $|M|$. Time reversal (3) is of little importance here; because of it, some separately degenerate representations can be considered to be degenerate. With an external magnetic field present, the time-reversal symmetry is broken, which leads to a splitting of the remaining double degeneracy for states with $M \neq 0$. The final invariant (6) need not be considered here in detail; it is taken care of by requiring the electronic wave function in (III.7) to be antisymmetric with respect to electron permutation (Pauli principle). The antisymmetry requirement, together with the specific electron spin coupling generally used when constructing an electronic wave function, leads to functions whose spatial and spin parts separately transform as conjugate pairs of irreducible representations of the permutation group, such that their product yields the antisymmetric irreducible representation. There is a subtle connection between the spin coupling and the permutation group, and the spin quantum number signifies the particular pair of conjugate irreducible representation chosen for spin and spatial part in (III.7).

The invariants (4) and (5) need to be explicated considerably. They are the basis for the conventional symmetry classification of electronic and vibrational states according to the molecular point group and of rotational states according to the four-group D_2 or D_∞ group for asymmetric- or symmetric-top rotors, respectively. However, the molecular point group and the internal rotation group are near-symmetry groups only, correct for rigid rotor molecules with small vibrational amplitudes, where rotational and rotation–vibration and electronic coupling can be neglected. Since we aim to deal just with these couplings, it is more appropriate to use the molecular symmetry (MS)[16] or extended molecular symmetry (EMS)[17] groups, based on the nuclear permutations (5) and coordinate inversion (4), which are fundamental invariants and permit a unifying classification of rotational, vibrational, and electronic (rovibronic) states. The nuclear permutation and inversion group has been introduced by Longuet-Higgins[18] to deal with nonrigid molecules, and Bunker has detailed the MS and EMS groups derived from it in an excellent review.[16]

The use of the MS or EMS group is complicated by the fact that nuclear permutations affect in general a rotation of the molecular fixed reference frame. The MS group, however, is frequently isomorphic with the molecular point group, so that the point group with its more accustomed and more easily visualized operations may be used in place of the MS group. For more details and the connection between MS and point group, we refer the reader to the review by Bunker[16] and confine our discussion to the essentials needed here.

To deal explicitly with the internal motion and the symmetry of a molecule, it is necessary to define a molecular fixed reference frame in terms of which the internal coordinates are described. We follow the standard procedure to define a nuclear (not molecular) fixed frame by first choosing some reference conformation $\boldsymbol{r}_\alpha^0$ for the position of nuclei measured relative to the center of mass, $\mathbf{0}$, of the reference; that is,

$$\sum_\alpha M_\alpha \boldsymbol{r}_\alpha^0 = \mathbf{0} \tag{IV.3}$$

This reference conformation is arbitrary and will in general (but not always) be chosen to correspond to an equilibrium conformation of the molecule. Defining now the instantaneous position vectors of the nuclei with respect to $\mathbf{0}$ as $\boldsymbol{r}_\alpha$, an optimal separation of external (translation and rotation) and internal motion is obtained by requiring the Eckart conditions[19]

$$\sum_\alpha M_\alpha (\boldsymbol{r}_\alpha - \boldsymbol{r}_\alpha^0) = \mathbf{0} \tag{IV.4}$$

and

$$\sum_\alpha M_\alpha (\boldsymbol{r}_\alpha - \boldsymbol{r}_\alpha^0) \times \boldsymbol{r}_\alpha^0 = \mathbf{0} \tag{IV.5}$$

to be satisfied. If internal torsion is to be considered, it is appropriate to impose in addition Sayvetz conditions[20] for the definition of torsional angles.

Condition (IV.4) ensures that for internal motions the center of nuclear mass remains $\mathbf{0}$, while (IV.5) keeps the reference frame fixed. This reference frame x, y, z with origin $\mathbf{0}$ is obtained by transforming the inertial tensor of the reference conformation

$$\mathbf{I} = \sum_\alpha M_\alpha (\boldsymbol{r}_\alpha^0 \boldsymbol{r}_\alpha^0 \mathbf{1} - \boldsymbol{r}_\alpha^0 \otimes \boldsymbol{r}_\alpha^0) \tag{IV.6}$$

into normal (diagonal) form.

We have used here the dyadic product

$$\boldsymbol{r}\otimes\boldsymbol{r}=\begin{pmatrix} xx & xy & xz \\ yx & yy & yz \\ zx & zy & zz \end{pmatrix} \tag{IV.7}$$

The translational motion is now the motion of **0** with respect to the origin of the laboratory fixed frame, and the external rotation is described in terms of the Euler angles that rotate the laboratory fixed frame (X,Y,Z) into the molecular fixed frame (x,y,z). Since (IV.5) implies

$$\sum_\alpha M_\alpha \boldsymbol{r}_\alpha^0 \times \dot{\boldsymbol{r}}_\alpha = 0 \tag{IV.8}$$

the Eckart conditions minimize the vibrational angular momentum and thus the coupling between rotation and vibration.

With the reference frame defined, we are in a position to specify the customarily used mass-scaled internal coordinates

$$S_k = \sum_\alpha \boldsymbol{d}_k(\alpha)\sqrt{M_\alpha}\,(\boldsymbol{r}_\alpha - \boldsymbol{r}_\alpha^0) \tag{IV.9}$$

for $k=1,2,\ldots,3N-6$, or $3N-5$ in case of linear molecules. The requirement to satisfy the Eckart conditions, (IV.4) and (IV.5), results in the following relations among the coefficients $\boldsymbol{d}_k(\alpha)$:

$$\sum_\alpha d_{k\mu}(\alpha)d_{l\mu}(\alpha) = \delta_{kl} \qquad \mu = x,y,z \tag{IV.10}$$

$$\sum_\alpha \sqrt{M_\alpha}\ \boldsymbol{d}_k(\alpha) = \boldsymbol{0} \tag{IV.11}$$

and

$$\sum_\alpha \sqrt{M_\alpha}\ \boldsymbol{r}_\alpha^0 \times \boldsymbol{d}_k(\alpha) = \boldsymbol{0} \tag{IV.12}$$

as well as the inverse relation

$$\boldsymbol{r}_\alpha = \boldsymbol{r}_\alpha^0 + \sqrt{M_\alpha}^{\,-1} \sum_k \boldsymbol{d}_k(\alpha) S_k \tag{IV.13}$$

and the orthogonality relation

$$\sum_k d_{k\mu}(\alpha)d_{k\nu}(\beta) = \delta_{\mu\nu}\left(\delta_{\alpha\beta} - \sqrt{M_\alpha M_\beta}\ /\ M\right)$$
$$- \sum{}_{\kappa\rho\lambda\sigma}\sqrt{M_\alpha M_\beta}\ e_{\mu\kappa\rho} e_{\nu\lambda\sigma} r_{\alpha\kappa}^0 (I^{-1})_{\rho\sigma} r_{\beta\lambda}^0 \tag{IV.14}$$

where $e_{\mu\kappa\rho}$ is the three-dimensional Levi-Civita antisymmetric pseudotensor, which has nonzero elements

$$e_{xyz} = e_{yzx} = e_{zxy} = 1 \tag{IV.15}$$

and

$$e_{yxz} = e_{xzy} = e_{zyx} = -1$$

By introducing linearly independent mass-scaled internal coordinates via a linear transformation in accordance with (IV.9) through (IV.14), the translational and rotational moments of inertia have been separated off. The next step is the transformation of the internal coordinates to symmetry coordinates, which imposes additional constraints on the coefficients $d_k(\alpha)$ such that the S_K's will transform as irreducible representations of the symmetry group of the system. This symmetry is determined through the topology of the potential surface $V(\mathfrak{R})$, resulting as a solution of (II.3). Its origin lies in the invariance of $\mathfrak{H}$ and $\mathfrak{H}_e$ with respect to (4) inversion of all spatial coordinates and (5) permutation of the coordinates, including spin, of identical nuclei. Thus we could classify the electronic wavefunction, the internal coordinates, and the rovibronic states according to the irreducible representations of the complete nuclear permutation and inversion (CNPI) group. The potential surface, corresponding to all electronic states, will always be totally symmetric in this group. With A an element of this group, $A \in G$, and P_A the corresponding operator, we have

$$P_A V(\mathfrak{R}) = V(A^{-1}\mathfrak{R}) = V(\mathfrak{R}) \tag{IV.16}$$

The number of elements in the CNPI group increases extremely rapidly with the number of identical nuclei; in ethane we would obtain, for example, a group with $6! \cdot 2! \cdot 2 = 2880$ elements, and using such a large group becomes laborious and unmanageable. Fortunately, we are most often not interested in all possible kinds of excursions of the nuclei, only in those which are feasible within a given, generally low energy. These "feasible" elements of the CNPI group introduced by Longuet-Higgins[18] form a subgroup of the CNPI group, the MS group.[16] The latter is a true symmetry group of the system, not a near-symmetry group like the point group. Hence it can be used for the classification of electronic, vibrational, and rotational states. For linear molecules, however, the MS needs to be extended to the EMS group,[17] where the infinite-order rotation around the internuclear axis is included appropriately.

For rigid molecules the MS group is isomorphic with the molecular point group, a situation obtained in most cases; thus the point group classification can be used. Even for some "nonrigid" cases, a higher-order point group can frequently be used. However, using the MS group, or the isomorphic point group in its place, has the advantage of permitting a symmetry classification, even though the nuclear conformation is not in or near its equilibrium. When classifying rotational states (rovibration or rovibronic states), though, care must be taken, since the nuclear permutations and the inversion affect the nuclear fixed reference frame. This is so because the reference frame is not completely determined by requesting the inertia tensor, (IV.6), to be diagonal, a phase (direction) must still be specified, and this specification fixes the axes as described by Bunker.[16] Confining ourselves to the classification of electronic and vibrational states and internal coordinates we need not concern ourselves here with the above-mentioned rotations or inversions of the reference frame.

We will give a few examples to get accustomed to the concepts stated above. For H_2O the elements of the CNPI and MS group are (all operations are feasible) E, the identity; (12), the exchange of H_1 with H_2; E^*, the coordinate inversion, and $(12)^* = E^* \cdot (12)$. The point group of the water molecule, C_{2v}, consists of the elements E, C_2, σ_{ab}, σ_{cb}, while the conventional rotation group for an asymmetric rotor is the four-group with the elements R_0, R_{2a}, R_{2b}, and R_{2c}, where a, b, and c are the principal axis forming a right-handed system with a and b in the molecular plane and b along the C_2 axis. To see the relations between these operations, consider C_2, which rotates the electronic wave function and the nuclear displacements by π around the b-axis, leaving the nuclear reference sites and thus the reference frame unaffected. On the other hand, (12) of the MS group permutes nuclei 1 and 2, leaving the nuclear displacements and the electronic wavefunction unaffected; however, the nuclear reference frame is rotated by π around b. Finally, R_{2b} rotates the entire molecule, nuclei, reference frame, nuclear displacements, and electronic wave function by π around the b-axis. Thus we have the relation $(12) = R_{2b}C_2$ and similarly for all the operations:

$$
\begin{aligned}
E &= R_0 E \\
(12) &= R_{2b} C_2 \\
E^* &= R_{2c} \sigma_{ab} \\
(12)^* &= R_{2a} \sigma_{cb}
\end{aligned}
\qquad \text{(IV.17)}
$$

The advantage of using the MS group over the isomorphic C_{2v} point group here is twofold: (1) the vibrational angular momentum, relative to the total angular momentum, and thus the Coriolis coupling, is unchanged by the operations of the MS group; and (2) the electronic wavefunctions, which depend parametrically on the internal coordinates, can be classified according to the irreducible representations of the MS group even for asymmetric nuclear conformations. The latter point permits the derivation of general selection rules; it allows a general symmetry classification of matrix elements integrated over the electronic coordinates but dependent on the internal nuclear coordinates, a classification we shall find extremely useful.

To illustrate the latter point, take an electronic wave function $\psi_e(S_1S_2S_3;\mathfrak{r})$, which depends parametrically on the internal symmetry coordinates (i.e., S_1—symmetric stretch, S_2—symmetric bend, and S_3—asymmetric stretch). Consider, for example, the operation (12) and we have

$$(12)\psi_e(S_1S_2S_3;\mathfrak{x}\mathfrak{y}\mathfrak{z})=\psi_e(S_1S_2-S_3;\mathfrak{x}-\mathfrak{y}\mathfrak{z})$$
$$=\pm\psi_e(S_1S_2S_3;\mathfrak{x}\mathfrak{y}\mathfrak{z}) \qquad \text{(IV.18)}$$

with the sign determined by the symmetry or antisymmetry of ψ with respect to (12), which is most easily seen when $S_3=0$, although (IV.17) holds for all values of $\mathfrak{S}$. To be sure, we could have used just as well the operations of the point group of the symmetric reference conformation here.

Generally, the action of an operator P_A corresponding to the operation A of the group of the system upon an electronic wavefunction gives for nondegenerate states

$$P_A\psi_\Gamma(\mathfrak{R},\mathfrak{r})=\psi_\Gamma(A^{-1}\mathfrak{R},A^{-1}\mathfrak{r})=x_\Gamma(A)\psi_\Gamma(\mathfrak{R},\mathfrak{r}) \qquad \text{(IV.19)}$$

where $x_\Gamma(A)$ is the character corresponding to A in the irreducible representation Γ. A point group and its classification can be used, which is just as well in most cases, provided that the group of the symmetric reference conformation isomorphic with the MS group is employed. Equation (IV.19) is general and relates the electronic wave functions of one nuclear conformation to those of another conformation. The characters are most easily ascertained in the symmetric reference conformation (i.e., when all displacements are zero), since now the relation is one of the electronic wave function to another with the same nuclear conformation used conventionally. To be sure, at a fixed displaced nuclear conformation, the representation of the electronic wave function will correspond to those of a lower (point) group. However, the representation Γ displayed in (IV.19) corresponding to the higher symmetry is still useful, and we propose the terminology “the state is of lineage Γ”.

To illustrate the usefulness of the lineage of an electronic state, consider an electronic matrix element, $M(\mathfrak{R})$, which is a function of the internal nuclear coordinates. If the operator $\hat{M}(\mathfrak{R},\mathrm{r})$ transforms as Γ_m of the MS-group, then from

$$M_\Gamma(\mathfrak{R}) = \langle \psi_{e\Gamma_1}(\mathfrak{R},\mathrm{r}) | \hat{M}_{\Gamma_m}(\mathfrak{R},\mathrm{r}) | \psi_{e\Gamma_2}(\mathfrak{R},\mathrm{r}) \rangle_\mathrm{r} \qquad \text{(IV.20)}$$

follows in general that $M_\Gamma(\mathfrak{R})$ will transform as the irreducible representation

$$\Gamma = \Gamma_1 \otimes \Gamma_m \otimes \Gamma_2 \qquad \text{(IV.21)}$$

of the MS-group, provided that all the irreducible representations are nondegenerate.

Clearly, the general relation (IV.20) includes the fact that $V(\mathfrak{R})$ will be always totally symmetric, for $\mathfrak{H}_e(\mathfrak{R},\mathrm{r})$ is totally symmetric and for nondegenerate irreducible representations $\Gamma_i \otimes \Gamma_i$ is totally symmetric; so Γ will be.

The concept of lineage, classifying the electronic wave functions throughout the nuclear configuration space according to the irreducible representations of the highly symmetric MS group permits us to relate the electronic wave function for less symmetric nuclear conformations to one corresponding to another equivalent conformation. With that we obtain information about the transformation properties of the hypersurface in nuclear coordinate space generated by a matrix element integrated over the electronic coordinates. Whether such a matrix element is zero or not for no nuclear displacements from the reference conformation, $\mathfrak{S} = 0$, is contained in (IV.20) and (IV.21) also. To obtain this information from group theory for other nuclear conformations, $\mathfrak{S} \neq 0$, it is still necessary to use the conventional symmetry classifications corresponding to the lower symmetry of the displaced nuclear conformation.

Before turning our attention to degenerate representations, let us look briefly at the aspect of "feasible" and "unfeasible" operations in the CNPI group. For this we consider as examples methyl chloride and ammonia. The CNPI group for both has the elements

$$E, (123), (132), (12)^*, (23)^*, (31)^*$$
$$E^*, (123)^*, (132)^*, (12), (23), (31)$$

and is isomorphic with D_{3h}. For methyl chloride, only the elements of the first row, isomorphic with C_{3v}, are feasible. The second row of elements are unfeasible, since for them an inversion of the molecule would be required, prohibited by an excessive energy barrier. The same holds in the case of ammonia for low energies and low spectral resolution. However, for higher energies, or for low energy, if the resolution is of the order of the state coupling through the barrier, it becomes necessary to use the full CNPI group, isomorphic with D_{3h}, as MS group. Thus, which operations are feasible and which are not depends on the energy barriers between nuclear conformations generated, the energy in the system, the coupling between localized states, and the spectral resolution.

For the EMS group or MS groups with cycles of order $n \geq 3$, equivalent to C_n with $n \geq 3$ for point groups, we have the possibility of degenerate irreducible representations. For electronic states transforming as degenerate representations, or if they have such a lineage, the important Jahn–Teller[21] and Renner–Teller[22] effects need to be considered. In general, for any conformation, where a degenerate electronic state is possible, there is always a nuclear displacement that will lift this degeneracy. Thus the energy surfaces corresponding to the degenerate lineage will either touch or intersect in the highly symmetric reference conformation, where they are energetically degenerate. In molecules with a linear reference conformation the electronic states with $|\Lambda| > 0$ will be doubly degenerate. This degeneracy will be split by a bending, S_b, the surfaces will touch at the linear conformation, but $\partial/\partial S_b V(\mathfrak{R}) = 0$ for $S_b = 0$. However, for the lower surface this may be a maximum, with additional minima at larger bending angles. For Π-states, $|\Lambda| = 1$, the energy splitting is second order in ΔS_b; for states with $|\Lambda| > 1$, it is of even higher order.[23] For molecules with nonlinear reference conformations in degenerate states the corresponding surfaces will always intersect at $\mathfrak{S} = 0$, and the highly symmetric reference conformation cannot be the equilibrium conformation since at $\mathfrak{S} = 0$ there will be at least one $\partial/\partial S_k V(\mathfrak{R}) \neq 0$.[24] The topology of these energy surfaces becomes rather complex in such cases and has been the subject of extensive study.[25] Particular care must be exercised, since at $\mathfrak{S} = 0$ the electronic wave function is not differentiable with respect to the nuclear coordinates.[26]

If the splittings are large, and the energy and nuclear excursions are such that degenerate points on the surface are not approached, then the lower MS group corresponding to the equilibrium conformation with no degeneracies would be appropriate, leaving only nondegenerate representations, as discussed above. However, if the higher-order MS group is appropriate, and we consider the nuclear motion in a state of degenerate lineage, then it will be necessary to consider simultaneously the two or more potentials and electronic wave functions arising from the degeneracy on an equal footing. Here we must expect a severe breakdown of the Born–Oppenheimer or adiabatic approximation. The meaning of a single-potential surface will lose its meaning, although the adiabatic separation indicated above and to be detailed below will still be viable.

Consider now a specific set of states of lineage Γ, a representation which is k-fold degenerate. We will obtain k energy surfaces and k electronic-state wave functions $\psi_{\Gamma a}(\mathfrak{R}, \mathfrak{r})$ of lineage Γ with subspecies a chosen such that, by definition, $\mathfrak{H}_e$ is diagonal in their representation. Since $\mathfrak{H}_e(\mathfrak{R}, \mathfrak{r})$ is invariant to the operations of the group and the $\psi_{\Gamma a}$'s are eigenfunctions of $\mathfrak{H}_e$, we can state in general that all k potential surfaces will be totally symmetric. Unfortunately, we are not able to prove such restrictive statements about the transformation properties for general matrix elements of the type

$$M(\mathfrak{R}) = \langle \psi_{e\Gamma_a}(\mathfrak{R}, \mathfrak{r}) | \hat{M}_{\Gamma_m}(\mathfrak{R}, \mathfrak{r}) | \psi_{e\Gamma_b}(\mathfrak{R}, \mathfrak{r}) \rangle_{\mathfrak{r}} \tag{IV.22}$$

In general, we can say only that $M(\mathfrak{R})$ must be a linear combination of parts transforming as the irreducible representations contained in the product $\Gamma \otimes \Gamma_m \otimes \Gamma$.

V. THE EQUATION FOR NUCLEAR MOTION

Thus far we have given a rather formal presentation of the adiabatic separation of electronic and nuclear motion. To consider the conventional adiabatic (II.4) in more detail and to carry out the necessary coordinate transformations from the laboratory fixed frame to the nuclear fixed frame, we write (II.4) in the more suggestive form

$$\sum_m \{X_{nm} + \delta_{nm}(V_n(\mathfrak{R}) - E_I)\} \theta_{mI}(\mathfrak{R}, \mathfrak{Q}) = 0 \tag{V.1}$$

which must be satisfied for all n with a given state energy E_I. This equation, where n and m refer to the collection of quantum numbers of the electronic states, is still exact and $\theta_{mI}(\mathfrak{R}, \mathfrak{Q})$ should be interpreted as the component of the nuclear motion function of state I to the electronic function ψ_m. To deal with this equation further, we want to eliminate the dependence on the coordinates of the laboratory frame of X_{nm}, (II.5), transforming it to the nuclear fixed frame. To this end we can separate from $\theta_{mI}(\mathfrak{R}, \mathfrak{Q})$ the functions $\Xi(\boldsymbol{Q}_0, E_T)$ describing the motion of the center of nuclear mass. These functions are eigenfunctions of the operator $\mathfrak{H}_{\text{trans}}$, defined in Appendix 1, (A.39). We can now integrate out the Q_0 dependence and obtain

$$\langle \Xi(\boldsymbol{Q}_0, E_T) | X_{nm} | \Xi(\boldsymbol{Q}_0, E_T') \rangle = \delta_{E_T E_T'}(E_T + \tilde{X}_{nm}) \tag{V.2}$$

with

$$\tilde{X}_{nm} = \langle \psi_n(\mathfrak{R}, \mathfrak{r}) | \mathfrak{H} - \mathfrak{H}_e - \mathfrak{H}_{\text{trans}} | \psi_m(\mathfrak{R}, \mathfrak{r}) \rangle_{\mathfrak{r}} \tag{V.3}$$

using the notation of (V.10) and (V.11),

$$|\phi(\Gamma_e a_e)NSJKM_J\rangle$$
$$= \sum_{M_S M_N} \langle SNM_S M_N | JM_J\rangle \, |\phi(\Gamma_e a_e)SM_S\rangle \, |NKM_N\rangle \tag{V.12}$$

for nonlinear molecules (K corresponds to P when J is replaced by N) and as

$$|\phi\Lambda NSJM_J\rangle$$
$$= \sum_{M_S M_N} \langle SNM_S M_N | JM_J\rangle \, |\phi\Lambda SM_S\rangle \, |N\Lambda M_N\rangle \tag{V.13}$$

for linear molecules. The general expression using the MS- or EMS-group classification is obtained by substituting for K the equivalent $\kappa = (\Gamma_N a_N r_N)$.

For the consideration of nuclear spins in strong magnetic fields an uncoupled basis is most appropriate:

$$|\gamma JM_J I_1 M_{I_1} \cdots I_N M_{I_N}\rangle = |\gamma JM_J\rangle \prod_{\alpha=1}^{N} |I_\alpha M_{I_\alpha}\rangle \tag{V.14}$$

In the weak-field case a basis coupled as

$$|\gamma JIFM_F\rangle = \sum_{M_J M_I} \langle JIM_J M_I | FM_F\rangle |\gamma JM_J\rangle |IM_I\rangle \tag{V.15}$$

is more convenient, where the nuclear spins are already coupled to each other in $|IM_I\rangle$.

Integrating the matrix elements X_{nm} with these rotational and nuclear spin basis functions over the Euler angles, the angles of orientation of the molecular fixed frame relative to the laboratory frame yields the matrix elements $\chi_{n_\alpha m_\beta}$, which are functions of the internal nuclear coordinates $\mathfrak{R}$ only. Here the indices n, m specify the electronic states as given by (V.4) and α and β represent the collection of substate quantum numbers of the basis, chosen as discussed above. After this reduction we may write (V.1) as

$$\sum_{n'\alpha'} \left[\chi_{n_\alpha n'_{\alpha'}}(\mathfrak{R}) + \delta_{n_\alpha n'_{\alpha'}} \left(V_n(\mathfrak{R}) - \bar{E}_I \right) \right] \Pi_{Iv, n'\alpha'}(\mathfrak{R}) = 0 \tag{V.16}$$

which determines the vibrational wave functions and depends only on the internal coordinates $\mathfrak{R}$ or what is equivalent $\mathfrak{S}$ (see the discussion in Section IV). In (V.16) the state energy is $\bar{E}_I = E_I - E_T$, and the nuclear motion functions describing the vibration must satisfy the orthonormality constraints.

$$\sum_{n_\alpha} \langle \Pi_{Iv,n_\alpha} | \Pi_{Jv',n_\alpha} \rangle = \delta_{IJ} \tag{V.17}$$

where $v \neq v'$ implies $I \neq J$. Again it should be emphasized that $\Pi_{Iv,m\beta}$ is the component to the basis function $|m_\beta\rangle$ of the vibrational state v belonging dominantly to state $|n_\alpha\rangle$ implicit in I.

In practice the infinite-order system of coupled differential equations, (V.16), must be truncated to a small, finite order. For example, in the usual Born–Oppenheimer approximation the sum is restricted in general to a single electronic state, n, and the substate dependence signified here by α is ignored. Such a drastic approximation is inadequate to describe a large class of more subtle molecular properties of interest here. For such a description it will be necessary to consider explicitly strongly coupled, degenerate, or near-degenerate electronic states $n_1 \ldots n_k$, while the effect of weakly coupled, energetically well separated electronic states can be treated perturbatively, in general to second order. This can be achieved using a van Vleck contact transformation[27] detailed in Section VII, (VII.1) to (VII.12).

The transformation matrix needed may be represented by $\mathbf{U}(\mathfrak{R})$, (VII.4), and we obtain formally, in symbolic matrix notation,

$$\tilde{\mathbf{X}} + \mathbf{V} = \mathbf{U}(\mathfrak{R})(\mathbf{X} + \mathbf{V})\mathbf{U}^\dagger(\mathfrak{R}) \tag{V.18}$$

where good through second order the operator $\mathbf{X} + \mathbf{V}$ is block-diagonal; that is, if we want to treat the coupling between the states $n_1 \cdots n_k$ explicitly only the leading $n_k \times n_k$ block of $\mathbf{X} + \mathbf{V}$ is of interest, while the weak coupling of the states $n_k + 1, \cdots$ has been folded in perturbatively. Thus (V.16) reduced to small, finite order becomes

$$\sum_{n'_{\alpha'}} \left[\tilde{\chi}_{n_\alpha n'_{\alpha'}}(\mathfrak{R}) + \delta_{n_\alpha n'_{\alpha'}} \left(V_n(\mathfrak{R}) - \bar{E}_I \right) \right] \tilde{\Pi}_{Iv,n'_{\alpha'}} = 0 \tag{V.19}$$

with the sum over $n'_{\alpha'}$ reduced to k terms. The relation of the eigenfunctions of (V.16) and (V.19) are given by

$$\tilde{\Pi}_{Iv,n_\alpha} = \sum_{m_\beta} U_{n_\alpha m_\beta} \Pi_{Iv,m_\beta} \tag{V.20}$$

In the conventional adiabatic approximation, only one term $n'=n$ is considered in (V.19) and the van Vleck transformation may or may not be included. Even with n restricted to one term, there remains the coupling between the rotational and fine-structure substates, α, corresponding to the electronic state n. Here we may use the freedom of a unitary transformation of the substate basis such that the angular momentum coupling between the various substates α and β is minimized. This would correspond to choosing, for example, for a given vibrational state an optimum intermediate coupling between Hund's cases (a) and (b), say. Using such an optimally coupled basis will lead to nearly decoupled equations, which, treated separately and ignoring the remaining coupling, yield solutions close to those of the fully coupled system. To find an approximation to such a basis, the following procedures may be used:

1. Solve the purely vibrational problem, neglecting all rotational, fine-structure, and so on, effects.
2. Using such an approximate vibrational wave function $|v\rangle$, say, compute the matrix elements

$$\mu_{n_\alpha, n_\beta} = \langle v | \tilde{\chi}_{n_\alpha n_\beta} + \delta_{\alpha\beta} V_n | v \rangle$$

3. Find the unitary matrix $\mathbf{U}$ which diagonalizes $\boldsymbol{\mu}$.
4. Obtain the optimally coupled operator for state n, v in matrix operator form as

$$\{\tilde{\mathbf{X}}(\mathfrak{R}) + \mathbf{V}(\mathfrak{R})\}_{\text{opt}} = \mathbf{U}\{\tilde{\mathbf{X}}(\mathfrak{R}) + \mathbf{V}(\mathfrak{R})\}\mathbf{U}^\dagger$$

A similar procedure could be followed if more than one electronic state n is considered explicitly.

VI. SPECTROSCOPIC CONSTANTS

In general, the observed spectroscopic data, transition frequencies, and line splittings (we are not concerned with transition probabilities here) are fitted to formulas containing parameters, the so-called spectroscopic constants. The formulas used are based and derived on the quantum mechanical description of the system, at least phenomenologically. Standard formulas for rovibronic fits are capable of representing regular systems as well as those which are irregular due to strong rovibronic perturbations, provided that sufficient states have been observed. Spectroscopic constants determined by such fits, which may be considered as a form of data reduction, are given in the tables of Landolt–Börnstein.[28] There, references to the formulas according to which such constants are evaluated are given as well.

Before we turn our attention to the detailed formalisms needed to determine these spectroscopic constants directly using *ab initio* quantum mechanical calculations, we feel it necessary to discuss some of the essential features of the spectroscopic data reduction. Certainly, it will not be possible here to give an extensive review of the full range of spectroscopic results and their interpretation. However, some general aspects need to be mentioned here.

The physical intuition inherent in analysis of spectra leads to an interpretation based on what we may call a phenomenological Hamiltonian useful for the assignment and analysis of spectra.

1. The adiabatic approximation defining a potential surface is assumed to hold generally. However, special attention is given to Renner–Teller[22] and Jahn–Teller[21] effects on the spectra.
2. The nuclear vibrational motion is treated in a perturbation expansion based on small displacements around well-defined potential minima.[29] For the consideration of large-amplitude nuclear motion, such as internal rotation, allowance is made by the introduction of special internal coordinates.[30]
3. The vibrational problem is solved in the harmonic oscillator limit[29] to the zeroth order, providing the vibrational function basis.
4. The rotational problem is formulated in the rigid-rotor approximation[29] leading to the rotational function basis. Vibrational angular momenta are included for degenerate vibrations in the same way as electronic momenta are introduced for degenerate electronic states of linear molecules.
5. The couplings in the higher-order terms of Nielson's[29] Hamiltonian are formulated in terms of the harmonic oscillator-rigid-rotor basis (e.g., nonrigidity effects).
6. Other internal magnetic interactions or external field effects are added and formulated in the appropriate angular momentum or spin-function basis leading to rovibronic eigenstates, including all possible splittings.

For steps (5) and (6) the effective operator technique is used throughout; that is, the total molecular Hamiltonian is replaced by some effective operators reproducing the correct matrix structure (see, e.g., Gerrat[31]). The effective Hamiltonians, each for different types of interaction, operate only on definite vibronic states, which we may label in analogy with our notation in the other chapters by ϕ, Γ_e, S, v, where ϕ stands for the electronic state of Γ-symmetry and spin S and v is the collection of vibrational state labels $v_1, v_2, \ldots,$ where each v_i is a particular harmonic oscillator quantum number. These vibrational states are classified with respect to the point group of the reference configuration. In case of the nonadiabatic vibronic states of the Renner–Teller effect in linear triatomics, only the vibronic species label should be used.

The dependence of the effective constants within the effective Hamiltonians (or operators) on the vibrational state number v_i is generally expanded into a series like

$$\tilde{A}_{v_i} = A_e + \sum_k \alpha_k (v_i + d_i/2)^k \tag{VI.1}$$

where A_e is the value at the equilibrium configuration, α_k are some parameters, and d_i is the degree of degeneracy of the vibration v_i in the harmonic oscillator limit.

If the variation of the fitted constants with the rotational quantum number J (or N) is accounted for, centrifugal distortion operators are added to the various terms, which are introduced at the places necessary.

For diatomics this problem may be formulated in a more closed form in terms of Dunham coefficients, which are defined by the series expansion

$$E_{vJ} = \sum_{ij} Y_{ij}\left(v + \tfrac{1}{2}\right)^i \left[J(J+1)\right]^j \tag{VI.2}$$

for the rovibrational energy E_{vJ} in one adiabatic state, where v is the vibrational and J the total angular momentum quantum number. The Dunham formula can be fitted for fixed J, in general the lowest $J = 0$, to the vibrational term values

$$E_{vJ}(J=0) = G_v + \omega_e\left(v + \tfrac{1}{2}\right) - \omega_e x_e\left(v + \tfrac{1}{2}\right)^2 + \omega_e y_e\left(v + \tfrac{1}{2}\right)^3 + \ldots \tag{VI.3}$$

whereas the purely rotational fit yields

$$E_{vJ} = B_v J(J+1) - D_v\left[J(J+1)\right]^2 + H_v\left[J(J+1)\right]^3 + \cdots \tag{VI.4}$$

where the B_v's can be fitted to

$$B_v = B_e - \alpha_e\left(v + \tfrac{1}{2}\right) + \gamma_e\left(v + \tfrac{1}{2}\right)^2 + \cdots \tag{VI.5}$$

with similar expressions for D_v, and so on. These constants are approximately related to the Dunham coefficients Y_{ij} according to the following table:

Y_{ij} j \ i	0	1	2	3		
0	Y_{oo}	ω_e	$-\omega_e x_e$	$\omega_e y_e$	$\rightarrow$	$T_e + G_v$
1	B_e	$-\alpha_e$	γ_e		$\rightarrow$	B_v
2	$-D_e$	β_e			$\rightarrow$	$-D_v$
3	H_e				$\rightarrow$	H_v

The Dunham constants are related to the potential curve of diatomics in accordance with Dunham's semiclassical analysis.[32] Following this line, all other effective constants of vibronic states can be expanded with respect to their vibrational and rotational dependence. In the following we will set up a list of such effective operators commonly used in spectroscopy without going into details.

First, we state that the energy of the vibronic state is given by

$$E = T_e + G_v \tag{VI.6}$$

where T_e is the electronic energy at the nuclear equilibrium conformation and G_v are the vibrational energies of the state designated by v as a collection of vibrational quantum numbers v_i, where each degenerate vibration is counted once. The angular momentum of degenerate vibronic states of linear molecules with respect to the symmetry axis is added to the rotational effective Hamiltonian. For the linear triatomic molecules we may extend the analysis by introducing the Renner–Teller splitting operator defined by[22]

$$\mathfrak{H}_{\mathrm{R-T}} = \varepsilon k_b \cos 2|\Lambda|(\varphi - \varphi_b)$$

where k_b is the mean force constant for the two potential curves (see Section VIII.A for more details). The fine structure of the levels is interpreted by the following Hamiltonian:

$$\tilde{\mathfrak{H}}_{fs} = \tilde{\mathfrak{H}}_{\mathrm{SO}} + \tilde{\mathfrak{H}}_{\mathrm{SS}} + \tilde{\mathfrak{H}}_{\mathrm{rot}} + \tilde{\mathfrak{H}}_{\mathrm{Sr}} + \tilde{\mathfrak{H}}_{d} \tag{VI.7}$$

The effective spin–orbit coupling operator $\tilde{\mathfrak{H}}_{\mathrm{SO}}$ for linear molecules is written as

$$\tilde{\mathfrak{H}}_{\mathrm{SO}} = \tilde{A} L_z S_z + \tilde{A}_D L_z S_z (\mathrm{N}^2 - N_z^2) + \cdots \tag{VI.8}$$

where $\tilde{A}$ is the effective spin–orbit constant.

$\tilde{A}_D$ is the centrifugal distortion constant; L_z, S_z are the electronic angular momentum and electronic spin operators, respectively, with respect to the symmetry axis; and $\mathbf{N}^2$, N_z are the angular momentum operators for the overall rotation. The operator L_z is considered to operate on the electronic function given at the equilibrium value (in the sense of the Herzberg–Teller approximation) yielding the expectation value Λ, also for the bending vibrational mode in linear triatomics. The effective spin–orbit coupling operator has to be included only in orbitally degenerate electronic states.

In nonlinear molecules this operator is taken into account for the metal complex ions of octahedral or related high-symmetry point groups, where it takes the effective form[21]

$$\tilde{\mathfrak{H}}_{SO} = \tilde{A}\mathbf{SL} \tag{VI.9}$$

The spin–spin coupling operator $\tilde{\mathfrak{H}}_{SS}$ is conveniently written for linear molecules as

$$\begin{aligned}\tilde{\mathfrak{H}}_{SS} = {} & \tfrac{2}{3}\lambda_0(3S_z^2 - \mathbf{S}^2) \\ & + \tfrac{1}{2}\lambda_{\perp\perp}(e^{-i2\varphi}S_+^2 + e^{i2\varphi}S_-^2)\end{aligned} \tag{VI.10}$$

where λ_0 is the spin–spin coupling constant and $\lambda_{\perp\perp}$ a spin–spin-Λ-doubling constant. The angle φ refers to angle of rotation of the whole electronic system. The second form in (VI.10) leads to a Λ-doubling effect in $\Omega = 0$ states of electronic Π-states in diatomics. (It separates the $\Sigma^+$and Σ^- representations contained in the product $\Pi \oplus \Pi$ of the electronic function times the molecular frame quantized spin function). For nonlinear molecules we write, in general,

$$\tilde{\mathfrak{H}}_{SS} = \sum_{\mu,\nu = x,y,z} \tfrac{1}{2}\tilde{\lambda}_{\mu\nu}(S_\mu S_\nu + S_\nu S_\mu) \tag{VI.11}$$

where symmetry is not considered.

The spin–spin coupling operator requires states of $S \geq 1$ to manifest itself. In general, only the diagonal elements of the property matrices, here $\boldsymbol{\lambda}$, are considered in spectroscopic analysis.

For linear molecules the rotational effective Hamiltonian is expressed by

$$\tilde{\mathfrak{H}}_{\mathrm{rot}} = \tilde{B}(\mathbf{N}^2 - N_z^2) - \tilde{D}(\mathbf{N}^2 - N_z^2)^2 + \tilde{H}(\mathbf{N}^2 - N_z^2)^3 + \cdots \tag{VI.12}$$

where $\tilde{B}$, $\tilde{D}$, $\tilde{H}$ are the rotational constant and the various centrifugal distortion constants, respectively, already mentioned above, and $\mathbf{N}$ is the angular momentum operator for the overall rotation:

$$\mathbf{N}=\mathbf{J}-\mathbf{S} \tag{VI.13}$$

where $\mathbf{J}$ is the total angular momentum operator and $\mathbf{S}$ the total electronic spin operator.

In order to give the correct form of $\tilde{\mathfrak{H}}_{\text{rot}}$, including all interactions in different function basis sets coupled according to Hund's case (a) or (b), we have to use $\mathbf{N}$ instead of $\mathbf{J}$ in (VI.12). In addition, we mention that N_z is given by $L_z + G_z$, where G_z is the angular momentum carried by the degenerate bending mode in triatomics.

For nonlinear molecules the rotational Hamiltonian $\mathfrak{H}_{\text{rot}}$ on the basis of the rigid-rotor approximation is given by

$$\tilde{\mathfrak{H}}^{(0)}_{\text{rot}}=\tilde{A}N_z^2+\tilde{B}N_y^2+\tilde{C}N_x^2 \tag{VI.14}$$

(the commonly used convention is $z \rightarrow a$, $y \rightarrow b$, $x \rightarrow c$ for $\tilde{A} \geq \tilde{B} \geq \tilde{C}$), where $\tilde{B}=\tilde{C}$ for symmetric-top molecules and $\tilde{A}=\tilde{B}=\tilde{C}$ for spherical-top molecules.

The centrifugal distortion is accounted for by adding to (VI.14)

$$\frac{1}{4}\sum_{\mu,\nu=x,y,z} \tau'_{\mu\mu\nu\nu} N_\mu^2 N_\nu^2$$

where the τ'-values are related to the potential surface in the harmonic limit according to the analysis given by several authors.[33] For a more refined theory, see Watson.[34]

The effective spin–rotational coupling operator for linear molecules is given by

$$\tilde{\mathfrak{H}}_{\text{Sr}}=\tilde{\gamma}(\mathbf{NS}-N_z S_z)+\text{centrifugal distortion terms} \tag{VI.15}$$

where $\tilde{\gamma}$ is the spin–rotational coupling constant. In molecules possessing a first-order spin–orbit effect, $\tilde{\gamma}$ is highly correlated to A_D and therefore cannot be determined accurately.[35] For nonlinear molecules we merely write

$$\mathfrak{H}_{\text{Sr}}=\sum_{\mu,\nu} \tilde{\gamma}_{\mu\nu}\left(\tfrac{1}{2}\right)(S_\mu N_\nu + N_\nu S_\mu) \tag{VI.16}$$

with a symmetric matrix $(\tilde{\gamma}_{\mu\nu})$. (If the matrix is considered to be nonsymmetric, the additional terms, linear in N, are properly added to the spin–orbit coupling operator.) There is no evidence for the necessity to include an effective rotation–electronic orbit interaction.

Under $\tilde{\mathfrak{H}}_{\Lambda-d}$ we summarize all the operators leading to Λ-doubling in vibronic Π-states of linear molecules which are essentially parts of the rotational operator $\tilde{\mathfrak{H}}_{\text{rot}}$. The influence of $\tilde{\mathfrak{H}}_{\Lambda-d}$ is best described by its matrix elements in a definite function basis or the appropriate operator equivalent. In electronic $^2\Pi$-states and vibrational Σ^+-states ($l=0$) we may write for the Hund's case (a) function basis,

$$\langle {}^c_d\Pi v\, l=0\, J\, \Omega=\tfrac{1}{2}\, S=\tfrac{1}{2}|\tilde{\mathfrak{H}}_{\Lambda-d}|{}^c_d\Pi v\, l=0\, J\, \Omega=\tfrac{3}{2}\, S=\tfrac{1}{2}\rangle$$
$$=\mp q_e\left(J+\tfrac{1}{2}\right)\sqrt{\left(J+\tfrac{1}{2}\right)^2-1} \qquad \text{(VI.17)}$$

and

$$\langle {}^c_d\Pi v\, l=0\, J\, \Omega=\tfrac{1}{2}\, S=\tfrac{1}{2}|\tilde{\mathfrak{H}}_{\Lambda-d}|\Pi v\, l=0\, J\, \Omega=\tfrac{1}{2}\, S=\tfrac{1}{2}\rangle$$
$$=\pm p_e\left(S+\tfrac{1}{2}\right)\left(J+\tfrac{1}{2}\right) \qquad \text{(VI.18)}$$

where c and d are the Kronig-symmetrized case (a) functions with $\Omega=|\Lambda| +\Sigma$ and l is the vibrational angular momentum quantum number ($l=0$ means vibrational Σ^+ states). From (VI.17) and (VI.18) we may express $\tilde{\mathfrak{H}}_{\Lambda-d}$ as

$$\tilde{\mathfrak{H}}_{\Lambda-d}=q_e\left(J_+^2 e^{-2i\varphi}+J_-^2 e^{2i\varphi}\right)$$
$$+p_e\left(J_+S_+e^{-2i\varphi}+J_-S_-e^{2i\varphi}\right) \qquad \text{(VI.19)}$$

In a more general form, valid for all cases, $J_\pm$ should be replaced by $N_\pm = J_\pm - S_\pm$.

In the same way, a vibrational Π-state ($|l|=1$) with $|K|=1$ in an electronic Σ-state is split according to $\pm\frac{1}{2}q_v\,(v+1)J(J+1)$, which results from the matrix element

$$\langle v\, l=\mp 1\, J\, K=\mp 1|\tilde{\mathfrak{H}}_{l-d}|v\, l=\pm 1\, J\, K=\pm 1\rangle$$
$$=\tfrac{1}{2}q_v(v+1)J(J+1) \qquad \text{(VI.20)}$$

where $K=\Lambda+l=1$ for Σ-states and $l=-v,-v+2,\ldots,v$. Therefore, $\tilde{\mathfrak{H}}_{l-d}$ may be expressed as

$$\tilde{\mathfrak{H}}_{l-d}=\tfrac{1}{2}q_v S_b^2\left(J_-^2 e^{2i\varphi_b}+J_+^2 e^{-2i\varphi_b}\right) \qquad \text{(VI.21)}$$

where S_2,φ_2 are the polar coordinates of the bending vibration. For the vibronic Renner–Teller states of Π-symmetry we will have the mixed form of the effective doubling operators (VI.20) and (VI.21).

For polyatomic nonlinear molecules we do not define here doubling operators, although in symmetric-top molecules a splitting equivalent to l-type doubling in linear molecules is present. We also refrain from giving a Δ-type doubling operator here, which would require higher-order perturbation theory.

Next we will introduce the hyperfine structure of the spectral lines produced by the interaction of the magnetic moments of the nuclei with spin $I_\alpha \neq 0$ with the magnetic moments of electron spins, electronic orbital motion, overall rotation, and other nuclear spins, or by the interaction of the nuclear charge distribution with the molecular charge, all contained in

$$\tilde{\mathfrak{H}}_{hfs} = \tilde{\mathfrak{H}}_{\mathrm{IO}} + \tilde{\mathfrak{H}}_{\mathrm{IS}} + \tilde{\mathfrak{H}}_{\mathrm{Ir}} + \tilde{\mathfrak{H}}_{\mathrm{II}} + \tilde{\mathfrak{H}}_{\mathrm{Q}} \tag{VI.22}$$

In (VI.22), $\tilde{\mathfrak{H}}_{\mathrm{IO}}$ and $\tilde{\mathfrak{H}}_{\mathrm{IS}}$ are the nuclear spin–electronic orbit and nuclear spin/electronic spin interactions, respectively, which are present only in radicals with an orbitally degenerate electronic state or for $S \neq 0$, respectively. The other terms are almost negligible in cases where the first two do not vanish, whereas in closed-shell molecules they are responsible for the hyperfine structures, which, can be resolved only in high-resolution spectroscopy in the microwave or radio-frequency region.

For linear molecules we may write

$$\tilde{\mathfrak{H}}_{\mathrm{IO}} = \sum_\alpha \tilde{\mathfrak{H}}_{\mathrm{IO}}(\alpha) = \sum_\alpha \tilde{a}_\alpha \mathsf{I}_\alpha \mathsf{L} \tag{VI.23}$$

The corresponding operator for nonlinear molecules is not investigated since the electronic angular momentum operator has no expectation values in nondegenerate electronic states. The nuclear spin electronic operator may be expressed for linear molecules separately for each nucleus as

$$\begin{aligned}\tilde{\mathfrak{H}}_{IS}(\alpha) = {} & \tilde{b}(\alpha)(\mathsf{S}\mathsf{I}_\alpha - S_z I_{\alpha z}) \\ & + (\tilde{c}(\alpha) + \tilde{b}(\alpha)) S_z I_{\alpha z} \\ & + \tilde{d}(\alpha)\left(S_+ I_{\alpha +} e^{-2i\varphi} + S_- I_{\alpha -} e^{2i\varphi}\right)\end{aligned} \tag{VI.24}$$

where the last term contributes only for $^2\Pi$-states.

For nonlinear molecules we may write

$$\tilde{\mathfrak{H}}_{IS}(\alpha) = \sum_{\mu,\nu} \tilde{b}_{\mu\nu}(\alpha) S_\mu I_{\alpha\nu} \tag{VI.25}$$

The quadrupole coupling operator $\mathfrak{H}_Q$ couples the nuclear spins with $I \geq 1$ to the rotation of the molecule. In the linear case we may write

$$\tilde{\mathfrak{H}}_Q = eQ_\alpha\left[q(\alpha)(3I_{\alpha z}^2 - \mathrm{I}^2)/4I_\alpha(2I_\alpha - 1)\right.$$
$$\left. + q_{\perp\perp}(\alpha)(I_{\alpha+}^2 e^{-2i\varphi} + I_{\alpha-}^2 e^{2i\varphi})/4I_\alpha(2I_\alpha - 1)\right] \qquad \text{(VI.26)}$$

for all α, where Q_α is the quadrupole moment of the nuclear charge distribution, $q(\alpha)$ the main quadrupole coupling constant, and $q_{\perp\perp}(\alpha)$ another constant, which contributes to the Λ-doubling in electronic Π-states.

For nonlinear molecules we may write in irreducible tensor notation

$$\tilde{\mathfrak{H}}_Q(\alpha) = \sum_{m=-2}^{2} (-1)^m \tilde{\mathrm{T}}_m^2(I_\alpha, I_\alpha)\tilde{c}_{-m}^2 \qquad \text{(VI.27)}$$

where

$$\tilde{c}_m^2 = \left[eQ_\alpha\sqrt{6}\,/2I_\alpha(2I_\alpha - 1)\right]\tilde{x}_m^{Q2}$$

and $\tilde{x}_m^{Q2}$ are the quadrupole coupling constants.

We should note here that all I_α-components are related to the molecular fixed frame.

The spin–rotation coupling may be expressed as

$$\tilde{\mathfrak{H}}_{\mathrm{I}r}(\alpha) = \tilde{C}_R(\alpha)\tfrac{1}{2}(\mathrm{I}_\alpha\mathrm{N} - I_{\alpha z}N_z + \mathrm{N}\mathrm{I}_\alpha - N_z I_{\alpha z}) \qquad \text{(VI.28)}$$

$$\tilde{\mathfrak{H}}_{\mathrm{I}r}(\alpha) = \sum_{\mu,\nu} \tilde{C}_{R\mu\nu}(\alpha)\tfrac{1}{2}(I_{\alpha\mu}N_\nu + N_\nu I_{\alpha\mu}) \qquad \text{(VI.29)}$$

for linear and for nonlinear molecules, respectively.

The nuclear spin–spin interactions are given by

$$\tilde{\mathfrak{H}}_{\mathrm{II}} = \sum_{\alpha\neq\beta}\sum \tilde{d}_{\alpha\beta}^s \mathbf{I}_\alpha \mathbf{I}_\beta + \tilde{d}_{\alpha\beta}^t(3I_{\alpha z}I_{\beta z} - \mathbf{I}_\alpha\mathbf{I}_\beta) \qquad \text{(VI.30)}$$

for linear molecules where $\tilde{d}^s$ and $\tilde{d}^t$ are the scalar and tensorial coupling constants, respectively,[36] and by

$$\tilde{\mathfrak{H}}_{\mathrm{II}} = \sum_{\alpha\neq\beta}\sum \sum_{\mu,\nu=x,y,z} \tilde{d}_{\mu\nu} I_{\alpha\mu} I_{\beta\nu} \qquad \text{(VI.31)}$$

for nonlinear molecules.

Applying external magnetic fields to the molecules will produce further splittings of the molecular levels, since the degeneracy of rotational states with respect to space rotations is lifted.

In open-shell molecules with an orbitally degenerate electronic state, a paramagnetic orbital Zeeman effect will arise. For linear molecules we have

$$\tilde{\mathfrak{H}}_H^L = \beta \tilde{g}_L L_z H_z \qquad \text{(VI.32)}$$

where β is the Bohr magneton ($\beta = \hbar e/2mc$) and H is the magnetic field measured in the molecular fixed frame. The orbital $\tilde{g}_L$-factor is different from 1 by a small amount because of molecular interactions, as described in Section VIII.E.

In states with total electronic spin $S \neq 0$, the paramagnetic Zeeman effect will exist. It may be expressed by

$$\tilde{\mathfrak{H}}_H^S = \beta \left[\tilde{g}_{S\perp}(\mathsf{SH} - S_z H_z) + \tilde{g}_{S\parallel} S_z H_z \right] \qquad \text{(VI.33)}$$

for linear molecules and by

$$\tilde{\mathfrak{H}}_H^S = \beta \sum_{\mu,\nu = x,y,z} \tilde{g}_{S\mu\nu} S_\mu H_\nu \qquad \text{(VI.34)}$$

for nonlinear molecules.

The diagonal spin g-factors $g_{S\mu\mu}$ or $g_{S\parallel}$ and $g_{S\perp}$ again differ by a small amount from the free spin value of about 2.0023. There are some more g-factors associated with the molecular rotation or the diamagnetic Zeeman effect.

For linear molecules we have

$$\begin{aligned} \tilde{\mathfrak{H}}_H^r = -\beta_N \Big[& \tilde{g}_R\left(\tfrac{1}{2}\right)(\mathsf{NH} - N_z H_z + \mathsf{HN} - H_z N_z) \\ & + \tilde{g}_{\perp\perp}\left(\tfrac{1}{2}\right)\left\{(N_+ H_+ + H_+ N_+)e^{-2i\varphi} \right. \\ & \left. + (N_- H_- + H_- N_-)e^{2i\varphi}\right\} \Big] \end{aligned} \qquad \text{(VI.35)}$$

with β_N = nuclear magneton ($\beta_N = \hbar e/2M_p c$, M_p = mass of the proton). The second term again contributes only in Π-states. For nonlinear molecules the appropriate form is

$$\tilde{\mathfrak{H}}_H^r = -\beta_N\left(\tfrac{1}{2}\right) \sum_{\mu,\nu = x,y,z} \tilde{g}_{R\mu\nu}(N_\mu H_\nu + H_\nu N_\mu) \qquad \text{(VI.36)}$$

The second-order Zeeman effect from the field-induced magnetic moment is introduced by the following effective operators:

$$\tilde{\mathfrak{H}}_x = \tfrac{1}{6}\{(\tilde{x}_{\parallel} + 2x_{\perp})\mathrm{H}^2 + (\tilde{x}_{\parallel} - \tilde{x}_{\perp})(3H_z^2 - \mathrm{H}^2) + \tilde{x}_{\perp\perp}(H_+^2 e^{-2i\varphi} + H_-^2 e^{2i\varphi})\} \tag{VI.37}$$

for linear molecules and

$$\tilde{\mathfrak{H}}_x = \tfrac{1}{2} \sum_{\mu,\nu = x,y,z} \tilde{x}_{\mu\nu} H_\mu H_\nu \tag{VI.38}$$

for nonlinear molecules.

The susceptibility values $x_{\parallel}$ and $x_{\perp}$ in (VI.38) are arranged to yield $(x_{\parallel} + 2x_{\perp})/3$ and $(x_{\perp} - x_{\parallel})/3$, which are the scalar and tensorial constants, respectively.[36]

The nuclear Zeeman effect is given by

$$\tilde{\mathfrak{H}}_H^I(\alpha) = -\beta_N g_\alpha\{[1 - (\tilde{\sigma}_{\parallel}(\alpha) + 2\tilde{\sigma}_{\perp}(\alpha))/3]\boldsymbol{I}_\alpha \mathrm{H} + \tfrac{2}{3}(\tilde{\sigma}_{\perp}(\alpha) - \sigma_{\parallel}(\alpha))(3I_{\alpha z}^2 H_z - \mathrm{I}_\alpha \mathrm{H})\} \tag{VI.39}$$

for linear molecules where g_α is the free nuclear g-factor and $\tilde{\sigma}_{\parallel}, \tilde{\sigma}_{\perp}$ are the diamagnetic screening constants.

For nonlinear molecules we have

$$\tilde{\mathfrak{H}}_H^I(\alpha) = -\beta_N g_\alpha\left(\mathrm{I}_\alpha \mathrm{H} - \sum_{\mu,\nu = x,y,z} \tilde{\sigma}_{\mu\nu}(\alpha) I_{\alpha\mu} H_\nu\right) \tag{VI.40}$$

The angular momentum carried by the degenerate vibrations may produce a further Zeeman effect, the vibrational Zeeman effect, which for a linear triatomic may be written as

$$\tilde{\mathfrak{H}}_H^l = -\beta_N \tilde{g}_l G_z H_z \tag{VI.41}$$

For the translating molecule the magnetic field produces a Stark effect which has the operator form

$$\tilde{\mathfrak{H}}_H^{\mathrm{tr}} = -\underline{\mu}(1/c)(\dot{\mathrm{Q}}_0 \times \mathrm{H}) \tag{VI.42}$$

or for linear molecules

$$\tilde{\mathfrak{H}}_H^{\mathrm{tr}} = -\underline{\mu_z}(1/c)(\dot{\mathrm{Q}}_0 \times \mathrm{H})_z \tag{VI.43}$$

with

$$\mathbf{U}=e^{i\lambda\mathbf{S}}=\mathbf{1}+i\lambda\mathbf{S}-\tfrac{1}{2}\lambda^2\mathbf{S}^2+\cdots \tag{VII.4}$$

where **1** is the unit matrix.

Inserting (VII.4) into (VII.3) and ordering according to powers of λ, we obtain

$$\overline{\mathbf{H}}_0=\mathbf{H}_0 \tag{VII.5}$$

$$\overline{\mathbf{H}}_1=\mathbf{H}'+i(\mathbf{H}_0\mathbf{S}-\mathbf{S}\mathbf{H}_0) \tag{VII.6}$$

$$\overline{\mathbf{H}}_2=\mathbf{S}\mathbf{H}_0\mathbf{S}+i(\mathbf{H}'\mathbf{S}-\mathbf{S}\mathbf{H}')-\tfrac{1}{2}(\mathbf{S}^2\mathbf{H}_0+\mathbf{H}_0\mathbf{S}^2) \tag{VII.7}$$

up to the second order.

We demand

$$\langle\gamma'|\langle n'|\overline{\mathfrak{H}}_1|n_k\rangle|\gamma\rangle=0 \tag{VII.8}$$

$$\langle\gamma''|\langle n''|S|n'\rangle|\gamma'\rangle=0 \tag{VII.9}$$

$$\langle\gamma'|\langle n_k|S|n_l\rangle|\gamma\rangle=0 \tag{VII.10}$$

for $n_k,n_l\in n_1,n_2$ and $n',n''\notin n_1,n_2$ to be fulfilled and obtain, from (VII.6),

$$\langle\gamma'|\langle n'|S|n_k\rangle|\gamma\rangle=-i\frac{\langle n'|\mathfrak{H}'|n_k\rangle}{V_{n_k}-V_{n'}}$$

$$\langle\gamma|\langle n_k|S|n'\rangle|\gamma'\rangle=i\frac{\langle n_k|\mathfrak{H}'|n'\rangle}{V_{n_k}-V_{n'}} \tag{VII.11}$$

Substituting the results of (VII.9) to (VII.11) into (VII.7) leads to

$$\begin{aligned}\langle\gamma'|\langle n_k|\overline{\mathfrak{H}}_2|n_l\rangle|\gamma\rangle &= \sum_{n''\gamma''}\langle\gamma'|\langle n_k|\mathfrak{H}'|n''\rangle|\gamma''\rangle\langle\gamma''|\langle n''|\mathfrak{H}'|n_l\rangle|\gamma\rangle\\ &\quad\times\tfrac{1}{2}\left[\left(V_{n_k}-V_{n''}\right)^{-1}+\left(V_{n_l}-V_{n''}\right)^{-1}\right]\end{aligned} \tag{VII.12}$$

For use in Section VIII we introduce the notation for the sum of the energy-difference denominators,

$$\overline{\Delta V}_{n''}^{-1}=\tfrac{1}{2}\left[\left(V_{n_k}-V_{n''}\right)^{-1}+\left(V_{n_l}-V_{n''}\right)^{-1}\right] \tag{VII.13}$$

As a result of the foregoing discussion we obtain, in addition to the perturbation corrections to the adiabatic potentials, the correction to the transition functions between the states to be treated explicitly. Such transition functions generally are important only for a certain domain in the nuclear configuration space, defined by the near degeneracies. The main nonadiabatic couplings are the kinetic vibrational operators (Section VIII.A) and the spin–orbit coupling (Section VIII.B). They may couple two states simultaneously or, because of their different symmetry, only one of them may couple the states. However, for states which are not coupled either by the vibrational operators or by the spin–orbit operator, other coupling operators, which are usually negligible, may become important, such as the spin–spin coupling or the rotational coupling.

A. Symmetry Notation

The more explicit presentation of various molecular property functions in Section VIII requires some explanations of the molecular symmetry to be used.

As stated in Section IV, the nuclear permutation-inversion group describes the key characteristics of molecular symmetry behavior. Therefore, in the formal approach we make use of the transformation properties of states and operators with respect to the MS group and of its subgroups.

The approach within the adiabatic separation (i.e., integrating the Hamiltonian with the adiabatic electronic functions only over the electronic coordinates) makes it necessary also to use the instantaneous point-group symmetry, which is defined as the group of transformations of internal coordinates and electronic and nuclear positions, which leave the momentary nuclear positions invariant. Thus we may use transformation properties with respect to electronic coordinates only, which determine the symmetry-required vanishing (or not) of integrals and also symmetry-required factors expressed by group coupling coefficients (see below). Therefore, in Section VIII we will classify the spatial part of the electronic state functions and operators with respect to the instantaneous point group $G(\mathfrak{R})$, which in general will be different from the MS group describing the global molecular symmetry. The symbol T_a^Γ is to denote the ath component of a basis corresponding to the irreducible representation Γ (if Γ is one-dimensional, a may be suppressed). The associated component in scalar products will be expressed as $(T_a^\Gamma)^\dagger$, which is given by T_a^Γ for a real basis and by $(-1)^a T_a^\Gamma$ for a complex basis of the irreducible representations. In general, we may write, therefore, $(T_a^\Gamma)^\dagger = \varepsilon(a) T_{\hat{a}}^\Gamma$ with $\varepsilon(a) = (-1)^a$ for a complex, $\varepsilon(a) = +1$ for a real basis and $\hat{a} = (-a)$ for a complex and $\hat{a} = a$ for a real basis. For nearly all nuclear configurations, real Cartesian component functions are adequate. In Appendix 3 we give some relationships between real (Cartesian) and complex basis functions classified according to the irreducible representations of the point groups and the irreducible tensor notation for the three-dimensional rotation group. The angular momentum and spin operators, which are classified with respect to finite or infinite point groups, can be identified with appropriate irreducible tensor components or linear combinations of them:

$$T_a^{\Gamma}(\mathrm{j})_r = S_{q\,(a,r)}^k(\mathrm{j}) = \sum_m c_{qm} T_m^k(\mathrm{j}) \tag{VII.14}$$

where j is taken as a representative of a general angular momentum operator.

In Section VIII we will prefer the S_q^k-notation for these operators. The associated operator in scalar products, $(S_q^k)^{\dagger}$, is denoted by $\varepsilon(q)S_{\hat{q}}^k$ with $\hat{q} = -q$ for a complex basis and $\hat{q} = q$ for a real basis and $\varepsilon = \pm 1$, as described above. In the same way we may define eigenfunctions of angular momentum and spin operators to form a basis for the irreducible representations of the point groups,

$$|j(\Gamma a)_r\rangle = \sum_m c_{qm}(r)|jm\rangle \tag{VII.15}$$

where $|jm\rangle$ is an eigenfunction of the angular momentum operators, transforming according to the irreducible representations of the three-dimensional rotation group SO(3) [or SU(2)], and r is a label to distinguish between functions of the same symmetry (see also Appendix 3). This procedure is well known, for example, when constructing the symmetrized functions of an asymmetric rigid rotor by a Wang transformation from the symmetric rotor basis, where the underlying point group is the four-group D_2.

In order to proceed within the manner described above, it will be necessary to resolve the irreducible representation of a higher group into those of the subgroup corresponding to the change in the nuclear configuration. (For the necessary correlation tables, see Herzberg.[39])

Although we need at times the instantaneous point group $G(\mathfrak{R})$, the higher-order molecular symmetry group (MS), which for "rigid molecules" is identical to the point group of highest symmetry, is generally useful. The information we get from the MS group is that the integrals as a function of the nuclear displacements coordinates may be classified with respect to its irreducible representations (see Section IV).

The integrals will transform as indicated by the direct product of the states and operators involved, all classified with respect to the MS group. The consequences are easily understood for situations where all states and operators are basis functions of a nondegenerate representation. Then we will get for a matrix element

$$\Gamma_{\text{int}} = \Gamma_{\text{st}_1} \otimes \Gamma_{\text{st}_2} \otimes \Gamma_{\text{op}}$$

and for a second-order perturbation sum

$$\Gamma_{\text{int}} = \Gamma_{\text{st}_1} \otimes \Gamma_{\text{st}_2} \otimes \Gamma_{\text{op}_1} \otimes \Gamma_{\text{op}_2}$$

where st stands for state and op for operator. The representations of the intermediate states over which the sum is performed are irrelevant here, since the sum is assumed to be complete. Examples: Let us consider the matrix elements of $\Sigma_i x_i$ for H_2O which has the equilibrium C_{2v}-configuration shown in Fig. 1. The operator $\Sigma_i x_i$ belongs to B_2-symmetry.

A matrix element $\langle \phi' A_2 | \Sigma_i x_i | \phi A_1 \rangle$ transforms according to $A_1 \otimes A_1 \otimes B_2 = B_2$ as a function of the nuclear coordinates. Thus we have only odd contributions of S_3 (see Fig. 1), which is compatible with B_2. In contrast, a matrix element $\langle \phi' A_1 | \Sigma_i x_i | \phi B_2 \rangle$ transforms according to $A_1 \otimes B_2 \otimes B_2 = A_1$, so that we can only have even contributions of S_3, which is compatible with A_1.

Unfortunately, for states belonging to degenerate species, we can state only that the integrals and also the second-order perturbation sums *may* contain all the different irreducible representations of the direct product of states and operators which are compatible with nuclear coordinate symmetry. One example is discussed in Section VIII.C.

If the totally symmetric representation is not contained in the direct product of the representations of the parts of the integrand, the integral will be zero at the reference conformation and small near it. Frequently, such integrals can be neglected, at least for small displacements. This is done in, and is the basis of, the Herzberg–Teller approximation.

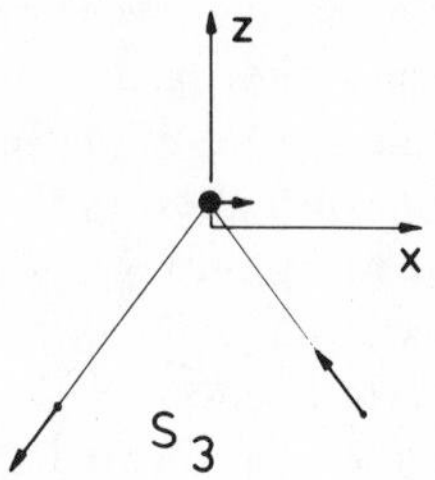

Fig. 1. S_3 coordinate shown for the H_2O molecule.

B. Coupling and Recoupling Coefficients for Symmetry Groups

In this work we use symmetry group coupling coefficients which are formally analogous to the Clebsch–Gordan coefficients of the three-dimensional rotation group.

Provided that $|u(\Gamma_1 a_1)\rangle$, $a_1=1,\ldots,n$, and $|v(\Gamma_2 a_2)\rangle$, $a_2=1,\ldots,m$, form a basis of the irreducible representations Γ_1 of dimension n and Γ_2 of dimension m of a group, then the product functions $|u(\Gamma_1 a_1)\rangle|v(\Gamma_2 a_2)\rangle$ with $a_1 \cdot a_2 = 1,\ldots,n\cdot m$, form a basis of a representation of dimension $n\cdot m$. If this representation is reducible, the reduction may be defined as[40]

$$|u(\Gamma_1 a_1)\rangle|v(\Gamma_2 a_2)\rangle = \sum_{\Gamma a} (\Gamma_1 a_1, \Gamma_2 a_2|\Gamma a)^* |(u\Gamma_1 v\Gamma_2)w(\Gamma a)\rangle \tag{VII.16}$$

where $(\Gamma_1 a_1, \Gamma_2 a_2|\Gamma a)$ are the symmetry-group coupling coefficients. The inverse relation of (VII.16) is given by

$$|(u\Gamma_1 v\Gamma_2)w(\Gamma a)\rangle = \sum_{a_1}\sum_{a_2} (\Gamma_1 a_1, \Gamma_2 a_2|\Gamma a)|u(\Gamma_1 a_1)\rangle|v(\Gamma_2 a_2)\rangle \tag{VII.17}$$

The values of the coupling coefficients depend on the choice of the basis of the irreducible representations and are defined except for a sign for each trio Γ_1, Γ_2, Γ_3 in $(\Gamma_1 a_1, \Gamma_2 a_2|\Gamma a)$. Griffith[40a] has constructed V-coefficients, which are analogous to the $3j$-symbols of Wigner[41] for the three-dimensional rotation group. Although the V-coefficients have a higher symmetry, we prefer using the coupling coefficients. For more details, see Appendix 2.

The Wigner–Eckart theorem,[42] which permits the extraction of subspecies-independent reduced matrix elements, allows us to define

$$\langle v(\Gamma_2 a_2)|T_a^\Gamma(\text{op})|u(\Gamma_1 a_1)\rangle = (\Gamma_1 a_1, \Gamma a|\Gamma_2 a_2)^* \langle v\Gamma_2\|T^\Gamma(\text{op})\|u\Gamma_1\rangle \tag{VII.18}$$

where $T_a^\Gamma(\text{op})$ is the operator classified according to symmetry species Γ and subspecies a and the double bars signify the reduced matrix element. This is analogous to the familiar use of the Wigner–Eckart theorem in the three-dimensional rotation group[43]

$$\langle j_2 m_2|\tilde{T}_m^k(\text{op})|j_1 m_1\rangle = \langle j_1 k m_1 m|j_2 m_2\rangle^* \langle j_2\|\tilde{T}^k(\text{op})\|j_1\rangle \tag{VII.19}$$

where $\tilde{T}^k_m(\text{op})$ is the mth compound of the irreducible tensor operator of rank k and $\langle j_1 k m_1 m | j_2 m_2 \rangle$ is a Clebsch–Gordan coefficient. In the standard basis used, the C-G coefficients are real and the * in (VII.19) is irrelevant. However, we will need at times nonstandard Cartesian basis functions, in this case the C-G coefficients become complex (see Appendix 3).

Another relation needed is for the recoupling of the product of two coupling coefficients. For this we use a definition of W-coefficients which is analogous to those of the Racah coefficients[41b,43,44] for the three-dimensional rotation group (see Appendix 2). Thus we get

$$(\Gamma_1 a_1, \Gamma_2 a_2 | \Gamma' a')(\Gamma' a', \Gamma_3 a_3 | \Gamma a) = \sum_{\Gamma'' a''} (\Gamma_2 a_2, \Gamma_3 a_3 | \Gamma'' a'')(\Gamma_1 a_1, \Gamma'' a'' | \Gamma a) \times W(\Gamma_1 \Gamma_2 \Gamma \Gamma_3; \Gamma' \Gamma'') \sqrt{\lambda(\Gamma')\lambda(\Gamma'')} \tag{VII.20}$$

with $\lambda(\Gamma)$ the dimension of the irreducible representation Γ. This is analogous to the standard definition in angular momentum theory given by

$$\langle j_1 j_2 m_1 m_2 | j' m' \rangle \langle j' j_3 m' m_3 | jm \rangle = \sum_{j'' m''} \langle j_2 j_3 m_2 m_3 | j'' m'' \rangle \langle j_1 j'' m_1 m'' | jm \rangle \times \tilde{W}(j_1 j_2 j j_3; j' j'') \sqrt{(2j'+1)(2j''+1)} \tag{VII.21}$$

The two relations above are needed to recouple the symmetry coupling coefficients in perturbation sums.

A more detailed exposition of these coupling coefficients and a comparison with other notations is presented in Appendix 2. For the evaluation of some matrix elements in Section VIII we use the fact that a space fixed component of an irreducible tensor operator $\tilde{T}^k_{m'}$ is related to the molecular fixed components $\tilde{T}^k_m$ by

$$\tilde{T}^k_{m'} = \sum_m \tilde{T}^k_m D^{k*}_{m'm}(\alpha, \beta, \gamma) \tag{VII.22}$$

where $D^k_{m'm}$ are the irreducible representations of the three-dimensional rotation group, parametrisized by the Euler angles α, β, γ. The rotational functions $|\text{NKM}_N\rangle$ are given by[1h]

$$|NKM_N\rangle = \sqrt{(2N+1)/8\pi^2}\, D^{N*}_{M_N K} \tag{VII.23}$$

Using the result for the integral over three $D^k_{m'm}$'s[41a] we obtain[1h]

$$\langle N'K'M'_N|D^{k*}_{m'm}|NKM_N\rangle = \sqrt{(2N'+1)/(2N+1)}\ \langle NkKm|N'K'\rangle \times \langle NkM_N m'|N'M'_N\rangle \tag{VII.24}$$

Similar equations hold for $|JPM_J\rangle$. Equations (VII.22) to (VII.24) may be generalized to the nonstandard components, which means that in (VII.24) the complex-conjugate coefficients have to be used. The formulas for the matrix elements of irreducible tensor operators in coupled basis functions or of coupled operators in an uncoupled basis used sometimes in Section VIII are presented in a convenient form by Carrington, Levy, and Miller.[1h]

VIII. EFFECTIVE OPERATORS AND EFFECTIVE CONSTANTS

It is well known that spectroscopic data are interpreted in terms of effective operators (see Section VI). They are designed to describe the molecular system effectively in the sense that their eigenvalue spectrum reproduces part of the spectrum of the total molecular Hamiltonian. These effective operators operate in the restricted space of orientational coordinates. In general, their eigenvalues are obtained by setting up the energy matrix in an appropriate basis of angular momentum and spin functions. This is a rather standard method and will not be treated here in detail. (For diatomics, see Carrington and Levy, and Miller.[1h]) As far as possible, we try to find effective operators which are analogs to the spectroscopic ones, but with effective constants as $\mathfrak{R}$-dependent.

This is not always possible, in particular if electronic transition matrix elements over spin operators are involved. Even here the concept of $\mathfrak{R}$-dependent constants remains useful. However, the matrix elements, which are in general a product of reduced matrix elements, independent of the angular momentum basis, and basis-dependent structure factors cannot be replaced here by $\mathfrak{R}$-dependent effective constants multiplied by integrals over basis-independent effective operators. For such transition matrix elements it is necessary to determine the basis-dependent structure coefficients explicitly, which is not required in other cases. (Examples of this are given in Sections VIII.B to VIII.D.)

We may summarize:

1. There are adiabatic state functions that can be expressed as effective operators acting within this electronic state.

2. There are nonadiabatic transition functions that cannot be expressed in terms of effective operator replacements.

Both contain an effective constant part and a structural part dependent on the symmetry considered.

A. Vibronic Interactions

In this first section of applying our ideas to the molecular system, we are concerned with the vibronic interactions. To separate them from the other interactions present in the molecule, we may introduce the hypothetical assumption that the molecule does not have any angular momentum or is in its ground state $J=0$ and that there are no couplings with the electronic spin, or $S=0$.

All those couplings that we do not consider here are treated in the following sections. Thus the remaining system to be described here is defined by the nonrelativistic Hamiltonian

$$\mathfrak{H} = -(\hbar^2/2m)\Sigma_i\nabla_i^2 - (\hbar^2/2M_N)(\Sigma_i\nabla_i)^2 + V_{\text{coul}}(\mathfrak{R}, \mathrm{r})$$
$$+\Sigma_k P_k^2 \qquad \text{(VIII.1)}$$

where k runs over the internal symmetry coordinates, $P_k = -i\hbar\partial/\partial S_k$ with S_k the kth symmetry coordinate; $\mathfrak{R}$ is the set of internal coordinates; and M_N is the total nuclear mass. All coordinates are molecular-frame-fixed. The second term is the mass-polarization correction to the electronic Hamiltonian.

From the adiabatic separation (see Sections II, III, and V) we get the potential surface $V(\mathfrak{R})$ as the solution of the electronic Schrödinger equation defined in (III.1). Clearly, the adiabatic potential function is the dominant part which enters into the nuclear motion problem.

We may define two types of vibronic interactions depending on the nuclear conformations considered:

1. The adiabatic approximation is valid for the electronic state, denoted by ϕ, Γ_e, S over the whole range of the nuclear conformation space considered. The vibronic state results just from the insertion of $V(\mathfrak{R})$ together with small, commonly neglected, first-order contributions from the mass-polarization correction and the electronic expectation value of the vibrational kinetic operator into the vibrational equation

$$(\Sigma_k P_k^2 + V(\mathfrak{R}) + \Delta V(\mathfrak{R}) - E)\Pi(\mathfrak{R}) = 0 \qquad \text{(VIII.2)}$$

The vibronic eigenfunction is given by

$$\Psi(\mathfrak{R}, \mathrm{r}) = \Pi(\mathfrak{R})\psi(\mathfrak{R}, \mathrm{r}) \tag{VIII.3}$$

The correction $\Delta V(\mathfrak{R})$ in (VIII.2) is given by

$$\Delta V(\mathfrak{R}) = \Sigma_k < \phi\Gamma_e a_e S | [P_k^2 | \phi\Gamma_e a_e S \rangle]$$
$$- (\hbar^2/2M_N)\langle\phi\Gamma_e a_e S | \Sigma_i \nabla_i^2 | \phi\Gamma_e a_e S\rangle \tag{VIII.4}$$

plus relativistic corrections considered in Section VIII.B.

In (VIII.4) and the following matrix elements the spin projection quantum numbers are omitted since they do not enter here. The square brackets denote a restricted operation on the parametric dependent electronic function. In principle, also in this approximation, second-order nonadiabatic terms from the coupling off-diagonal in the electronic states, given by

$$\sum_k \left(\langle\phi'\Gamma_e' a_e' S | [P_k | \phi\Gamma_e a_e S\rangle] P_k + \langle\phi'\Gamma_e' a_e' S | [P_k^2 | \phi\Gamma_e a_e S\rangle]\right) \tag{VIII.5}$$

could be introduced as an additional small correction to $V(\mathfrak{R})$ with a modification of the kinetic vibrational operator $\Sigma_k P_k^2$ (see Bunker and Moss[45]). This case is the one adopted in nearly all spectroscopic investigations.

2. The adiabatic approximation does not hold in cases of near degeneracy of electronic eigenvalues $V(\mathfrak{R})$ over a certain region of $\mathfrak{R}$. Then we have to consider the potential-coupling terms explicitly. These couplings will be dominated by matrix elements of the type

$$\sum_k \langle\phi'\Gamma_e' a_e' | [P_k | \phi\Gamma_e a_e\rangle] P_k \tag{VIII.5$'$}$$

Since the vibrational normal coordinates possess a definite symmetry with respect to the molecular symmetry group as well as to the instantaneous point groups, there are selection rules for the vanishing or not of the term (VIII.5). These couplings will give rise to coupled equations (compare Section V) yielding the vibronic eigenstates.

We do not consider an intermediate case between (1) and (2). If the couplings are strong enough to give an appreciable influence, we use case (2).

We should note some "strange" properties of the integral involving the kinetic vibrational operators:

(i)
$$\langle\psi'|\mathfrak{H}_{v-kin}|\psi\rangle_{\mathrm{r}} = \langle\psi'|\sum_k P_k^2|\psi\rangle_{\mathrm{r}}$$
$$= \sum_k \{\delta_{\psi'\psi}P_k^2 + 2\langle\psi'|[P_k|\psi\rangle_{\mathrm{r}}]P_k + \langle\psi'|[P_k^2|\psi\rangle_{\mathrm{r}}]\} \quad \text{(VIII.6)}$$

(ii)
$$\langle\psi'|[P_k|\psi\rangle_{\mathrm{r}}] = \langle\psi|[P_k|\psi'\rangle_{\mathrm{r}}^*] \quad \text{(VIII.7)}$$

where * denotes the complex-conjugate value:

(iii)
$$\langle\psi'|[P_k^2|\psi\rangle_{\mathrm{r}}] = \langle\psi|[P_k^2|\psi'\rangle_{\mathrm{r}}] + 2[P_k\langle\psi'|[P_k|\psi\rangle_{\mathrm{r}}]] \quad \text{(VIII.8)}$$

Equation (VIII.8) tells us that P_k^2 is not a Hermitian operator in the electronic function basis alone. Only after integrating with a nuclear coordinate function over the nuclear coordinates is the Hermiticity of the operator P_k^2 also regained. Property (iii) will introduce some complications in the formulations of the couplings. However, the couplings (iii) between different states are of minor interest, compared to the couplings (ii). In addition, we note that property (iii) is also present if we consider integrals of the type

$$\langle\psi'|P_kA|\psi\rangle_{\mathrm{r}} \quad \text{(VIII.9)}$$

where A is a pure electronic operator.

Then we have

$$\langle\psi'|[P_kA|\psi\rangle_{\mathrm{r}}] = \langle\psi|[P_kA|\psi'\rangle_{\mathrm{r}}] + [P_k\langle\psi'|A|\psi\rangle_{\mathrm{r}}] \quad \text{(VIII.10)}$$

Before going into more detail, it is worth noting that the couplings defined in this section are not sufficient to determine the vibronic states fully. In general, this is possible only after introducing th relativistic spin–orbit and possibly some other couplings to be considered in the next sections.

For polyatomic molecules whose molecular symmetry groups have only nondegenerate irreducible representations, there are no symmetry-induced degeneracies; other degeneracies may take place by chance. Nevertheless, conditions for near-degeneracies, intersections of the Jahn–Teller type,[46] or crossings of surfaces of different symmetry should be quite common, especially in excited states.

The symmetry condition for vibronic coupling will be: The direct product of the irreducible representations of the functions and the coupling normal coordinate must contain the totally symmetric representation Γ_0:

$$\Gamma_e' \otimes \Gamma_e \otimes \Gamma(S_k) \supset \Gamma_0 \qquad \text{(VIII.11)}$$

More interesting are the molecules of higher symmetry, with at least threefold axial, octrahedral, or tetahedral symmetry. Here we have in addition to the already mentioned cases the possibility of symmetry-required degeneracies of electronic state function. These degeneracies will be lifted by non-totally symmetric nuclear displacements. This behavior is well known for the triatomic linear molecule in the electronic Π-state, which is split by the degenerate Π-bending mode (i.e., the Renner–Teller effect). In nonlinear molecules we have the Jahn–Teller effect. In Figs. 2 and 3 we present examples of potential surfaces of the Renner–Teller and Jahn–Teller types of potentials for triatomic molecules, where the nonadiabatic behavior of the system, at least near the reference configuration, is quite obvious. The following discussion is restricted to the triatomic linear molecule of the Renner–Teller type (Fig. 2).

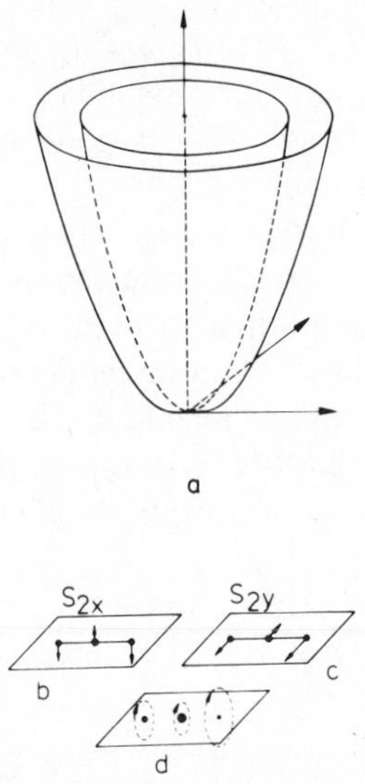

Fig. 2. (*a*) Adiabatic potential surfaces with a Renner–Teller touching in the linear conformation for a linear triatomic molecule in the space of the two degenerate symmetry coordinates S_{2x} and S_{2y}, the bending vibrational coordinates. In (*b*) and (*c*) the coordinates S_{2x} and S_{2y} are explained. The internal angular momentum associated with the degenerate vibration is demonstrated in (*d*).

1. Triatomic Linear Molecule

The normal coordinates of the triatomic linear molecule are $S_1(\Sigma^+)$, S_{2x}, $S_{2y}(\Pi)$, and $S_3(\Sigma^+)$. The coordinates S_{2x} and S_{2y}, forming a basis for the Π irreducible representation, may conveniently be transformed to cylindrical polar coordinates S_2, φ_2 by

$$S_2 e^{\pm i\varphi_2} = S_{2x} \pm iS_{2y} \tag{VIII.12}$$

This choice is very helpful regarding the invariance of the potential function $V(\mathfrak{R})$ with respect to φ_2, which follows from the cylindrical symmetry of the linear molecule in its reference configuration. At this point we should give some explanation of the angles involved in linear molecules, although we do not consider rotational dependence in this section. We define the angles φ and φ_2, where φ is the angle of the whole electronic system and φ_2 the coordinate from (VIII.12). Both φ and φ_2 are with respect to an arbitrary plane containing the internuclear axis in the reference conformation. This plane defines the third Euler angle γ with respect to the laboratory fixed frame.

This enables us to introduce the factor $(1/\sqrt{2\pi})e^{iK\gamma}$ as a pure rotational function part, where K is the projection of $\mathbf{N}$, the angular momentum of the overall rotation, onto the molecular fixed symmetry axis z and N_z is given by $N_z = L_z + G_z$, where L_z and G_z are the z-components of the electronic and vibrational angular momentum operators, respectively (see also Section VIII.C). Although K is a good quantum number for pure vibronic interactions, Λ is not.

As shown in Fig. 2, an electronic degenerate state, without loss of generality a Π-state, is split by the degenerate bending mode. However, it is not necessary for the absolute minima to lie at $S_2 = 0$. The lower curve may have deeper minima outside.[1c] If these minima are fairly deep, the molecule should not be considered as linear; it is a bent molecule of C_s or C_{2v} point group symmetry. If the barrier between the minima are small compared to vibrational energies, we speak of quasi-linear molecules of $C_{\infty v}(D_{\infty h})$ symmetry.

Let us denote the two potential functions of Fig. 2 by $V_1(S_1S_2S_3)$ and $V_2(S_1S_2S_3)$, both independent of φ_2. The integrals $\langle\psi_p|\mathfrak{H}_{v\text{-kin}}|\psi_q\rangle$ for $p, q = 1,2$ and

$$\mathfrak{H}_{v\text{-kin}} = -\hbar^2\left(\sum_{\mu=1}^{3}\frac{\partial^2}{\partial S_\mu^2} + \frac{1}{S_2}\frac{\partial}{\partial S_2} + \frac{1}{S_2^2}\frac{\partial^2}{\partial\varphi_2^2}\right) \tag{VIII.13}$$

may be evaluated, yielding

$$
\begin{aligned}
\langle\psi_p|\mathfrak{H}_{v\text{-kin}}|\psi_q\rangle_{\mathfrak{r}} = \delta_{pq}\mathfrak{H}_{v\text{-kin}}\\
-\hbar^2\Bigg\{ 2\sum_\mu \langle\psi_p|\left[\frac{\partial}{\partial S_\mu}|\psi_q\rangle_{\mathfrak{r}}\right]\frac{\partial}{\partial S_\mu} + \frac{1}{S_2}\langle\psi_p|\frac{\partial}{\partial S_2}|\psi_q\rangle_{\mathfrak{r}}\\
+\sum_\mu \langle\psi_p|\left[\frac{\partial^2}{\partial S_\mu^2}|\psi_q\rangle_{\mathfrak{r}}\right]\delta_{pq}\\
+2\frac{1}{S_2^2}\langle\psi_p|\left[\frac{\partial}{\partial\varphi^2}|\psi_q\rangle_{\mathfrak{r}}\right]\frac{\partial}{\partial\varphi_2}\\
+\frac{1}{S_2^2}\langle\psi_p|\left[\frac{\partial^2}{\partial\varphi_2^2}|\psi_q\rangle_{\mathfrak{r}}\right]\delta_{pq}\Bigg\}
\end{aligned}
\tag{VIII.14}
$$

The last two δ_{pq}-terms arise because a totally symmetric operator may not combine the functions ψ_1 and ψ_2, which are symmetric and antisymmetric (or vice versa) with respect to the reflection at the plane containing the linear reference configuration and the displaced nuclei. On the other hand $\langle\psi_p|[\partial/\partial S_\mu|\psi_p\rangle_{\mathfrak{r}}]$ for $\mu=1, 2, 3$ and $\langle\psi_p|[\partial/\partial\varphi_2|\psi_p\rangle_{\mathfrak{r}}]$ for $p=1, 2$ vanish. All these results remain valid for $S_2\to 0$, where ψ_1 and ψ_2 become degenerate.

The vibronic eigenfunctions

$$
\Psi_t(\mathfrak{R},\mathfrak{r}) = \sum_{s=1}^{2} \pi_s^t(\mathfrak{R})\psi_s(\mathfrak{R},\mathfrak{r}) \tag{VIII.15}
$$

and the eigenvalues E_r are obtained from the coupled equation (neglecting spin–orbit effects for the moment)

$$
\begin{bmatrix} \langle\psi_1|\mathfrak{H}_{v\text{-kin}}|\psi_1\rangle_{\mathfrak{r}} + V_1 - E & \langle\psi_1|\mathfrak{H}_{v\text{-kin}}|\psi_2\rangle_{\mathfrak{r}} \\ \langle\psi_2|\mathfrak{H}_{v\text{-kin}}|\psi_1\rangle_{\mathfrak{r}} & \langle\psi_2|\mathfrak{H}_{v\text{-kin}}|\psi_2\rangle_{\mathfrak{r}} + V_2 - E \end{bmatrix} \begin{bmatrix} \Pi_1 \\ \Pi_2 \end{bmatrix} = \begin{bmatrix} 0 \\ 0 \end{bmatrix} \tag{VIII.16}
$$

using (VIII.14).

Although we do not want to restrict ourselves to small nuclear displacements, we may introduce the "spectroscopic approximations" to give some feeling for the solutions in the case of small oscillations. These approximations are

(a) $$V_s(S_1,S_2,S_3) = \tfrac{1}{2}\left(k_1S_1^2 + (k_2 \pm \delta)S_2^2 + k_3S_3^2\right) \tag{VIII.17}$$

with $s=1,2$ and $+/-$ for $s=1/2$;

(b)
$$\langle\psi_1|\frac{\hbar}{i}\frac{\partial}{\partial\varphi_2}|\psi_2\rangle_{\mathrm{r}}=-\langle\psi_2|\frac{\hbar}{i}\frac{\partial}{\partial\varphi_2}|\psi_1\rangle_{\mathrm{r}}=i\Lambda \qquad \text{(VIII.18)}$$

where Λ is the expectation value of L_z in the linear configuration [note that $(\hbar/i)(\partial/\partial\varphi_2)\equiv L_z$ in the linear conformation]. Within these approximations (VIII.16) is decoupled into normal one-dimensional harmonic oscillation problems for S_1 and S_3 and into the coupled equation for S_2 and φ_2 alone. The latter becomes

$$\begin{bmatrix} d_+-E & a \\ a & d_--E \end{bmatrix}\begin{bmatrix} \chi_1(S_2,\varphi_2) \\ \chi_2(S_2,\varphi_2) \end{bmatrix}=\begin{bmatrix} 0 \\ 0 \end{bmatrix} \qquad \text{(VIII.19)}$$

with

$$d_\pm=-\hbar^2\left[\frac{\partial^2}{\partial S_2^2}+\frac{1}{S_2}\frac{\partial}{\partial S_2}+\frac{1}{S_2^2}\left(\frac{\partial^2}{\partial\varphi_2^2}-\Lambda^2\right)\right]+\tfrac{1}{2}(k_2\pm\delta)S_2^2$$

and

$$a=-2\hbar^2\frac{\Lambda}{S_2^2}\frac{\partial}{\partial\varphi_2}$$

Since the Hamiltonian (VIII.1) is invariant with respect to the symmetry operations of the point group $C_{\infty v}$, the vibronic functions Ψ_t form a basis for its irreducible representations. Therefore, we may adopt for the angular dependence the form $\Psi_t\sim\exp(iK\varphi_2)$, with K a good vibronic quantum number, so that the wave function, (VIII.15), can be expressed as

$$\Psi_t(S_1,S_2,S_3,\varphi_2,\mathrm{r})=\sum_{s=1,2} f(S_1)h(S_3)g_s(S_2)e^{iK\varphi_2}\psi_s \qquad \text{(VIII.20)}$$

The functions $g_s(S_2)$, $s=1,2$, are conveniently replaced by real functions η_+ and η_- through

$$g_1=(\eta_++\eta_-)/\sqrt{2} \qquad g_2=i(\eta_+-\eta_-)/\sqrt{2} \qquad \text{(VIII.21)}$$

Using the relations (VIII.20) with $\chi_s=g_s\exp(iK\varphi_2)$ and (VIII.21) the system of coupled equations, VIII.19) reduces to

$$\begin{bmatrix} d'_+-E & a' \\ a' & d'_--E \end{bmatrix}\begin{bmatrix} \eta_+(S_2) \\ \eta_-(S_2) \end{bmatrix}=\begin{bmatrix} 0 \\ 0 \end{bmatrix} \qquad \text{(VIII.22)}$$

with

$$d'_{\pm} = -\hbar^2\left(\frac{\partial^2}{\partial S_2^2} + \frac{1}{S_2}\frac{\partial}{\partial S_2} - \frac{l_{\pm}^2}{S_2^2}\right) + \tfrac{1}{2}k_2 S_2^2$$

and

$$a' = \tfrac{1}{2}\delta S_2^2$$

where $l_{\pm} = K \pm \Lambda$.

The coupled system (VIII.22) is equivalent to the formulation already presented by Renner in 1934.[22] The Renner splitting parameter ε used in spectroscopy is defined as

$$\varepsilon = \frac{\delta}{k_2} \qquad \text{(VIII.23)}$$

for $K=0$ we have $l_{\pm} = \pm\Lambda$ and the diagonal components of the matrix in (VIII.21) are the same. If (VIII.21) is retransformed with B, we get two decoupled equations, one for the lower and one for the upper curve of Fig. 2. These two solutions are labeled by Σ^+ and Σ^-, respectively.[22]

It is obvious from the list of approximations above that the theory applied usually to the Renner–Teller splitting is not suitable for an *ab initio* calculation, but is appropriate for a phenomenological description. Provided that the potential functions V_1 and V_2 are given together with their eigenfunctions ψ_1 and ψ_2, the coupled equations (VIII.16) should be solved directly.

In group theoretical language the vibronic functions with $K = 0, 1, 2, \ldots$ belong to the irreducible representations $\Sigma^{\pm}$, Π, $\Delta, \ldots$. In the same way, the electronic functions and vibrational basis functions form a basis of the irreducible representations of $C_{\infty v}$ dependent on their Λ and l values given at the reference configuration. Their direct product determines the species of the vibronic functions. For example, $|\Lambda| = 1$ and $|l| = 1$ will lead through $\Pi \otimes \Pi = \Sigma^+ + \Sigma^- + \Delta$ to three vibronic states, where two are nondegenerate and one is degenerate. The classification of the vibronic levels remains valid also for the bent configuration, where Λ and l are no longer good quantum numbers.

2. *Nonlinear Molecules*

In the case of electronically degenerate states of nonlinear molecules, there are no high-resolution spectroscopic studies, as far as we know, which reveal unambiguously the nonadiabatic effects expected due to Jahn–Teller splitting. The Jahn–Teller effect will always produce an equilibrium conformation that is not electronically degenerate. However,

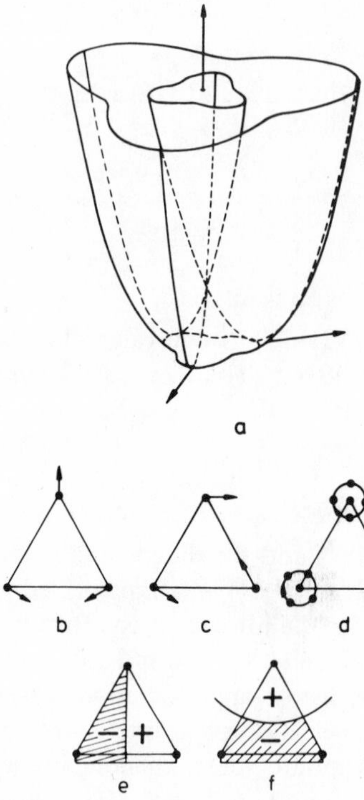

Fig. 3. (*a*) Adiabatic potential surface for a triatomic nonlinear molecule with a Jahn–Teller conical intersection in the space of the two degenerate symmetry coordinates S_{2x} and S_{2y}. These two coordinates $S_{2x}S_{2y}$ are defined in (*b*) and (*c*). They may be combined to yield internal angular momentum, as shown in (*d*). The possible symmetries of two functions leading to the degeneracy at $S_2=0$ are shown in (*e*) and (*f*).

for high-resolution studies it is necessary to consider both branches of the intersecting surfaces (Fig. 3). In this case a coupled description of the nuclear motion analogous to (VIII.16) is necessary and possible. However, at the conical intersection of the potential surfaces, care is required since the derivative of the electronic functions ψ_1 and ψ_2 with respect to $\mathfrak{R}$ may be noncontinuous.[26]

B. Relativistic Effects (Spin–Orbit, Spin–Spin Coupling)

In this second section we include the interactions resulting from the relativistic corrections within the molecular Hamiltonian. These effects are included in the Pauli-type approximation, where all two particle interactions are described using the Breit operator for two electrons (see Howard and Moss[47] and literature cited therein). We focus our attention on molecules with light nuclei, where the relativistic effects, mainly the spin–orbit coupling, may be considered as a small perturbation relative to the electrostatic energies of the nonrelativistic Hamiltonian. We discuss here the following parts of the molecular Hamiltonian:

$$\mathfrak{H}_{\text{rel}} = \mathfrak{H}_{\text{SO}} + \mathfrak{H}_{\text{SS}}^{(0)} + \mathfrak{H}_{\text{SS}}^{(2)} \quad \text{(VIII.24a)}$$

$$+ \mathfrak{H}_{rm} + \mathfrak{H}_{\text{D}} + \mathfrak{H}_{\text{OO}} \quad \text{(VIII.24b)}$$

$$+ \mathfrak{H}_{\text{S}v} + \mathfrak{H}_{\text{O}v} \quad \text{(VIII.24c)}$$

$$+ \mathfrak{H}_{\text{S}r} + \mathfrak{H}_{\text{O}r} \quad \text{(VIII.24d)}$$

$$+ \mathfrak{H}_{\text{SI}} + \mathfrak{H}_{\text{OI}} \quad \text{(VIII.24e)}$$

The individual terms are detailed in Appendix 1, (A.46) through (A.57). The most important operators are given by (VIII.24a), that is, the spin–orbit, spin–other-orbit, and the spin–spin coupling operators.

The operators in (VIII.24a) and (VIII.24b) are purely electronic and thus have nothing to do with the separation of nuclear and electronic motions. This effect on the electronic eigenfunctions and eigenvalues is usually considered by perturbations to the zeroth-order solution of the nonrelativistic electronic equation, (III.1). Only for cases of surface crossing or near-degeneracies, which fulfill specific symmetry conditions for the spin quantum numbers S and the spatial electronic wavefunction, will the coupling of adiabatic states require a nonadiabatic explicit solution. For the vibronic interactions discussed in the preceding section, the spin–orbit interaction also has to be introduced into the coupling elements between the equations needed to generate the vibronic states. The operators in (VIII.24b), that is, $\mathfrak{H}_{\text{rm}}$, $\mathfrak{H}_{\text{D}}$, and $\mathfrak{H}_{\text{OO}}$, [see also Appendix 1, (A.50) to (A.52)] are all totally symmetric and small; thus they can be considered in first-order perturbation theory. They give relativistic corrections to the adiabatic potential $\Delta V_{\text{rel}}(\mathfrak{R})$ only, which may be added to the potential surface for very accurate calculations. The vibrational interactions with the electronic spins, $\mathfrak{H}_{\text{S}v}$, and with the electronic orbital motion, $\mathfrak{H}_{\text{O}v}$, are of little interest, since they do not have any first-order effect. They are omitted from the following discussion. Heretofore they have not been considered in analyzing experimental results. The relativistic interactions between electronic motion and molecular rotation of (VIII.24d) are considered in Section VIII.C, and the couplings with nuclear spins (VIII.24e) leading to hyperfine structure are treated in Section VIII.D.

Let us come back to the terms of (VIII.24a). The spin–orbit operator $\mathfrak{H}_{\text{SO}}$ may be written in abbreviated form, (A.46):

$$\mathfrak{H}_{\text{SO}} = \sum_i \mathsf{u}_i \cdot \mathsf{s}_i \quad \text{(VIII.25)}$$

where it is important to realize that u_i transforms as the electronic angular momentum operator l_i of the ith electron under the symmetry operations of the molecular symmetry group. The spin–spin operator $\mathfrak{H}_{SS}$ may be written in irreducible tensor notation[1h] as follows [see also (A.48) and (A.49)]:

$$\mathfrak{H}_{SS} = \mathfrak{H}_{SS}^{(0)} + \mathfrak{H}_{SS}^{(2)}$$

with

$$\mathfrak{H}_{SS}^{(0)} = -\sum_{i<j} \sqrt{3}\, \tilde{T}^0(\mathbf{s}_i, \mathbf{s}_j) D^0(i,j) \quad \text{(VIII.26a)}$$

$$\mathfrak{H}_{SS}^{(2)} = \sum_{m=-2}^{2} (-1)^m \sum_{i<j} \tilde{T}_m^2(\mathbf{s}_i, \mathbf{s}_j) D_{-m}^2(i,j) \quad \text{(VIII.26b)}$$

where $\tilde{T}_m^k$ are the irreducible tensor components built from the electronic spins and the D_m^k's transform as the irreducible tensor components built from electronic coordinate functions $[D^0(i,j) = (\beta^2 g^2 \pi/3)\delta(\mathbf{r}_{ij}),\ D_m^2(i,j) = \beta^2 g^2 \sqrt{6}\ C_m^2(\omega_{ij}) r_{ij}^{-3}]$.

The spin–orbit operator gives a first-order effect only for orbitally degenerate or at least near-degenerate electronic states. In addition, it may contribute significantly to other effects through its second-order terms, as will be demonstrated here and the following sections. In experimental work and theoretical analysis the first-order terms are considered only for linear molecules and for transition metal complexes in ligand field theory.[48]

In order to present the matrix elements of the spin–orbit and spin–spin operators, we have to choose a definite basis in one of the possible coupling cases. We will use for our formulas Hund's case (a), which is adequate for strong spin–orbit coupling relative to the rotational energy. After that we will discuss briefly the analogous formulas for Hund's case (b), which are generally better for states without a first-order spin–orbit effect or for light molecules with high rotational energies and small spin–orbit coupling energy. For matrix elements diagonal in the spin quantum number S, one may form operator replacements, which allow a description independent of the function basis coupling. However, this is no longer possible for matrix elements off-diagonal in S.

The second-order perturbation sums due to $\mathfrak{H}_{SO}$ make significant contributions to the effective spin–spin coupling term, whereas their contributions to the effective spin–orbit terms may be neglected for most calculations.

The spin–orbit coupling operator (VIII.25) may be written in symmetry notation as

$$\mathfrak{H}_{SO} = \sum_{\tilde{\Gamma}\tilde{a}r} \varepsilon(\tilde{a}) \sum_i T_{\tilde{a}}^{\tilde{\Gamma}}(\mathbf{u}_i)_r T_{\tilde{a}}^{\tilde{\Gamma}}(\mathbf{s}_i)_r \quad \text{(VIII.27)}$$

where $\varepsilon = (-)^{\tilde{a}}$ for complex representations and $+1$ otherwise. Writing the spin operator in terms of irreducible SO(3) tensor components yields

$$\mathfrak{H}_{\mathrm{SO}} = \sum_{\tilde{\Gamma}\tilde{a}r} \varepsilon(\tilde{a}) \sum_{i} T_{\tilde{a}}^{\tilde{\Gamma}}(\mathrm{u}_i)_r S_{\hat{q}(\tilde{a},r)}^{1}(\mathrm{s}_i) \qquad \text{(VIII.28)}$$

where $S_{\hat{q}}$ is identical to $T_{\tilde{a}}^{\tilde{\Gamma}}$ with $\hat{q} = -q$ for a complex basis and $\hat{q} = q$ for a real basis and $\varepsilon(\tilde{a})$ explained after (VIII.27).

Examples for (VIII.28) are, for $C_{\infty v}$,

$$\begin{aligned}\mathfrak{H}_{\mathrm{SO}} = \sum_{i} \big[& T^{\Sigma^-}(\mathrm{u}_i) \cdot S_0^1(\mathrm{s}_i) \\ & - T_{+}^{\Pi}(\mathrm{u}_i) \cdot S_{-1}^{1}(\mathrm{s}_i) \\ & - T_{-}^{\Pi}(\mathrm{u}_i) \cdot S_{+1}^{1}(\mathrm{s}_i) \big]\end{aligned}$$

with

$$T^{\Sigma^-}(\mathrm{u}_i) = u_{iz} \qquad S_0^1(\mathrm{s}_i) = s_{iz}$$

$$T_{\pm}^{\Pi}(\mathrm{u}_i) = \mp \frac{1}{\sqrt{2}} (u_{ix} \pm i u_{iy})$$

$$S_{\pm 1}^{1}(\mathrm{s}_i) = \mp \frac{1}{\sqrt{2}} (s_{ix} \pm i s_{iy})$$

and for 0 or T_d,

$$\mathfrak{H}_{\mathrm{SO}} = \sum_{\tilde{a} = x,y,z} T_{\tilde{a}}^{T_1}(\mathrm{u}_i) \cdot S_{q(\tilde{a})}^{1}(\mathrm{s}_i)$$

with

$$T_{x,y,z}^{T_1}(\mathrm{u}_i) = u_{ix,yz} \qquad S_{q(x,y,z)}^{1}(\mathrm{s}_i) = s_{ix,y,z}$$

Analogously, we have in symmetry notation for the components of the spin–spin operator

$$\mathfrak{H}_{\mathrm{SS}}^{(0)} = \sum_{i<j} \sqrt{3}\; T^{\Gamma_0}(D^0(i,j))\, T^{\Gamma_0}(\mathrm{s}_i, \mathrm{s}_j) \qquad \text{(VIII.29)}$$

$$= -\sum_{i<j} \sqrt{3}\; T^{\Gamma_0}(D^0(i,j))\, S_0^0(\mathrm{s}_i, \mathrm{s}_j) \qquad \text{(VIII.30)}$$

where Γ_0 is the totally symmetric irreducible representation and

$$\mathfrak{H}_{SS}^{(2)} = \sum_{\bar{\Gamma}\bar{a}r} \varepsilon(\bar{a}) \sum_{i<j} T_{\bar{a}}^{\bar{\Gamma}}(D^2(i,j))_r T_{\bar{a}}^{\bar{\Gamma}}(\mathbf{s}_i, \mathbf{s}_j)_r \tag{VIII.31}$$

$$= \sum_{\bar{\Gamma}\bar{a}r} \varepsilon(\bar{a}) \sum_{i<j} T_{\bar{a}}^{\bar{\Gamma}}(D^2(i,j))_r S_{\hat{q}(\bar{a},r)}^2(\mathbf{s}_i, \mathbf{s}_j) \tag{VIII.32}$$

with $\hat{q}$ as explained above and $\varepsilon(\bar{a}) = (-1)^{q(\bar{a})}$ for a complex basis and $+1$ for a real basis.

Examples for (VIII.32) are, for $C_{\infty v}$,

$$\begin{aligned}\mathfrak{H}_{SS}^{(2)} = \sum_{i<j} \big[& T^{\Sigma^+}(D^2 i,j)) S_0^2(\mathbf{s}_i, \mathbf{s}_j) \\ & - T_+^{\Pi}(D^2(i,j)) S_{-1}^2(\mathbf{s}_i, \mathbf{s}_j) \\ & - T_-^{\Pi}(D^2(i,j)) S_{+1}^2(\mathbf{s}_i, \mathbf{s}_j) \\ & + T_+^{\Delta}(D^2(i,j)) S_{-2}^2(\mathbf{s}_i, \mathbf{s}_j) \\ & + T_-^{\Delta}(D^2(i,j)) S_{+2}^2(\mathbf{s}_i, \mathbf{s}_j) \big]\end{aligned}$$

with

$$\begin{aligned} T^{\Sigma^+}(D^2(i,j)) &= D_0^2(i,j) \\ T_\pm^{\Pi}(D^2(i,j)) &= D_{\pm 1}^2(i,j) \\ T_\pm^{\Delta}(D^2(i,j)) &= D_{\pm 2}^2(i,j) \\ S_q^2(\mathbf{s}_i, \mathbf{s}_j) &= \tilde{T}_q^2(\mathbf{s}_i, \mathbf{s}_j) \qquad q = 0, \pm 1, \pm 2 \end{aligned}$$

and for 0 or T_d,

$$\begin{aligned}\mathfrak{H}_{SS}^{(2)} = \sum_{i<j} \big[& T_\theta^E(D^2(i,j)) S_0^2(\mathbf{s}_i, \mathbf{s}_j) \\ & + T_\varepsilon^E(D^2(i,j)) S_{(x^2-y^2)}^2(\mathbf{s}_i, \mathbf{s}_j) \\ & + T_\xi^{T_2}(D^2(i,j)) S_{xz}^2(\mathbf{s}_i, \mathbf{s}_j) \\ & + T_\eta^{T_2}(D^2(i,j)) S_{yz}^2(\mathbf{s}_i, \mathbf{s}_j) \\ & + T_\zeta^{T_2}(D^2(i,j)) S_{xy}^2(\mathbf{s}_i, \mathbf{s}_j) \big]\end{aligned}$$

with

$$T_{\theta}^{E}(D^2(i,j)) = D_0^2(i,j)$$

$$T_{e}^{E}(D^2(i,j)) = (1/\sqrt{2})(D_{+2}^2(i,j) + D_{-2}^2(i,j))$$

$$T_{\xi}^{T_2}(D^2(i,j)) = (1/\sqrt{2})(-D_{+1}^2(i,j) + D_{-1}^2(i,j))$$

$$T_{\eta}^{T_2}(D^2(i,j)) = (i/\sqrt{2})(D_{+1}^2(i,j) + D_{-1}^2(i,j))$$

$$T_{\zeta}^{T_2}(D^2(i,j)) = -(i/\sqrt{2})(D_{+2}^2(i,j) - D_{-2}^2(i,j))$$

$$S_0^2(\mathbf{s}_i, \mathbf{s}_j) = \tilde{T}_0^2(\mathbf{s}_i, \mathbf{s}_j)$$

$$S_{(x^2-y^2)}^2(\mathbf{s}_i, \mathbf{s}_j) = (1/\sqrt{2})(\tilde{T}_{+2}^2(\mathbf{s}_i, \mathbf{s}_j) + \tilde{T}_{-2}^2(\mathbf{s}_i, \mathbf{s}_j))$$

$$S_{xz}^2(\mathbf{s}_i, \mathbf{s}_j) = (1/\sqrt{2})(-\tilde{T}_{+1}^2(\mathbf{s}_i, \mathbf{s}_j) + \tilde{T}_{-1}^2(\mathbf{s}_i, \mathbf{s}_j))$$

$$S_{yz}^2(\mathbf{s}_i, \mathbf{s}_j) = (i/\sqrt{2})(\tilde{T}_{+1}^2(\mathbf{s}_i, \mathbf{s}_j) + \tilde{T}_{-1}^2(\mathbf{s}_i, \mathbf{s}_j))$$

$$S_{xy}^2(\mathbf{s}_i, \mathbf{s}_j) = -(i/\sqrt{2})(\tilde{T}_{+2}^2(\mathbf{s}_i, \mathbf{s}_j) - \tilde{T}_{-2}^2(\mathbf{s}_i, \mathbf{s}_j))$$

(See also Appendix 3.)
Using these operators, the effective matrix elements for spin–orbit and spin–spin coupling, including second-order contributions, may be expressed as follows for Hund's case (a) coupled functions (i.e., $|S\Sigma\rangle|JPM_J\rangle$ basis, where in general we omit $|JPM_J\rangle$, since here only diagonal elements are needed).

For spin–orbit coupling, we get

$$\begin{aligned}
&\langle \phi'(\Gamma_e' a_e') S'\Sigma' | \tilde{\mathfrak{H}}_{\mathrm{SO}} | \phi(\Gamma_e a_e) S\Sigma \rangle \\
&= \sum_{\tilde{\Gamma}\tilde{a}r} (\Gamma_e a_e, \tilde{\Gamma}\tilde{a}_e | \Gamma_e' a_e')^* \langle S 1 \Sigma \hat{q}(\tilde{a}, r) | S'\Sigma' \rangle^* \\
&\times \Bigg\{ \varepsilon(\tilde{a}) \langle \phi' \Gamma_e' S' \| \sum_i T^{\tilde{\Gamma}}(\mathbf{u}_i)_r S^1(\mathbf{s}_i) \| \phi \Gamma_e S \rangle + \sum_{\tilde{\Gamma}_1 \tilde{a}_1 r_1} \varepsilon(\tilde{a}_1) \sum_{\tilde{\Gamma}_2 \tilde{a}_2 r_2} \varepsilon(\tilde{a}_2) \\
&\times \sum_{\phi''\Gamma_e'' S''} \overline{\Delta V}^{-1}_{\phi''\Gamma_e'' S''} \langle \phi' \Gamma_e' S' \| \sum_i T^{\tilde{\Gamma}_1}(\mathbf{u}_i)_{r_1} S^1(\mathbf{s}_i) \| \phi'' \Gamma_e'' S'' \rangle \\
&\times \langle \phi'' \Gamma_e'' S'' | \sum_i T^{\tilde{\Gamma}_2}(\mathbf{u}_i)_{r_2} S'(\mathbf{s}_i) \| \phi \Gamma_e S \rangle (\tilde{\Gamma}_2 \tilde{a}_2, \tilde{\Gamma}_1 \tilde{a}_1 | \tilde{\Gamma} \tilde{a})^* \\
&\times W(\Gamma_e \tilde{\Gamma}_2 \Gamma_e' \tilde{\Gamma}_1 ; \Gamma_e'' \tilde{\Gamma}) \sqrt{\lambda(\Gamma_e'')\lambda(\tilde{\Gamma})} \, \langle 1 1 \hat{q}_2(\tilde{a}_1, r_1) \hat{q}_1(\tilde{a}_2, r_2) | 1 \hat{q}(\tilde{a}, r) \rangle^* \\
&\times \tilde{W}(S 1 S' 1 ; S'' 1) \sqrt{3(2S''+1)} \Bigg\}
\end{aligned} \tag{VIII.33}$$

where the sums $\Sigma_{\tilde{\Gamma}}$, $\Sigma_{\tilde{\Gamma}_1}$, $\Sigma_{\tilde{\Gamma}_2}$ run over all spin–orbit symmetries of (VIII.28); $\hat{q}$, $\hat{q}_1$, and $\hat{q}_2$ are related to $\tilde{a}$, $\tilde{a}_1$, $\tilde{a}_2$, respectively, as indicated in (VIII.28); r, r_1, r_2 are additional labels to specify components unambiguously:

$$\overline{\Delta V}^{-1}_{\phi''\Gamma''_e S''} = \tfrac{1}{2}\left(\left(V_{\phi\Gamma_e S} - V_{\phi''\Gamma''_e S''}\right)^{-1} + \left(V_{\phi'\Gamma'_e S'} - V_{\phi''\Gamma''_e S''}\right)^{-1}\right) \tag{VIII.34}$$

W is the group recoupling coefficient, and $\lambda(\Gamma)$ denotes the degree of the irreducible representation Γ. $\tilde{W}$ is the Racah coefficient for the three-dimensional rotation group (see Section VII and Appendix 2). The term in braces may be considered as an effective spn–orbit constant.

For spin–spin coupling, we get the two components

$$\begin{aligned}
\langle \phi'(\Gamma'_e a'_e) S'\Sigma' | \tilde{\mathfrak{H}}^{(0)}_{SS} | \phi(\Gamma_e a_e) S\Sigma \rangle &= \delta_{\Gamma'_e\Gamma_e}\delta_{a'_e a_e}\delta_{S'S} \\
&\times \Bigg\{ \langle \phi'\Gamma_e S \| \sum_{i<j} D^0(i,j)(\mathbf{s}_i \cdot \mathbf{s}_j) \| \phi\Gamma_e S \rangle \\
&+ \sum_{\tilde{\Gamma}_1 \tilde{a}_1 r_1} \varepsilon(\tilde{a}_1) \sum_{\tilde{\Gamma}_2 \tilde{a}_2 r_2} \varepsilon(\tilde{a}_2) \sum_{\phi''\Gamma''_e S''} \overline{\Delta V}^{-1}_{\phi''\Gamma''_e S''} \\
&\times \langle \phi'\Gamma_e S \| \sum_i T^{\tilde{\Gamma}_1}(\mathbf{u}_i)_{r_1} S^1(\mathbf{s}_i) \| \phi''\Gamma''_e S'' \rangle \\
&\times \langle \phi''\Gamma''_e S'' \| \sum_i T^{\tilde{\Gamma}_2}(\mathbf{u}_i)_{r_2} S^1(\mathbf{s}_i) \| \phi\Gamma_e S \rangle \\
&\times (\tilde{\Gamma}_2\tilde{a}_2, \tilde{\Gamma}_1\tilde{a}_2 | \Gamma_0)^* \, W(\Gamma_e\tilde{\Gamma}_2\Gamma_e\tilde{\Gamma}_1; \Gamma''_e\Gamma_0)\sqrt{\lambda(\Gamma''_e)} \\
&\times \tilde{W}(S1S1; S''0)\sqrt{2S''+1} \Bigg\}
\end{aligned} \tag{VIII.35}$$

with Γ_0 the totally symmetric representation, and

$$\langle \phi'(\Gamma_e' a_e') S'\Sigma' | \tilde{\mathfrak{H}}_{SS}^{(2)} | \phi(\Gamma_e a_e) S\Sigma \rangle$$

$$= \sum_{\bar{\Gamma}\bar{a}r} (\Gamma_e a_e, \bar{\Gamma}\bar{a} | \Gamma_e' a_e')^* \langle S 2 \Sigma \hat{q}(\bar{a}, r) | S'\Sigma' \rangle$$

$$\times \Bigg\{ \varepsilon(a) \langle \phi' \Gamma_e' S' \| \sum_{i<j} T^{\bar{\Gamma}}(D^2(i,j))_r S^2(\mathbf{s}_i, \mathbf{s}_j) \| \phi \Gamma_e S \rangle$$

$$+ \sum_{\tilde{\Gamma}_1 \tilde{a}_1 r_1} \varepsilon(\tilde{a}_1) \sum_{\tilde{\Gamma}_2 \tilde{a}_2 r_2} \varepsilon(\tilde{a}_2) \sum_{\phi'' \Gamma_e'' S''} \overline{\Delta V}_{\phi'' \Gamma_e'' S''}^{-1}$$

$$\times \langle \phi' \Gamma_e' S' \| \sum_i T^{\tilde{\Gamma}_1}(\mathbf{u}_i)_{r_1} S^1(\mathbf{s}_i) \| \phi'' \Gamma_e'' S'' \rangle$$

$$\times \langle \phi'' \Gamma_e'' S'' \| \sum_i T^{\tilde{\Gamma}_2}(\mathbf{u}_i)_{r_2} S^1(\mathbf{s}_i) \| \phi \Gamma_e S \rangle$$

$$\times (\tilde{\Gamma}_2 a_2, \tilde{\Gamma}_1 \tilde{a}_1 | \bar{\Gamma}\bar{a})^* W(\Gamma_e \tilde{\Gamma}_2 \Gamma_e' \tilde{\Gamma}_1 \cdot \Gamma_e'' \bar{\Gamma}) \sqrt{\lambda(\Gamma_e'')\lambda(\bar{\Gamma})}$$

$$\times \langle 1 1 \hat{q}_2(\tilde{a}_1, r_1) \hat{q}_1(\tilde{a}_2, r_2) | 2 \hat{q}(\bar{a}, r) \rangle \tilde{W}(S 1 \dot{S}' 1; S'' 2) \sqrt{5(2S''+1)} \Bigg\}$$

(VIII.36)

where the sum $\Sigma_{\bar{\Gamma}}$ runs over the symmetries of the spatial part of the spin–spin operator, (VIII.31). The term in braces in (VIII.35) may be considered as the effective isotropic spin–spin coupling constant, whereas those in (VIII.36) are the effective tensorial constants.

As mentioned in Section VII, it is useful to consider the matrix elements, (VIII.33) to (VIII.36), after a unitary transformation of the $|S\Sigma\rangle$-functions such that the transformed functions $|S\sigma\rangle$ span a basis of the irreducible representation of the molecular symmetry group $|S(\Gamma_S a_S r_S)\rangle$, where r_S is a number to separate different functions of the same symmetry.

We will get as many splittings from a state Γ_e, Γ_S as there are irreducible representations in the direct product $\Gamma_e \otimes \Gamma_S$. For example, for diatomics,

Γ_e and Γ_S are represented by Λ and Σ and the direct products are labeled by Ω with the additional parity labeling for Λ, Σ, $\Omega=0$ corresponding to $\Sigma^{\pm}$, so that we have $\Omega=|\Lambda-\Sigma|,\ldots,|\Lambda+\Sigma|$, where all $\Omega=0$ are split into 0^+ and 0^-. (The splitting of $\Omega=0$ for $|\Lambda|=|\Sigma|>1$ requires a higher-order perturbation than second order or, equivalently, the splittings become smaller the larger $|\Lambda|$ is.)

In the transformed matrix elements the Clebsch–Gordan coefficients $\langle Sk\sigma q|S'\sigma'\rangle$ are replaced by internal symmetry factors $(\Gamma_S a_S, \tilde{\Gamma}\tilde{a}|\Gamma'_S a'_S)$, which yields additional information. This is so because $\mathfrak{H}_{SO}$ and $\mathfrak{H}_{SS}$ are no longer invariant with respect to spin permutations; thus S is no longer a good quantum number, while the MS group remains good, and thus it is preferable to characterize the spin functions with respect to it.

Tables for the coefficients entering the matrix elements of (VIII.33) to (VIII.36) are given in Appendix 2 for the point group $C_{\infty v}$, as are references to similar tables for other groups.

Let us apply the formalism to an electronic Π-state of diatomics with the appropriate point group $C_{\infty v}$. (For $D_{\infty h}$ only g or u labels have to be added.)

In (VIII.33) we have to consider the diagonal matrix elements

$$\langle \phi(\Pi a'_e)S\Sigma'|\tilde{\mathfrak{H}}_{SO}|\phi(\Pi a_e)S\Sigma\rangle$$

with

$$a_e, a'_e = \pm$$

$$\Sigma, \Sigma' = -S, -S+1, \ldots, S$$

The coefficient $(\Gamma_e a_e, \tilde{\Gamma}\tilde{a}|\Gamma'_e a'_e)$ for $\Gamma_e=\Gamma'_e=\Pi$ is zero unless $\tilde{\Gamma}=\Sigma^-$ and $a_e = a'_e$; then we obtain $(\Pi\pm, \Sigma^-|\Pi\pm)=\pm 1$. The appropriate C-G coefficient is $\langle S1\Sigma 0|S\Sigma\rangle$, which gives $\Sigma/\sqrt{S(S+1)}$.

In the summation $\Sigma_{\tilde{\Gamma}_1\tilde{a}_1}\Sigma_{\tilde{\Gamma}_2\tilde{a}_2}$, only the terms with $\tilde{\Gamma}_1=\tilde{\Gamma}_2=\Pi$ survive, yielding $(\Pi\pm, \Pi\mp|\Sigma^-)=\pm\sqrt{1/2}$. The recoupling coefficient $W(\Pi\Pi\Pi\Pi;\Gamma''_e\Sigma^-)$ is zero unless $\Gamma''_e=\Sigma^+,\Sigma^-$, or Δ, giving

$$W(\Pi\Pi\Pi\Pi;\Delta\Sigma^-)=\tfrac{1}{2}$$

or

$$W(\Pi\Pi\Pi\Pi;\Sigma^{\pm}\Sigma^-)=\mp\tfrac{1}{2}$$

(Compare the definitions of W's in Appendix 2 and the values for $C_{\infty v}$ in Table II of Appendix 2.) The appropriate C-G coefficient of the three-dimensional rotation group is $\langle 11 \pm 1 \mp 1|10\rangle = \pm 1/\sqrt{2}$; thus we get

$$\langle \phi(\Pi\pm)S\Sigma|\tilde{\mathfrak{H}}_{SO}|\phi(\Pi\pm)S\Sigma\rangle = \pm\Sigma\sqrt{S(S+1)}^{\,-1}$$
$$\Bigg\{\langle\phi\Pi S\|\sum_i T^{\Sigma^-}(\mathbf{u}_i)S^1(\mathbf{s}_i)\|\phi\Pi S\rangle$$
$$+\sum_{\phi''S''}\Bigg[-\frac{1}{2\sqrt{2}}\overline{\Delta V}^{-1}_{\phi''\Delta S''}|\langle\phi\Pi S\|\sum_i T^{\Pi}(\mathbf{u}_i)S^1(\mathbf{s}_i)\|\phi''\Delta S''\rangle|^2$$
$$+\frac{1}{4}\sum_{s=\pm}(-1)^s\overline{\Delta V}^{-1}_{\phi''\Sigma^s S''}|\langle\phi\Pi S\|\sum_i T^{\Pi}(\mathbf{u}_i)S^1(\mathbf{s}_i)\|\phi''\Sigma^s S''\rangle|^2\Bigg]$$
$$\times\tilde{W}(S1S1;S''1)\sqrt{3(2S''+1)}\Bigg\} \qquad \text{(VIII.37)}$$

with $(-1)^s = +1$ for $s = +$ and -1 for $s = -$.

For (VIII.35), the only possibilities for $\tilde{\Gamma}_1\tilde{a}_1/\tilde{\Gamma}_2\tilde{a}_2$ are given by $\Pi_+/\Pi_-, \Pi_-/\Pi_+$, and Σ^-/Σ^-. The needed W-coefficients are

$$W(\Pi\Pi\Pi\Pi;\Delta\Sigma^+) = \tfrac{1}{2}$$
$$W(\Pi\Pi\Pi\Pi;\Sigma^{\pm}\Sigma^+) = \pm\tfrac{1}{2}$$
$$W(\Pi\Sigma^-\Pi\Sigma^-;\Pi\Sigma^+) = 1/\sqrt{2}$$

so that we get

$$\langle\phi(\Pi\pm)S\Sigma|\tilde{\mathfrak{H}}^{(0)}_{SS}|\phi(\Pi\pm)S\Sigma\rangle$$
$$=\langle\phi\Pi S\|\sum_{i<j}D^0(i,j)(\mathbf{s}_i\cdot\mathbf{s}_j)\|\phi\Pi S\rangle$$
$$+\Bigg[\sum_{\phi''S''}\Bigg(\tfrac{1}{2}\overline{\Delta V}^{-1}_{\phi''\Delta S''}|\langle\phi\Pi S\|\sum_i T^{\Pi}(\mathbf{u}_i)S^1(\mathbf{s}_i)\|\phi''\Delta S''\rangle|^2$$
$$+\sum_{s=\pm}(-1)^s\tfrac{1}{2}\overline{\Delta V}^{-1}_{\phi''\Sigma^s S''}|\langle\phi\Pi S\|\sum_i T^{\Pi}(\mathbf{u}_i)S^1(\mathbf{s}_i)\|\phi''\Sigma^s S''\rangle|^2$$
$$+\frac{1}{\sqrt{2}}\overline{\Delta V}^{-1}_{\phi''\Pi S''}|\langle\phi\Pi S\|\sum_i T^{\Sigma^-}(\mathbf{u}_i)S^1(\mathbf{s}_i)\|\phi''\Pi S''\rangle|^2\Bigg)$$
$$\times\tilde{W}(S1S1;S''0)\sqrt{(2S''+1)}\Bigg] \qquad \text{(VIII.38)}$$

For (VIII.36) the following relations hold: the coefficient $(\Gamma_e a_e, \bar{\Gamma}\bar{a}|\Gamma_e a_e')$ for $\Gamma_e = \Pi$ is not equal to zero for $\bar{\Gamma}\bar{a} = \Sigma^+$, yielding $(\Pi\pm,\Sigma^+|\Pi\pm)=1/\sqrt{2}$, and for $\bar{\Gamma}\bar{a}=\Delta_\pm$, yielding $(\Pi\pm,\Delta\mp|\Pi\mp)=1$.

The appropriate C-G coefficients $\langle S2\Sigma q|S\Sigma'\rangle$ are

$$\langle S2\Sigma 0|S\Sigma\rangle\delta_{\Sigma'\Sigma}=(3\Sigma^2-S(S+1))\left[(2S+3)(2S-1)S(S+1)\right]^{-1/2}$$

for $\bar{\Gamma}=\Sigma^+$, and

$$\langle S2\Sigma\mp 2|S(\Sigma\mp 2)\rangle\delta_{\Sigma'(\Sigma\mp 2)}$$
$$=\left[3(S\mp\Sigma-1)(S\pm\Sigma+1)(S\pm\Sigma+2)/2(2S+3)(2S-1)S(S+1)\right]^{1/2}$$

for $\bar{\Gamma}\bar{a}=\Delta_\pm$.

In the summation $\Sigma_{\tilde{\Gamma}_1\tilde{a}_1}\Sigma_{\tilde{\Gamma}_2\tilde{a}_2}$ the following terms are not equal to zero:
(a) $\tilde{\Gamma}_1=\tilde{\Gamma}_2=\Sigma^-$, yielding

$$(\Sigma^-,\Sigma^-|\Sigma^+)=-1$$

(b) $\tilde{\Gamma}_1=\tilde{\Gamma}_2=\Pi$, yielding

$$(\Pi\pm,\Pi\mp|\Sigma^+)=\frac{1}{\sqrt{2}} \qquad \text{and} \qquad (\Pi\pm,\Pi\pm|\Delta\pm)=1$$

The W-coefficients needed are
(a) $W(\Pi\Sigma^-\Pi\Sigma^-;\Gamma_e''\Sigma^+)$, which requires $\Gamma_e''=\Pi$ with

$$W(\Pi\Sigma^-\Pi\Sigma^-;\Pi\Sigma^+)=\frac{1}{\sqrt{2}}$$

(b) $W(\Pi\Pi\Pi\Pi;\Gamma_e''\Sigma^+)$, which requires $\Gamma_e''=\Delta$ with

$$W(\Pi\Pi\Pi\Pi;\Delta\Sigma^+)=\tfrac{1}{2}$$

or $\Gamma_e''=\Sigma^\pm$ with

$$W(\Pi\Pi\Pi\Pi;\Sigma^\pm\Sigma^+)=\pm\tfrac{1}{2}$$

and $W(\Pi\Pi\Pi\Pi;\Gamma_e''\Delta)$ requires $\Gamma_e''=\Sigma^\pm$ with

$$W(\Pi\Pi\Pi\Pi;\Sigma^\pm\Delta)=\tfrac{1}{2}$$

The appropriate C-G coefficients are

$$\langle 1 1 \pm 1 \mp 1 | 2 0 \rangle = 1/\sqrt{6}$$

$$\langle 1 1 0 0 | 2 0 \rangle = -2/\sqrt{6}$$

$$\langle 1 1 \pm 1 \pm 1 | 2 \pm 2 \rangle = \sqrt{2}$$

Thus we get

$$\begin{aligned}
&\langle \phi(\Pi\pm) S\Sigma | \tilde{\mathfrak{H}}^{(2)}_{SS} | \phi(\Pi\pm) S\Sigma \rangle \\
&= (3\Sigma^2 - S(S+1)) \sqrt{2(2S+3)(2S-1)S(S+1)}^{\,-1} \\
&\times \Bigg\{ \langle \phi\Pi S \| \sum_{i<j} T^{\Sigma^+}(D^2(i,j)) S^2(\mathrm{s}_i, \mathrm{s}_j) \| \phi\Pi S \rangle \\
&\quad + \Bigg(\sum_{\phi'' S''} \Bigg[\tfrac{1}{2} \overline{\Delta V}^{-1}_{\phi''\Delta S''} |\langle \phi\Pi S \| \sum_i T^{\Pi}(\mathrm{u}_i) S^1(\mathrm{s}_i) \| \phi''\Delta S'' \rangle|^2 \\
&\qquad + \sum_{s=\pm} (-1)^s \frac{1}{2\sqrt{2}} \overline{\Delta V}^{-1}_{\phi''\Sigma^s S''} |\langle \phi\Pi S \| \sum_i T^{\Pi}(\mathrm{u}_i) S^1(\mathrm{s}_i) \| \phi''\Sigma^s S'' \rangle|^2 \\
&\qquad + 2\overline{\Delta V}^{-1}_{\phi''\Pi S''} |\langle \phi\Pi S \| \sum_i T^{\Sigma^-}(\mathrm{u}_i) S^1(\mathrm{s}_i) \| \phi''\Pi S'' \rangle|^2 \Bigg] \Bigg) \Bigg\} \\
&\times \frac{1}{\sqrt{6}} \tilde{W}(S1S1; S''2) \sqrt{5(2S''+1)}
\end{aligned} \tag{VIII.39}$$

and

$$\begin{aligned}
&\langle \phi(\Pi\pm) S\Sigma | \tilde{\mathfrak{H}}^{(2)}_{SS} | \phi(\Pi\mp) S(\Sigma\pm 2) \rangle \\
&= \big[(3(S\mp\Sigma-1)(S\mp\Sigma)(S\pm\Sigma+1)(S\pm\Sigma+2))/2S(S+1)(2S+3)(2S-1) \big] \\
&\times \Bigg\{ \sum_{s=\pm} \sum_{\phi'' S''} \overline{\Delta V}^{-1}_{\phi''\Sigma^s S''} |\langle \phi\Pi S \| \sum_i T^{\Pi}(\mathrm{u}_i) S^1(\mathrm{s}_i) \| \phi''\Sigma^s S'' \rangle|^2 \\
&\quad \times \tilde{W}(S1S1; S''2) \sqrt{5(2S''+1)} \Bigg\}
\end{aligned} \tag{VIII.40}$$

For convenience we present the $\tilde{W}$-coefficients here for $S'' = S$:

$$\tilde{W}(S1S1; S0)\sqrt{(2S+1)} = -1/\sqrt{3}$$

$$\tilde{W}(S1S1; S1)\sqrt{3(2S+1)} = 1/\sqrt{2S(S+1)}$$

$$\tilde{W}(S1S1; S2)\sqrt{5(2S+1)} = \sqrt{(2S+3)(2S-1)/6S(S+1)}$$

For a Hund's case (b) basis [see (V.6)], we get analogous results by replacing the C-G coefficients $\langle Sk\Sigma q|S'\Sigma'\rangle$ for $k=0, 1, 2$ by

$$(-1)^{J-S'-N}\tilde{W}(S'SN'N;kJ)\delta_{J'J}\delta_{M'_JM_J}\sqrt{(2N+1)(2S'+1)}$$
$$\times\langle NkKq|N'K'\rangle$$

in the matrix elements of (VIII.33) to (VIII.36) with the bras or kets $|\phi\Gamma_e NSJKM_J\rangle$ of the basis.

The matrix elements diagonal in ϕ, Γ_e, S are independent of a definite basis and operator equivalents can be formed. For $\mathfrak{H}_{SO}$ one obtains in symmetry notation

$$\tilde{\mathfrak{H}}_{SO}=\sum_{\tilde{\Gamma}\tilde{a}r}\varepsilon(\tilde{a})\tilde{A}_r^{\tilde{\Gamma}}T_{\tilde{a}}^{\tilde{\Gamma}}(\mathsf{L})_r S^1_{\hat{q}(\tilde{a},r)}(\mathsf{S}) \tag{VIII.41}$$

with

$$\tilde{A}_r^{\tilde{\Gamma}}=\left[\langle\phi\Gamma_e\|T^{\tilde{\Gamma}}(\mathsf{L})_r\|\phi\Gamma_e S\rangle\langle S\|S^1(\mathsf{S})\|S\rangle\right]^{-1}$$
$$\times\Bigg\{\langle\phi\Gamma_e S\|\sum_i T^{\tilde{\Gamma}}(\mathsf{u}_i)_r S^1(\mathsf{s}_i)\|\phi\Gamma_e S\rangle+\varepsilon(\tilde{a})\sum_{\tilde{\Gamma}_1\tilde{a}_1r_1}\varepsilon(\tilde{a}_1)\sum_{\tilde{\Gamma}_2\tilde{a}_2r_2}\varepsilon(\tilde{a}_2)$$
$$\times\sum_{\phi''\Gamma''_e S''}\overline{\Delta V}^{-1}_{\phi''\Gamma''_e S''}\langle\phi\Gamma_e S\|T^{\tilde{\Gamma}_1}(\mathsf{u}_i)_{r_1}S^1(\mathsf{s}_i)\|\phi''\Gamma''_e S''\rangle\langle\phi''\Gamma''_e S''\|$$
$$\times\sum_i T^{\tilde{\Gamma}_2}(\mathsf{u}_i)_{r_2}S^1(\mathsf{s}_i)\|\phi\Gamma_e S\rangle(\tilde{\Gamma}_2\tilde{a}_2,\tilde{\Gamma}_1\tilde{a}_1|\tilde{\Gamma}\tilde{a})^* W(\Gamma_e\tilde{\Gamma}_2\Gamma'_e\tilde{\Gamma}_1;\Gamma''_e\tilde{\Gamma})$$
$$\times\sqrt{\lambda(\Gamma''_e)\lambda(\tilde{\Gamma})}\,\langle 11\hat{q}_2(\tilde{a}_1,r_1)\hat{q}_1(\tilde{a}_2,r_2)|1\hat{q}(\tilde{a},r)\rangle^*\tilde{W}(S1S1;S''1)$$
$$\times\sqrt{3(2S''+1)}\Bigg\}$$

For molecules in the linear conformation the only nonvanishing term in (VIII.41) is given by $\tilde{A}^{\Sigma^-}\cdot L_z\cdot S_z$, where we used that $T^{\Sigma^-}(\mathsf{L})=L_z$. For octahedral and tetahedral molecules, one gets $\tilde{A}^{T_1}\cdot\mathsf{L}\cdot\mathsf{S}$ for (VIII.41).

In the same way, we get

$$\tilde{\mathfrak{H}}^{(0)}_{SS}=\tilde{\lambda}^0\mathsf{S}^2 \tag{VIII.42}$$

where $\tilde{\lambda}^0$ is the value in braces in (VIII.35) divided by $\sqrt{S(S+1)}$.

For $\tilde{\mathfrak{H}}^{(2)}_{SS}$, we can write

$$\tilde{\mathfrak{H}}^{(2)}_{SS}=\sum_{\bar{\Gamma}\bar{a}r}\varepsilon(\bar{a})\left(\tilde{\lambda}^{\bar{\Gamma}}_{\bar{a}}\right)_r S^2_{\hat{q}(\bar{a},r)}(\mathsf{S},\mathsf{S}) \tag{VIII.43}$$

with

$$(\lambda^{\bar{\Gamma}}_{\bar{a}})_r = (\Gamma_e a_e, \bar{\Gamma}\bar{a}|\Gamma_e a'_e)^* [\langle S \| S^2(\mathsf{S},\mathsf{S}) \| S\rangle]^{-1}$$

$$\times \Bigg\{ \langle \phi\Gamma_e S \| \sum_{i<j} T^{\bar{\Gamma}}(D^2(i,j))_r S^2(\mathsf{s}_i, \mathsf{s}_j) \| \phi\Gamma_e S\rangle + \varepsilon(\bar{a}) \sum_{\tilde{\Gamma}_1\tilde{a}_1 r_1} \varepsilon(\tilde{a}_1)$$

$$\times \sum_{\tilde{\Gamma}_2\tilde{a}_2 r_2} \varepsilon(\tilde{a}_2) \sum_{\phi''\Gamma''_e S''} \overline{\Delta V}^{-1}_{\phi''\Gamma''_e S''} \langle \phi\Gamma_e S \| \sum_i T^{\tilde{\Gamma}_1}(\mathsf{u}_i)_{r_1} S^1(\mathsf{s}_i) \| \phi''\Gamma''_e S''\rangle$$

$$\times \langle \phi''\Gamma''_e S'' \| T^{\tilde{\Gamma}_2}(\mathsf{u}_i) S^1(\mathsf{s}_i) \| \phi\Gamma_e S\rangle (\tilde{\Gamma}_2\tilde{a}_2, \tilde{\Gamma}_1\tilde{a}_1|\bar{\Gamma}\bar{a})^*$$

$$\times W(\Gamma_e\tilde{\Gamma}_2\Gamma'_e\tilde{\Gamma}_1; \Gamma''_e\bar{\Gamma}) \sqrt{\lambda(\Gamma''_e)\lambda(\bar{\Gamma})} \langle 11\hat{q}_2(\tilde{a}_1, r_1)\hat{q}_1(\tilde{a}_2, r_2)|2\hat{q}(\bar{a}, r)\rangle$$

$$\times {}^*\tilde{W}(S1S1; S''2)\sqrt{5(2S''+1)} \Bigg\}$$

C. Rotation and Rotation–Electronic Couplings

In this section we discuss the fine-structure effects resulting from the molecular rotation and the coupling of rotation with electronic spin and orbital motion. The rotation–electronic orbit interaction is of minor importance and is included here only for completeness. At times it will be necessary to add second-order electronic terms already dealt with in the previous section. The operators yielding the first-order contributions to the effective property functions discussed in this section are presented in Appendix 1, (A.20) to (A.37). The field-free rotation operator is expressed as

$$\mathfrak{H}_{\text{rot}} = \tfrac{1}{2}\hbar^2 \mathsf{R}\cdot\boldsymbol{\mu}\mathsf{R} \qquad \text{(VIII.44)}$$

where R is the angular momentum operator for the nuclei defined by

$$\mathsf{R} = \mathsf{N} - \mathsf{L} - \mathsf{G} \qquad \text{(VIII.45)}$$

In (VIII.45), N, L, and G are the angular momentum operators for overall rotation, electronic orbital motion, and vibration, respectively, all in units of $\hbar$. The matrix $\boldsymbol{\mu}$, the inverse inertia moment dyadic, contains the main, nuclear contribution to the rotational constants as functions of the internal nuclear coordinates. The elements of the $\boldsymbol{\mu}$-dyadic possess some interesting symmetry properties:

1. In all nuclear conformations belonging to the point group of highest symmetry (i.e., the reference configuration and all those reached by a totally symmetric displacement), only those elements are nonzero which transform totally symmetric. For example, in C_{3v}-symmetric molecules only μ_{zz} and $\mu_{xx}+\mu_{yy}$ are nonzero in the reference configuration, whereas for T_d-symmetric molecules only one value exists, given by $\mu_{xx}+\mu_{yy}+\mu_{zz}$, and for C_{2v}-symmetric molecules three different values exist, μ_{xx}, μ_{yy}, and μ_{zz}.
2. In nuclear configurations related by a nontotally symmetric displacement to the reference, only those elements of $\boldsymbol{\mu}$ may become unequal zero which transform according to the irreducible representation of the nuclear displacement considered. The transformation properties of the elements of the $\boldsymbol{\mu}$-dyadic may be found in nearly all character tables for point groups.

As a consequence of point 2, we will get more coupling terms within the rotational operator (VIII.44) for conformations of the lower instantaneous point group symmetry. The additional terms may lead to rotational line splittings. For instance, in a molecule with C_{3v}-symmetry, a vibrational displacement coordinate transforming according to the degenerate E-representation may produce nonvanishing values for the anisotropy constant $\mu_{xx}-\mu_{yy}$ and for $\mu_{xy}+\mu_{yx}$. However, for triatomic molecules with D_{3h} MS-symmetry, there is no normal coordinate which may cause such an anisotropy $\mu_{xx}-\mu_{yy}$. This may be present, nevertheless, as a result of second-order electronic contributions.

For molecules with a linear reference configuration (i.e., for linear molecules), the only nonvanishing elements of $\boldsymbol{\mu}$ are given by $\mu_{xx}=\mu_{yy}=2\cdot B$ for all nuclear displacements, so that (VIII.44) reduces to

$$\mathfrak{H}_{\text{rot}}=(\mathbf{R}-R_z\mathbf{e}_z)\cdot B(\mathbf{R}-R_z\mathbf{e}_z) \qquad \text{(VIII.46)}$$

where $\mathbf{e}_z$ is the unit vector in the direction of the linear reference axis denoted by z. For linear molecules the existence of the anisotropy value $\mu_{xx}-\mu_{yy}$ is known to produce part of the Λ-doubling effect. Another part of the Hamiltonian we investigate here is the spin–rotation coupling operator $\mathfrak{H}_{\mathrm{S}r}$, (A.58), which we write as

$$\mathfrak{H}_{\mathrm{S}r}=\tfrac{1}{2}\sum_i\left(\mathbf{s}_i\cdot\boldsymbol{\gamma}_i I''^{1/2}\mathbf{R}I''^{-1/2}+I''^{1/2}\mathbf{R}I''^{-1/2}\cdot\boldsymbol{\gamma}_i\mathbf{s}_i\right) \qquad \text{(VIII.47)}$$

where

$$\gamma_i = -(\hbar g\beta e/c)\sum_{\alpha} Z_\alpha r_{i\alpha}^{-3}(\mathbf{r}_{i\alpha}\cdot\mathbf{r}_\alpha^0\mathbf{1} - \mathbf{r}_{i\alpha}\otimes\mathbf{r}_\alpha^0)\mathbf{I}''^{-1} \qquad \text{(VIII.48)}$$

For linear molecules γ_i reduces to

$$\gamma_i = -(\hbar g\beta e/c)\sum_{\alpha} Z_\alpha r_{i\alpha}^{-3} z_\alpha^0 I''^{-1}\begin{bmatrix} z_{i\alpha} & 0 & 0 \\ 0 & z_{i\alpha} & 0 \\ -x_{i\alpha} & -y_{i\alpha} & 0 \end{bmatrix} \qquad \text{(VIII.49)}$$

and for diatomics with $I''^{-1} = (\mu R R^0)^{-1}$, where $\mu = (M_\alpha M_\beta)/(M_\alpha + M_\beta)$ and $R = z_\alpha - z_\beta$, we get

$$\gamma_i = -(g\beta e\hbar/cR)\left[\frac{Z_\alpha}{M_\alpha} r_{i\alpha}^{-3}\begin{bmatrix} z_{i\alpha} & 0 & 0 \\ 0 & z_{i\alpha} & 0 \\ -x_i & -y_i & 0 \end{bmatrix} - \frac{Z_\beta}{M_\beta} r_{i\beta}^{-3}\begin{bmatrix} z_{i\beta} & 0 & 0 \\ 0 & z_{i\beta} & 0 \\ -x_i & y_i & 0 \end{bmatrix}\right] \qquad \text{(VIII.50)}$$

where α and β denote the two nuclei.

In the following we look for the resultant effective functions in nuclear configuration space of various types:

1. Effective rotational constant matrix for the adiabatic states.
2. Effective spin–rotational constant matrix for the adiabatic states.
3. Transition functions between two or more nearly degenerate states contributing to the rotational constants.
4. The same as type 3 for spin–rotation constants.

The effective adiabatic functions, as well as the transition functions, will include second-order contributions from all other electronic states sufficiently far away, which are not treated explicitly in the coupled equation formulation.

In order to introduce the molecular symmetry, we write $\mathfrak{H}_{\text{rot}}$, (VIII.44), in the irreducible tensor form

$$\mathfrak{H}_{\text{rot}} = \tfrac{1}{2}\sum_{km}(-1)^m \sum_{m_1 m_2}\langle 11 m_1 m_2 | km\rangle \tilde{T}^1_{m_1}(\mathsf{R})\tilde{T}^k_{-m}(\boldsymbol{\mu})\tilde{T}^1_{m_2}(\mathsf{R})$$

(VIII.51)

where

$$\sum_{m_1 m_2} \langle 11 m_1 m_2 | km \rangle \tilde{T}^1_{m_1}(\mathsf{R}) \tilde{T}^1_{m_2}(\mathsf{R}) = \tilde{T}^k_m(\mathsf{R},\mathsf{R}) \qquad \text{(VIII.52)}$$

For (VIII.51) we may write in the more general case (see Section VII),

$$\mathfrak{H}_{\mathrm{rot}} = \tfrac{1}{2} \sum_{kq} \varepsilon(q) \sum_{q_1 q_2} \langle 11 q_1 q_2 | kq \rangle S^1_{q_1}(\mathsf{R}) \mu^k_{\hat{q}} S^1_{q_2}(\mathsf{R}) \qquad \text{(VIII.53)}$$

where $\varepsilon(q) = (-1)^q$ for complex notation when (VIII.53) becomes identical to (VIII.51) and $+1$ otherwise, and $\hat{q} = -q$ for complex and $\hat{q} = q$ for real notation. The numbers q and the relation between μ^k_q and $\boldsymbol{\mu}$ are presented in Appendix 3.

The k-summation goes only over $k = 0$ and 2, since $\boldsymbol{\mu}$ is a symmetric dyadic. For linear molecules we get for (VIII.53),

$$\begin{aligned} \mathfrak{H}_{\mathrm{rot}} = &\sum_q \varepsilon(q) S^1_q(\mathsf{R}) B S^1_{\hat{q}}(\mathsf{R}) \\ &- S^1_0(\mathsf{R}) B S^1_0(\mathsf{R}) \end{aligned} \qquad \text{(VIII.54)}$$

Before we go on, we introduce a formal notation for the vibrational angular momentum operator G. Since G operates not only on nuclear coordinate functions but also on the electronic functions, we use the formal substitution of G by

$$\mathsf{G} = \mathsf{G}^{\mathfrak{R}} + \mathsf{G}^{\mathrm{el}} \qquad \text{(VIII.55)}$$

from the definition

$$\langle \psi' | \mathsf{G} | \psi \rangle_{\mathrm{r}} = \delta_{\psi'\psi} \mathsf{G}^{\mathfrak{R}} + \langle \psi' | [\mathsf{G}^{\mathrm{el}} | \psi \rangle_{\mathrm{r}}]$$

where the square brackets indicate a restricted operation only onto the electronic function part denoted by ψ.

This enables us to find the following decomposition of $\mathfrak{H}_{\mathrm{rot}}$:

$$\begin{aligned} \mathfrak{H}_{\mathrm{rot}} = \tfrac{1}{2}\{ &(\mathsf{N} - \mathsf{G}^{\mathfrak{R}}) \cdot \boldsymbol{\mu} (\mathsf{N} - \mathsf{G}^{\mathfrak{R}}) && \text{(VIII.56a)} \\ &+ (\mathsf{L} + \mathsf{G}^{\mathrm{el}}) \cdot \boldsymbol{\mu} (\mathsf{L} + \mathsf{G}^{\mathrm{el}}) && \text{(VIII.56b)} \\ &- (\mathsf{N} - \mathsf{G}^{\mathfrak{R}}) \cdot \boldsymbol{\mu} (\mathsf{L} + \mathsf{G}^{\mathrm{el}}) && \text{(VIII.56c)} \\ &- (\mathsf{L} + \mathsf{G}^{\mathrm{el}}) \cdot \boldsymbol{\mu} (\mathsf{N} - \mathsf{G}^{\mathfrak{R}}) \} && \text{(VIII.56c)} \end{aligned}$$

Even more explicitly, (VIII.56) may be written for linear molecules as

$$\mathfrak{H}_{\text{rot}} = B(\mathbf{N}^2 - N_z^2) \qquad \text{(VIII.57a.1)}$$

$$+\mathbf{G}^{\mathfrak{R}} B \mathbf{G}^{\mathfrak{R}} - G_z^{\mathfrak{R}} B G_z^{\mathfrak{R}} \qquad \text{(VIII.57a.2)}$$

$$-\tfrac{1}{2}\big(N_+ B G_-^{\mathfrak{R}} + G_-^{\mathfrak{R}} B N_+$$

$$+ N_- B G_+^{\mathfrak{R}} + G_+^{\mathfrak{R}} B N_-\big) \qquad \text{(VIII.57a.3)}$$

$$+ B\big\{(\mathbf{L}^2 - L_z^2) \qquad \text{(VIII.57b.1)}$$

$$+ (\mathbf{G}^{\text{el}2} - G_z^{\text{el}2}) \qquad \text{(VIII.57b.2)}$$

$$+ L_+ G_-^{\text{el}} + L_- G_+^{\text{el}}\big\} \qquad \text{(VIII.57b.3)}$$

$$- B(N_+ L_- + N_- L_+) \qquad \text{(VIII.57c.1)}$$

$$+\tfrac{1}{2}\big(G_+^{\mathfrak{R}} B L_- + G_-^{\mathfrak{R}} B L_+ \qquad \text{(VIII.57c.2)}$$

$$+ L_+ B G_-^{\mathfrak{R}} + L_- B G_+^{\mathfrak{R}}$$

$$- B\big(N_+ G_-^{\text{el}} + N_- G_+^{\text{el}}\big) \qquad \text{(VIII.57c.3)}$$

$$-\tfrac{1}{2}\big(G_+^{\mathfrak{R}} B G_-^{\text{el}} + G_-^{\mathfrak{R}} B G_+^{\text{el}}$$

$$+ G_+^{\text{el}} B G_-^{\mathfrak{R}} + G_-^{\text{el}} B G_+^{\mathfrak{R}}\big) \qquad \text{(VIII.57c.4)}$$

Note that $\mathbf{G}^{\mathfrak{R}}$ does not commute with $B(\mathfrak{R})$.

Since parts (a) of (VIII.56) and (VIII.57) are not affected by electronic integration, giving only a Kronecker-δ-type contribution $\delta_{\Phi'\Phi}\delta_{\Gamma_e'\Gamma_e}\delta_{a_e'a_e}$, they need not to be considered further. Parts (b) are totally symmetric electronic operators, which will yield only negligible modifications of the adiabatic potential energies. Parts (c) are responsible for the first- and second-order contributions to the various properties. We will restrict the following discussion to the term involving the coupling between $\mathbf{N}$ and $\mathbf{L}$ (VIII.57c.1). We do this because the other $\mathbf{G}^{\mathfrak{R}}$ with $\mathbf{L}$ and $\mathbf{N}$, $\mathbf{G}^{\mathfrak{R}}$ with $\mathbf{G}^{\text{el}}$ are of the same structure, and in addition they will give only negligible contributions.

In order to introduce the instantaneous point-group symmetry, we start, using (VIII.51), with the terms of interest;

$$\mathfrak{H}'_{\text{rot}} = -\sum_{kq} \varepsilon(q) \sum_{q_1 q_2} \langle 11 q_1 q_2 | kq \rangle S_{q_1}^1(\mathbf{N}) S_{q_2}^1(\mathbf{L}) \mu_{\bar{q}}^k \qquad \text{(VIII.58)}$$

and set

$$S^1_{q_2}(\mathsf{L}) \equiv T^{\tilde{\Gamma}}_{\tilde{a}}(\mathsf{L})_r$$

where r is the additional index to specify the identification precisely. The summation Σ_{q_2} is replaced by $\Sigma_{\tilde{\Gamma}\tilde{a}r}$, and therefore we get

$$\mathfrak{H}'_{\mathrm{rot}} = -\sum_{kq} \varepsilon(q) \sum_{q_1} \sum_{\tilde{\Gamma}\tilde{a}r} \langle 11 q_1 q_2(\tilde{a},r)|kq\rangle$$

$$\times S^1_{q_1}(\mathsf{N}) T^{\tilde{\Gamma}}_{\tilde{a}}(\mathsf{L})_r \mu^k_{\hat{q}} \qquad \text{(VIII.59)}$$

For a linear molecule, this is

$$\mathfrak{H}'_{\mathrm{rot}} = +2B\left(T^{\Pi}_{+}(\mathsf{L}) S^1_{-1}(\mathsf{N}) + T^{\Pi}_{-}(\mathsf{L}) S^1_{+1}(\mathsf{N})\right) \qquad \text{(VIII.60)}$$

with

$$T^{\Pi}_{\pm}(\mathsf{L}) = \mp(L_x \pm iL_y)/\sqrt{2}$$

$$S^1_{\pm 1}(\mathsf{N}) = \mp(N_x \pm iN_y)/\sqrt{2}$$

By analogy with $\mathfrak{H}_{\mathrm{rot}}$ we consider the coupling of N with s_i in $\mathfrak{H}_{\mathrm{S}r}$, (VIII.47), and therefore may write in irreducible tensor notation

$$\mathfrak{H}'_{\mathrm{S}r} = \tfrac{1}{2} \sum_{k=0,2q} \varepsilon(q) \sum_{q_1 q_2} \sum_{i} \langle 11 q_1 q_2 | k\hat{q}\rangle \gamma^{(k)}_{iq}$$

$$\times\left(S^1_{q_1}(\mathsf{N}) S^1_{q_2}(\mathsf{s}_i) + S^1_{q_2}(\mathsf{s}_i) S^1_{q_1}(\mathsf{N})\right) \qquad \text{(VIII.61)}$$

Translation of γ^k_{iq} into the symmetry notation of the instantaneous point group yields

$$\mathfrak{H}'_{\mathrm{S}r} = \tfrac{1}{2} \sum_{k=0,2} \sum_{\Gamma ar} \varepsilon(a) \sum_{q_1 q_2} \sum_{i} \langle 11 q_1 q_2 | k\hat{q}(a,r)\rangle$$

$$\times T^{\Gamma}_{a}(\gamma^k_i)_r \left(S^1_{q_1}(\mathsf{N}) S^1_{q_2}(\mathsf{s}_i) + S^1_{q_2}(\mathsf{s}_i) S^1_{q_1}(\mathsf{N})\right) \qquad \text{(VIII.62)}$$

For a linear molecule one adds the $q=0$ terms for $k=0$ and 2 in (VIII.61):

$$\begin{aligned}\mathfrak{H}_{Sr}=\tfrac{1}{2}\sum_i \{ & T^{\Sigma^+}(\gamma_i)[S^1_{+1}(\mathbf{s}_i)S^1_{-1}(\mathbf{N}) \\ & + S^1_{-1}(\mathbf{s}_i)S^1_{+1}(\mathbf{N}) + S^1_{+1}(\mathbf{N})S^1_{-1}(\mathbf{s}_i) \\ & + S^1_{-1}(\mathbf{N})S^1_{+1}(\mathbf{s}_i)] \\ & + T^{\Pi}_{+}(\gamma_i^{(2)})[S^1_0(\mathbf{s}_i)S^1_{-1}(\mathbf{N}) + S^1_{-1}(\mathbf{N})S^1_0(\mathbf{s}_i)] \\ & + T^{\Pi}_{-}(\gamma_i^{(2)})[S^1_0(\mathbf{s}_i)S^1_{+1}(\mathbf{N}) + S^1_{+1}(\mathbf{N})S^1_0(\mathbf{s}_i)]\} \end{aligned} \qquad \text{(VIII.63)}$$

with

$$T^{\Sigma^+}(\gamma_i)=(ge\beta\hbar/c)\sum_\alpha Z_\alpha r_{i\alpha}^{-3} z_\alpha^0 z_{i\alpha} I''^{-1}$$

$$T^{\Pi}_{\pm}(\gamma_i^{(2)})=(ge\beta\hbar/c)\sum_\alpha Z_\alpha r_{i\alpha}^{-3} z_\alpha^0 I''^{-1}[\mp(x_{i\alpha}\pm iy_{id})/\sqrt{2}\,]$$

$$S^1_{\pm 1}(\mathbf{s}_i)=\mp(s_{ix}\pm is_{iy})/\sqrt{2}$$

$$S^1_0(\mathbf{s}_i)=s_{iz}$$

$$S^1_{\pm 1}(\mathbf{N})=\mp(N_x\pm iN_y)/\sqrt{2}$$

$$S^1_0(\mathbf{N})=N_z$$

where $T^{\Sigma^+}(\gamma)=-\frac{1}{2}(\gamma_{ixx}+\gamma_{iyy})$, which is equal to $(\sqrt{3}/2)\gamma_{io}^{(0)}$ and $(\sqrt{6}/2)\gamma_{io}^{(2)}$.

A problem arises due to the anomalous commutation relations of the molecular fixed components of the angular momentum operators **N**, **J**, and **F**.[49] Following a suggestion by Carrington, Levy, and Miller,[1h] we introduce a redefinition of the components

$$\begin{aligned} S^1_{\pm 1}(\mathbf{J}') &= \mp(J_x \mp iJ_y) = -S^1_{+1}(\mathbf{J}) \\ S^1_{x/y}(\mathbf{J}') &= \pm J_{x/y} = \pm S^1_{x/y}(\mathbf{J}) \\ S^1_0(\mathbf{J}') &= S^1_z(\mathbf{J}') = J_z \end{aligned} \qquad \text{(VIII.64)}$$

such that the new **J**′ or **N**′ operators fulfill the normal commutation relations and can be used for the standard application of the irreducible tensor formulas.

However, we have to be aware that all $S^1_q(\mathbf{N})$, and so on [e.g., in (VIII.54) and (VIII.57)] have to be replaced by their analogous forms. For the general case, $S^1_q(\mathbf{N})\rightarrow\tilde{\varepsilon}(q)S^1_{q^+}(\mathbf{N}')$, where $q^+=-q$, $\tilde{\varepsilon}(q)=(-1)^q$ for the complex basis, and $q^+=q$, $\tilde{\varepsilon}(q)=-1$ for the y-component, $+1$ otherwise, for the real basis, when standard formulas are used.

Now we look for all the contributions to the effective rotational matrix $\tilde{\mu}(\phi'\Gamma'_e,\phi\Gamma_e)$, that is, the adiabatic terms and the coupling functions in

$$\tilde{\mathfrak{H}}_{\text{rot}}=\tfrac{1}{2}(\mathbf{N}-\mathbf{G}^R)\cdot\tilde{\mu}(\phi'\Gamma'_e,\phi\Gamma_e)(\mathbf{N}-\mathbf{G}^R) \tag{VIII.65}$$

Instead of $(\mathbf{N}-\mathbf{G}^R)$, we now use $\mathbf{N}$, for reasons given above. We get, for a general unspecified rotational basis $|\lambda\rangle$,

$$\begin{aligned}
&\langle\phi'(\Gamma'_e a'_e)\lambda'|\tilde{\mathfrak{H}}_{\text{rot}}|\phi(\Gamma_e a_e)\lambda\rangle\\
&=\delta_{\phi'\phi}\delta_{\Gamma'_e\Gamma_e}\delta_{a'_e a_e}\tfrac{1}{2}\sum_{k=0,2}\sum_{q}\varepsilon(q)\mu_q^k\langle\lambda'|S_q^k(\mathbf{N},\mathbf{N})|\lambda\rangle\\
&\quad+\sum_{k=0,2}\sum_{q}\sum_{k_1q_1}\sum_{k_2q_2}\varepsilon(q_1)\varepsilon(q_2)\sum_{q_3q_5}\\
&\quad\times\sum_{\bar{\Gamma}\bar{a}r}(\Gamma_e a_e,\bar{\Gamma}\bar{a}|\Gamma'_e a'_e)^*\langle\lambda'|S_q^k(\mathbf{N},\mathbf{N})|\lambda\rangle\sum_{\tilde{\Gamma}_1\tilde{a}_1r_1}\varepsilon(\tilde{a}_1)\sum_{\tilde{\Gamma}_2\tilde{a}_2r_2}\varepsilon(\tilde{a}_2)\\
&\quad\times\sum_{\phi''\Gamma''_e}\overline{\Delta V}^{-1}_{\phi''\Gamma''_e}\langle\phi'\Gamma'_e\|T^{\tilde{\Gamma}_1}(\mathbf{L})_{r_1}\|\phi''\Gamma''_e\rangle\langle\phi''\Gamma''_e\|T^{\tilde{\Gamma}_2}(\mathbf{L})_{r_2}\|\phi\Gamma_e\rangle\\
&\quad\times(\tilde{\Gamma}_2\tilde{a}_2,\tilde{\Gamma}_1\tilde{a}_1|\bar{\Gamma}\bar{a})^*W(\Gamma_e\tilde{\Gamma}_2\Gamma'_e\tilde{\Gamma}_1;\Gamma''_e\bar{\Gamma})\sqrt{\lambda(\Gamma''_e)\lambda(\bar{\Gamma})}\\
&\quad\times\langle 11q_3q_5|kq\rangle^*\langle 11q_3q_4(\tilde{a}_1r_1)|k_1q_1\rangle\langle 11q_5q_6(\tilde{a}_2,r_2)|k_2q_2\rangle\,\mu_{\hat{q}_1}^{k_1}\mu_{\hat{q}_2}^{k_2}
\end{aligned} \tag{VIII.66}$$

If we introduce $S_q^k(\mathbf{N}',\mathbf{N}')$, we have to multiply (VIII.61) with $\tilde{\varepsilon}(q_3)\cdot\tilde{\varepsilon}(q_5)$. Additional matrix elements linear in $\mathbf{N}$ are

$$\begin{aligned}
&\langle\phi'(\Gamma'_e a'_e)\lambda'|\tilde{\mathfrak{H}}'_{\text{rot}}|\phi(\Gamma_e a_e)\lambda\rangle\\
&=\sum_{\tilde{\Gamma}\tilde{a}r}\sum_{q'}(\Gamma_e a_e,\tilde{\Gamma}\tilde{a}|\Gamma'_e a'_e)^*\langle\lambda'|S_{q_1}^1(\mathbf{N})|\lambda\rangle\\
&\quad\times\Bigg\{-\sum_{kq}\varepsilon(q)\langle 11q'q_2(\tilde{a},r)|kq\rangle\mu_q^k\langle\phi'\Gamma'_e\|T^{\tilde{\Gamma}}(\mathbf{L})_r\|\phi\Gamma_e\rangle\\
&\quad\times-\sum_{k_1q_1}\varepsilon(q_1)\sum_{k_2q_2}\varepsilon(q_2)\sum_{q_3q_4}\sum_{\tilde{\Gamma}_1\tilde{a}_1r_1}\varepsilon(\tilde{a}_1)\sum_{\tilde{\Gamma}_2\tilde{a}_2r_2}\varepsilon(\tilde{a}_2)\\
&\quad\times\sum_{\phi''\Gamma''_e}\overline{\Delta V}^{-1}_{\phi''\Gamma''_e}\langle\phi'\Gamma'_e\|T^{\tilde{\Gamma}_1}(\mathbf{L})_{r_1}\|\phi''\Gamma''_e\rangle\langle\phi''\Gamma''_e\|T^{\tilde{\Gamma}_2}(\mathbf{L})_{r_2}\|\phi\Gamma_e\rangle\\
&\quad\times\langle 11q_3q_5|1q'\rangle^*\langle 11q_3q_4(\tilde{a}_1,r_1)|k_1q_1\rangle\langle 11q_5q_6(\tilde{a}_2r_2)|k_2q_2\rangle\mu_{\hat{q}_1}^{k_1}\mu_{\hat{q}_2}^{k_2}\Bigg\}
\end{aligned} \tag{VIII.67}$$

where we have used $S_q^1(\mathsf{N},\mathsf{N}) = -S_q^1(\mathsf{N})$.

Equations (VIII.66) and (VIII.67) are rather difficult to comprehend in detail. Therefore, we present some applications to triatomic linear molecules in their linear as well as bent nuclear conformation. For the bent case it is advantageous to use the real notation; that is, we write for (VIII.61),

$$\mathfrak{H}'_{\mathrm{rot}} = -2B\left[T^{A'}(\mathsf{L})S_x^1(\mathsf{N}) + T^{A''}(\mathsf{L})S_y^1(\mathsf{N})\right] \qquad \text{(VIII.68)}$$

where A' and A'' are the species of the C_s point group. For molecules with $D_{\infty h}$ symmetry, the reference configuration C_{2v}-symmetry is also possible, and we would have A_2 for A' and B_1 for A'' for the axis identification $z \to x$, $y \to z$, $x \to y$.

For the linear nuclear arrangement, there are no expectation values for the terms of (VIII.57c), since they are of Π-symmetry and the representation Π is not contained in the direct product $\Gamma_e \otimes \Gamma_e$ for any $\Gamma_e = \Sigma^{\pm}, \Pi, \Delta, \ldots$. This operator may couple states of Λ-values different by one. In the bent conformation the possibly degenerate electronic states are split. For nondegenerate states there is no expectation value of electronic angular momentum operators. The couplings between the split states, which are nearly degenerate for small displacements, are small for reasons of continuity and may be neglected. For large displacements the coupling may be strong; however, the state will be split so far that second-order perturbation theory will be sufficient in all nuclear configurations.

Equation (VIII.65) gives for linear molecules in their linear conformation,

$$\langle \phi'(\Gamma'_e a'_e)\lambda' | \tilde{\mathfrak{H}}_{\mathrm{rot}} | \phi(\Gamma_e a_e)\Lambda \rangle = \langle \lambda' | \mathsf{N}^2 - N_z^2 | \lambda \rangle$$

$$\times \Big\{ B\delta_{\phi'\phi}\delta_{\Gamma'_e\Gamma_e}\delta_{a'_e a_e} + 2B^2 \sum_{\phi''\Gamma''_e} S_{\phi''\Gamma''_e,\Gamma'_e\Gamma_e,\Pi\Pi}\,(\Pi+,\Pi-|\Sigma^+)^*$$

$$\times (\Gamma_e a_e, \Sigma^+ | \Gamma_e a_e)\delta_{\Gamma'_e\Gamma_e}\delta_{a'_e a_e} W(\Gamma_e\Pi\Gamma_e\Pi;\Gamma''_e\Sigma^+)\sqrt{\lambda(\Gamma''_e)} \Big\} \qquad \text{(VIII.69a)}$$

$$+2B^2 \sum_{\phi''\Gamma''_e} S_{\phi''\Gamma''_e,\Gamma'_e\Gamma_e,\Pi\Pi}$$

$$\times \Big\{ (\Pi+,\Pi+|\Delta+)^*(\Gamma_e a_e,\Delta+|\Gamma'_e a'_e)^*$$

$$\times W(\Gamma_e\Pi\Gamma'_e\Pi;\Gamma''_e\Delta)\sqrt{\lambda(\Gamma''_e)2}\,\langle\lambda'|\tfrac{1}{2}N_-^2|\lambda\rangle + (\Pi-,\Pi-|\Delta-)^*$$

$$\times (\Gamma_e a_e,\Delta-|\Gamma'_e a'_e)^*\, W(\Gamma_e\Pi\Gamma'_e\Pi;\Gamma''_e\Delta)\sqrt{\lambda(\Gamma''_e)2}\,\langle\lambda'|\tfrac{1}{2}N_+^2|\lambda\rangle \Big\}$$

$$\text{(VIII.69b)}$$

with

$$S_{\phi''\Gamma_e'',\Gamma_e'\Gamma_e,\Pi\Pi}$$
$$=\overline{\Delta V}_{\phi''\Gamma_e''}^{-1}\langle\phi'\Gamma_e'\|T^{\Pi}(\mathrm{L})\|\phi''\Gamma_e''\rangle\langle\phi''\Gamma_e''\|T^{\Pi}(\mathrm{L})\|\phi\Gamma_e\rangle$$

The second-order term in (VIII.64a) gives a small electronic correction to the rotational constant, modifying $B(\mathfrak{R})$ to $\tilde{B}(\mathfrak{R})$.

The two terms in (VIII.69b) contribute to the Λ-doubling. For expectation values, $\Gamma_e'=\Gamma_e$, the group coupling coefficient $(\Gamma_e a_e,\Delta\pm|\Gamma_e a_e')$ requires that $\Gamma_e=\Pi$ and $a_e/a_e'\mp/\pm$. The $W(\Pi\Pi\Pi\Pi;\Gamma_e''\Delta)$ coefficients are $+\frac{1}{2}$ for $\Gamma_e''=\Sigma^+$, $-\frac{1}{2}$ for $\Gamma_e''=\Sigma^-$, and zero otherwise.

Therefore, (VIII.69b) may be replaced by

$$\tfrac{1}{2}\tilde{q}\left(N_+^2e^{-i2\varphi}+N_-^2e^{+i2\varphi}\right) \tag{VIII.70}$$

where

$$\tilde{q}(R)=B^2\Sigma_{\phi''}(S_{\phi''\Sigma^+,\Pi\Pi,\Pi\Pi}-S_{\phi''\Sigma^-,\Pi\Pi,\Pi\Pi})$$

For electronic $\Sigma^{\pm}$-states, only the first term in (VIII.69a) will contribute to the expectation value.

Equation (VIII.62) has the following terms for the linear nuclear conformation:

$$\begin{aligned}\langle\phi'(\Gamma_e'a_e')\lambda'|\tilde{\mathfrak{H}}_{\mathrm{rot}}'|\phi(\Gamma_e a_e)\lambda\rangle=&-B\langle\phi'\Gamma_e'\|T^{\Pi}(\mathrm{L})\|\phi\Gamma_e\rangle\\&\{(\Gamma_e a_e,\Pi+|\Gamma_e'a_e')^*\langle\lambda'|N_-/\sqrt{2}\,|\lambda\rangle\\&+(\Gamma_e a_e,\Pi-|\Gamma_e'a_e')^*\langle\lambda'|N_+/\sqrt{2}\,|\lambda\rangle\}\end{aligned} \tag{VIII.71a}$$

$$\begin{aligned}&+2B^2\sum_{\phi''\Gamma_e''}S_{\phi''\Gamma_e'',\Gamma_e'\Gamma_e,\Pi\Pi}\\&\times(\Pi+,\Pi-|\Sigma^-)^*(\Gamma_e a_e,\Sigma^-|\Gamma_e a_e')^*\delta_{\Gamma_e'\Gamma_e}\\&\times W(\Gamma_e\Pi\Gamma_e\Pi;\Gamma_e''\Sigma^-)\sqrt{\lambda(\Gamma_e'')}\,\langle\lambda'|N_z|\lambda\rangle\end{aligned} \tag{VIII.71b}$$

Whereas the right-hand side of (VIII.79a) does not contribute to expectation values, the second term, (VIII.71b), does so only for degenerate electronic states.

For the bent triatomic conformation, we have to formulate the problem in the instantaneous point group $C_s(C_{2v})$. We will get two types of matrix elements for those states which are electronically degenerate in the linear conformation:

1. Expectation values for the two degenerate components.
2. Transition values between the two components.

Clearly, there must be a smooth continuation from the linear to the bent case.

The expectation values are for the state $|\phi_1 A'\rangle$ and $|\phi_2 A''\rangle$:

$$\langle\phi_1 A'\lambda'|\tilde{\mathfrak{H}}_{\text{rot}}|\Phi_1 A'\lambda\rangle = \langle\lambda'|\mathbf{N}^2 - N_z^2|\lambda\rangle \times \left\{B + 2B^2 \sum_{\phi''} \left(S_{\phi''A',A'A',A'A'} + S_{\phi''A'',A'A',A''A''}\right)\right\} + \langle\lambda'|N_x^2 - N_y^2|\lambda\rangle 2B^2 \sum_{\phi''} \left(S_{\phi''A',A'A',A'A'} - S_{\phi''A'',A'A',A''A''}\right) \tag{VIII.72}$$

$$\langle\phi_2 A''\lambda'|\tilde{\mathfrak{H}}_{\text{rot}}|\phi_2 A''\lambda\rangle = \langle\lambda'|\mathbf{N}^2 - N_z^2|\lambda\rangle \times \left\{B + 2B^2 \sum_{\phi''} \left(S_{\phi''A'',A''A'',A'A'} + S_{\phi''A',A''A'',A''A''}\right)\right\} + \langle\lambda'|N_x^2 - N_y^2|\lambda\rangle 2B^2 \sum_{\phi''} \times \left(S_{\phi''A'',A''A'',A'A'} - S_{\phi''A',A''A'',A''A''}\right) \tag{VIII.73}$$

$$\langle\phi_2 A''\lambda'|\tilde{\mathfrak{H}}_{\text{rot}}|\phi_1 A'\lambda\rangle = B^2\left[\sum_{\phi''} \left(S_{\phi''A',A''A',A'A''} + S_{\phi''A'',A''A',A''A'}\right) \times \langle\lambda'|N_x N_y + N_y N_x|\lambda\rangle - i\sum_{\phi''} \left(S_{\phi''A',A''A',A'A''} - S_{\phi''A'',A''A',A''A'}\right)\langle\lambda'|N_z|\lambda\rangle\right] \tag{VIII.74}$$

with $S_{\phi''\Gamma_e'',\Gamma_e'\Gamma_e,\tilde{\Gamma}_1\tilde{\Gamma}_2}$ as defined for (VIII.69). The matrix element $\langle\phi_1 A'\lambda|\tilde{\mathfrak{H}}_{\text{rot}}|\phi_2 A''\lambda'\rangle$ is just the complex conjugate of (VIII.74). Whereas the first second-order terms in (VIII.72) and (VIII.73) yield the electronic contribution to B, the second term in (VIII.72) and (VIII.73) and the first in (VIII.74) are responsible for the Λ-doubling. The last term in (VIII.74) again gives a very small contribution to a matrix element linear in N_z and L_z.

In Fig. 4 we give a qualitative picture of the nuclear coordinate dependence in the (S_{2x}, S_{2y})-space of the electronic second-order contributions.

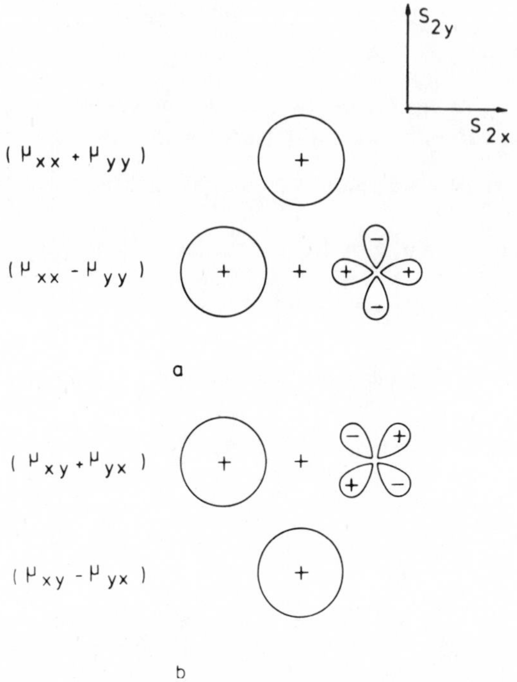

Fig. 4. Qualitative angular dependence in the S_2-conformation space of the second-order contributions to the rotational constants μ_{xx}, μ_{yy}, μ_{xy}, and μ_{yx} for a molecule with a linear reference conformation. The effective rotational operator is given by $(\mu_{xx}+\mu_{yy})(N_x^2+N_y^2)/4+(\mu_{xx}-\mu_{yy})(N_x^2-N_y^2)/4+\mu_{xy}N_xN_y/2+\mu_{yx}N_yN_x/2$. The first-order contribution is given by $B(N_x^2+N_y^2)$, where B is larger than $(\mu_{xx}+\mu_{yy})$ by some orders and is obviously spherically symmetric in the S_2-conformation space. In the upper part are shown the expectation values for the two near-degenerate states, which have a similar dependence; in the lower part, the transition function part between these two states. For $S_2=0$ the expectation values for the two states coincide and $\mu_{xx}-\mu_{yy}=\mu_{xy}+\mu_{yx}$, as can be seen from (VIII.70).

With this figure the overall symmetry of these functions in the molecular symmetry group $C_{\infty v}$ is demonstrated.

All these results may now be transferred to the terms with $\mathbf{G}^{\mathfrak{R}}$ instead of $\mathbf{N}$ [see (VIII.60)], leading essentially to the same expressions. Terms with $G_+^{\mathfrak{R}2}$ and $G_-^{\mathfrak{R}2}$ or, equivalently, $G_x^{\mathfrak{R}2}-G_y^{\mathfrak{R}2}$ and $G_x^{\mathfrak{R}}G_y^{\mathfrak{R}}+G_y^{\mathfrak{R}}G_x^{\mathfrak{R}}$ would contribute to the l-type doubling or for rovibronic states to K-type doubling. However, these terms may be neglected compared to the dominant effect resulting from the pure rovibrational interaction. The rovibrational interaction is hidden in the formulation of coupled equations for the nuclear motion problem and can be recovered only using a harmonic oscillator (or other appropriate) function basis for the vibrational eigenfunctions. In addition to the terms $\mathsf{G}^{\mathfrak{N}}\cdot\mathsf{G}^{\mathfrak{N}}$, we get as the appropriate cross terms $\mathsf{G}^{\mathfrak{N}}\cdot\mathsf{N}$.

For the spin–rotational coupling we have to consider the first-order terms of the spin–rotation Hamiltonian (VIII.42) and the second-order terms due to the perturbational coupling of $\mathfrak{H}_{\text{rot}}$ with $\mathfrak{H}_{\text{SO}}$ [see (VIII.2)]. Here the second-order contributions are generally not smaller than the first-order terms.

Allowing for matrix elements off-diagonal in the spin quantum numbers, we have to specify the spin–rotational function basis. Again we consider only the terms $N_\mu \cdot L_\nu$ of $\mathfrak{H}_{\text{rot}}$ in the perturbation sums and $N_\mu \cdot s_{i\nu}$ in $\mathfrak{H}_{\text{Sr}}$ for the first-order matrix elements. We get, for Hund's case (b),

$$
\begin{aligned}
&\langle \phi'(\Gamma'_e a'_e) N'S'J'K'M'_J | \tilde{\mathfrak{H}}_{\text{Sr}} | \phi(\Gamma_e a_e) NSJKM_J \rangle \\
&\quad = \tfrac{1}{2} \sum_{\bar{\Gamma}\bar{a}r} \left(\Gamma_e a_e, \bar{\Gamma}\bar{a} | \Gamma'_e a'_e\right)^* \sum_{k=0,2} \Bigg[\varepsilon(\bar{a}) \langle \phi'\Gamma'_e S' \| T^{\bar{\Gamma}}(\gamma_i^{(k)})_r S^1(\mathsf{s}_i) \| \phi\Gamma_e S \rangle \\
&\qquad - \sum_{\tilde{\Gamma}_1 \tilde{a}_1 r_1} \varepsilon(\tilde{a}_1) \sum_{\tilde{\Gamma}_2 \tilde{a}_2 r_2} \varepsilon(\tilde{a}_2) \sum_{\phi''\Gamma''_e S''} \overline{\Delta V}^{-1}_{\phi''\Gamma''_e S} \\
&\qquad \times \langle \phi'\Gamma'_e S' \| \sum_i T^{\tilde{\Gamma}_1}(\mathsf{u}_i)_{r_1} S^1(\mathsf{s}_i) \| \phi''\Gamma''_e S \rangle \\
&\qquad \times \langle \phi''\Gamma''_e S \| T^{\tilde{\Gamma}_2}(\mathsf{L})_{r_2} \| \phi\Gamma_e S \rangle \sum_{k'q'} \varepsilon(q') \sum_{q_1} \langle 11 q_1 q_2(\tilde{a}_2, r_2) | k'q' \rangle \mu^{k'}_{q'} \\
&\qquad \times \langle 11 \hat{q}_3(\tilde{a}_1, r_1) q_1 | k\hat{q}(\bar{a}, r) \rangle \left(\tilde{\Gamma}_2 \tilde{a}_2, \tilde{\Gamma}_1 \tilde{a}_1 | \tilde{\Gamma}\bar{a}\right)^* W\left(\Gamma_e \tilde{\Gamma}_2 \Gamma'_e \tilde{\Gamma}_1; \Gamma''_e \bar{\Gamma}\right) \\
&\qquad \times \sqrt{\lambda(\Gamma'_e)\lambda(\bar{\Gamma})} \Big\{ (-1)^{J-s'-N} \sqrt{(2N+1)(2S'+1)} \\
&\qquad\qquad \times \tilde{W}(S'SN'N; 1J) \delta_{J'J} \delta_{M'_J M_J} \\
&\qquad\qquad \times \langle NkKq | N'K' \rangle^* \tilde{W}(N'Nk1; 1N) \\
&\qquad\qquad \times \langle N \| S^1(\mathsf{N}) \| N \rangle \cdot \left((-1)^{k-1} \sqrt{(2k+1)/3}\,\right) \Big\} \Bigg]
\end{aligned}
\tag{VIII.75}
$$

plus the terms with the reversed order of s_i and N, where the factors $\langle 11 q_1 q_3 | kq \rangle$, $\sqrt{3(2N'+1)}$, $\tilde{W}(N'N1k; 1N')$, and $\langle N' \| S^1(\mathsf{N}) \| N' \rangle = \sqrt{N'(N'+1)}$, respectively, have to be substituted for the equivalent factors above.

For Hund's case (a) (i.e., $\mathbf{N}$ replaced by $\mathbf{J}-\mathbf{S}$), we get

$$\langle\phi'(\Gamma_e'a_e')S'\Sigma'J'P'M_J'|\tilde{\mathfrak{H}}_{\mathrm{Sr}}|\phi(\Gamma_e a_e)S\Sigma JPM_J\rangle$$

$$=\tfrac{1}{2}\sum_{\bar{\Gamma}\bar{a}r}\left(\Gamma_e a_e,\bar{\Gamma}\bar{a}|\Gamma_e'a_e'\right)^*\delta_{J'J}\delta_{M_J'M_J}\sum_{q_1q_2}$$

$$\left\{\sum_{k=0,2}\varepsilon(\bar{a})\langle\phi'\Gamma_e'S'\|\sum_i T^{\bar{\Gamma}}(\gamma_i^{(k)})_r S^1(\mathbf{s}_i)\|\phi\Gamma_e S\rangle\langle 11q_1q_2|k\hat{q}(\bar{a},r)\rangle\right.$$

$$-\sum_{\tilde{\Gamma}_1\tilde{a}_1r_1}\varepsilon(\tilde{a}_1)\delta_{q_1,\hat{q}_1'(\tilde{a}_1,r_1)}\sum_{\tilde{\Gamma}_2\tilde{a}_2r_2}\varepsilon(\tilde{a}_2)\sum_{\phi''\Gamma_e''S}\overline{\Delta V}^{-1}_{\phi''\Gamma_e''}$$

$$\times\langle\phi'\Gamma_e'S\|\sum_i T^{\tilde{\Gamma}_1}(\mathbf{u}_i)_{r_1}S^1(\mathbf{s}_i)\|\phi''\Gamma_e''S\rangle\langle\phi''\Gamma_e''S\|T^{\tilde{\Gamma}_2}(\mathbf{L})_{r_2}\|\phi\Gamma_e S\rangle$$

$$\times\left(\tilde{\Gamma}_2\tilde{a}_2,\tilde{\Gamma}_1\tilde{a}_1|\bar{\Gamma}\bar{a}\right)^* W\left(\Gamma_e\tilde{\Gamma}_2\Gamma_e'\tilde{\Gamma}_1;\Gamma_e''\bar{\Gamma}\right)\sqrt{\lambda(\Gamma_e'')\lambda(\bar{\Gamma})}$$

$$\times\sum_{k'q'}\varepsilon(q')\langle 11q_2q_3(\tilde{a}_2,r_2)|k'q'\rangle\mu_{\hat{q}'}^{k'}$$

$$\times\left[\langle s1\Sigma q_1|S'\Sigma'\rangle^*\langle J1Pq_2^+|J'P'\rangle^*\tilde{\varepsilon}(q_2)\langle J\|S^1(\mathbf{J})\|J\rangle\right.$$

$$-\sum_{k''q''}\langle 11q_2q_1|k''q''\rangle^*\langle Sk''\Sigma q''|S'\Sigma'\rangle^*$$

$$\left.\left.\times\tilde{W}(S1S'1;S'k'')\sqrt{(2S'+1)(2k''+1)}\,\langle S\|S^1(\mathbf{S})\|S\rangle\right]\right\}$$

(VIII.76)

plus the terms with reserved order of $\mathbf{s}_i$ and $\mathbf{N}$ with the factors $W(1S'1S;S'k'')$, $\langle S'\|S^1(S)\|S'\rangle$ instead of the equivalent factors above. Index q_2^+ and sign factor $\tilde{\varepsilon}(q_2)$ follow both from the replacement of components of the operator $\mathbf{J}$ by the appropriate components of $\mathbf{J}'$, already described [see (VIII.64)] and the text which follows. For diatomics, K in (VIII.75) and P in (VIII.76) have to be replaced by Λ and Ω, respectively.

In the same way as described for the rotational operator, we now investigate (VIII.71) and (VIII.72) for the linear molecule, but confine ourselves to the linear nuclear arrangement. To avoid the complication of matrix elements off-diagonal in S, we consider only expectation values in degenerate or nondegenerate states. In this case we may use the effective operator replacement and use a general spin–rotational function basis. We get for (VIII.75) and (VIII.76) using (VIII.63):

$$\langle\phi(\Gamma_e a_e')\lambda'|\tilde{\mathfrak{H}}_{\mathrm{Sr}}|\phi(\Gamma_e a_e)\lambda\rangle$$
$$=(\Gamma_e a_e,\Sigma^+|\Gamma_e a_e)^*\delta_{a_e'a_e}\langle\lambda'|(\mathbf{N}\cdot\mathbf{S}-N_zS_z)|\lambda\rangle/\langle S\|S^1(\mathbf{s})\|S\rangle$$
$$\times\Big\{\langle\phi\Gamma_e S\|\sum_i T^{\Sigma^+}(\gamma_i)S^1(\mathbf{s}_i)\|\phi\Gamma_e S\rangle$$
$$-2B\sum_{\phi''\Gamma_e''}S'_{\phi''\Gamma_e'',\Gamma_e\Gamma_e,\Pi\Pi}(\Pi+,\Pi-|\Sigma^+)^*$$
$$\times W(\Gamma_e\Pi\Gamma_e\Pi;\Gamma_e''\Sigma^+)\sqrt{\lambda(\Gamma_e'')}\Big\} \qquad \text{(VIII.77a)}$$

$$-2B\sum_{\phi''\Gamma_e''}S'_{\phi''\Gamma_e'',\Gamma_e\Gamma_e,\Pi\Pi}$$
$$\times\Big\{(\Gamma_e a_e,\Sigma^-|\Gamma_e a_e')^*\langle\lambda'|S_z|\lambda\rangle/\langle S\|S^1(\mathbf{S})\|S\rangle$$
$$\times(\Pi+,\Pi-|\Sigma^-)^*W(\Gamma_e\Pi\Gamma_e\Pi;\Gamma_e''\Sigma^-)\sqrt{\lambda(\Gamma_e'')} \qquad \text{(VIII.77b)}$$
$$+(\Gamma_e a_e,\Delta+|\Gamma_e a_e')^*\langle\lambda'|N_-S_-|\lambda\rangle$$
$$\times(\Pi+,\Pi+|\Delta+)^*W(\Gamma_e\Pi\Gamma_e\Pi;\Gamma_e''\Delta)\sqrt{2\lambda(\Gamma_e'')} \qquad \text{(VIII.77c)}$$
$$+(\Gamma_e a_e,\Delta-|\Gamma_e a_e')^*\langle\lambda'|N_+S_+|\lambda\rangle$$
$$\times(\Pi-,\Pi-|\Delta-)^*W(\Gamma_e\Pi\Gamma_e\Pi,\Gamma_e''\Delta)\sqrt{2\lambda(\Gamma_e'')}\Big\} \qquad \text{(VIII.77d)}$$

with

$$S'_{\phi''\Gamma_e'',\Gamma_e\Gamma_e,\Pi\Pi}$$
$$=\overline{\Delta V}^{-1}_{\phi''\Gamma_e''}\langle\phi\Gamma_e S\|\sum_i T^{\Pi}(\mathbf{u}_i)S^1(\mathbf{s}_i)\|\phi''\Gamma_e''S\rangle$$
$$\times\langle\phi''\Gamma_e''S\|T^{\Pi}(\mathbf{L})\|\phi\Gamma_e S\rangle$$

In this approximation the Π-type operators are not considered, since they are off-diagonal in Γ_e.

In (VIII.77a), we find the "main" spin–rotational effect. Equations (VIII.77c) and (VIII.77d) give the contribution to the Λ-doubling due to this perturbation. The term (VIII.77b) is not so important; it is proportional to $L_z\cdot S_z$ (i.e., is a spin–orbit splitting term) and may be neglected compared to the main spin–orbit effect presented in Section VIII.B.

The last term to be discussed is the rotation–orbit operator $\mathfrak{H}_{\mathrm{Or}}$ [see Appendix 1, (A.66)]. Its influence in spectroscopy is negligible since it has no diagonal elements in the adiabatic states. It can only couple states of different symmetry (i.e., states with Λ-values different by one). Moss and Howard assume that the condition for an orbit–rotation effect can be fulfilled in degenerate states of linear molecules. However, since it is an electronic Π-operator, it cannot have expectation values in the adiabatic approximation. We assume that the orbit–rotational effect, as well as the orbit–vibration interaction, will have no practical significance. If the conditions for coupling of near-degenerate states with the correct symmetry are fulfilled, even the first-order effect in the resulting vibronic states will be negligible, since it will produce nondiagonal matrix elements in the rotational function basis.

D. Nuclear Spin Interactions (Hyperfine Structure)

The nuclear spins may interact with the spins of the electrons and of other nuclei and with the motions of electrons and nuclei. Here we consider the magnetic dipole and the electrostatic quadrupole interactions producing the hyperfine structure of molecular states.

The strongest effects are caused by the nuclear spin–electron spin and nuclear spin–electronic orbit operator [see Appendix 1, (A.68)] in open-shell states. Operators $\mathfrak{H}_{\mathrm{SI}_\alpha}$ and $\mathfrak{H}_{\mathrm{OI}_\alpha}$ both correspond with respect to their symmetry to the spin–spin and spin–orbit coupling operators discussed in Section VIII.B. In the abbreviated form analogous to (VIII.28), (VIII.30), and (VIII.32), we may write in symmetry notation for the field-free case:

$$\mathfrak{H}_{\mathrm{OI}_\alpha} = \sum_{\tilde{\Gamma}\tilde{a}r} \varepsilon(\tilde{a}) \sum_i T_{\tilde{a}}^{\tilde{\Gamma}}(\mathbf{a}_{i\alpha})_r S_{\hat{q}(\tilde{a},r)}^1(\mathbf{I}_\alpha) \tag{VIII.78}$$

$$\mathfrak{H}_{\mathrm{SI}_\alpha}^{(0)} = \sum_i b_{i\alpha}^{(0)}(\mathbf{s}_i \cdot \mathbf{I}_\alpha) \tag{VIII.79}$$

$$\mathfrak{H}_{\mathrm{SI}_\alpha}^{(2)} = \sum_{\bar{\Gamma}\bar{a}r} \varepsilon(\bar{a}) \sum_i T_{\bar{a}}^{\bar{\Gamma}}(b_{i\alpha}^{(2)})_r S_{\hat{q}(\bar{a},r)}^2(\mathbf{s}_i, \mathbf{I}_\alpha) \tag{VIII.80}$$

In (VIII.78), $\mathbf{a}_{i\alpha}$ transforms in the same way as the electronic angular momentum l_i. Since the second-order contributions for the effective operators are generally negligible, the effective matrix elements are easily written down. We again use the symbol λ for the set of rotational and spin quantum numbers of various choices of angular momentum couplings if this set needs not to be specified. Thus we have for $\mathfrak{H}_{\mathrm{OI}_\alpha}$,

$$\begin{aligned}\langle\phi'(\Gamma_e' a_e')\lambda'|\tilde{\mathfrak{H}}_{\mathrm{OI}_\alpha}|\phi(\Gamma_e a_e)\lambda\rangle = \sum_{\tilde{\Gamma}\tilde{a}r} \varepsilon(\tilde{a})\langle\phi'\Gamma_e'\|\sum_i T^{\tilde{\Gamma}}(\mathbf{a}_{i\alpha})_r\|\phi\Gamma_e\rangle \\ \times\left(\Gamma_e a_e, \tilde{\Gamma}\tilde{a}|\Gamma_e' a_e'\right)^*\langle\lambda'|S_{\hat{q}(\tilde{a},r)}^1(\mathbf{I}_\alpha)|\lambda\rangle\end{aligned} \tag{VIII.81}$$

plus higher-order terms. For the nuclear spin–electronic spin coupling operator, we define a function basis allowing for matrix elements nondiagonal in S. For Hund's case (a), we use

$$|S\Sigma J I_\alpha F P M_F\rangle = |S\Sigma\rangle \sum_{M_J M_{I_\alpha}} \langle J I M_J M_{I_\alpha}|F M_F\rangle |J P M_J\rangle |I_\alpha M_{I_\alpha}\rangle \tag{VIII.82}$$

for one nuclear spin coupled to the total angular momentum F.

For Hund's case (b), we use

$$|S N J I_\alpha F K M_F\rangle = \sum_{M_N M_S} \sum_{M_J M_{I_\alpha}} \langle N S M_N M_S|J M_J\rangle \langle J I_\alpha M_J M_{I_\alpha}|F M_F\rangle |S M_S\rangle |N K M_N\rangle |I_\alpha M_{I_\alpha}\rangle \tag{VIII.83}$$

(see also Section V).

In addition, we make deliberate use of the decomposition

$$S_q^k(\mathsf{l}_\alpha, \mathsf{s}_i) = \sum_{q_1 q_2} \langle 1 1 q_1 q_2|k q\rangle S_{q_1}^1(\mathsf{l}_\alpha) S_{q_2}^1(\mathsf{s}_i) \tag{VIII.84}$$

For Hund's case (a), we obtain the Fermi contact term

$$\begin{aligned}
&\langle \phi'(\Gamma_e' a_e') S' \Sigma' J' I_\alpha' F' P' M_F' |\tilde{\mathfrak{H}}^{(0)}_{\mathrm{SI}_\alpha}| \phi(\Gamma_e a_e) S \Sigma J I_\alpha F P M_F\rangle \\
&\quad = \langle \phi' \Gamma_e S' \| \sum_i b_{i\alpha}^{(0)} S^1(\mathsf{s}_i) \| \phi \Gamma_e S\rangle \delta_{\Gamma_e' \Gamma_e} \delta_{a_e' a_e} \delta_{F' F} \delta_{M_F' M_F} \\
&\qquad \times \sum_{q_1} \langle S 1 \Sigma q_1 | S' \Sigma'\rangle^* \langle J 1 P q_1 | J' P'\rangle^* (-1)^{F - I_\alpha - J'} \sqrt{(2I_\alpha + 1)(2J' + 1)} \\
&\qquad \times \tilde{W}(I_\alpha I_\alpha J' J; 1F) \langle I_\alpha \| S^1(\mathsf{l}_\alpha) \| I_\alpha\rangle \delta_{I_\alpha' I_\alpha}
\end{aligned} \tag{VIII.85}$$

and

$$\begin{aligned}
&\langle \phi'(\Gamma_e' a_e') S' \Sigma' J' I_\alpha' F' P' M_F' |\tilde{\mathfrak{H}}^{(2)}_{\mathrm{SI}_\alpha}| \phi(\Gamma_e a_e) S \Sigma J I_\alpha F P M_F\rangle \\
&\quad = \sum_{\bar{\Gamma} \bar{a} r} \varepsilon(\bar{a}) \langle \phi' \Gamma_e' S' \| \sum_i T^{\bar{\Gamma}}(b_{i\alpha}^{(2)})_r S^1(\mathsf{s}_i) \| \phi \Gamma_e S\rangle \\
&\qquad \times \left(\Gamma_e a_e, \bar{\Gamma} \bar{a} | \Gamma_e' a_e'\right)^* \sum_{q_1 q} \langle 1 2 q_1 q_2 (\bar{a}, r) | 1 q\rangle \left(-\sqrt{5/3}\,\right) \\
&\qquad \times \langle S 1 \Sigma q_1 | S' \Sigma'\rangle^* \langle J 1 P q | J' P'\rangle^* \\
&\qquad \times (-1)^{F - I_\alpha - J'} \sqrt{(2I_\alpha + 1)(2J' + 1)}\; \tilde{W}(I_\alpha I_\alpha J' J; 1F) \\
&\qquad \times \langle I_\alpha \| S^1(\mathsf{l}_\alpha) \| I_\alpha\rangle \delta_{I_\alpha' I_\alpha} \delta_{F' F} \delta_{M_F' M_F}
\end{aligned} \tag{VIII.86}$$

For Hund's case (b), we get

$$\langle\phi'(\Gamma_e'a_e')S'N'J'I_\alpha'F'K'M_F'|\tilde{\mathfrak{H}}^{(0)}_{\mathrm{SI}_\alpha}|\phi(\Gamma_e a_e)SNJI_\alpha FKM_F\rangle$$
$$=\langle\phi'\Gamma_e S'\|\sum_i b^{(0)}_{i\alpha}S^1(\mathbf{s}_i)\|\phi\Gamma_e S\rangle\delta_{\Gamma_e'\Gamma_e}\delta_{a_e'a_e}\delta_{F'F}\delta_{M_F'M_F}$$
$$\times(-1)^{F-I_\alpha-J'}\sqrt{(2I_\alpha+1)(2J'+1)}\;\tilde{W}(I_\alpha I_\alpha J'J;1F)$$
$$\times(-1)^{S'-N+J-1}\sqrt{(2J+1)(2S'+1)}\;\tilde{W}(S'SJ'J;1N)$$
$$\times\langle I_\alpha\|S^1(\mathbf{I}_\alpha)\|I_\alpha\rangle\delta_{I_\alpha'I_\alpha} \qquad \text{(VIII.87)}$$

and

$$\langle\phi'(\Gamma_e'a_e')S'N'J'I_\alpha'F'K'M_F'|\tilde{\mathfrak{H}}^{(2)}_{\mathrm{SI}_\alpha}|\phi(\Gamma_e a_e)SNJI_\alpha FKM_F\rangle$$
$$=\sum_{\bar{\Gamma}\bar{a}r}\varepsilon(\bar{a})\langle\phi'\Gamma_e'S'\|\sum_i T^{\bar{\Gamma}}(b^{(2)}_{i\alpha})_r S^1(\mathbf{s}_i)\|\phi\Gamma_e S\rangle$$
$$\times\left(\Gamma_e a_e,\bar{\Gamma}\bar{a}|\Gamma_e'a_e'\right)^*\langle N2Kq(\bar{a},r)|NK\rangle^*$$
$$\times(-1)^{F-J'-I_\alpha}\sqrt{(2I_\alpha+1)(2J'+1)}\;\tilde{W}(I_\alpha I_\alpha J'J;1F)$$
$$\times\sqrt{(2S'+1)(2J+1)(2N+1)3}\begin{Bmatrix} J' & J & 1\\ S' & S & 1\\ N' & N & 2\end{Bmatrix}\left(-\sqrt{5/3}\right)$$
$$\times\langle I_\alpha\|S^1(\mathbf{I}_\alpha)\|I_\alpha\rangle\delta_{I_\alpha'I_\alpha}\delta_{F'F}\delta_{M_F'M_F} \qquad \text{(VIII.88)}$$

where the term in braces is the 9*j*-symbol of Wigner.[41]

For the decoupled basis $|JM_J\rangle|I_\alpha M_{I_\alpha}\rangle$, we only have to replace

$$(-1)^{F-J'-I_\alpha}\sqrt{(2J'+1)(2I_\alpha+1)}\;\tilde{W}(I_\alpha I_\alpha J'J;kF)$$

by

$$\sum_{q_1q_2}\langle JkM_Jq_1|J'M_J'\rangle\langle I_\alpha kM_{I_\alpha}q_2|I_\alpha M_I'\rangle\langle 11q_1q_2|00\rangle\sqrt{3} \qquad \text{(VIII.89)}$$

As already mentioned, the interpretation of these equations is quite similar to those of Section VIII.B. For convenience, we again discuss some simple results for diatomics.

The nuclear spin-orbit coupling, (VIII.78), may contribute only for $\tilde{\Gamma}=\Sigma^-$, for such expectation values $\mathfrak{H}_{\mathrm{OI}}$ may be replaced by $\tilde{a}(R)\cdot L_z\cdot I_{\alpha z}$ with

$$\tilde{a}(R)=\langle\phi\Gamma_e\|\sum_i T^{\Sigma^-}(\mathbf{a}_i)\|\phi\Gamma_e\rangle/|\Lambda| \qquad \text{(VIII.90)}$$

and

$$T^{\Sigma^-}(\mathbf{a}_i)=(g_\alpha\beta_N e/mc)r_{i\alpha}^{-3}l_{iz}(\alpha) \tag{VIII.91}$$

The nuclear spin–electronic spin couplings, (VIII.78) to (VIII.80), contain species Σ^+ for $\mathfrak{H}_{\mathrm{SI}_\alpha}^{(0)}$ and $\mathfrak{H}_{\mathrm{SI}_\alpha}^{(2)}$, Π and Δ for $\mathfrak{H}_{\mathrm{SI}\alpha}^{(2)}$. The Π-species do not contribute to expectation values, whereas the Δ-term gives rise to Λ-doubling elements. Thus we get three effective constants in the following effective operators from (VIII.85) or (VIII.87):

(a) $\tilde{c}(R)(\mathbf{S}\cdot\mathbf{I}_\alpha)$, with

$$\tilde{c}(R)=\langle\phi\Gamma_e S\|\sum_i b_{i\alpha}^{(0)}S^1(\mathbf{s}_i)\|\phi\Gamma_e S\rangle/\langle S\|S^1(\mathbf{S})\|S\rangle$$

and

$$b_{i\alpha}^{(0)}=(\beta g\beta_N g_\alpha 8\pi/3)\delta(\mathbf{r}_{i\alpha})$$

and from (VIII.86) or (VIII.88).

(b) $\tilde{b}(R)S_0^2(\mathbf{S},\mathbf{I}_\alpha)$, with

$$\tilde{b}(R)=\langle\phi\Gamma_e S\|\sum_i T^{\Sigma+}(b_{i\alpha}^{(2)})S^1(\mathbf{s}_i)\|\phi\Gamma_e S\rangle/\langle S\|S^1(\mathbf{S})\|S\rangle$$

and

$$T^{\Sigma^+}(b_{i\alpha}^{(2)})=\beta g\beta_N g_\alpha r_{i\alpha}^{-3}(3\cos^2\theta_{i\alpha}-1)/\sqrt{6}$$

with the addition for Π-states.

(c) $\tilde{d}(R)\frac{1}{2}(S_+I_{\alpha+}e^{-i2\phi}+S_-I_{\alpha-}e^{+i2\phi})$, with

$$\tilde{d}(R)=\langle\phi\Pi S\|\sum_i T^{\Delta}(b_{i\alpha}^{(2)})S^1(\mathbf{s}_i)\|\phi\Pi S\rangle/\langle S\|S^1(\mathbf{S})\|S\rangle$$

and

$$T_{\pm}^{\Delta}(b_{i\alpha}^{(2)})=\beta g\beta_N g_\alpha r_{i\alpha}^{-3}\sin^2\theta_{i\alpha}e^{\pm i2\phi_i}$$

where $\theta_{i\alpha}$ and ϕ_i define the orientation of the vector $\mathbf{r}_i-\mathbf{r}_\alpha$ in spherical coordinates.

In closed-shell states the hyperfine splittings are smaller, since there are no first-order effects of $\mathfrak{H}_{\mathrm{OI}_\alpha}$ and $\mathfrak{H}_{\mathrm{SI}_\alpha}$. In those states, the nuclear quadrupole coupling operator will, in general, have the strongest influence [Appendix 1, (A.44)].

For the dominant part (i.e., the part containing electronic coordinates) we may write for the quadrupole coupling operator, (A.44):

$$\mathfrak{H}_Q = \sum_{\Gamma \bar{a} r} \varepsilon(\bar{a}) \sum_i T_a^{\Gamma}(D^2(i,\alpha))_r S^2_{\hat{q}(\bar{a},r)}(\mathsf{I}_\alpha, \mathsf{I}_\alpha) \qquad \text{(VIII.92)}$$

Neglecting all additional pseudoquadrupole coupling and second-order contribution (see, e.g., Mizushima[50] for diatomics), we get for the general case with an unspecified rotational basis,

$$\langle \phi'(\Gamma'_e a'_e)\lambda' | \tilde{\mathfrak{H}}_Q | \phi(\Gamma_e a_e)\lambda \rangle = \sum_{\bar{\Gamma}\bar{a}r} \varepsilon(\bar{a}) \langle \phi' \Gamma'_e \| \sum_i T^{\bar{\Gamma}}(D^2(i,\alpha))_r \| \phi \Gamma_e \rangle$$

$$\times \left(\Gamma_e a_e, \bar{\Gamma}\bar{a} | \Gamma'_e a'_e\right)^* \langle \lambda' | S^2_{\hat{q}(\bar{a},r)}(\mathsf{I}_\alpha, \mathsf{I}_\alpha) | \lambda \rangle$$

(VIII.93)

For linear molecules we again have $\bar{\Gamma} = \Sigma^+$ and $\bar{\Gamma}\bar{a} = \Delta\pm$ contributions, the latter for Π-states only.

Another very small effect is caused by the coupling of the nuclear spins with molecular rotation [Appendix 1, (A.71) to (A.79)]. Since it is very weak, it is considered only for very high resolution rotational spectroscopy in the microwave region. Since in electronic open-shell states other couplings are much stronger, we may restrict ourselves to closed-shell states (i.e., we may neglect first-order matrix elements of l_i and s_i). The operator, presented in Appendix 1, also contains the Thomas correction of Ramsey,[51] which accounts for the relativistic corrections analogous to the spin–orbit operator of Section VIII.B.

In abbreviated form we may write

$$\mathfrak{H}_{I_\alpha r} = -I''^{1/2}(\mathsf{I}_\alpha \cdot \mathbf{m}\mathsf{R} + \mathsf{R} \cdot \mathbf{m}\mathsf{I}_\alpha) I''^{-1/2} \qquad \text{(VIII.94)}$$

with

$$\mathbf{m} = (\hbar e \beta_N / 2c)\mathbf{I}''^{-1} \sum_{\beta(\neq\alpha)} Z_\beta \left[(\mathsf{r}^0_\beta \cdot \mathsf{r}_{\alpha\beta})\mathbf{1} - \mathsf{r}^0_\beta \otimes \mathsf{r}_{\alpha\beta} \right] r_{\alpha\beta}^{-3}$$

$$+ (e g_\alpha \beta_N / 2c)(1 - \gamma_\alpha) \sum_{\beta(\neq\alpha)} Z_\beta \left[(\mathsf{r}_\alpha \cdot r_{\beta\alpha})\mathbf{1} - \mathsf{r}_\alpha \otimes \mathsf{r}_{\beta\alpha} \right] r_{\alpha\beta}^{-3}$$

$$- (e g_\alpha \beta_N / 2c)(1 - \gamma_\alpha) \sum_i \left[(\mathsf{r}_\alpha \cdot \mathsf{r}_{i\alpha})\mathbf{1} - \mathsf{r}_\alpha \otimes \mathsf{r}_{i\alpha} \right] r_{i\alpha}^{-3} \qquad \text{(VIII.95)}$$

and $\mathsf{R} = \mathsf{N} - \mathsf{G}$. The effective nuclear spin–rotation coupling operator contains second-order contributions of the same size due to the coupling of $\mathfrak{H}_{I_\alpha r}$ with $\mathfrak{H}_{\text{rot}}$.

Before we use symmetry notation, let us consider what we get in a nondegenerate state with the adiabatic approximation valid for a general polyatomic molecule without any symmetry. The effective operator will be

$$\tilde{\mathfrak{H}}_{I_\alpha r} = -I''^{1/2}\left[\mathsf{I}_\alpha \cdot \mathbf{M}(\mathsf{N}-\mathsf{G}) + (\mathsf{N}-\mathsf{G})\cdot \mathbf{M}^{+}\mathsf{I}_\alpha\right] I''^{-1/2} \qquad \text{(VIII.96)}$$

with

$$\mathbf{M} = \langle \phi | \mathbf{m} | \phi \rangle + \mathbf{n}$$

and

$$\mathbf{n} = (e g_\alpha \beta_N / mc) \sum_{\phi''}{}' \frac{\langle \phi | \sum_i \boldsymbol{l}_i(\alpha) r_{i\alpha}^{-3} | \phi'' \rangle \otimes \langle \phi'' | \mathsf{L} | \phi \rangle}{V_\phi - V_{\phi''}} \boldsymbol{\mu}$$

where $\otimes$ denotes the dyadic product. By a change of the origin to the nucleus α it is possible to derive a more symmetrical expression. It should be mentioned, however, that the Thomas corrections in $\mathbf{M}$ will not vanish in this case as claimed by Flygare[52]; however, its contribution will be rather small. Making the transformation we get nearly the same expressions as before, with r_β^0 in (VIII.94) replaced by $\mathsf{r}_{\beta\alpha}^0$ and L in (VIII.96) replaced by $\mathsf{L}(\alpha)$ (i.e., referring to the nucleus α), plus a correction term that cancels parts of the two Thomas correction terms, leaving only the $(-\gamma_\alpha)$ of the $(1-\gamma_\alpha)$-factor. Clearly, using symmetry will give some more information on the elements of the $\mathbf{m}$-dyadic.

For a more detailed investigation we express the nuclear spin-rotation interaction operator, (VIII.88), as

$$\begin{aligned}\mathfrak{H}_{I_\alpha r} &= \left(-\mathrm{Tr}(\mathbf{m})/\sqrt{3}\,\right) S_0^0(\mathsf{I}_\alpha, \mathsf{N}-\mathsf{G}) \\ &\quad + \sum_{\Gamma a r} \varepsilon(a)\left[\, T_a^\Gamma(m^{(2)}),\, S_{\hat{q}(a,r)}^2(\mathsf{I}_\alpha, \mathsf{N}-\mathsf{G}) \right. \\ &\qquad \left. + S_{\hat{q}(a,r)}^2(\mathsf{I}_\alpha, \mathsf{N}-\mathsf{G})\, T_a^\Gamma(m^{(2)}) \right] \qquad \text{(VIII.97)}\end{aligned}$$

The operators leading to the second-order contributions are $\mathfrak{H}_{\mathrm{OI}_\alpha}$, (VIII.78), and the relevant parts of $\mathfrak{H}_{\mathrm{rot}}$, which we write in the form

$$\mathfrak{H}'_{\mathrm{rot}} = \sum_{\tilde{\Gamma}\tilde{a}r} \varepsilon(\tilde{a})\, T^{\tilde{\Gamma}}\tilde{a}(\mathsf{L}),\, S_{\hat{q}(\tilde{a},r)}^1(\boldsymbol{\mu}(\mathsf{N}-\mathsf{G})) \qquad \text{(VIII.98)}$$

where

$$\begin{aligned}S_q^1(\boldsymbol{\mu}(\mathsf{N}-\mathsf{G})) &= \mu_0^{(0)} S_q^1(N-G) \\ &\quad - \left(\sqrt{5/3}\,\right) \sum_{q_2 q_1} \langle 2 1 q_2 q_1 | 1 q \rangle \mu_{q_2}^{(2)} S_{q_1}^1(\mathsf{N}-\mathsf{G}) \qquad \text{(VIII.99)}\end{aligned}$$

From eqs. (VIII.78), (VIII.97), and (VIII.98) we get for the expectation value in a non-degenerate electronic state for general rotational-spin functions

$$\langle\phi\Gamma_e\lambda'|\tilde{\mathfrak{H}}_{I_\alpha r}|\phi\Gamma_e\lambda\rangle$$

$$=\langle\Lambda'|-S_0^0(\mathsf{I}_\alpha,\mathsf{N}-\mathsf{G})|\lambda\rangle$$

$$\times\Bigg\{\langle\phi\Gamma_e\|Tr(\mathbf{m})/\sqrt{3}\,\|\phi\Gamma_e\rangle+\sum_{\tilde{\Gamma}_1\tilde{a}_1r_1}\varepsilon(\tilde{a}_1)\sum_{\tilde{\Gamma}_2\tilde{a}_2r_2}\varepsilon(\tilde{a}_2)\sum_{\phi''\Gamma_e''}{}'\,\overline{\Delta V}_{\phi''\Gamma_e''}^{-1}$$

$$\times\langle\phi\Gamma_e\|\Sigma_i T^{\tilde{\Gamma}_i}(\mathsf{a}_i)_{r_1}\|\phi''\Gamma_e''\rangle\langle\phi''\Gamma_e''\|T^{\tilde{\Gamma}_2}(\mathsf{L})_{r_2}\|\phi\Gamma_e\rangle(\tilde{\Gamma}_2\tilde{a}_2,\tilde{\Gamma}_1\tilde{a}_1|\Gamma_0)$$

$$\times W(\Gamma_e\tilde{\Gamma}_2\Gamma_e\tilde{\Gamma}_1;\Gamma_e''\Gamma_0)\sqrt{\lambda(\Gamma_e'')}$$

$$\times\Bigg[\langle 11\hat{q}_1(\hat{a}_1,r_2)\hat{q}_2(\tilde{a}_2,r_2)|00\rangle^*\mu_0^{(0)}+\sum_{q_3q_4}\langle 11\hat{q}_1(\hat{a}_1,r_1)q_3|00\rangle^*$$

$$\times\langle 21q_4q_3|1\hat{q}_2(\tilde{a}_2,r_2)\rangle\mu_{q_4}^{(2)}\Big(-\sqrt{5/3}\Big)\Bigg]\Bigg\}$$

$$+\sum_r\langle\lambda'|S_{q(r)}^2(\mathsf{I}_\alpha,\mathsf{N}-\mathsf{G})|\lambda\rangle$$

$$\times\Bigg\{\langle\phi\Gamma_e\|T^{\Gamma_0}(m^{(2)})_r\|\phi\Gamma_e\rangle+\sum_{\tilde{\Gamma}_1\tilde{a}_1r_1}\varepsilon(\tilde{a}_1)\sum_{\tilde{\Gamma}_2\tilde{a}_2r_2}\varepsilon(\tilde{a}_2)\sum_{\phi''\Gamma_e''}{}'\,\overline{\Delta V}_{\phi''\Gamma_e''}^{-1}$$

$$\times\langle\phi\Gamma_e\|\Sigma_i T^{\tilde{\Gamma}_1}(\mathsf{a}_i)_{r_1}\|\phi''\Gamma_e''\rangle\langle\phi''\Gamma_e''\|T^{\tilde{\Gamma}_2}(\mathsf{L})_{r_2}\|\phi\Gamma_e\rangle(\tilde{\Gamma}_2\tilde{a}_2,\tilde{\Gamma}_1\tilde{a}_1|\Gamma_0)^*$$

$$\times W(\Gamma_e\tilde{\Gamma}_2\Gamma_e\tilde{\Gamma}_1;\Gamma_e''\Gamma_0)\sqrt{\lambda(\Gamma_e'')}$$

$$\times\Bigg[\langle 11\hat{q}_1(\tilde{a}_1,r_1)\hat{q}_2(\tilde{a}_2,r_2)|2\hat{q}(r)\rangle^*\mu_0^{(0)}+\sum_{q_3q_4}\langle 11\hat{q}_1(\tilde{a}_1,r_1)q_3|2q(r)\rangle^*$$

$$\times\langle 21q_4q_3|1\hat{q}_2(\tilde{a}_2,r_2)\rangle\,\mu_{q_4}^{(2)}\Big(-\sqrt{5/3}\Big)\Bigg]\Bigg\}\qquad\text{(VIII.100)}$$

plus all terms with I_α and $(\mathsf{N}-\mathsf{G})$ reversed.

In (VIII.96), a nuclear spin–vibration coupling is already present through the vibrational angular momentum G. From the nuclear spin–

vibrational operators [see Appendix 1, (A.78) to (A.80)] we get an additional contribution of the same kind, which may contribute in degenerate excited vibrational modes. This can be deduced from the operator (A.79) by replacing $\mathbf{r}_{\beta\alpha}$ with $\mathbf{r}_\beta^0 - \mathbf{r}_\alpha + (\mathbf{r}_\beta - \mathbf{r}_\beta^0)$ and introducing $(1/\sqrt{M\beta})\Sigma_k l_k(\beta) S_k$ for $(\mathbf{r}_\beta - \mathbf{r}_\beta^0)$. Thus we get in (A.79) terms of the sort

$$\sum_{\beta(\neq\alpha)} (Z_\beta / M_\beta) r_{\alpha\beta}^{-3} \zeta_{lk} S_l P_k \tag{VIII.101}$$

which give rise to a vibrational angular momentum-type operator with a first-order effect for degenerate vibrational modes, which we will not consider in detail; however, see the vibrational Zeeman effect in the next section.

The last coupling to be discussed is the nuclear spin–spin coupling, which again is quite small. Its importance lies mainly in NMR spectroscopy, but spin–spin coupling may also be observed in high-resolution microwave spectra. Whereas in NMR spectroscopy only the rotational averaged values (i.e., the traces of the spin–spin coupling tensors) can be measured, rotational spectroscopy will give information about the anisotropies of the diagonal elements. There are a large number of first- and second-order contributions to the effective spin–spin coupling constants which are defined through $\mathbf{I}_\alpha \cdot \mathbf{J} \mathbf{I}_\beta$.

First-order contributions come from the operators, given in Appendix 1, (A.69), which we may express as

$$\begin{aligned}\mathfrak{H}_{I_\alpha I_\beta} = & \sum_{m=-2}^{2} (-1)^m \tilde{T}_m^2(\mathbf{I}_\alpha, \mathbf{I}_\beta) D_{-m}^2(\alpha,\beta) \\ & + \sum_i \left(\tilde{T}_0^0(\mathbf{I}_\alpha, \mathbf{I}_\beta) Tr(\boldsymbol{\rho}_i) / \sqrt{3} \right. \\ & + \sum_{m=-2}^{2} (-1)^m \tilde{T}_m^2(\mathbf{I}_\alpha, \mathbf{I}_\beta) \tilde{T}_{-m}^2(\boldsymbol{\rho}_i) \end{aligned} \tag{VIII.102}$$

where

$$D_m^2(\alpha,\beta) = \beta_N^2 g_\alpha g_\beta \sqrt{6}\, C_m^2(\omega_{\alpha\beta}) r_{\alpha\beta}^{-3} \tag{VIII.103}$$

and

$$\begin{aligned}\boldsymbol{\rho}_i = & \left(\beta_N^2 g_\alpha g_\beta e^2 / 2mc^2 \right) r_{i\alpha}^{-3} r_{i\beta}^{-3} \\ & \left[(\mathbf{r}_{i\alpha} \cdot \mathbf{r}_{i\beta}) \mathbf{1} - \mathbf{r}_{i\alpha} \otimes \mathbf{r}_{i\beta} \right] \end{aligned} \tag{VIII.104}$$

The second-order contributions arise from the perturbational coupling of the operators discussed in the beginning of this section [i.e., $\mathfrak{H}_{\mathrm{OI}_\alpha}$, (VIII.78); and $\mathfrak{H}_{\mathrm{SI}_\alpha}$, (VIII.79) and (VIII.80)]. Expressing the coordinate function in $\mathfrak{H}_{I_\alpha I_\beta}$ in group-symmetrized form and using (VIII.78) to

(VIII.80), we get for a nondegenerate electronic state with $S=0$ in the adiabatic approximation, the following matrix elements with the rotational spin function basis $|\lambda\rangle$ unspecified:

$$
\begin{aligned}
&\langle\phi\Gamma_e\lambda'|\tilde{\mathfrak{H}}_{I_\alpha I_\beta}|\phi\Gamma_e\lambda\rangle=\langle\lambda'|S_0^0(\mathbf{I}_\alpha,\mathbf{I}_\beta)|\lambda\rangle\\
&\times\Big\{\langle\phi\Gamma_e\|-\sum_i Tr(\boldsymbol{\rho}_i)/\sqrt{3}\,\|\phi\Gamma_e\rangle+\sum_{\tilde{\Gamma}_1\tilde{a}_1r_1}\varepsilon(\tilde{a}_1)\sum_{\tilde{\Gamma}_2\tilde{a}_2r_2}\varepsilon(\tilde{a}_2)\sum_{\phi''\Gamma_e''}\overline{\Delta V}^{-1}_{\phi''\Gamma_e''}\\
&\quad\times\langle 11\hat{q}_2(\tilde{a}_2,r_2)\hat{q}_1(\tilde{a}_1,r_1)|00\rangle\langle\phi\Gamma_e\|\sum_i T^{\tilde{\Gamma}_1}(\mathbf{a}_i)_{r_1}\|\phi''\Gamma_e''\rangle\\
&\quad\times\langle\phi''\Gamma_e''\|\sum_i T^{\tilde{\Gamma}_2}(\mathbf{a}_i)_{r_2}\|\phi\Gamma_e\rangle(\tilde{\Gamma}_2\tilde{a}_2,\tilde{\Gamma}_1\tilde{a}_1|\Gamma_0)^*\delta_{\tilde{\Gamma}_1\tilde{\Gamma}_2}W(\Gamma_e\tilde{\Gamma}_1\Gamma_e\tilde{\Gamma}_1;\Gamma_e''\Gamma_0)\\
&\quad\times\sqrt{\lambda(\Gamma_e'')}+\sum_{\phi''}\overline{\Delta V}^{-1}_{\phi''\Gamma_e}|\langle\phi\Gamma_eS\|\sum_i c_iS^1(\mathbf{s}_i)\|\phi\Gamma_eS+1\rangle|^2\\
&\quad+\sum_{\bar{\Gamma}_1\bar{a}_1r_1}\varepsilon(\bar{a}_1)\sum_{\bar{\Gamma}_2\bar{a}_2r_2}\varepsilon(\bar{a}_2)\sum_{\phi''\Gamma_e''}\overline{\Delta V}^{-1}_{\phi''\Gamma_e''S+1}\langle 11\hat{q}_2(\bar{a}_2,r_2)\hat{q}_1(\bar{a}_1,r_1)|00\rangle\\
&\quad\times\langle\phi\Gamma_eS\|\sum_i T^{\bar{\Gamma}_1}(\mathbf{b}_i)_{r_i}S^1(\mathbf{s}_i)\|\phi''\Gamma_e''S+1\rangle\\
&\quad\times\langle\phi''\Gamma_e''S+1\|\sum_i T^{\bar{\Gamma}_2}(\mathbf{b}_i)_{r_2}S^1(\mathbf{s}_i)\|\phi\Gamma_eS\rangle(\bar{\Gamma}_2\bar{a}_2,\bar{\Gamma}_1\bar{a}_1|\Gamma_0)^*\\
&\quad\times\delta_{\bar{\Gamma}_1\bar{\Gamma}_2}W(\Gamma_e\bar{\Gamma}_1\Gamma_e\bar{\Gamma}_1;\Gamma_e''\Gamma_0)\sqrt{\lambda(\Gamma_e'')}+\sum_{\Gamma r}\delta_{\Gamma\Gamma_0}\langle\lambda'|S^2_{q(r)}(\mathbf{I}_\alpha,\mathbf{I}_\beta)|\lambda\rangle\\
&\quad\times\Big\{T^{\Gamma_0}(\tilde{\mathbf{C}}^2(\mathbf{r}_\alpha,\mathbf{r}_\beta))_r+\langle\phi\Gamma_e\|T^{\Gamma_0}(\boldsymbol{\rho}_i^{(2)})_r\|\phi\Gamma_e\rangle+\sum_{\tilde{\Gamma}_1\tilde{a}_1r_1}\varepsilon(\tilde{a}_1)\sum_{\tilde{\Gamma}_2\tilde{a}_2r_2}\varepsilon(\tilde{a}_2)\\
&\quad\times\sum_{\phi''\Gamma_e''}\overline{\Delta V}^{-1}_{\phi''\Gamma_e''}\langle 11\hat{q}_2(\hat{a}_2,r_2)q_1(\tilde{a}_1,r_1)|2\hat{q}(r)\rangle^*\langle\phi\Gamma_e\|\sum_i T^{\tilde{\Gamma}_1}(\mathbf{a}_i)_{r_1}\|\phi''\Gamma_e''\rangle\\
&\quad\times\langle\phi''\Gamma_e''\|\sum_i T^{\tilde{\Gamma}_2}(\mathbf{a}_i)_{r_2}\|\phi\Gamma_e\rangle\,(\tilde{\Gamma}_2\tilde{a}_2,\tilde{\Gamma}_1\tilde{a}_1|\Gamma_0)^*\delta_{\tilde{\Gamma}_1\tilde{\Gamma}_2}W(\Gamma_e\tilde{\Gamma}_1\Gamma_e\tilde{\Gamma}_1;\Gamma_e''\Gamma_0)\\
&\quad\times\sqrt{\lambda(\Gamma_e'')}+\sum_{\bar{\Gamma}_1\bar{a}_1r_1}\varepsilon(\bar{a}_1)\sum_{\bar{\Gamma}_2\bar{a}_2r_2}\varepsilon(\bar{a}_2)\sum_{\phi''\Gamma_e''}{}'\,\overline{\Delta V}^{-1}_{\phi''\Gamma_e''S+1}\\
&\quad\times\langle 11\hat{q}_2(\bar{a}_2,r_2)\hat{q}_1(\bar{a}_1,r_1)|2\hat{q}(r)\rangle^*\langle\phi\Gamma_eS\|\sum_i T^{\bar{\Gamma}_1}(\mathbf{b}_i)_{r_1}S^1(\mathbf{s}_i)\|\phi''\Gamma_e''S+1\rangle\\
&\quad\times\langle\phi''\Gamma_e''S+1\|\sum_i T^{\bar{\Gamma}_2}(\mathbf{b}_i)_{r_2}S^1(\mathbf{s}_i)\|\phi\Gamma_eS\rangle\,(\bar{\Gamma}_2\bar{a}_2,\bar{\Gamma}_1\bar{a}_1|\Gamma_0)^*\\
&\quad\times\delta_{\bar{\Gamma}_1\bar{\Gamma}_2}W(\Gamma_e\bar{\Gamma}_1\Gamma_e\bar{\Gamma}_1;\Gamma_e''\Gamma_0)\sqrt{\lambda(\Gamma_e'')}\Big\}
\end{aligned}
\tag{VIII.105}
$$

E. Zeeman and Stark Effects

Finally, we consider static external magnetic and electric fields. They remove the isotropy of the space and thus lead to splittings of the $(2J+1)$ (or $2F+1$) degenerate rovibronic states. Although the quantum number $J(F)$ is no longer good, it remains nearly good in many cases, since the effects due to the external fields are generally small compared to the separations of the field-free states.

Depending on the strength of the applied fields and on the internal couplings, an angular momentum function basis selection different from the field-free case may be appropriate. Often a decoupled basis (i.e., for nuclear spins in the Zeeman effect) is more suitable than the coupled one. In this section we will not consider the matrix elements in the angular momentum function basis, as they can be derived by the standard techniques of angular momentum theory. For the effective Hamiltonian operators, akin to those given in Section VI, an effective magnetic moment dyadic is associated with angular momentum operators, represented by j, according to

$$\underline{\mu}_m = -\beta\tilde{\mathbf{g}}\mathbf{j} \tag{VIII.106}$$

where β is the Bohr magneton (for nuclear rotational and spin operators β is replaced by $-\beta_N$ the nuclear magneton). The $\tilde{\mathbf{g}}$-dyadics, as functions of the nuclear coordinates rather than vibronic constants, may be evaluated for the different types of magnetic interaction. In addition, the magnetic field induces a magnetic moment

$$\underline{\mu}_{\text{ind}} = -\tilde{\mathbf{x}}\mathbf{H} \tag{VIII.107}$$

where $\tilde{\mathbf{x}}$ is the effective susceptibility dyadic. The Hamiltonian operators for these interactions are given by $\underline{\mu}_m H$ and $-\frac{1}{2}\mathbf{H}\cdot\tilde{\mathbf{x}}\mathbf{H}$, respectively.

The Stark effect describes the interaction of the electric dipole moment $\underline{\mu}$ with the electric field given by $\underline{\mu}\cdot E$. Equivalent to the induced magnetic moment, one has an effective induced electric dipole moment

$$\underline{\mu}_{\text{ind}} = -\boldsymbol{\alpha}\mathbf{E} \tag{VIII.108}$$

where $\boldsymbol{\alpha}$ is the effective polarizability dyadic, leading to the interaction operator $-\frac{1}{2}\mathbf{E}\cdot\boldsymbol{\alpha}\mathbf{E}$.

Particular attention must be paid to the gauge invariance[53] of some of the effective constants presented in this section (i.e., the magnetic susceptibility values and the magnetic shielding constants). In the following paragraphs we will discuss the various Zeeman effects before we come to the Stark effects of first and second order in E.

The strongest magnetic effects are present for open-shell states with either nonzero spin or an nonzero orbital angular momentum leading to paramagnetic dipole moments. With the spin of a free electron a magnetic moment of $\underline{\mu}_m = -\beta g \mathbf{s}_i$ is associated, with g a number close to 2 [see (A.52)], which may be calculated from relativistic quantum field theory.[54] Thus, without any further interaction, we would expect a molecular magnetic moment of

$$\underline{\mu}_m = -\beta g \mathbf{S} \tag{VIII.109}$$

where S is the total electronic spin. However, owing to additional intramolecular interactions, a modified molecular magnetogyric ratio dyadic must be used

$$\tilde{\mathbf{g}}_S = g\mathbf{1} + \tilde{\mathbf{g}}'_S \tag{VIII.110}$$

where the elements of $\tilde{\mathbf{g}}'_S$ are small compared to g. Nevertheless, they can be measured, at least the diagonal values, and therefore should also be calculated.

The following operators contribute to the paramagnetic spin Zeeman effect in first order:

$$\mathfrak{H}_1 = (g\hbar^2/4m^3c^2)\sum_i \nabla_i^2(\mathbf{s}_i\cdot\mathbf{H}) \tag{VIII.111}$$

from (A.51), and

$$\mathfrak{H}_2 = -(\beta e^2 g/4mc^2)\sum_i \mathbf{s}_i\left\{\sum_{j\neq i} r_{ij}^{-3}\mathbf{r}_{ij}\times\left[(\mathbf{H}\times\mathbf{r}_i)-2(H\times\mathbf{r}_j)\right]\right.$$
$$\left.-\sum_\alpha Z_\alpha r_{i\alpha}^{-3}\left[\mathbf{r}_{i\alpha}\times(\mathbf{H}\times\mathbf{r}_i)\right]\right\} \tag{VIII.112}$$

from (A.47).

In second order the perturbational coupling of

$$\mathfrak{H}_3 = \beta\mathbf{L}\cdot\mathbf{H} \tag{VIII.113}$$

the leading term of (A.6), with the spin–orbit operator

$$\mathfrak{H}_{SO} = \sum_i \mathsf{u}_i \cdot \mathsf{s}_i \qquad \text{(VIII.114)}$$

[see (A.48) for the definition of u_i] will be dominant. The second-order terms dominate the first-order terms due to the operator of (VIII.111) and (VIII.112). For a shorter notation, we introduce for $\mathfrak{H}_1$, (VIII.111),

$$o_i = (g\hbar^2/4m^3c^2)\nabla_i^2 \qquad \text{(VIII.115)}$$

and for $\mathfrak{H}_2$, (VIII.112),

$$\boldsymbol{\kappa}_i = -(\beta e^2 g/4mc^2)\left[\sum_{j\neq i} r_{ij}^{-3}\{[\mathsf{r}_{ij}\cdot\mathsf{r}_i\mathbf{1} - \mathsf{r}_{ij}\otimes\mathsf{r}_i] - 2[(\mathsf{r}_{ij}\cdot\mathsf{r}_j)\mathbf{1} - \mathsf{r}_{ij}\otimes\mathsf{r}_j]\}\right.$$

$$\left. - \sum_\alpha Z_\alpha r_{i\alpha}^{-3}[(\mathsf{r}_{i\alpha}\cdot\mathsf{r}_i)\mathbf{1} - \mathsf{r}_{i\alpha}\otimes\mathsf{r}_i]\right]$$

$$\text{(VIII.116)}$$

For the elements of $\boldsymbol{\kappa}_i$ we form the irreducible tensor components $S_q^k(\boldsymbol{\kappa}_i) = \kappa_{iq}^{(k)}$, $k = 0, 2$, as described in Table III of Appendix 3.

Thus in our symmetry notation for the operator $\mathfrak{H}_1$, (VIII.111), and $\mathfrak{H}_2$, (VIII.112),

$$\mathfrak{H}_1' = \sum_i (\mathsf{s}_i\cdot\mathsf{H})\left(o_i - \tfrac{1}{3}\mathrm{Tr}(\boldsymbol{\kappa}_i)\right) \qquad \text{(VIII.117)}$$

$$\mathfrak{H}_2' = \sum_{\overline{\Gamma}\bar{a}r} \varepsilon(\bar{a}) \sum_i T_{\bar{a}}^{\overline{\Gamma}}(\kappa_i^{(2)})_r S^2_{\bar{q}(\bar{a},r)}(\mathsf{s}_i, \mathsf{H}) \qquad \text{(VIII.118)}$$

where we have omitted the symbol T^{Γ_0} to indicate that o_i and $\mathrm{Tr}(\kappa_i)$ are totally symmetric, and the S_0^0-symbol for $\mathsf{s}_i\cdot\mathsf{H}$ to indicate a zero-rank operator.

The effective matrix elements, diagonal in S, are given by

$$\langle\phi'\Gamma_e'a_e'\lambda'|\tilde{\mathfrak{H}}_H^S|\phi\Gamma_e a_e\lambda\rangle = \delta_{\Gamma_e'\Gamma_e}\delta_{a_e'a_e}\langle\lambda'|S_0^0(\mathsf{S},\mathsf{H})|\lambda\rangle$$
$$\times\Bigg\{-\beta g + \langle\phi'\Gamma_e'S\|\sum_i(o_i + Tr(\kappa_i)/3)S^1(\mathsf{s}_i)\|\phi\Gamma_e S\rangle\langle S\|S^1(\mathsf{S})\|S\rangle^{-1}$$
$$+\sum_{\tilde{\Gamma}_1\tilde{a}_1r_1}\varepsilon(\tilde{a}_1)\sum_{\tilde{\Gamma}_2\tilde{a}_2r_2}\varepsilon(\tilde{a}_1)\sum_{\phi''\Gamma_e''}\overline{\Delta V}^{-1}_{\phi''\Gamma_e''S}$$
$$\times\langle\phi'\Gamma_e'S\|\sum_i T^{\tilde{\Gamma}_1}(\mathsf{u}_i)_{r_1}S^1(\mathsf{s}_i)\|\phi''\Gamma_e''S\rangle$$
$$\times\langle\phi''\Gamma_e''S\|T^{\tilde{\Gamma}_2}(\mathsf{L})_{r_2}\|\phi\Gamma_e S\rangle\langle S\|S^1(\mathsf{S})\|S\rangle^{-1}(\tilde{\Gamma}_2\tilde{a}_2,\tilde{\Gamma}_1\tilde{a}_1|\Gamma_0)^*$$
$$\times\delta_{\tilde{\Gamma}_1\tilde{\Gamma}_2}W(\Gamma_e\tilde{\Gamma}_1\Gamma_e\tilde{\Gamma}_1;\Gamma_e''\Gamma_0)\sqrt{\lambda(\Gamma_e'')}\,\langle 11\hat{q}_2(\tilde{a}_2,r_2)\hat{q}_1(\tilde{a}_1,r_1)|00\rangle^*\Bigg\}$$
$$+\sum_{\bar{\Gamma}\bar{a}r}\varepsilon(\bar{a})\langle\lambda'|S^2_{\bar{q}(\bar{a},r)}(\mathsf{S},\mathsf{H})|\lambda\rangle$$
$$\times\Bigg\{(\Gamma_e a_e,\bar{\Gamma}\bar{a}|\Gamma_e'a_e')\langle\phi'\Gamma_e'S\|\sum_i T^{\bar{\Gamma}}(\kappa_i)_r S^1(\mathsf{s}_i)\|\phi\Gamma_e S\rangle\langle S\|S^1(\mathsf{S})\|S\rangle^{-1}$$
$$+\sum_{\tilde{\Gamma}_1\tilde{q}_1r_1}\varepsilon(\tilde{a}_1)\sum_{\tilde{\Gamma}_2\tilde{a}_2r_2}\varepsilon(\tilde{a}_2)\sum_{\phi''\Gamma_e''}{}'\,\overline{\Delta V}^{-1}_{\phi''\Gamma_e''S}$$
$$\times\langle\phi'\Gamma_e'S\|\sum_i T^{\tilde{\Gamma}_1}(\mathsf{u}_i)_{r_1}S^1(\mathsf{s}_i)\|\phi''\Gamma_e''S\rangle$$
$$\times\langle\phi''\Gamma_e''S\|T^{\tilde{\Gamma}_2}(\mathsf{L})_{r_2}\|\phi\Gamma_e S\rangle\langle S\|S^1(\mathsf{S})\|S\rangle^{-1}\,(\tilde{\Gamma}_2\tilde{a}_2,\tilde{\Gamma}_1\tilde{a}_1|\bar{\Gamma}\bar{a})^*$$
$$\times W(\Gamma_e\tilde{\Gamma}_2\Gamma_e'\tilde{\Gamma}_1;\Gamma_e''\bar{\Gamma})\sqrt{\lambda(\Gamma_e'')\lambda(\bar{\Gamma})}\,\langle 11\hat{q}_2(\tilde{a}_2,r_2)\hat{q}_1(\tilde{a}_1,r_1)|2\hat{q}(\bar{a},r)\rangle^*\Bigg\}$$

(VIII.119)

plus the terms with reversed order of operators in the perturbation sums.

For molecules in the linear conformation and diatomics, we get only two nonvanishing expectation values, $g'_{S\parallel}$ and $g'_{S\perp}$, where $\parallel$ and $\perp$ signify the components parallel and perpendicular to the internuclear axis, respectively. For Π-states we have also a Λ-doubling type of matrix element with $g'_{S\perp\perp}$.

With the interaction of the magnetic moment due to the electronic angular momentum with the external magnetic field we associate the effective operator

$$\tilde{\mathfrak{H}}_H^L = \beta \mathsf{H} \cdot \tilde{\mathbf{g}}_l \mathsf{L} \tag{VIII.120}$$

where $\tilde{\mathbf{g}}_L$ may be expressed as

$$\tilde{\mathbf{g}}_L = \mathbf{1} + \tilde{\mathbf{g}}'_L \tag{VIII.121}$$

with elements of $\tilde{\mathbf{g}}'_L$ small compared to 1. The dominant part, $\beta \mathsf{H} \cdot \mathsf{L}$, is derived from (A.6) with the leading term in (A.9). As is clear from (VIII.120), the orbital Zeeman effect requires an orbitally degenerate or at least near-degenerate state.

The following operators are expected to yield the most important first-order contributions to $\tilde{\mathbf{g}}'_L$:

$$\mathfrak{H}_1 = (\beta\hbar/2m^2c^2) \sum_i \nabla_i^2 \mathsf{l}_i \cdot \mathsf{H} \tag{VIII.122}$$

from (A.51)

$$\mathfrak{H}_2 = -(\beta e^2/4mc^2\hbar)\left\{ \sum_{i\neq j} r_{ij}^{-1}\left[(\mathsf{r}_j \times \mathsf{p}_i) + (\mathsf{r}_i \times \mathsf{p}_j)\right] + r_{ij}^{-3}\left[(\mathsf{r}_i \times \mathsf{r}_{ij})(\mathsf{r}_{ij}\cdot \mathsf{p}_j)\right.\right.$$

$$\left.\left. + (\mathsf{r}_j \times \mathsf{r}_{ij})(\mathsf{r}_{ij}\cdot \mathsf{p}_i)\right]\right\}\mathsf{H} \tag{VIII.123}$$

from (A.54), and

$$\mathfrak{H}_3 = -(m/M_N)\beta \mathsf{L}\cdot\mathsf{H} \tag{VIII.124}$$

These are the most important terms of (A.6) using (A.9), and of (A.12). In addition, we get

$$\mathfrak{H}_4 = \mathsf{L}\cdot \mathbf{I}''^{-1}\mathsf{t}\mathsf{H} \tag{VIII.125}$$

due to (A.29), using (A.30) and (A.31).

The contributions from (VIII.124) and (VIII.125) to the $\mathbf{g}_H^L$ do not contain electronic operators and can be evaluated with the nuclear masses and coordinates. The operators in (VIII.122) and (VIII.123) can be expressed as

$$\mathfrak{H}_1+\mathfrak{H}_2=\sum_i \mathsf{f}_i\cdot\mathsf{H} \tag{VIII.126}$$

where f_i transforms as the electronic angular momentum operator $\boldsymbol{l}_i$.

Second-order contributions are expected to arise due to $\beta\mathsf{L}\cdot\mathsf{H}$ and that part of the rotational Hamiltonian which couples the angular momentum operator N with the electronic angular momentum L (see Section VIII.C), that is, $-\mathsf{N}\cdot\boldsymbol{\mu}\mathsf{L}$, using the fact that the commutator is

$$[\mathsf{N},\mathsf{H}]=i\mathsf{H} \tag{VIII.127}$$

in the molecular fixed frame.

With this we obtain

$$\begin{aligned}
\langle\phi'\Gamma_e'a_e'\lambda'|\tilde{\mathfrak{H}}_H^L|\phi\Gamma_e a_e\lambda\rangle &= \sum_{\tilde{\Gamma}\tilde{a}r}\langle\lambda'|S^1_{\hat{q}(\tilde{a},r)}(\mathsf{H})|\lambda\rangle(\Gamma_e a_e,\tilde{\Gamma}\tilde{a}|\Gamma_e'a_e')^*\\
&\times\Big\{\varepsilon(\tilde{a})\langle\phi'\Gamma_e'\|\,\beta(1-m/M_N)T^{\tilde{\Gamma}}(\mathsf{L})_r+\sum_i T^{\tilde{\Gamma}}(\mathsf{f}_i)_r\|\phi\Gamma_e\rangle\\
&+i\sum_{\tilde{\Gamma}_1\tilde{a}_1r_1}\varepsilon(\tilde{a}_1)\sum_{\tilde{\Gamma}_2\tilde{a}_2r_2}\varepsilon(\tilde{a}_2)\sum_{\phi''\Gamma_e''}{}'\,\overline{\Delta V}^{-1}_{\phi''\Gamma_e''}\langle\phi'\Gamma_e'\|T^{\tilde{\Gamma}_1}(\mathsf{L})_{r_1}\|\phi''\Gamma_e''\rangle\\
&\times\langle\phi''\Gamma_e''\|\,T^{\tilde{\Gamma}_2}(\mathsf{L})_{r_2}\|\phi\Gamma_e\rangle\,(\tilde{\Gamma}_2\tilde{a}_2,\tilde{\Gamma}_1\tilde{a}_1|\tilde{\Gamma}\tilde{a})^*\\
&\times W(\Gamma_e\tilde{\Gamma}_2\Gamma_e'\tilde{\Gamma};\Gamma_e''\tilde{\Gamma})\sqrt{\lambda(\Gamma_e'')\lambda(\tilde{\Gamma})}\,\langle 11\hat{q}_2(\tilde{a}_2,r_2)\hat{q}_1(\tilde{a}_1,r_1)|1\hat{q}(\tilde{a},r)\rangle^*\\
&\times\sum_{k_1q_3q_4}\sum_{k_2q_5q_6}\varepsilon(q_3)\varepsilon(q_5)\mu^{k_1}_{\hat{q}_3}\mu^{k_2}_{\hat{q}_5}\langle 11q_4q_1(\tilde{a}_1,r_1)|k_1q_3\rangle\\
&\times\langle 11q_6q_2(\tilde{a}_2,r_2)|k_2q_5\rangle\Big\}
\end{aligned} \tag{VIII.128}$$

where we used (VIII.54) for the rotational coupling operator. For linear conformations an expectation value is not equal to zero only for $\tilde{\Gamma}=\Sigma^-$.

Rather small in size but investigated extensively for closed-shell molecules is the rotational Zeeman effect, which is the leading term for diamagnetic molecules. It is by a factor of m/M_p smaller than the paramagnetic electronic spin and orbital Zeeman effect. From the combination of rotational g-factors and susceptibility values, an experimental determination of molecular quadrupole charge distribution is possible.[37a] The effective operator for the rotational Zeeman effect in one adiabatic state may be written as

$$\tilde{\mathfrak{H}}_H^r = -\beta_N \tfrac{1}{2}\{(\mathsf{N}-\mathsf{G}^{\mathfrak{R}})\cdot\tilde{\mathbf{g}}_r\mathsf{H}+\mathsf{H}\cdot\tilde{\mathbf{g}}_r(\mathsf{N}-\mathsf{G}^{\mathfrak{R}})\} \qquad \text{(VIII.129)}$$

with β_N the nuclear magneton and $\tilde{\mathbf{g}}_r$ the effective rotational g-factor dyadic. The rotational $\tilde{g}$-factors have a first-order nuclear contribution and a second-order electronic contribution. The first is contained in (A.29) and may be expressed by

$$\mathbf{g}_r^{\text{nuc}} = M_p\mathbf{I}''^{-1}\mathbf{t} \qquad \text{(VIII.130)}$$

For diatomic we get

$$\mathbf{g}_r^{\text{nuc}} = \frac{Z_\alpha M_\beta^2 + Z_\beta M_\alpha^2}{M_\alpha M_\beta(M_\alpha + M_\beta)}\begin{bmatrix}1 & 0 & 0\\ 0 & 1 & 0\\ 0 & 0 & 1\end{bmatrix} \qquad \text{(VIII.131)}$$

The electronic contributions have their origin in the perturbational coupling of the rotational operator

$$\mathfrak{H}'_{\text{rot}} = -\tfrac{1}{2}(\mathsf{L}\cdot\underline{\mu}(\mathsf{N}-\mathsf{G}^{\mathfrak{R}})+(\mathsf{N}-\mathsf{G}^{\mathfrak{R}})\cdot\underline{\mu}\mathsf{L}) \qquad \text{(VIII.132)}$$

with

$$\mathfrak{H}_H^L = \beta\mathsf{L}\cdot\mathsf{H} \qquad \text{(VIII.133)}$$

We use the following notation for (VIII.132) and (VIII.133), respectively:

$$\mathfrak{H}'_{\text{rot}} = -\tfrac{1}{2}\sum_{\tilde{\Gamma}\tilde{a}r}\varepsilon(\tilde{a})\Big(T_{\tilde{a}}^{\tilde{\Gamma}}(\mathsf{L})_r S_{\hat{q}(\tilde{a},r)}^{1}(\boldsymbol{\mu}(\mathsf{N}-\mathsf{G}^{\mathfrak{R}}))$$

$$+ S_{\hat{q}(\tilde{a},r)}^{1}(\boldsymbol{\mu}(\mathsf{N}-\mathsf{G}^{\mathfrak{R}}))\,T_{\tilde{a}}^{\tilde{\Gamma}}(\mathsf{L})_r \qquad \text{(VIII.134)}$$

and

$$\mathfrak{H}_H^L = \beta\sum_{\tilde{\Gamma}\tilde{a}r}\varepsilon(\tilde{a})\,T_{\tilde{a}}^{\tilde{\Gamma}}(\mathsf{L})_r S_{\hat{q}(\tilde{a},r)}^{1}(\mathsf{H}) \qquad \text{(VIII.135)}$$

With this the following effective matrix elements are obtained for the electronic contribution to the rotational Zeeman effect.

$$\langle \phi'\Gamma_e' a_e' \lambda' | \tilde{\mathfrak{H}}_H^{r\,\mathrm{el}} | \phi \Gamma_e a_e \lambda \rangle =$$

$$\begin{aligned} &-\sum_{\bar{\Gamma}\bar{a}} \left(\Gamma_e a_e, \bar{\Gamma}\bar{a} | \Gamma_e' a_e'\right)^* \sum_{kq} \sum_{\tilde{\Gamma}_1 \tilde{a}_1 r_1} \varepsilon(\tilde{a}_1) \sum_{\tilde{\Gamma}\tilde{a}_2 r_2} \varepsilon(\tilde{a}_2) \sum_{\phi''\Gamma_e''}{}' \\ &\times \left[f(\tilde{\Gamma}_2\tilde{a}_2 r_2; \tilde{\Gamma}_1 \tilde{a}_1 r_1; \phi''\Gamma_e'') \langle 11 \hat{q}_2(\tilde{a}_2, r_2) \hat{q}_1(\tilde{a}_1, r_1) | kq \rangle \right. \\ &\quad \times \langle \lambda' | \tfrac{1}{2} S_q^k(\mathsf{H}, \boldsymbol{\mu}(\mathsf{N} - \mathsf{G}^{\mathfrak{R}})) | \lambda \rangle + \langle 11 \hat{q}_1(\tilde{a}_1, r_1) \hat{q}_2(\tilde{a}_2, r_2) | kq \rangle \\ &\quad \left. \times \langle \lambda' | \tfrac{1}{2} S_q^k((\mathsf{N} - \mathsf{G}^{\mathfrak{R}})\boldsymbol{\mu}, \mathsf{H}) | \lambda \rangle\, f(\tilde{\Gamma}_1 \tilde{a}_1 r_1; \tilde{\Gamma}_2 \tilde{a}_2 r_2; \phi''\Gamma_e'') \right] \end{aligned} \qquad \text{(VIII.136)}$$

with

$$\begin{aligned} f(\tilde{\Gamma}_2\tilde{a}_2 r_2; \tilde{\Gamma}_1\tilde{a}_1 r_1; \phi''\Gamma_e'') = {} & i\overline{\Delta V}_{\phi''\Gamma_e''}^{-1} \langle \phi'\Gamma_e' \| T^{\tilde{\Gamma}_1}(\mathsf{L})_{r_1} \| \phi''\Gamma_e'' \rangle \\ & \times \langle \phi''\Gamma_e'' \| T^{\tilde{\Gamma}_2}(\mathsf{L})_{r_2} \| \phi\Gamma_e \rangle \left(\tilde{\Gamma}_2\tilde{a}_2, \tilde{\Gamma}_1\tilde{a}_1 | \bar{\Gamma}\bar{a}\right)^* \\ & \times W\left(\Gamma_e \tilde{\Gamma}_2 \Gamma_e' \tilde{\Gamma}_1; \Gamma_e'' \bar{\Gamma}\right) \sqrt{\lambda(\Gamma_e'')\lambda(\bar{\Gamma})} \end{aligned}$$

and $k = 0, 2$.

The $k = 1$ contributions have already been accounted for in the orbital Zeeman effect, (VIII.128). For $\phi' = \phi$, $\Gamma_e' = \Gamma_e$, the $\tilde{\mathbf{g}}_r^{\mathrm{el}}$-electronic contributions to $\mathbf{g}_r$ can be identified in (VIII.136).

As stated in Section VI, (VI.32), also purely vibrational Zeeman effects not contained in the rotational Zeeman effect, (VIII.129), can be observed for excited degenerate vibrational modes.[55] This effect is due to part of the vibrational operator, (A.13),

$$\begin{aligned} \mathfrak{H}_H^v &= -\tfrac{1}{2} \sum_k (P_k A_k + A_k P_k) \\ &= -(e/4c) \sum_k \sum_\alpha \left(Z_\alpha / \sqrt{M_\alpha}\,\right) (P_k \mathsf{d}_k(\alpha) \cdot (\mathsf{H} \times \mathsf{r}_\alpha) \\ &\qquad + \mathsf{d}_k(\alpha) \cdot (H \times \mathsf{r}_\alpha) P_k) \end{aligned} \qquad \text{(VIII.137)}$$

in first order and the coupling of $\frac{1}{2}\sum_k P_k^2$ with $\beta \mathsf{L} \cdot \mathsf{H}$ in second order.

Equation (VIII.137) may be written as

$$\mathfrak{H}_H^v = -(e/2c) \sum_k \sum_\alpha \left(Z_\alpha / \sqrt{M_\alpha}\,\right) (\mathsf{r}_\alpha^0 \times \mathsf{d}_k(\alpha)) \mathsf{H} P_k \qquad \text{(VIII.138a)}$$

$$-(e/4c) \sum_k \sum_l \sum_\alpha (Z_\alpha / M_\alpha)(\mathsf{d}_k(\alpha) \times \mathsf{d}_l(\alpha)) \mathsf{H} (P_k S_l + S_l P_k) \qquad \text{(VIII.138b)}$$

The term (VIII.138a) can be neglected for spectroscopic investigations since it cannot have first-order matrix elements in the harmonic oscillator approximation. The second term (VIII.138b), however, has first-order matrix elements in the harmonic oscillator basis.

The electronic second-order terms can be treated best if we resolve the kinetic vibrational operator $\frac{1}{2}\Sigma_k P_k^2$ analogous to (VIII.55) for G into

$$\frac{1}{2}\sum_k P_k^2 = \frac{1}{2}\sum_k \left(P_k^{\mathfrak{R}2} + P_k^{\text{el}2} + 2P_k^{\mathfrak{R}}P_k^{\text{el}}\right) \tag{VIII.139}$$

The last term can be transformed to

$$\sum_k P_k^{\mathfrak{R}}P_k^{\text{el}} = \sum_k \sum_{\Gamma(k)a} T_a^{\Gamma}(P_k^{\text{el}})P_k^{\mathfrak{R}} \tag{VIII.140}$$

for which a coordinate transformation among the S_k may be necessary. As an example for a linear triatomic molecule, we may write

$$\frac{1}{2}\sum_k P_k^2 = -\frac{\hbar^2}{2}\left[\sum_{\mu=1,2,3}\frac{\partial^2}{\partial S_\mu^2} + \frac{1}{S_2}\frac{\partial}{\partial S_2} + \frac{1}{S_2^2}\frac{\partial^2}{\partial \phi_2^2}\right] \tag{VIII.141}$$

(see Section VIII.A) and we get

$$\sum_k P_k^{\text{el}}P_k^{\mathfrak{R}} = -\hbar^2\left[\sum_{\mu=1,2,3}\left(\frac{\partial}{\partial S_\mu}\right)^{\text{el}}\left(\frac{\partial}{\partial S_\mu}\right)^{\mathfrak{R}} + \frac{1}{S_2^2}\left(\frac{\partial}{\partial \phi_2}\right)^{\text{el}}\left(\frac{\partial}{\partial \phi_2}\right)^{\mathfrak{R}}\right]$$

$$= -\hbar^2\left[\sum_{\mu=1,2,3} T^{\Sigma^+}\left(\left(\frac{\partial}{\partial S_\mu}\right)^{\text{el}}\right)\left(\frac{\partial}{\partial S_\mu}\right)^{\mathfrak{R}} + \frac{1}{S_2^2}T^{\Sigma^-}\left(\left(\frac{\partial}{\partial \phi_2}\right)^{\text{el}}\right)\left(\frac{\partial}{\partial \phi_2}\right)^{\mathfrak{R}}\right] \tag{VIII.142}$$

For the vibrational Zeeman effect in linear molecules it is the term with $T^{\Sigma^-}((\partial/\partial\phi_2)^{\text{el}})$ which is most important.

The second-order perturbation gives the following effective matrix elements:

$$\langle\phi'\Gamma_e' a_e'\lambda'|\tilde{\mathfrak{H}}_H^{\mathfrak{v}}|\phi\Gamma_e a_e\lambda\rangle = \beta\sum_{\bar{\Gamma}\bar{a}r}(\Gamma_e a_e, \Gamma a|\Gamma_e' a_e')^* \times \sum_k \sum_{\Gamma a r_1} \sum_{\tilde{\Gamma}\tilde{a}r_2} \varepsilon(\tilde{a}_1) \sum_{\phi''\Gamma_e''}{}' \{ f(\Gamma a r_1; \tilde{\Gamma}\tilde{a}r_2; \phi''\Gamma_e'') P_k^{\mathfrak{R}} + P_k^{\mathfrak{R}} f'(\tilde{\Gamma}\tilde{a}r_2; \Gamma a r_1; \phi''\Gamma_e'')\} \langle\lambda'|S^{1}_{\hat{q}(\bar{a},r_2)}|\lambda\rangle \tag{VIII.143}$$

with

$$f(\Gamma a r_1; \tilde{\Gamma}\tilde{a}r_2; \phi''\Gamma_e'') = \overline{\Delta V}^{-1}_{\phi''\Gamma_e''} \langle\phi'\Gamma_e'\| T^{\Gamma}(P_k^{\mathrm{el}})_{r_1}\|\phi''\Gamma_e''\rangle\langle\phi''\Gamma_e''\| T^{\tilde{\Gamma}}(\mathsf{L})_{r_2}\|\phi\Gamma_e\rangle \times (\tilde{\Gamma}\tilde{a}, \Gamma a|\bar{\Gamma}\bar{a})^* W(\Gamma_e\tilde{\Gamma}_2\Gamma_e'\Gamma; \Gamma_e''\bar{\Gamma})\sqrt{\lambda(\Gamma_e'')\lambda(\bar{\Gamma})}$$

For the description of the second-order Zeeman effect (which should not be confused with the second-order perturbation contributions due to the first-order Zeeman operator resulting in matrix elements off-diagonal in the rotational basis), we have to consider the following operators. In first order we obtain from (A.6),

$$\mathfrak{H}_x = -(e^2/8mc^2)\sum_i \mathsf{H}\cdot(\mathsf{r}_i^2\mathbf{1} - \mathsf{r}_i\otimes\mathsf{r}_i)\mathsf{H} = -(e^2/8mc^2)\sum_i \mathsf{H}\cdot\mathsf{t}_{ei}\mathsf{H}$$
$$= -(e^2/8mc^2)\sum_{k=0,2}\sum_{m=-k}^{k}(-1)^m\sum_i \tilde{T}_m^k(\mathsf{t}_{ei})\tilde{T}_{-m}^k(\mathsf{H},\mathsf{H})$$
$$= -(e^2/8mc^2)\Big[(1/3)\sum_i Tr(\mathsf{t}_{ei})\mathsf{H}^2 + \sum_{\bar{\Gamma}\bar{a}r}\varepsilon(\bar{a})\sum_i T_{\bar{a}}^{\bar{\Gamma}}(t_{ei}^{(2)})_r S^{2}_{\hat{q}(\bar{a},r)}(\mathsf{H},\mathsf{H})\Big] \tag{VIII.144}$$

with many smaller contributions neglected. The dominant second-order contribution comes from the perturbational coupling of the operator $\beta\mathsf{L}\cdot\mathsf{H}$ with itself.

The following effective matrix is obtained using an undefined rotational function basis $|\lambda\rangle$:

$$\langle\phi'\Gamma'_e a'_e\lambda'|\tilde{\mathfrak{H}}_x|\phi\Gamma_e a_e\lambda\rangle = -(e^2/8mc^2)$$

$$\times\Big[\delta_{\Gamma_e\Gamma'_e}\delta_{a_e a'_e}\Big\{\langle\phi'\Gamma_e\|\sum_i -\mathrm{Tr}(\mathbf{t}_{ei})/\sqrt{3}\ \|\phi\Gamma_e\rangle + 2(\hbar^2/m)\sum_{\tilde{\Gamma}_1\tilde{a}_1 r_1}\varepsilon(\tilde{a}_1)$$

$$\times\sum_{\tilde{\Gamma}_2\tilde{a}_2 r_2}\varepsilon(\tilde{a}_2)\sum_{\phi''\Gamma''_e}{}'\ \overline{\Delta V}^{-1}_{\phi''\Gamma''_e}\langle\phi'\Gamma'_e\|T^{\tilde{\Gamma}_1}(\mathsf{L})_{r_1}\|\phi''\Gamma''_e\rangle$$

$$\times\langle\phi''\Gamma''_e\|T^{\tilde{\Gamma}_2}(\mathsf{L})_{r_2}\|\phi\Gamma_e\rangle\big(\tilde{\Gamma}_2\tilde{a}_2,\tilde{\Gamma}_1\tilde{a}_1|\Gamma_0\big)^*$$

$$\times W\big(\Gamma_e\tilde{\Gamma}_2\Gamma'_e\tilde{\Gamma}_1;\Gamma''_e\Gamma_0\big)\sqrt{\lambda(\Gamma''_e)}\ \langle 11\hat{q}_2(\tilde{a}_2,r_2)\hat{q}_1(\tilde{a}_1,r_1)|00\rangle\Big\}$$

$$\times\langle\lambda'|S_0^0(H,H)|\lambda\rangle + \sum_{\bar{\Gamma}\bar{a}r}\big(\Gamma_e a_e,\bar{\Gamma}\bar{a}|\Gamma'_e a'_e\big)^*$$

$$\times\Big\{\varepsilon(\bar{a})\langle\phi'\Gamma'_e\|\sum_i T^{\bar{\Gamma}}(t_{ei}^{(2)})_r\|\phi\Gamma_e\rangle + 2(\hbar^2/m)\sum_{\tilde{\Gamma}_1\tilde{a}_1 r_1}\varepsilon(\tilde{a}_1)\sum_{\tilde{\Gamma}_2\tilde{a}_2 r_2}\varepsilon(\tilde{a}_2)$$

$$\sum_{\phi''\Gamma''_e}\overline{\Delta V}^{-1}_{\phi''\Gamma''_e}BX\langle\phi'\Gamma'_e\|T^{\tilde{\Gamma}_1}(\mathsf{L})_{r_1}\|\phi''\Gamma''_e\rangle\langle\phi''\Gamma''_e\|T^{\tilde{\Gamma}_2}(\mathsf{L})_{r_2}\|\phi\Gamma_e\rangle$$

$$\times\big(\tilde{\Gamma}_2\tilde{a}_2,\tilde{\Gamma}_1\tilde{a}_1|\bar{\Gamma}\bar{a}\big)^* W\big(\Gamma_e\tilde{\Gamma}_2\Gamma'_e\tilde{\Gamma}_1;\Gamma''_e\bar{\Gamma}\big)\sqrt{\lambda(\Gamma''_e)\lambda(\bar{\Gamma})}$$

$$\times\langle 11\hat{q}_2(\tilde{a}_2,r_1)\hat{q}_1(\tilde{a}_1,r_1)|2\hat{q}(\bar{a},r)\rangle^*\Big\}\langle\lambda'|S^2_{\hat{q}(\bar{a},r)}(\mathsf{H},\mathsf{H})|\lambda\rangle\Big]$$

(VIII.145)

In (VIII.145), we have separated the isotropic and second-rank contributions. There is no $k=1$ contribution, since the commutator $[\mathsf{H},\mathsf{H}]$ vanishes. In the adiabatic approximation (i.e., $\phi=\phi'$, $\Gamma_e=\Gamma'_e$) we may identify the elements of the effective susceptibility tensor from (VIII.145), which is defined by $-\frac{1}{2}\mathsf{H}\cdot\tilde{\chi}\mathsf{H}$ in Cartesian notation. In this approximation the $\tilde{\chi}$-values are independent of the origin, which gives an opportunity to test *ab initio* calculations.

The parts of the molecular Hamiltonian involving the magnetic dipoles of the nuclei are expressed by

$$\mathfrak{H}_{\mathrm{ZI}} = -\beta g_\alpha \mathsf{I}_\alpha\cdot\mathsf{H} \qquad \text{(VIII.146)}$$

However, the effective magnetic field at the nuclear positions is reduced due to the electrons of the molecule. The magnetic nuclear shielding contains first- and second-order contributions of the same size. In first order the operator

$$\begin{aligned}\mathfrak{H}_d^{\text{sh}} &= (e^2/2mc^2)\mathsf{I}_\alpha \sum_i r_{i\alpha}^{-3}\left[(\mathsf{r}_i\cdot\mathsf{r}_{i\alpha})\mathbf{1} - \mathsf{r}_i\otimes\mathsf{r}_{i\alpha}\right]\mathsf{H}\\ &= (e^2/2mc^2)\mathsf{I}_\alpha \sum_i \boldsymbol{\sigma}_i \mathsf{H}\\ &= (e^2/2mc^2)\sum_i \sum_{k=0,2} \sum_{m=-k}^{k} (-1)^m \tilde{T}_m^k(\boldsymbol{\sigma}_i)\tilde{T}_{-m}^k(\mathsf{I}_\alpha,\mathsf{H})\\ &= (e^2/2mc^2)\sum_i \left[(1/3)Tr(\boldsymbol{\sigma}_i)\mathsf{I}_\alpha\cdot\mathsf{H} + \sum_{\bar{\Gamma}\bar{a}r} \varepsilon(\bar{a})T^{\bar{\Gamma}}\bar{a}(\sigma_i^{(2)})_r S^2_{\hat{q}(\bar{a},r)}(\mathsf{I}_\alpha,\mathsf{H})\right]\end{aligned} \tag{VIII.147}$$

from (A.68) has to be considered, whereas the second-order contributions arise from the perturbational coupling of $\beta \mathsf{L}\cdot\mathsf{H}$ with $\mathfrak{H}_{\mathrm{OI}_\alpha}$, (A.78).

The effective matrix elements in a general rotational basis are given by

$$\begin{aligned}&\langle\phi'\Gamma_e' a_e'\lambda'|\tilde{\mathfrak{H}}_H^{\text{sh}}|\phi\Gamma_e a_e\lambda\rangle = (e^2/2mc^2)\\ &\times\Bigg[\delta_{\Gamma_e'\Gamma_e}\delta_{a_e'a_e}\Bigg\{\langle\phi'\Gamma_e\|\sum_i - Tr(\boldsymbol{\sigma}_i)/\sqrt{3}\,\|\phi\Gamma_e\rangle + (\hbar c/e)\sum_{\tilde{\Gamma}_1\tilde{a}_1r_1}\varepsilon(\tilde{a}_1)\\ &\qquad\times\sum_{\tilde{\Gamma}_2\tilde{a}_2r_2}\varepsilon(\tilde{a}_2)\sum_{\phi''\Gamma_e''}\overline{\Delta V}_{\phi''\Gamma_e''}^{-1}\langle\phi'\Gamma_e'\|\sum_i T^{\tilde{\Gamma}_1}(\mathsf{a}_i)_{r_1}\|\phi''\Gamma_e''\rangle\\ &\qquad\times\langle\phi''\Gamma_e''|T^{\tilde{\Gamma}_2}(\mathsf{L})_{r_2}\|\phi\Gamma_e\rangle\,(\tilde{\Gamma}_2\tilde{a}_2,\tilde{\Gamma}_1\tilde{a}_1|\Gamma_0)^*\\ &\qquad\times W(\Gamma_e\tilde{\Gamma}_2\Gamma_e'\tilde{\Gamma}_1;\Gamma_e''\Gamma_0)\sqrt{\lambda(\Gamma_e'')}\,\langle 11\hat{q}_2(\tilde{a}_2,r_2)\hat{q}_1(\tilde{a}_1,r_1)|00\rangle^*\Bigg\}\\ &\quad\times\langle\lambda'|S_0^0(\mathsf{I}_\alpha,\mathsf{H})|\lambda\rangle + \sum_{\bar{\Gamma}\bar{a}r}(\Gamma_e a_e,\bar{\Gamma}\bar{a}|\Gamma_e'a_e')^*\\ &\quad\times\Bigg\{\langle\phi'\Gamma_e'\|\sum_i T^{\bar{\Gamma}}(\sigma_i^{(2)})\|\phi\Gamma_e\rangle + (\hbar c/e)\sum_{\tilde{\Gamma}_1\tilde{a}_1r_1}\varepsilon(\tilde{a}_1)\sum_{\tilde{\Gamma}_2\tilde{a}_2r_2}\varepsilon(\tilde{a}_2)\\ &\qquad\times\sum_{\phi''\Gamma_e''}{}'\,\overline{\Delta V}_{\phi''\Gamma_e''}^{-1}\langle\phi'\Gamma_e'\|\sum_i T^{\tilde{\Gamma}_1}(\mathsf{a}_i)_{r_1}\|\phi''\Gamma_e''\rangle\langle\phi''\Gamma_e''|T^{\tilde{\Gamma}_2}(\mathsf{L})_{r_2}\|\phi\Gamma_e\rangle\\ &\qquad\times(\tilde{\Gamma}_2\tilde{a}_2,\tilde{\Gamma}_1\tilde{a}_1|\bar{\Gamma}\bar{a})^*\,W(\Gamma_e\tilde{\Gamma}_2\Gamma_2'\tilde{\Gamma}_1;\Gamma_e''\bar{\Gamma})\sqrt{\lambda(\Gamma_e'')\lambda(\bar{\Gamma})}\\ &\qquad\times\langle 11\hat{q}_1(\tilde{a}_1,r_1)\hat{q}_2(\tilde{a}_2,r_2)|2\hat{q}(\bar{a},r)\rangle^*\Bigg\}\langle\lambda'|S^2_{\hat{q}(\bar{a},r)}(\mathsf{I}_\alpha,\mathsf{H})|\lambda\rangle\Bigg]\end{aligned} \tag{VIII.148}$$

plus the terms with reversed order of operators within the perturbation sums.

If the adiabatic approximation is valid, an effective operator may be deduced from (VIII.148) for the electronic state by identifying the appropriate terms with the effective form

$$\tilde{\mathfrak{H}}_H^{\mathrm{sh}} = \beta_N g_\alpha \mathsf{I}_\alpha \cdot \tilde{\sigma} \mathsf{H} \qquad \text{(VIII.149)}$$

with Cartesian notation for the $\tilde{\sigma}$-dyadic. The $\tilde{\sigma}$-values here are independent of the origin; therefore, for a more symmetrical formula, r_i in (VIII.148) may be replaced by $\mathsf{r}_{i\alpha}$ and L by $\mathsf{L}(\alpha)$.

Finally, we consider a mixed Zeeman–Stark term, the translational Zeeman effect due to the operator (A.40). In the classical limit of treating the translational motion, this operator can be written as

$$\mathfrak{H}_H^{\mathrm{tr}} = -\underline{\mu}(1/c)(\dot{\mathsf{Q}}_0 \times \mathsf{H}) \qquad \text{(VIII.150)}$$

where $\dot{\mathsf{Q}}_0$ is the velocity of the molecule in the space fixed system. If the scalar product of (VIII.151) is expressed in the molecular fixed frame the effective matrix is given by

$$\langle \phi'\Gamma_e' a_e' \lambda' | \tilde{\mathfrak{H}}_H^{\mathrm{tr}} | \phi \Gamma_e a_e \lambda \rangle = -\sum_{\Gamma ar} \varepsilon(a)(\Gamma_e a_e, \Gamma a | \Gamma_e' a_e')^* \langle \phi' \Gamma_e' \| T^{\Gamma}(\underline{\mu})_r \| \phi \Gamma_e \rangle$$
$$\times \langle \lambda' | (1/c) S^1_{\hat{q}(a,r)} (\dot{\mathsf{Q}}_0 \times \mathsf{H}) | \lambda \rangle \qquad \text{(VIII.151)}$$

with no second-order contributions.

For the discussion of the Stark effect it is sufficient to consider the operator

$$\mathfrak{H}_E = -\underline{\mu} \cdot \mathsf{E} \qquad \text{(VIII.152)}$$

with $\underline{\mu}$ the electric dipole moment, (A.3). The operator $\mathfrak{H}_E$ may be written in irreducible tensor notation

$$\mathfrak{H}_E = -\sum_m (-1)^m \tilde{T}^1_m(\underline{\mu}) \tilde{T}^1_{-m}(\mathsf{E}) \qquad \text{(VIII.153)}$$

in the molecular fixed frame. The appropriate symmetry notation is

$$\mathfrak{H}_E = -\sum_{\Gamma ar} \varepsilon(a) T^{\Gamma}_a(\underline{\mu})_r S^1_{\hat{q}(a,r)}(\mathsf{E}) \qquad \text{(VIII.154)}$$

The effective matrix of the first- and second-order terms is given by

$$\langle \phi'\Gamma_e' a_e' \lambda' | \tilde{\mathfrak{H}}_E | \phi \Gamma_e a_e \lambda \rangle = -\sum_{\Gamma a r} (\Gamma_e a_e, \Gamma a | \Gamma_e' a_e')^* \langle \phi' \Gamma_e' \| T^{\Gamma}(\underline{\mu})_r \| \phi \Gamma_e \rangle \times \langle \lambda' | S^1_{\hat{q}(a,r)}(\mathsf{E}) | \lambda \rangle \tag{VIII.155}$$

and

$$\begin{aligned} \langle \phi'\Gamma_e' a_e' \lambda' | \tilde{\mathfrak{H}}_\alpha | \phi \Gamma_e a_e \lambda \rangle &= \sum_{\Gamma a r} (\Gamma_e a_e, \Gamma a | \Gamma_e' a_e')^* \langle \lambda' | S^{k}_{\hat{q}(a,r)}{}^{(\Gamma,r)}(\mathsf{E},\mathsf{E}) | \lambda \rangle \\ &\times \sum_{\Gamma_1 a_1 r_1} \varepsilon(a_1) \sum_{\Gamma_2 a_2 r_2} \varepsilon(a_2) \sum_{\phi'' \Gamma_e''}{}' \overline{\Delta V}^{-1}_{\phi''\Gamma_e''} \\ &\times \langle \phi' \Gamma_e' \| T^{\Gamma_1}(\underline{\mu})_{r_1} \| \phi'' \Gamma_e'' \rangle \langle \phi'' \Gamma_e'' \| T^{\Gamma_2}(\underline{\mu})_{r_2} \| \phi \Gamma_e \rangle \\ &\times (\Gamma_2 a_2, \Gamma_1 a_1 | \Gamma a)^* W(\Gamma_e \Gamma_2 \Gamma_e' \Gamma_1; \Gamma_e'' \Gamma) \sqrt{\lambda(\Gamma_e'')\lambda(\Gamma)} \\ &\times \langle 11 \hat{q}_2(a_1, r_1) \hat{q}_1(a_2, r_2) | k(\Gamma_1 r) \hat{q}(a,r) \rangle^* \end{aligned} \tag{VIII.156}$$

with an undefined rotational function basis $|\lambda\rangle$.

Confining ourselves to expectation values (i.e., $\phi' = \phi$, $\Gamma_e' = \Gamma_e$) we get the identification of the effective dipole moment constant

$$T_a^{\Gamma}(\tilde{\underline{\mu}})_r = \langle \phi \Gamma_e \| T^{\Gamma}(\underline{\mu})_r \| \phi \Gamma_e \rangle \tag{VIII.157}$$

and the polarizability constants

$$\begin{aligned} T_a^{\Gamma}(\tilde{\alpha})_r = -2 \Bigg[& \sum_{\Gamma_1 a_1 r_1} \varepsilon(a_1) \sum_{\Gamma_2 a_2 r_2} \varepsilon(a_2) \sum_{\phi'' \Gamma_e''} \overline{\Delta V}^{-1}_{\phi''\Gamma_e''} \\ &\times \langle \phi \Gamma_e \| T^{\Gamma_1}(\underline{\mu})_{r_1} \| \phi'' \Gamma_e'' \rangle \langle \phi'' \Gamma_e'' \| T^{\Gamma_2}(\underline{\mu})_{r_2} \| \phi \Gamma_e \rangle \\ &\times (\Gamma_2 a_2, \Gamma_1 a_1 | \Gamma a)^* W(\Gamma_e \Gamma_2 \Gamma_e \Gamma_1; \Gamma_e'' \Gamma) \sqrt{\lambda(\Gamma_e'')\lambda(\Gamma)} \\ &\times \langle 11 \hat{q}_2(a_2, r_2) \hat{q}_1(a_1, r_1) | k(\Gamma, r) \hat{q}(a,r) \rangle \Bigg] \end{aligned} \tag{VIII.158}$$

with $k = 0, 2$ and $\overline{\Gamma} = \Gamma_0$ for $k = 0$.

IX. COMPUTATION OF MOLECULAR PROPERTIES

In the preceding sections we have outlined the theory underlying the definition of the various molecular properties as determined experimentally, and we have given explicit formulas useful for the detailed computation of these properties using *ab initio* methods. Here in the final section we want to review the state of the art in carrying out such computations. One way to achieve this would be to compile a complete annotated bibliography, critically evaluating work done and research in progress concerning the direct computation of molecular properties; a valuable though monumental task, viewing, evaluating, and referencing several thousand research papers. Fortunately, such compilations exist,[2] and hopefully the effort of updating them continues. We will, instead, restrict ourselves here to some general comments about the *ab initio* computation of molecular properties, the possibility, reliability, effectiveness, and efficiency of such computations, pointing to some difficulties and problems as we see them, hopefully stimulating some theoretical research in those directions where knotty, unsolved problems remain. To be sure, such an assessment is strongly influenced by our own personal bias and should be seen as such.

In the present discussion we could use a classification of molecular properties paralleling that used in Section VIII:

1. Coarse structure; electronic, vibrational, and rotational energies linked directly to the structure of adiabatic potential surfaces.
2. Fine structure; spin–orbit and spin–spin coupling as well as Λ-type doubling and spin–rotation coupling.
3. Hyperfine structure; couplings of nuclear moments with electronic and molecular motion.
4. Structure and effects observed in the presence of external static electric or magnetic fields; electric moments and polarizability, magnetic screening and susceptibility.

It seems, however, more appropriate here to use a classification according to the type of computational procedure required for the evaluation of the molecular properties considered:

1. Properties obtained directly as expectation values over the electronic wave function.
2. Properties requiring the evaluation of perturbation expressions over electronic states.
3. Properties that require the explicit solution of equations for the nuclear motion with and without detailed couplings.

A. Properties as Expectation Values

Potential curves or surfaces as defined by adiabatic separation, (III.5 and III.6), with or without the diagonal part of the adiabatic correction, (VIII.4), are an obvious direct result of conventional electronic wave function calculations. From such potentials the dissociation energies, D_e, and, if several states are computed, the excitation energies, T_e, are obtained directly. Just as direct is the determination of the equilibrium nuclear position vectors, r_e, as the positions of the potential minima. From these zeroth-order approximations to the inertial tensor, $\mathbf{I}_e$ (B_e in diatomics), result; see (A.27) and (A.33). Computing, in general numerically, the second derivation of the potential surfaces at the minima with respect to the normal coordinates yields zeroth-order approximations to the ω_e's. Experience shows that well-correlated electronic wave functions derived from carefully balanced computations (careful basis set and configuration mixing selection) yield a reliability of ± 0.2 eV for D_e's and T_e's. The local consistency of such potentials is in general better; thus ω_e's are obtained with uncertainties of about 2%. The reliability achievable for the positions of the potential minima (i.e., the r_e's or the molecular conformation) is even better; the uncertainty for bond angles is generally less than 2° and for bond distances less than 1 pm, values that seem to be quite insensitive to the quality of the electronic wave functions; even Hartree–Fock functions seem to give reliable molecular conformations. To be sure, for small molecules, less than 20 electrons, and especially for diatomics, better reliability and accuracy can be and has been achieved. Here it is advisable to solve the nuclear motion problem also and determine term values, or D_0, T_{00}, ΔG_v's (ν_0's), B_v's, or B_0, and so on, directly, values that are experimentally more accessible. The determination of equilibrium properties and potential curve (surface) characteristics from experiment is fraught with fitting and inversion problems which should not be underestimated.[3a]

In Section VIII we presented four more properties that can be computed directly as expectation values over the electronic wave function as a function of the nuclear positions. To be sure, the classification of properties into some which are first order and others also requiring second-order contributions is somewhat arbitrary and depends on the neglect of higher-order terms that are small compared to the leading ones. For example, we discuss spin–orbit coupling in the next section as requiring a second-order contribution, even though this correction is in general by orders of magnitude smaller than the first-order term, while we have classified as purely first order the following four properties:

1. The nuclear spin–electronic orbit interaction in the case of nondegenerate electronic states can be determined using (VIII.81) and (VIII.68).

All that is necessary is to compute the one-electron integrals with the operator $\mathbf{r}_{i\alpha} \times \mathbf{p}_i / r_{i\alpha}^3 = \mathbf{l}_{i\alpha} / r_{i\alpha}^3$. The second term in (A.68), requiring two-electron integrals, corresponds to a mass polarization correction and can be ignored in general.

2. The nuclear spin–electronic spin interaction is given by (VIII.85), (VIII.86), and (A.64). For its calculation the value of the electronic wave function (spin density) at the nuclei is required plus the one-electron integrals with the operator $Y_{2m}(\omega_{i\alpha}) / r_{i\alpha}^3$, where $\omega_{i\alpha}$ is the solid angle between the vectors for the positions of electron i and nucleus α measured from some center, the center of nuclear mass or one nucleus, say.
3. Integrals of the same operator as above are needed to evaluate the nuclear quadrupole coupling constants as given in detail by (VIII.93) and (A.44).
4. The electric dipole or multipole moments given in (VIII.155), (VIII.157), and (A.43) are obtained directly from the one-electron integrals over the operator $\mathbf{r}_i$ or its appropriate tensor products.

Because the properties discussed above can be computed directly as expectation values over the electronic wave function and because the integrals needed are of the one-electron type, which are evaluated readily, there are no difficulties in calculating these properties. This is and should be done routinely whenever electronic state functions are computed explicitly; the extra effort is minimal.

Unfortunately, the expectation values depending on operators containing angular dependence times $r_{i\alpha}^{-3}$ will not be reliable in general. These values will depend strongly on an angular polarization of the K-shells, the possibility of which is rarely included in the basis sets chosen. Equally problematic is the calculation of the $\delta_{i\alpha}$ term in the nuclear spin-electronic spin coupling. To compute such a point-type property reliably places severe demands on the accuracy of the wave function.[56] More experience exists, and reliable values (± 0.1 D) can be obtained, for the calculations of dipole or higher electric moments. Some difficulties are known here for calculations with Gaussian-type basis functions.[57] These functions drop off rapidly as the $r_{i\alpha}$'s get large, which can lead to significant errors in the calculation of moments, especially higher moments, of the electronic charge distribution.

B. Properties Requiring Perturbation Expressions

While the calculation of the properties listed in Section IX.A can be considered to be rather simple and routine, provided that good electronic wave functions are available, direct calculation of the properties to be considered here is difficult, and additional theoretical work is required. The evaluation of the perturbation sums requires significant extra computational effort. Furthermore, there is frequently the need to compute additional two-electron integrals over operators other than $1/r_{ij}$, a nontrivial task. Provided that these extra integrals are available (for these integrals, see Ref. 58), three different approaches can be followed to determine the perturbation contributions.

1. Finite perturbations[59]; that is, the perturbing operators are included explicitly into the total Hamiltonian and with it a conventional electronic wave function and energy calculation is carried out. This approach is cumbersome at best, since a full energy computation has to be performed for each property required, a computation more difficult than a standard energy-wave-function calculation, because of lower symmetry and the need of additional one- and two-electron integrals.
2. Using the coupled[60] or uncoupled[61] Hartree–Fock or MCSCF[62] theory for the perturbation calculations would require a similar effort, as outlined above for calculations using finite perturbations. Here the orbital equations, or equations for the corrections to the orbitals, would have to be solved for each type of property (perturbation operator). In addition, the results would be limited to the Hartree–Fock model or, if coupled perturbation MCSCF equations were derived and used, to the MCSCF model. The former and the latter (in the form it can be used efficiently) model are frequently not adequate for an accurate energy and wave-function calculation.
3. The direct summation of the perturbation expressions appears to be theoretically and computationally the easiest procedure. In one-particle functions, with orbitals determined as is necessary for a conventional configuration mixing calculation, it should present little difficulty to determine the CSF's and the matrix elements of the terms in the perturbation sums. The sums could be evaluated readily up to the limit imposed by the finite number of basis functions chosen. The calculation would be no more difficult than a standard configuration mixing calculation. To be sure, the orbital basis may have to be augmented to account for the different symmetry required for some of the CSF's in the perturbations. To what degree the values of the perturbation sums depend on the basis and orbitals chosen should be investigated.

The general expression to be evaluated for spin–orbit interaction is given in (VIII.33) with u_i defined through (A.48). Here the second term, containing three-electron operators, is the mass-polarization correction, which is negligible in general. The perturbation sum in (VIII.33) will be small compared to the expectation value term except in those cases where two states of adjacent multiplicity are nearly energy-degenerate. A major difficulty in the calculation of the spin–orbit interaction for molecules is the evaluation of the two-electron integrals with the operator $\mathbf{r}_{ij} \times \mathbf{p}_i / r_{ij}^3$. Significant effort has gone into developing appropriate algorithms for the calculation of these integrals over STF's for diatomics[63] and over GTF in general;[64] however, efficient programs for these integrals do not seem to be readily available. Fortunately, there is an indication that for these integrals the zero differential overlap approximation seems to be adequate[65] in many cases. This approximation, or a Mulliken-type approximation,[66] is expected to be good because of the r_{ij}^3 term in the denominator.

Spin–spin coupling is detailed in (VIII.35) and (VIII.36) with (A.48) to (A.50). For the explicit evaluation of these expressions the comments made above apply as well; however, here the expectation-value terms will in general not dominate the perturbation sums. Furthermore, it is to be expected that the expectation value for the $\delta(\mathbf{r}_{ij})$ operator of (A.49) will be obtained accurately only for the very best, highly correlated electronic wave functions, which include r_{ij}'s explicitly. In addition to the two-electron integrals needed for spin–orbit interaction, those over the operator $Y_{2m}(\omega_{ij})/r_{ij}^3$ are required also.

To evaluate spin–rotation coupling explicitly, (VIII.75) should be used with the argument of the tensor defined through (A.58). The calculation of these terms should be quite similar in complexity to the full calculation of spin–orbit couplings.

It is not possible in general to give a simple, compact formula for the direct calculation of Λ-doubling constants. The origin of Λ-doubling can be traced to the electronic contribution to the rotational constant [i.e., the perturbation sum in (VIII.66) detailed more explicitly in (VIII.69b)]. From these the Λ-doubling constant q for linear molecules is given by (VIII.70). The corresponding p constant originates from spin-rotation coupling.

The calculation of nuclear spin–rotation coupling requires only one-electron integrals; the detailed formula is given by (VIII.100), with (VIII.95) defining the tensorial argument $\mathbf{m}$. The origin for these expressions is explicated in (A.70) and equations following. Even though these expressions appear formidable, detailed evaluation should present little difficulty, since no new two-electron integrals are involved. One electron integrals are required here for the operators $\mathbf{r}_{i\alpha}$, $\mathbf{r}_{i\alpha}/r_{i\alpha}^3$, and $\mathbf{r}_{i\alpha} \otimes \mathbf{r}_{i\beta} / r_{i\alpha}^3 r_{i\beta}^3$.

There are four contributions to nuclear spin–nuclear spin coupling, which are given explicitly by (VIII.105) with $\boldsymbol{\rho}_i$ defined by (VIII.104). Again, only one-electron integrals are needed in the evaluation, and these integrals are of the same type as for nuclear spin–electron spin and rotation coupling. Note that in spherically averaged NMR spectra, only the traces of the tensorial expressions are observed.

There are a number of magnetogyric ratios. The general expressions for the electronic spin, g_s, value is given by (VIII.119), with o_i defined by

(VIII.115) and $\boldsymbol{\kappa}_i$ defined through (VIII.116). For the calculation of g_s, several new types of integrals are required, one-electron integrals with the operator $\mathbf{r}_{i\alpha} \otimes \mathbf{r}_i / r_{i\alpha}^3$, and two-electron integrals with $\mathbf{r}_{ij} \otimes \mathbf{r}_i / r_{ij}^3$.

The orbital value, g_L, is given in detail through (VIII.128). For the evaluation of this expression, new integrals are again required, with the operators $\nabla_i^2 \cdot \mathbf{l}_i$, $\mathbf{r}_j \times \mathbf{p}_i / r_{ij}$, and $(\mathbf{r}_i \times \mathbf{r}_{ij})(\mathbf{r}_{ij} \cdot \mathbf{p}_j) / r_{ij}^3$. Little effort has gone thus far into developing formulas and algorithms for the calculation of these integrals and those required for g_s.

Easier is the direct computation of the rotational, vibrational, and translational g-values with the detailed expressions given by (VIII.136), (VIII.143), and (VIII.151), respectively. Difficult here are the integrals with the operator P_k, the vibrational nuclear momentum along a normal coordinate. Procedures to calculate those integrals have been developed, however.[67]

There exists a plethora of literature on the calculation of magnetic susceptibility,[3b–d] the general expression for which is given by (VIII.145). Frequently, the attempt is made to obtain magnetic susceptibility by just evaluating the expectation-value term with the coordinate origin chosen "optimally."[68] Such as optimal origin choice is possible in principle, since the origin dependence of the first- and second-order terms are reversed. However, finding it in practice requires the evaluation of both contributions. With normal limited basis sets, even the sum of the first- and second-order contributions exhibits a strong origin dependence. This basis-set inadequacy is believed to be remedied by using gauge-invariant basis functions,[69] which guarantee an origin-independent susceptibility. Using such basis functions, which really depend on the gauge chosen,[70] does not overcome, nor does it give information about, errors in the susceptibility values obtained due to an inadequate basis. More promising appear to be recent developments, where for normal basis sets an optimal origin for the evaluation of the first- and second-order contributions is chosen.[71] The integrals over the operator $\mathbf{r}_i \otimes \mathbf{r}_i$ needed for such calculations present little difficulty.

For the calculation of diamagnetic screening constants, given by (VIII.153), the same arguments concerning the origin dependence apply as given above. For the expectation-value term, additional integrals over $\mathbf{r}_i \times \mathbf{r}_{i\alpha} / r_{i\alpha}^3$ are required.

Finally, the polarizability tensor, given by (VIII.158), requires evaluation of a perturbation term only, with the integrals over the operator $\mathbf{r}_i$.

C. Properties and Nuclear Motion

We have discussed above direct evaluation of various molecular properties from the electronic wave function of the molecule. These values will depend on the nuclear conformation chosen, and we will get, for a molecule with N nuclei for each electronic state $3N-6$ (or -5 for linear cases), dimensional hypersurfaces for these properties, akin to the potential-energy hypersurface. These hypersurfaces are never directly observable. What is seen are these quantities for specific rovibronic states or combinations thereof. In the case of properties for low rotational and vibrational states, the values computed for the equilibrium conformation will in general provide good approximations to the observed quantities. Using R-centroids[72] for the rovibrational state considered should provide an even better approximation. It is, however, better to solve the equation for the wave functions of the nuclear motion, (V.16), and average the properties, which are obtained as functions of $\mathfrak{R}$, appropriately. An approximate rovibrational wave function is already required to determine R-centroids.

Solving the nuclear motion equation presents little difficulty for diatomics, since here the equation is one-dimensional and can be integrated numerically,[73] or a basis-function expansion technique[74] can be used. Here it is even feasible to treat sets of coupled equations necessary if coupling terms, which explicitly connect electronic with nuclear motion, are to be treated adequately.[75] The problem of solving the nuclear motion equation accurately becomes significantly more complicated for triatomics[76] and almost hopeless for polyatomics. For triatomics the potential is a three-dimensional hypersurface, which is six-dimensional for four atomics, and so on. For a five-atomic molecule the mapping of the complete nine-dimensional potential-energy hypersurface appears to be prohibitive. Finding simplified and adequate representations for such potentials in terms of internuclear distances or generalized normal coordinates or the like is a problem that needs to be solved before the nuclear motion equation can be treated for such systems in general. In the case of triatomics, several promising procedures have been developed and applied successfully.[76,77] Using a Hartree-type product representation for the nuclear motion wave function, and treating correlation with CM-type methods leads to gratifying results, indeed.[78] However, even for triatomics, the required computational effort for an accurate calculation becomes large, comparable to that of an electronic wave-function calculation. It is therefore unlikely that these methods can be extended readily to polyatomic systems. Here, significant additional research and new ideas seem to be required.

APPENDIX 1. THE MOLECULAR HAMILTONIAN

In this appendix we present the full molecular Hamiltonian, which contains relativistic corrections in the Pauli-type approximation and two-particle interactions on the basis of the Breit operator for two electrons. We use a Hamiltonian correct to order c^{-2} akin to the one transformed by Howard and Moss[47] from a space fixed to a molecular fixed frame for nonlinear[47a] and linear molecules.[47b] The molecular fixed frame has its origin in the center of mass of the nuclei. We introduce the following modifications with respect to the work of Howard and Moss:

1. For accomplishing the gauge transformation, we prefer to use a modified scalar function

$$F=\{[\mathsf{H}\times(\mathsf{Q}_0+\mathsf{r}_0)]\cdot\underline{\mu}-q[\mathsf{H}\times\mathsf{Q}_0]\cdot\mathsf{r}_0\}/2c \qquad \text{(A.1)}$$

   where H is the external magnetic field, Q_0 the position vector of the center of molecular mass in the space fixed coordinate system, and r_0 the position vector of the center of molecular mass in the molecular fixed coordinate system given by

$$\mathsf{r}_0=-(m/M)\sum_i \mathsf{r}_i \qquad \text{(A.2)}$$

   where m is the mass of an electron, M the total molecular mass, r_i the position vector of electron i in the molecular fixed frame, and the sum Σ_i runs over all electrons. $\underline{\mu}$ is the dipole moment given by

$$\underline{\mu}=e\left(\sum_\alpha Z_\alpha \mathsf{r}_\alpha-\sum_i \mathsf{r}_i\right) \qquad \text{(A.3)}$$

   where Z_α is the atomic number of nucleus α, e the absolute value of the electronic charge, and the sum Σ_α runs over all nuclei. With this choice of gauge it is possible to eliminate the electronic coordinate dependence out of the vector potential due to the moving molecular frame.
2. We have retained the nuclear spin–rotation and nuclear spin–vibration interactions as well as nuclear spin–nuclear spin interactions, even though they are formally of order smaller than c^{-2}, since they lead to observable effects.

In the following we present the various terms of the molecular Hamiltonian separated into nonrelativistic terms, electronic relativistic contributions, interaction operators which all have their origin in the relativistic corrections, and the operator for the nuclear spin Zeeman effect:

$$\mathfrak{H}=\mathfrak{H}_{\text{nonrel}}+\mathfrak{H}_{\text{rel-el}}+\mathfrak{H}_{\text{rel-W}}+\mathfrak{H}_{\text{ZI}} \qquad \text{(A.4)}$$

The nonrelativistic parts of the Hamiltonian are given by the expression

$$\begin{aligned}\mathfrak{H}_{\text{nonrel}} = \mathfrak{H}_e^0 + \mathfrak{H}_{\text{el}}^{\text{mp}} \\ + \mathfrak{H}_{v-\text{kin}} + U_W \\ + \mathfrak{H}_{\text{rot}} \\ + \mathfrak{H}_{\text{trans}} \\ + \mathfrak{H}_{\text{ST}} \\ + \mathfrak{H}_Q\end{aligned} \tag{A.5}$$

whose individual terms are explained below.

The zeroth-order electronic Hamiltonian in the nonrelativistic limit is given by [see also Section III, (III.1)]

$$\mathfrak{H}_e^0 = (1/2m)\sum_i \underline{\pi}_i^2 + V_{\text{Coul}} \tag{A.6}$$

with

$$\underline{\pi}_i = \mathbf{p}_i + (e/c)\mathbf{a}_i \tag{A.7}$$

$$\mathbf{p}_i = -i\hbar\nabla_i \tag{A.8}$$

where the gradient ∇ has to be taken in the molecular fixed frame,

$$\mathbf{a}_i = \tfrac{1}{2}(\mathbf{H}\times\mathbf{r}_i) + (m/Me)(\mathbf{H}\times\underline{\mu}) - (mq/2Me)(\mathbf{H}\times\mathbf{r}_0) \tag{A.9}$$

where q is the molecular charge,

$$V_{\text{Coul}} = e^2\left[\sum_{i<j} r_{ij}^{-1} + \sum_{\alpha<\beta} Z_\alpha Z_\beta r_{\alpha\beta}^{-1} - \sum_i \sum_\alpha Z_\alpha r_{i\alpha}^{-1}\right] \tag{A.10}$$

Here r_{ij}, $r_{i\alpha}$, and $r_{\alpha\beta}$ are the distances between the particles $i, j, \ldots$; that is,

$$r_{ij} = |\mathbf{r}_i - \mathbf{r}_j| \tag{A.11}$$

and so on.

The mass polarization correction to the electronic operator is given by

$$\mathfrak{H}_{\text{el}}^{\text{mp}} = \left(\frac{1}{2M_N}\right)\left(\sum_i \underline{\pi}_i\right)^2 \tag{A.12}$$

with M_N the total nuclear mass.

The kinetic-energy operator for the vibration is

$$\mathfrak{H}_{v-\mathrm{kin}}=\tfrac{1}{2}\sum_k \underline{\Pi}_k^2 \tag{A.13}$$

with

$$\underline{\Pi}_k = P_k - A_k, \tag{A.14}$$

$$P_k = -i\hbar\frac{\partial}{\partial S_k} \tag{A.15}$$

where S_k is the normal coordinate,

$$A_k=(e/2c)\sum_\alpha \left(Z_\alpha/\sqrt{M_\alpha}\,\right)(\mathsf{d}_k(\alpha)\cdot(\mathsf{H}\times\mathsf{r}_\alpha)) \tag{A.16}$$

where M_α is the mass of nucleus α and $\mathsf{d}_k(\alpha)$ is defined in Section IV, (IV.9), by

$$S_k=\sum_\alpha \mathsf{d}_k(\alpha)\sqrt{M_\alpha}\,(\mathsf{r}_\alpha-\mathsf{r}_\alpha^0) \tag{A.17}$$

The Watson term U_W is given by

$$U_W=-\hbar^2/8\,\mathrm{Tr}(\boldsymbol{\mu}) \tag{A.18}$$

with $\boldsymbol{\mu}$ given below in (A.25) and where Tr indicates the trace of a matrix.

The rotational Hamiltonian is expressed as

$$\mathfrak{H}_{\mathrm{rot}}=\tfrac{1}{2}\underline{\Pi}_r\cdot\boldsymbol{\mu}\underline{\Pi}_r \tag{A.19}$$

with

$$\underline{\Pi}_r=\hbar\mathsf{R}-\mathsf{A}_r \tag{A.20}$$

$$\mathsf{R}=\mathsf{N}-\mathsf{L}-\mathsf{G} \tag{A.21}$$

where N is the molecular rotational angular momentum operator, L the electronic orbital angular momentum operator, and G the vibrational angular momentum operator which is given by

$$\mathsf{G}=\sum_{kl}\underline{\zeta}_{kl}S_kP_l \tag{A.22}$$

The Coriolis coupling coefficient $\underline{\zeta}_{kl}$ is defined by

$$\underline{\zeta}_{kl} = \sum_{\alpha} \mathsf{d}_k(\alpha) \times \mathsf{d}_l(\alpha) \tag{A.23}$$

$$\mathsf{A}_r = (e/2c) \sum_{\alpha} Z_\alpha \left[\mathsf{r}_\alpha \times (\mathsf{H} \times \mathsf{r}_\alpha) - \left(1/\sqrt{M_\alpha}\,\right) \sum_{kl} \underline{\zeta}_{kl} S_k \mathsf{d}_l(\alpha) \cdot (\mathsf{H} \times \mathsf{r}_\alpha) \right] \tag{A.24}$$

$$\boldsymbol{\mu} = \mathbf{I}'^{-1} \tag{A.25}$$

where $\mathbf{I}'^{-1}$ is given by

$$\begin{aligned} \mathbf{I}' &= \mathbf{I} - \sum_{klm} \underline{\zeta}_{km} \otimes \underline{\zeta}_{lm} S_k S_l \\ &= \mathbf{I}^{(0)} + \sum_{k} \mathbf{a}_k S_k + \tfrac{1}{4} \sum_{kl} \mathbf{a}_k \mathbf{I}^{(0)-1} \mathbf{a}_l S_k S_l \end{aligned} \tag{A.26}$$

where the inertia moment tensor $\mathbf{I}$ is given by

$$\mathbf{I} = \sum_{\alpha} M_\alpha (r_\alpha^2 \mathsf{1} - \mathsf{r}_\alpha \otimes \mathsf{r}_\alpha) \tag{A.27}$$

$\mathbf{I}^{(0)}$ is the inertia moment tensor in the reference conformation, and

$$\mathbf{a}_k = \left. \frac{\partial \mathbf{I}}{\partial S_k} \right|_0 \tag{A.28}$$

where the derivative has to be taken at the nuclear reference conformation; $\otimes$ denotes a dyadic product.

$\mathfrak{H}_{\text{rot}}$ may be rearranged into the form

$$\begin{aligned} \mathfrak{H}_{\text{rot}} &= (\hbar^2/2) \mathsf{R} \cdot \boldsymbol{\mu} \mathsf{R} + \tfrac{1}{2} \{ \mathsf{R} \cdot \mathbf{I}''^{-1} \mathsf{t} \mathsf{H} + \mathsf{H}^{+} \mathbf{t} \mathbf{I}''^{-1} \mathsf{R} \} \\ &\quad + \tfrac{1}{2} \mathsf{A}_r \cdot \boldsymbol{\mu} \mathsf{A}_r \end{aligned} \tag{A.29}$$

with

$$\mathbf{I}'' = \mathbf{I}^{(0)} + \tfrac{1}{2} \sum_{k} \mathbf{a}_k S_k \tag{A.30}$$

$$\mathsf{t} = (\hbar e/c) \sum_{\alpha} Z_\alpha (\mathsf{r}_\alpha^0 \cdot \mathsf{r}_\alpha \mathbf{1} - \mathsf{r}_\alpha^0 \otimes \mathsf{r}_\alpha) \tag{A.31}$$

For linear molecules (A.29) can be written as

$$\mathfrak{H}_{\text{rot}} = \mathsf{R}' \cdot B\mathsf{R}' + B(\mathsf{R}' \cdot \mathsf{A}_r + \mathsf{A}_r \cdot \mathsf{R}') + B\mathsf{A}_r^2 \tag{A.32}$$

with

$$\mathsf{R}' = \mathsf{N} - \mathsf{L} - \mathsf{G} - (N_z - L_z - G_z)\mathsf{e}_z$$

where e_z is the unit vector in the direction of the symmetry axis

$$B = (\hbar^2/2)I'^{-1} \tag{A.33}$$

$$I' = I''^2/I^{(0)}$$

$$\mathsf{A}_r = -(e/2\hbar c)I''I^{(0)-1}\sum_\alpha Z_\alpha z_\alpha^0(\mathsf{e}_z \times (\mathsf{H} \times \mathsf{r}_\alpha)) \tag{A.34}$$

$$I'' = I^{(0)} + (1/2)\sum_k a_k S_k \tag{A.35}$$

$$I^{(0)} = \sum_\alpha M_\alpha z_\alpha^{02} \tag{A.36}$$

$$a_k = \left.\frac{\partial I_{xx}}{\partial S_k}\right|_0 = \left.\frac{\partial I_{yy}}{\partial S_k}\right|_0 \tag{A.37}$$

For diatomic molecules only one internal nuclear coordinate exists. Instead of a normal coordinate S, we use the internuclear distance R directly. Denoting the two nuclei by α and β, we have

$$R = z_\alpha - z_\beta \tag{A.38}$$

With this restriction we get

$$\begin{aligned}\mathfrak{H}_{\text{rot}} = {} & B\big(\mathsf{N}^2 - N_z^2 + \mathsf{L}^2 - L_z^2 - 2(\mathsf{NL} - N_z L_z)\big) \\ & - (\hbar e/2c)\big[(Z_\alpha M_\beta^2 + Z_\beta M_\alpha^2)/(M_\alpha + M_\beta)M_\alpha M_\beta\big] \\ & \times\big[(\mathsf{N} - \mathsf{L})\mathsf{H} - (N_z - L_z)H_z\big] + (e^2/8c^2)\big[(Z_\alpha M_\beta^2 + Z_\beta M_\alpha^2)^2/ \\ & \times (M_\alpha + M_\beta)^3 M_\alpha M_\beta\big] R^2\big[\mathsf{H}^2 - H_z^2\big]\end{aligned} \tag{A.39}$$

where B, (A.33), is now given by $(\hbar^2(M_\alpha + M_\beta)/2M_\alpha M_\beta R^2)$.

The translational kinetic energy is given by

$$\mathfrak{H}_{\text{trans}} = (1/2M)\underline{\Pi}_0^2 \tag{A.40}$$

with

$$\underline{\Pi}_0 = \mathbf{P}_0 - \mathbf{A}_0 = -i\hbar\nabla_0 - \mathbf{A}_0 \tag{A.41}$$

where ∇_0 is the gradient with respect to the space-fixed frame,

$$\mathbf{A}_0 = (1/c)(\mathbf{H}\times\underline{\mu}) + (q/2c)(\mathbf{H}\times\mathbf{Q}_0) - (q/2c)(\mathbf{H}\times\mathbf{r}_0) \tag{A.42}$$

The Stark effect operators are given by

$$\mathfrak{H}_{\mathrm{ST}} = -q\mathbf{Q}_0\cdot\mathbf{E} - \underline{\mu}\cdot\mathbf{E} \tag{A.43}$$

where $\mathbf{E}$ is the external electric field.

The nuclear quadrupole coupling operator is

$$\mathfrak{H}_Q = -\sum_{\alpha}\left(eQ_\alpha\sqrt{6}\,/2I_\alpha(2I_\alpha-1)\right)\sum_{m}(-1)^m\tilde{T}^2_m(\mathbf{I}_\alpha,\mathbf{I}_\alpha)$$
$$\times\left[\sum_i C^2_{-m}(\omega_{i\alpha})r_{i\alpha}^{-3} - \sum_{\beta(\neq\alpha)} Z_\beta C^2_{-m}(\omega_{\beta\alpha})r_{\beta\alpha}^{-3}\right] \tag{A.44}$$

with $\tilde{T}^2_m(\mathbf{I}_\alpha,\mathbf{I}_\alpha)$ the irreducible tensor operator of rank 2 formed by the components of $\mathbf{I}_\alpha$, where $\mathbf{I}_\alpha$ is the spin operator for nucleus α and I_α the appropriate spin quantum number. The Racah spherical harmonics are defined as

$$C^2_m(\omega_{i\alpha}) = \sqrt{4\pi/5}\;Y^2_m(\omega_{i\alpha}) \tag{A.45}$$

where the Y^k_m are the standard spherical harmonics and $\omega_{i\alpha}$ specifies the orientation of the vector $(\mathbf{r}_i - \mathbf{r}_\alpha)$ in a spherical coordinate system, with a similar definition for $\omega_{\beta\alpha}$. We denote by Q_α the quadrupole moment of the nuclear charge distribution. For diatomic molecules the nuclear contribution (Σ_β) vanishes for $m\neq 0$.

The electronic relativistic corrections are given by

$$\mathfrak{H}_{\text{rel-el}} = \mathfrak{H}_{\mathrm{SO}} + \mathfrak{H}^{(0)}_{\mathrm{SS}} + \mathfrak{H}^{(2)}_{\mathrm{SS}} + \mathfrak{H}_{rm} + \mathfrak{H}_{\mathrm{D}} + \mathfrak{H}_{\mathrm{OO}} \tag{A.46}$$

whose individual terms are given by the following expressions. The spin–orbit, and spin–other-orbit interaction, and its mass polarization corrections are included in

$$\mathfrak{H}_{\mathrm{SO}} = \sum_i \mathbf{u}_i\mathbf{s}_i \tag{A.47}$$

with $\mathbf{s}_i$ the spin operator of electron i in units of $\hbar$ and

$$\mathbf{u}_i = -(\beta g e/2mc)\left\{ \sum_{j(\neq i)} r_{ij}^{-3}(\mathbf{r}_{ij}\times(\underline{\pi}_i - 2\underline{\pi}_j)) - \sum_{\alpha} Z_\alpha r_{i\alpha}^{-3}(\mathbf{r}_{i\alpha}\times\underline{\pi}_i)\right\}$$

$$-(\beta g e/2M_N c)\sum_j \left\{ \sum_{l(\neq i)} r_{il}^{-3}(\mathbf{r}_{il}\times\underline{\pi}_j) - \sum_{\alpha} Z_\alpha r_{i\alpha}^{-3}(\mathbf{r}_{i\alpha}\times\underline{\pi}_j)\right\} \quad \text{(A.48)}$$

where the first term contains the spin–orbit and spin–other-orbit interaction and the two last terms are the mass polarization corrections.

The spin–spin coupling operators are separated into zero- and second-rank tensor contributions with

$$\mathfrak{H}_{SS}^{0} = -(\beta^2 g^2 8\pi/3)\sum_{i<j} \mathbf{s}_i\cdot\mathbf{s}_j\delta(\mathbf{r}_{ij}) \quad \text{(A.49)}$$

where δ is the Dirac δ-function and

$$\mathfrak{H}_{SS}^{2} = -\beta^2 g^2\sqrt{6}\sum_{i<j} r_{ij}^{-3}\sum_{m=-2}^{2}(-1)^m \tilde{T}_m^2(\mathbf{s}_i,\mathbf{s}_j)C_{-m}^2(\omega_{ij}) \quad \text{(A.50)}$$

The relativistic correction to the kinetic energy is accounted for by

$$\mathfrak{H}_{rm} = (1/8m^3c^2)\sum_i(\underline{\pi}_i^2 + g\mathbf{s}_i)^2 \quad \text{(A.51)}$$

where g is the free-electron spin g-factor,[54]

$$g \approx 2.023 \quad \text{(A.52)}$$

The relativistic Darwin corrections are given by

$$\mathfrak{H}_D = -(e^2\hbar^2\pi/2m^2c^2)\left\{\sum_{i\neq j}\delta(\mathbf{r}_{ij}) - \sum_i\sum_\alpha Z_\alpha\delta(\mathbf{r}_{i\alpha})\right\} \quad \text{(A.53)}$$

The orbit-orbit interaction is

$$\mathfrak{H}_{OO} = -\sum_{i\neq j}(e^2/4m^2c^2)r_{ij}^{-1}\{\underline{\pi}_i\cdot\underline{\pi}_j + r_{ij}^{-2}\mathbf{r}_{ij}\cdot(\mathbf{r}_{ij}\cdot\underline{\pi}_i)\underline{\pi}_j\} \quad \text{(A.54)}$$

neglecting small mass polarization corrections presented in the work of Howard and Moss.

The Zeeman effect of the electronic spin is given by

$$\mathfrak{H}_{ZS} = g\beta\,\mathbf{S}\mathbf{H} \quad \text{(A.55)}$$

where β is the Bohr magneton,

$$\beta = \frac{\hbar e}{2mc} \tag{A.56}$$

and S is the total electronic spin operator in units of $\hbar$.

The operators for the interaction of electronic spins and orbital motion with rotation, vibration, and nuclear spins and of nuclear spins with rotation, vibration, and other nuclear spins are

$$\mathfrak{H}_{\text{rel-}W} = \mathfrak{H}_{\text{S}r} + \mathfrak{H}_{\text{S}v} + \mathfrak{H}_{\text{SI}} + \mathfrak{H}_{\text{O}r} + \mathfrak{H}_{\text{O}v} + \mathfrak{H}_{\text{O}I} + \mathfrak{H}_{\text{II}} + \mathfrak{H}_{\text{I}r} + \mathfrak{H}_{\text{I}v} \tag{A.57}$$

The individual terms are detailed below.

The electronic spin–rotation interaction operator is

$$\begin{aligned}\mathfrak{H}_{\text{S}r} = \tfrac{1}{2}\Big\{ & \sum_i \sum_\alpha Z_\alpha (g\beta e/c) I''^{1/2} \\ & \times \left[\mathsf{s}_i \cdot (\mathsf{r}_{i\alpha} \times (\mathsf{r}_\alpha^0 \times \mathbf{I}''^{-1}\mathsf{R})) \right. \\ & \left. + ((\mathsf{R}\mathbf{I}''^{-1} \times \mathsf{r}_\alpha^0) \times \mathsf{r}_{i\alpha}) \cdot \mathsf{s}_i \right] I''^{-1/2}\end{aligned} \tag{A.58}$$

where

$$I'' = \det(\mathbf{I}'') \tag{A.59}$$

[see (A.35)].

For linear molecules (A.58) can be simplified to the expression

$$\begin{aligned}\mathfrak{H}_{\text{S}r} = \tfrac{1}{2}\Big\{ & \sum_i \sum_\alpha Z_\alpha (g\beta e/c) I''^{-1} r_{i\alpha}^{-3} Z_\alpha^0 \\ & \times \left[\mathsf{s}_i \cdot (\mathsf{r}_{i\alpha} \times (\mathsf{e}_z \times \mathsf{R})) + (\mathsf{r}_{i\alpha} \times (\mathsf{e}_z \times \mathsf{R})) \cdot \mathsf{s}_i \right] \Big\}\end{aligned} \tag{A.60}$$

For diatomics we get an additional simplification by

$$I'' = R^0 R M_\alpha M_\beta / (M_\alpha + M_\beta) \tag{A.61}$$

and

$$z_\alpha^0 = R^0 M_\beta / (M_\alpha + M_\beta) \qquad z_\beta^0 = -R^0 M_\alpha / (M_\alpha + M_\beta) \tag{A.62}$$

The electronic spin–vibration interaction is given by

$$\mathfrak{H}_{Sv} = -\sum_i \sum_\alpha \sum_k Z_\alpha \left(g\beta e / c\sqrt{M_\alpha} \right) I''^{1/2} \times r_{i\alpha}^{-3} \mathbf{s}_i (\mathbf{r}_{i\alpha} \times \mathbf{d}_k(\alpha)) P_k I''^{-1/2} \tag{A.63}$$

The electronic spin–nuclear spin interactions are

$$\mathfrak{H}_{SI} = \beta g (8\pi/3) \sum_\alpha \sum_i g_\alpha \mathbf{s}_i \cdot \mathbf{I}_\alpha \delta(\mathbf{r}_{i\alpha}) - \beta g \beta_N \sqrt{6} \sum_\alpha \sum_i g_\alpha r_{i\alpha}^{-3} \sum_{m=-2}^{2} (-1)^m \tilde{T}_m^2(\mathbf{s}_i, \mathbf{I}_\alpha) C_{-m}^2(\omega_{i\alpha}) \tag{A.64}$$

with the nuclear magneton β_N defined as

$$\beta_N = \hbar e / 2M_p c \tag{A.65}$$

where M_p is the mass of the proton.

The orbit–rotation interaction is given by

$$\mathfrak{H}_{Or} = -(e^2/2mc^2) I''^{1/2} \sum_i \sum_\alpha Z_\alpha r_{i\alpha}^{-1} \times \left[\underline{\pi}_i \cdot (\mathbf{r}_\alpha^0 \times \mathbf{I}''^{-1} \underline{\Pi}_r) + r_{i\alpha}^{-2} \mathbf{r}_{i\alpha} \cdot (\mathbf{r}_{i\alpha} \cdot \underline{\pi}_i)(\mathbf{r}_\alpha^0 \times \mathbf{I}''^{-1} \underline{\Pi}_r) \right] I''^{-1/2} \tag{A.66}$$

The orbit–vibration interaction is

$$\mathfrak{H}_{Ov} = -(e^2/2mc^2) I''^{1/2} \sum_i \sum_\alpha \sum_k \left(Z_\alpha / \sqrt{M_\alpha} \right) \times r_{i\alpha}^{-1} \left[(\underline{\pi}_i \mathbf{d}_k(\alpha) \underline{\Pi}_k) + r_{i\alpha}^{-2} r_{i\alpha} (r_{i\alpha} \underline{\pi}_i)(\mathbf{d}_k(\alpha) \underline{\Pi}_k) \right] I''^{-1/2} \tag{A.67}$$

The orbit–nuclear spin interaction is given by

$$\mathfrak{H}_{OI} = (e\beta_N / mc) \sum_\alpha \sum_i g_\alpha r_{i\alpha}^{-3} \mathbf{I}_\alpha (\mathbf{r}_{i\alpha} \times \underline{\pi}_i) + (e\beta_N / M_N c) \sum_\alpha \sum_i \sum_j g_\alpha r_{i\alpha}^{-3} \mathbf{I}_\alpha (\mathbf{r}_{ij} \times \underline{\pi}_j) \tag{A.68}$$

where the second term is a mass polarization correction.

The nuclear spin–spin interaction operator has the following contributions:

$$
\begin{aligned}
\mathfrak{H}_{\mathrm{II}} = {} & \beta_N^2 \sum_{\alpha<\beta} g_\alpha g_\beta \sum_{m=-2}^{2} (-1)^m \tilde{T}_m^2(\mathbf{I}_\alpha, \mathbf{I}_\beta) C_{-m}^2(\omega_{\alpha\beta}) \\
& + (\beta_N^2 e^2/2mc^2) \sum_\alpha \sum_\beta g_\alpha g_\beta \\
& \times \mathbf{I}_\alpha \cdot \left(\sum_i (\mathbf{r}_{i\alpha}\mathbf{r}_{i\beta}\mathbf{1} - \mathbf{r}_{i\alpha} \otimes \mathbf{r}_{i\beta}) r_{i\alpha}^{-3} r_{i\beta}^{-3} \right) \mathbf{I}_\beta
\end{aligned}
\tag{A.69}
$$

The nuclear spin–rotation interaction is given by

$$
\mathfrak{H}_{\mathrm{I}r} = \mathfrak{H}_{\mathrm{I}r}^0 + \mathfrak{H}_{\mathrm{I}r}^T \tag{A.70}
$$

where $\mathfrak{H}_{\mathrm{I}r}^0$ is the nuclear spin–rotation interaction operator resulting from the interaction of the magnetic field produced by the nuclear spins with the rotation, and $\mathfrak{H}_{\mathrm{I}r}^T$ is the Thomas correction introduced by Ramsey,[51] which is analogous to the spin–orbit interaction of electrons. Since a quantum mechanical relativistic invariant theory for nuclear particles is lacking, it cannot be derived in the same way as the spin–orbit effect and has to be considered as a classical effect.

In detail, we have

$$
\begin{aligned}
\mathfrak{H}_{\mathrm{I}r}^0 = {} & -(\hbar e \beta_N/2c) I''^{1/2} \sum\sum_{\alpha \neq \beta} Z_\alpha \\
& \times \{ \mathbf{R} \cdot \mathbf{I}''^{-1} r_{\alpha\beta}^{-3} (\mathbf{r}_\alpha^0 \times (\mathbf{I}_\beta \times \mathbf{r}_{\alpha\beta})) \\
& + r_{\alpha\beta}^{-3} (\mathbf{r}_\alpha^0 \times (\mathbf{I}_\beta \times \mathbf{r}_{\alpha\beta})) \cdot \mathbf{I}''^{-1} \mathbf{R} \} I''^{-1/2}
\end{aligned}
\tag{A.71}
$$

$$
\begin{aligned}
\mathfrak{H}_{\mathrm{I}r}^T = {} & -(\beta_N/2c) \sum_\alpha g_\alpha (1-\gamma_\alpha) \\
& \times \{ \mathbf{I}_\alpha \cdot (\mathbf{E}^i(\mathbf{r}_\alpha) \mathbf{r}_\alpha \mathbf{1} - \mathbf{E}^i(\mathbf{r}_\alpha) \otimes \mathbf{r}_\alpha) \boldsymbol{\mu} \mathbf{R} \\
& + \mathbf{R} \cdot \boldsymbol{\mu} (\mathbf{E}^i(\mathbf{r}_\alpha) \cdot \mathbf{r}_\alpha \mathbf{1} - \mathbf{E}^i(\mathbf{r}_\alpha) \otimes \mathbf{r}_\alpha) \mathbf{I}_\alpha \}
\end{aligned}
\tag{A.72}
$$

with

$$
\mathbf{E}^i(\mathbf{r}_\alpha) = e \left[\sum_{\beta(\neq\alpha)} Z_\beta \mathbf{r}_{\beta\alpha} r_{\beta\alpha}^{-3} - \sum_i \mathbf{r}_{i\alpha} r_{i\alpha}^{-3} \right] \tag{A.73}
$$

$$
\gamma_\alpha = Z_\alpha M_p / g_\alpha M_\alpha \tag{A.74}
$$

For molecules with a linear reference conformation, we get for $\mathfrak{H}^0_{\mathrm{I}r}$, (A.71):

$$\mathfrak{H}^0_{\mathrm{I}r} = -(\hbar e\beta_N/2c)\sum_{\alpha\neq\beta}\sum Z_\alpha \times\{\mathsf{R}I''^{-1}r_{\alpha\beta}^{-3}Z_\alpha^0(Z_{\alpha\beta}\mathbf{1}-\mathsf{e}_z\otimes\mathsf{r}_{\alpha\beta})\mathsf{I}_\beta + \mathsf{I}_\beta r_{\alpha\beta}^{-3}z_\alpha^0(z_{\alpha\beta}\mathbf{1}-\mathsf{e}_z\otimes\mathsf{r}_{\alpha\beta})I''^{-1}\mathsf{R}\} \tag{A.75}$$

For diatomics $\mathfrak{H}^0_{\mathrm{I}r}$ is simplified further. For the two nuclei denoted by α and β, we get for $\mathfrak{H}^0_{\mathrm{I}r}$:

$$\begin{aligned}\mathfrak{H}^0_{\mathrm{I}r} = &-(\hbar e\beta_N/2c)R^{-3}\\ &\times\{(Z_\beta/M_\beta)[(\mathsf{I}_\alpha\cdot\mathsf{R}-I_{\alpha z}R_z)+(\mathsf{R}\cdot\mathsf{I}_\alpha-R_zI_{\alpha z})]\\ &-(Z_\alpha.M_\alpha)[(\mathsf{I}_\beta\cdot\mathsf{R}-I_{\beta z}R_z)+(\mathsf{R}\cdot\mathsf{I}_\beta-R_zI_{\beta z})]\end{aligned} \tag{A.76}$$

and for $\mathfrak{H}^T_{\mathrm{I}r}$:

$$\begin{aligned}\mathfrak{H}^T_{\mathrm{I}r} = &-(\beta_N/2c)\{-g_\alpha(1-\gamma_\alpha)M_\beta E_z^i(z_\alpha)\\ &\times[(\mathsf{I}_\alpha\cdot\mathsf{R}-I_{\alpha z}R_z)+(\mathsf{R}\cdot\mathsf{I}_\alpha-R_zI_{\alpha z})]-g_\beta(1-\gamma_\beta)M_\alpha E_z^i(Z_\beta)\\ &\times[(\mathsf{I}_\beta\cdot\mathsf{R}-I_{\beta z}R_z)+(\mathsf{R}\cdot\mathsf{I}_\beta-R_zI_{\beta z})]\}(R/(M_\alpha+M_\beta))\end{aligned} \tag{A.77}$$

In the same way we describe the nuclear spin–vibration interactions:

$$\mathfrak{H}_{\mathrm{I}v} = \mathfrak{H}^0_{\mathrm{I}v} + \mathfrak{H}^T_{\mathrm{I}v} \tag{A.78}$$

$$\mathfrak{H}^0_{\mathrm{I}v} = -(e\beta_N/c)\sum_{\alpha\neq\beta}\sum\sum_k g_\alpha\left(Z_\beta/\sqrt{M_\beta}\right) \times\{P_k r_{\alpha\beta}^{-3}(\mathsf{r}_{\beta\alpha}\times\mathsf{d}_k(\alpha))\mathsf{I}_\alpha + \mathsf{I}_\alpha r_{\alpha\beta}^{-3}\cdot(\mathsf{r}_{\beta\alpha}\times\mathsf{d}_k(\alpha))P_k\} \tag{A.79}$$

$$\mathfrak{H}^T_{\mathrm{I}v} = -(\beta_N/2c)\sum_\alpha\sum_k g_\alpha(1-\gamma_\alpha)\sqrt{M_\alpha}^{\,-1} \times\{\mathsf{I}_\alpha\cdot(\mathsf{E}^i(\mathsf{r}_\alpha)\times\mathsf{d}_k(\alpha))P_k + P_k(\mathsf{E}^i(\mathsf{r}_\alpha)\times\mathsf{d}_k(\alpha))\cdot\mathsf{I}_\alpha\} \tag{A.80}$$

Finally, the nuclear Zeeman effect operator is given by

$$\mathfrak{H}_{\mathrm{ZI}} = -\beta_N\sum_\alpha g_\alpha\mathsf{I}_\alpha\cdot\mathsf{H} \tag{A.81}$$

APPENDIX 2. COUPLING AND RECOUPLING COEFFICIENTS

For a real basis of the irreducible representations of the symmetry group, Griffith[40a] has constructed real V-coefficients, which are related to coupling coefficients by

$$V_{\text{Gr}}\begin{pmatrix} \Gamma_3 & \Gamma_1 & \Gamma_2 \\ a_3 & a_1 & a_2 \end{pmatrix}\sqrt{\lambda(\Gamma_3)} = (\Gamma_1 a_1, \Gamma_2 a_2 | \Gamma_3 a_3) \tag{A.82}$$

Griffith defines, in addition, a complex basis for doubly and triply degenerate irreducible representations by

$$\begin{aligned} f_{\pm 1} &= (-if_x + f_y)/\sqrt{2} \\ f_0 &= if_z \end{aligned} \tag{A.83}$$

For this basis the V_{Gr} are related to now-complex coupling coefficients by

$$[-1]^{\Gamma_3}(-1)^{a_3} V_{\text{Gr}}\begin{pmatrix} \Gamma_3 & \Gamma_1 & \Gamma_2 \\ -a_3 & a_1 & a_2 \end{pmatrix}\sqrt{\lambda(\Gamma_3)} = (\Gamma_1 a_1, \Gamma_2 a_2 | \Gamma_3 a_3) \tag{A.84}$$

with $[-1]^{\Gamma_3} = -1$ for triply degenerate representations and $+1$ otherwise.

The V-coefficients are even or odd with respect to the permutation of their columns, depending on whether the representations Γ_1, Γ_2, Γ_3 contain the antisymmetric products of degenerate representations. For an odd permutation the V-coefficients are multiplied by $(-1)^{\Gamma_1+\Gamma_2+\Gamma_3}$, where $(-1)^{\Gamma}$ is -1 for any Γ, which is contained in the antisymmetric product of any irreducible representation with itself; it is $+1$ otherwise.

For many purposes it is convenient to use the definitions of Griffith, for example in the case of a real basis for the irreducible representations. However, the convention adopted by Griffith is not compatible with the standard convention used for the three-dimensional rotation group. This becomes particularly apparent in the case of degenerate representations with a complex basis. In order to use a phase convention consistent with the three-dimensional rotation group as far as possible, we define with real coupling coefficients

$$V\begin{pmatrix} \Gamma_1 & \Gamma_2 & \Gamma_3 \\ a_1 & a_2 & -a_3 \end{pmatrix}\sqrt{\lambda(\Gamma_3)} = (-1)^{\Gamma_1+\Gamma_2}(\Gamma_1 a_1, \Gamma_2 a_2 | \Gamma_3 a_3) \tag{A.85}$$

where the permutation symmetry for the V's is as described above. To avoid ambiguity we specify the complex basis functions such that they transform in the same way as the spherical harmonics, Y^l_m, which form a basis for these representations. This means that for $C_{\infty v}$:

$$
\begin{aligned}
\Sigma^+:&\quad Y^l_0, \qquad l=0,1,2,\ldots\\
\Sigma^-:&\left(Y^l_m(1)\,Y^l_{-m}(2)-Y^l_{-m}(1)\,Y^l_m(2)\right)/\sqrt{2}\,, \qquad l\geq m\\
\Pi_\pm:&\quad Y^l_{\pm 1}, \qquad l\geq 1\\
\Delta_\pm:&\quad Y^l_{\pm 2}, \qquad l\geq 2
\end{aligned}
$$

etc. .

With this definition the meaning of a corresponding "real" basis is spoiled, in as much as some functions of nondegenerate species (i.e., Σ^- in $C_{\infty v}$) will be purely imaginary. The real basis is connected to the complex one by

$$
\begin{aligned}
f_x &= ((-1)^m f_{+1} + f_{-1})/\sqrt{2}\\
f_y &= -i((-1)^m f_{+1} - f_{-1})/\sqrt{2}\\
f_z &= f_0
\end{aligned}
\tag{A.86}
$$

with m given by the appropriate Y^l_m forming the complex basis.

The V-coefficients for this "real" basis are given by

$$
V\begin{pmatrix}\Gamma_1 & \Gamma_2 & \Gamma_3\\ a_1 & a_2 & a_3\end{pmatrix}\sqrt{\lambda(\Gamma_3)} = \{-1\}^{\Gamma_3}(\Gamma_1 a_1, \Gamma_2 a_2 | \Gamma_3 a_3) \tag{A.87}
$$

For instance, for $C_{\infty v}$, $\{-1\}^{\Gamma}=-1$ for $\Gamma=\Sigma^-, \Pi, \Phi, \ldots$ and $+1$ for $\Gamma=\Sigma^+, \Delta, \ldots$. Again the V-coefficients defined in (A.87) have the same permutation symmetry as specified above. In Table I we give a list of V-coefficients for $C_{\infty v}$ with our phase definition.

The Wigner–Eckart theorem, expressed in terms of V-coefficients, takes the following form for a real basis:

$$
\begin{aligned}
&\langle v(\Gamma_2 a_2)|\,T^{\Gamma}_a(\mathrm{op})\,|u(\Gamma_1 a_1)\rangle\\
&\qquad = \{-1\}^{\Gamma_2} V\begin{pmatrix}\Gamma_2 & \Gamma & \Gamma_1\\ a_2 & a & a_1\end{pmatrix}\langle v\Gamma_2\|T^{\Gamma}(\mathrm{op})\|u\Gamma_1\rangle
\end{aligned}
\tag{A.88}
$$

TABLE I
V-Coefficients for $C_{\infty v}$[a]

Complex basis	Real basis
$V\begin{pmatrix} \Sigma^- & \Sigma^- & \Sigma^{\pm} \\ \cdot & \cdot & \cdot \end{pmatrix} = -1$	$V\begin{pmatrix} \Sigma^- & \Sigma^- & \Sigma^{+} \\ \cdot & \cdot & \cdot \end{pmatrix} = -1$
$V\begin{pmatrix} E_n & E_n & \Sigma^+ \\ + & - & \cdot \end{pmatrix} = \frac{1}{\sqrt{2}}$	$V\begin{pmatrix} E_n & E_n & \Sigma^+ \\ a & a & \cdot \end{pmatrix} = \frac{1}{\sqrt{2}}, a = x,y$
$V\begin{pmatrix} E_n & E_n & \Sigma^- \\ + & - & \cdot \end{pmatrix} = -\frac{1}{\sqrt{2}}$	$V\begin{pmatrix} E_n & E_n & \Sigma^- \\ x & y & \cdot \end{pmatrix} = -\frac{i}{\sqrt{2}}$
	$V\begin{pmatrix} E_n & E_m & E_p \\ x & x & x \end{pmatrix} = \frac{1}{\sqrt{2}}$
$V\begin{pmatrix} E_n & E_m & E_p \\ \pm & \pm & \mp \end{pmatrix} = 1$	$V\begin{pmatrix} E_n & E_m & E_p \\ y & y & x \end{pmatrix} = \frac{1}{\sqrt{2}}$
	$V\begin{pmatrix} E_n & E_m & E_p \\ x & y & y \end{pmatrix} = \frac{1}{\sqrt{2}}$
with $p > m \geq n, p = n + m$	

[a]They are related to the coupling coefficients as indicated by (A.85) and (A.87). The E_n with $n = 1, 2, \ldots$ mean $\Pi, \Delta, \ldots$. The single component of a one-dimensional representation is represented by $\cdot$.

or for a complex basis,

$$\langle v(\Gamma_2 a_2)| T_a^{\Gamma}(\text{op})|u(\Gamma_1 a_1)\rangle = V\begin{pmatrix} \Gamma_2 & \Gamma & \Gamma_1 \\ -a_2 & a & a_1 \end{pmatrix} \langle v\Gamma_2 \| T^{\Gamma}(\text{op}) \| u\Gamma_1 \rangle \tag{A.89}$$

With Griffith's definitions we could omit $\{-1\}^{\Gamma_2}$ in (A.88) and would have to multiply $[-1]^{\Gamma_2}(-1)^{a_2}$ in (A.89). This may be compared with (VII.18). Clearly, the definition of the Wigner–Eckart theorem by (VII.18) and (A.88) and (A.89) are different insofar as the reduced matrix elements in (A.88) and (A.89) contain the factor $\sqrt{\lambda(\Gamma_2)}$. This is fully analogous to the difference of reduced matrix elements for the three-dimensional rotation group using the $3j$-symbols or Clebsch–Gordan coefficients.

We define the recoupling coefficients (Table II) analogous to the definition of Racah coefficients,[41a,43,44] that is,

$$\begin{aligned} &W(\Gamma_1\Gamma_2\Gamma\Gamma_3; \Gamma'\Gamma'')\sqrt{\lambda(\Gamma')\lambda(\Gamma'')} \\ &\quad = \Sigma_{a_1,a_2,a',a'',a}(\Gamma_1 a_2, \Gamma_2 a_2|\Gamma' a')(\Gamma' a', \Gamma_3 a_3|\Gamma a) \\ &\quad\quad \times (\Gamma_1 a_1, \Gamma'' a''|\Gamma a)^*(\Gamma_2 a_2, \Gamma_3 a_3|\Gamma'' a'')^* \end{aligned} \tag{A.90}$$

TABLE II
Recoupling Coefficients for $C_{\infty v}$[a]

$$W_{\mathrm{Gr}}\begin{pmatrix}\Sigma^+ & \Gamma_1 & \Gamma_2\\ \Gamma & \Gamma_3 & \Gamma_4\end{pmatrix}=(-1)^{\Gamma_1+\Gamma_3+\Gamma}\delta_{\Gamma_1\Gamma_2}\delta_{\Gamma_3\Gamma_4}\delta(\Gamma_1,\Gamma_3,\Gamma)\frac{1}{\sqrt{\lambda(\Gamma_1)\lambda(\Gamma_3)}}$$

$$\text{with } \delta(\Gamma_1,\Gamma_3,\Gamma)=\begin{cases}1 & \text{if } \Gamma \text{ is contained in } \Gamma_1\otimes\Gamma_3\\ 0 & \text{otherwise}\end{cases}$$

$$W_{\mathrm{gr}}\begin{pmatrix}\Sigma^- & E_n & E_n\\ \Sigma^- & E_n & E_n\end{pmatrix}=\tfrac{1}{2}$$

$$W_{\mathrm{Gr}}\begin{pmatrix}\Sigma^- & E_m & E_m\\ E_p & E_n & E_n\end{pmatrix}=\begin{cases}-\tfrac{1}{2} & \text{for } p=m+n\\ \tfrac{1}{2} & \text{for } n=m+p \text{ or } m=n+p\end{cases}$$

$$W_{\mathrm{Gr}}\begin{pmatrix}E_m & E_n & E_p\\ E_q & E_r & E_s\end{pmatrix}=\begin{cases}\tfrac{1}{2} & \text{if } m=n+p=r+s=p+q+s\\ 0 & \text{otherwise}\end{cases}$$

[a]The W-coefficients used in the text are related to the listed W_{Gr}-coefficients by (A.91).

independent of a_3. For a standard basis with real coupling coefficients, the complex conjugate becomes irrelevant.

Our W-coefficients are related to those defined by Griffith,[40a] W_{Gr}, by the relation

$$W(\Gamma_1\Gamma_2\Gamma\Gamma_3;\Gamma'\Gamma'')=(-1)^{\Gamma_1+\Gamma_2+\Gamma_3+\Gamma}W_{\mathrm{Gr}}\begin{pmatrix}\Gamma_1 & \Gamma_2 & \Gamma'\\ \Gamma_3 & \Gamma & \Gamma''\end{pmatrix} \tag{A.91}$$

Equation (A.91) is again similar to the relation between Racah coefficients and the $6j$-symbols of Wigner.[44] The W_{Gr}-coefficients possess a permutation symmetry analogous to that of $6j$-symbols, which is advantageous for tabulation (i.e., they are invariant with respect to the permutation of the columns and with respect to turning any pair of columns upside down).

APPENDIX 3. RELATIONS BETWEEN REAL AND COMPLEX NOTATION USED HEREIN

First we consider scalar products of the types

$$\underline{\gamma}\cdot \mathbf{A} \tag{A.92}$$

and

$$\mathbf{A}\cdot\gamma\mathbf{B} \tag{A.93}$$

where A and B are angular momentum and spin operators and operate on the electronic spatial coordinates. In irreducible tensor notation we write for (A.92)

$$\underline{\gamma}\cdot\mathsf{A}=\sum_{m=-1}^{1}(-1)^{m}\tilde{T}_{m}^{1}(\underline{\gamma})\tilde{T}_{-m}^{1}(\mathsf{A}) \tag{A.94}$$

with $\tilde{T}_{m}^{1}(\mathsf{A})=\mp(A_x+iB_y)/\sqrt{2}$, and so on.

It is always possible to find unitary transformation for the $\tilde{T}_{m}^{1}(\underline{\gamma})$ to yield components that transform according to the irreducible representations of the symmetry group considered:

$$T_{a}^{\Gamma}(\underline{\gamma})_{r}=\sum_{m}c_{am}(\Gamma,r)\tilde{T}_{m}^{1}(\underline{\gamma}) \tag{A.95}$$

where Γ is an irreducible representation, a is one component of Γ, and r is a label used to distinguish between the same representations. The c_{am} (Γ,r)'s are coefficients of the unitary transformation matrix.

The inverse relation for (A.95) is

$$\tilde{T}_{m}^{1}(\underline{\gamma})=\sum_{\Gamma ar}c_{ma}^{*}(\Gamma,r)T_{a}^{\Gamma}(\underline{\gamma})_{r} \tag{A.96}$$

Insertion of (A.96) into (A.94) gives the following relations:

$$\sum_{m=-1}^{1}(-1)^{m}\tilde{T}_{m}^{1}(\underline{\gamma})\tilde{T}_{-m}^{1}(\mathsf{A})=\sum_{\Gamma ar}\sum_{m}(-1)^{m}c_{ma}(\Gamma,r)T_{a}^{\Gamma}(\underline{\gamma})_{r}\tilde{T}_{-m}^{1}(\mathsf{A})$$

$$=\sum_{\Gamma ar}\varepsilon(a)T_{a}^{\Gamma}(\underline{\gamma})_{r}\cdot S_{\hat{q}(a,r)}^{1}(\mathsf{A}) \tag{A.97}$$

with

$$S_{\hat{q}(a,r)}^{1}(\mathsf{A})=\sum_{m=-1}^{1}c_{ma}^{*}(\Gamma,r)\tilde{T}_{-m}^{1}(\mathsf{A}) \tag{A.98}$$

and

$$\varepsilon(a)=\begin{cases}(-1)^{q(a)} & \text{for a complex basis}\\ +1 & \text{for a real basis}\end{cases}$$

We distinguish between $q(a,r)$ and $\hat{q}(a,r)$ by $\hat{q}=-q$ for the complex basis and $\hat{q}=q$ for the real basis. By (A.98) real components of the irreducible tensor operator may be generated.

In the same way we may treat the tensorial scalar product

$$\mathsf{A}\cdot\gamma\mathsf{B} \tag{A.93}$$

where for simplicity A may commute with γ. We get, in irreducible tensor rotation,

$$\mathsf{A}\cdot\gamma\mathsf{B}=\sum_{k=0}^{2}\sum_{m=-2}^{2}(-1)^m\tilde{T}_m^k(\gamma)\tilde{T}_{-m}^k(\mathsf{A},\mathsf{B}) \tag{A.99}$$

where

$$\tilde{T}_m^k(\mathsf{A},\mathsf{B})=\sum_{k_1k_2m_1m_2}\langle k_1k_2m_1m_2|km\rangle\tilde{T}_{m_1}^1(\mathsf{A})\tilde{T}_{m_2}^1(\mathsf{B}) \tag{A.100}$$

and $\tilde{T}_m^k(\gamma)$ are the irreducible tensor components of the tensor (see Table III). We get for (A.99):

$$\mathsf{A}\cdot\gamma\mathsf{B}=\sum_{k=0}^{2}\sum_{\Gamma ar}\varepsilon(a)T_a^\Gamma(\gamma^k)_r S_{\hat{q}(a,r)}^k(\mathsf{A},\mathsf{B}) \tag{A.101}$$

with

$$S_q^k(\mathsf{A},\mathsf{B})=\sum_{m=-k}^{k}c_{ma}^*(\Gamma,r)\tilde{T}_{-m}^k(\mathsf{A},\mathsf{B})$$

Let us apply (A.97) and (A.101) to the complex and real notation, respectively.

As an example we use the spin–orbit operator $\Sigma_i\mathsf{u}_i\cdot\mathsf{s}_i$ for the first case where u_i operates only on the spatial electronic coordinates and s_i on the electronic spin coordinates. First we consider those symmetry groups for which it is advantageous to use complex components for degenerate representations. The $\tilde{T}_m^1(\mathsf{u}_i)$, $m=\pm 1$, form a basis of a degenerate representation E (e.g., Π in $C_{\infty v}$). Then we have the identification

$$T_\pm^E(\mathsf{u}_i)=\tilde{T}_{\pm 1}^1(\mathsf{u}_i)$$

and

$$T^A(\mathsf{u}_i)=\tilde{T}_0^1(\mathsf{u}_i) \tag{A.102}$$

TABLE III

Components $\gamma_q^k = S_q^k(\gamma)$ of a Dyadic γ for Real and Complex Notation, Respectively

		Complex		Real
γ_0^0		$(-1/\sqrt{3})\mathrm{Tr}(\gamma)$		$(-1/\sqrt{3})\mathrm{Tr}(\gamma)$
γ_q^1	0:	$(-i/\sqrt{2})(\gamma_{xy}-\gamma_{yx})$	z:	$(-1/\sqrt{2})(\gamma_{xy}-\gamma_{yx})$
			x:	$(-i/\sqrt{2})(\gamma_{yz}-\gamma_{zy})$
	± 1:	$(1/2)[(\gamma_{xz}-\gamma_{zx})\pm i(\gamma_{yz}-\gamma_{zy})]$	y:	$(-i/\sqrt{2})(\gamma_{zx}-\gamma_{xz})$
γ_q^2	0:	$(1/\sqrt{6})(2\gamma_{zz}-\gamma_{xx}-\gamma_{yy})$	0:	$(1/\sqrt{6})(2\gamma_{zz}-\gamma_{xx}-\gamma_{yy})$
			xz:	$(1/\sqrt{2})(\gamma_{xz}+\gamma_{zx})$
	± 1:	$(\mp 1/2)[(\gamma_{xz}+\gamma_{zx})\pm i(\gamma_{yz}+\gamma_{zy})]$	yz:	$(1/\sqrt{2})(\gamma_{yz}+\gamma_{zy})$
			(x^2-y^2):	$(\gamma_{xx}-\gamma_{yy})$
	± 2:	$(1/2)[(\gamma_{xx}-\gamma_{yy})\pm i(\gamma_{xy}+\gamma_{yx})]$	xy:	$(\gamma_{xy}+\gamma_{yx})$

with an appropriate nondegenerate representation A (e.g., $A-\Sigma^-$ for $C_{\infty v}$, A_2 for C_{3v}, etc.). The $S_q^1(\mathbf{s}_i)$ are identical to the $\tilde{T}_m^1(\mathbf{s}_i)$; that is,

$$S^1_{q\,(E_\pm)} = \tilde{T}^1_{\pm 1}(\mathbf{u}_i)$$

and

$$S^1_{q\,(A)} = \tilde{T}^1_0(\mathbf{u}_i) \tag{A.103}$$

Frequently, it is more convenient to use real basis functions, for instance if the $\tilde{T}^1_{\pm 1}(\mathbf{u}_i)$'s belong to one-dimensional representations only. Then we would form u_{ix}, u_{iy}, and u_{iz}, which may be classified with respect to the symmetry group. The appropriate $S_q^1(\mathbf{s}_i)$ are given by

$$S_x^1(\mathbf{s}_i) = (-\tilde{T}^1_{+1}(\mathbf{s}_i) + T^1_{-1}(\mathbf{s}_i))/\sqrt{2}$$

$$S_y^1(\mathbf{s}_i) = (\tilde{T}^1_{+1}(\mathbf{s}_i) + \tilde{T}^1_{-1}(\mathbf{s}_i))(i/\sqrt{2})$$

$$S_z^1(\mathbf{s}_i) = \tilde{T}^1_0(\mathbf{s}_i) \tag{A.104}$$

For the second case we present the different forms of $\gamma_q^k = S_q^k(\gamma)$ and $S_q^k(\mathbf{A},\mathbf{B})$ in Tables III and IV, respectively. From Table III the symmetry-group representation of the γ_q^k are easily derived for the different point groups.

TABLE IV

Components $S_q^k(\mathsf{A},\mathsf{B})$ for Real and Complex Notation, Respectively

	Complex		Real
S_0^0	$(-1/\sqrt{3})\mathsf{A}\cdot\mathsf{B}$		$(-1/\sqrt{3})\mathsf{A}\cdot\mathsf{B}$
S_q^1	0: $(-1/\sqrt{2})(A_xB_y-A_yB_x)$	z:	$(-i/\sqrt{2})(A_xB_y-A_yB_x)$
		x:	$(-i/\sqrt{2})(A_zB_x-A_xB_z)$
	±1: $(1/2)[(A_xB_z-A_zB_x)\pm i(A_yB_z-A_zB_y)]$	y:	$(-i/\sqrt{2})(A_xB_y-A_yB_x)$
S_q^2	0: $(1/\sqrt{6})(2A_zB_z-A_xB_x-A_yB_y)$	0:	$(1/\sqrt{6})(2A_zB_z-A_xB_x-A_yB_y)$
		xz:	$(1/\sqrt{2})(A_xB_z+A_zB_x)$
	±1: $(\mp1/2)(A_xB_z+A_zB_x)\pm i(A_yB_z+A_zB_y)$	yz:	$(1/\sqrt{2})(A_yB_z+A_zB_y)$
		(x^2-y^2):	$(A_xB_x-A_yB_y)$
	±2: $(1/2)[(A_xB_x-A_yB_y)\pm i(A_xB_y+A_yB_x)]$	xy:	$(A_xB_y+A_yB_x)$

Since we deal with irreducible tensor operators with a nonstandard basis, it is convenient to present some appropriate formulas for this basis. The equations

$$\tilde{T}_m^k(\mathsf{A},\mathsf{B})=\sum_{k_1k_2m_1m_2}\langle k_1k_2m_1m_2|km\rangle\tilde{T}_{m_1}^{k_1}(\mathsf{A})\tilde{T}_{m_2}^{k_2}(\mathsf{B})$$

$$\tilde{T}_{m_1}^{k_1}(\mathsf{A})\tilde{T}_{m_2}^{k_2}(\mathsf{B})=\sum_{km}\langle k_1k_2m_1m_2|km\rangle\tilde{T}_m^k(\mathsf{A},\mathsf{B}) \tag{A.105}$$

have to be replaced by the more general equations

$$S_q^k(\mathsf{A},\mathsf{B})=\sum_{k_1k_2q_1q_2}\langle k_1k_2q_1q_2|kq\rangle^*S_{q_1}^{k_1}(\mathsf{A})S_{q_2}^{k_2}(\mathsf{B})$$

$$S_{q_1}^{k_1}(\mathsf{A})S_{q_2}^{k_2}(\mathsf{B})=\sum_{kq}\langle k_1k_2q_1q_2|kq\rangle S_q^k(\mathsf{A},\mathsf{B}) \tag{A.106}$$

where the $\langle k_1k_2m_1m_2|km\rangle$ are the standard Clebsch–Gordan coefficients and $\langle k_1k_2q_1q_2|kq\rangle$ are the generalized Clebsch–Gordan coefficients. The latter coincide with the standard coefficients for a complex basis, where for a real basis they are certain complex linear combinations of the standard coefficients.

In Table V we give a list of those Clebsch–Gordan coefficients for a real basis, which are frequently used within the text. In Table VI we show how the Clebsch–Gordan coefficients for a real basis are related to the standard coefficients, when irreducible tensor operators of rank 0, 1, and 2 are considered.

TABLE V
Clebsch–Gordan Coefficients $\langle j_1 j_2 q_1 q_2 | jq\rangle$ with $j_1 = j_2 = 1, j = 0, 1, 2$ for the Nonstandard Components given by x, y, z for $j = 1$ and by 0, $(x^2 - y^2), xy, xz, yz$ for $j = 2$, Which Define Our "Real" Basis

$$\langle 11xx|00\rangle = \langle 11yy|00\rangle = \langle 11zz|00\rangle = -1/\sqrt{3}$$
$$\langle 11xy|1z\rangle = \langle 11zx|1y\rangle = \langle 11yz|1x\rangle =$$
$$-\langle 11yx|1z\rangle = -\langle 11xz|1y\rangle = -\langle 11zy|1x\rangle = -i/\sqrt{2}$$
$$\langle 11xx|20\rangle = \langle 11yy|20\rangle = -1/\sqrt{6}$$
$$\langle 11xx|2(x^2-y^2)\rangle = -\langle 11yy|2(x^2-y^2)\rangle = 1$$
$$\langle 11xy|2xy\rangle = \langle 11yx|2xy\rangle = 1$$
$$\langle 11xz|2xz\rangle = \langle 11zx|2xz\rangle = 1/\sqrt{2}$$
$$\langle 11yz|2yz\rangle = \langle 11zy|2yz\rangle = 1/\sqrt{2}$$

TABLE VI
Clebsch–Gordan Coefficients $\langle j_1 k m_1 q | j_2 m_2\rangle$ with $k = 0, 1, 2$ for the Nonstandard Real q-Components Given in Terms of the Standard Clebsch–Gordan Coefficients

$$\langle j_1 0 m_1 0 | j_2 m_1\rangle = \delta_{j_1 j_2}$$
$$\langle j_1 1 m_1 x | j_2 m_1 \pm 1\rangle = \mp(1/\sqrt{2}\,)\langle j_1 1 m_1 \pm 1 | j_2 m_1 \pm 1\rangle$$
$$\langle j_1 1 m_1 y | j_2 m_1 \pm 1\rangle = -(i/\sqrt{2}\,)\langle j_1 1 m_1 \pm 1 | j_2 m_1 \pm 1\rangle$$
$$\langle j_1 1 m_1 z | j_2 m_1\rangle = \langle j_1 1 m_1 0 | j_2 m_1\rangle$$
$$\langle j_1 2 m_1 0 | j_2 m_1\rangle = \langle j_1 2 m_1 0 | j_2 m_1\rangle$$
$$\langle j_1 2 m_1 (x^2-y^2) | j_2 m_1 \pm 2\rangle = \langle j_1 2 m_1 \pm 2 | j_2 m_1 \pm 2\rangle$$
$$\langle j_1 2 m_1 xy | j_2 m_1 \pm 2\rangle = \pm i\langle j_1 2 m_1 \pm 2 | j_2 m_1 \pm 2\rangle$$
$$\langle j_1 2 m_1 xz | j_2 m_1 \pm 1\rangle = \mp(1/\sqrt{2}\,)\langle j_1 2 m_1 \pm 1 | j_1 m_1 \pm 1\rangle$$
$$\langle j_1 2 m_1 yz | j_2 m_1 \pm 1\rangle = -(i/\sqrt{2}\,)\langle j_1 2 m_1 \pm 1 | j_2 m_1 \pm 1\rangle$$

Finally, we introduce angular momentum and spin-basis functions classified according to the symmetry groups. They are related to the basis functions of the infinite three-dimensional rotation group by the same type of unitary transformations as given by (A.95):

$$|j(\Gamma ar)\rangle = \sum_m c_{am}(\Gamma, r)|jm\rangle \tag{A.107}$$

where, again, r specifies the functions unambiguously.

For nondegenerate representations we have linear combinations of the type

$$|j(\Gamma r)\rangle = \{|jm\rangle \pm |j-m\rangle\}/\sqrt{2} \tag{A.108}$$

which may be classified dependent on j, m, and the sign in (A.108). For two-dimensional degenerate irreducible representations, two functions of the type $|jm\rangle$, $|j-m\rangle$ may form a basis. For three-dimensional representations we get more complicated relations. For the octahedral group, 0, they are given for $j=0$ up to $j=6$ by Watanabe.[40b] The functions are conveniently expressed as $|jq\rangle$ with q as a label equivalent to the m. The $(2j+1)$ functions $|jm\rangle$ are transformed into $(2j+1)$ functions $|jq\rangle$, which can be classified with respect to the different symmetry groups.

APPENDIX 4. LIST OF SYMBOLS

$\mathsf{A}_0, A_k, \mathsf{A}_r, \mathsf{a}_i$	generalized vector potential functions for the molecular translation, (A.42); the kth vibrational mode, (A.16); the molecular rotation, (A.16); and the ith electron, (A.9), respectively
a	designates one component of a basis for irreducible representations of molecular symmetry groups
$\mathsf{a}_{i\alpha}$	spatial function part of the nuclear spin–orbit operator defined in (VIII.78) to (A.68)
A	one-dimensional irreducible representation
$\tilde{A}$	effective rotational constant, (VI.12)
$\tilde{A}$	effective spin–orbit constants, (VI.8) and (VIII.41)
$\tilde{A}_D$	centrifugal distortion spin–orbit constant, (VI.8)
α, β	index for nuclei ranging from 1 up to N
α, β	set of quantum numbers in Section V
α, β, γ	Euler angles
$\tilde{\alpha}$	effective polarizability values
$\tilde{\mathbf{b}}$	effective hyperfine structure constants, (VI.25)
$\tilde{b}$	one of the effective hyperfine-structure constants for linear molecules, (VI.24)
$b^{(k)}_{i\alpha_q}, k=0,2$	zero- and second-rank irreducible tensor components of the spatial function parts of the electronic spin–nuclear spin coupling operator defined in (VIII.79) and (VIII.80), through (A.64)
B	one-dimensional irreducible representation
$\tilde{B}$	effective rotational constant, (VI.12)
B	rotational constant of linear molecules, (A.33)

β	Bohr magneton, (A.56)
β_N	nuclear magneton, (A.65)
c	speed of light
$\tilde{c}_r$	effective rotational–nuclear spin coupling constants, (VI.28) and (VI.29)
$\tilde{c}$	one of the effective hyperfine structure constants for linear molecules, (VI.24)
$\tilde{c}_m^2$	effective quadrupole coupling constants defined in (VI.27)
C_m^k	spherical harmonics as defined in (A.45)
$D_q^k(i,j),\ k=0,2$	kth-rank irreducible tensor components of the spatial-function part of the electronic spin–spin coupling operator defined in (VIII.26a) and (VIII.26b)
$D_q^2(\alpha,\beta)$	second-rank irreducible tensor components of the spatial-function part of the nuclear spin–spin coupling operator defined in (VIII.103)
$D_q^2(i,\alpha)$	second-rank irreducible tensor operator components of the spatial-function part of the quadrupole coupling operator defined in (VIII.92) to (A.44)
$D_{m'm}^k$	elements of the irreducible representation of the three-dimensional group
d_k	linear transformation coefficients relating the internal symmetry coordinates to the nuclear Cartesian displacement coordinates, (IV.9) or (A.17)
$\tilde{d}$	one of the hyperfine structure constants for linear molecules, (VI.24)
$\tilde{\mathbf{d}}$	matrix of hyperfine-structure constants defined in (VI.31)
$\tilde{d}_{\alpha\beta}$	effective nuclear spin coupling constants, (VI.30) and (VI.31)
D	centrifugal distortion constant for diatomic molecules
δ_{ij}	Kronecker symbol: 1 for $i=j$, 0 otherwise
$\delta(\cdot)$	Dirac δ-function
Δ	irreducible representation in $C_{\infty v}$
$\overline{\Delta V}^{-1}$	see (VIII.13)

∇	gradient operator
e	absolute value of the electronic charge
$\mathbf{e}_\mu$	unit vector in direction of the molecular fixed axes $\mu = x,y,z$
$e_{\lambda\mu\nu}$	antisymmetric Levi-Civita tensor, (VI.15)
E	doubly degenerate representation
$\mathbf{E}$	external electric field
ε	electronic energy in Section III
$\varepsilon(q), \varepsilon(a)$	sign value, ± 1, as defined in Section VII
ε	Renner splitting parameter, (VIII.22)
$\mathbf{1}$	unit matrix
$\mathbf{F}, F$	total angular momentum operator, including nuclear spins and the appropriate quantum number, respectively
φ	angle of orientation of the electronic system in linear molecules
ϕ	parameter to signify the electronic adiabatic functions
$\mathbf{G}$	vibrational angular momentum operator, (A.22)
$\mathbf{G}^{\mathfrak{R}}, \mathbf{G}^{el}$	defined in (VIII.55)
$G_\pm$	$G_x \pm iG_y$
g	free electron g-factor
g_α	g-factor of nucleus α
$\tilde{g}_S$	effective electronic spin g-factors, (VI.33) and (VI.34)
$\tilde{g}_L$	effective electronic orbital g-factors, (VI.32)
$\tilde{g}_r$	effective rotational g-factors, (VI.35) and (VI.36)
$\tilde{g}_l$	effective vibrational g-factor, (VI.41)
γ	set of quantum numbers in Sections III and VII
$\tilde{\gamma}$	effective spin–rotational coupling constants
$\gamma_i, \gamma_{i_q}^{(k)}, k=0,2$	dyadic and appropriate irreducible tensor components of the spatial function part of the spin–rotation coupling operator, (VIII.49) and (VIII.61)
Γ	irreducible representation of a general symmetry group
$\Gamma_e, \Gamma_S, \Gamma_N, \Gamma_J$	irreducible representations signifying the spatial electronic function, the electronic spin function, and the rotational functions, respectively

$\hbar$	Planck's constant divided by 2π
H	third-order centrifugal distortion constant for diatomic molecules, (VI.4)
H	external magnetic field
i,j	index for electrons ranging from 1 up to n
i	$\sqrt{-1}$
$\mathsf{I}_\alpha, I_\alpha$	operator for the spin of nucleus α and the appropriate quantum number
$I_{\alpha\pm}$	$I_{\alpha x} \pm i I_{\alpha y}$
I, I	operator for the total nuclear spin and the appropriate quantum number
$\mathbf{I}, \mathbf{I}^0, \mathbf{I}', \mathbf{I}''$	inertia moment tensors as defined in (A.26), (A.27), and (A.30)
I''	determinant of $\mathbf{I}''$
I, I^0, I', I''	inertia moments of linear molecules as defined in (A.33), (A.35), and (A.36)
J, J	total angular momentum operator, excluding nuclear spins and the appropriate quantum number
J'	redefined $\boldsymbol{J}$, see Section VIII.C
$J_\pm$	$J_x \pm iJ_y$
Θ	function depending on all nuclear coordinates
k	rank of irreducible tensor operators
k	force constants
K	quantum number for the projection of N onto one of the molecular fixed frame axes, in general z
κ	generalized K; see Section V
$\boldsymbol{\kappa}_i, \kappa_{iq}^{(k)}, k=0,2$	dyadic and appropriate irreducible tensor components of the spatial-function part of the electronic spin Zeeman effect defined in (VIII.116)
l	vibrational quantum number for the degenerate bending mode of the linear molecules
l_i	orbital angular momentum operator of electron i
L	total electronic orbital angular momentum operator
$L_\pm$	$L_x \pm iL_y$
λ	degree of the irreducible representation Γ
λ	angular momentum function basis in Section VIII

$\tilde{\lambda}$	effective spin–spin coupling constants, (VI.11), (VIII.42), and (VIII.43)
Λ	quantum number for the projection of L onto the internuclear axis of linear molecules
m	mass of an electron
$\mathbf{m}$	part of the nuclear spin–rotational coupling operator defined in (VIII.94) and (VIII.95)
M	total molecular mass
M_N	total nuclear mass
M_p	mass of a proton
M_α	mass of nucleus α
$M_J, M_N, M_S, M_{I_\alpha}, M_I, M_F$	quantum numbers for the projection of the angular momentum operators onto one of the laboratory fixed axes, in general Z
$\underline{\mu}$	electric dipole moment operator
$\underline{\mu}_m$	magnetic dipole moment
$\boldsymbol{\mu}, \mu_q^{(k)}, k=0,2$	inverse inertia moment dyadic (tensor), (A.25), and the appropriate irreducible tensor components
μ, ν	index for the molecular frame fixed coordinates x,y,z
n	number of electrons in the molecule
N	number of nuclei in the molecule
N, N	rotational angular momentum operator and the appropriate quantum number
N'	redefined N, see Section VIII.C
$N_\pm$	$N_x \pm iN_y$
Ω	quantum number for the projection of J onto the internuclear axes in linear molecules
P	quantum number for the projection of J onto one of the molecular fixed frame axis, in general z, for nonlinear molecules
π	generalized P; see Section V
$\mathsf{P}_0, \underline{\Pi}_0$	translational momentum operator, (A.41)
P_k, Π_k	vibrational momentum operators, (A.14) and (A.15)
$\underline{\Pi}_r$	nuclear rotational angular momentum operator, (A.20)

$\mathbf{p}_i, \underline{\pi}_i$	electronic momentum operators, (A.7) and (A.8)
p	Λ-doubling parameter, (VI.18)
Π	irreducible representation in $C_{\infty v}$
π	real number
Π	function depending on the nuclear displacement coordinates
q	Λ-doubling parameter, (VI.17); l-type doubling parameter, (VI.20)
$\tilde{q}$	quadrupole coupling constants
q	designates a component of an irreducible tensor operator S_q^k
$\mathbf{Q}$	position vector in the laboratory system
Q_α	quadrupole moment of the charge distribution in nucleus α
$\mathfrak{q}$	set of electronic coordinates, including spin in the laboratory system
$\mathfrak{Q}$	set of nuclear coordinates, including spin in the laboratory system
Ξ	translational function in Section V
$\mathbf{r}$	position vector in the molecular fixed frame
$\mathbf{r}_\alpha^0$	nuclear positions within the reference conformation
$\hat{\mathbf{r}}_i$	electronic molecular fixed frame coordinates, including spin used in Section III
$\mathbf{r}_{ab}$	distance vector $\mathbf{r}_a - \mathbf{r}_b$
r_{ab}	absolute value of $\mathbf{r}_{ab}$
$\mathfrak{r}$	set of electronic coordinates $\hat{\mathbf{r}}_i$
r	additional index defined in Section VII
R	internuclear distance in diatomic molecules
$\mathfrak{R}$	set of nuclear coordinates, including spin in the molecular fixed system
$\boldsymbol{\rho}_i, \rho_{i_q}^{(k)}, k=0,2$	spatial-function part of the nuclear spin interaction operator in dyadic and irreducible tensor form defined in (VIII.102) and (VIII.104)
$\boldsymbol{\rho}$	matrix involving derivatives with respect to nuclear coordinates in Section V
s	parity label for Σ-representations in $C_{\infty v}$ with $(-1)^s$ equals $+1$ for $s=+$ and -1 for $s=-$

$\mathbf{s}_i$	spin operator of electron i
$\mathbf{S}, S$	total electronic spin operator and the appropriate quantum number
$S_{\pm}$	$S_x \pm iS_y$
$\mathfrak{S}$	set of internal symmetry coordinates
S_k	internal symmetry coordinates defined in (IV.9) and (A.17)
S_q^k	generalized irreducible tensor components
σ	generalized Σ for nonlinear molecules; see Section V
$\tilde{\sigma}$	effective magnetic shielding values, (VI.39) and (VI.40)
$\boldsymbol{\sigma}$	general property matrix in Section V
$\boldsymbol{\sigma}_i, \sigma_{iq}^{(k)}, k=0,2$	spatial function part of the magnetic shielding operator defined in (VIII.147) in dyadic and irreducible tensor form
$\Sigma^{\pm}$	one-dimensional representations in $C_{\infty v}$
Σ	quantum number for the projection of S onto one molecular frame fixed axis, in general z
$\mathbf{t}$	rotational Zeeman effect matrix, defined in (A.31)
$\tilde{T}_m^k$	standard irreducible tensor operator components
T_a^{Γ}	designates an operator transforming according the component a of the irreducible representation Γ
T	triply degenerate representations
T_e	electronic energy of the equilibrium conformation, (VI.6)
Tr	trace of a matrix
τ'	centrifugal distortion coefficients; see Section VI
$\mathbf{u}_i$	spatial-function part of the spin–orbit and spin–other-orbit operator defined in (A.48)
$\mathbf{U}$	unitary transformation in Section V
V	adiabatic electronic energies
V	coupling coefficients for point groups, Appendix 2
v	vibrational quantum number
ψ	electronic function within the adiabatic approximation

Ψ	total molecular function
$\tilde{W}$	Racah recoupling coefficients
W	recoupling coefficients for point groups, Section VII and Appendix 2
x,y,z	molecular fixed frame Cartesian coordinates
X,Y,Z	laboratory fixed frame Cartesian coordinates
$\tilde{X}$	effective susceptibility values
Y_m^k	spherical harmonics
Y_{ij}	Dunham coefficients, (VI.2)
Z_α	atomic number of nucleus α
$\underline{\zeta}_{kl}$	Coriolis coupling coefficients, (A.23)

Special Symbols

$\otimes$	direct product of representations
$\otimes$	dyadic product, (IV.7)
$\times$	vector product
$\cdot$	scalar product
$\hat{a},\hat{q}$	associated components in scalar products; see Section VII
$\dagger$	Hermitian conjugation
$*$	complex conjugation
$a^\dagger, q^\dagger$	redefined molecular fixed components of a rotational angular momentum operator obeying normal commutation relations
$\tilde{\Gamma},\tilde{a}$	$\tilde{}$ designates the transformation property of operators, which transform in the same way as **L**
$\bar{\Gamma},\bar{a}$	$\bar{}$ designates the transformation property of operators, which transform in the same way as the Y_m^2
$\sim$	indicates effective operators and constants
$\tilde{T},\tilde{W}$	$\tilde{}$ signifies standard irreducible tensor notation
$\cdot$	derivative with respect to the time

REFERENCES

1. (a) G. Herzberg, *Spetra of Diatomic Molecules*, Vol. 1 of *Molecular Spectra and Molecular Structure*, Van Nostrand Reinhold, New York, 1950; (b) G. Herzberg, *Infrared and Raman Spectra*, Vol 2 of *Molecular Septra and Molecular Structure*, Van Nostrand Reinhold, New York, 1964; (c) G. Herzberg, *Electronic Spetra of Polyatomic Molecules*, Vol 3 of *Molecular Spetra and Molecular Structure*, Van Nostrand Reinhold, New York, 1966; (d) W. Gordy and R. L. Cock, *Microwave Molecular Spetra*, Interscience, New York, 1970; (e) G. H. Townes and A. L. Shawlow, *Microwave Sepctroscopy*, McGraw-Hill, New York, 1955; (f) A. Abragam, *The Principles of Nuclear Magnetism*, Oxford University Press, New York, 1961; (g) J. A. Pople, W. G. Schneider, and H. I. Bernstein, *High-*

resolution Nuclear Magnetic Resonance, McGraw-Hill, New York, 1959; (h) A. Carrington, D. H. Levy, and T. A. Miller, *Adv. Chem. Phys.*, **18**, 149 (1970). (i) I. Kovacs, *Rotational Structure in the Spectra of Diatomic Molecules*, Hilger, London, 1969; (j) M. Mizushima, *The Theory of Rotating Diatomic Molecules*, Wiley, New York, 1975.

2. (a) H. F. Schäfer, *The Electronic Structure of Atoms and Molecules*, Addison Wesley, Reading, Mass, 1972; (b) H. F. Schäfer, *Annu. Rev. Phys. Chem.*, **27**, 269 (1976); (c) W. G. Richards, T. E. H. Walker, and R. K. Hinkley, *Bibliography of ab initio Molecular Wavefunctions*, Clarendon Press, Oxford, 1971; (d) W. G. Richards, T. E. H. Walker, L. Farnell, and P. R. Scott, *Bibliography of ab initio Molecular Wavefunctions: Supplement for* 1970–1973, Clarendon Press, Oxford, 1974.
3. (a) W. G. Richards, I. Raftery, and R. K. Hinkley, *Spec. Period. Rep. Theor. Chem.*, **1**, 1 (1974) and references therein; (b) W. N. Lipscomb, *MTP Int. Rev. Sci. Phys. Chem. Ser.* 1, **1**, 167 (1972) and references therein; (c) R. Ditchfield, *MTP Int. Rev. Sci. Phys. Chem. Ser.* 1, **2**, 91 (1972) and references therein; (d) W. N. Lipscomb, *Adv. Magn. Reson.*, **2**, 137 (1966); (e) D. W. Davies, *The Theory of the Electric and Magnetic Properties of Molecules*, Wiley, New York, 1967, and references therein; (f) A. J. Sadley, *Mol. Phys.*, **34**, 731 (1977). (g) F. D. Wayne and E. A. Colbourn, *Mol. Phys.*, **34**, 1141 (1977).
4. (a) M. Born, *Nachr. Akad. Wiss. Göttingen, Math.-Phys. Kl.*, (1951); (b) M. Born and K. Huang, *Dynamical Theory of Crystal Lattices*, Oxford University Press, New York, 1956.
5. J. Hinze, *Adv. Chem. Phys.*, **26**, 213 (1974).
6. (a) J. H. Wilkinson, *The Algebraic Eigenvalue Problem*, Oxford University Press, London, 1965; (b) J. H. Wilkinson and C. Reinsch, *Linear Algebra*, Vol 2 of *Handbook for Automatic Computation*, Springer-Verlag, New York, 1971.
7. (a) R. K. Nesbet, *J. Chem. Phys.*, **43**, 311 (1965); (b) C. F. Bender, R. P. Hosteny, A. Pipano, and I. Shavitt, *J. Comp. Phys.*, **11**, 90 (1973); (c) E. Davidson, *J. Comp. Phys.*, **17**, 87 (1975).
8. (a) B. O. Roos, *Chem. Phys. Lett.*, **15** (1978); (b) R. F. Hausman Jr., S. D. Bloom, and C. F. Bender, *Chem. Phys. Lett.*, **32**, 483 (1975); (c) B. O. Roos, *Mod. Theor. Chem.*, **3**, 277 (1977).
9. (a) C. C. J. Roothaan, *Rev. Mod. Phys.*, **23**, 69 (1951); (b) C. C. J. Roothaan, *Rev. Mod. Phys.*, **32**, 179 (1960).
10. (a) J. Hinze, *J. Chem. Phys.*, **59**, 6424 (1973); (b) G. Das and A. C. Wahl, *J. Chem. Phys.*, **44**, 87 (1966); **56**, 1765 (1972); (c) A. C. Wahl and G. Das, *Mod. Theor. Chem.*, **3**, 51 (1977).
11. (a) I. Shavitt, *Bull. Am. Phys. Soc.*, **19**, 195 (1974); (b) I. Shavitt, *Mod. Theor. Chem.*, **3**, 189 (1977).
12. R. J. Bunker and S. A. Peyerimhoff, *Theor. Chim. Acta*, **35**, 33 (1974).
13. E. R. Davidson, *Rev. Mod. Phys.*, **44**, 451 (1972).
14. See Refs. 2(a) and (b).
15. G. C. Lie, J. Hinze, and B. Liu, *J. Chem. Phys.*, **59**, 1872 (1973).
16. P. R. Bunker, "Practically Everything You Ought to Know about the Molecular Symmetry Group," in James R. Durig, ED., *Vibrational Spetra and Structure*, Dekker, New York, 1975.
17. P. R. Bunker and D. Papausek, *J. Mol. Spectrosc.*, **32**, 419 (1969).
18. H. C. Longuet-Higgins, *Mol Phys.*, **6**, 445 (1963).
19. (a) C. Eckart, *Phys. Rev.*, **47**, 552 (1935); (b) E. B. Wilson Jr., J. C. Decius, and P. C. Cross, *Molecular Vibrations*, McGraw-Hill, New York, 1955; (c) J. D. Louck and H. W. Galbraith, *Rev. Mod. Phys.*, **48**, 69 (1976).
20. A. Saynetz, *J. Chem. Phys.*, **7**, 383 (1939).
21. H. A. Jahn and E. Teller, *Proc. R. Soc. Lond. Ser. A*, **161**, 220 (1937).
22. (a) G. Herzberg and E. Teller, *Z. Phys. Chem. B*, **21**, 410 (1933); (b) R. Renner, *Z. Phys.*, **92** 172 (1934).

23. J. A. Pople and H. C. Longuet-Higgins, *Mol. Phys.*, **1**, 372 (1958).
24. S. Aronowitz, *Phys. Rev. A*, **14**, 1319 (1976).
25. A. D. Liehr, *J. Phys. Chem.*, **67**, 389, 471 (1963).
26. E. R. Davidson, *J. Am. Soc.*, **99** 397 (1977).
27. J. H. Van Vleck, *Phys. Rev.*, **33**, 467 (1929).
28. Landolt-Börustein, *Numerical Data and Functional Relationships in Science and Technology*, new ser., Group II, Vols. 1, 4, 6, and 9a, Springer-Verlag, New York.
29. H. N. Nielsen, "The Vibration-Rotation Energies of Molecules and Their Spectra in the Infra-Red," *Handbuch der Physik*, Vol. 37/1, Atome III, Moleküle I, Springer-Verlag, New York, 1959.
30. (a) C. C. Lin and J. D. Swalen, *Rev. Mod. Phys.*,**31**, 841 (1959); (b) R. Meyer and Hs. H. Günthard, *J. Chem. Phys.*, **49**, 1510 (1968) and literature cited therein; (c) H. M. Picket, *J. Chem. Phys.* 56, 1715 (1971) and literature cited therein; (d) H. Bauder, R. Meyer, and Hs. H. Günthard, *Mol. Phys.*, **28**, 1305 (1974); (e) H. Frei, R. Meyer, A. Bander, and Hs. H. Günthard, *Mol. Phys.*, **32**, 443 (1976); **34**, 1198 (1977).
31. J. Gerrat, *Annu. Rep. Prog. Chem. A*, **65**, 3 (1968).
32. J. L. Dunham, *Phys. Rev.*, **41**, 721 (1932); (b) I. Sandeman, *Proc. R. Soc. Edinb.*, **54**, 1, 130 (1930); **60**, 210 (1940).
33. E. B. Wilson, Jr., *J. Chem. Phys.*, **4**, 526 (1936); (b) E. B. Wilson, Jr. and J. B. Howard, *J. Chem. Phys.*, **4**, 260 (1936); (c) D. Kivelson and E. B. Wilson, *J. Chem. Phys.*, **20**, 1575 (1952); (d) D. Kivelson and E. B. Wilson, *J. Chem. Phys.*, **21**, 1229 (1953); (e) J. K. G. Watson, *J. Chem. Phys.*, **46**, 1935 (1967).
34. J. K. G. Watson "Aspects of Quantic and Sextic Centrifugal Effects on Rotational Energy Levels," in James R. Durig, Ed., *Vibrational Spectra and Structure*, Vol. 6, Dekker, 1972.
35. J. M. Brown and J. K. G. Watson, *J. Mol. Spectrosc.*, **65**, 65 (1977).
36. C. Schlier, *Fortschr. Phys.*, **9**, 455 (1961).
37. (a) D. H. Sutter and W. H. Flygare, *The Molecular Zeeman Effect*, Vol. 63 of *Topics in Current Chemistry*, Springer-Verlag, New York, 1976. (b) L. Engelbrecht, dissertation, University of Kiel (1975).
38. See Ref. 1(d), p. 650.
39. See Ref. 1(c), Appendix IV, Tables 59 and 60.
40. (a) J. S. Griffith, *The Irreducible Tensor Method for Molecular Symmetry Groups*, Prentice-Hall, Englewood Cliffs, N.J., 1962; (b) H. Watanabe, *Operator Methods in Ligand Field Theory*, Prentice-Hall, Englewood Cliffs, N.J., 1966.
41. (a) A. R. Edmonds, *Angular Momentum in Quantum Mechanics*, Princeton University Press, Princeton, N.J., 1957; (b) E. P. Wigner, "On Matrices Which Reduce Kronecker Products of Representations of S.R. Groups," 1951, unpublished.
42. (a) E. P. Wigner, *Gruppentheorie*, Vieweg, Braunschweig, 1931; (b) C. Eckart, *Rev. Mod. Phys.*, **2**, 305 (1930).
43. M. E. Rose, *Elementary Theory of Angular Momentum*, Wiley, New York, 1957.
44. G. Racah, *Phys. Rev.*, **62**, 438 (1942).
45. P. R. Bunker and R. E. Moss, *Mol. Phys.*, **33**, 417 (1977).
46. G. Herzberg and H. C. Longuet-Higgins, *Discuss. Faraday Soc.*, **35**, 77 (1963); (b) H. C. Longuet-Higgins, *Proc. R. Soc. Lond. Ser. A*, **344**, 147 (1975).
47. (a) B. J. Howard and R. E. Moss, *Mol. Phys.*, **19**, 433 (1970); (b) B. J. Howard and R. E. Moss, *Mol. Phys.*, **20**, 147 (1971).
48. (a) C. J. Ballhausen, *Introduction to Ligand Field Theory*, McGraw-Hill, New York, 1962; (b) J. S. Griffith, *The Theory of Transition Metal Ions*, Cambridge University Press, New York, 1971; (c) R. Engelman, *The Jahn-Teller Effect in Molecules and Crystals*, Wiley, New York, 1972.

49. (a) J. H. Van Vleck, *Rev. Mod. Phys.*, **23**, 213 (1951); (b) K. F. Freed, *J. Chem. Phys.*, **45**, 4214 (1966).
50. See Ref. 1(j), pp. 130, 131.
51. N. F. Ramsey, *Phys. Rev.*, **90**, 232 (1951).
52. W. H. Flygare, *J. Chem Phys.*, **41**, 206 (1964).
53. J. H. Van Vleck, *The Theory of Electric and Magnetic Susceptibilities*, Oxford University Press, New York, 1966.
54. M. J. Levine and R. Roskies, *Phys. Rev. Lett.*, **30**, 772 (1973).
55. W. Hüttner and K. Morgenstern, *Z. Naturforsch.*, **A25**, 547 (1970).
56. L. B. Redei, *Phys. Rev.*, **130**, 420 (1963).
57. For example, J. D. Petke and J. L. Whitten, *J. Chem. Phys.*, **59**, 4885 (1973).
58. (a) A. C. Wahl, P. E. Cade, and C. C. J. Roothaan, *J. Chem. Phys.*, **41**, 2578 (1964); (b) R. L. Matcha, C. W. Kern, and D. M. Schrader, *J. Chem. Phys.*, **51**, 2152 (1969); (c) R. L. Matcha and C. W. Kern, *J. Chem. Phys.*, **51**, 3434 (1969); *Phys. Rev. Lett.*, **25**, 981 (1970); *J. Chem. Phys.*, **55**, 469 (1971); (d) M. B. Milleur and R. L. Schrader, *J. Chem. Phys.*, **57**, 3029 (1972); (e) R. H. Pritchard and C. W. Kern, *J. Chem. Phys.*, **57**, 2590 (1972); (f) O. Matsyoka, *Int. J. Quantum Chem.*, **5**, 1 (1971); **7**, 365 (1973); (g) H. Ito and Y. J. I'Haya, *Mol. Phys.*, **24**, 1103 (1972); *Chem. Phys. Lett.*, **17**, 16 (1972); (h) B. W. N. Lo, *Int. J. Quantum Chem.*, **9**, 1055 (1975).
59. (a) H. D. Cohen and C. C. J. Roothaan, *J. Chem. Phys.*, **43**, S34 (1965); (b) R. E. Stanton, *J. Chem. Phys.*, **36**, 1298 (1962); (c) H.-J. Werner and W. Meyer, *Mol. Phys.*, **31**, 855 (1976).
60. (a) A. Dalgarno, *Adv. Phys.*, **11**, 281 (1962); (b) P. W. Langhoff, M. Karplus, and R. P. Hurst, *J. Chem. Phys.*, **44**, 505 (1966); (c) T. C. Caves and M. Karplus, *J. Chem. Phys.*, **50**, 3649 (1969); (d) A. J. Sadlej, *Chem. Phys. Lett.*, **8**, 100 (1971).
61. (a) H. Peng, *Proc. R. Soc. Lond. Ser A*, **178**, 499 (1941); (b) L. C. Allen, *Phys. Rev.*, **118**, 167 (1960); (c) R. M. Stevens, R. M. Pitzer, and W. N. Lipscomb, *J. Chem. Phys.*, **38**, 550 (1963); (d) G. H. F. Diercksen and R. Mc Weeny, *J. Chem. Phys.*, **44**, 3354 (1966); (e) G. P. Arrighini, M. Maestro, and R. Moccia, *Chem. Phys. Lett.*, **1**, 242 (1967); *J. Chem. Phys.*, **49**, 882 (1968).
62. M. Jaszunski and A. J. Sadlej, *Theor. Chim. Acta*, **40**, 157 (1975); *Int. J. Quantum Chem.*, **11**, 233 (1977).
63. See Refs. 58(b) to (e).
64. See Refs. 58(f) and (g).
65. (a) T. E. H. Walker and W. G. Richards, *Symp. Faraday Soc.*, **2**, 64 (1968); *J. Chem. Phys.*, **52**, 1311 (1970); (b) J. A. Hall, J. Schamps, J. M. Robbe, and H. Lefebvre-Brion, *J. Chem. Phys.*, **59**, 3271 (1973); however, see also (c) R. Mc Weeny and B. T. Sufcliffe, *Methods of Molecular Quantum Mechanics*, Academic, London, 1969.
66. R. S. Mulliken, *J. Chem. Phys.*, **46**, 497 (1949).
67. (a) M. Desouter-Lecomte and J. C. Lorquet, *J. Chem. Phys.*, **66**, 4006 (1977) and literature cited herein; (b) C. Galloy and J. C. Lorquet, *J. Chem. Phys.*, **67**, 4672 (1977).
68. (a) S. Fraga and G. Malli, *Many-Electron Systems and Interactions*, Saunders, Philadelphia, 1968; for optimal gauge origin, see (b) S. I. Chan and T. P. Das, *J. Chem. Phys.*, **7** 383 (1975); and (c) R. Moccia, *Chem. Phys. Lett.*, **5**, 260, 265 (1970).
69. (a) F. London, *J. Phys. Radium (Paris)*, **8**, 397 (1937); (b) H. F. Hameka, *Rev. Mod. Phys.*, **34**, 87 (1962); (c) R. Ritchfield, *J. Chem. Phys.*, **56**, 5688 (1972) and references therein; (d) M. Zauces, D. Pumpernik, M. Hladnik, and A. Azman, *Chem. Phys. Lett.*, **48**, 139 (1977).
70. S. T. Epstein, *J. Chem. Phys.*, **58**, 1952 (1972).
71. A. J. Sadlej, *Acta Phys. Pol. A*, **49**, 667 (1978); (b) A. J. Sadlej and W. T. Raynes, *Chem. Phys.*, **7**, 383 (1975); (c) P. A. Braun and T. K. Rebane, *Chem. Phys. Lett.*, **49** 765 (1977).

72. (a) W. R. Jarmain and P. A. Frazer, *Proc. Phys. Soc. Lond. Sect. A.*, **66**, 1153 (1953); (b) P. A. Frazer, *Can. J. Phys.*, **32**, 515 (1954).
73. (a) J. W. Cooley, *Math. Comput.*, **15**, 363 (1961); (b) D. G. Truhlar and W. P. Tarara, *J. Chem. Phys.*, **64**, 237 (1976); (c) B. R. Johnson, *J. Chem. Phys.*, **67**, 4086 (1977).
74. (a) C. S. Lin and G. W. F. Drake, *Chem Phys. Lett.*, **16**, 35 (1972); (b) R. Gordan, *J. Chem. Phys.*, **51**, 14 (1969); (c) P. Julienne, *J. Mol. Spectrosc.*, **48**, 508 (1973).
75. F. H. Mies, *J. Mol. Spectrosc.*, **53**, 150 (1974).
76. G. D. Carney, L. L. Sprandel, and C. W. Kern, *Adv. Chem. Phys.*, **37**, 305 (1978).
77. (a) M. G. Bucknell, N. C. Handy, and S. F. Boys, *Mol. Phys.*, **28**, 759 (1973); **28**, 777 (1973); (b) R. J. Whitehead and N. C. Handy, *J. Mol. Spectrosc.*, **55**, 356 (1975); (c) G. D. Carney and R. N. Porter, *J. Chem. Phys.*, **65**, 3547 (1976); (d) G. D. Carney, L. A. Curtiss, and S. R. Langhoff, *J. Mol. Spectrosc.*, **61**, 371 (1976); *Appl. Spectrosc.*, **30**, 453 (1976); (e) G. D. Carney, S. R. Langhoff, and L. A. Curtiss, *J. Chem. Phys.*, **66**, 3724 (1977); (f) S. M. Calwell and N. C. Handy, *Mol. Phys.*, **35**, 1183 (1978) and references therein.
78. J. M. Bowman, *J. Chem. Phys.*, **68**, 608 (1977).

GENERALIZED LANGEVIN EQUATIONS AND MANY-BODY PROBLEMS IN CHEMICAL DYNAMICS

S. A. ADELMAN*

Department of Chemistry
Purdue University
West Lafayette, Indiana

CONTENTS

I. INTRODUCTION

This review deals with new theoretical methods[1] which are designed to handle the many-body problems arising in molecular treatments of condensed-phase chemistry. The new methods developed as a response from theory to the challenge posed by the emergence of experimental techniques for probing condensed-phase chemical dynamics on *molecular timescales*. The new probes which particularly stimulated the development of the theory are molecular beam collision methods for study of solid-surface-catalyzed chemical reactions[2] and laser spectroscopic techniques for picosecond time-scale investigations of chemical dynamics in liquids and solids.[3]

The new theoretical techniques, as one might expect, draw heavily on ideas from both gas-phase and condensed-phase theory. The methods, in fact, are best regarded as a union of the classical trajectory method of gas-phase chemical kinetics[4,5] with the theory of generalized Brownian motion,[6,7] one of the cornerstones of modern nonequilibrium statistical mechanics.[6–14]

*Alfred P. Sloan Foundation Fellow.

Before beginning a detailed discussion of the new methods, we first summarize some of the previous work upon which they are based.

The classical trajectory method was developed into a powerful tool for simulating small-molecule gas-phase chemistry by Karplus, Porter, and Sharma[4] as well as many others.[4] The trajectory method has been reviewed elsewhere.[5]

The theory of generalized Brownian motion, which we believe provides the most powerful approach available to many-body dynamical problems, emerged from the modern time correlation function theory of irreversible phenomena. The correlation function approach was developed by many people, including Green,[8] Callen,[9] Zwanzig,[10] Mori,[6,11] Kadanoff and Martin,[12] and particularly Kubo.[7,13] Various aspects of time correlation function methods have been reviewed by many authors.[14] The ideas of generalized Brownian motion theory, while implicit in many early papers, were first laid down in an explicit and unified form by Mori[6] and Kubo.[7]

The first indication of a connecting link between the chemical dynamical and statistical mechanical theories appeared in an early paper by Zwanzig,[15] who dealt with a model for atomic scattering off a collinear harmonic chain. Zwanzig's treatment was generalized and applied by other workers[16] to a variety of gas–solid phenomena.

Zwanzig's formulation, while providing an excellent beginning, is deficient for a number of practical and conceptual reasons. Most obviously, it provided no clean way to treat gas scattering off finite-temperature solids.

Our own efforts,[1,17] carried on partially in collaboration with J. D. Doll, amounted in essence to fusing Zwanzig's ideas with the methods of generalized Brownian motion theory and with gas-phase classical trajectory methodology.[4,5] This fusion required a *reshaping* of generalized Brownian motion theory. The Brownian motion theory was originally developed for macroscopic transport rather than molecular dynamical problems, and the reshaping was required to make the theory convenient for molecular time-scale phenomena.

The modified formulation of generalized Brownian motion theory (which we believe actually provides the most symmetric form of the theory), together with algorithms for applying the theory to chemical dynamics, were first presented for molecule scattering off *perfectly harmonic* solids.[1] In the present treatment and elsewhere[17] we lift the harmonic restriction and formulate the theory in general terms. Thus the present version of the theory is applicable to liquid-state as well as solid-state chemical phenomena.

The present theory has two very attractive features:

1. It provides a general and systematic method for reducing many-body chemical dynamical problems to *effective* few-body problems. These effective few-body problems may be solved in a straightforward manner by adaptations of gas-phase classical trajectory methods. *Thus the theory provides a* practical *generalization of the gas-phase classical trajectory method to the many-body problems characteristic of condensed-phase chemistry*.
2. The theory provides a *conceptual framework* for *understanding* as well as computing condensed-phase chemical dynamics. It provides a *language* for thinking about and discussing complex many-body phenomena in simple qualitative terms. We believe that these *conceptual* aspects of the theory will eventually prove to be as useful as the computing power it supplies.

The plan of this review is as follows. Section II develops generalized Brownian motion theory, formulated so as to be useful in chemical dynamics applications, in a unified and comprehensive manner. Our discussion is restricted to the approximation of classical mechanics but is otherwise general. In Section III we present some simple applications of the theory to gas–solid collision dynamics. These applications are designed to illustrate the conceptual content of the theory rather than to demonstrate its use as a tool for large-scale numerical simulation. We expect that results of large-scale simulations, which are now becoming available,[18] will soon be reviewed elsewhere.

II. THEORY OF GENERALIZED BROWNIAN MOTION

Our approach to many-body problems in chemical dynamics draws heavily on modern ideas concerning the treatment of general many-body dynamical problems. We work, in particular, within the framework of the effective equation-of-motion approach to many-body problems. The most general formulation of this approach constitutes the subject matter of generalized Brownian motion theory. The first comprehensive and unified treatments of generalized Brownian motion are due to Mori[6] and Kubo.[7] We will develop generalized Brownian motion theory in this section, in a form that will lend itself readily to application to chemical dynamical problems. Our overall viewpoint is that of our earlier work.[1] That work, however, was restricted to perfectly harmonic many-body systems. Our analysis here may be regarded as an extension[17] of our earlier work to general many-body systems. We begin with a discussion of general concepts important in many-body dynamical problems.

A. General Concepts in Many-Body Dynamical Problems

In order to focus our general discussion and as preparation for our development of generalized Brownian motion theory, we will consider the dynamics of a macroscopic (M) particle (e.g., ball bearing) moving in a liquid. This "trivial" problem displays many features common to many-body problems and will provide an opportunity to introduce simple concepts of general importance. The M-particle problem is simply stated: calculate the observed classical trajectory for the M-particle moving in a prescribed external force field for a given set of M-particle initial conditions.

To solve this problem one may:

1. Solve the coupled set of Newton's equations of motion for the M-particle and the $\sim 10^{23}$ to 10^{24} liquid molecules. This solution will depend on the liquid-molecule initial conditions as well as the initial conditions of the M-particle and the prescribed external force.
2. Find the observable or macroscopic trajectory of the M-particle by averaging the M-particle trajectories determined in (1) over the thermal distribution of liquid molecule initial conditions.

This "straightforward" approach to our simply stated problem thus requires numerical solution of vast numbers of coupled nonlinear differential equations. These equations, moreover, must be solved a vast number of times in order to carry out the thermal average. The brute-force procedure, moreover, produces an immense amount of information about the trajectories of liquid molecules which is not observable and thus not directly relevant to the problem at hand.

The situation just outlined is characteristic of many-body dynamical problems. The remedy is invariably the same. One mentally divides the many-body system into two (or more) parts. The first part, which we will call the primary system or primary zone, is comprised of those *few* particles whose dynamics one wishes to follow in detail (the primary "particles" may actually be fictitious particles: e.g., perturbed normal modes, plasma oscillations, etc.). The second part, which we call the heatbath, consists of the remainder of the many-body system. In the simple example under discussion, the M- particle is the primary system and the liquid molecules comprise the heatbath.

Given this mental division of the system:

1. One writes down or derives *effective equations of motion* involving only the primary system particles explicitly. These effective equations of motion contain only that information about the heatbath that is relevant to the motion of the primary system. We call this *limited* information the heatbath influence.
2. One next explicitly determines the heatbath influence. This typically involves the calculation of certain time correlation functions (see below) describing those aspects of the heatbath motion relevant to the dynamics of the primary system. The heatbath is a many-body system and thus, in general, one cannot make exact calculations of the heatbath influence. One must rather construct a simplified and tractable model for the heatbath. This model heatbath must simulate the true heatbath influence to some desired accuracy but in other respects may bear little resemblance to the true heatbath. The true heatbath influence actually contains very detailed information about its complex many-body dynamics [see (II.132) below]. Thus for the model heatbath method to be useful, the fine detail in the true heatbath influence must be of minor consequence for measurements on the primary system. This, in fact, often turns out to be the case. Highly simplified heatbath models which account for only the grosser features of the true heatbath influence often are useful in interpreting and predicting measurements of primary system dynamics. This fact is basic to the real power of the effective equation-of-motion method. If only exact or nearly exact heatbath influences were useful, the effective equation-of-motion method would provide an elegant formal language but not a practical computational tool.
3. One finally solves the effective equations of motion and calculates the dynamics of the primary system. Once these dynamics are known, it is more or less straightforward to calculate such observables as diffusion coefficients, NMR relaxation times, infrared line shapes, cross-sections for reactive scattering at solid surfaces, and so on.

In traditional applications of the effective equations-of-motion method,[6,12] solving the equations is straightforward because of their linear character. For the newer applications, which are a main concern of this review (i.e., gas–solid collisions and related problems), step (3) is much less straightforward since the effective equations of motion are nonlinear. In fact, we will see that for these latter problems, steps (2) and (3), are strongly coupled. That is, the solvability of the effective equations of motion (by adaptations of traditional few-body techniques) hinges on the development of appropriate heatbath models. This will become clearer below.

Now let us see how these general concepts apply to the simple problem of M-particle dynamics in a liquid. For this problem, we will simply write down the effective equation of motion (step 1 above), (II.2) below, on empirical and intuitive grounds and make no attempt to derive it from microscopic dynamics.

Let the mass of the M-particle be M and let its coordinate be $\vec{r}(t)$. From the standpoint of the M-particle the liquid simply serves as a source of frictional resistance to its motion. The frictional force $\vec{F}_f$ is found experimentally to be proportional to the velocity $\dot{\vec{r}}(t)$ of the particle and to act in a direction opposite to the velocity:

$$\vec{F}_f = -M\beta\dot{\vec{r}}(t) \tag{II.1}$$

where β, the heatbath influence, is the coefficient of friction of the particle. The effective equation of motion for the M-particle is thus

$$M\ddot{\vec{r}}(t) = -M\beta\dot{\vec{r}}(t) + \vec{F}(\vec{r}) \tag{II.2}$$

where $\vec{F}(\vec{r})$ is an arbitrary external nondissipative force acting on the M-particle (e.g., the force of gravity).

Notice that the effective (II.2) is vastly simpler than the exact set of $\sim 10^{24}$ coupled Newton equations for the M-particle and the liquid molecules. The potential-energy functions coupling the motion of the liquid molecules and the M-particle never appear and the complexities of many-body dynamics are compressed into a single parameter β. To calculate β exactly, of course, requires solution of the full many-body problem. This unpleasant problem may be circumvented by introducing an appropriate heatbath model. The most sensible model is based on the observation that from the standpoint of the M-particle the liquid appears to be a structureless continuum rather than a collection of molecules. Thus we may model the true heatbath, whose dynamics is governed by $\sim 10^{24}$ coupled differential equations, by a hydrodynamic continuum whose dynamics is determined by a single differential equation (the Navier–Stokes equation) for the velocity field in the liquid. With this heatbath modeling scheme, β is readily found analytically as[19]

$$M\beta = 6\pi\eta R \tag{II.3}$$

Equation (II.3) is the celebrated Stokes law, which determines the friction β in terms of the *readily measurable* liquid viscosity η and the M-particle radius R.

To complete the effective equation of motion program (points 1, 2, 3 above) for the M-particle case, (II.2) must be solved. For arbitrary $\vec{F}(\vec{r})$ this may be accomplished by straightforward numerical integration. Integration of (II.2) may be accomplished analytically if $\vec{F}(\vec{r})=\vec{0}$, $\vec{F}(\vec{r})=\vec{F}$ (a constant), or $\vec{F}(\vec{r})=-k\vec{r}$. This last case of linear forces or damped harmonic motion will be studied in detail since it will lead us naturally into the general theory of the effective equation-of-motion method. Our analysis of damped harmonic M-particle motion will be cast int the language of correlation and response functions. While this language is not really needed for this special case, it will prove essential when we generalize the theory in Section II.C.

B. Response Functions and Damped Macroscopic Harmonic Motion

For damped harmonic motion the force $\vec{F}(\vec{r})$ may be written as

$$\vec{F}(\vec{r}) = -M\Omega_M^2\vec{r}$$

where Ω_M is the effective circular frequency of the oscillator. Combining the foregoing force law with (II.2) yields the following equation of damped harmonic motion:

$$\ddot{\vec{r}}(t) = -\Omega_M^2\vec{r}(t) - \beta\dot{\vec{r}}(t) \tag{II.4}$$

We solve (II.4) by the Laplace transform technique. We define the Laplace transform $\hat{f}(z)$ of an arbitrary function of time $f(t)$ by

$$\hat{f}(z) = \int_0^\infty e^{-zt} f(t)\,dt \equiv \mathcal{L} f(t) \tag{II.5a}$$

The inverse relationship is

$$f(t) = \int_C e^{zt}\hat{f}(z)\,dz \equiv \mathcal{L}^{-1}\hat{f}(z) \tag{II.5b}$$

where C denotes the Laplace inversion contour.

Solving (II.4) by Laplace transforms yields for the Laplace transform $\hat{\vec{r}}(z)$ of the trajectory $\vec{r}(t)$,

$$\hat{\vec{r}}(z) = \hat{\chi}_M(z)\dot{\vec{r}}(0) + \hat{\phi}_M(z)\vec{r}(0) \tag{II.6}$$

where

$$\hat{\chi}_M(z) = \left[z^2 + \Omega_M^2 + z\beta\right]^{-1} \tag{II.7a}$$

and

$$\hat{\phi}_M(z) = (z+\beta)\hat{\chi}_M(z) \tag{II.7b}$$

The quantities $\hat{\chi}_M(z)$ and $\hat{\phi}_M(z)$ are *response functions* for the harmonically bound M-particle. Their time-domain transforms, which we will also call response functions, are given by

$$\chi_M(t) = \mathcal{L}^{-1}\hat{\chi}_M(z) = \omega_1^{-1} e^{-(1/2)\beta t} \sin \omega_1 t \tag{II.8a}$$

$$\phi_M(t) = \mathcal{L}^{-1}\hat{\phi}_M(z) = e^{-(1/2)\beta t}\left[\cos \omega_1 t + \frac{\beta}{2\omega_1} \sin \omega_1 t\right] \tag{II.8b}$$

where $\omega_1 = (\Omega_M^2 - \frac{1}{4}\beta^2)^{1/2}$. Notice that the response functions decay to zero as time increases:

$$\lim_{t\to\infty} \chi_M(t) = \lim_{t\to\infty} \phi_M(t) = 0 \tag{II.9}$$

The time-domain trajectory $\vec{r}(t) = \mathcal{L}^{-1}\hat{r}(z)$ of the M-particle is determined from (II.6) as

$$\vec{r}(t) = \chi_M(t)\dot{\vec{r}}(0) + \phi_M(t)\vec{r}(0) \tag{II.10}$$

Thus the response function $\chi_M(t)$ determines the possible M-particle trajectories since $\phi_M(t)$ and $\chi_M(t)$ are related by (II.7b) [one may show that $\phi_M(t) = \dot{\chi}_M(t) + \beta\chi_M(t)$]. The response function $\chi_M(t)$, moreover, also determines the oscillator trajectory if it is driven by an external time-dependent force $M\vec{f}_e(t)$. For this case, the effective equation of motion is

$$\ddot{\vec{r}}(t) = -\Omega_M^2\vec{r}(t) - \beta\dot{\vec{r}}(t) + \vec{f}_e(t) \tag{II.11}$$

The solution of (II.11) may be found by Laplace transforms. The result is

$$\vec{r}(t) = \chi_M(t)\dot{\vec{r}}(0) + \phi_M(t)\vec{r}(0) + \int_0^t \chi_M(\tau)\vec{f}_e(t-\tau)\,d\tau \tag{II.12}$$

The trajectory of the driven oscillator, (II.12), is a superposition of its motion in the absence of $\vec{f}_e(t)$, (II.10), and of a perturbation on this "free" motion which is linear in $\vec{f}_e(t)$. For $t \gg \beta^{-1}$ the free motion decays to zero and the oscillator trajectory becomes independent of its initial conditions. For $\vec{f}_e(t)=\vec{f}$, a constant, $\vec{r}(t)$ approaches a steady-state value $\vec{r}(\infty)$ as $t \to \infty$, where

$$\vec{r}(\infty)=\hat{\chi}_M(z=0)\vec{f}=\Omega_M^{-2}\vec{f} \tag{II.13}$$

Thus Ω_M^{-2} is the steady-state extension of the oscillator induced by a unit external force and thus functions as a susceptibility or generalized polarizability for the M-particle.

Equations (II.10) and (II.12) show that $\chi_M(t)$ determines individual M-particle trajectories. It also governs trajectory-averaged M-particle dynamics. Thermally averaging (II.12) over oscillator initial conditions $\dot{\vec{r}}(0)$ and $\vec{r}(0)$ and then differentiating the resulting equation with respect to time yields

$$\langle \dot{\vec{r}}(t) \rangle = \int_0^t \dot{\chi}_M(\tau)\vec{f}_e(t-\tau)\,d\tau \tag{II.14}$$

where $\langle \cdots \rangle$ denotes a thermal average and where we have used $\langle \vec{r}(0) \rangle = \langle \dot{\vec{r}}(0) \rangle = \vec{0}$. Equation (II.14) shows that the *average* velocity of the M-particle *induced* by $\vec{f}_e(t)$ is determined by the response function $\dot{\chi}_M(t)$. The response function also determines the decay of spontaneous fluctuations of the velocity of the M-particle from its equilibrium average value $\langle \dot{\vec{r}} \rangle = \vec{0}$. The temporal development of spontaneous fluctuations is described mathematically by equilibrium time correlation functions (tcf). Thus the response function $\chi_M(t)$ is simply related to an M-particle tcf. This relationship is readily derived from (II.10). Multiplying (II.10) by $\dot{\vec{r}}(0)$ and taking a thermal average over M-particle initial conditions yields

$$\langle \vec{r}(t)\cdot\dot{\vec{r}}(0) \rangle = \chi_M(t)\langle \dot{\vec{r}}(0)\cdot\dot{\vec{r}}(0) \rangle + \phi_M(t)\langle \vec{r}(0)\cdot\dot{\vec{r}}(0) \rangle \tag{II.15}$$

Within the approximation of classical statistical mechanics, position and velocity are statistically independent and the average kinetic energy in each degree of freedom is $\frac{1}{2}k_BT$, where k_BT is Boltzmann's constant times Kelvin temperature. Thus

$$\langle \dot{\vec{r}}(0)\dot{\vec{r}}(0) \rangle = \frac{k_BT}{M}\overleftrightarrow{1} \tag{II.16a}$$

and

$$\langle \vec{r}(0)\dot{\vec{r}}(0) \rangle = \overleftrightarrow{0} \tag{II.16b}$$

where $\overleftrightarrow{0}$ and $\overleftrightarrow{1}$ are, respectively, the null and unit Cartesian tensors. Equations (II.15) and (II.16) yield the following basic relationship between response and correlation functions:

$$\chi_M(t) = \frac{M}{3k_BT}\langle \vec{r}(t)\cdot\dot{\vec{r}}(0)\rangle \tag{II.17a}$$

or, equivalently,

$$\dot{\chi}_M(t) = \frac{M}{3k_BT}\langle \dot{\vec{r}}(t)\cdot\dot{\vec{r}}(0)\rangle \tag{II.17b}$$

Thus $\dot{\chi}_M(t)$ is proportional to the velocity autocorrelation function of the M-particle. Equations (II.17) are an example of the first fluctuation–dissipation theorem. The fluctuation–dissipation theorem is a basic result in nonequilibrium statistical mechanics and is an essential component of the effective equation-of-motion method.

The significance of the first fluctuation–dissipation theorem becomes clearer if we combine (II.14) and (II.17b). This yields

$$\langle \dot{\vec{r}}(t)\rangle = \frac{M}{3k_BT}\int_0^t \langle \dot{\vec{r}}(\tau)\cdot\dot{\vec{r}}(0)\rangle \vec{f}_e(t-\tau) \tag{II.18}$$

Equation (II.18) is an example of the central result of linear response theory; that is, it is a relation between the average value of a dynamical variable in the presence of an external force and the correlation of fluctuations in the system in the *absence* of that force.

An equation of motion for the response function $\chi_M(t)$ is obtained by combining (II.4) and (II.17). The result is

$$\ddot{\chi}_M(t) = -\Omega_M^2\chi_M(t) - \beta\dot{\chi}_M(t) \tag{II.19}$$

Thus $\chi_M(t)$ satisfies the same equation of motion as the trajectory $\vec{r}(t)$. The initial conditions on $\chi_M(t)$ are found from (II.16) and (II.17) to be

$$\begin{aligned}\chi_M(0) &= 0\\ \dot{\chi}_M(0) &= 1\end{aligned} \tag{II.20}$$

So far our discussion has been based on the special problem of M-particle dynamics. We assumed on empirical grounds that these dynamics are governed by (II.2) which becomes (II.4) for harmonic restoring forces. This assumed dynamics has led us to the key results of this section, (II.17) to (II.20). We now undertake a more general analysis of the response function. This analysis shows that the fluctuation–dissipation results, (II.17) and (II.18), have general validity but that the assumed effective equation of motion, (II.4) or, equivalently, (II.19), suffers from inconsistencies. These inconsistencies are of little practical consequence for macroscopic motion but often lead to serious error in application of the effective equation-of-motion method to many-body *molecular* dynamical problems.

C. Response and Correlation for Microscopic Harmonic Motion

So far our discussion of the effective equation-of-motion method has been limited to the treatment of macroscopic particle dynamics. We begin, in this section, generalization of the theory to permit treatment of motion of particles of molecular size. Our analysis is based on classical mechanics but is otherwise quite general.

We consider as our primary system an arbitrary harmonically bound particle of mass m coupled to an arbitrary heatbath. We will call such a system an m-oscillator. Two familiar examples are provided by a diatomic molecule dissolved in a liquid (in this case the m-oscillator is the relative coordinate of the molecule and its mass is the reduced mass of the molecule) and also by an atom bound to a solid (this atom may either be part of the pure solid or it may be an impurity: e.g., a chemisorbed species bound to the solid surface). Since unbound damped motion is a limiting case of harmonic damped motion, most of our discussion also holds for this case.

We begin by noting that the linear response results, (II.17) and (II.18), hold for m-oscillators. That is, the average induced trajectory of an m-oscillator $\langle \vec{r}(t) \rangle$ is related to the external force $\vec{f}_e(t)$ by

$$\langle \vec{r}(t) \rangle = \int_0^t \dot{\chi}_m(t-\tau)\vec{f}_e(\tau)\,d\tau = \int_0^t \dot{\chi}_m(\tau)\vec{f}_e(t-\tau)\,d\tau \tag{II.21}$$

The response function $\chi_m(t)$ of the m-oscillator is related to its velocity autocorrelation function by

$$\dot{\chi}_m(t) = \frac{m}{3k_BT}\langle \dot{\vec{r}}(t)\cdot\dot{\vec{r}}(0) \rangle \tag{II.22a}$$

Equivalently,

$$\chi_m(t) = \frac{m}{3k_BT}\langle \vec{r}(t)\cdot\dot{\vec{r}}(0) \rangle = -\frac{m}{3k_BT}\langle \dot{\vec{r}}(t)\cdot\vec{r}(0) \rangle \tag{II.22b}$$

Equations (II.21) and (II.22) require several comments. First, (II.21) with (II.22a) may be derived by a straightforward application of Kubo's linear response method.[13] To keep this article self-contained, we present the derivation in Appendix 1. To derive (II.22b) from (II.22a), we have used the fact that $\langle \vec{r}(0)\cdot\dot{\vec{r}}(0)\rangle = 0$, (II.16b). Equation (II.16) holds for general classical thermal systems. The second equality in (II.22b) is a consequence of the fact that equilibrium time correlation functions describe fluctuations from a stationary situation (i.e., equilibrium) and thus cannot depend on which time one chooses to call $t=0$. Thus for an arbitrary dynamical variable $A(t)$,

$$\langle A(t+\tau)A(\tau)\rangle = \langle A(t)A(0)\rangle \tag{II.23a}$$

which is independent of τ. Differentiating (II.23) with respect to τ and then setting $\tau=0$ yields

$$\langle \dot{A}(t)A(0)\rangle = -\langle A(t)\dot{A}(0)\rangle \tag{II.23b}$$

The second equality in (II.22b) is a special case of (II.23b).

Setting $\tau=-t$ in (II.23a) demonstrates that the autocorrelation function $\langle A(t)A(0)\rangle$ is an even function of time:

$$\langle A(t)A(0)\rangle = \langle A(0)A(-t)\rangle = \langle A(-t)A(0)\rangle \tag{II.23c}$$

A number of important properties of $\chi_m(t)$ may be derived from (II.21) to (II.23) and (II.16).

First the initial values of $\chi_m(t)$ and $\dot{\chi}_m(t)$ are given by [cf. (II.20)]

$$\begin{aligned} \chi_m(0) &= 0 \\ \dot{\chi}_m(0) &= 1 \end{aligned} \tag{II.24}$$

From (II.22a) and (II.23c) we see that

$$\dot{\chi}_m(-t) = \dot{\chi}_m(t) \tag{II.25}$$

That is, $\dot{\chi}_m(t)$ is an even function of time and hence $\chi_m(t)$ is odd in time.

We next introduce the *spectral density* $\rho(\omega)$ of $\dot{\chi}_m(t)$ defined by the Fourier transform relationship

$$\rho(\omega) = \frac{1}{\pi}\int_{-\infty}^{\infty} e^{i\omega t}\dot{\chi}_m(t)\,dt \tag{II.26}$$

Because $\dot{\chi}_m(t)$ is an even function, (II.26) may be rewritten as the cosine transform relationship

$$\rho(\omega) = \frac{2}{\pi}\int_0^{\infty} \cos\omega t\,\dot{\chi}_m(t)\,dt \tag{II.27a}$$

The inverse relationship is

$$\dot{\chi}_m(t) = \int_0^\infty \cos\omega t \rho(\omega)\, d\omega \tag{II.27b}$$

The spectral density is simply related to the Laplace transform $\hat{\chi}_m(z) = \mathfrak{L}\chi_m(t)$. From (II.24) and (II.27a) and the fact that $\lim_{t\to\infty}\chi_m(t) = 0$ for a many-body heatbath, one finds that

$$\rho(\omega) = -\frac{2\omega}{\pi}\mathrm{Im}\,\hat{\chi}_m(i\omega) \tag{II.27c}$$

From the linear response result (II.21) and the fact that motion of the m-oscillator leads to energy dissipation (on the average) if the heatbath is a true many-body system, one may prove that $\rho(\omega)$ is nonnegative:

$$\rho(\omega) \geq 0 \tag{II.28}$$

The proof that is due to Kubo[13] is reviewed in Appendix 2. Equation (II.28) is the basis of the interpretation of $\rho(\omega)$ as a spectral *density*.

From (II.24) and (II.27b) show that $\rho(\omega)$ is unit-normalized:

$$\int_0^\infty \rho(\omega)\, d\omega = 1 \tag{II.29}$$

The second moment of $\rho(\omega)$,

$$\omega_e^2 = \int_0^\infty \omega^2 \rho(\omega)\, d\omega \tag{II.30}$$

will be of particular interest below. We will call ω_e the Einstein frequency of the m-oscillator. From (II.27b) we see that the short-time behavior of $\dot{\chi}_m(t)$ is determined by the Einstein frequency:

$$\lim_{t\to 0}\dot{\chi}_m(t) = 1 - \tfrac{1}{2}\omega_e^2 t^2 \tag{II.31}$$

and further that

$$\ddot{\chi}_m(0) = 0 \tag{II.32a}$$

and

$$\dddot{\chi}_m(0) = -\omega_e^2 \tag{II.32b}$$

Let us now compare (II.31) with the short-time behavior of the response function $\chi_0(t)$ for an *isolated* (i.e., gas phase) harmonic oscillator of mass m_0, coordinate $\vec{r}_0$, and frequency ω_0. It is straightforward to show [e.g., set $\beta=0$ in (II.8)] that

$$\dot{\chi}_0(t)=\frac{m_0}{3k_BT}\langle\vec{r}_0(t)\cdot\vec{r}_0(0)\rangle=\cos\omega_0 t=1-\tfrac{1}{2}\omega_0^2t^2+\cdots \tag{II.33}$$

Comparing (II.31) and (II.33) shows that for *short times*, the m-oscillator response function is that of an isolated or Einstein oscillator with frequency ω_e; hence our designation of ω_e as the Einstein frequency of the m-oscillator. Physically, this means that for sufficiently short times, the heatbath cannot respond to the m-oscillator motion and thus for short times the m-oscillator dynamics is that of an effective isolated harmonic system (we will modify this last statement in Section II.I when we introduce heatbath fluctuations; i.e., thermal noise, into the theory). We emphasize that ω_e is *not*, in general, the frequency that the m-oscillator would have if it was truly isolated since ω_e depends on the exact many-body spectral density, via (II.30).

The Einstein frequency ω_e governs the short-time response of the m-oscillator. A second harmonic frequency Ω also plays a key role in our analysis. The frequency Ω determines the long-time or more correctly the static or zero-frequency response of the oscillator. The frequency Ω also governs very slow oscillator motion and thus we will call it the adiabatic frequency. Since the heatbath can rearrange in response to slow oscillator motion, Ω includes the effects of this heatbath response. Thus, in general, $\Omega\neq\omega_e$. In fact, $\Omega\leq\omega_e$, as will be discussed below.

The adiabatic frequency may be determined by calculating the response of the m-oscillator to a static external force $\vec{f}_e(t)=\vec{f}_e$. We consider, in particular, the mean steady-state extension $\langle\vec{r}(\infty)\rangle$ of this oscillator induced by $\vec{f}_e$. From (II.21), we see that this extension is [cf. (II.13)]

$$\langle\vec{r}(\infty)\rangle=\left[\int_0^\infty \chi_m(\tau)\,d\tau\right]\vec{f}_e=\hat{\chi}_m(0)\vec{f}_e \tag{II.34a}$$

Using (II.22b), the equation above may be rewritten as

$$\langle\vec{r}(\infty)\rangle=\frac{m}{3k_BT}\langle r^2\rangle\vec{f}_e \tag{II.34b}$$

Comparing (II.34b) with the corresponding result for an isolated oscillator,

$$\langle\vec{r}_0(\infty)\rangle=\frac{m_0}{3k_BT}\langle r_0^2\rangle\vec{f}_e=\frac{1}{\omega_0^2}\vec{f}_e \tag{II.35}$$

shows that the frequency Ω defined by

$$\Omega^2 = \frac{3k_B T}{m\langle r^2 \rangle} \tag{II.36}$$

functions as an inverse polarizability which controls the static response of the oscillator. For most systems, as mentioned above, $\Omega \neq \omega_e$. The isolated oscillator is an obvious exception since it is not subject to heatbath effects. Equations (II.33) and (II.35) indeed show, as expected, that for this case $\Omega = \omega_e = \omega_0$.

In addition to determining the static response of the m-oscillator, the frequency Ω also fully governs its dynamics in the quasistatic limit (i.e., when the oscillator is moving so slowly that the heatbath is able to perfectly follow its motion). For the quasistatic case, the m-oscillator motion does not lead to energy dissipation, hence the term adiabatic for the frequency Ω.

The adiabatic frequency may be expressed in terms of the spectral density $\rho(\omega)$. The result, which may be derived by integrating (II.27b) twice with respect to time using (II.16) and (II.22), is

$$\Omega^{-2} = \int_0^\infty \omega^{-2} \rho(\omega)\, d\omega \tag{II.37}$$

Since $\rho(\omega)$ is nonnegative, (II.28), comparison of the definition of ω_e, (II.30), with (II.37), shows that

$$\omega_e \geq \Omega \tag{II.38}$$

The reason for this inequality is that Ω has a larger contribution from low-frequency regions of the integrand in the integration over $\rho(\omega)$ than does ω_e.

The distinction between Einstein and adiabatic frequencies is often not emphasized in discussions of the correlation function approach to nonequilibrium statistical mechanics. This is perhaps because earlier applications of the theory were mainly concerned with *macroscopic* transport phenomena (e.g., heat flow, macroscopic fluid flow, etc.). For macroscopic phenomena, time scales are long compared to molecular motional time scales, and thus Ω is invariably the relevant frequency. For the applications of the theory to chemical dynamics, our concern here, the relevant time scales are molecular. Thus ω_e is the more appropriate frequency and the distinction between short and long time frequencies is of practical as well as conceptual importance. We will see, for example, how the distinction plays a key role in our formulation of the effective equation of motion for the m-oscillator in Sections II.D and II.E. We will see its practical importance when we discuss energy transfer in gas–solid collisions in Section III.

So far our discussion in this section has been general and formal. In order to implement the formalism practically, some type of modeling of the many-body system is always necessary (see Section II.A). We close this section by developing a simple model for the spectral density and response function. For the case that the many-body system is a solid, our model is that employed by Debye in his heat capacity theory. Thus we will call the model the Debye model. The Debye model is conveniently developed in terms of $\rho(\omega)$.

After a brief calculation, we may rewrite (II.27a) for $\rho(\omega)$ as

$$\rho(\omega)=\frac{2m}{\pi 3k_BT}\omega^2\int_0^\infty\langle\vec{r}(t)\cdot\vec{r}(0)\rangle\cos\omega t\,dt \tag{II.39}$$

From (II.39) we see that $\rho(0)=0$ if the integral exists. It is tempting to try to develop $\rho(\omega)$ in a power series in ω^2 by expanding $\cos\omega t$ in powers of ω^2. This is only valid if the integrals so obtained exist. In particular, the following equality is true:

$$\lim_{\omega\to 0}\rho(\omega)=\frac{2m\omega^2}{3\pi k_BT}\int_0^\infty dt\langle\vec{r}(t)\cdot\vec{r}(0)\rangle \tag{II.40}$$

only if $\langle\vec{r}(t)\cdot\vec{r}(0)\rangle$ decays sufficiently rapidly for large t. This is true, for example, for three-dimensional solids. It is not true, however, for liquids due to molecular diffusion.[20] If (II.40) is valid, we may make the Debye approximation. We assume that the low-frequency behavior $\rho(\omega)\sim\omega^2$ holds for all ω less than a cutoff frequency, or Debye frequency, denoted by ω_D. For $\omega>\omega_D$, we assume that $\rho(\omega)=0$. Since $\rho(\omega)$ must be unit-normalized, (II.29), the Debye spectrum is

$$\rho(\omega)=\frac{3}{\omega_D^3}\omega^2\eta(\omega-\omega_D) \tag{II.41}$$

where $\eta(x)$ is the unit step function. Equations (II.40) and (II.41) are in harmony if

$$\omega_D=\left\{\frac{9\pi k_BT}{2m}\left[\int_0^\infty dt\langle\vec{r}(t)\cdot\vec{r}(0)\rangle\right]^{-1}\right\}^{1/3} \tag{II.42}$$

Equation (II.42) provides an expression for ω_D in terms of the exact dynamics of the many-body system.

Using the Debye model, (II.41), we may evaluate the Einstein and adiabatic frequencies from (II.30) and (II.37). The results are

$$\omega_e = \sqrt{\tfrac{3}{5}}\,\omega_D \qquad \text{(II.43)}$$

and Debye

$$\Omega = \sqrt{\tfrac{1}{3}}\,\omega_D \qquad \text{(II.44)}$$

Note that for the Debye model $\omega_e > \Omega$, in agreement with (II.38).

Finally, we may evaluate $\dot{\chi}_m(t)$ for the Debye model using (II.27b) and (II.41). The result is

$$\dot{\chi}_m(t) = 3\left(\frac{\sin x}{x} + \frac{2\cos x}{x^2} - \frac{2\sin x}{x^3}\right) \qquad \text{Debye} \qquad \text{(II.45)}$$

where $x = \omega_D t$. Notice that the response function $\dot{\chi}_m(t)$ differs markedly from the M-particle response function given in (II.8); in particular, $\dot{\chi}_m(t)$ decays asymptotically as $\sim t^{-1}\sin\omega_D t$ rather than exponentially. This suggests that our intuitive guess for the effective equation of motion, (II.2), must be regarded with caution (see below).

We now turn to the formulation of an exact effective equation of motion for the response function.

D. Effective Equation of Motion for the Response Function

So far we have found that for either very high frequency (short time scale) or very low frequency processes the m-oscillator responds like an isolated oscillator. The appropriate frequency of the effective isolated oscillator is the Einstein frequency ω_e, (II.30), for fast processes and the adiabatic frequency Ω, (II.37), for very slow processes. Our results may be summarized by the effective equations of motion

$$\ddot{\chi}_m(t) = -\omega_e^2 \chi_m(t) \qquad \text{short times} \qquad \text{(II.46a)}$$

$$\hat{\chi}_m(z) = (z^2 + \Omega^2)^{-1} \qquad \text{low frequencies} \qquad \text{(II.46b)}$$

Most processes occur on intermediate time scales, so we require an *exact* effective equation of motion which interpolates between the limiting results of (II.46). For processes on intermediate time scales the heatbath has time to partially adjust to the m-oscillator motion (thus the Einstein limit does not obtain), but it cannot respond rapidly enough to *perfectly* follow the oscillator motion (as in the adiabatic limit). The result is that the heatbath

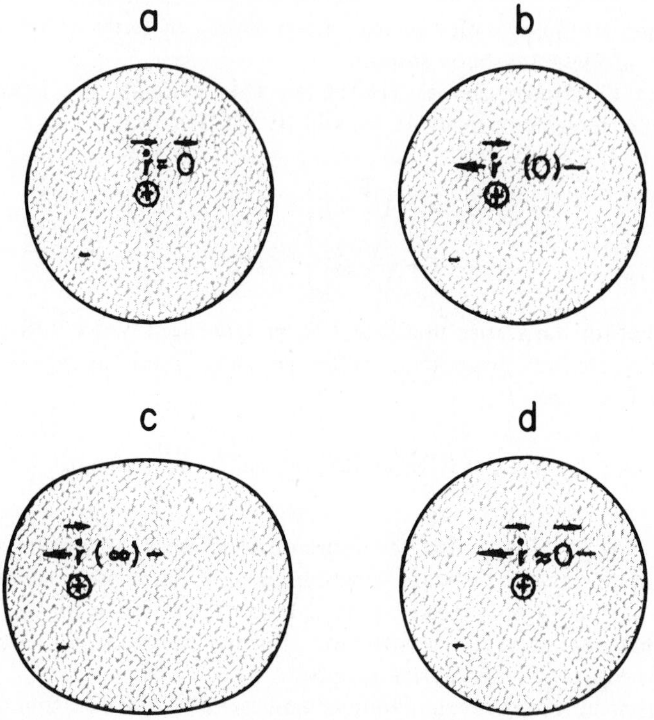

Fig. 1. Electronic drag on a moving metal ion—a prototype of generalized friction. (*a*) Spherically symmetric Debye cloud is induced in the metal's electron gas (Heatbath) by a stationary metal cation (primary system). By symmetry the cloud exerts zero force on the ion. (*b*) Ion commences motion with initial velocity $\dot{\vec{r}}(0)$. Cloud initially remains stationary, producing an instantaneous restoring force on the ion $= -\omega_{e_0}^2$ (ion displacement). (*c*) Ion attains a steady velocity $\dot{\vec{r}}(\infty)$, thus inducing a steady-state Debye cloud which is ellipsoidal due to heatbath lag. Ellipsoidal cloud exerts a constant frictional force on the ion $= -\beta\dot{\vec{r}}(\infty)$. (*d*) Quasistatic ion motion, $\dot{\vec{r}} \approx \vec{0}$; ellipsoidal cloud becomes nearly spherical, energy dissipation becomes negligible, and the adiabatic limit obtains.

motion lags behind the oscillator motion. This lag leads to energy dissipation or generalized frictional damping of the *m*-oscillator. The lag effect is easy to visualize physically for the case of damping of the motion of a metal ion (primary system) due to energy dissipation into the electron gas of the metal (heatbath). (The energy dissipation mechanism we have in mind involves mainly single-particle excitations of the electron gas rather than excitations of the plasma oscillations of the gas; see, e.g., the discussion of Pines.[21]) For this reason the lag effect in this system is as illustrated in Fig. 1.

Frictional damping *is* included in the equation of motion for the macroscopic oscillator response function, (II.19), through the friction coefficient β. We will now show, however, that this equation is inadequate since it is incorrect in the Einstein limit.

The difficulties with the short-time behavior of (II.19) are readily established by comparing the exact short-time results

$$\begin{aligned} \dot{\chi}_m(0) &= 0 \\ \ddot{\chi}_m(0) &= -\omega_e^2 \end{aligned} \tag{II.47}$$

derived in Section II.C with the corresponding M-particle results

$$\begin{aligned} \dot{\chi}_M(0) &= -\beta \\ \ddot{\chi}_M(0) &= -\Omega_M^2 + \beta^2 \end{aligned} \tag{II.48}$$

which may be readily found from (II.8a) or equivalently from (II.19) to (II.20). Equations (II.47) and (II.28) clearly disagree because (II.48) incorrectly include friction effects which should vanish in the short-time regime. Moreover, (II.48) has an additional defect. We will see below that $\Omega_M = \Omega$, the adiabatic frequency, and thus even for the case $\beta = 0$, (II.47) and (II.48) differ.

One way to formulate an effective equation of motion which is free of the deficiencies of (II.19) is to replace the friction parameter β by a time nonlocal *friction kernel* $\beta(t)$. With this replacement, we have as our effective equation of motion.

$$\ddot{\chi}_m(t) = -\Omega^2 \chi_m(t) - \int_0^t \beta(t-\tau)\dot{\chi}_m(\tau)\,d\tau \tag{II.49a}$$

The Laplace space form of (II.49a) is [recall the initial conditions, (II.24)]

$$\hat{\chi}_m(z) = \left[z^2 + \Omega^2 + z\hat{\beta}(z) \right]^{-1} \tag{II.49b}$$

Equations (II.49) *define* $\beta(t)$ in terms of $\chi_m(t)$; for example, from (II.49b),

$$z\hat{\beta}(z) = \hat{\chi}_m^{-1}(z) - z^2 - \Omega^2 \tag{II.50}$$

From (II.49a) and (II.24) one immediately deduces that

$$\begin{aligned} \dddot{\chi}_m(0) &= 0 \\ \ddot{\chi}_m(0) &= -\Omega^2 - \beta(0) \end{aligned} \tag{II.51}$$

Equations (II.51) and (II.47) are in agreement and, in fact, demonstrate that the initial value of the friction kernel $\beta(0)$ must be given by

$$\beta(0)=\omega_e^2-\Omega^2\geq 0 \tag{II.52}$$

In order that the representation of (II.49a) be consistent, it must also reduce properly in the adiabatic limit to (II.46b). That it does is apparent from (II.49b), which reduces to (II.46b) in the low-frequency $z\to 0$ limit.

Using the representation of (II.49b), we may readily derive a *microscopic* expression for the phenomenological friction coefficient β. Comparing the response function $\hat{\chi}_M(z)$, (II.7a), and the corresponding microscopic function $\hat{\chi}_m(z)$ demonstrates that the two forms agree for low frequency (low z) if β is defined as

$$\beta=\hat{\beta}(0)=\int_0^\infty \beta(\tau)\,d\tau \tag{II.53a}$$

The friction coefficient β may alternatively be expressed microscopically as

$$\beta=\left.\frac{d\hat{\chi}_m^{-1}(z)}{dz}\right|_{z=0}=\Omega^4\int_0^\infty \tau\chi_m(\tau)\,d\tau \tag{II.53b}$$

The first form of (II.53b) is obtained by differentiating (II.50), letting $z\to 0$, and comparing the result with (II.53a); the second form is found by evaluating the derivative explicitly using (II.59) below.

The local friction or "Markovian" approximation is defined by the replacement $\hat{\beta}(z)=\beta$ for all z in (II.49b). This then gives, in the time domain, the M-particle effective equation of motion, (II.19). The Markovian approximation must always be used with caution in the time domain even if β, as defined by (II.53), exists. The existence of β merely implies that the low-z or low-frequency limit of $\hat{\chi}_m(z)$ has the M-particle form, (II.7a). It *does not* imply that $\lim_{t\to\infty}\chi_m(t)$ is given by the M-particle form, (II.8a).

For the Debye model, for example, the time-domain response function decays asymptotically as $\sim t^{-1}\sin\omega_D t$, (II.45). This differs markedly from the exponential decay of the response function [e.g., (II.8)], which is the signal of a true Markovian heatbath.[22] Yet β calculated from (II.53b) for the Debye model (II.45) is finite and, in fact, is given by

$$\beta=\frac{\pi}{6}\omega_D \qquad \text{Debye} \tag{II.54}$$

The effective equation of motion (II.49a) has often been discussed in the literature since it provides a formally exact and quite convenient representation of the *heatbath-averaged* dynamics of the oscillators. When one generalizes the theory to include heatbath fluctuations or thermal noise, the analog of (II.49a) [see (II.148) below], has an awkward behavior at short times. We will see that the thermal noise cannot be approximated as a simple random process but rather contains systematic transient terms. If one ignores these transients, then one incorrectly predicts that Ω rather than ω_e is the correct short-time-scale frequency [see (II.149) below]. This deficiency is very serious in applications of the effective equation-of-motion method to molecular dynamics since molecular processes occur on a time scale comparable to the decay time of the transients. Gas-solid energy transfer, for example, depends strongly on the frequency of the solid atom(s) struck by the incident gas particle (see, e.g., Fig. 5). If the transient frequency of the struck solid atom is taken as Ω rather than ω_e (recall that $\Omega=\sqrt{\frac{5}{9}}\ \omega_e$ for the Debye model), then serious error is incurred in calculation of sticking probabilities (overestimation often by a factor of 2 or more) and other processes (e.g., chemical reactions), influenced by gas-solid energy exchange.

For this reason we will work here with an alternative *exact* representation of the dynamics of the m-oscillator. This representation is free of the difficulties just described. The alternative form for the effective equation of motion is

$$\ddot{\chi}_m(t) = -\omega_e^2\chi_m(t) + \int_0^t \Theta(t-\tau)\chi_m(\tau)\,d\tau \tag{II.55a}$$

The quantity $\Theta(t)$, which we will call the *damping kernel*, is *defined* by (II.55a). The damping kernel will be regarded throughout the remainder of this article as the *heatbath influence*, discussed in general terms in Section II.A.

The Laplace domain form of (II.55a) is [cf. (II.49b)]

$$\hat{\chi}_m(z) = \left[z^2 + \omega_e^2 - \hat{\Theta}(z)\right]^{-1} \tag{II.55b}$$

An explicit defining relationship for the damping kernel in terms of the primary system response function is found by rearranging (II.55b) to yield

$$\hat{\Theta}(z) = z^2 + \omega_e^2 - \hat{\chi}_m^{-1}(z) \tag{II.56}$$

Using the initial values given in (II.24), we may readily verify from (II.55a) that

$$\dddot{\chi}_m(0)=0$$
$$\ddot{\chi}_m(0)=-\omega_e^2 \tag{II.57}$$

which are identical to (II.47). Thus (II.55a) provides a representation for $\chi_m(t)$ which is consistent at short times. A related point is that the Einstein limit, (II.46a), is apparent from (II.55a), since the integral or heatbath response term vanishes at very short times. Notice that for the representation of (II.49a), emergence of the Einstein limit is not as clear; rather, the vanishing of the integral term at $t=0$ *appears* to show that Ω governs the short-time dynamics. This is, of course, not actually true since (II.49a) is exact. [Equation (II.51) shows that Ω cannot govern the short-time dynamics unless $\beta(0)=0$, which is not true since $\omega_e > \Omega$.] This "appearance," however, is the origin of the awkwardness of the representation of (II.49a) when heatbath noise is introduced.

The effective equation of motion, (II.55), properly reduces to the adiabatic limit, (II.46b), as well as the Einstein limit. To show this [see (II.55b)] we must demonstrate that

$$\hat{\Theta}(0)=\omega_e^2-\Omega^2 \tag{II.58}$$

since if (II.58) is true, (II.55b) and (II.46b) are identical as $z \to 0$.

Equation (II.58) may be verified by combining (II.56), with the result

$$\hat{\chi}(0)=\Omega^{-2} \tag{II.59}$$

obtained by comparing (II.34) and (II.36).

The relationship between the representations of (II.49) and (II.55) may readily be established by determining the relationship between the friction kernel $\beta(t)$ and the damping kernel $\Theta(t)$. A straightforward calculation yields

$$\beta(t)=\int_t^{\infty}\Theta(\tau)\,d\tau \tag{II.60}$$

Note that (II.58) and (II.60) reproduce (II.52). This provides a consistency check on the whole theory.

Before closing this section, we summarize the essentials of what has been done in Sections II.B to II.D. Starting with the macroscopic effective equation of motion, (II.4), which was adopted on intuitive and empirical grounds, we derived as rigorous consequences of (II.4) the first fluctuation–dissipation theorem, (II.17b), and the linear response result (II.18).

We next found that these two results were generally correct within classical statistical mechanics (quantum generalizations, of course, exist[13]) and hold for arbitrary m-oscillators, (II.21) and (II.22). That is, the fluctuation–dissipation results are not tied to the phenomenological model of (II.4), but may be derived in the general case from rigorous microscopic considerations[13] (Appendix 1).

The remainder of our derivation consisted in essence of reversing our M-particle analysis, that is, going from the fluctuation–dissipation results, (II.21) and (II.22), to the exact equations of motion, (II.49), or, equivalently, (II.55). Our development of the exact equations of motion was guided by the realization that the m-oscillator acts like an isolated oscillator at short times and at low frequencies. The oscillator frequencies associated with the two regimes ω_e and Ω are in general different; the physical origin of the difference is illustrated in Fig. 1. The existence of a distinct short-time frequency is closely related to the stationary property of equilibrium tcf, (II.23a), and in fact our whole analysis depended heavily on this property and its consequence $\dot{\chi}_M(t)=\dot{\chi}_M(-t)$.

Finally, we note that our program of reversing the M-particle analysis to obtain the exact effective equation of motion is not yet complete, since so far we only have the equation of motion for the response function, the analog of (II.19), rather than the effective equation of motion for the trajectory, the analog of (II.4). Before turning to this problem in Section II.I, we continue our analysis of the response function.

E. The Damping Kernel and Heatbath Response

We have seen so far that exact effective equations of motion may be formulated for the m-oscillator response or correlation functions. These equations incorporate the *systematic* effects of energy transfer between m-oscillator and heatbath through the generalized friction kernel $\beta(t)$ or equivalently through the damping kernel $\Theta(t)$. As discussed above (see Fig. 1), the damping effect arises from delayed heatbath response to the m-oscillator. Thus we may view the damping kernel from an alternative standpoint. Namely, we may regard it as a heatbath response function and subject it to the same analysis we employed in Sections II.C and II.D for m-oscillator response function $\chi_m(t)$. This simple shift in viewpoint will lead to remarkable results. The most fundamental and far-reaching result is the second fluctuation–dissipation theorem, (II.74) below. We will see that this result is a cornerstone of the effective equation-of-motion method.

We begin our analysis by establishing that $\dot{\Theta}(t)$ is even in time; that is, that

$$\dot{\Theta}(-t)=\dot{\Theta}(t) \tag{II.61}$$

Equation (II.61) may be proven in several different ways. For example, we may rewrite (II.55a) as

$$\ddot{\chi}_m(t) = -\omega_e^2\chi_m(t) + \int_0^t \chi_m(t-\tau)\Theta(\tau)\,d\tau \qquad \text{(II.55c)}$$

Making the substitution $t \to -t$ in (II.55c) and using the evenness of $\dot{\chi}(t)$ yields

$$\ddot{\chi}_m(t) = -\omega_e^2\chi_m(t) + \int_0^{-t} \chi_m(t+\tau)\Theta(\tau)\,d\tau$$

The previous equation may be rewritten as (let $\tau \to -\tau$)

$$\ddot{\chi}_m(t) = -\omega_e^2\chi_m(t) - \int_0^t \chi_m(t-\tau)\Theta(-\tau)\,d\tau$$

Finally, comparing the equation above with (II.55c) shows that

$$\Theta(-\tau) = -\Theta(\tau)$$

which is equivalent to (II.61).

Equation (II.61) is a key result, since much of our earlier analysis of $\chi_m(t)$ relied only on the evenness of $\dot{\chi}_m(t)$. In fact, the whole theory of $\chi_m(t)$ (Sections II.C and II.D) applies also to $\Theta(t)$ [or rather $\theta(t)$, (II.64) below].

We begin our discussion by determining the initial values of the damping kernel. From (II.61) we have immediately

$$\Theta(0) = 0 \qquad \text{(II.62a)}$$

The initial value of $\dot{\Theta}(t)$ is determined by differentiating (II.55a) three times and using the spectral density representation, (II.27b), to express $d^5\chi(0)/dt^5$ in terms of the fourth moment of $\rho(\omega)$. The result is

$$\dot{\Theta}(0) = \omega_c^4 \geq 0 \qquad \text{(II.62b)}$$

The quantity ω_c^4 is mean-square *fluctuation* from the Einstein force constant ω_e^2 as calculated from the distribution $\rho(\omega)$:

$$\omega_c^4 = \int_0^\infty \rho(\omega)\left[\omega^2 - \omega_e^2\right]^2 d\omega \qquad \text{(II.63)}$$

We will see below that ω_c^2 acts as an effective force constant coupling the m-oscillator and the heatbath. It vanishes only for isolated primary system oscillators.

We now introduce a normalized heatbath response function $\theta(t)$ defined by

$$\theta(t)=\omega_c^{-4}\Theta(t) \tag{II.64}$$

The normalized response function is obviously even

$$\dot{\theta}(t)=\dot{\theta}(-t) \tag{II.65}$$

and satisfies the initial conditions

$$\begin{aligned}\theta(0)&=0\\ \dot{\theta}(0)&=1\end{aligned} \tag{II.66}$$

These are exactly the conditions satisfied by $\chi_m(t)$, (II.24) and (II.25). Thus one may immediately deduce the following results concerning $\theta(t)$ and its spectral density $\sigma(\omega)$ [cf. (II.27) and (II.29)].

$$\sigma(\omega)=\sigma(-\omega)=\frac{2}{\omega}\int_0^\infty \cos\omega t\dot{\theta}(t)\,dt \tag{II.67a}$$

$$\dot{\theta}(t)=\int_0^\infty \cos\omega t\sigma(\omega)\,d\omega \tag{II.67b}$$

$$\sigma(\omega)=-\frac{2\omega}{\pi}\operatorname{Im}\hat{\theta}(iz) \tag{II.67c}$$

$$\int_0^\infty \sigma(\omega)\,d\omega=1 \tag{II.68}$$

Further combining (II.27c), (II.64), and (II.67c) with the effective equation of motion in the form of (II.56) yields

$$\sigma(\omega)=\omega_c^{-4}\frac{\rho(\omega)}{|\hat{\chi}_m(i\omega)|^2} \tag{II.69}$$

Equation (II.69) is the most succinct statement of the content of the effective equation of motion, (II.55). From (II.69) and (II.28), one may further see that $\sigma(\omega)$ is nonnegative-definite [cf. (II.28)]:

$$\sigma(\omega)\geq 0 \tag{II.70}$$

Thus the second moment of $\sigma(\omega)$, $\omega_{e_1}^2$, where [cf. (II.30)]

$$\omega_{e_1}^2\equiv\int_0^\infty \omega^2\sigma(\omega)\,d\omega \tag{II.71}$$

is nonnegative. This means that ω_{e_1} is real and may be interpreted as the Einstein frequency for the heatbath [cf. (II.31)]:

$$\lim_{t\to 0}\dot{\theta}(t)=1-\tfrac{1}{2}\omega_{e_1}^2 t^2 \tag{II.72}$$

and thus [cf. (III.32)]

$$\omega_{e_1}^2=-\ddot{\theta}(0) \tag{II.73a}$$

The quantity $\ddot{\theta}(0)$ may be expressed in terms of moments of $\rho(\omega)$ by a continuation of the procedure used to derive (II.62b) and (II.63). Thus ω_{e_1} may be expressed in terms of moments of $\rho(\omega)$ as well as moments of $\sigma(\omega)$. The result is

$$\omega_{e_1}^2=\omega_c^{-4}\left[\omega_e^6-2\omega_e^2\int_0^\infty \omega^4\rho(\omega)\,d\omega+\int_0^\infty \omega^6\rho(\omega)\,d\omega\right] \tag{II.73b}$$

The preceding analysis strongly suggests that the heatbath and m-oscillator response functions share the same formal properties. This leads to the *conjecture* that heatbath response function $\dot{\theta}(t)$, like $\dot{\chi}_m(t)$, may be represented as an autocorrelation function of a dynamical variable. We call this variable $\dot{\vec{R}}(t)$. Thus we conjecture [cf. (II.22)]

$$\dot{\theta}(t)=\frac{m}{3k_BT}\langle\dot{\vec{R}}(t)\cdot\dot{\vec{R}}(0)\rangle \tag{II.74a}$$

or, equivalently,

$$\theta(t)=\frac{m}{3k_BT}\langle\vec{R}(t)\cdot\dot{\vec{R}}(0)\rangle=-\frac{m}{3k_BT}\langle\dot{\vec{R}}(t)\cdot\vec{R}(0)\rangle \tag{II.74b}$$

An alternative form of (II.74) involving the friction kernel $\beta(t)$ follows from (II.60), (II.64), and (II.74). It is

$$\beta(t)=\frac{m\omega_c^4}{3k_BT}\langle\vec{R}(t)\cdot\vec{R}(0)\rangle \tag{II.74c}$$

[Note: In deriving (II.47c), we assume that $\langle\vec{R}(\infty)\cdot\vec{R}(0)\rangle=0$; this is true for many-body heatbaths (no Poincaré recurrences).]

Equations (II.74) have so far not been justified. They will be used without justification in this section and will be established rigorously in Section II.I. Equations (II.74) are a statement of the second fluctuation–dissipation theorem, a major result in nonequilibrium statistical mechanics.

The heatbath dynamical variable $\vec{R}(t)$ has a fundamental physical interpretation. The quantity $\omega_c^2\vec{R}(t)$ is the stochastic or random force on the primary system arising from thermal fluctuations in the heatbath. This stochastic force is the generalization of the random force appearing in phenomenological Brownian motion theory.[22] Thus the second fluctuation–dissipation theorem[6,7] has the deep physical significance of relating the two types of force the heatbath exerts on the primary system, the *systematic* dissipative force linear in $\Theta(t)$ and the random or stochastic force. The physical origin of this relationship is that both dissipative and random forces arise from the same mechanism, heatbath interaction with the primary system. The exact form of the relationship of (II.74) is a consequence of the fact that the dissipative force, which removes energy from the primary system (spontaneous emission), must on the average be in balance with the random force, which may either remove energy from or provide energy to the primary system (induced emission or absorption), in order that the equilibrium distribution of energies in the primary system be maintained or restored.

An important practical consequence of the second fluctuation–dissipation theorem is that all information about random (as well as systematic) heatbath influence is contained in $\Theta(t)$ [or $\theta(t)$]. Hence this function provides *complete* information about the influence of the heatbath on primary system dynamics and explicit determination of $\theta(t)$ (point 2 of Section II.A) is tantamount to solving the many-body aspects of the primary system dynamical problem.

We now return to the analysis of $\Theta(t)$. Immediate consequences of (II.66) and (II.74) are

$$\tfrac{1}{2}m\langle \dot{R}^2\rangle = \tfrac{3}{2}k_BT \tag{II.75a}$$

and

$$\langle \vec{R}(0)\cdot\dot{\vec{R}}(0)\rangle = 0 \tag{II.75b}$$

An adiabatic frequency of the heatbath Ω_1 may be defined by [cf. (II.59)]

$$\Omega_1^{-2} = \hat{\theta}(0) = \omega_c^{-4}\hat{\Theta}(0) \tag{II.76a}$$

The heatbath adiabatic frequency describes the quasi-static or low-frequency response of the heatbath. From (II.58) and (II.76a) we see that the heatbath adiabatic frequency is related to primary system frequencies by

$$\Omega_1^2 = \frac{\omega_c^4}{\omega_e^2 - \Omega^2} = \frac{\omega_c^4}{\beta(0)} \tag{II.76b}$$

One may further derive from (II.74) to (II.76) that [cf. (II.36)]

$$\Omega_1^2 = \frac{3k_BT}{m\langle R^2\rangle} \tag{II.77}$$

[Note; Our derivation of (II.77) requires that $\langle \vec{R}(\infty)\cdot\vec{R}(0)\rangle = 0$, which is true for many-body heatbaths.] An alternative expression for Ω_1 follows from (II.67b), (II.74), and (II.77). It is [cf. (II.37)]

$$\Omega_1^{-2} = \int_0^\infty \omega^{-2}\sigma(\omega)\,d\omega \tag{II.78}$$

Finally, we may express $\sigma(\omega)$ as [cf. (II.39)]

$$\sigma(\omega) = \frac{2m\omega^2}{3\pi k_BT}\int_0^\infty \langle \vec{R}(t)\cdot\vec{R}(0)\rangle \cos\omega t\,dt \tag{II.79}$$

and we may define a heatbath Debye model by [cf. (II.41)]

$$\sigma(\omega) = \frac{3}{\omega_{D_1}^3}\omega^2\eta(\omega-\omega_{D_1}) \tag{II.80}$$

where the heatbath Debye frequency [cf. (II.42)] is defined by

$$\omega_{D_1} = \left[\frac{9\pi k_BT}{2m}\left[\int_0^\infty dt\langle \vec{R}(t)\cdot\vec{R}(0)\rangle\right]^{-1}\right]^{1/3} \tag{II.81}$$

if the integral exists. The integral exists if the local friction parameter $\beta \equiv \hat{\beta}(0)$ (II.53a) exists. In fact, (II.74c) shows that

$$\int_0^\infty \langle \vec{R}(t)\cdot\vec{R}(0)\rangle dt \equiv \frac{3k_BT\beta}{m\omega_c^4} \tag{II.82}$$

and hence

$$\omega_{D_1} = \left(\frac{3\omega_c^4\pi}{2\beta}\right)^{1/3} \tag{II.83}$$

Finally, we note that heatbath response function may be represented by an exact equation of motion identical in form to (II.55a) for $\chi_m(t)$; that is,

$$\ddot{\theta}(t) = -\omega_{e_1}^2\theta(t) + \int_0^t \Theta_2(t-\tau)\theta(\tau)\,d\tau \tag{II.84a}$$

Equation (II.84) *defines* $\Theta_2(t)$. It is a *damping kernel for the heatbath.*

Equivalently, it is the response function (unnormalized) of the *heatbath's heatbath*. The Laplace space form of (II.84a) is

$$\hat{\theta}(z)=\left[z^2+\omega_{e_1}^2-\hat{\Theta}_2(z)\right] \qquad \text{(II.84b)}$$

We may verify that (II.84) describes the heatbath response in both Einstein (ω_{e_1}) and adiabatic (Ω_1) limits by following the analogous analysis for $\chi_m(t)$ [(II.55) to (II.60)]. Thus (II.84) provide a fully consistent effective equation of motion for the heatbath response.

The results of this section may be summarized by the following statement. *The heatbath is an m-oscillator-characterized by the normalized response function* $\theta(t)$. The quantitative form of this statement is the effective equation of motion (II.84a) with the initial conditions (II.66).

Notice that our concept of an m-oscillator has become much more abstract. We began with the analysis of a concrete system, a macroscopic harmonic oscillator suspended in a liquid. Our generalization of the macroscopic (M) theory to microscopic (m) oscillators has lead us to the conclusion that the heatbath coordinate $\vec{R}(t)$ is an m-oscillator. The heatbath coordinate is not, however, the coordinate of a physical particle. It is not even the coordinate of a conventional collective variable or fictitious particle (e.g., a normal mode). [An important exception is the case of perfectly harmonic heatbaths; for this case, $\vec{R}(t)$ is a linear combination of the coordinates of the real particles in the heatbath or, equivalently, a linear combination of the normal-mode coordinates of the heatbath.] The heatbath coordinate $\vec{R}(t)$ in the general case is a complex function of the coordinates of all particles in the many-body system, *including the primary system coordinate*. The abstract character of the heatbath coordinate $\vec{R}(t)$ is evident from, say, (II.75a). This equation *formally* resembles the classical equipartition law for kinetic energy if one interprets $\dot{\vec{R}}(t)$ as the "velocity" associated with the heatbath "position" $\vec{R}(t)$. The formal character of (II.75a) is evident, however, since the "heatbath mass" m appearing in the equation is the primary system m-oscillator mass.

Our discussion thus leads us to the following abstract definition of an m-oscillator: An m-oscillator is a variable $\vec{R}(t)$ whose response function $\theta(t)$ satisfies an equation of motion identical in form to (II.84) with the initial conditions given in (II.66). The relationship between the *statistical* time evolution of $\vec{R}(t)$ and the response function $\theta(t)$ is given by a fluctuation–dissipation theorem of the form of (II.74).

This abstract definition of generalized Brownian motion, which arises naturally from our correlation function analysis of m-oscillator motion, explains why Mori[6] was able, through an explicit microscopic analysis, based on projection operator[10] methods, to obtain effective equations of motion for *arbitrary dynamical variables* of generalized Langevin form.

Mori's important result seems natural, rather than surprising, when the underlying structure of the effective equation-of-motion theory is exposed by a correlation function analysis of the type presented by us here or by Kubo[7] elsewhere.

F. Hierarchy of Heatbaths

Given the discussion in Section II.E, it is now straightforward to develop the general structure of the effective equation-of-motion theory for correlation functions. We have found that $\theta(t)$, the heatbath response function, is an m-oscillator (more correctly its associated dynamical variable is the m-oscillator) and thus shares the formal properties of $\chi_m(t)$, the primary system response function. Included is the property that the damping kernel $\Theta_2(t)$ is also an m-oscillator (i.e., the heatbath's heatbath is an m-oscillator). The damping kernel of the heatbath's heatbath is, in turn, an m-oscillator, and so on. Thus the formal structure of the effective equation-of-motion theory is now apparent. The theory is an infinite hierarchy of coupled effective equations of motion for a sequence of (normalized) response functions $\theta_n(t)$. These effective equations of motion are identical in *form* to (II.84). The initial conditions on each response function are identical in form to those specified in (II.66). The theory is closed exactly by specification of $\theta_0(t) \equiv \chi_m(t)$. The practical, as opposed to formal, value of the theory derives, in part, from the fact that even if only limited information about $\chi_m(t)$ is known (e.g., if only the first few positive moments of $\rho(\omega)$ are known), the theory may be closed approximately by making a plausible guess for one of the $\theta_n(t)$.

Our plan in the next three sections is as follows. In this section we will first explicitly write down the coupled hierarchy of effective equations of motion. We next develop a continued-fraction representation for the Laplace domain response function $\hat{\theta}_n(z)$. Finally, we briefly discuss the moment analysis of response functions. The continued-fraction representation is in essence a graphic display of the structure of the coupled hierarchy, and moment analysis is one of its useful practical applications. We examine in Section II.G what is, in essence, the time-domain analog of the continued-fraction representation. This analog is an *exact* representation of many-body dynamics in terms of the dynamics of a fictitious nearest-neighbor collinear harmonic chain. This *exact* effective harmonic chain representation will suggest *approximate* heatbath models[1b] (see Section II.A) which are of key importance in application of the effective equation-of-motion theory to many-body problems in chemical dynamics. Moreover, the effective harmonic chain representation will lead us naturally into the concept of exact "normal modes" for general (i.e., nonharmonic) many-body systems. This concept is explored in Section II.H.

We begin with the coupled hierarchy of effective equations of motion.

1. *Heatbath Effective Equations of Motion*

The general structure of the effective equation-of-motion theory as just discussed is based on the idea that one may prove by iteration that the many-body system is described by an infinite hierarchy of coupled heatbaths. Each heatbath is an *m*-oscillator and thus shares the formal properties of the *zeroth heatbath*, which is the primary system. Thus one has immediately that $\dot{\theta}_n(t)$, the *normalized* response function of the *n*th heatbath, is even in time [cf. (II.25)]

$$\dot{\theta}_n(t)=\dot{\theta}_n(-t) \tag{II.85}$$

and satisfies the initial conditions [cf. (II.24)]

$$\begin{aligned}\theta_n(0)&=0\\ \dot{\theta}_n(0)&=1\end{aligned} \tag{II.86}$$

Equations (II.85) and (II.86) immediately demonstrate that a spectral density $\sigma_n(\omega)$ is associated with $\theta_n(t)$ and is defined by [cf. (II.27)]

$$\sigma_n(\omega)=\frac{2}{\pi}\int_0^\infty \cos\omega t\dot{\theta}_n(t)\,dt \tag{II.87a}$$

The inverse relationship is

$$\dot{\theta}_n(t)=\int_0^\infty \cos\omega t\sigma_n(\omega)\,d\omega \tag{II.87b}$$

The spectral density is nonnegative and unit-normalized [cf. (II.28) and (II.29)]:

$$\sigma_n(\omega)\geq 0 \tag{II.88}$$

and

$$\int_0^\infty \sigma_n(\omega)\,d\omega=1 \tag{II.89}$$

Thus $\sigma_n(\omega)$ may be consistently interpreted as a probability density. The odd moments of $\sigma_n(\omega)$ vanish and the even moments are positive definite. In particular, the second moment of $\sigma_n(\omega)$, $\omega_{e_n}^2$, given by [cf. (II.30)]

$$\omega_{e_n}^2\equiv\int_0^\infty \omega^2\sigma_n(\omega)\,d\omega \tag{II.90}$$

is positive definite and hence ω_{e_n} may be interpreted as an Einstein frequency which governs the short-time behavior of $\theta_n(t)$. Note that [cf. (II.32b)]

$$\omega_{e_n}^2 = -\ddot{\theta}_n(0) \tag{II.91}$$

The response function $\theta_n(t)$ is related to a coordinate $\vec{R}_n(t)$ by the nth fluctuation–dissipation theorem. This theorem may be written in the following three equivalent forms [cf. (II.74)]:

$$\dot{\theta}_n(t) = \frac{m}{3k_BT}\langle \dot{\vec{R}}_n(t)\cdot\dot{\vec{R}}_n(0)\rangle \tag{II.92a}$$

$$\theta_n(t) = \frac{m}{3k_BT}\langle \vec{R}_n(t)\cdot\dot{\vec{R}}_n(0)\rangle = -\frac{m}{3k_BT}\langle \dot{\vec{R}}_n(t)\cdot\vec{R}_n(0)\rangle \tag{II.92b}$$

$$\beta_n(t) \equiv \omega_{c_n}^4 \int_t^\infty \theta_n(\tau)\,d\tau = \frac{m}{3k_BT}\omega_{c_n}^4 \langle \vec{R}_n(t)\cdot\vec{R}_n(0)\rangle \tag{II.93}$$

In (II.93) $\omega_{c_n}^2$ is an effective force constant coupling the nth and $(n-1)$th heatbath. It is given in terms of the spectral density of the $(n-1)$th heatbath by [cf. (II.63)]

$$\omega_{c_n}^2 = \int_0^\infty \sigma_{n-1}(\omega)\left[\omega^2 - \omega_{e_{n-1}}^2\right]^2 d\omega \tag{II.94}$$

Finally, the adiabatic frequency of the nth heatbath Ω_n is given by [cf. (II.36)]

$$\Omega_n^{-2} = \hat{\theta}_n(0) = \frac{3k_BT}{m\langle R_n^2\rangle} \tag{II.95}$$

The nth heatbath is directly coupled to the $(n-1)$th and $(n+1)$th heatbath by the following pair of effective equations of motion [cf. (II.53a) and (II.64)]

$$\ddot{\theta}_{n-1}(t) = -\omega_{e_{n-1}}^2\theta_{n-1}(t) + \omega_{c_n}^4 \int_0^t \theta_n(t-\tau)\theta_{n-1}(\tau)\,d\tau \tag{II.96a}$$

and

$$\ddot{\theta}_n(t) = -\omega_{e_n}^2\theta_n(t) + \omega_{c_{n+1}}^4 \int_0^t \theta_{n+1}(t-\tau)\theta_n(\tau)\,d\tau \tag{II.96b}$$

Note that our present definitions are consistent with those of Sections II.C to II.E if $\chi_m(t) \equiv \theta_0(t)$, $\theta(t) \equiv \theta_1(t)$, $\omega_e \equiv \omega_{e_0}$, $\omega_{c_1} \equiv \omega_c$, and $\Omega_0 \equiv \Omega$.

The formal content of the theory is contained in (II.85), (II.86), and (II.96). This content is summarized by the statement: the transformation $\theta_n(t) \to \theta_{n+1}(t)$ defined by (II.96b) converts an m-oscillator into an m-oscillator.

Finally, we emphasize that although the analysis of *physical* microscopic damped harmonic motion (e.g., the dynamics of a chemisorbed species or the motion of a diatomic molecule dissolved in a liquid) motivated the development of the effective equation of motion theory, the theory is now of abstract character and is defined independently of any physical interpretation of the m-oscillator. The abstract character is an important reason for the extremely broad regime of applicability of the effective equation-of-motion method.

We develop the continued-fraction representation of $\hat{\theta}_n(z)$.

2. *Continued-Fraction Representation for* $\hat{\theta}_n(z)$

We begin development of the continued-fraction representation by writing down the Laplace transform of (II.96b). This is

$$\hat{\theta}_n(z) = \left[z^2 + \omega_{e_n}^2 - \omega_{c_{n+1}}^4 \hat{\theta}_{n+1}(z) \right]^{-1} \tag{II.97}$$

Similarly,

$$\hat{\theta}_{n+1}(z) = \left[z^2 + \omega_{e_{n+1}}^2 - \omega_{c_{n+2}}^4 \hat{\theta}_{n+2}(z) \right]^{-1} \tag{II.98}$$

Using (II.98) in (II.97) to eliminate $\hat{\theta}_{n+1}(z)$ gives

$$\hat{\theta}_n(z) = \left[z^2 + \omega_{e_n}^2 - \omega_{c_{n+1}}^4 \left[z^2 + \omega_{e_{n+1}}^2 - \omega_{c_{n+2}}^4 \hat{\theta}_{n+2}(z) \right]^{-1} \right]^{-1} \tag{II.99}$$

We may similarly eliminate $\hat{\theta}_{n+2}(z)$ from (II.99). Continuing this procedure then gives the continued-fraction representation

$$\hat{\theta}_n(z) = \cfrac{1}{z^2 + \omega_{e_n}^2 - \cfrac{\omega_{c_{n+1}}^4}{z^2 + \omega_{e_{n+1}}^2 - \cfrac{\omega_{c_{n+2}}^4}{z^2 + \omega_{e_{n+2}}^2 + \cdots}}} \tag{II.100a}$$

An important special case of (II.100) is the case $n=0$. This gives the continued-fraction representation of $\hat{\chi}_m(z)$ as [recall, $\omega_{e_0}^2 \equiv \omega_e^2$ and $\omega_{c_1}^2 \equiv \omega_c^2$]

$$\hat{\chi}_m(z) = \cfrac{1}{z^2+\omega_e^2 - \cfrac{\omega_c^4}{z^2+\omega_{e_1}^2 - \cfrac{\omega_{c_2}^4}{z^2+\omega_{e_2}^2+\cdots}}} \tag{II.100b}$$

The continued-fraction representations in (II.100) are equivalent but not identical to those presented by Mori.[6b] A special case of (II.100) has been derived elsewhere.[1a]

Notice that the continued-fraction representation of, say, $\hat{\chi}_m(z)$ involves the Einstein frequencies $\omega_e \equiv \omega_{e_0}$, ω_{e_1}, and so on, and the coupling force constants $\omega_c^2 \equiv \omega_{c_1}^2$, $\omega_{c_2}^2$, and so on. These frequencies and force constants may all be expressed in terms of moments of the primary system spectral density $\rho(\omega)$ (see below). These moments, in turn, are related to the odd time derivatives of $\chi_m(t)$, $\dddot{\chi}_m(0)$, $d^5\chi_m(0)/dt^5$, and so on [cf. (II.32b)]; that is, the moments are determined by the short-time expansion of $\chi_m(t)$ in powers of t^2. If one, however, truncates the continued-fraction representation of $\hat{\chi}_m(z)$ at the nth level, by setting $\hat{\theta}_n(z)=0$, computes those moments of $\rho(\omega)$ sufficient to evaluate the truncated representation explicitly, and then inverts the Laplace transform to obtain $\chi_m^{(n)}(t)$, one finds that $\chi_m^{(n)}(t)$ may be written as [cf. (II.123) below]

$$\chi_m^{(n)}(t) = \sum_{\lambda=1}^{n} U_{0\lambda}^2 \cos\omega_\lambda t \tag{II.101}$$

The quantities ω_λ are the roots of the equation

$$\left[\chi_m^{(n)}(i\omega)\right]^{-1} = 0 \tag{II.102}$$

and are approximations to the "normal-mode" frequencies of the many-body system. The $U_{0\lambda}$, as we will see in Section II.G, are elements of a unitary matrix which transforms between an "atom" and a "mode" representation of the exact many-body dynamics of the full system (see Section II.H). Notice that (II.101) involves information about $\chi_m(t)$ obtained from its power series expansion in t^2, yet the representation of (II.101) is formally infinite-order in t^2 [since $\cos\omega_\lambda t = \sum_{n=0}^{\infty} (-1)^n (t^{2n}/(2n)!)$]. Thus the continued-fraction representation of response or correlation functions allows one to utilize short-time information about the correlation function to obtain a representation of its behavior valid at considerably longer times (see Section II.H); that is, it provides an automatic and sometimes optimal[23a] algorithm for rearranging and resuming power series in t^2. This algorithm has a simple physical interpretation (see Section II.H). Continued fraction methods have been used in a manner similar to that just described to calculate tcf for harmonic lattices[23] and for *anharmonic* oscillators subject to simple frictional damping.[24]

The ability of the continued-fraction representation to automatically resume complex power series and give usable results for time correlation functions is one (rather small) illustration of the great power of the effective equation-of-motion method that underlies it.

In order to employ the continued-fraction representation to, say, calculate tcf, one must express the Einstein and coupling force constants $\omega_{e_n}^2$ and $\omega_{c_n}^2$ in terms of moments of $\rho(\omega)$. We now briefly indicate how this is done.

3. *The Moment Analysis*

Comprehensive discussions of the moment analysis and its application for a variety of physical problems are available in the literature.[23] For this reason, and also because the moment analysis is of computational rather than conceptual interest in many-body chemical dynamics, our discussion here will be brief. We will sketch

1. How the quantities $\omega_{e_n}^2$ and $\omega_{c_n}^2$ may be related to moments of $\rho(\omega)$.
2. How moments of $\rho(\omega)$ may be expressed as *equilibrium* properties of the many-body system, which may be computed from its partition function.

The reduction of the problem of computing moments to a problem in *equilibrium* statistical mechanics represents important progress, since as we have just seen, (II.101), *dynamic* properties may be computed from the moments. Even equilibrium properties, however, cannot be computed exactly for all but the simplest many-body systems (e.g., harmonic lattices[25]), and this limits the method of moments as a practical computational tool. In favorable cases, however, the moments may be determined from experiment.[26]

To understand how, say, $\omega_{e_n}^2$ is related to moments of $\rho(\omega)$ [point (1) above], recall that

$$\omega_{e_n}^2 = -\ddot{\theta}_n(0)$$

which is (II.91). The quantity $\ddot{\theta}_n(0)$, however, may be related to *initial* time derivatives of $\theta_{n-1}(t)$ via (II.96a). These initial time derivatives may, in turn, be related to initial time derivatives of $\theta_{n-2}(t)$, and so on. Finally, one obtains a relationship between $\ddot{\theta}_n(0)$ and the initial time derivatives of $\chi_m(t)$ or, equivalently, the moments of $\rho(\omega)$. By this recursion procedure, $\omega_{e_n}^2$ for arbitrary n may be expressed in terms of $\rho(\omega)$; a similar argument holds for $\omega_{c_n}^2$. A simple example of the type of computation just described is given by (II.73b), which relates $\omega_{e_1}^2$ to the second, fourth, and sixth moments of $\rho(\omega)$.

Once the relationship between, say, a particular $\omega_{e_n}^2$ and moments of $\rho(\omega)$ has been worked out, one must next express the moments as equilibrium properties of the many-body system. A simple example is provided by the second moment of $\rho(\omega)$:

$$\omega_e^2 = -\ddot{\chi}_m(0) = \int_0^\infty \omega^2 \rho(\omega)\, d\omega = -\frac{m}{3k_B T}\langle \ddot{r}(0)\cdot\dot{r}(0)\rangle$$

$$= +\frac{m}{3k_B T}\langle \dot{r}^2\rangle = \frac{1}{3mk_B T}\langle F^2\rangle \qquad \text{(II.103)}$$

where $\langle F^2\rangle$ is the *equilibrium* mean-square force acting on the primary system m-oscillator. To obtain (II.103), we have used (II.22a), (II.23b), and (II.32b). Higher-order moments of $\rho(\omega) \sim (d^5/dt^5)\chi(0)$, and so on, may be computed as equilibrium properties by extensions of the procedure used in (II.103).

The last step in the moment analysis evaluating the equilibrium properties explicitly is the only step that is not, in general, straightforward. For harmonic systems, however, special techniques are available[25] which make the calculation of moments relatively straightforward.

Summarizing, we have shown in this section that the essence of the formal structure of the effective equation of motion theory is the fact that the effective equation of motion transforms m-oscillators into m-oscillators. Second, we have discussed one practical application of this structure, namely, the computation of dynamic properties (i.e., tcf) in terms of equilibrium properties, the moments of the spectral density of the primary system $\rho(\omega)$. The bridge linking the dynamic and equilibrium properties is the continued fraction representation, (II.100).

G. Many-Body Dynamics in Harmonic Chain Form and Heatbath Modeling

Four closely related topics are developed in this section: (1) the nearest-neighbor fictitious harmonic chain representation of *exact* many-body dynamics, (2) the concept of equivalent heatbaths, (3) the "physical" interpretation of the harmonic chain representation, and (4) the methodology of heatbath modeling.

The heatbath models[1b,17] we develop in this section are of particular importance. They are the basis of the *practical* applicability of the effective equation of motion method to many-body problems in chemical dynamics.

We first develop the harmonic chain representation of the many-body problem.

1. Nearest-Neighbor Harmonic Chain Representation

We show here that the effective equation of motion for the primary system m-oscillator may be rigorously recast as a set of $N+1$ coupled equations of mition, (II.109), for a fictitious nearest-neighbor collinear harmonic chain composed of $N+1$ "atoms" numbered $0, 1, \ldots, N$. The trajectory of atom 0 for the initial conditions of (II.111) is the response function $\chi_m(t)$; thus solution of the harmonic eigenvalue problem to determine the normal modes of chain is sufficient to determine $\chi_m(t)$. The number of chain atoms $N+1$ must be at least 2, but is otherwise arbitrary. Thus the chain representation of many-body dynamics is really a class of representations for $N=1,2,3,\ldots$.

We first develop the nearest-neighbor chain representation for the case $N=1$ and then extend the derivation to arbitrary N. We begin with (II.55a), rewritten as

$$\ddot{\chi}_m(t) = -\omega_{e_0}^2 \chi_m(t) + \omega_{c_1}^4 \int_0^t \theta_1(t-\tau)\chi_m(\tau)\,d\tau \tag{II.104}$$

[Recall that $\omega_{e_0}^2 \equiv \omega_e^2$, $\omega_{c_1}^2 \equiv \omega_c^2$, $\theta_1(t) \equiv \theta(t)$, and $\Theta(t) = \omega_{c_1}^4 \theta(t)$.] We next introduce the "atom" coordinates $r_0(t)$ and $r_1(t)$ defined by

$$r_0(t) = \chi_m(t) \tag{II.105a}$$

and

$$r_1(t) = \omega_{c_1}^2 \int_0^t \theta_1(t-\tau)\chi_m(\tau)\,d\tau \tag{II.105b}$$

Using (II.86), one verifies that the "atom" coordinates satisfy the initial conditions

$$r_0(0) = 0 \tag{II.106a}$$
$$\dot{r}_0(0) = 1$$

and

$$r_1(0)=\dot{r}_1(0)=0 \tag{II.106b}$$

The Laplace transform of (II.105b) gives

$$\hat{r}_1(z)=\omega_{c_1}^2\hat{\theta}_1(z)\hat{\chi}_m(z)=\omega_{c_1}^2\hat{\theta}_1(z)\hat{r}_0(z)$$

or, equivalently [use (II.84b)],

$$\left[z^2+\omega_{e_1}^2-\omega_{c_2}^4\hat{\theta}_2(z)\right]\hat{r}_1(z)=\omega_{c_1}^2\hat{r}_0(z) \tag{II.107}$$

Using (II.105) in (II.104) gives the following result for the equation of motion of $r_0(t)$:

$$\ddot{r}_1(t)=-\omega_{e_0}^2 r_0(t)+\omega_{c_1}^2 r_1(t) \tag{II.108a}$$

Laplace inversion of (II.107) gives with the use of (II.106b) the following result for the effective equation of motion for $r_1(t)$:

$$\begin{aligned}\ddot{r}_0(t)=&-\omega_{e_1}^2 r_1(t)+\omega_{c_1}^2 r_0(t)\\&+\omega_{c_2}^4\int_0^t\theta_2(t-\tau)r_1(\tau)\,d\tau\end{aligned} \tag{II.108b}$$

Equations (II.108) provide an *exact* representation of the primary system many-body dynamics in terms of a fictitious two-atom nearest-neighbor coupled harmonic chain. The self-frequencies for atoms 0 and 1 are, respectively, ω_{e_0} and ω_{e_1} and $\omega_{c_1}^2$ acts as a nearest-neighbor force constant which couples the atoms. Notice that atom 2 is coupled to a heatbath characterized by the influence $\theta_2(t)$, and thus is subject to direct generalized damping. Atom 1, however, is not directly damped and experiences energy dissipation only because it is coupled to atom 2 by $\omega_{c_1}^2$.

Equations (II.108) amount to replacing the true many-body heatbath by a single m-oscillator with coordinate r_1. This generalized oscillator is coupled linearly to the primary system by the force constant $\omega_{c_1}^2$. The fact that the true many-body heatbath, perhaps a liquid, may be rigorously replaced by a single generalized harmonic oscillator illustrates a simple but valuable principle of the effective equation-of-motion method: since the effective equation of motion for an m-oscillator involves its heatbath only through the heatbath influence $\theta(t)$, *the true many-body heatbath may be rigorously replaced by an* equivalent heatbath *which duplicates its influence but which is otherwise arbitrary*. The statement above is, of course, only true if one is concerned solely with primary system dynamics. Useful equivalent heatbaths usually bear little resemblance to the many-body system they simulate, so their dynamics is of no independent interest.

The practical value of the equivalent heatbath method is that equivalent heatbaths are often much more readily modeled than real heatbaths. For example, a simple heatbath model is readily derived by dropping the damping term in (II.108b). This heatbath model is an isolated harmonic oscillator of frequency ω_{e_1}; it is equivalent to the following continued-fraction representation of $\hat{\chi}_m(z)$ [cf. (II.100b)]:

$$\hat{\chi}_m^{(2)}(z)=\cfrac{1}{z^2+\omega_{e_0}^2-\cfrac{\omega_{c_1}^4}{z^2+\omega_{e_1}^2}}$$

The equation above may be readily derived from the Laplace transform of (II.108). This simplest heatbath model yields a two-"mode" approximation for $\chi_m(t)$; cf. (II.101).

Equations (II.108) provide the nearest-neighbor $(N+1)$-atom chain representation of many-body dynamics for the case $N=1$. The general result may be derived by extension of the preceding argument. One finds the following equation of motion:

$$\ddot{r}_0(t)=-\omega_{e_0}^2 r_0(t)+\omega_{c_1}^2 r_1(t) \tag{II.109a}$$

and

$$\begin{aligned}
\ddot{r}_1(t)&=-\omega_{e_1}^2 r_1(t)+\omega_{c_1}^2 r_0(t)+\omega_{c_2}^2 r_2(t)\\
\ddot{r}_2(t)&=-\omega_{e_2}^2 r_2(t)+\omega_{c_2}^2 r_1(t)+\omega_{c_3}^2 r_3(t)\\
&\vdots\\
\ddot{r}_N(t)&=-\omega_{e_N}^2 r_N(t)+\omega_{c_N}^2 r_{N-1}(t)+\omega_{c_{N+1}}^4\int_0^t \theta_{N+1}(t-\tau)r_N(\tau)\,d\tau
\end{aligned} \tag{II.109b}$$

where the coordinate $r_0(t)=\chi_m(t)$ and the remaining coordinates are given by [cf. (II.105b)]

$$r_n(t)=\omega_{c_n}^2\int_0^t \theta_n(t-\tau)r_{n-1}(\tau)d\tau \qquad n\geq 1 \tag{II.110}$$

The initial conditions are [cf. (II.106)]

$$\begin{aligned} r_n(0) &= 0 \\ \dot{r}_0(0) &= 1 \\ \dot{r}_n(0) &= 0 \qquad n \geq 1 \end{aligned} \tag{II.111}$$

The equivalent heatbath corresponding to (II.109) is the dynamical system defined by (II.109b) with $r_0(t)=0$. The simplest heatbath model derivable from this equivalent heatbath is found by setting $\theta_{N+1}(t)=0$ in (II.109b). This approximation models the heatbath as an isolated (no-damping) N-atom harmonic chain.

Our ability to *formally* reduce the complexities of many-body dynamics to the trivialities of nearest-neighbor harmonic chain motion depends on the recursive structure (m-oscillator$\rightarrow$$m$-oscillator) of the effective equation-of-motion method.

The practical value of this formal reduction will become apparent in Section III. There we show that the equivalent chain representation of the full many-body problem allows one to rigorously generalize the classical trajectory method of gas-phase chemical dynamics[4,5] to many-body or condensed-phase chemical dynamics. This generalization is not merely formal but is rather a practical computational scheme whose value is now solidly documented.[1,18a,b]

2. *"Physical" Interpretation of Effective Chain Dynamics*

The effective chain representation of many-body dynamics leads to a quasiphysical interpretation of energy dissipation in the primary system. This interpretation yields insight into the role of many-body effects (i.e., breakdown of the Einstein model) in condensed-phase processes (see Section III). The "physical" interpretation is based on the fact that $\chi_m(t)$ is the trajectory of chain atom 0 for the initial conditions of (II.111).

These initial conditions may be interpreted in the following manner. Suppose that the chain is initially at temperature $T=0$ K, but that it obeys the laws of *classical* statistical mechanics [i.e., it is initially quiescent and thus $r_n(0)=\dot{r}_n(0)=0$]. (Real chains at $T=0$ K, of course, have zero-point oscillations and hence do not satisfy classical statistics. The artificial case of *classical* $T=0$ K many-body systems arises because the present formulation is based on the approximation of classical mechanics.) Next, suppose that at $t=0$ atom 0 is subjected to a force pulse which instantaneously increments its velocity from $\dot{r}_0(0)=0$ to $\dot{r}_0(0)=1$. Immediately after the pulse has acted, the chain's phase-space coordinate is described by the initial conditions of (II.111). We now follow the trajectory of atom 0. The initial kinetic energy of atom 0, equal to $\frac{1}{2}m\dot{r}^2(0)=\frac{1}{2}m$, begins to flow out of atom 0 and into atom 1. This initial energy flow occurs roughly on a time scale $\tau_{0\rightarrow 1}$, where

$$\tau_{0\rightarrow 1} \sim \left[\frac{\omega_{e_1}^2 + \omega_{e_2}^2}{2} + \omega_{c_1}^2 \right]^{-1/2}$$

if $\omega_{e_1} \approx \omega_{e_2}$ and $\omega_{c_1} \neq 0$. This energy flow leads to damping of the motion of atom 0. The motion of atom 1, in turn, is damped since it feeds energy to atom 2 and since it also reflects some energy back to atom 0. This process is continued down the $(N+1)$-atom chain until the energy finally reaches atom N, where it is dissipated into the $(N+1)$th heatbath. Thus the trajectory $r_0(t) = \chi_m(t)$ damps out in time except for roughly periodic "recharges" due to the reflection effect (e.g., $0 \rightarrow 1 \rightarrow 0$). On this damped motion of atom 0 must be superimposed on an overall periodic motion governed by the Einstein frequency ω_{e_0}. Thus we expect $\chi_m(t)$, in general, to display damped oscillatory behavior (an exception is the case of overdamped motion for which the damping is complete before atom 0 can execute a single cycle).

The detailed behavior of the response function, of course, depends on the values of ω_{e_n} and ω_{c_n}. We will present particular examples in Section III. We now turn to the problem of constructing heatbath models which are useful in many-body chemical dynamics.

3. *Heatbath Models*

The isolated chain heatbath models $[\omega_{c_{N+1}}^4 \theta_{N+1}(t) \rightarrow 0]$ discussed above are *not* useful in chemical dynamics applications of effective equation-of-motion methods. The reason is that these are finite systems and are thus subject to Poincaré recurrences. In the "physical" language of Section II.G.2, a compressional pulse initiated at atom 0 in an isolated chain would return to atom 0 after reflection off the spurious chain boundary at atom N. The Poincaré recurrence time τ_R depends on the detailed values of the ω_{e_n} and ω_{c_n}; for the case $\omega_{e_n} = \omega_{c_n} = \omega_0$, all n, $\tau_R \sim 4\pi N \omega_0^{-1}$.

Recurrences are disastrous for chemical dynamics applications of the theory.[16] Recurrences, for example, may "stimulate" unphysical desorptions in computer studies of the dynamics of molecules adsorbed on solids. Such desorptions would make the results of a computer study of, say, heterogeneous reaction dynamics meaningless.

Thus heatbath models which are useful in chemical dynamics work (Fig. 2) must satisfy the following criteria. They must:

1. Be free from recurrences.
2. Be equivalent to the true many-body heatbath to some specified accuracy.
3. Involve a relatively small number of fictitious atoms N.

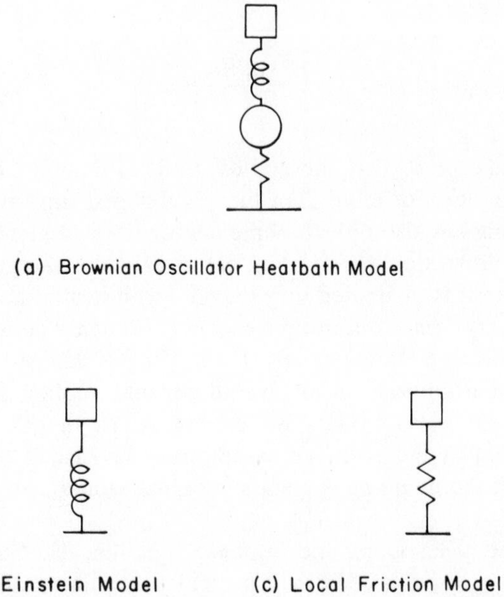

Fig. 2. Simplest models for many-body dynamics. (*a*) Brownian oscillator heatbath model. Primary system, the cubic atom, is harmonically bound with spring constant $\omega_{e_0}^2$ and coupled by a force constant $\omega_{c_1}^2$ to a simple Brownian oscillator, the atom bound by a zigzag spring, which models the true many-body heatbath. Friction coefficient β_2 and frequency Ω_1 [see (II.105b)] of the Brownian oscillator are chosen to optimally mimic the response of the true many-body heatbath. This gives the simplest model which is correct in the short-time limit and which lacks Poincaré recurrences. (*b*) Einstein model. Full many-body system is replaced by an isolated harmonic oscillator with frequency ω_{e_0}. This model reproduces the true response of the many-body system only at very short times and suffer recurrences for $t \gtrsim \omega_{e_0}^{-1}$. (*c*) Friction model. Full many-body system is replaced by a single Brownian oscillator with a time local friction coefficient $\beta = \hat{\beta}(0)$ and restoring frequency = adiabatic frequency Ω. The friction model is always incorrect on short time scales and probably also fails for long time scales for most molecular systems.

Point (3) follows because once one couples the chemical part of the problem (Section III) to the many-body system (via a nonharmonic potential) the effective equations of motion are nonlinear and must be solved numerically. Thus if N is large, the computer work becomes very costly.

The isolated chain heatbath models do not satisfy point (1) for times $t \gtrsim \tau_R$. The recurrence time τ_R, of course, may always be increased by increasing N, but eventually the dimensionality of the problem, point (3), [and also the difficulty of computing high-order moments of $\rho(\omega)$, Section II.F.3 make this remedy impractical.

The fact that the effective equations of motion involve nonlinear forces in chemical dynamics applications leads to a fourth criterion.

4. The effective equations of motion must contain no retarded kernel (i.e., trajectory-history-dependent) terms.

This criterion follows because if such terms appear, the effective equations of motion are integrodifferential equations and are nearly intractable on a computer. (The reason is that each history-dependent term may be "unfolded" into an infinite effective harmonic chain, and thus history-dependent terms, if treated exactly, are equivalent to an infinite set of coupled differential equations.)

Criteria 1 to 4 may be satisfied by replacing the history-dependent integral term in the last member of (II.109b) by a simple friction term. We may accomplish this by rewriting the history-dependent term as

$$\omega_{c_{N+1}}^4 \int_0^t \theta_{N+1}(t-\tau) r_N(\tau)\, d\tau = \beta_{N+1}(0) r_N(t) - \int_0^t \beta_{N+1}(t-\tau) \dot{r}_N(\tau)\, d\tau \qquad \text{(II.112)}$$

To obtain (II.112), we have used the definition of $\beta_n(t)$, (II.93), and the initial value $r_n(0)=0$, (II.111).

The quantity $\beta_{N+1}(0)$ is, however, related to the Einstein and adiabatic frequencies of the Nth heatbath by [cf. (II.52)]

$$\beta_{N+1}(0) = \omega_{e_N}^2 - \Omega_N^2 \qquad \text{(II.113)}$$

Finally, we make the local friction approximation

$$\beta_{N+1}(t) = 2\beta_{N+1}\delta(t) \qquad \text{(II.114)}$$

where [cf. (II.53)]

$$\beta_{N+1} \equiv \int_0^\infty \beta_{N+1}(\tau)\, d\tau = \frac{d}{dz}\hat{\theta}_N^{-1}(z)\bigg|_{z=0} = \Omega_N^4 \int_0^\infty \tau\theta_N(\tau)\, d\tau. \qquad \text{(II.115)}$$

Using (II.114) to (II.115) in (II.109) yields the following *model* effective equations of motion:

$$\ddot{r}_0(t) = -\omega_{e_0}^2 r_0(t) + \omega_{c_1}^2 r_1(t) \qquad \text{(II.116a)}$$

and

$$\ddot{r}_1(t) = -\omega_{e_1}^2 r_1(t) + \omega_{c_1}^2 r_0(t) + \omega_{c_2}^2 r_2(t)$$

$$\vdots$$

$$\ddot{r}_N(t) = -\Omega_N^2 r_N(t) + \omega_{c_N}^2 r_{N-1}(t) - \beta_{N+1}\dot{r}_N(t) \qquad \text{(II.116b)}$$

Equations (II.116) amount to replacing the true many-body heatbath by an N-atom nearest-neighbor chain whose dynamics is governed by (II.116b) with $r_0(t) \to 0$. This model heatbath is then linearly coupled to the primary system (atom 0) via the force constant $\omega_{c_1}^2$.

The model heatbath has several very desirable features. First, it is free of recurrences due to the frictional damping of atom N. While the damping is not treated in a precise manner because of the local friction approximation, *the error is of negligible consequence for the motion of atom 0* [i.e., $\chi_m(t)$] *if N is sufficiently large*. This is because for large N, $\chi_m(t)$ is negligible by the time atom N is stimulated. Thus the damping is simply a device to smoothly cutoff $\chi_m(t)$ or equivalently $\Theta(t)$ at large times. This cutoff is essential, however, since otherwise recurrences, and their consequences mentioned above, would occur (Fig. 3).

From our discussion it is clear that the friction parameter β_{N+1} need not be calculated rigorously from (II.115); any β_{N+1} which is large enough to prevent a nonnegligible energy flow back to atom 0 suffices. In fact, in practical applications, we wish to keep N as small as possible (see point (3) above), and so β_{N+1}, and also Ω_{N+1}, are typically chosen so that the response of the model heatbath, (II.116b), optimally fits the response of the true many-body heatbath $\Theta(t) = \omega_{c_1}^4 \theta_1(t) = \omega_c^4 \theta(t)$ at short times. [Recall that

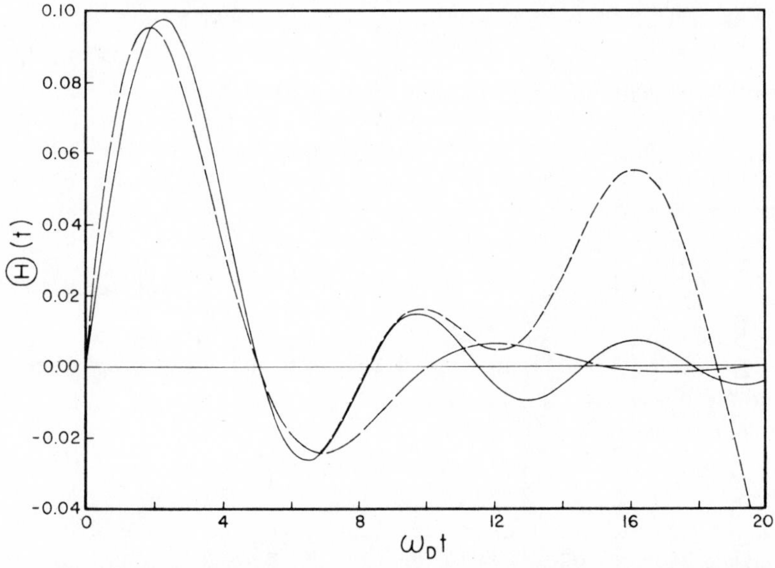

Fig. 3. Damping kernel $\Theta(t)$ for the Debye model. $\Theta(t)$ is plotted vs. $\omega_D t$: (—), exact Debye result; (– – –), four-atom *undamped* harmonic chain approximant displaying a recurrence; (– –), $N=1$ *damped* harmonic chain approximant or Brownian oscillator heatbath model of Fig. 2*a*.

the short-time behavior of $\Theta(t)$ may be accurately calculated by combining the continued fraction and moment techniques of Section II.F.] Error is of course incurred at longer times, say $t > \tau_E$, but this error is of no consequence if $\Theta(\tau_E) \ll \Theta(0)$.

The case $N = 1$ leads to the simplest model heatbath. This model heatbath amounts the replacing the true heatbath by a *single* damped harmonic oscillator; that is, the heatbath coordinate $r_1(t)$ satisfies

$$\ddot{r}_1(t) = -\Omega_1^2 r_1(t) - \beta_2 \dot{r}_1(t) \tag{II.117}$$

The full many-body system thus is replaced by a two-atom harmonic chain with damping on the second atom:

$$\ddot{r}_0(t) = -\omega_{e_0}^2 r_0(t) + \omega_{c_1}^2 r_1(t) \tag{II.118a}$$

and

$$\ddot{r}_1(t) = -\Omega_1^2 r_1(t) + \omega_{c_1}^2 r_0(t) - \beta_2 \dot{\vec{r}}_1(t) \tag{II.118b}$$

Equations (II.118) provide the simplest model of the full many-body system which is free of gross qualitative error and which is thus useful in computer simulations of many-body chemical dynamics. The great value of this simplest model is that it reduces, say, the many-body problem of atom–solid collisions to an effective *three-body* classical trajectory problem involving the incident gas atom plus two fictitious particles, atoms 0 and 1. Unlike the simple damped oscillator model, (II.19), the model (II.118) correctly describes short-time-scale (Einstein-limit) processes. This, as mentioned, is an essential feature of any qualitatively correct theory for molecular time-scale processes. Moreover, the model of (II.118), which is illustrated in Fig. 2a, is free of recurrences.

The weakness of the model is, of course, that it is only roughly equivalent (point 2) to the true many-body system. This weakness, of course, may be systematically corrected by increasing N. The labor involved in computer simulations, however, increases as N increases. In practice, a compromise between precision (large N) and computational convenience (small N) is always made.

In summary, the recursive structure of the effective equation-of-motion method has lead us to the nearest-neighbor harmonic chain representation of many-body dynamics. This representation, which is exact, yields equivalent heatbaths which are easily modeled. The model heatbaths are N-atom nearest-neighbor harmonic chains with atom N subject to simple frictional damping. These heatbaths are very useful in chemical dynamics applications of the theory since they are free from Poincaré recurrences, they correctly describe $\chi_m(t)$ at short times, and they lead to effective equations of motion which may be easily solved by standard classical trajectory methods on a computer.

H. "Normal Modes" of the Many-Body System

Section II.G dealt with an "atom" picture of exact many-body dynamics. Here we discuss the corresponding "normal-mode" representation. The "modes" of a (nonharmonic) many-body system, like the "atoms" of the chain representation, are fictitious. They merely provide a convenient language for discussing many-body dynamics and a rigorous framework for introducing "physically" motivated approximations.

We begin the "mode" analysis by defining the force constant matrix $\underline{\omega}^2$ of the full many-body system as [cf. (II.109)]

$$[\underline{\omega}^2]_{nm} = \omega_{e_n}^2 \delta_{n,m} - \omega_{c_n}^2 \delta_{n-1,m} - \omega_{c_{n+1}}^2 \delta_{n+1,m} \qquad n,m \geq 0 \qquad \text{(II.119)}$$

This matrix is symmetric and thus may be diagonalized by an orthogonal matrix U:

$$\mathsf{U}^T \underline{\omega}^2 \mathsf{U} = \begin{bmatrix} \omega_0^2 & 0 & 0 & \cdots \\ 0 & \omega_1^2 & 0 & \cdots \\ 0 & 0 & \omega_3^2 & 0 \\ \vdots & & \ddots & \end{bmatrix} \qquad \text{(II.120a)}$$

where

$$\mathsf{U}^T \mathsf{U} = \mathsf{U}\mathsf{U}^T = 1 \qquad \text{(II.120b)}$$

The frequencies $\omega_\lambda = \omega_0, \omega_1, \ldots,$ in (II.120a) are the "normal-mode" frequencies of the many-body system.

The "atom" coordinates $r_n(t)$ and "mode" coordinates $\zeta_\lambda(t)$ are related by

$$r_n(t) = \sum_{n=0}^{\infty} U_{n\lambda} \zeta_\lambda(t) = \sum_{\lambda=0}^{\infty} U_{n\lambda} \left[\zeta_\lambda(0) \cos \omega_\lambda t + \dot{\zeta}_\lambda(0) \frac{\sin \omega_\lambda t}{\omega_\lambda} \right] \qquad \text{(II.121a)}$$

Using the initial conditions of (II.111) and (II.120b), we find that the initial "position" and "velocity" of mode λ are given by

$$\zeta_\lambda(0) = 0$$
$$\dot{\zeta}_\lambda(0) = U_{\lambda 0} \qquad \text{(II.122)}$$

Thus

$$r_n(t) = \sum_{\lambda=0}^{\infty} U_{n\lambda} U_{\lambda 0} \frac{\sin \omega_\lambda t}{\omega_\lambda} \tag{II.121b}$$

From (II.121b), we see that the expansion of $\chi_m(t)$ in normal modes is

$$\dot{\chi}_m(t) = \dot{r}_0(t) = \sum_{\lambda=0}^{\infty} U_{0\lambda} U_{\lambda 0} \cos \omega_\lambda t \tag{II.123}$$

Introducing the local density of states at atom 0 by

$$\rho_0(\omega) = \sum_{\lambda=0}^{\infty} U_{0\lambda} U_{\lambda 0} \delta(\omega - \omega_\lambda) \tag{II.124}$$

we may rewrite $\chi_m(t)$ as

$$\chi_m(t) = \int_0^{\infty} \rho_0(\omega) \cos \omega t \, d\omega \tag{II.125}$$

Comparing (II.27b) and (II.123) shows that the local density of states $\rho_0(\omega)$ is just the spectral density $\rho(\omega)$:

$$\rho(\omega) = \sum_{\lambda=0}^{\infty} U_{0\lambda} U_{\lambda 0} \delta(\omega - \omega_\lambda) \tag{II.126}$$

Equation (II.126) provides a "physical" interpretation of $\rho(\omega)$.

The Laplace transform $\hat{\chi}_m(z)$ may be written in the mode representation as

$$\hat{\chi}_m(z) = \sum_{\lambda=0}^{\infty} \frac{U_{0\lambda} U_{\lambda 0}}{z^2 + \omega_\lambda^2} \tag{II.127a}$$

Equations (II.120b) and (II.127a) show that

$$\lim_{z \to \infty} \hat{\chi}_m(z) = z^{-2} \tag{II.128}$$

In fact [compare (II.126) and (II.127a)],

$$\hat{\chi}_m(z) = z^{-2}\left[1 - z^{-2}\int_0^{\infty} \rho(\omega)\omega^2 \, d\omega + z^{-4}\int_0^{\infty} \rho(\omega)\omega^4 \, d\omega + \cdots \right] \tag{II.129}$$

and thus the moments of $\rho(\omega)$ may be computed from the high-z expansion of $\hat{\chi}_m(z)$ as well as the short-time expansion of $\chi_m(t)$.

Equation (II.127a) may be rewritten as

$$\hat{\chi}_m(i\omega) = -\frac{1}{2\omega}\sum_{\lambda=0}^{\infty} U_{0\lambda}U_{\lambda 0}\left[\frac{1}{\omega-\omega_\lambda}+\frac{1}{\omega+\omega_\lambda}\right] \qquad \text{(II.127b)}$$

Equation (II.127b) shows that $\hat{\chi}_m(i\omega)$ has poles at the "mode" frequencies ω_λ.

An analysis of the heatbath modes analogous to that just presented for the primary system shows that [cf. (II.121b)

$$\dot{\theta}(t) = \sum_{\lambda=1}^{\infty} U_{1\lambda}^{(1)}U_{\lambda 1}^{(1)}\cos\omega_\lambda^{(1)}t \qquad \text{(II.130)}$$

where $\mathsf{U}^{(1)}$ is the unitary transformation which diagonalizes the heatbath frequency matrix given by [cf. (II.119)]

$$\left[\underline{\omega}_1^2\right]_{nm} = \omega_{e_n}^2\delta_{n,m} - \omega_{e_n}^2\delta_{n-1,m} - \omega_{c_{n+1}}^2\delta_{n+1,m} \qquad n,m\geq 1 \qquad \text{(II.131)}$$

and $\omega_\lambda^{(1)}$ are the normal modes of the heatbath. The heatbath spectral density $\sigma(\omega)$ may be interpreted as a local density of heat bath states at atom 1 [cf. (II.126)]:

$$\sigma(\omega) = \sum_{\lambda=1}^{\infty} U_{1\lambda}^{(1)}U_{\lambda 1}^{(1)}\delta\left[\omega-\omega_\lambda^{(1)}\right] \qquad \text{(II.132)}$$

The heatbath response function $\hat{\theta}(i\omega)$ has the spectral expansion [cf. (II.127b)]

$$\hat{\theta}(i\omega) = -\frac{1}{2\omega}\sum_{\lambda=1}^{\infty} U_{1\lambda}^{(1)}U_{\lambda 1}^{(1)}\left[\frac{1}{\omega-\omega_\lambda^{(1)}}+\frac{1}{\omega+\omega_\lambda^{(1)}}\right] \qquad \text{(II.133)}$$

which shows that $\hat{\theta}(i\omega)$ has poles at the heatbath "mode" frequencies.

From (II.69) and (II.127b), one sees that

$$\sigma(\omega_\lambda) = 0 \qquad \lambda = 0,1,2,\ldots \qquad \text{(II.134a)}$$

and hence that

$$\rho(\omega_\lambda^{(1)}) = 0 \qquad \lambda = 1,2,3,\ldots \qquad \text{(II.143b)}$$

Thus the heatbath and primary system normal-mode frequencies are always unequal (in fact, they interleave). Finally, we see from (II.127b), (II.133), and (II.56) that

$$\hat{\chi}_m(i\omega_\lambda^{(1)})=0 \tag{II.135a}$$

and

$$\hat{\theta}(i\omega_\lambda)=\omega_{e_0}^2-\omega_\lambda^2 \tag{II.135b}$$

This completes our brief discussion of the "mode" picture of many-body dynamics.

I. The Generalized Langevin Equation

We develop in this and succeeding sections a generalization of the effective equation-of-motion method which permits one to compute *individual* primary system trajectories as solutions of an effective equation of motion. This equation of motion is known as the generalized Langevin equation (GLE). Since application of the effective equation-of-motion method to condensed phase chemical processes requires computation of trajectories (as in the gas phase), the present generalization is essential to such applications.

The development of the GLE theory will require a qualitative discussion of heatbath fluctuations from equilibrium and the related concepts of stochastic or random forces (such forces are also known as thermal noise sources). We present such a discussion in Section II.I.1 and follow in Section II.I.2 with a general formulation of the GLE theory. In Section II.J we develop the theory in fuller detail for heatbaths whose fluctuations are distributed according to Gaussian statistics (Gaussian noise approximation). The Gaussian case is of particular importance since application of the theory for this case is straightforward. Section II.K deals with the generalization of the nearest-neighbor harmonic chain representation of many-body dynamics, Section II.G, to include heatbath fluctuations. Finally, in Section II.L, we generalize the theory to the case of a primary system composed of N coupled m-oscillators.

1. Random Forces and Heatbath Fluctuations

The heatbath (see Section II.D) exerts a *systematic* dissipative force on the primary system. This dissipative force is *induced* by the primary system motion and arises from a *lag* (Fig. 1*c*) in the *average* local density of heatbath molecules in the neighborhood of the primary system. In addition to this dissipative force, the heatbath also exerts a random or stochastic force, $\vec{f}(t)$, on the primary system. This random force arises from *fluctuations* in the local density of heatbath molecules about the average value. These fluctuations in local density arise from the random thermal motions of the heatbath molecules. Since the magnitude of thermal kinetic energy increases with temperature T, the random force becomes increasingly important as T increases. For *classical* $T = 0°K$ systems, thermal motion ceases, and the random force should vanish since (unlike the systematic force) the random force is *not* induced by primary system motion but rather is due to intrinsic motion of the heatbath.

The random force produces random momentum transfers to the primary system. For *M-particle* primary systems the random momentum transfers have only negligible influence because the momentum transfers are of microscopic magnitude and, moreover, are of random direction and hence cancel on macroscopic time scales. A corollary is that macroscopic motion in a heatbath *ceases* shortly after external driving forces are switched off.

Motion of *microscopic particles* (macromolecular size or smaller), in contrast, is *perpetual* even in the absence of external forces. The driving force for this perpetual or Brownian motion is the internal or random force $\vec{f}(t)$. Thus the random force is responsible for the chaotic nature of individual molecular trajectories.

Since classical mechanics is strictly deterministic (i.e., a complete set of initial conditions for the full many-body system *precisely* determines the trajectory of the primary system), the concept of a "random" force requires further clarification. The random character of $\vec{f}(t)$ arises fundamentally from imprecise knowledge of initial conditions of the heatbath molecules. Suppose, for example, that one starts off N microscopic particles with *identical* initial conditions in N *macroscopically identical* heatbaths. The particles will all execute different trajectories. While it is convenient to ascribe the different trajectories to the effects of a "random force," at the microscopic level the differences arise simply because the initial conditions of each heatbath are different.

Let us illustrate these qualitative points with a simple example. Consider as our "many-body" system a two-atom harmonic chain. Let atom 0 be the primary system and atom 1 the heatbath. The equation of motion for the chain is [cf. (II.108)]

$$\ddot{r}_0(t) = -\omega_{e_0}^2 r_0(t) + \omega_{c_1}^2 r_1(t) \tag{II.136a}$$

and

$$\ddot{r}_1(t) = -\omega_{e_1}^2 r_1(t) + \omega_{c_1}^2 r_0(t) \tag{II.136b}$$

The Laplace transform of (II.136b) may be written as

$$\hat{r}_1(z)=\omega_{c_1}^2\hat{\theta}_1(z)r_0(z)+\omega_{c_1}^{-2}\hat{f}(z) \tag{II.137a}$$

where the damping kernel $\hat{\theta}_1(z)$ is defined by

$$\hat{\theta}_1(z)=\left[z^2+\omega_{e_1}^2\right]^{-1} \tag{II.138a}$$

and where the random force $\hat{f}(z)$ is defined by

$$\hat{f}(z)=\omega_{c_1}^2\hat{\theta}_1(z)\left[\dot{r}_1(0)+zr_1(0)\right] \tag{II.139a}$$

In the time domain, the trajectory of atom 1 is found by Laplace inversion of (II.137a) to be

$$r_1(t)=\omega_{c_1}^2\int_0^t\theta_1(t-\tau)r_0(\tau)\,d\tau+\omega_{c_1}^{-2}f(t) \tag{II.137b}$$

and the random force is

$$f(t)=\omega_{c_1}^2\left[\theta_1(t)\dot{r}_1(0)+\dot{\theta}(t)r_1(0)\right] \tag{II.138b}$$

Using (II.137b) to eliminate $r_1(t)$ from (II.136a) yields the following *generalized Langevin equation* for $r_0(t)$:

$$\ddot{r}_0(t)=-\omega_{e_0}^2r_0(t)+\omega_{c_1}^4\int_0^t\theta_1(t-\tau)r_0(\tau)\,d\tau+f(t) \tag{II.139b}$$

Equation (II.139b) nicely illustrates the general points just discussed. First notice that the random force $f(t)$ depends on the initial conditions of the heatbath, (II.138b). Moreover, $f(t)$ is proportional to the trajectory of the heatbath for the case that the primary system is *clamped* at $r_0(t)=0$. This illustrates that the random force is due to intrinsic heatbath motion and is *not* induced by primary system motion. The *induced* part of the trajectory of the heatbath $\sim\int_0^t\theta_1(t-\tau)r_0(\tau)\,d\tau$ becomes the systematic dissipative part of the heatbath force on the primary system. Since at $T=0°K$, the intrinsic motion of the heatbath vanishes; $f(t)=0$ at $T=0°K$. The stochastic character of $f(t)$ arises because $r_1(0)$ and $\dot{r}_1(0)$ are not known precisely but are, rather, distributed thermally at finite temperature. Thus only at $T=0°K$, where the classical thermal distribution narrows to a δ-function in position and momentum, does the dynamics of the primary system become deterministic or, equivalently, $f(t)$ vanishes. Finally, for harmonic chains (and harmonic systems in general) the initial conditions of the heatbath are distributed Gaussianly and thus $f(t)$ is rigorously a Gaussian stochastic process for such systems.

Most of our comments about the simple two-atom chain example are generally valid. We now turn to the general formulation of the theory.

2. *Generalized Langevin Equation and the Second Fluctuation–Dissipation Theorem*

Our development of the GLE theory begins with (II.55a), the effective equation of motion for the primary system response function $\chi_m(t)$. This may be rewritten as

$$\langle \ddot{\vec{r}}(t)\cdot\dot{\vec{r}}(0)\rangle = -\omega_e^2\langle \vec{r}(t)\cdot\dot{\vec{r}}(0)\rangle + \omega_c^4\int_0^t \theta(t-\tau)\langle \vec{r}(\tau)\cdot\dot{\vec{r}}(0)\rangle\, d\tau \qquad \text{(II.140)}$$

To obtain (II.140) from (II.55a) we have used (II.22) and (II.64).

We now note that the effective equation of motion for the correlation function $\langle \vec{r}(t)\cdot\dot{\vec{r}}(0)\rangle$, (II.140), is equivalent to the following GLE for the trajectory $\vec{r}(t)$ [cf. (II.139b)]:

$$\ddot{\vec{r}}(t) = -\omega_e^2\vec{r}(t) + \omega_c^4\int_0^t \theta(t-\tau)\vec{r}(\tau)\, d\tau + \vec{f}(t) \qquad \text{(II.141)}$$

if the random force $\vec{f}(t)$ satisfies the following constraint:

$$\langle \vec{f}(t)\cdot\dot{\vec{r}}(0)\rangle = 0 \qquad \text{(II.142a)}$$

Equation (II.141) with (II.142a) *defines* the random force $\vec{f}(t)$. Notice that (II.142a) is necessary in order that (II.140) may be derived from (II.141).

A second constraint on $\vec{f}(t)$ holds for thermal heatbaths. For such heatbaths the kinetic energy of the primary system must approach its equilibrium value as $t\to\infty$:

$$\lim_{t\to\infty} \tfrac{1}{2}m\langle \dot{r}^2(t)\rangle = \tfrac{3}{2}k_B T \qquad \text{(II.143)}$$

Equation (II.143) places the following additional constraint on $f(t)$:

$$\langle \vec{f}(t)\cdot\vec{f}(0)\rangle = \frac{3k_B T}{m}\omega_c^4\dot{\theta}(t) \qquad \text{(II.142b)}$$

Equation (II.142b), which is derived in Appendix 3, is the second fluctuation–dissipation theorem *conjectured* in (II.74). Comparison of (II.74a) and (II.142b) shows that

$$\vec{f}(t)=\omega_c^2\vec{R}(t) \tag{II.144}$$

Thus, as discussed in Section II.E, the heatbath dynamical variable $\vec{R}(t)$ is proportional to the random force $\vec{f}(t)$. Moreover, our proof of the second fluctuation–dissipation theorem (Appendix 3) clearly shows that the connection between the dissipative and random forces is intimately related to the requirement that the system relax to thermal equilibrium. This was discussed in Section II.E.

Notice that at $T=0°\mathrm{K}$, we expect $\vec{f}(t)=\vec{0}$. Thus at $T=0°\mathrm{K}$,

$$\ddot{\vec{r}}(t)=-\omega_e^2\vec{r}(t)+\omega_c^4\int_0^t\theta(t-\tau)\vec{r}(\tau)\,d\tau \qquad T=0°\mathrm{K} \tag{II.145}$$

Equation (II.145) is, however, identical to the effective equation of motion for $\chi_m(t)$, (II.55a). In fact, for the special initial conditions $r_1(0)=0$ and $\dot{r}_1(0)=1$, the $T=0°\mathrm{K}$ trajectory is *identical* to $\chi_m(t)$. Moreover, if we average $\vec{r}(t)$ over heatbath initial conditions (such an average is denoted by $\langle\ \rangle_B$) and we see that the $T=0°\mathrm{K}$ equation of motion also holds for the average trajectory:

$$\langle\ddot{\vec{r}}(t)\rangle_B=-\omega_e^2\langle\vec{r}(t)\rangle_B+\omega_c^4\int_0^t\theta(t-\tau)\langle\vec{r}(\tau)\rangle_B\,d\tau \tag{II.146}$$

Equation (II.9.18) follows since

$$\langle\vec{f}(t)\rangle_B=\vec{0} \tag{II.147}$$

if $\vec{f}(t)$ is truly random.

Thus the following qualitative picture of primary system dynamics emerges. The $T=0°\mathrm{K}$ trajectory is the average trajectory at finite T. The spread about this average trajectory grows (from zero at $T=0°\mathrm{K}$) as T increases. The increase in the spread with T reflects the increasing importance of intrinsic thermal motion of the bath [i.e., $\vec{f}(t)$]. Equivalently, it reflects our increasingly imprecise knowledge of heatbath initial conditions as T increases.

This simple qualitative picture *does not* hold at short times for the conventional GLE.[6,7] This may be seen by rewriting (II.141) in friction kernel form as [cf. (II.112)]

$$\ddot{\vec{r}}(t)=-\Omega^2\vec{r}(t)-\int_0^t\beta(t-\tau)\dot{\vec{r}}(\tau)\,d\tau+\vec{f}_\beta(t) \tag{II.148}$$

where $\beta(t)$ is defined in terms of $\Theta(t)=\omega_c^4\theta(t)$ in (II.60) and where

$$\vec{f}_\beta(t)=\vec{f}(t)-\beta(t)\vec{r}(0) \tag{II.149}$$

Since $\lim_{t\to\infty}\beta(t)=0$ [see (II.60)], for long times $\vec{f}_\beta(t)=\vec{f}(t)$. For short times, $\vec{f}_\beta(t)$ contains the *systematic* transient term (see discussion in Section II.D) $\beta(t)\vec{r}(0)$. This term spoils the interpretation of $\vec{f}_\beta(t)$ as a random force.[27b] For example [cf. (II.147)],

$$\langle\vec{f}(t)\rangle_B=\beta(t)\vec{r}(0)$$

and at $T=0$ K, $\vec{f}(t)=\beta(t)\vec{r}(0)$. Thus the simple qualitative picture of stochastic dynamics sketched above does not hold at short times if $\vec{f}_\beta(t)$ is interpreted as the random force. A closely related point is that Ω^2 rather than ω_e^2 appears explicitly in (II.148). [In fact, the role of the systematic transient term is to convert the adiabatic restoring force $-\Omega^2\vec{r}(t)$ to the corresponding Einstein force $-\omega_e^2\vec{r}(t)$ as $t\to 0$.] For these reasons (II.141) is preferred over (II.148) for molecular-time-scale processes.

We finally note that within the local friction or Markovian approximation [cf. (II.114)],

$$\beta(t)=2\beta\delta(t) \tag{II.150a}$$

the GLE takes the form

$$\ddot{\vec{r}}(t)=-\Omega^2\vec{r}(t)-\beta\dot{\vec{r}}(t)+\vec{f}(t) \tag{II.150b}$$

[the Markovian approximation, if it is ever valid at all, is only valid as $t\to\infty$ and thus the distinction between $\vec{f}(t)$ and $\vec{f}_\beta(t)$ is not relevant for this case] and the fluctuation–dissipation theorem becomes

$$\langle\vec{f}(t)\cdot\vec{f}(0)\rangle=\frac{6k_BT}{m}\beta\delta(t) \tag{II.151}$$

Equations (II.150) and (II.151), together with the assumption that the random force $\vec{f}(t)$ is distributed Gaussianly, comprise the content of the phenomenological theory of Brownian motion.[22] We now turn to the special case of Gaussian heatbaths.

J. Gaussian Heatbaths and the Probability Distribution Function

The GLE, (II.141), yields for $T>0$°K an ensemble of trajectories $\vec{r}(t)$ for each set of primary system initial conditions. That is, each of the vast number of possible random forces or, equivalently, each of the thermally accessible sets of heatbath initial conditions determines a unique trajectory. One, however, is almost never interested in individual trajectories.

Rather observables such as gas–solid adsorption probabilities or time resolved reaction rates in liquids are determined by averages over swarms of trajectories. The relevant theoretical problem is thus to compute the probability distribution function (pdf) for trajectories rather than individual trajectories. Observables may then be expressed as averages over the pdf.

The pdf may always be constructed by Monte Carlo sampling of $\vec{f}(t)$ and explicit calculation of trajectories. This brute-force method, in fact, is *currently* the most useful one for treating the nonlinear stochastic processes arising on chemical dynamics applications of the GLE theory (Section III). For linear GLE's, e.g. (II.141), and for the case of Gaussian random forces, the pdf may be constructed much more conveniently using the techniques of linear Gaussian fluctuation theory.

We develop the linear fluctuation theory in this section. Our motivation is twofold. First, the theory is of considerable interest in its own right; for example, it illustrates the fundamental character of the second fluctuation–dissipation theorem in a clear manner. Second, the theory is the logical departure point for nonlinear fluctuation methods. If practical nonlinear methods could be developed which improve upon brute-force Monte Carlo sampling, application of the GLE theory to many-body chemical processes would be considerably facilitated. Attempts in this direction have already been made.[28]

Our Gaussian fluctuation treatment of the GLE, (II.141), employs certain elementary results in Gaussian probability theory. For completeness these results are reviewed in Appendix 4.

We now turn to the linear fluctuation calculation of the pdf for trajectories.

1. Probability Functions for Generalized Brownian Oscillators

The trajectory of the primary system at $T=0°\text{K}$, denoted by $\vec{z}(t)$, is readily found by solving the GLE, (II.145), for the case $\vec{f}(t)=\vec{0}$. The result is

$$\vec{z}(t)=\chi_m(t)\vec{r}(0)+\dot{\chi}_m(t)\vec{r}(0) \qquad \text{(II.152)}$$

The $T=0°\text{K}$ trajectory, $\vec{z}(t)$, is also the average or most probable trajectory for finite T. The trajectories $\vec{r}(t)$ for $T>0°\text{K}$ *fluctuate* about $\vec{z}(t)$ due to thermal noise (i.e., the random force). The trajectory fluctuation $\vec{y}(t)=\vec{r}(t)-\vec{z}(t)$ is found from (II.145) to be

$$\vec{y}(t)=\vec{r}(t)-\vec{z}(t)=\int_0^t \chi_m(t-\tau)\vec{f}(\tau)\,d\tau \qquad \text{(II.153a)}$$

Similarly, the fluctuation $\vec{y}(t)=\vec{r}(t)-\vec{z}(t)$ of velocity about the average velocity $\dot{\vec{z}}(t)$ may be written as

$$\vec{y}(t)=\vec{r}(t)-\vec{z}(t)=\int_0^t \dot{\chi}_m(t-\tau)\vec{f}(\tau)\,d\tau \tag{II.153b}$$

We now assume that $\vec{f}(t)$ obeys Gaussian statistics. This means that the probability that $\vec{f}(t)$ has the value $\vec{f}$ at an arbitrary time t is described by the unit normalized Gaussian pdf [cf. (A.54)]

$$P[\vec{f}\,]=\left(\frac{3}{2\pi\langle f^2\rangle}\right)^{3/2} e^{-(3/2)(f^2/\langle f^2\rangle)} \tag{II.154}$$

where $\langle f^2\rangle$ is the second moment of the distribution [cf. (A.38b)]. The second moment $\langle f^2\rangle$ is $\sim T$, (II.156) below, and is a measure of the thermal spread in heatbath initial conditions. The quantity $\langle f^2\rangle$ may be expressed in terms of known quantities using the second fluctuation–dissipation theorem which we write as [combine (II.47c) and (II.144)]

$$\beta(t)=\frac{m}{3k_BT}\langle\vec{f}(t)\cdot\vec{f}(0)\rangle \tag{II.155}$$

Using (II.155), (II.52), and (II.76b), one finds for $\langle f^2\rangle$ the following expressions:

$$\langle f^2\rangle=\langle\vec{f}(0)\cdot\vec{f}(0)\rangle=\frac{3k_BT}{m}\beta(0)=\frac{3k_BT}{m}\left[\omega_e^2-\Omega^2\right]=\frac{3k_BT}{m}\frac{\omega_c^4}{\Omega_1^2} \tag{II.156}$$

The last result in (II.156) shows that $\langle f^2\rangle$, which measure the importance of the random force, increases with T, with the coupling constant ω_c^4, and with the heatbath polarizability Ω_1^{-2}. This makes good physical sense.

The fluctuations $\vec{y}(t)$ and $\dot{\vec{y}}(t)$ are Gaussian random variables since they are sums (integrals) of Gaussian random variables $\vec{f}(t_1)$, $\vec{f}(t_2)$, and so on. This follows since linear combinations of Gaussian random variables are themselves Gaussian random variables (Appendix 4).

Thus the pdf for $\vec{y}(t)$ is determined by the second moment $\langle y^2(t)\rangle$ as

$$P[\vec{y}(t)]=\left(\frac{3}{2\pi\langle y^2(t)\rangle}\right)^{1/2}\exp\left[-\frac{3}{2}\frac{y^2(t)}{\langle y^2(t)\rangle}\right] \tag{II.157}$$

Notice that knowledge of $\vec{y}(t)$ is sufficient to determine $\vec{r}(t)$ if the initial conditions $\vec{r}(0)$ and $\dot{\vec{r}}(0)$ are known [see (II.152) and (II.153b)]. Thus $P[\vec{y}(t)]$ may be interpreted as the pdf. $P[\vec{r},\vec{r}(0);\dot{\vec{r}}(0);t]$ for the position $\vec{r}$ at time t given the initial position $\vec{r}(0)$ and the initial velocity $\dot{\vec{r}}(0)$. Thus

$$P\left[\vec{r},\vec{r}(0);\dot{\vec{r}}(0);t\right]=\left(\frac{3}{2\pi\langle y^2(t)\rangle}\right)^{3/2}\exp\left[-\frac{3}{2}\frac{y^2(t)}{\langle y^2(t)\rangle}\right] \quad \text{(II.158a)}$$

Similarly, the pdf for velocity is

$$P\left[\dot{\vec{r}},\vec{r}(0);\dot{\vec{r}}(0);t\right]=\left(\frac{3}{2\pi\langle \dot{y}^2(t)\rangle}\right)^{3/2}\exp\left[-\frac{3}{2}\frac{\dot{y}^2(t)}{\langle \dot{y}^2(t)\rangle}\right] \quad \text{(II.158b)}$$

and the *joint* pdf for position and velocity is [cf. (A.44)]

$$\left(\frac{3}{2\pi}\right)^{6/2}\left(\frac{1}{\det\langle \mathbf{y}\mathbf{y}^T\rangle}\right)^{3/2} e^{-(3/2)\mathbf{y}^T\cdot\langle \mathbf{y}\mathbf{y}^T\rangle^{-1}\cdot \mathbf{y}} \quad \text{(II.158c)}$$

where

$$\mathbf{y}=\begin{pmatrix}\vec{y}(t)\\ \dot{\vec{y}}(t)\end{pmatrix} \quad \text{and} \quad \mathbf{y}^T=(\vec{y}(t)\dot{\vec{y}}(t))$$

and where the *correlation matrix* $\langle \mathbf{y}\cdot\mathbf{y}^T\rangle$ is given by [cf. (A.46)]

$$\langle \mathbf{y}\cdot\mathbf{y}^T\rangle=\begin{pmatrix}\langle y^2(t)\rangle & \langle \vec{y}(t)\cdot\dot{\vec{y}}(t)\rangle\\ \langle \dot{\vec{y}}(t)\cdot\vec{y}(t)\rangle & \langle \dot{y}^2(t)\rangle\end{pmatrix}\vec{1} \quad \text{(II.159)}$$

For $T=0°$K, the fluctuations $\langle y^2(t)\rangle$, $\langle \dot{y}^2(t)\rangle$, and $\langle \dot{\vec{y}}(t)\cdot\vec{y}(t)\rangle$ vanish and the pdf's in (II.158) reduce to δ-functions centered at the average trajectory; that is, the dynamics is deterministic, like gas-phase dynamics. (Unlike gas-phase motion, however, classical condensed-phase motion damps out at $T=0°$K, because of the dissipative forces.)

Equations (II.158c) show that *all* information about the primary system dynamics is carried by the second moments $\langle y^2(t)\rangle$, $\langle \dot{y}^2(t)\rangle$, and $\langle \vec{y}(t)\cdot\dot{\vec{y}}(t)\rangle$. These may be readily calculated from (II.153). For example, $\langle y^2(t)\rangle$ is given by

$$\begin{aligned}\langle y^2(t)\rangle&=\int_0^t d\tau\int_0^t d\tau'\chi_m(t-\tau)\langle\vec{f}(\tau)\cdot\vec{f}(\tau')\rangle\chi_m(t-\tau')\\ &=\int_0^t d\tau\int_0^t d\tau'\chi_m(\tau)\langle\vec{f}(\tau-\tau')\cdot\vec{f}(0)\rangle\chi_m(\tau')\end{aligned} \quad \text{(II.160)}$$

where we have used (II.23a) to obtain the second equality in (II.160).

To evaluate (II.160) explicitly, an expression for $\langle \vec{f}(t)\cdot\vec{f}(0)\rangle$ is required. The second fluctuation–dissipation theorem, (II.155), provides the required expression. Using (II.155) in (II.160) yields

$$\langle y^2(t)\rangle = \frac{3k_BT}{m}\int_0^t d\tau\int_0^t d\tau' \chi_m(\tau)\beta(\tau-\tau')\chi_m(\tau') \qquad \text{(II.161)}$$

Equation (II.161) illustrates the pivotal role played by the second fluctuation–dissipation theorem in the effective equation-of-motion theory. The theorem permits one to determine the second moments (e.g., $\langle y^2(t)\rangle$) and hence *all* observable properties of a finite-temperature Gaussian primary system (i.e., the pdf's) in terms of the response function $\chi_m(t)$ [recall that $\beta(t)$ is determined by $\chi_m(t)$ via (II.49)]. This is a remarkable result since the response function originally had only the limited significance of describing the velocity response of the primary system in the presence of a weak external force, (II.21).

Moreover, this result has great practical value since $\dot{\chi}_m(t)$ may be usefully approximated from its first few moments (Section II.F) using the heatbath modeling techniques of Section II.G.3.

Thus the formal apparatus of the effective equation-of-motion theory often allows one to calculate a remarkable amount of information about the dynamic behavior of the primary system from a small number of moments of $\rho(\omega)$ or equivalently from the short-time behavior of $\dot{\chi}_m(t)$. We will see applications of this idea in Section III.

We now recast (II.160) in a much more useful form. Differentiating (II.160) with respect to time yields

$$\begin{aligned}\frac{d}{dt}\langle y^2(t)\rangle &= \frac{6k_BT}{m}\chi_m(t)\int_0^t \beta(t-\tau)\chi_m(\tau)\,d\tau\\ &= \frac{6k_BT}{m}\chi_m(t)\mathcal{L}^{-1}\left[\hat{\beta}(z)\hat{\chi}_m(z)\right] \qquad \text{(II.162)}\end{aligned}$$

Using (II.50) to eliminate $\hat{\beta}(z)$ from the equation above and then inverting the Laplace transform gives

$$\frac{d}{dt}\langle y^2(t)\rangle = \frac{6k_BT}{m}\chi_m(t)\left[1-\dot{\chi}_m(t)-\Omega^2\phi_m(t)\right] \qquad \text{(II.163)}$$

where

$$\phi_m(t) = \int_0^t \chi_m(\tau)\,d\tau \qquad \text{(II.164)}$$

Finally, integrating (II.163) and using (II.164) and also $\langle y^2(0)\rangle = 0$, which follows from (II.153a), yields our final result for the second moment $\langle y^2(t)\rangle$:

$$\langle y^2(t)\rangle = \frac{6k_BT}{m}\left[\phi_m(t) - \frac{1}{2}\chi_m^2(t) - \frac{\Omega^2}{2}\phi_m^2(t)\right] \qquad \text{(II.165a)}$$

Notice that (II.165) depends only on $\chi_m(t)$, as just discussed. Equations (II.158a) and (II.165) thus provide a closed result for the position distribution of the primary system at any time t, for any temperature T, and for any set of initial conditions $\vec{r}(0)$ and $\dot{\vec{r}}(0)$. This result may be evaluated explicitly if $\chi_m(t)$ is known [e.g., if the Debye approximation for $\chi_m(t)$, (II.45) is used]. For the case of Markovian heatbaths, (II.165) with (II.8), reduces to the *phenomenological* Brownian motion results of Ornstein and Uhlenbeck,[22] Chandrasekhar,[22] Wang and Uhlenbeck,[22] and others.

The $t\to\infty$ limit of (II.165) is informative. Using the fact that $\lim_{t\to\infty}\chi_m(t) = 0$ and using (II.164) and (II.59) yields

$$\lim_{t\to\infty}\langle y^2(t)\rangle = \frac{3k_BT}{m\Omega^2} \qquad \text{(II.166a)}$$

Since the $T=0°\text{K}$ motion damps out at long times [see (II.152)],

$$\lim_{t\to\infty}\vec{z}(t) = 0$$

we see that [see (II.153a)]

$$\lim_{t\to\infty}\vec{y}(t) = \vec{r}(t) \qquad \text{(II.167)}$$

Thus the pdf (II.158a) becomes independent of $\vec{r}(0)$ and $\dot{\vec{r}}(0)$ as $t\to\infty$ and, in fact, (II.158a), (II.166a), and (II.167) show that

$$\lim_{t\to\infty} P\left[\vec{r},\vec{r}(0);\dot{\vec{r}}(0);t\right] = P_{\text{eq}}(\vec{r}) = \left(\frac{m\Omega^2}{2\pi k_BT}\right)^{3/2}\exp\left(-\frac{1}{2}\frac{m\Omega^2r^2}{k_BT}\right) \qquad \text{(II.168a)}$$

Equation (II.168) is the *equilibrium* distribution function for primary system coordinates. Notice that $P_{\text{eq}}(\vec{r})$ properly involves the adiabatic or *long-time* frequency Ω. The theory thus properly predicts relaxation to thermal equilibrium of the initial nonequilibrium pdf $P[\vec{r},0;\vec{r}(0),\dot{\vec{r}}(0)] = \delta[\vec{r} - \vec{r}(0)]$. The reason is that regression to thermal equilibrium is built into the theory through the second fluctuation–dissipation theorem.

Notice that (II.168a) implies that

$$\langle r^2\rangle = \int r^2 P_{\text{eq}}(\vec{r})\,d\vec{r} = \frac{3k_BT}{m\Omega^2} \tag{II.169}$$

Equation (II.169) was derived earlier, (II.36), by a completely different argument. The present derivation demonstrates the internal consistency of the theory.

We may now analogously derive the second moments required to explicitly evaluate the pdf for velocity and the joint pdf for position and velocity, (II.158b) and (II.158c). The results for these second moments are [cf. (II.165a)]

$$\langle \dot{y}^2(t)\rangle = \frac{3k_BT}{m}\left[1-\dot{\chi}_m^2(t)-\Omega^2\chi_m^2(t)\right] \tag{II.165b}$$

$$\langle \vec{y}(t)\cdot\dot{\vec{y}}(t)\rangle = \frac{3k_BT}{m}\chi_m(t)\left[1-\Omega^2\phi_m(t)-\dot{\chi}_m(t)\right] \tag{II.165c}$$

Notice that

$$\lim_{t\to\infty}\langle \dot{y}^2(t)\rangle = \frac{3k_BT}{m} \tag{II.166b}$$

and

$$\lim_{t\to\infty}\langle \vec{y}(t)\cdot\dot{\vec{y}}(t)\rangle = 0 \tag{II.166c}$$

Thus

$$\lim_{t\to\infty} P\left[\dot{\vec{r}},\dot{\vec{r}}(0);\vec{r}(0);t\right] = P_{\text{eq}}\left[\dot{\vec{r}}\right] \tag{II.168b}$$

and

$$\lim_{t\to\infty} P\left[\vec{r},\vec{r}(0);\dot{\vec{r}},\dot{\vec{r}}(0);t\right] = P_{\text{eq}}\left[\vec{r}\right]P_{\text{eq}}\left[\dot{\vec{r}}\right] \tag{II.168c}$$

where $P_{\text{eq}}(\vec{r})$ is given by (II.168a) and where $P_{\text{eq}}[\dot{\vec{r}}]$ is the Maxwellian velocity distribution

$$P_{\text{eq}}\left[\dot{\vec{r}}\right] = \left(\frac{m}{2\pi k_BT}\right)^{3/2}\exp\left(-\frac{1}{2}\frac{m\dot{r}^2}{k_BT}\right) \tag{II.170}$$

Equations (II.168) and (II.170) show that all pdf's for the primary system properly go over to the appropriate equilibrium pdf's as $t\to\infty$.

So far we have dealt with the generalized Langevin representation of primary system dynamics. We now turn to a corresponding generalized Fokker–Planck representation.

2. *The Generalized Fokker–Planck Representation*

While the generalization of the phenomenological Langevin equation, (II.150b), the GLE of (II.148) has been known for some time, the corresponding generalized Fokker–Planck representation of many-body dynamics has only been arrived at more recently.[27]

In order to make the content of the generalized Fokker–Planck theory transparent, it will first be helpful to recast the GLE, (II.141), in "time-dependent transport coefficient" form.[27] For simplicity we will restrict ourselves to the $T=0°\text{K}$ case, $\vec{f}(t)=\vec{0}$.

For this case the solution to the GLE is (II.152), which we rewrite as

$$\vec{r}(t)=\chi_m(t)\dot{\vec{r}}(0)+\dot{\chi}_m(t)\vec{r}(0) \qquad T=0\text{ K} \tag{II.171a}$$

Taking the first and second derivatives of (II.171a) gives

$$\dot{\vec{r}}(t)=\dot{\chi}_m(t)\dot{\vec{r}}(0)+\ddot{\chi}_m(t)\vec{r}(0) \tag{II.171b}$$

and

$$\ddot{\vec{r}}(t)=\ddot{\chi}_m(t)\dot{\vec{r}}(0)+\dddot{\chi}_m(t)\vec{r}(0) \tag{II.171c}$$

We may now solve (II.171a) for $\dot{\vec{r}}(0)$ and $\vec{r}(0)$ in terms of $\vec{r}(t)$ and $\dot{\vec{r}}(t)$ and use the result to eliminate $\dot{\vec{r}}(0)$ and $\vec{r}(0)$ from (II.171c). The result is the following effective equation of motion for the trajectory $\vec{r}(t)$,

$$\ddot{\vec{r}}(t)=-\omega^2(t)r(t)-\tilde{\beta}(t)\dot{\vec{r}}(t) \tag{II.172}$$

where the time-dependent frequency

$$\omega^2(t)=\frac{\ddot{\chi}_m^2(t)-\dot{\chi}_m(t)\dddot{\chi}_m(t)}{\det \mathsf{T}(t)} \tag{II.173a}$$

and the time-dependent friction coefficient

$$\tilde{\beta}(t)=\frac{-d\ln\det \mathsf{T}(t)}{dt} \tag{II.173b}$$

where

$$\mathsf{T}(t)=\begin{pmatrix}\dot{\chi}_m(t) & \chi_m(t)\\ \ddot{\chi}_m(t) & \dot{\chi}_m(t)\end{pmatrix} \tag{II.174}$$

Equations (II.20) and (II.47) show that

$$\mathsf{T}(0)=\begin{pmatrix}1 & 0\\ 0 & 1\end{pmatrix} \tag{II.175}$$

Equations (II.173) and (II.175) with (II.147) yield the initial values

$$\lim_{t\to 0}\omega^2(t)=\omega_e^2 \tag{II.176a}$$

and

$$\lim_{t\to 0}\tilde{\beta}(t)=0 \tag{II.176b}$$

Thus

$$\ddot{\vec{r}}(t)=-\omega_e^2\vec{r}(t) \qquad t\to 0 \tag{II.177}$$

Thus for short times, frictional effects properly disappear from (II.172), (II.176b), and the short-time dynamics is governed, as expected, by ω_e, (II.177).

For Markovian heatbaths, the response function is given by (II.8a). Use of this response function in (II.173) and (II.174) yields

$$\left.\begin{aligned}\omega^2(t)&=\Omega_M^2\\ \tilde{\beta}(t)&=\beta\end{aligned}\right\} \quad \text{Markov} \tag{II.178}$$

Thus the effective equation of motion, (II.172), reduces to the following result:

$$\ddot{r}(t)=-\Omega^2\vec{r}(t)-\beta\dot{\vec{r}}(t) \qquad \text{Markov limit} \tag{II.179}$$

This is just the Langevin equation (II.150b) with $\vec{f}(t)$ set equal to zero.

The phase-space pdf $P[\vec{r},\vec{r}(0);\dot{\vec{r}},\dot{\vec{r}}(0);t]\equiv P$ for the phenomenological Brownian oscillator satisfies the following generalized phase-space equation:

$$\left[\frac{\partial}{\partial t}+\dot{\vec{r}}\cdot\frac{\partial}{\partial\vec{r}}-\Omega_M^2\vec{r}\cdot\frac{\partial}{\partial\dot{\vec{r}}}\right]P=\beta\frac{\partial}{\partial\dot{\vec{r}}}\left[\dot{\vec{r}}P\right] \tag{II.180}$$

Equation (II.180) becomes the ordinary Liouville equation for a gas-phase oscillator of frequency Ω_M if $\beta=0$. The equivalence of (II.179) and (II.180) may be established by noting that the solution of (II.180) is

$$P\left[\vec{r},r(0);\dot{\vec{r}},\dot{\vec{r}}(0);t\right]=\delta\left[\vec{r}-\vec{r}(t)\right]\delta\left[\dot{\vec{r}}-\dot{\vec{r}}(t)\right] \tag{II.181}$$

where $\vec{r}(t)$ is the solution of (II.179) for the initial conditions $\vec{r}(0)$ and $\dot{\vec{r}}(0)$.

Thus one might expect that the phase-space equation corresponding to (II.172) is [cf. (II.181)]

$$\left[\frac{\partial}{\partial t}+\dot{\vec{r}}\frac{\partial}{\partial \vec{r}}-\omega^2(t)\vec{r}\cdot\frac{\partial}{\partial \dot{\vec{r}}}\right]P=\tilde{\beta}(t)\frac{\partial}{\partial \dot{\vec{r}}}[\dot{r}P] \qquad \text{(II.182)}$$

That this is indeed the case may be readily verified by an argument that parallels the verification of (II.180).

So far our phase-space theory does not yet include temperature or, equivalently, heatbath fluctuations. The effect of these fluctuations in the Markov case is well known[22] and gives the following *Fokker–Planck* equation for the phase-space pdf:

$$\left[\frac{\partial}{\partial t}+\dot{\vec{r}}\frac{\partial}{\partial \vec{r}}-\Omega_M^2\vec{r}\cdot\frac{\partial}{\partial \dot{\vec{r}}}\right]P=\beta\frac{\partial}{\partial \dot{\vec{r}}}[\dot{\vec{r}}P]+\frac{k_BT}{M}\beta\frac{\partial^2 P}{\partial \dot{r}^2} \qquad \text{(II.183)}$$

Notice that (2.183) reduces to (II.180) for $T=0°$K. The *diffusive term*$\sim \partial^2P/\partial\dot{r}^2$ causes the phase-space pdf of the phenomenological Brownian oscillator to broaden Gaussianly about the $T=0°$K result of (II.181). Thus the diffusive term is the Fokker–Planck analog of the random force appearing in the Langevin equation (II.150b). The diffusive term also has the second fluctuation–dissipation result, (II.151), built in since (II.183) properly predicts that P relaxes to the equilibrium distribution

$$P_{\text{eq}}[\vec{r},\dot{\vec{r}}]\sim\exp\left(-\frac{1}{2}\frac{M\dot{r}^2}{k_BT}\right)\exp\left(-\frac{1}{2}\frac{M\Omega_M^2r^2}{k_BT}\right) \qquad \text{(II.184a)}$$

as $t\to\infty$. A related point is that (II.184) is a solution of (II.183).

The complete equivalence of the Fokker–Planck representation, (II.183), and the Langevin representation, (II.150b) and (II.151), may be proven by noting that the phase-space pdf of (II.158c) with the second moments of (II.165) specialized to the Markov approximation, (II.8a), is the solution of (II.183) with the initial value

$$\lim_{t\to 0}P=\delta(\vec{r}-\vec{r}(0))\delta(\dot{\vec{r}}-\dot{\vec{r}}(0)) \qquad \text{(II.184b)}$$

The exact phase space (II.182) may also be generalized to finite temperature. The derivation[27] involves a lengthy calculation. We omit the derivation here and pass to the result, which is[27b]

$$\left[\frac{\partial}{\partial t}+\dot{\vec{r}}\cdot\frac{\partial}{\partial \vec{r}}-\omega^2(t)\vec{r}\cdot\frac{\partial}{\partial \dot{\vec{r}}}\right]P=\tilde{\beta}(t)\frac{\partial}{\partial \dot{\vec{r}}}[\dot{r}P]+\frac{k_BT}{m}\left[A(t)\frac{\partial^2P}{\partial \dot{\vec{r}}^2}+B(t)\frac{\partial^2P}{\partial\dot{\vec{r}}\,\partial\vec{r}}\right] \qquad \text{(II.185)}$$

where

$$A(t)=\tilde{\beta}(t)+\chi_m(t)\omega^2(t)\left[1-\Omega^2\phi_m(t)\right]$$
$$-\Omega^2\chi_m(t)\left[\dot{\chi}_m(t)+\chi_m(t)\tilde{\beta}(t)\right] \qquad \text{(II.186a)}$$

and

$$B(t)=\left\{\frac{\omega^2(t)-\Omega^2}{\Omega^2}-\frac{\omega^2(t)}{\Omega^2}\left[1-\phi_m(t)\Omega^2\right]^2\right\}$$
$$+\left(1-\Omega^2\phi_m(t)\right)\left(\dot{\chi}_m(t)+\chi_m(t)\tilde{\beta}(t)\right) \qquad \text{(II.186b)}$$

where $\phi_m(t)=\int_0^t \chi_m(\tau)\,d\tau$.

Equation (II.185) with (II.186) is completely equivalent to the GLE theory of (II.141) and (II.142b) for the case of Gaussian random forces; that is, the solution of (II.184) for the initial conditions given by (II.184b) generates the phase-space pdf (II.158c) with the second moments given by (II.165). For $T=0°\mathrm{K}$, (II.185) reduces to (II.182). For short times, (II.185) reduces to

$$\left[\frac{\partial}{\partial t}+\dot{\vec{r}}\cdot\frac{\partial}{\partial \vec{r}}-\omega_e^2\vec{r}\cdot\frac{\partial}{\partial \dot{\vec{r}}}\right]P=0 \qquad \text{short times} \qquad \text{(II.187)}$$

This is the Liouville equation for an isolated oscillator; as expected, the frequency of this oscillator is the Einstein frequency. For times sufficiently long that the systematic transient term $\beta(t)\vec{r}(t)$ in $\vec{f}_\beta(t)$ [see (II.149)] has relaxed away, (II.185) reduces to the much more appealing form [cf. (II.183)]

$$\left[\frac{\partial}{\partial t}+\dot{\vec{r}}\cdot\frac{\partial}{\partial \vec{r}}-\omega^2(t)\vec{r}\cdot\frac{\partial}{\partial \dot{\vec{r}}}\right]P=\beta(t)\frac{\partial}{\partial \dot{\vec{r}}}\left[\dot{\vec{r}}P\right]$$
$$+\frac{k_BT}{m}\left[\tilde{\beta}(t)\frac{\partial^2 P}{\partial \dot{\vec{r}}^2}+\frac{\omega^2(t)-\Omega^2}{\Omega^2}\frac{\partial}{\partial \vec{r}}\cdot\frac{\partial}{\partial \dot{\vec{r}}}\right]P \qquad \text{(II.188)}$$

Notice that in the Markovian limit, $\tilde{\beta}(t)=\beta$ and $\omega^2(t)=\Omega$, (II.188) reduces to the phenomenological Fokker–Planck equation, (II.183). For many systems the phenomenological theory is *not* approached at long times. An ex-

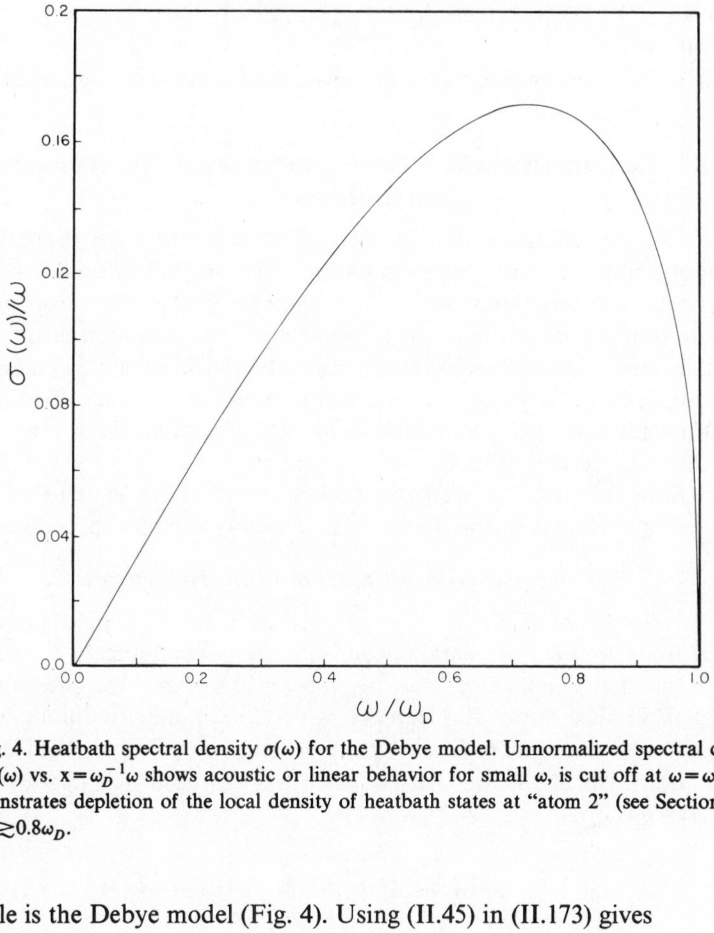

Fig. 4. Heatbath spectral density $\sigma(\omega)$ for the Debye model. Unnormalized spectral density $\omega^{-1}\sigma(\omega)$ vs. $x=\omega_D^{-1}\omega$ shows acoustic or linear behavior for small ω, is cut off at $\omega=\omega_D$, and demonstrates depletion of the local density of heatbath states at "atom 2" (see Section II.G) for $\omega \gtrsim 0.8\omega_D$.

ample is the Debye model (Fig. 4). Using (II.45) in (II.173) gives

$$\tilde{\beta}(t)=\frac{2\omega_D}{x}\left[1+\frac{\sin x\cos x}{x}-\frac{2\sin^2 x}{x}\right]\left[1-\frac{\sin^2 x}{x^2}\right]^{-1} \tag{II.189a}$$

$$\omega^2(t)=\omega_D^2\left[1-\frac{1+\cos^2 x}{x^2}+\frac{2\sin 2x}{x^3}-\frac{2\sin^2 x}{x^4}\right]\left[1-\frac{\sin^2 x}{x^2}\right]^{-1} \quad \text{Debye} \tag{II.189b}$$

where $x=\omega_D t$. Note that the Debye results *never* go over to the Markov approximation, (II.178). Despite this, (II.188) properly describes relaxation to equilibrium at long times since the fluctuation–dissipation theorem is built into (II.158) [e.g., the equilibrium pdf (II.184a) is a solution to (II.188)].

This concludes our discussion of generalized Gaussian fluctuation theory.

K. Heatbath Hierarchy at Finite Temperature and the Harmonic Chain Representation

We do three things in this section. First we generalize the nearest-neighbor harmonic chain representation of Section II.G to finite-temperature many-body dynamics. Second, we indicate here and in Appendix 5 how the recursive structure of the effective equation of motion theory, Section II.F, easily generalizes to finite temperature. The harmonic chain representation, recall, is a consequence of this recursive structure. Finally, we develop a finite-temperature heatbath modeling scheme which generalizes the scheme of Section II.G.3.

The finite-temperature heatbath models are what are actually used in practical applications of the theory to many-body chemical processes.

1. Nearest-Neighbor Chains at Finite Temperature

Our development of the $(N+1)$-atom nearest-neighbor chain representation at finite temperature parallels our analogous development for the response function $\chi_m(t)$, or equivalently for $T=0°\text{K}$ dynamics, presented in Section II.G. The essential difference is that the initial conditions of the chain are now distributed thermally rather than according to (II.111).

We begin with the GLE, (II.141), rewritten as [recall that $\omega_e=\omega_{e_0}$, $\omega_c=\omega_{c_1}$, $\theta(t)=\theta_1(t)$]

$$\ddot{\vec{r}}_0(t)=-\omega_{e_0}^2\vec{r}_0(t)+\omega_{c_1}^4\int_0^t d\tau\theta_1(t-\tau)r_0(\tau)+\vec{f}_1(t) \qquad \text{(II.190)}$$

where $\vec{r}_0(t)$, the coordinate of "atom" 0, is the primary system coordinate:

$$\vec{r}_0(t)=\vec{r}(t) \qquad \text{(II.191a)}$$

and where

$$\vec{f}_1(t)\equiv\vec{f}(t) \qquad \text{(II.191b)}$$

We now define the coordinate of "atom" 1, $\vec{r}_1(t)$, by

$$\vec{r}_1(t)=\omega_{c_1}^2\int_0^t\theta_1(t-\tau)\vec{r}_0(\tau)\,d\tau+\vec{R}_1(t) \qquad \text{(II.192)}$$

where

$$\vec{r}_1(t) = \vec{R}(t) = \omega_{c_1}^{-2}\vec{f}(t) \equiv \omega_{c_1}^{-2}\vec{f}_1(t) \tag{II.193}$$

Thus $\vec{R}_1(t)$ is the heatbath dynamical variable introduced in Section II.E. Equation (II.192) shows that $\vec{R}_1(t)$ may be interpreted as the trajectory of atom 1 if atom 0 is *clamped* at its equilibrium position [i.e., if $\vec{r}_0(t) = \vec{0}$]. Equivalently, $\vec{R}_1(t)$ is the part of the trajectory of atom 1 which is *not* induced by the motion of atom 0, but rather arises from the *intrinsic* motion of atom 1 [see related discussion after (II.139b)].

The Laplace transform of (II.192) is

$$\hat{\vec{r}}_1(z) = \omega_{c_1}^2 \hat{\theta}_1(z) \hat{\vec{r}}_0(z) + \hat{\vec{R}}_1(z)$$

Using (II.84b), the equation above may be rewritten as [cf. (II.107)]

$$\left[z^2 + \omega_{e_1}^2 - \omega_{c_2}^4 \hat{\theta}_2(z)\right] \hat{\vec{r}}_1(z) - \left[z\vec{r}_1(0) + \dot{\vec{r}}_1(0)\right] = \omega_{c_1}^2 \hat{\vec{r}}_0(z) + \omega_{c_2}^2 \hat{\vec{R}}_2(z) \tag{II.194}$$

where we *define* the coordinate of the heatbath's heatbath, $\hat{\vec{R}}_2(z)$, by

$$\omega_{c_2}^2 \hat{\vec{R}}_2(z) = -\left[z\vec{r}_1(0) + \dot{\vec{r}}_1(0)\right] + \left[z^2 + \omega_{e_1}^2 - \omega_{c_2}^4 \hat{\theta}_2(z)\right] \hat{\vec{R}}_1(z) \tag{II.195}$$

Using (II.192) in (II.190) gives the following effective equation of motion for atom 0,

$$\ddot{\vec{r}}_0(t) = -\omega_{e_0}^2 \vec{r}_0(t) + \omega_{c_1}^2 \vec{r}_1(t) \tag{II.196a}$$

Inverting the Laplace transform in (II.194) gives the following effective equation of motion for atom 1:

$$\ddot{\vec{r}}_1(t) = -\omega_{e_1}^2 \vec{r}_1(t) + \omega_{c_1}^2 \vec{r}_0(t) + \omega_{c_2}^4 \int_0^t \theta_2(t-\tau) \vec{r}_1(\tau)\, d\tau + \omega_{c_2}^2 \vec{R}_2(t) \tag{II.196b}$$

Equations (II.196) are the effective $(N+1)$-atom nearest-neighbor chain representation of many-body dynamics for the case $N = 1$. They generalize (II.108) to finite-temperature dynamics and they reduce to the $T = 0°\mathrm{K}$ representation, (II.108), if the random force acting on atom 1,

$$\vec{f}_2(t) \equiv \omega_{c_2}^2 \vec{R}_2(t) \tag{II.197}$$

is set equal to zero.

Equations (II.196) amount to replacing the true heatbath by an equivalent heatbath whose coordinate $R_1(t)$ satisfies the generalized Langevin equation

$$\ddot{\vec{R}}_1(t) = -\omega_{e_1}^2 \vec{R}_1(t) + \omega_{c_2}^4 \int_0^t \theta_2(t-\tau)\vec{R}_1(\tau)\,d\tau + \vec{f}_2(t) \qquad \text{(II.198)}$$

The equivalent heatbath is then linearly coupled to the primary system via $\omega_{c_1}^2$.

Equation (II.198) is identical in form to the original GLE, (II.141), and one may show (Appendix 5) that $\vec{f}_2(t)$ satisfies the constraint [cf. (II.142a)]

$$\langle \vec{f}_2(t) \cdot \dot{\vec{R}}_1(0) \rangle = 0 \qquad \text{(II.199)}$$

and also the (third) fluctuation–dissipation theorem [cf. (II.92) and (II.93)]

$$\dot{\theta}_2(t) = \frac{m}{3k_B T} \langle \dot{\vec{R}}_2(t) \cdot \dot{\vec{R}}_2(0) \rangle \qquad \text{(II.200a)}$$

or, equivalently,

$$\langle \vec{f}_2(t) \cdot \vec{f}_2(0) \rangle = \frac{3k_B T}{m} \beta_2(t) \qquad \text{(II.200b)}$$

Equations (II.198) to (II.200) illustrate that the recursive structure (m-oscillator→m-oscillator) of the effective equation-of-motion theory generalized to finite temperature [i.e., the heatbath coordinate $\vec{R}(t) = \vec{R}_1(t)$] is an m-oscillator and thus its time development is governed by the GLE, (II.198), with the fluctuation–dissipation theorem of (II.200). The coordinate of the heatbath's heatbath $\vec{R}_2(t)$ is also an m-oscillator and is thus governed by the GLE

$$\ddot{\vec{R}}_2(t) = -\omega_{e_2}^2 \vec{R}_2(t) + \omega_{c_3}^4 \int_0^t \theta_3(t-\tau)\vec{R}_2(\tau)\,d\tau + \vec{f}_3(\tau) \qquad \text{(II.201)}$$

with an associated fluctuation–dissipation theorem, and so on.

Development of the $(N+1)$-atom nearest-neighbor chain representation is straightforward given this recursive structure. The result of the coordinate of its heatbath is an m-oscillator, and so on. This point was discussed in Section II.F.

Development of the general $(N+1)$-atom nearest-neighbor chain representation is now straightforward. The result is [cf. (II.109)]

$$\ddot{\vec{r}}_0(t) = -\omega_{e_0}^2 \vec{r}_0(t) + \omega_{c_1}^2 \vec{r}_1(t) \qquad \text{(II.202a)}$$

and

$$\ddot{\vec{r}}_1(t) = -\omega_{e_1}^2 \vec{r}_1(t) + \omega_{c_1}^2 \vec{r}_0(t) + \omega_{c_2}^2 \vec{r}_2(t)$$

$$\ddot{\vec{r}}_2(t) = -\omega_{e_2}^2 \vec{r}_2(t) + \omega_{c_2}^2 \vec{r}_1(t) + \omega_{c_3}^2 \vec{r}_3(t)$$

$$\vdots$$

$$\ddot{\vec{r}}_N(t) = -\omega_{e_N}^2 \vec{r}_N(t) + \omega_{c_N}^2 \vec{r}_{N-1}(t) + \omega_{c_{N+1}}^4 \int_0^t \theta_{N+1}(t-\tau) \vec{r}_N(\tau)\, d\tau + \vec{f}_{N+1}(t)$$

(II.202b)

The coordinate of atom 0 is, as before, (II.191a), $\vec{r}_0(t) = \vec{r}(t)$. The coordinate of the nth atom is defined recursively as [cf. (II.192) and (II.110)]

$$\vec{r}_n(t) = \omega_{c_n}^2 \int_0^t \theta_n(t-\tau) \vec{r}_{n-1}(\tau) + \vec{R}_n(t) \qquad \text{(II.203)}$$

where the coordinate of the nth heatbath $\vec{R}_n(t)$ is related to the corresponding random force $\vec{f}_n(t)$ by [cf. (II.193)]

$$\vec{R}_n(t) = \omega_{c_n}^{-2} \vec{f}_n(t) \qquad \text{(II.204)}$$

The nth random force (Appendix 5) is completely specified by the constraint

$$\langle \vec{f}_n(t) \cdot \vec{R}_{n-1}(0) \rangle = 0 \qquad \text{(II.205)}$$

and the fluctuation–dissipation theorem [cf. (II.199) and (II.200)]

$$\dot{\theta}_n(t) = \frac{m}{3k_B T} \langle \dot{\vec{R}}_n(t) \cdot \vec{R}_n(0) \rangle \qquad \text{(II.206a)}$$

or, equivalently,

$$\langle \vec{f}_n(t) \cdot \vec{f}_n(0) \rangle = \frac{3k_B T}{m} \beta_n(t) \qquad \text{(II.206b)}$$

Equations (II.202) to (II.206) summarize the nearest-neighbor effective harmonic chain representation of many-body dynamics at finite temperature. It amounts to replacing the true heatbath by the fictitious dynamical system found by setting $\omega_{c_1}^2 = 0$ in (II.202b). This equivalent heatbath is coupled linearly to atom 0, the primary system, by the force constant $\omega_{c_1}^2$. The advantage of the finite-temperature chain representation, like its $T = 0°K$ analog, is that it leads to a useful heatbath modeling scheme which we now develop.

2. *Heatbath Modeling at Finite Temperature*

The heatbath models at finite temperature are derived by replacing the generalized damping on the Nth chain atom by the local frictional damping of (II.114). This leads to the following effective equation of motion for the primary system:

$$\ddot{\vec{r}}_0(t) = -\omega_{e_0}^2 \vec{r}_0(t) + \omega_{c_1}^2 \vec{r}_1(t) \tag{II.207a}$$

and

$$\begin{aligned} \ddot{\vec{r}}_1(t) &= -\omega_{e_1}^2 \vec{r}_1(t) + \omega_{c_1}^2 \vec{r}_0(t) + \omega_{c_2}^2 \vec{r}_2(t) \\ &\vdots \\ \ddot{\vec{r}}_N(t) &= -\Omega_N^2 \vec{r}_N(t) + \omega_{c_N}^2 \vec{r}_{N-1}(t) - \beta_{N+1} \dot{\vec{r}}_N(t) + \vec{f}_{N+1}(t) \end{aligned} \tag{II.207b}$$

The statistics of the random force $\vec{f}_N(t)$ are given by the fluctuation–dissipation theorem [cf. (II.151)]

$$\langle \vec{f}_{N+1}(t) \cdot \vec{f}_{N+1}(0) \rangle = \frac{6k_B T}{m} \beta_{N+1} \delta(t) \tag{II.208}$$

Notice (II.207) and (II.116) are identical at $T=0$, where the random force $\vec{f}_N(t) = \vec{0}$. The simplest model heatbath ($N=1$) case is a single Brownian oscillator whose Langevin equation [cf. (II.117)]

$$\ddot{\vec{r}}_1(t) = -\Omega_1^2 \vec{r}_1(t) - \beta_2 \dot{\vec{r}}_1(t) + \vec{f}_2(t) \tag{II.196c}$$

is of phenomenological *form* [cf. (II.150b)]; recall, however, that the parameter Ω_1 and β_2 may be computed from *exact* many-body dynamics via (II.95) and (II.115). Finally, the heatbath model of (II.209) corresponds to replacing the full many-body system by a two-atom harmonic chain whose effective equation of motion is given by [cf. (II.118)]

$$\ddot{\vec{r}}_0(t) = -\omega_{e_0}^2 \vec{r}_0(t) + \omega_{c_1}^2 \vec{r}_1(t) \tag{II.209a}$$

and

$$\ddot{\vec{r}}_1(t) = -\Omega_1^2 \vec{r}_1(t) + \omega_{c_1}^2 \vec{r}_0(t) - \beta_2 \dot{\vec{r}}_1(t) + \vec{f}_2(t) \tag{II.209b}$$

The value of this model has been discussed at the end of Section II.G.3. It is illustrated pictorially in Fig. 2*a*.

L. Generalized Brownian Motion Theory for *n* Coupled *m*-oscillators

So far we have considered as our primary system a single microscopic or *m*-oscillator. All other particles in the many-body system comprise the heatbath, which influences the dynamics of the primary system through stochastic and dissipative terms in the effective equation of motion, the GLE, for the primary system. Often, however, it will be more convenient to *include* several particles, say n, in the primary system. This may occur, for example, if the dynamics of all n particles is of simultaneous interest. More frequently, however, the dynamics of only one (or a few $<n$) particle(s) is of experimental concern. This dynamics is so strongly influenced by coupling to $n-1$ other particles, however, that all n particles must be treated explicitly. In other words, useful heatbath models can only be constructed if $n-1$ particles are removed from the heatbath and included in the primary system.

A prototypical example is provided by gas–solid reaction dynamics as probed by, say, thermal molecular beam scattering. Only scattered gas molecules may be detected. If one is interested, however, in performing a computer simulation of a heterogeneous reaction, it is insufficient to include only the gas molecules in the primary system. This is because the potential energy surface governing the chemical reaction strongly couples the gas molecules to those solid atoms in the reaction or *primary* zone. The effects of this potential-energy surface cannot be accounted for by simple models of the heatbath influence. Thus all strongly interacting particles, gas molecules plus primary zone solid atoms, must be included in the primary system.

We now briefly outline the most salient results of the effective equation of motion theory for an *n*-particle primary system. This outline contains no new results or concepts, but rather is simply a restatement of some of the results of the previous sections in generalized notation. It is included for the sake of completeness and may be skipped with little loss. For simplicity of notation we assume that all primary system particles have the same mass $=m$.

The key quantity in the analysis is the response matrix $\underline{\chi}_m(t)$ defined by its elements

$$\dot{\chi}_{m_{\alpha\beta}}(t)=\frac{m}{k_B T}\langle\dot{\vec{r}}_\alpha(t)\vec{r}_\beta(0)\rangle \tag{II.210a}$$

All matrices here and below, like $\underline{\chi}_m(t)$, have as elements three-dimensional

Cartesian tensors. In matrix notation

$$\dot{\underline{\chi}}_m(t)=\frac{m}{k_BT}\langle\dot{\mathbf{r}}(t)\dot{\mathbf{r}}^T(0)\rangle \tag{II.210b}$$

where

$$\mathbf{r}(t)=\begin{bmatrix}\vec{r}_1(t)\\ \vec{r}_2(t)\\ \vdots\\ \vec{r}_n(t)\end{bmatrix} \tag{II.211a}$$

and where (note that T denotes a matrix transpose)

$$\mathbf{r}^T(t)=(\vec{r}_1(t)\vec{r}_2(t)\cdots r_n(t)) \tag{II.211b}$$

The coordinates $\vec{r}_\alpha$ and velocities $\dot{\vec{r}}_\alpha$ have the following thermal correlations:

$$\langle\dot{\vec{r}}_\alpha(0)\dot{\vec{r}}_\beta(0)\rangle=\frac{k_BT}{m}\overleftrightarrow{1}\delta_{\alpha\beta} \tag{II.212a}$$

and

$$\langle\dot{\vec{r}}_\alpha(0)\vec{r}_\beta(0)\rangle=\vec{0} \tag{II.212b}$$

Equations (II.210) and (II.212) yield the initial conditions

$$\underline{\chi}_m(0)=\mathsf{0}$$
$$\dot{\underline{\chi}}_m(0)=\mathsf{1} \tag{II.213}$$

where $\mathsf{0}$ and $\mathsf{1}$ are the direct product of the null and unit $n\times n$ matrices with the unit Cartesian tensor $\overleftrightarrow{1}$. Further, because of microscopic reversibility of molecular trajectories and the stationary character of the thermal distribution, we have

$$\dot{\underline{\chi}}_m(-t)=\dot{\underline{\chi}}_m(t)=\dot{\underline{\chi}}_m^T(t) \tag{II.214a}$$

and

$$\underline{\chi}_m(-t)=-\underline{\chi}_m(t)=-\underline{\chi}_m^T(t) \tag{II.2148b}$$

where

$$\underline{\chi}_m(t) = \frac{m}{k_B T}\langle \mathbf{r}(t)\dot{\mathbf{r}}^T(0)\rangle = -\frac{m}{k_B T}\langle \dot{\mathbf{r}}(t)\mathbf{r}^T(0)\rangle \tag{II.215}$$

These results imply that the spectral density matrix

$$\underline{\rho}(\omega) = \frac{2}{\pi}\int_0^\infty \cos\omega t \dot{\underline{\chi}}_m(t)\, dt \tag{II.216a}$$

or

$$\dot{\underline{\chi}}_m(t)0 = \int_0^\infty \cos\omega t \underline{\rho}(\omega)\, d\omega \tag{II.216}$$

is even and symmetric:

$$\underline{\rho}(-\omega) = \underline{\rho}(\omega) = \underline{\rho}^T(\omega)$$

and is also unit-normalized:

$$\int_0^\infty \underline{\rho}(\omega)\, d\omega = 1 \tag{II.217}$$

The second moment of $\underline{\rho}(\omega)$ is the symmetric Einstein frequency matrix given by

$$\underline{\omega}_e^2 = \underline{\omega}_{e_0}^2 = \int_0^\infty \omega^2 \underline{\rho}(\omega)\, d\omega = -\ddot{\underline{\chi}}_m(0) \tag{II.218}$$

The adiabatic frequency matrix is given by

$$\underline{\Omega}^{-2} = \hat{\underline{\chi}}_m(0) = \int_0^\infty \omega^{-2}\underline{\rho}(\omega)\, d\omega \tag{II.219}$$

The response matrix $\underline{\chi}_m(t)$ obeys the following effective equation of motion:

$$\ddot{\underline{\chi}}_m(t) = -\underline{\omega}_e^2 \underline{\chi}_m(t) + \int_0^t \underline{\Theta}_1(t-\tau)\underline{\chi}_m(\tau)\, d\tau \tag{II.220}$$

The damping matrix $\underline{\Theta}_1(t)$ is odd:

$$\underline{\Theta}_1(-t) = -\underline{\Theta}(-t) \tag{II.221a}$$

and its time derivative is thus even:

$$\underline{\dot{\Theta}}_1(-t) = \underline{\dot{\Theta}}_1(t) \tag{II.221b}$$

It satisfies the following initial conditions:

$$\begin{aligned} \underline{\Theta}_1(0) &= 0 \\ \underline{\dot{\Theta}}_1(0) &= \underline{\omega}_c^4 = \underline{\omega}_{c_1}^4 \end{aligned} \tag{II.222}$$

where the coupling constant matrix $\underline{\omega}_{c_1}^2$ is given by

$$\underline{\omega}_{c_1}^4 = \int_0^\infty \underline{\rho}(\omega)\left[\omega^2 1 - \underline{\omega}_{e_0}^2\right]^2 d\omega \tag{II.223}$$

Thus the normalized heatbath response matrix $\underline{\theta}(t)$ defined by

$$\underline{\Theta}_1(t) = \underline{\omega}_{c_1}^2 \underline{\theta}_1(t) \underline{\omega}_{c_1}^2 \tag{II.224}$$

satisfies

$$\begin{aligned} \underline{\theta}_1(t) &= -\underline{\theta}_1(t) \\ \underline{\dot{\theta}}_1(t) &= \underline{\dot{\theta}}_1(-t) \end{aligned} \tag{II.225}$$

with the initial conditions

$$\begin{aligned} \underline{\theta}_1(0) &= 0 \\ \underline{\dot{\theta}}_1(0) &= 1 \end{aligned} \tag{II.226}$$

Hence we may deduce that $\underline{\theta}(t)$ has an effective equation of motion of the form

$$\underline{\theta}_1(t) = -\underline{\omega}_{e_1}^2 \underline{\theta}(t) + \int_0^t \underline{\Theta}_2(t-\tau)\underline{\theta}_1(\tau)\, d\tau \tag{II.227}$$

where $\underline{\omega}_{e_1}^2$ is the second moment of the heatbath spectral density matrix $\underline{\sigma}_1(\omega)$:

$$\underline{\omega}_{e_1}^2 = \int_0^\infty \omega^2 \underline{\sigma}_1(\omega)\, d\omega = -\underline{\ddot{\theta}}_1(0) \tag{II.228a}$$

where

$$\underline{\sigma}_1(\omega)=\underline{\sigma}_1(-\omega)=\underline{\sigma}_1^T(\omega)=\frac{2}{\pi}\int_0^\infty \cos\omega t\,\dot{\underline{\theta}}_1(t)\,dt \tag{II.229}$$

We may write

$$\underline{\Theta}_2(t)=\underline{\omega}_{c_2}^2\underline{\theta}_2(t)\underline{\omega}_{c_2}^2 \tag{II.230}$$

where the second coupling constant matrix $\underline{\omega}_{c_2}^2$ is defined by

$$\underline{\omega}_{c_2}^4=\int_0^\infty \underline{\sigma}_1(\omega)\left[\omega^2\mathbf{1}-\underline{\omega}_{e_1}^2\right]^2 d\omega \tag{II.228b}$$

Finally, we may show that

$$\dot{\underline{\theta}}_2(-t)=\dot{\underline{\theta}}_2(t)$$

and thus establish that $\underline{\theta}_2(t)$ is a coupled set of m-oscillators, and so on. Continuing this process, we find that the nth heatbath response matrix is directly coupled to the $(n-1)$th and $(n+1)$th heatbath response matrices by

$$\ddot{\underline{\theta}}_{n-1}(t)=-\underline{\omega}_{e_{n-1}}^2\underline{\theta}_{n-1}(t)+\int_0^t \underline{\Theta}_n(t-\tau)\underline{\theta}_{n-1}(\tau)\,d\tau \tag{II.231a}$$

$$\ddot{\underline{\theta}}_n(t)=-\underline{\omega}_{e_n}^2\underline{\theta}_n(t)+\int_0^t \underline{\Theta}_{n+1}(t-\tau)\theta_n(\tau)\,d\tau \tag{II.231b}$$

where

$$\underline{\omega}_{e_n}^2=-\ddot{\underline{\theta}}_n(0)=\int_0^\infty \omega^2\underline{\sigma}_n(\omega)\,d\omega \tag{II.232}$$

and where

$$\underline{\omega}_{c_n}^2=\int_0^\infty \underline{\sigma}_{n-1}\left[\omega^2\mathbf{1}-\underline{\omega}_{e_{n-1}}^2\right]^2 d\omega \tag{II.233}$$

The unit-normalized spectral density matrix of the nth heatbath $\underline{\sigma}_n(\omega)$ is given by

$$\underline{\sigma}_n(\omega)=\frac{2}{\pi}\int_0^\infty \cos\omega t\,\dot{\underline{\theta}}_n(t)\,dt \tag{II.234}$$

From the recursive structure of (II.231) one may develop the generalization of the continued fraction representation, the moment analysis, the $t=0°\mathrm{K}$ harmonic chain representation and the "mode" analysis of Sections II.F to II.H.

At finite temperature, the effective equation of motion becomes the generalized Langevin equation

$$\ddot{\mathbf{r}}(t)=-\underline{\omega}_{e_0}^2\mathbf{r}(t)+\int_0^t\underline{\Theta}(t-\tau)\mathbf{r}(\tau)\,d\tau+\mathbf{f}(t) \tag{II.235a}$$

where

$$\mathbf{f}(t)=\begin{bmatrix}\vec{f}_1(t)\\ \vec{f}_2(t)\\ \vdots\\ \vec{f}_n(t)\end{bmatrix}$$

Written in more explicit notation, the GLE for particle α, $1\leq\alpha\leq n$, is

$$\ddot{\vec{r}}_\alpha(t)=-\sum_\gamma\left[\overleftrightarrow{\omega}_{e_0}^2\right]_{\alpha\gamma}\cdot\vec{r}_\gamma(t)+\int_0^t\overleftrightarrow{\Theta}_{\alpha\gamma'}(t-\tau)\cdot\vec{r}_{\gamma'}(\tau)\,d\tau+\vec{f}_\alpha(t) \tag{II.235b}$$

The random force $\mathbf{f}(t)$ is related to the damping matrix by the second fluctuation–dissipation theorem,

$$\langle\dot{\mathbf{f}}(t)\dot{\mathbf{f}}^T(0)\rangle=\frac{k_BT}{m}\underline{\omega}_{c_1}^2\underline{\theta}_1(t)\underline{\omega}_{c_1}^2=\frac{3k_BT}{m}\dot{\underline{\Theta}}_1(t) \tag{II.236a}$$

or, equivalently

$$\langle\mathbf{f}(t)\mathbf{f}^T(0)\rangle=\frac{k_BT}{m}\underline{\beta}_1(t) \tag{II.236}$$

where the friction matrix $\underline{\beta}_1(t)$ is defined by

$$\underline{\beta}_1(t)=\int_t^\infty\underline{\Theta}_1(\tau)\,d\tau \tag{II.237}$$

Defining the heatbath dynamical variables $\mathbf{R}_1(t)$ by

$$\mathbf{f}_1(t)=\underline{\omega}_{c_1}^2\mathbf{R}_1(t) \tag{II.238}$$

may show that $\mathbf{R}(t)$ satisfies the following GLE:

$$\ddot{\mathbf{r}}_1(t) = -\underline{\omega}_{e_1}^2 \mathbf{r}_1(t) + \underline{\omega}_{c_2}^2 \int_0^t \underline{\theta}_2(t-\tau) \underline{\omega}_{c_2}^2 \mathbf{R}_1(\tau)\, d\tau + \underline{\omega}_{c_2}^2 \mathbf{R}_2(t) \quad \text{(II.239)}$$

with the fluctuation–dissipation theorem

$$\langle \underline{\omega}_{c_2}^2 \dot{\mathbf{R}}_2(t) \dot{\mathbf{R}}_2^T(0) \left[\underline{\omega}_{c_2}^2\right]^T \rangle = \frac{k_B T}{m} \underline{\omega}_{c_2}^2 \dot{\underline{\theta}}_2(t) \underline{\omega}_{c_2}^2 = \frac{k_B T}{m} \underline{\Theta}_2(t) \quad \text{(II.240)}$$

Continuing this process gives the general recursive structure of the effective equation of motion theory at finite temperature [cf. (II.231)], the GLE

$$\ddot{\mathbf{R}}_{n-1}(t) = -\underline{\omega}_{e_{n-1}}^2 \mathbf{R}_{n-1}(t) + \underline{\omega}_{c_n}^2 \int_0^t \underline{\theta}_n(t-\tau) \underline{\omega}_{c_n}^2 \mathbf{R}_{n-1}(\tau)\, d\tau + \underline{\omega}_{c_n}^2 R_n(t) \quad \text{(II.241)}$$

with the $(n+1)$th fluctuation–dissipation theorem

$$\langle \underline{\omega}_{c_n}^2 \mathbf{R}_n(t) \mathbf{R}_n^T(t) \underline{\omega}_{c_n}^2 \rangle = \frac{k_B T}{m} \underline{\omega}_{c_n}^2 \dot{\underline{\theta}}_n(t) \underline{\omega}_{c_n}^2 = \frac{k_B T}{m} \underline{\Theta}_n(t) \quad \text{(II.242)}$$

This concludes our discussion of generalized Brownian motion theory.

III. GAS–SOLID COLLISIONS: A CASE STUDY IN MANY-BODY CHEMICAL DYNAMICS

A general formalism for many-body problems in chemical dynamics, based on classical mechanics but otherwise unrestricted, was developed in Section II. This formalism amounts to a *reshaping*[1,17] of the modern correlation function approach[6–14] to general many-body dynamical problems. This reshaping was aimed at converting the effective equation of motion theory from a rather formal approach to macroscopic transport problems[12] (e.g., diffusion, fluid flow, heat transfer, electrical conduction, etc.) into a flexible and convenient theoretical framework for *molecular-time-scale* dynamics. The value of the new theory is twofold. First, it allows one to *compute* condensed phase chemical dynamics from *small-scale* classical trajectory simulations. This is an important development, since previously classical trajectory studies were limited to small-molecule gas-phase processes[4,5] or else involved massive-scale molecular dynamics calculations.[29] Second, and equally important, is that the theory provides a *language* which helps one think about and understand how the *collective* behavior of many interacting heatbath molecules influence the primary or chemical system dynamics. Thus the *theoretical* framework of Section II provides a *practical* tool which allows one to both compute and understand the subtle effects arising because of interplay between many-body and chemical interactions in real condensed-phase processes.

We believe that this formal framework will be of wide applicability and will enable one to study via computer simulations a broad range of molecular-time-scale physical and chemical processes. These include, for example:

1. Vibrational energy relaxation in liquids and solids.
2. Time-resolved chemical reaction dynamics in liquids, including free-radical and ion-recombination dynamics, rates of conformation change, and so on.
3. Fundamental gas–solid energy exchange and absorption processes.
4. Heterogeneous chemical reaction dynamics at both the gas–solid and liquid-solid interfaces.
5. Molecular-time-scale polymer and bipolymer dynamics.

The purpose of this article is to develop and clearly present the concepts, methods, and models which are likely to be of widest applicability in many-body chemical problems. For this reason, we present in this section a study of what is probably the simplest example of a many-body chemical process, namely nonreactive gas–solid surface collision dynamics. This simple (and very important) prototype illustrates the new types of behavior one sees in condensed-phase processes. It clearly shows, moreover, how the formal framework of Section II allows one to both compute this behavior and understand it in qualitative physical language.

The plan of this section is as follows. The nonlinear generalized Langevin equation for gas–solid collisions is introduced in Section III.A. In Sections III.B and III.c we discuss the heatbath response function for the simplest model of use in gas–solid dynamics, the Debye model of (II.41) and we also deal with numerical methods for solving the GLE based on the effective harmonic chain representation of many-body dynamics, Section II.K. Section III.D deals with model computer simulations of gas–solid energy-exchange processes based on the Debye model and two more realistic theories. The results in Section III.D underline the importance of many-body effects in gas–solid energy transfer and further demonstrate how physical effects built into the response function are mirrored in computed gas–solid scattering attributes. Thus the model calculations show how the GLE framework allows one to understand the many-body effects in simple terms. Finally, in Section III.E we close with a summary.

A. Generalized Langevin Equation for Gas–Solid Dynamics

The new qualitative features which emerge in many-body chemical dynamics are strikingly illustrated by the simplest treatment of the simplest many-body chemical process, atom–solid surface energy-exchange processes in which the incident gas atom directly and collinearly strikes a single solid atom. For this case, the primary system is comprised of the incident gas atom and the single surface atom which is directly struck during the collision. If we let $\vec{R}$ and M and $\vec{r}$ and m be, respectively, the coordinates and masses of the incident gas atom and the struck surface atom, and let $W(\vec{r},\vec{R})$ be the atom–solid potential-energy function, then the GLE for atom–solid collisions becomes

$$M\ddot{\vec{R}}(t) = -\frac{\partial W(\vec{r},R)}{\partial \vec{R}} \tag{III.1a}$$

and

$$\ddot{\vec{r}}(t) = -m^{-1}\frac{\partial W}{\partial \vec{r}}(\vec{r},\vec{R}) - \omega_e^2\vec{r}(t) + \int_0^t \Theta(t-\tau)\dot{\vec{r}}(\tau)\,d\tau + \vec{f}(t) \tag{III.1b}$$

with $\Theta(t) = \omega_c^4\theta(t) = \omega_{c_1}^4\theta_1(t)$ and with the statistics of the random force given by the second fluctuation dissipation theorem, (II.74), (II.142b), or (II.155).

Equation (III.1) requires several comments.

1. The GLE for the "free" surface atom, given by (II.141), has simply been coupled to the gas atom equation of motion, (III.1a), in the manner familiar from ordinary (nondissipative) mechanics; that is, the heatbath response $\Theta(t)$ used in (III.1b) is assumed to be the response in the *absence* of the gas atom. This may be rigorously justified if the solid is perfectly harmonic but is otherwise an (essential) approximation.
2. The potential-energy function $W[\vec{r},\vec{R}]$ can include the *static* influence of the heatbath solid atoms (i.e., it can depend on the *equilibrium* position of *all* atoms in solid). In this manner, the geometry of, say, the crystal face struck by the solid atom may be built into the theory.[18a] We stress, however, that all atoms whose *instantaneous* coordinate is included in the gas–solid potential must be treated explicitly in the GLE (see Section II.L).
3. The damping kernel $\Theta(t)$ contains the following effects:
   a. The influence of the nuclear motion of all heatbath solid atoms on the dynamics of the primary system; this nuclear motion may be either harmonic or anharmonic.
   b. Any influence of nonnuclear degrees of freedom,[30] for example electronic or spin degrees of freedom, on atomic motion. An example of

such influence [breakdown of Born–Oppenheimer approximation] is depicted in Fig. 1.

c. Any anharmonicities in the primary system motion. Often these anharmonicities are sufficiently specific and important that they cannot be usefully lumped into the response kernel but rather should be introduced explicitly as nonlinear forces in (III.2) (Fig. 5).

4. The equation of motion of the gas atom, (III.1a), may also be written as an effective equation of motion involving a damping kernel and an associated random force. This could account for the neglect of heatbath fluctuations in $W(\vec{r},\vec{R})$, and interaction with nonnuclear degrees of freedom.[30] Such refinements will not be considered here.

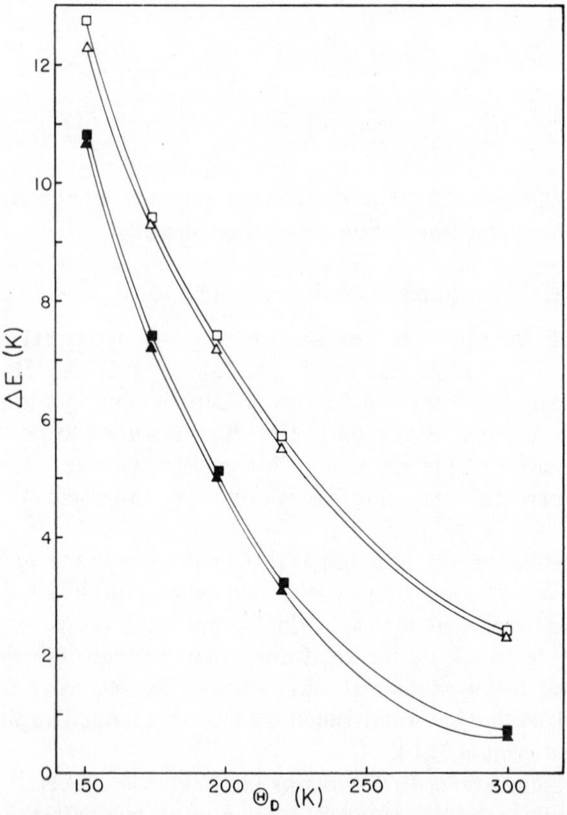

Fig. 5. Energy transfer vs. Debye temperature for ne–metal collisions. A Morse potential $W(\vec{r},\vec{R})=D[e^{-2\alpha(|\vec{R}-\vec{r}|)}-2e^{-\alpha(|\vec{R}-\vec{r}|)}]$ is employed. Morse parameters are $D_e=112°K$, $\alpha=1.8\ Å^{-1}$. Incident gas energy is $E_i=296°K$. Single Brownian oscillator approximant to Debye heatbath with (□) and without (△) explicit inclusion of an anharmonic restoring force on the primary solid atom; corresponding Einstein model results are indicated by (■) and (▲).

B. Heatbath Response for the Debye Model

The heatbath response function $\Theta(t)$ for an arbitrary system may be conveniently computed from its spectral density $\sigma(\omega)$ using (II.64) and (II.67b). The heatbath spectral density, in turn, may be computed from the primary system spectral density $\rho(\omega)$ using (II.69). The result for $\dot{\Theta}(t)$ is

$$\dot{\Theta}(t)=\int_0^{\infty}\frac{\rho(\omega)\cos\omega t}{|\hat{\chi}_m(i\omega)|^2}\,d\omega \tag{III.2}$$

For the Debye model $\rho(\omega)$ is given by (II.41) and $|\hat{\chi}_m(i\omega)|^2$ is readily computed from

$$\hat{\chi}_m(z)=\frac{3}{\omega_D^3}\left[\omega_D-z\tan^{-1}\left(\frac{\omega_D}{z}\right)\right] \qquad \text{Debye} \tag{III.3}$$

found from the Laplace transform of (II.45). The Debye model $\Theta(t)$ found from (III.2) and (III.3) is plotted in Fig. 3. Notice that it displays damped oscillatory behavior. This may be explored in the "physical" language of Section II.G.2, since $\Theta(t)$ like $\chi_m(t)$ has a simple interpretation within the effective harmonic chain representation. Specifically, $\Theta(t)$ is the trajectory of atom 1 of the fictitious harmonic chain for the case that atom 0 is *clamped* at its equilibrium position and for the special initial conditions $r_n(0)=0$, $\dot{r}(0)=0$, $n>1$, and $\dot{r}(0)=\omega_c^4$.

If we define the *many-body effect* on computed gas–solid scattering attributes as the difference between the results calculated within the GLE theory, (III.1), and within the *Einstein model* defined by dropping the dissipative and random terms in (III.1b) [i.e., set $\Theta(t)=0$ and $\vec{f}(t)=\vec{0}$], then we see that the magnitude of the many-body effect depends strongly on $\dot{\Theta}(0)=\omega_c^4$. Within the Debye model, (II.41), (II.43), and (II.63) yield for the coupling constant

$$\omega_c^2=\sqrt{\tfrac{12}{175}}\,\omega_D^2 \tag{III.4}$$

Thus the magnitude of the many-body effect increases with Debye frequency. This is physically very sensible; the larger the characteristic frequency of the solid, the greater is the speed of sound in the solid and the more quickly energy in the primary system may be dissipated.

A second important parameter in governing the magnitude of the many-body effect is the Einstein frequency of the heatbath ω_{e_1}. The importance of the many-body effect *decreases* with increasing ω_{e_1} if ω_c^2 is kept constant. This may be readily understood in the effective harmonic chain language. If atom 1 is restored by a "stiff" spring (i.e., if ω_{e_1} is large), then the energy from the primary system is not easily pumped into atom 1. Alternatively, ω_{e_1}, roughly speaking, governs the oscillation period of $\Theta(t)$ (see Fig. 3), and if this period is short, the response integrand in the GLE, (III.1b), is highly oscillatory and the many-body effects tend to cancel, particularly when one averages over trajectories. For the Debye model, ω_{e_1} is given by [use (II.41), (II.43), and (II.73b)]

$$\omega_{e_1} = 0.71\omega_D \tag{III.5}$$

Comparing (III.5) with the Debye-model result for ω_{e_0}, (II.43),

$$\omega_{e_0} = \sqrt{\tfrac{3}{5}}\ \omega_D = 0.77\omega_D \tag{III.6}$$

we see that the restoring frequency of the heatbath is comparable to that for the primary system.

Notice that $\omega_{c_1}^2$ and ω_{e_1} are not independent, since both are proportional to ω_D. That these very different quantities are so closely coupled is probably due to weakness in the Debye model.

The heatbath spectral density $\sigma(\omega)$ provides another way to look at the many-body system. We plot $\omega^{-1}\sigma(\omega)$ vs. $x = \omega_D^{-1}\omega$ in Fig. 4 for the Debye model. Note that $\sigma(\omega)$ like $\rho(\omega)$ is cutoff at $x = 1$. This follows from the interleaving property of heatbath and full system "normal modes" (Section II.H). The low-frequency part of $\omega^{-1}\sigma(\omega)$ shows typical "acoustic" behavior in accord with (II.79). At high frequencies ($x \gtrsim 0.8$), $\sigma(\omega)$ quickly falls to zero [i.e., the local density of heatbath modes at atom 1, (II.130), is depleted at higher frequencies].

C. Numerical Solution of the GLE and Heatbath Models

Heatbath modeling is necessary at two levels in our approach to many-body chemical dynamics. First, the true system must be modeled since the *exact* response characteristics are not known for any real many-body system. Second, even with model response functions, the nonlinear GLE [e.g., (III.1)] is numerically intractable (see the discussion in Section II.G) unless the heatbath modeling scheme of Sections II.G and II.K (or its equivalent[1b]) is employed. An example of the first type of modeling is our use of the Debye approximation for $\rho(\omega)$. The second type of modeling was first alluded to in Section II.A.

We now discuss numerical solution of the GLE with emphasis on the heatbath modeling techniques of Sections II.G and II.K. Within the framework of harmonic chain representation, Section II.K, the GLE (III.1) becomes [cf. (II.202)]

$$M\ddot{\vec{R}}(t) = -\frac{\partial W[\vec{r}_0, \vec{R}]}{\partial \vec{R}} \tag{III.7a}$$

$$\ddot{\vec{r}}_0(t) = -\frac{1}{m}\frac{\partial W[\vec{r}_0, \vec{R}]}{\partial \vec{r}} - \omega_{e_0}^2 \vec{r}_0(t) + \omega_{c_1}^2 \vec{r}_1(t) \tag{III.7b}$$

$$\ddot{\vec{r}}_1(t) = -\omega_{e_1}^2 \vec{r}_1(t) + \omega_{c_1}^2 \vec{r}_0(t) + \omega_{c_2}^2 \vec{r}_2(t)$$

$$\vdots$$

$$\ddot{\vec{r}}_N(t) = -\Omega_N^2 \vec{r}_N(t) + \omega_{c_N}^2 \vec{r}_{n-1}(t) - \beta_{N+1}\dot{\vec{r}}_N(t) + \vec{f}_{N+1}(t) \tag{III.7c}$$

Equation (III.7) is to be supplemented by the fluctuation–dissipation theorem of (II.208).

Even with the fluctuation–dissipation theorem, (III.7) is incomplete at finite T unless the statistics of $\vec{f}(t)$ and the initial values $\vec{r}_n(0)$ and $\dot{\vec{r}}_n(0)$ are specified. We assume that these are distributed *Gaussianly*. The Gaussian assumption is exact only for perfectly harmonic many-body systems. For other systems little is known about the exact statistics (which depend ultimately on the true many-body partition function) and the Gaussian approximation will be adopted pragmatically. Once Gaussian statistics are assumed, Monte Carlo sampling of $\vec{r}_n(0)$, $\dot{\vec{r}}_n(0)$, and $\vec{f}(t)$ is relatively straightforward.

Let us illustrate the Gaussian sampling methods for the initial velocities of the chain atoms, $\dot{\vec{r}}_n(0)$. On intuitive grounds, one would expect that the p.d.f. for $\dot{\vec{r}}_n(0)$ is the Maxwellian distribution

$$P[\dot{r}_n] \sim \exp\left(-\frac{1}{2}\frac{m\dot{r}_n^2}{k_B T}\right) \tag{III.8}$$

Actually, the "obvious" result of (III.8) must be *proven* since the chain atoms are fictitious. [In fact, (III.8) is *not* true in general but rather depends on assumption of Gaussian statistics.] The proof is based on the recursive *definition* of the coordinates $\vec{r}_n(t)$, (II.203). From (II.203) it follows that

$$\dot{\vec{r}}_n(t) = \omega_{c_n}^2 \theta_n(0)\vec{r}_{n-1}(t) + \omega_{c_n}^2 \int_0^t \theta_{n-1}(t-\tau)\dot{\vec{r}}_n(\tau)\,d\tau + \dot{\vec{R}}_n(t) \tag{III.9}$$

Using the initial condition $\theta_n(0) = 0$, (III.9) gives at $t = 0$,

$$\dot{\vec{r}}_n(0) = \dot{\vec{R}}_n(0) \tag{III.10}$$

Thus because of the assumed Gaussian statistics and also (III.10), the p.d.f. for $\dot{r}_n$ is

$$P[\dot{r}_n] \sim \exp\left(-\frac{3}{2}\frac{\dot{r}_n^2}{\langle \dot{R}_n^2 \rangle}\right) \tag{III.11}$$

But from (II.92a),

$$\langle \dot{R}_n^2 \rangle = \langle \vec{\dot{R}}_n(0) \cdot \vec{\dot{R}}_n(0) \rangle = \frac{3k_B T}{m} \dot{\theta}_n(0) = \frac{3k_B T}{m} \tag{III.12}$$

Combining (III.11) and (III.12) gives the desired result, (III.8).

Sampling of the atom positions $\vec{r}_n(0)$ is more involved. The positions and velocities are statistically uncorrelated:

$$\langle \vec{r}_n(0) \cdot \vec{\dot{r}}_n(0) \rangle = 0 \tag{III.13}$$

This result may be derived from (II.203) to (II.205). The coordinates themselves, however, are correlated [i.e., $\langle \vec{r}_n(0) \cdot \vec{r}_m(0) \rangle \neq 0$ if $n \neq m$]. "Physically" the correlation is due to the connecting "springs" linking the "atoms," and thus one must sample from an $(N+1) \times (N+1)$ Gaussian pdf [cf. (A.44) to (A.46)]. This sampling problem is not all serious (since typically $N = 1$ to 4) and can be overcome by, say, moving to the normal-mode representation of Section II.H. The modes are statistically independent, as expected on "physical" grounds. Similar considerations are involved in sampling the random force $\vec{f}_{N+1}(t)$. We will not pursue details of the sampling further here except to mention that at $T = 0°$K, one may show that

$$\vec{r}_n(0) = \vec{\dot{r}}(0) = \vec{f}_{N+1}(t) = \vec{0} \qquad T = 0\text{ K} \tag{III.14}$$

and thus there is no sampling problem at $T = 0$K.

Once the random variables have been sampled, (III.7) may be solved by standard gas-phase trajectory methodology (e.g., Runge-Kutta integration). The elimination of history-dependent terms in the effective equations of motion is crucial to the solvability of (III.7) by standard techniques as discussed in Section II.G.3. In fact, the requirement of numerical solvability of the GLE is what motivated our development of the nearest-neighbor chain representation and associated models in Sections II.G and II.K.

The key point of (III.7) is that the *original* $\sim 10^{24}$*-body problem has been reduced to an* effective *few-body problem involving* $N+1$ *ficitious particles*. The reduction is completely rigorous for perfectly harmonic many-body systems; that is, as $N \to \infty$, the scattering attributes as computed from (III.7) will approach the exact scattering attributes obtained by performing the full $\sim 10^{24}$ body dynamical calculation. This claim *cannot* be made for nonharmonic systems. The method, however, is physically very plausible and appealing for even nonharmonic systems. The practical value method derives from the fact that excellent results may be obtained even if N in (III.7) is small. The case $N=1$, for example, which requires only an effective *three-body* collision calculation, gives a good semiquantitative account of many gas-solid energy-transfer processes.[1b] Convergence studies of atom-solid sticking probabilities show[1b] that results predicted by models for which $N \sim 3$ to 5 are converged to well within the accuracy of the classical trajectory method.

The qualitatively desirable features of the heatbath modeling scheme used in (III.7) is depicted pictorially in Fig. 3, where we plot the exact Debye model $\Theta(t)$ and two approximants. The approximants are the single Brownian oscillator heatbath model ($N=1$) (Fig. 2a) and the $N=4$ response function computed with damping suppressed [$\beta_5=0$ and $\vec{f}_5(t)=\vec{0}$ in (III.7c)]. Ignoring damping leads to unphysical behavior in the response function for $\omega_D t > 10$. This behavior is a reflection of the Poincaré recurrences occurring in the ($N=4$) undamped model heatbath. The disastrous consequences of these recurrences for chemical dynamics simulations were extensively discussed in Section II.G.3 and are apparent from Fig. 3. The damped heatbath models are free from recurrences. This is clear for the $N=1$ model from Fig. 3.

They also have a second essential property. They give correct results for short-time-scale processes. This has been mentioned in Section II.G.3. This is particularly easy to see at $T=0°\text{K}$, where $\vec{f}(t)=\vec{0}$, since the heatbath response integral in (III.1b) does not contribute at $t \to 0$. The models are also exact on short time scales at finite T as long as the *initial* value of $\vec{f}(t)$ is chosen from the correct distribution function. This is guaranteed [see (II.154) and (II.156)] if the correct values of ω_{c_1} and Ω_1 are built into the models. This is easily accomplished.

Thus the model dynamical problem of (III.7) gives correct results for very short time-scale processes (Einstein limit), is free from recurrence problems at longer times, and is very easy to treat computationally. Even the simplest model ($N=1$) is free of qualitative error. Moreover, at any level the error in calculated dynamics due to modeling may be assessed by a convergence study and the error may be systematically reduced by increasing N. (The price paid, of course, is that computational labor increases with N.)

D. Numerical Implementation of the Generalized Langevin Theory

The generalized Langevin theory reviewed in this section provides a computationally practical generalization of the gas-phase classical trajectory theory which makes the trajectory method applicable to the many-body problems characteristic of condensed-phase chemistry. The theory may be used to simulate a wide variety of condensed-phase processes.

The required input into the calculations is a potential-energy surface, as in gas-phase studies, and additionally a response function for the many-body system. If the potential-energy surface and response function accurately represent a real system, then the theory should give results in good accord with experiments on that system. Useful response functions may be constructed by a variety of methods. Thus the main difficulty in applying the GLE theory to real systems lies in the potential-energy surface problem. This obstacle also is a central problem in gas-phase dynamics work.

While fairly large scale finite-temperature GLE simulations of gas–solid reaction dynamics[18a] and vibrational relaxation in solids[18b] have been carried out using model potential-energy surfaces, this work will not be reviewed here. In keeping with the spirit of this chapter, we restrict ourselves to the simplest model calculations which display the essential new features arising in condensed-phase dynamics due to many-body effects. Since the random force complicates discussion of these features, we will present only $T=0°$K results here.

Figure 5 gives a first illustration of the role of many-body effects on gas–solid energy transfer. The many-body effect is assessed by comparing energy transfer ΔE computed from the Brownian oscillator approximant to the exact Debye-model *heatbath* [Fig. 2*a* or $N=1$ in (III.7c)] with the ΔE value calculated from the Einstein model (Fig. 2*b*) with the Debye $\omega_{e_0}=\sqrt{\frac{3}{5}}\ \omega_D$. Recall that the many-body effect is defined as the difference between GLE and Einstein results and thus the calculation of Fig. 5 gives a direct measure of the effect. The results for energy transfer ΔE are plotted as a function of Debye temperature Θ_D [defined by $k_B\Theta_D=\hbar\omega_D$] for Ne-metal scattering. (Potential parameters, solid atom masses, and so on, are here and below relegated to the figure captions.) Figure 5 clearly shows that many-body effects *enhance* energy transfer at $T=0°$K. While the overall magnitude of the energy transfer *decreases* with Θ_D, the percent contribution of the many-body effect to energy transfer *increases* with Θ_D. The reason is that increase in Θ_D increases the Einstein frequency ω_{e_0}, (III.6), and hence decreases the *initial* energy transfer into the solid. Increase in Θ_D, however, *increases* the speed of sound in the solid. This increase in sound velocity, in turn, enhances the efficiency of removal of energy from the primary system and thus increases the many-body influence. Equivalently, the primary system/heatbath coupling constant ω_c^2 increases with ω_D, (III.4), and this increase enhances the many-body contribution.

Trapping thresholds, defined as the maximum gas atom incident energy which will permit trapping, are plotted as a function of Θ_D in Fig. 6. The trapping thresholds, which provide a measure of the ease of trapping, are

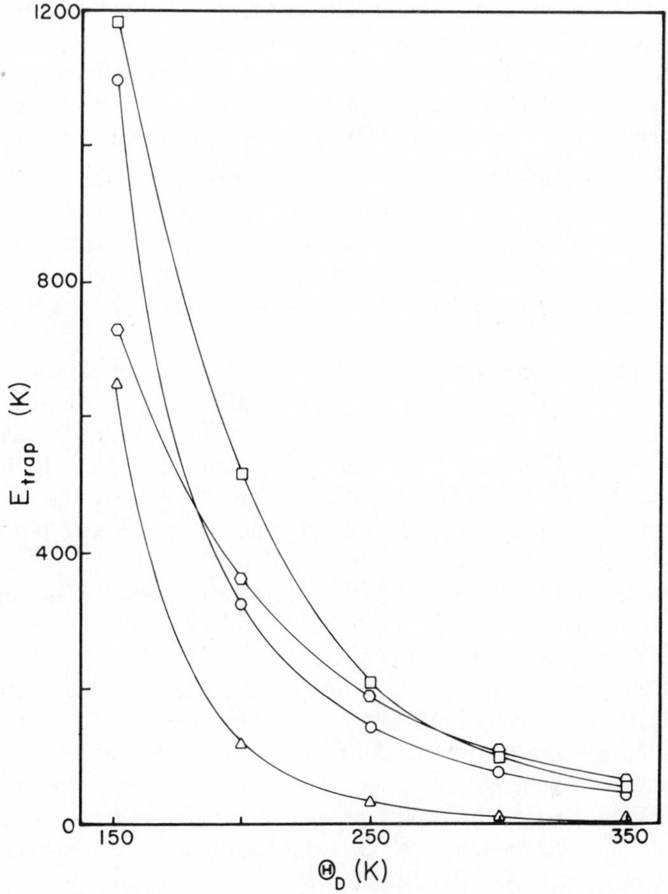

Fig. 6. Trapping threshold vs. Θ_D. Incident gas energy $E_i = 300°K$. Morse parameters $D_e = 417.8°K$, $\alpha = 1.69$ Å^{-1} are chosen to simulate Ar–Ag scattering. (○) Single Brownian oscillator approximant to Debye heatbath; (△) Einstein model with $\omega_e = \sqrt{\frac{3}{5}}\ \omega_D$; (□) Local friction model with Debye friction coefficient, $\beta_D = (\pi/6)\omega_D$; (⬡) Local friction model with β chosen to give optimal fit to Debye $\chi_m(t)$, $\beta_F = 0.268\omega_D$. For both local friction models the Debye adiabatic frequency $\Omega = \sqrt{\frac{1}{3}}\ \omega_D$ is employed.

plotted as a function of Θ_D for a model designed to simulate Ar–Ag scattering. Results for the $(N = 1)$ Brownian heatbath and Einstein approximants to the exact Debye response function are presented. Also plotted are results for scattering off a phenomenological Brownian oscillator model for the *full* many-body system, (II.150b). The Brownian oscillator or local friction model for the full system is illustrated in Fig. 2*c*.

Figure 6 shows that trapping efficiency as measured by the trapping threshold is extremely sensitive to the Einstein frequency, $\sim\Theta_D$, and is greatly enhanced by many-body effects. The sensitivity to Einstein frequency underlines the importance of using ω_e rather than the adiabatic frequency Ω in the simulations. This point was stressed in Section II. Since for the Debye model $\Omega=\sqrt{\frac{5}{9}}\,\omega_e=0.75\omega_e$, we see from Fig. 6 that if Ω rather than ω_e is used as the basic frequency, the trapping threshold at $\Theta_D=225°K$ is increased (within the isolated oscillator approximation for the many-body system) from $\Delta E\cong 80°K$ to $\Delta E\cong 300°K$. Thus use of Ω rather than ω_e would lead to predicted trapping efficiencies which are much too large.

Figure 6 additionally shows that the Einstein model underestimates trapping efficiency. For $\Theta_D=150°K$, for example, trapping occurs at gas energies less than $T\sim550°K$ for the Einstein model and for gas energies less than $T\sim1000°K$ for the GLE theory (implemented within the Brownian oscillator heatbath approximation). Thus the "agreement" between single oscillator (soft cube[31]) models and experiment for sticking probabilities, which has been claimed in the literature, is undoubtedly spurious. Data may always be fit by juggling system parameters; unless the dynamical theory upon which the data fit is based is reliable, however, "agreement" between theory and experiment is meaningless.

The trapping efficiency (Fig. 6) follows the overall pattern of the energy transfer (Fig. 5). Overall efficiency is greatest at small Θ_D because of the initial energy transfer effect; the percent many-body contribution, however, increases with Θ_D due to the sound-velocity effect. For example, at $\Theta_D=300°K$ the Einstein model predict essentially no trapping, while the GLE model, by providing a mechanism for dissipating collisional energy, leads to a trapping threshold of $T\sim90°K$.

The results for the local friction model (Fig. 2*c*) for two reasonable choices of β disagree with one another and show no clear pattern. The reason is that these models are invalid in all time-scale regimes for the Debye model of the full many-body system.

The important role of time-scale effects in many-body chemical dynamics is clearly evident from the "ratio plots"[1b] presented in Fig. 7. We plot there the ratios $R_E=\Delta E_E/\Delta E_{\mathrm{GLE}}$ of Einstein to GLE energy transfer as a function of the reduced collision time $\omega_D\tau_c$ (where τ_c is defined in the figure caption). Figure 7 is again based on the Debye solid model and ΔE_{GLE} is the energy transfer computed using the Brownian oscillator heatbath model of Fig. 2*a*.

The reduced collision time is a direct measure of the importance of many-body effects on energy transfer since if the primary solid atom can execute one or more oscillations during a collision time (i.e., if $\omega_D\tau_c\gtrsim1$),

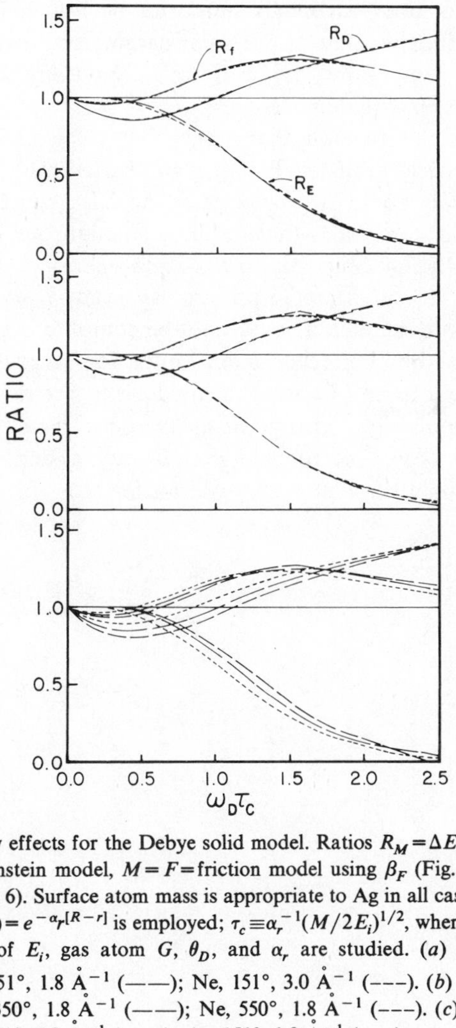

Fig. 7. Many-body effects for the Debye solid model. Ratios $R_M = \Delta E_M/\Delta E$ as a function of $\omega_D\tau_c$; $M = E =$ Einstein model, $M = F =$ friction model using β_F (Fig. 6), $M = D =$ friction model using β_D (Fig. 6). Surface atom mass is appropriate to Ag in all cases, and an exponential potential $W(r,R) = e^{-\alpha_r[R-r]}$ is employed; $\tau_c \equiv \alpha_r^{-1}(M/2E_i)^{1/2}$, where $E_i =$ incident beam energy. Variations of E_i, gas atom G, θ_D, and α_r are studied. (*a*) (G, θ_D, α_r)Ne, 151°, 1.0 Å^{-1} (—); Ne, 151°, 1.8 Å^{-1} (——); Ne, 151°, 3.0 Å^{-1} (---). (*b*) (G, θ_D, α_r) Ne, 151°, 1.8 Å^{-1}, (—); Ne, 350°, 1.8 Å^{-1} (——); Ne, 550°, 1.8 Å^{-1} (---). (*c*) (G, θ_D, α_r)He, 151°, 1.8 Å^{-1} (—); Ne, 151°, 1.8 Å^{-1} (——); Ar, 151°, 1.8 Å^{-1} (---).

then collision energy may be transferred from the primary system into the solid *before the gas atom departs*. (Energy dissipation occurring after the gas atom departs, of course, can have no effect on measurable scattering attributes.) Since the ratio R_E is a direct measure of the many-body effect, it is reasonable to surmise that it depends mainly on $\omega_D\tau_c$ and that it is *nearly independent* of other variables, such as gas mass, solid atom mass, Θ_D, potential parameters, incident gas energy, and so on. The absolute energy transfer, of course, is very sensitive to all these parameters (see, e.g.,

Fig. 5). Our conjecture is strikingly supported by Fig. 7. The ratios R_E are almost perfectly independent of potential parameters, incident gas energy, and solid Debye temperature Θ_D and are only weakly dependent on gas atom mass in the He–Ar mass range.

The ratio plots clearly show that many-body effects are very important for $\omega_D \tau_c \gtrsim 1$ for energy transfer. For $\omega_D \tau_c = 2$, for example, the Einstein energy transfer is only about 20% as large as the GLE energy transfer. Comparing Figs. 5 and 6 indicates that trapping fractions are even more sensitive to many-body influence than is simple energy transfer. Residence times, which are not presented here, display many-body influence in an even more dramatic manner. They cannot be computed even semiquantitatively unless many-body effects are accounted for because of the spurious recurrences present in any finite chain model (see Section II.G).

The ratio plots of Fig. 7 also point up the deficiencies of the Brownian oscillator model (Fig. 1*c*) for the many-body systems. The ratios of Einstein to Brownian oscillator energy transfer (R_f and R_d in Fig. 8) are

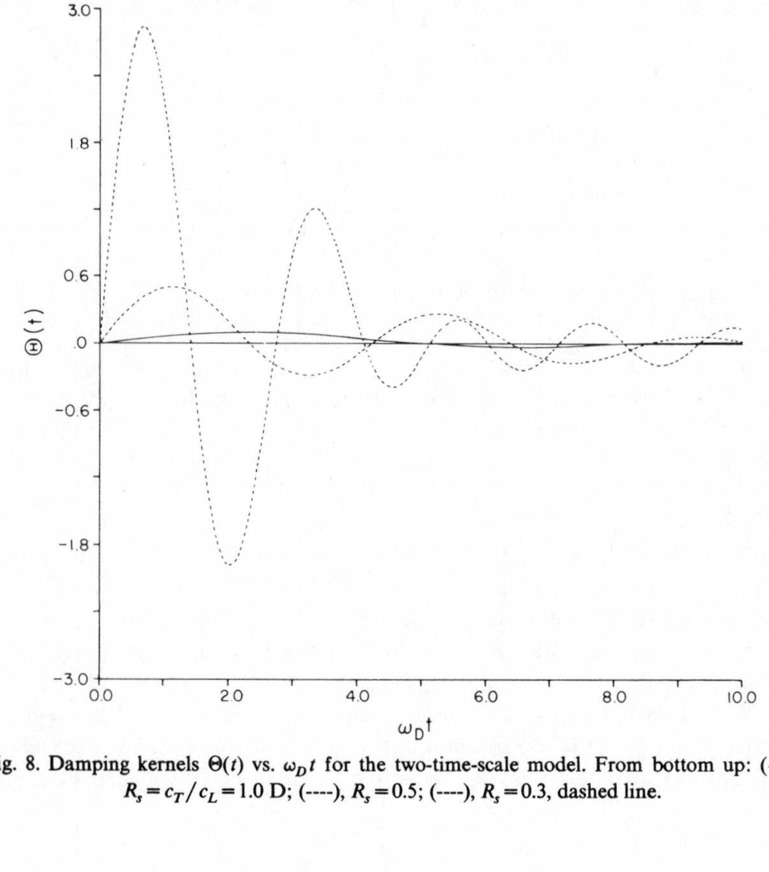

Fig. 8. Damping kernels $\Theta(t)$ vs. $\omega_D t$ for the two-time-scale model. From bottom up: (—), $R_s = c_T/c_L = 1.0$ D; (----), $R_s = 0.5$; (----), $R_s = 0.3$, dashed line.

unphysical both for fast processes $\omega_D \tau_c \ll 1$ and for slow processes $\omega_D \tau_c \gg 1$. For $\omega_D \tau_c = 2$, for example, the Brownian oscillator model predicts that Einstein energy transfer is *greater than* many-body transfer. This is clearly incorrect for the Debye model of a many-body system. [Such behavior ($\Delta E_E > \Delta E_{\text{GLE}}$) could occur for *some* many-body systems for *certain* $\omega_D \tau_c$. This would in no way, however, justify the simple friction model since it is *always* incorrect for $\omega_D \tau_c \ll 1$.]

We now turn to two models for the many-body system, in this case a solid, which include important physics present for real solids but absent in the Debye model. Our motivation is to show how inclusion of new physical features modifies ω_{e_0} and $\Theta(t)$ in a clearly understandable way and how, moreover, these modifications are mirrored in computed gas–solid scattering attributes. Thus our discussion will exemplify how the language provided by our theory allows one to understand as well as compute many-body influence on condensed-phase dynamics.

The Debye model involves only a single time scale $\sim \omega_D^{-1} \sim ac^{-1}$, where a is a lattice spacing and c is the average speed of sound in the solid. The many-body response of real solids cannot be described in terms of sound velocities except at low frequencies and for acoustic modes. Moreover, even in this limit the sound velocities are in general anisotropic (i.e., depend on the direction in $\vec{k}$ space of the wave vector of the acoustic mode). For isotropic continuum solid models, the full response of the system *is* describable in terms of *isotropic* sound velocities. Even for this simplest case, however, the simple Debye model, (II.45), for the response is inadequate. The reason is that *longitudinal* and *transverse* normal modes have distinct sound velocities denoted respectively by c_L and c_T. For real solids the anisotropy averaged sound velocity ratio $R_s = c_T / c_L = 0.1 - 0.7$. Given the fact that each sound velocity is associated with a different time scale for many-body motion and given the importance of time-scale effects on primary system/heatbath energy exchange (see, e.g., Fig. 7), it is clear that a solid model incorporating the feature of distinct longitudinal and transverse sound velocities might display scattering attributes quite different from that of the Debye model.

We thus propose an augmented Debye model, which we call the two-time-scale model (TTM) described by the *unit-normalized* spectral density $\rho(\omega)$, where [32]

$$\rho(\omega) = \frac{2\omega^2}{\omega_T^2}\eta(\omega - \omega_T) + \frac{\omega^2}{\omega_L^2}\eta(\omega - \omega_L) \qquad \text{TTM} \qquad \text{(III.15)}$$

The longitudinal and transverse cutoff frequencies ω_L and ω_T are defined in terms of the sound velocities c_L and c_T and a characteristic lattice spacing a by

$$\omega_L = c_L a^{-1}\pi\left(\frac{6}{\pi}\right)^{1/3} \tag{III.16a}$$

and

$$\omega_T = c_T a^{-1}\pi\left(\frac{6}{\pi}\right)^{1.3} \tag{III.16b}$$

The response function $\chi_m(t)$ for the TTS is readily obtained from (II.27b) and (III.15) as [cf. (II.45)]

$$\dot{\chi}(t) = 2\left[\frac{\sin x}{x} + \frac{2\cos x}{x^2} - \frac{2\sin x}{x^3}\right] + \left[\frac{\sin y}{y} + \frac{2\cos y}{y^2} - \frac{2\sin y}{y^3}\right] \quad \text{TTM} \tag{III.17}$$

where

$$x = \omega_T t$$
$$y = \omega_L t \tag{III.18}$$

The Debye frequency ω_D is defined in terms of ω_c and ω_L by

$$\omega_D = \left[\tfrac{1}{3}\omega_L^{-3} + \tfrac{2}{3}\omega_T^{-3}\right]^{-1/3} \tag{III.19}$$

For $c_T = c_L$, the equations above show that the TTM reduces to the Debye model, (II.41).

From (III.17) one may readily compute an Einstein frequency and a damping kernel $\Theta(t)$ for the TTM; these are required input for the GLE, (III.1). The Einstein frequency is [cf. (II.43)]

$$\omega_{e_0}^2 = \tfrac{1}{5}\left[2\omega_T^2 + \omega_L^2\right] \tag{III.20}$$

The quantities ω_e and $\Theta(t)$ may be evaluated as a function of c_T/c_L for *fixed* ω_D. The results for ω_e are

$$\omega_{e_0} = 1.42\omega_D \qquad c_T/c_L = 0.3 \qquad \text{TTM} \tag{III.21a}$$
$$\omega_{e_0} = 0.98\omega_D \qquad c_T/c_L = 0.5 \tag{III.21b}$$

This is to be compared with the Debye result, (II.43),

$$\omega_{e_0} = 0.77\omega_D \tag{III.21c}$$

Equations (III.21) show that the introduction of a second time-scale greatly changes ω_{e_0} from that predicted by the Debye model. Given the great sensitivity of gas–solid energy-exchange processes to ω_{e_0}, which recall governs the *initial* exchange, we expect that the introduction of a second sound velocity will lead to a *decreased efficiency* of gas–solid energy transfer over that predicted by the Debye model, at least for short-time-scale processes.

The many-body effect, which is important for all but the shortest time-scale processes, depends on $\Theta(t)$, which is plotted vs. $\omega_D t$ for the TTM in Fig. 8 for $c_T/c_L=0.3$, 0.5, and 1.0 D. The results are striking. The magnitude of $\Theta(t)$ (and hence ω_c^2) increases enormously as c_T/c_L decreases. Thus the many-body effect will be much more important within the TTM for physical values of $c_T/c_L \sim 0.3$ to 0.6 than for the Debye value $c_T/c_L=1$. For solids at $T=0°\text{K}$, the change in Einstein frequency and the change in $\Theta(t)$ due to taking $c_T \neq c_L$, will compete. The net effect is usually to greatly enhance gas–solid energy exchange and sticking or trapping probabilities. The following discussion, which may be made much more detailed if one computes $\omega_{c_1}^2, \omega_{e_1}, \omega_{c_2}^2, \ldots$, show how the GLE *formalism* yields insight, *even before dynamical calculations are made*, into the effects of many-body motion on the chemical part of the dynamics.

The physics behind the results predicted by the formalism is as follows. Equations (III.16) and (III.19) show that as c_T/c_L is decreased from unity to physical values ~ 0.3 to 0.6 for *fixed* ω_D, ω_L, and hence c_L must *increase* over their values when $c_T=c_L$. The increase in longitudinal sound velocity means that energy is more quickly removed from the primary solid atom and dissipated into the heatbath within the TTM than within the Debye theory. Thus many-body effects predicted by the TTM are larger than those predicted by the Debye theory. The increase in Einstein frequency within the TTM arises because ω_L increases for fixed ω_D, (III.19) (i.e., $\omega_T < \omega_D$ and $\omega_L > \omega_D$ for $R_s < 1$).

The TTM should provide a considerable improvement over the Debye model for processes which take place *inside* a solid (e.g., vibrational energy and phase[18b] relaxation of a diatomic impurity embedded in a solid matrix). The TTM, however, like the bulk Debye model, does not include surface effects and thus is suspect for gas–solid scattering.

Surface effects may be included within the continuum elastic framework by solving the equation of elasticity with surface boundary conditions.[32] We will call the resulting theory the surface elastic model (SEM). The SEM like the TTM depends on two parameters $R_s = c_T/c_L$ and ω_D.

The normal modes of the SEM may be divided into surface or Rayleigh modes, which are confined to the surface layers, and bulk modes which are linear combinations of the TTM modes. Motion (i.e., response functions) of *bulk* solid atoms within the SEM is *identical* to the motion predicted by the TTM. The dynamics of *surface layer* atoms, however, predicted by the SEM is completely different than TTM dynamics. The reason is that the Rayleigh modes, absent in the TTM, contribute importantly to surface-layer dynamics in the SEM. The frequency spectrum of the Rayleigh modes, moreover, is very different from that of the bulk modes. The Rayleigh mode spectrum, for example, is *insensitive* to the sound-velocity ratio $R_s = c_T/c_L$, while the TTM (Fig. 8) spectrum is extremely sensitive to R_s.

The Einstein frequencies calculated within the SEM (the Einstein frequency within the SEM is an anisotropic tensor; we here give the zz component, where z is the direction perpendicular to the surface plane) are, for example,

$$\omega_e = 0.65\omega_D \qquad c_T/c_L = 0.3 \qquad \text{SEM} \tag{III.22a}$$

$$\omega_e = 0.63\omega_D \qquad c_T/c_L = 0.5 \tag{III.22b}$$

Comparison of (III.21) and (III.22) shows that the SEM results are much less sensitive to R_s than the TTM results and, moreover, are much closer to the Debye results than the corresponding TTM results.

Most of the contribution to ω_e comes from the Rayleigh modes. The Rayleigh contribution to ω_e is, for example, given by

$$\omega_{e_0}^{\text{Rayleigh}} = 0.587\omega_D \qquad c_T/c_L = 0.3 \tag{III.23a}$$

$$\omega_{e_0}^{\text{Rayleigh}} = 0.586\omega_D \qquad c_T/c_L = 0.5 \tag{III.23b}$$

Note that $\omega_{e_0}^{\text{Rayleigh}}$ is insensitive to R_s.

The reason is that the Rayleigh modes have their own characteristic sound velocity c_R which describes surface dissipation. This sound velocity is[33]

$$c_R = \xi c_T \tag{III.24}$$

where ξ is the physical solution of the following algebraic equation:

$$\xi^6 - 8\xi^4 + 8\xi^2(3 - 2R_s^2) - 16(1 - R_s^2) = 0 \tag{III.25}$$

As R_s varies from 0.0 to 0.5, ξ varies from 0.87 to 0.96. Thus c_R is insensitive to R_s.

Moreover, it is clear from the discussion above that $c_R = \xi c_T$ is less than either c_T or c_L. Thus we expect many-body effects to be *less important* in the SEM than in the Debye model and much less important than in the TTM.

This is illustrated in Fig. 9, where we compare the (zz component) of $\Theta(t)$ computed from (1) the SEM with $R_s = 0.3$, (2) the pure Rayleigh approximation to the SEM (bulk mode contributions neglected) and (3) the Debye model. Notice that Debye and SEM response functions are similar in magnitude (cf. Fig. 8) with the Debye $\Theta(t)$ larger as predicted. Further notice the importance of the Rayleigh contributions to $\Theta(t)$.

We may summarize the results of the foregoing analysis as follows. The Debye model, by ignoring the difference between longitudinal and transverse sound velocities, grossly underestimates the efficiency of energy transfer for many bulk processes. The rates of such processes are likely to be extremely sensitive to *both* ω_D and R_s. For surface processes the Debye model *fortuitously* will probably give more reasonable results. The reason is that surface processes are strongly influenced by the Rayleigh modes,

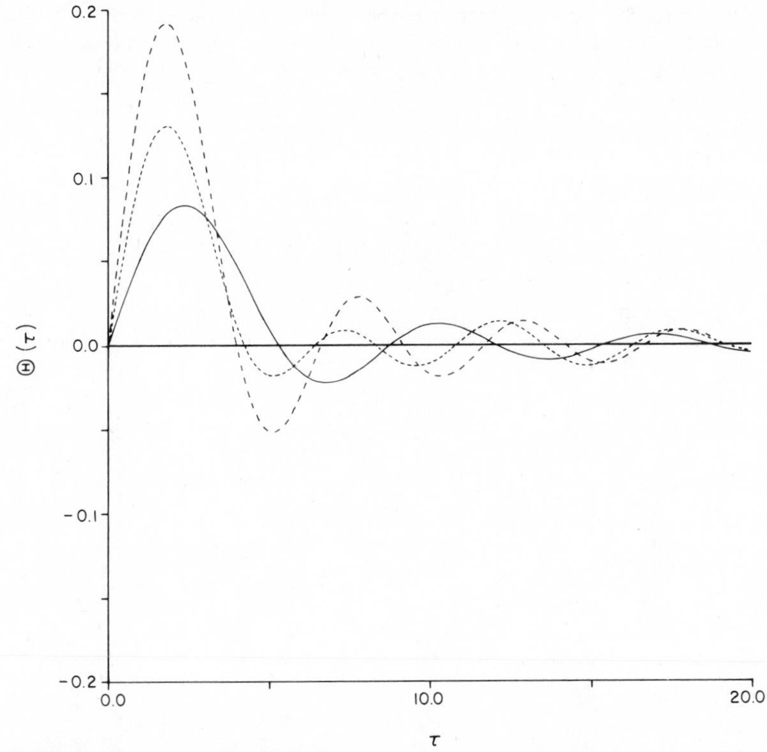

Fig. 9. Damping kernel $\Theta(t)$ for the surface elastic model (SEM) $\Theta(t)$ (zz component) vs. reduced time $\tau = 2\sqrt{\pi}\, a^{-1} c_T t$. (---), SEM result for $R_s = c_T/c_L = 0.3$; (—), Rayleigh mode approximation to SEM result for $R_s = 0.3$. For comparison the Debye result (———) is presented.

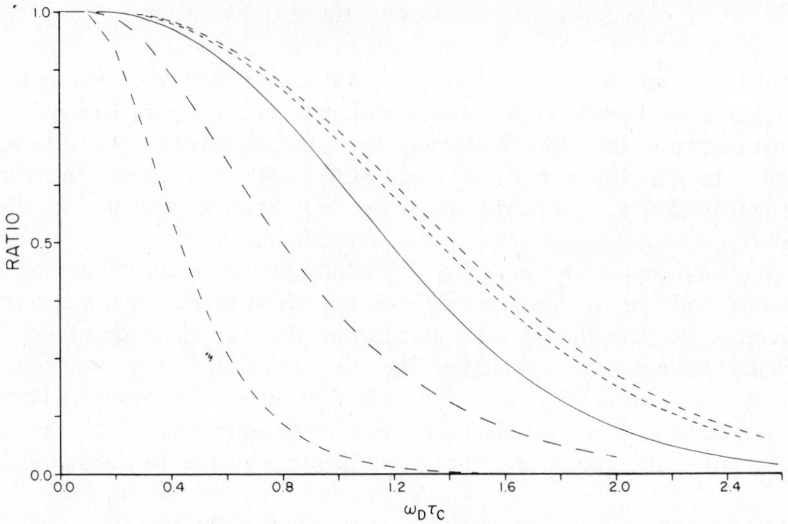

Fig. 10. Many-body effects for continuum elastic solid models. Ratio plots (see text) are, from bottom: (----), TTM $R_s=0.3$; (———), $R_s=0.5$; (—), Debye model; (---), SEM $R_s=0.3$; (---), SEM $R_s=0.5$.

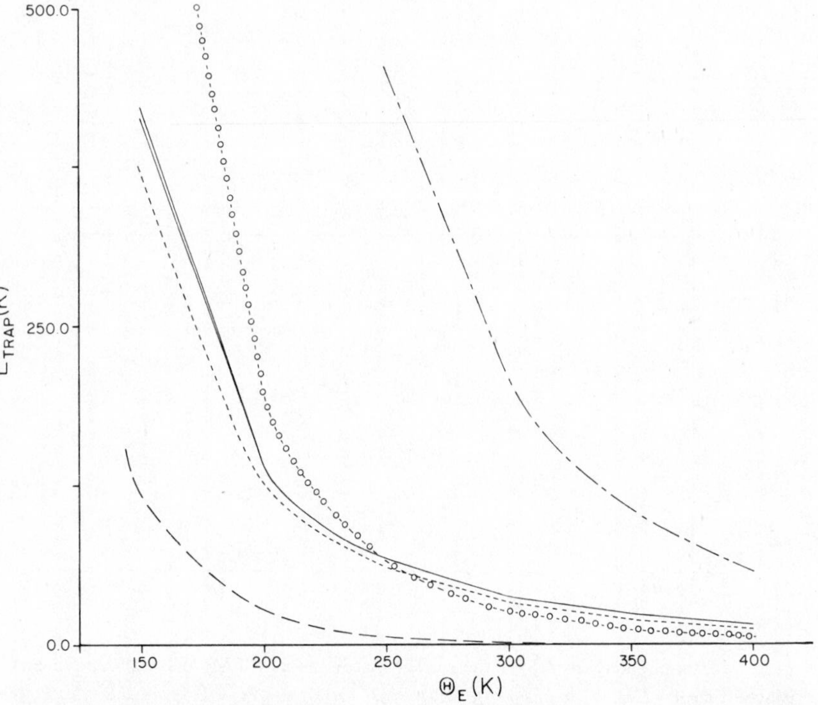

Fig. 11. Trapping thresholds $E_{\text{trap}}(K)$ vs. Einstein temperature $\Theta_E(K)=h\omega_e/k_B$. Incident gas energy $E_i=300°$K; Morse parameters $D_e=417.8°$K, $\alpha=1.69$ Å^{-1}, are chosen to simulate Ar–Ag scattering. Results, from bottom up: (---), Einstein; (---),-(----) Debye; (—), SEM $R_s=0.5$; (—), $R_s=0.3$; (0-0-0), TTM $R_s=0.5$; (---), $R_s=0.3$.

which are insensitive to R_s and which yield ω_e and $\Theta(t)$ results which are similar in magnitude to the Debye results. The fortuitous agreement between surface and Debye results in only, however, very rough. The initial energy transfer governed by ω_{e_0}, for example, should be larger in the SEM than in the Debye model while the many-body effect shows a reverse order [cf. (III.21c), (III.22), and Fig. 9].

The foregoing qualitative conclusions, derived without the help of simulations but rather via physical reasoning guided by the GLE formalism, are confined by the numerical simulation results displayed in Figs. 10 and 11. Figure 10 emphasizes via a ratio plot (cf. Fig. 7) the different behavior of many-body effects for the different models. Figure 11, which displays the dependence of trapping thresholds on Einstein frequency, provides an alternative way of looking at the many-body influence on atom–solid dynamics. Notice that GLE results computed using the Debye, TTM, and SEM models are intercompared and are also compared with the corresponding Einstein model results. This last comparison demonstrates in a striking manner the dominant importance of many-body influence on condensed-phase dynamics.

E. Summary and Concluding Remarks

We have presented in this article a comprehensive and general theoretical framework for treating many-body problems in chemical dynamics within the approximation of classical mechanics. The theory is based on the modern correlation function formulation of generalized Brownian motion theory. Generalized Brownian motion theory was originally developed[6,7] to provide a microscopic foundation for Onsager's theory of irreversibility,[34] and thus most applications were directed toward *macroscopic* transport problems. Our own efforts involved *reshaping* the theory so as to make it applicable to molecular-time-scale processes. The departure point for our work was an important early paper by Zwanzig[15] on atom–solid scattering. The Zwanzig work and the first version of our own theory[1] were restricted in that the heatbath was required to be perfectly harmonic. This restriction has now been lifted and the present theory, while only *exact* for harmonic many-body systems, is generally applicable. Thus while the first applications of the theory dealt with gas–atom collisions with solid surfaces, the theory is also useful for treating liquid-state chemical phenomena. The present theory has a number of very attractive features. These include:

1. It provides a *general* and *comprehensive* framework for reducing the many-body dynamical problems characteristic of condensed-phase chemical processes to *equivalent* effective few-body problems. These equivalent few-body problems may be *modeled* in a systematic and convergent

manner. *The combination of the reduction and modeling techniques allow one to rigorously generalize the gas-phase classical trajectory method to condensed-phase chemical dynamics*. This generalization is not merely formal but is rather a practical algorithm whose computational value is now solidly documented.[1, 18a,b] Thus the GLE methods solve a fundamental problem in molecular collision theory.

2. The theory provides a *language* (Einstein frequencies, equivalent harmonic chains, heatbath response functions, etc.) for qualitative discussion of many-body influence on condensed-phase dynamics. This language provides new and simple ways of thinking about many-body dynamics. We believe that the new language will eventually prove to be as useful as the computing power supplied by the theory.

3. The theory cleanly separates chemical and statistical mechanical aspects of the full many-body problem. By isolating the statistical mechanical aspects of the problem in the heatbath response function, the theory shows that only *limited information* about the heatbath's complex dynamics is relevant and, moreover, precisely *defines* the relevant information. This feature allows one to sharply focus one's attack on the heatbath many-body problem toward development of specific techniques (e.g., the moment analysis) aimed at calculation of *only* the *relevant* heatbath information. The contrast with brute-force molecular dynamics is striking; there all heatbath information, including much that is irrelevant, is computed. An important related point is that since *all* (systematic *and* random) many-body influence is compressed into the heatbath response function $\Theta(t)$, the many-body problem of computing $\Theta(t)$ has to be solved only *once*. The response function may then be repeatedly used in a variety of chemical dynamics studies.

The contrast between GLE methods and conventional molecular dynamics on this point is again striking. In the brute-force approach, the (full) many-body dynamics of the heatbath is recomputed in every run.

The main weakness of the present theory is one it shares with all molecular dynamical theories. Potential-energy functions for real systems not readily accessible. This obstacle is particularly severe in condensed-phase problems.

Despite this we believe that the present theory, both its practical and conceptual aspects, when used in conjunction with modern experimental probes such as picosecond laser spectroscopy[3] and molecular beam scattering,[2] will lead to an increased understanding of condensed-phase processes.

APPENDIX 1

We here prove (II.21) with (II.22a) via an application of Kubo's linear response technique.[13]

Let $i\mathfrak{L}_0$ be the Liouville operator governing the dynamics of the *full* many-body system in the *absence of external potentials*. For an N-particle many-body system composed of spherical particles each with mass m interacting via pair potentials $u(r)$, the Liouville operator is

$$i\mathfrak{L}_0 = \sum_{\alpha=1}^{N} \frac{\vec{p}_a}{m} \cdot \frac{\partial}{\partial \vec{r}_\alpha} - \sum_{\alpha=1}^{N} \sum_{\beta>\alpha} \frac{\partial u}{\partial \vec{r}_\alpha}(\vec{r}_{\alpha\beta}) \cdot \frac{\partial}{\partial \vec{p}_\alpha} \tag{A.1}$$

The unperturbed time development of any dynamical variable $A^{(0)}(t)$ is governed by the following equation of motion:

$$\frac{dA^{(0)}(t)}{dt} = i\mathfrak{L}_0 A^{(0)}(t) \qquad A^{(0)}(0) = A \equiv A(0) \tag{A.2}$$

In particular,

$$\frac{d\vec{r}_\alpha(t)}{dt} = i\mathfrak{L}_0 \vec{r}_\alpha(t) = \frac{\vec{p}_\alpha(t)}{m} \tag{A.3a}$$

$$\frac{d\vec{p}_\alpha(t)}{dt} = i\mathfrak{L}_0 \vec{p}_\alpha(t) = -\sum_{\beta\neq\alpha} \frac{\partial u}{\partial \vec{r}_\alpha}(r_{\alpha\beta}) \tag{A.3b}$$

Notice that (A.3) are just Hamilton's equations for the system, familiar from classical trajectory work.

The solution to (A.3b) may be formally written as

$$\vec{p}_\alpha(t) = e^{i\mathfrak{L}_0 t}\vec{p}(0) \tag{A.4}$$

since if (A.4) is correct,

$$\dot{p}_\alpha(t) = i\mathfrak{L}_0 e^{i\mathfrak{L}_0 t} p_\alpha(0) = i\mathfrak{L}_0 p_\alpha(t) \tag{A.5}$$

which is identical to (A.3b).

We now apply an external potential $\phi_e(\vec{r}, t)$ given by

$$\phi_e(\vec{r}, t) = -m\vec{f}_e(t)\vec{r} \tag{A.6}$$

The external potential produces an arbitrary force $m\vec{f}_e(t)$ on the primary system (coordinate $\vec{r}$) but does not directly couple to the other system particles. The Liouville operator in the presence of the external potential $i\mathfrak{L}(t)$ is given by

$$i\mathfrak{L}(t) = i\mathfrak{L}_0 + i\mathfrak{L}_1(t) \tag{A.7}$$

where $$i\mathcal{L}_1(t) = -\frac{\partial}{\partial \vec{r}}\phi_e(\vec{r},t)\cdot\frac{\partial}{\partial \vec{p}} = m\vec{f}_e(t)\cdot\frac{\partial}{\partial \vec{p}} \tag{A.8}$$

We now analyze the motion of the primary system in the presence of the external force $m\vec{f}_e(t)$. Our argument is general and does not depend on the specific form of $i\mathcal{L}_0$ given in (A.1).

The equation of motion for arbitrary system dynamical variable $A(t)$ is, in the presence of the external force [cf. (A.2)],

$$\frac{dA(t)}{dt} = i\mathcal{L}(t)A(t) \tag{A.9}$$

The propagator for motion of the system dynamical variables $U(t,t_0)$ satisfies the equation of motion

$$\frac{dU(t,t_0)}{dt} = i\mathcal{L}(t)U(t,t_0) \tag{A.10}$$

Comparing (A.9) and (A.10) shows that

$$A(t) = U(t,t_0)A(t_0) \tag{A.11}$$

Hence the term propagator. We will now assume that $\vec{f}_e(t)$ is a weak perturbation and calculate the propagator by the methods of time-dependent perturbation theory just as is done in quantum mechanics.[(35)] [Note that (A.10) is formally identical for the equation of motion for the quantum propagation.] Following the methods of quantum mechanics,[35] one may derive the following *exact* integral equation for $U(t,t_0)$:

$$U(t,t_0) = e^{i\mathcal{L}_0(t-t_0)} + \int_{t_0}^{t} e^{i\mathcal{L}_0(t-\tau)}i\mathcal{L}_1(\tau)U(\tau,t_0) \tag{A.12}$$

Combining (A.11) and (A.12) yields the following exact integral equation for $A(t)$,

$$A(t) = A^{(0)}(t) + \int_{t_0}^{t} e^{i\mathcal{L}_0(t-\tau)}i\mathcal{L}_1(\tau)A(\tau)\,d\tau \tag{A.13}$$

where $A^{(0)}(t)$ is the unperturbed development of A from $A(0)\equiv A$.

We next thermally average $A(t)$ over the equilibrium distribution of initial conditions $\vec{r}_\alpha(0),\vec{p}_\alpha(0)$ in the system. The equilibrium distribution function is

$$P(\mathbf{r},\mathbf{p})=\frac{1}{Z}\exp\left\{-\beta\left[\sum_{\alpha=1}^{M}\left(\frac{p_\alpha^2}{2m}+\sum_{\beta\neq\alpha}u(r_{\alpha\beta})\right)\right]\right\}=\frac{1}{Z}e^{-\beta H_0(\mathbf{r},\mathbf{p})} \tag{A.14}$$

where $Z=\int d\mathbf{r}\,d\mathbf{p}\,e^{-\beta H_0[\mathbf{r},\mathbf{p}]}$ is the partition function. Denoting this thermal average by $\langle\cdots\rangle$, (A.13) becomes for the case $\langle A^{(0)}(t)\rangle=0$,

$$\langle A(t)\rangle=\int_{t_0}^{t}\langle e^{i\mathfrak{L}_0(t-\tau)}i\mathfrak{L}_1(\tau)A(\tau)\rangle d\tau \tag{A.15}$$

We now specialize to the case of a weak perturbation $i\mathfrak{L}_1(\tau)$. To linear order in the perturbation, (A.8) and (A.15) yield

$$\begin{aligned}\langle A(t)\rangle&=m\int_{t_0}^{t}\langle e^{i\mathfrak{L}_0(t-\tau)}\vec{f}_e(\tau)\cdot\frac{\partial}{\partial\vec{p}}A^{(0)}(\tau)\rangle d\tau\\&=m\int_{t_0}^{t}\vec{f}_e(\tau)\cdot\langle\frac{\partial}{\partial\vec{p}}A^{(0)}(t-\tau)\rangle d\tau\end{aligned} \tag{A.16}$$

where we have used the fact that $\vec{f}_e(t)$ is independent of $\mathbf{r},\mathbf{p}$ to obtain the second equality in (A.16). Finally,

$$\begin{aligned}\langle\frac{\partial}{\partial\vec{p}}A^{(0)}(t-\tau)\rangle&=\frac{1}{Z}\int d\mathbf{r}\,d\mathbf{p}\,e^{-\beta H(\mathbf{r},\mathbf{p})}\frac{\partial}{\partial\vec{p}}A^{(0)}(t-\tau)\\&=-\frac{1}{Z}\int d\mathbf{r}\,d\mathbf{p}\,A^{(0)}(t-\tau)\frac{\partial}{\partial\vec{p}}e^{-\beta H_0(\mathbf{r},\mathbf{p})}\\&=\frac{\beta}{m}\langle A^{(0)}(t-\tau)\vec{p}(0)\rangle\end{aligned}$$

Combining the result above and (A.16) gives the following linear response result:

$$\langle A(t)\rangle=\beta\int_{t_0}^{t}\vec{f}_e(\tau)\cdot\langle A^{(0)}(t-\tau)\vec{p}(0)\rangle\,d\tau \tag{A.17}$$

relating the *induced* value of $A(t)$ to an equilibrium tcf. Finally, specializing to the case $A(t)=\vec{r}(t)$ in (A.17) [assuming that different Cartesian coordinates of $\vec{r}(t)$ are uncoupled] yields (II.21) with (II.22).

APPENDIX 2

We here prove that $\rho(\omega)$ is nonnegative, (II.28). The key point is that if $\rho(\omega)$ were negative, primary system motion could lead to energy being pumped into the primary system (on the average) from the *systematic* part of the primary system/heatbath interaction $[\theta(t)]$ rather than being dissipated. This could lead to *negative* friction coefficients for macroscopic particles in clear violation of the second law of thermodynamics.

To prove (II.28), it is most convenient to consider an external force $\vec{f}_e(t)$ which "turned on" at $t=-\infty$. Then (II.21) is modified as follows:

$$\langle\vec{r}(t)\rangle=\int_{-\infty}^{t}\dot{\chi}_m(t-\tau)\vec{f}_e(\tau)\,d\tau \tag{A.18}$$

The total average work w done *on* the m-oscillator by the external force is

$$w=\int_{-\infty}^{\infty}\vec{f}_e(t)\cdot\langle\vec{r}(t)\rangle\,dt=\int_{-\infty}^{\infty}dt\,\vec{f}_e(t)\cdot\int_{-\infty}^{t}\dot{\chi}_m(t-\tau)\vec{f}_e(\tau)\,d\tau \tag{A.19}$$

Defining the Fourier transform $\vec{F}_e(\omega)$ by

$$\vec{f}_e(t)=\int_{-\infty}^{\infty}\vec{F}_e(\omega)e^{i\omega t}\,d\omega \tag{A.20}$$

and introducing (A.20) into (A.19) yields

$$w=\int_{-\infty}^{\infty}d\omega\vec{F}_e(\omega)\cdot\int_{-\infty}^{\infty}d\omega'\vec{F}_e(\omega')\int_{-\infty}^{\infty}dt\,e^{i\omega t}\int_{-\infty}^{t}d\tau\,e^{i\omega'\tau}\dot{\chi}_m(t-\tau) \tag{A.21}$$

Equation (A.21) may be rewritten as [let $u=t-\tau$]

$$w=2\pi\int_{-\infty}^{\infty}d\omega\vec{F}_e(\omega)\cdot\vec{F}_e(-\omega)\int_{0}^{\infty}du\,e^{i\omega u}\dot{\chi}_m(u) \tag{A.22}$$

Since $\vec{f}_e(t)$ is real, $\vec{F}_e^*(-\omega)=\vec{F}_e(\omega)$. Further, the work w must be real. Thus

$$w=2\pi\int_{-\infty}^{\infty}d\omega|F_e(\omega)|^2\rho(\omega) \tag{A.23}$$

where we have used (II.27b).

Finally, in order that energy be *dissipated*, the work w *on* the m-oscillator must be *positive* or at least not negative. This can only occur for arbitrary $\vec{f}_e(t)$ if $\rho(\omega)\geq 0$.

APPENDIX 3

We here prove the second fluctuation–dissipation theorem in the form (II.115)

$$\beta(t)=\frac{m}{3k_BT}\langle\vec{f}(t)\cdot\vec{f}(0)\rangle \tag{A.24}$$

Our proof will parallel our derivation of (II.165). We begin with the requirement that the mean kinetic energy of the primary system approach its equilibrium value, (II.143), or equivalently,

$$\lim_{t\to\infty}\langle\dot{r}^2(t)\rangle=\frac{3k_BT}{m} \tag{A.25}$$

As $t\to\infty$, the velocity of the primary system $\dot{\vec{r}}(t)$ and $\dot{\vec{y}}(t)$, the fluctuation about the $T=0°$K velocity, are identical, (II.167). Thus it is sufficient to prove that

$$\lim_{t\to\infty}\langle\dot{y}^2(t)\rangle=\frac{3k_BT}{m} \tag{A.26}$$

Using (II.153b), it is straightforward to show that [cf. (II.161)]

$$\langle\dot{y}^2(t)\rangle=\frac{3k_BT}{m}\int_0^t d\tau\int_0^t d\tau'\dot{\chi}_m(\tau)\gamma(\tau-\tau')\dot{\chi}_m(\tau') \tag{A.27}$$

where $\gamma(t)$ is *defined* by

$$\gamma(t)=\frac{m}{3k_BT}\langle\vec{f}(t)\cdot\vec{f}(0)\rangle \tag{A.28}$$

To prove the second fluctuation–dissipation theorem, we thus must show that

$$\gamma(t)=\beta(t) \tag{A.29}$$

Differentiating (A.27) with respect to time yields [cf. (II.162)]

$$\frac{d}{dt}\langle\dot{y}^2(t)\rangle=\frac{6k_BT}{m}\dot{\chi}_m(t)\mathfrak{L}^{-1}[z\hat{\gamma}(z)\hat{\dot{\chi}}_m(z)] \tag{A.30}$$

We write $\hat{\gamma}(z)$ as [cf. (II.50)]

$$z\hat{\gamma}(z)=\hat{\chi}_m^{-1}(z)-z^2-\Omega^2-\hat{\psi}(z) \tag{A.31}$$

Thus $\hat{\gamma}(z)=\hat{\beta}(z)$ if $\hat{\psi}(z)=0$. Combining (A.30) and (A.31) yields

$$\frac{d}{dt}\langle\dot{y}^2(t)\rangle=-\frac{6k_BT}{m}\dot{\chi}_m(t)\left[\ddot{\chi}_m(t)+\Omega^2\chi_m(t)+\int_0^t\psi(t-\tau)\chi_m(\tau)\,d\tau\right] \tag{A.32}$$

or, since $\langle\dot{y}^2(0)\rangle=0$,

$$\langle\dot{y}^2(t)\rangle=\frac{3k_BT}{m}\left[1-\dot{\chi}_m^2(t)-\Omega^2\chi_m^2(t)-2\int_0^t d\tau'\,\dot{\chi}_m(\tau')\int_0^{\tau'}d\tau\,\psi(\tau'-\tau)\chi_m(\tau)\right] \tag{A.33}$$

Thus since $\chi_m(\infty)=0$ for many-body systems,

$$\lim_{t\to\infty}\langle\dot{y}^2(t)\rangle=\frac{3k_BT}{m}\left[1-2\int_0^\infty d\tau'\int_0^{\tau'}d\tau\,\dot{\chi}_m(\tau')\psi(\tau-\tau')\chi_m(\tau)\right] \tag{A.34}$$

Equations (A.26) and (A.34) are only consistent if

$$\int_0^\infty d\tau'\int_0^{\tau'}d\tau\,\dot{\chi}_m(\tau')\psi(\tau-\tau')\chi_m(\tau)=0 \tag{A.35}$$

Using Fourier transform, the integral above may be transformed into a Fourier-space integral with a nonnegative integrand. This integral can only vanish if the integrand vanishes at each point in Fourier space or, equivalently, if $\psi(t)=0$. This is sufficient to prove (A.24).

APPENDIX 4. GAUSSIAN PROBABILITY DISTRIBUTION FUNCTIONS

We here briefly review elementary properties of Gaussian probability distribution functions (pdf). A random variable x is said to be Gaussian if the probability of choosing x is described by the pdf $P(x)$, where

$$P(x)=\left\{\frac{1}{2\pi\sigma^2}\right\}^{1/2}\exp\left[-\frac{1}{2}\frac{(x-x_0)^2}{\sigma^2}\right] \tag{A.36}$$

Note that $P(x)$ is unit normalized:

$$\int_{-\infty}^{\infty} P(x)\,dx = 1 \tag{A.37}$$

and its first and second moments are [$\langle\cdot\rangle$ now denotes an average over the distribution $P(x)$]

$$\langle x\rangle = \int_{-\infty}^{\infty} xP(x)\,dx = x_0 \tag{A.38a}$$

and

$$\langle x^2\rangle = \int_{-\infty}^{\infty} x^2P(x)\,dx = \sigma^2 + x_0^2 \tag{A.39b}$$

Equations (A.38) show that x_0 is the mean value of x and σ is the standard deviation of the pdf:

$$\sigma^2 = \langle[x - \langle x\rangle]^2\rangle$$

We will be concerned below with Gaussian random variables x whose mean is zero (i.e., $\langle x\rangle = x_0 = 0$). The pdf for such a variable is

$$P(x) = \left(\frac{1}{2\pi\langle x^2\rangle}\right)^{1/2} \exp\left(-\frac{1}{2}\frac{x^2}{\langle x^2\rangle}\right) \tag{A.39}$$

Alternatively, one may deal with the Fourier transform of $P(x), M(q)$, where $M(q)$ is defined by

$$M(q) = \int_{-\infty}^{\infty} e^{iqx}P(x)\,dx \tag{A.40}$$

Note that for the Gaussian pdf of (A.39),

$$M(q) = e^{-(1/2)q^2\langle x^2\rangle} \tag{A.41}$$

Note that from (A.40), we see that the power series expansion of $M(q)$ has as its coefficients the nonvanishing (even) moments of $P(x)$:

$$M(q) = 1 - \tfrac{1}{2}q^2\langle x^2\rangle + \tfrac{1}{24}q^4\langle x^4\rangle + \cdots \tag{A.42}$$

Comparing (A.42) with the power-series expansion of (A.41) shows that

$$\begin{aligned}\langle x^4\rangle &= 3\langle x^2\rangle^2\\ \langle x^6\rangle &= 15\langle x^2\rangle^3\end{aligned} \tag{A.43}$$

Equation (A.43) illustrates that arbitrary even moments of a Gaussian distribution may always be expressed as the sum of all distinct products of second moments. This is a generally correct result.

So far we have considered the case of a single Gaussian random variable x. Often we must simultaneously consider n Gaussian random variables $x_1, x_2, \ldots, x_n$. The unit-normalized *joint* Gaussian pdf for these random variables may be written as

$$P(\mathbf{x})=\left(\frac{1}{2\pi}\right)^{n/2}\left(\frac{1}{\det\langle\mathbf{x}\mathbf{x}^T\rangle}\right)^{1/2} e^{-(1/2)\mathbf{x}^T\langle\mathbf{x}\mathbf{x}^T\rangle^{-1}\mathbf{x}} \tag{A.44}$$

where

$$\mathbf{x}=\begin{bmatrix} x_1 \\ x_2 \\ \vdots \\ x_n \end{bmatrix} \quad \text{and} \quad \mathbf{x}^T=(x_1, x_2, \ldots, x_n) \tag{A.45}$$

and where $\langle\mathbf{x}\mathbf{x}^T\rangle$ is the correlation matrix for the random variable:

$$\langle\mathbf{x}\mathbf{x}^T\rangle=\begin{bmatrix} \langle x_1^2\rangle & \langle x_1x_2\rangle & \cdots & \cdot & \langle x_1x_n\rangle \\ \vdots & \langle x_2^2\rangle & \ddots & & \vdots \\ \langle x_nx_1\rangle & \cdot\cdot & \cdots & \cdot & \langle x_n^2\rangle \end{bmatrix} \tag{A.46}$$

Note that $P(\mathbf{x})d\mathbf{x}$ is the *simultaneous* probability that x_1 will be found between x_1 and x_1+dx_1, x_2 will be found between x_2 and x_2+dx_2, and so on. The correlation functions or second moments $\langle x_\alpha x_\beta\rangle$ are given by [cf. (A.38b)]

$$\langle x_\alpha x_\beta\rangle=\int P(x)x_\alpha x_\beta\, dx \tag{A.47}$$

We will often be interested in a second set of variables,

$$\mathbf{y}=\begin{bmatrix} y_1 \\ \vdots \\ y_m \end{bmatrix} \tag{A.48}$$

which are linearly related to the first set $\mathbf{x}$:

$$y_\alpha=\sum_{\beta=1}^{n} L_{\alpha\beta}x_\beta \tag{A.49a}$$

and

$$x_\beta = \sum_{\alpha=1}^{m} (L^{-1})_{\beta\alpha} y_\alpha \tag{A.49b}$$

The matrix L must be nonsingular but is otherwise arbitrary (e.g., $n \neq m$ in general, and thus L is not necessarily orthogonal).

An important property of Gaussian random variables is that if $\mathbf{x}$ is distributed Gaussianly, then $\mathbf{y}$ is also distributed Gaussianly. This follows because

$$\mathbf{x}^T\langle\mathbf{x}\mathbf{x}^T\rangle^{-1}\mathbf{x} = \mathbf{y}^T(\mathsf{L}^T)^{-1}\langle\mathsf{L}^{-1}\mathbf{y}\mathbf{y}^T(\mathsf{L}^T)^{-1}\rangle^{-1}\mathsf{L}^{-1}\mathbf{y} = \mathbf{y}^T\langle\mathbf{y}\mathbf{y}^T\rangle^{-1}\mathbf{y}$$

Comparing the equation above with (A.44) shows that $\mathbf{y}$ is distributed Gaussianly. An important special case of a linear transform is the orthogonal transform U which diagonalizes the (symmetric) correlation matrix. The variables

$$\mathbf{z} = \mathsf{U}\mathbf{x} \tag{A.50}$$

have a diagonal correlation matrix

$$\langle z_\alpha z_\beta\rangle = \langle z_\alpha^2\rangle\delta_{\alpha\beta} \tag{A.51}$$

(i.e., they are statistically independent). For this special case the pdf becomes a product of one-dimensional Gaussian pdf [cf. (II.154)]:

$$P(z) = \prod_{\lambda=1}^{n}\left(\frac{1}{2\pi\langle z_\lambda^2\rangle}\right)^{1/2}\exp\left(-\frac{1}{2}\frac{z^2}{\langle z_\lambda^2\rangle}\right) \tag{A.52}$$

An important special case of (A.52) is the pdf for a three-dimensional Gaussian random vector variable which has the property

$$\langle\vec{v}\vec{v}\rangle = \tfrac{1}{3}\langle v^2\rangle\overleftrightarrow{1} \tag{A.53}$$

The pdf for this variable is

$$P(\vec{v}) = \left(\frac{3}{2\pi\langle v^2\rangle}\right)^{3/2}\exp\left(-\frac{3}{2}\frac{v^2}{\langle v^2\rangle}\right) \tag{A.54}$$

The moment generating function

$$M(\vec{q}) = \int e^{i\vec{q}\cdot\vec{v}}P(\vec{v})\,d(\vec{v}) \tag{A.55a}$$

is given by

$$M(\vec{q}) = e^{-(1/6)q^2v^2} \tag{A.55b}$$

This concludes our brief summary of the properties of Gaussian pdf.

APPENDIX 5

We here prove (II.199), (II.200), (II.205), and (II.206). We begin with the GLE for the heatbath coordinate, (II.198). Taking the dot product of (II.198) with $\dot{\vec{R}}_1(0)$ and then performing a thermal average yields

$$\langle \ddot{\vec{R}}_1(t) \cdot \dot{\vec{R}}_1(0) \rangle = -\omega_{e_1}^2 \langle \vec{R}_1(t) \cdot \dot{\vec{R}}_1(0) \rangle + \omega_{c_2}^4 \int_0^t \theta_2(t-\tau) \langle \vec{R}_1(\tau) \cdot \dot{\vec{R}}_1(0) \rangle \, d\tau + \langle \vec{f}_2(t) \cdot \dot{\vec{R}}_1(0) \rangle \tag{A.56}$$

But using (II.193) and (II.142b) shows that

$$\langle \dot{\vec{R}}_1(t) \cdot \dot{\vec{R}}_1(0) \rangle = \frac{3k_BT}{m} \dot{\theta}(t) \tag{A.57}$$

Thus (A.56) may be rewritten as

$$\ddot{\theta}(t) = -\omega_{e_1}^2 \theta(t) + \omega_{c_2}^4 \int_0^t \theta_2(t-\tau) \, d\tau + \langle \vec{f}_2(t) \cdot \dot{\vec{R}}_1(0) \rangle \tag{A.58}$$

Comparing (A.58) and (II.64) and (II.84) shows that $\langle \vec{f}_2(t) \cdot \dot{\vec{R}}_1(0) \rangle = 0$ which is (II.199).

To prove the fluctuation–dissipation result, (II.200), we note that the solution to the heatbath GLE, (II.198), is

$$\vec{R}_1(t) = \dot{\theta}_1(t)\vec{R}_1(0) + \theta_1(t)\dot{\vec{R}}_1(0) + \int_0^t \theta(t-\tau)\vec{f}_2(\tau) \, d\tau \tag{A.59}$$

Thus since $\theta_1(\infty) = 0$,

$$\lim_{t\to\infty} \vec{R}_1(t) = \int_0^t \theta(t-\tau)\vec{f}_2(\tau) \, d\tau$$

This shows that $\lim_{t\to\infty} \dot{\vec{R}}_1(t)$ is independent of the initial conditions. Hence since the many body system is stationary (thermal equilibrium) in the *statistical* sense, it follows that

$$\lim_{t\to\infty} \langle \dot{R}_1^2(t) \rangle = \langle \dot{R}_1^2(0) \rangle = \langle \dot{\vec{R}}_1(0) \cdot \dot{\vec{R}}_1(0) \rangle \tag{A.60}$$

Combining (II.142b), (II.193), and (A.60) and using $\dot{\theta}(0)=1$ gives

$$\lim_{t\to\infty} \langle \dot{R}_1^2(t) \rangle = \frac{3k_BT}{m} \tag{A.61}$$

This, however, is identical in form to (A.25), and thus the third fluctuation–dissipation theorem of (II.200) may be proven using a replica of the argument in Appendix 3. Equations (II.205) and (II.206) may be proven by recursively continuing the argument above.

ACKNOWLEDGEMENT

Support of this work by the National Science Foundation under Grant No. 768113 and through the Purdue University NSF-MRL program under DMR 76-00889 is gratefully acknowledged. I greatly benefited from many valuable conversations with Professors J. D. Doll and B. J. Garrison and Drs. Y.-W. Lin, M. Berkowitz, C. Y. Mou, and A. Diebold concerning various aspects of the work described above.

REFERENCES

1. (a) S. A. Adelman and J. D. Doll, *J. Chem. Phys.*, **61,** 4242 (1974); 2374 (1976); *Acc. Chem. Res.* **10,** 378 (1977); (b) S. A. Adelman and B. J. Garrison, *J. Chem. Phys.*, **65,** 3571 (1976); B. J. Garrison and S. A. Adelman, *Surf. Sci.*, **66,** 253 (1977); (c) J. D. Doll and D. R. Dion, *J. Chem. Phys.*, **65,** 3762 (1976).
2. For recent reviews of experimental work in gas-solid dynamics with many references, see, for example, W. H. Weinberg, *Adv. Colloid Interface Sci.*, **4,** 301 (1975); R. J. Madix and J. Benzinger, *Annu. Rev. Phys. Chem.*, to be published. S. T. Ceyer and G. Samorjai, *Annu. Rev. Phys. Chem.*, **28,** 447 (1977). Also see, for example, M. Balooch, M. J. Cardillo, D. R. Miller, and R. E. Stickney, *Surf. Sci.*, **46,** 358 (1974); D. Auerbach, C. Becker, J. Cowin, and L. Wharton, *Appl. Phys.*, **14,** 411 (1977).
3. For recent reviews with many references, see C. B. Moore, Ed., *Chemical and Biochemical Applications of Lasers*, Vols. 1–3, Academic, New York, 1974. For reviews slanted toward liquid-state applications, see, for example, K. B. Eisenthal, *Annu. Rev. Phys. Chem.*, **28,** 207 (1977); D. Oxtoby, *Adv. Chem. Phys.*, to be published.
4. M. Karplus, R. N. Porter, and R. D. Sharma, *J. Chem. Phys.*, **43,** 3259 (1965). Also see, for example, D. L. Bunker, *J. Chem. Phys.*, **40,** 1946 (1964); J. C. Polanyi, *Acc. Chem Res.*, **5,** 161 (1972); J. T. Muckerman, *J. Chem. Phys.*, **54,** 1155 (1971); N. C. Blais and D. G. Truhlar, *J. Chem. Phys.*, **58,** 1090 (1973); D. L. Bunker and W. Hase, *J. Chem. Phys.*, **59,** 4621 (1973); J. C. Tully and R. Preston, *J. Chem. Phys.*, **55,** 562 (1971).
5. For reviews of classical trajectory methods, see, for example, D. L. Bunker, in B. Alder, S. Fernbach, and M. Rotenberg, Eds., *Methods in Computational Physics*, Academic, New York, 1971; R. N. Porter, *Annu. Rev. Phys. Chem.*, **25,** 317 (1974).
6. (a) H. Mori, *Prog. Theo. Phys.*, **33,** 423 (1965); (b) H. Mori, *Prog. Theo. Phys.*, **34,** 399 (1965).
7. R. Kubo, *Rep. Prog. Theor. Phys.*, **33,** 425 (1965).
8. M. S. Green, *J. Chem. Phys.*, **20,** 1281 (1952); **22,** 398 (1952).
9. H. B. Callen and T. A. Welton, *Phys. Rev.*, **83,** 34 (1951); H. B. Callen and R. F. Green, *Phys. Rev.*, **86,** 702 (1952).
10. R. Zwanzig, *J. Chem. Phys.*, **33,** 1338 (1960); *Lectures in Theoretical Physics*, Vol. 3 (W. E. Brittin, B. W. Downs, and J. Downs, Eds.), Interscience, New York, 1961; *Phys. Rev.*, **124,** 983 (1961).

11. H. Mori, *Phys. Rev.*, **112,** 1829 (1958); *Prog. Theo. Phys.*, **27,** 529 (1962).
12 L. P. Kadanoff and P. C. Martin, *Ann. Phys..*, **24,** 419 (1963).
13. R. Kubo, *J. Phys. Soc. Jap.*, **12,** 570 (1957); *Lec. Theor. Phys.*, (Boulder), **1,** 120 (1958); R. Kubo and K. Tomita, *J. Phys. Soc. Jap.*, **9,** 888 (1954); R. Kubo in D. ter Haar, Ed., *Fluctuation, Dissipation, and Resonance in Magnetic Systems*, Plenum, New York, 1962.
14. For reviews of the correlation function approach to various problems in nonequilibrium statistical mechanics with many references, see, for example, R. Zwanzig, *Annu. Rev. Phys. Chem.*, **16,** 67 (1965); B. J. Berne and G. H. Harp, *Adv. Chem. Phys.*, **63,** (1970).
15. R. Zwanzig, *J. Chem. Phys.*, **32,** 1173 (1960).
16. F. O. Goodman, *J. Phys. Chem. Solids*, **23,** 1269, 1491 (1962); B. McCarroll and G. Ehrlich, *J. Chem. Phys.*, **39,** 1317 (1963).
17. S. A. Adelman, *J. Chem. Phys.*, **71,** (1979).
18. For recent applications of the present formalism, see (a) A. Diebold and G. Wolken, *Sur. Sci.*, to be published; (b) M. Shugard, J. Tully, and A. Nitzan, *J. Chem. Phys.*, **66,** 2345 (1977); **69,** 336 (1978); A. Nitzan, M. Shugard, and J. C. Tully, *J. Chem. Phys.*, **69,** 2525 (1978). For a review of other approaches to gas-solid energy-transfer processes, see, for example, F. O. Goodman and H. Y. Wachman, *Dynamics of Gas-Surface Scattering*, Academic, New York, 1976. This book contains many references to earlier work. Also see, for example, J. D. McClure, *J. Chem. Phys.*, **51,** 1687 (1969); **57,** 2810 (1972); **57,** 2823 (1972); J. Lorenzen and L. M. Raff, *J. Chem. Phys.*, **49,** 1165 (1968); R. A. Oman, *J. Chem. Phys.*, **48,** 2919 (1968); F. O. Goodman, *Surf. Sci.*, **30,** 1 (1972); J. H. Weare, *J. Chem. Phys.*, **61,** 2900 (1974); J. L. Beeby, *J. Phys.*, **C5,** 3438 (1972); Y.-W. Lin and G. Wolken, *J. Chem. Phys.*, **65,** 2634, 3729 (1976); H. Metiu, *J. Chem. Phys.*, **67,** 5456 (1977); S. Mukamel, *J. Chem. Phys.*, to be published. For a recent approach to quantum scattering theory based on the classical GLE formalism, see Y.-W. Lin and S. A. Adelman, *J. Chem. Phys.*, **68,** 9 (1978). For alternative approaches to many-body problems in chemical dynamics, see, for example, K. Kitakara, H. Metiu, J. Ross, and R. Silbey, *J. Chem. Phys.*, **69,** 2871 (1976).
19. L. Landau and E. M. Lifshitz, *Fluid Mechanics*, Addison-Wesley, Reading, Mass., 1959.
20. See, for example, B. J. Alder and T. E. Wainwright, *Phys. Rev. A*, **1,** 18 (1970); R. Zwanzig and M. Bixon, *Phys. Rev. A*, **2,** 2005 (1970); T. Keyes and I. Oppenheim, *Phys. Rev. A*, **7,** 1384 (1973).
21. D. Pines, *Elementary Excitations in Solids*, W. A. Benjamin, New York, 1964.
22. S. Chandrasekhar, *Rev. Mod. Phys.*, **15,** 1 (1943); M. C. Wang and G. E. Uhlenbeck, *Rev. Mod. Phys.*, **17,** 323 (1945); G. E. Uhlenbeck and L. S. Ornstein, *Phys. Rev.*, **36,** 823 (1930). Also see, for example, M. Lax, *Rev. Mod. Phys.*, **38,** 541 (1966).
23. See, for example, (a) R. I. Cukier and J. C. Wheeler, *J. Chem. Phys.*, **60,** 4629 (1974); O. Platz and R. G. Gordon, *Phys. Rev. Lett.*, **20,** 264 (1973); (b) R. J. Rubin, *J. Math. Phys.*, **1,** 309 (1960); **2,** 373 (1961); R. Mazur and E. Montroll, *J. Math. Phys.*, **1,** 701 (1960); G. W. Ford, M. Kac, and P. Mazur, *J. Math Phys.*, **6,** 504 (1965). Other related work includes J. M. Deutch and R. Silbey, *Phys. Rev. A*, **3,** 2049 (1971); K. Wada and J. Hori, *Prog. Theor. Phys.*, **49,** 129 (1973); T. Munakata, *Prog. Theor. Phys.*, **48,** 1173 (1972); E. L. Chang, R. M. Mazo, and J. T. Hynes, *Mol. Phys.*, **28,** 997 (1974).
24. M. Bixon and R. Zwanzig, *J. Stat. Phys.*, **3,** 245 (1971).
25. For a review of work on harmonic systems with many references, see, A. A. Maradudin, E. W. Montroll, G. H. Wriss, and I. P. Ipatova, *Theory of Lattice Dynamics in the Harmonic Approximation*, 2nd ed., Academic, New York, 1971.
26. See, for example, J. H. van Vleck, *Phys. Rev.*, **74,** 1168 (1948); A. D. Buckingham, *Trans. Faraday Soc.*, **56,** 753 (1960); R. G. Gordon, *J. Chem. Phys.*, **39,** 2728 (1963).
27. (a) S. A. Adelman, *J. Chem. Phys.*, **64,** 124 (1976); (b) S. A. Adelman, *Mol. Phys.*, **33,** 1171 (1977). Also see R. F. Fox, *J. Stat. Phys.*, to be published.

28. J. D. Doll, L. E. Meyers, and S. A. Adelman, *J. Chem. Phys.*, **63,** 4908 (1975).
29. See, for example, B. J. Alder, *J. Chem. Phys.*, **31,** 459 (1959); **33,** 1439 (1960); A. Rahman, *Phys. Rev.*, **136,** A405 (1964); A. Rahman and F. Stillinger, *J. Chem. Phys.*, **55,** 3336 (1971).
30. E. G. d'Agliano, P. Kumar, W. Schaich, and H. Suhl, *Phys. Rev. B*, **11,** 2122 (1975).
31. R. Logan and R. Stickney, *J. Chem. Phys.*, **44,** 195 (1966); **49,** 860 (1968). For related work, see P. J. Pagni and J. C. Keck, *J. Chem. Phys.*, **58,** 1162 (1973).
32. A. Diebold, S. A. Adelman, and C. Y. Mou, *J. Chem. Phys.*, **71,** (1979).
33. L. Landau and E. M. Lifshitz, *Theory of Elasticity*, Addison-Wesley, Reading, Mass., 1959.
33. L. Onsager, *Phys. Rev.*, **37,** 405 (1931); **38,** 2265 (1931); L. Onsager and S. Machlup, *Phys. Rev.*, **91,** 1505, 1512 (1953).
34. See, for example, A. Messiah, *Quantum Mechanics*, North-Holland, Amsterdam, 1966, pp. 722–724.

EXPERIMENTAL AND THEORETICAL STUDIES OF ROTOTRANSLATIONAL CORRELATION FUNCTIONS

MYRON EVANS AND GARETH EVANS

Department of Chemistry
University College of Wales
Aberystwyth, Wales

RUSSELL DAVIES

Department of Applied Mathematics
University College of Wales
Aberystwyth, Wales

CONTENTS

Synopsis

This article is concerned with the experimental and theoretical attempts which have recently been made to solve the molecular dynamical problem of how systems of N ($\approx 10^{23}$) interacting molecules evolve in fluids and mesophases such as plastic and liquid crystals. The development in the 1960s of computerized Michelson interferometry has closed the gap between the microwave and mid-infrared region of the electromagnetic spectrum. The region 1 to 300 cm^{-1} is of particular interest to molecular dynamicists since the power absorption coefficient, $\alpha(\omega)$, observed here in liquids as a broad band (the Poley absorption) is the high-frequency adjunct of the dielectric loss, $\varepsilon''(\omega)$, observed in the microwave and lower-frequency regions. As a result of the relation

$$\alpha(\omega) = \frac{\omega\varepsilon''(\omega)}{n(\omega)c}$$

where $n(\omega)$ is the refractive index and c the velocity of light, high-frequency absorptions which are barely visible in measurements of $\varepsilon''(\omega)$ appear in great detail when $\alpha(\omega)$ is measured because of the frequency multiplication. By observing the absorption and dispersion of a given N-particle system using measurements both on $\varepsilon''(\omega)$ and $\alpha(\omega)$, we can obtain the dipole correlation function, $C_\mu(t)$, and its second derivative, $\ddot{C}_\mu(t)$, which has essentially the same form as the angular velocity correlation function $C_\omega(t)$. Measurements of both $C_\mu(t)$ and $\ddot{C}_\mu(t)$ provide very stringent tests for models of liquid or mesophase dynamics which are apparently successful in other fields, such as depolarized Rayleigh scattering, infrared and Raman broadening, and classical dielectric spectroscopy. One way in which a particular model may be tested is via the orientational autocorrelation function that it generates. We thus define the relation between this function and $C_\mu(t)$, and also the role played by the dynamic internal field in this problem. In addition, since nondipolar molecules display a rotational-type absorption from static to THz frequencies, we discuss the methods used to distinguish between this interaction-induced absorption and the permanent dipolar absorption. We show that both types of bands may be reproduced theoretically within the framework of the Mori continued-fraction representation of the Liouville equation for motions of permanent or induced (temporary) dipoles.

Many, if not all, the dynamical models describing the liquid state may be classified as low-order approximants of the (in general) infinite Mori continued fraction. Since this truncation process necessitates a loss of statistical information, the representation of the intermolecular potential contribution to the system Hamiltonian may be unreliable. On the other hand, higher-order approximants can be used only with the introduction of too many phenomenological parameters. We describe the advantages and disadvantages of various low-order approximant models in different situations.

Having discussed the continued fraction for column vectors of linearly independent dynamical variables, we proceed to discuss its effectiveness in describing the features of experimental static-THz broad-band spectra. First, we attempt to define the role of translation–rotation coupling. This has a direct effect on neutron time-of-flight spectra, but its role in absorption spectra is subtler. The continued fraction may be used for the column vector $\begin{bmatrix} \mathbf{v} \\ \boldsymbol{\omega} \end{bmatrix}$, where $\boldsymbol{\omega}$ is the molecular angular velocity and $\mathbf{v}$ the linear velocity, to show how the correlation matrices $\langle \mathbf{v}(t)\boldsymbol{\omega}^T(0)\rangle$ and $\langle \boldsymbol{\omega}(t)\mathbf{v}^T(0)\rangle$ affect the zero-THz absorption, the Fourier transform of purely orientational correlations. The same approach may be used to determine rototranslational correlation functions which appear in the theory of thermal neutron scattering, *without* assuming the decorrelation of rotation from translation. After dealing with this fundamental question we proceed to evaluate the various models of the Liouville equation, using different experimental approaches to the liquid state.

Approach via Compressed Gas

In this regime we might expect the extended diffusion models to be adequate, but this is clearly not so when measurements on $\alpha(\omega)$ are used. The most characteristic feature of the power absorption is its large shift in peak frequency to higher values as the number density of the absorber is increased. This cannot be followed by any simple variant of J-diffusion or m-diffusion theory, where the angular momentum is randomized at each collision according to a Poisson distribution. These models merely broaden asymmetrically the gas phase ($J \rightarrow J+1$) distribution. Efforts at evaluating extended diffusion in fields such as NMR relaxation and infrared/Raman broadening indicate that it is not widely known how simple features of the zero-THz absorption may facilitate the assessment of these models. For example, a feature of this type of absorption is that autocorrelation functions should possess Taylor expansions in time t which are even up to t^4, at least. The orientational correlation functions for extended diffusion models, however, expand evenly up to t^2.

Some excellent zero-THz data for the strongly dipolar species CH_2Cl_2 are now available, and these are used to evaluate the following models in addition to extended diffusion.

1. An approximant of the Mori continued fraction resulting from an exponential second memory for the orientational correlation function.
2. The approximant equivalent to the model of itinerant libration. Mathematically, this results from an exponential second memory for the angular velocity autocorrelation function. Physically, the situation is described equally well by the Brownian motion of two interacting dipoles.
3. As liquid densities are attained, the Langevin equation for the inertia-corrected rotational diffusion of the asymmetric top (developed within the last two or three years). This is a zeroth-order approximant, and *even with correction for inertial effects* is unrealistic at THz frequencies. Without the inclusion of inertial effects, the power absorption becomes a plateau at high frequency so that the theoretical zero-THz band is effectively infinitely wide.

Techniques involving Raman/infrared band broadening, light scattering, NMR relaxation, and classical dielectric loss measurements are relatively insensitive to the short-time details of molecular rotational dynamics. Thus data over an extended range of temperature and number density are often needed before departures from inertialess rotational diffusion may be observed with any precision. We develop this point by comparing results for the rotational dynamics of CH_2Cl_2 liquid obtained from the different techniques available. In contrast to the confused and often contradictory deductions drawn from some other fields, the zero-THz profile discriminates much more readily between models, and is thus at the very least a useful source of complementary information. A self-consistent appraisal is attempted within the framework constructed by Mori. This aims at a satisfactory evaluation of such quantities as the mean-square torque and its derivatives, so that some statistical assessment may be made of the potential part of the total N-particle Hamiltonian. At present, features of observed spectra are reflected in model correlation times which are often physically meaningless (e.g., the Debye relaxation time or the time between elastic collisions of extended diffusion) and sometimes directly contradictory.

Approach to the Liquid State via Mesophases (Liquid and Plastic Crystals) and Glasses

These mesophases restrict the degree of rototranslational freedom available in the isotropic fluid. For example, some solid lattices of dipolar molecules with pseudospherical van der Waals contours are rotationally disordered but translationally ordered. Liquid crystals in the smectic, cholesteric, and nematic phases are composed of molecules which are rotationally hindered (about their short axes) but retain a considerable degree of translational freedom. This is especially so in the nematic phase, where, in the absence of external magnetic and electric fields, the director axes meander in space. One of the most far-reaching consequences of zero-THz spectroscopy of these systems is the realization that the whole loss or absorption profile must be regarded as a continuous function of frequency, arising from an ensemble dynamical process which itself evolves continuously in time from the initial $t=0$. In this context we develop some model calculations of the absorption expected in the nematic phase on the basis of restricted rotational freedom. Preliminary calculations suggest that the potential well experienced by a nematogen, for example, in the field of its neighbors, needs to be considerably narrower and steeper than that of isotropic dipolar liquids. The well depth estimated from best fit to the zero-THz data agrees surprisingly well with a rough calculation using a potential of the form

$$V=a\exp(-br)-\frac{c}{r^6}$$

in which the only intermolecular interactions considered are those between the benzene rings of the nematogen.

In supercooled solutions and glasses of some small, dipolar molecules (such as CH_2Cl_2) it is remarkable that the evolution of the rotational dynamics of the dipolar solute extends over as much as 10 to 15 decades of frequency. This is a direct consequence again of the result that for a well-behaved orientational correlation function, low-frequency losses must extend analytically to the THz region, where, as power absorptions, they rise above the Debye plateau and regain transparency typically at 200 to 300 cm^{-1}. Thus any viable model of the rotations in glasses, liquid crystals, and disordered or plastic solids must be capable of linking loss peaks separated in this way. For example, the dielectric loss observed in a glassy solution of CH_2Cl_2 in decalin has been measured recently in the kHz and THz frequency regions at 107 to 148°K. The low-frequency part of the loss curve exhibits a peak which shifts upward about two decades with a 4°K increase in temperature, and at the glass-to-liquid transition temperature moves almost immediately out of the audiofrequency range to the microwave. The far-infrared (or THz) part of the loss is displaced by 50 to 116 cm^{-1} in the glass as compared with the liquid at 293°K. In addition, therefore, to the well-documented primary and secondary losses in glasses and polymers there exists in general a tertiary process at far-infrared frequencies analogous to the Poley absorption in liquids. Using the itinerant librator model, features described above can be reproduced, and quite satisfactorily so in such specialized cases as fluorobenzene dissolved in a plastic crystal of benzene.

Molecular Dynamics Simulations

Probability density functions are simulated with an atom–atom potential and compared with ones calculated analytically from the itinerant librator model discussed above. The former are non-Gaussian and the latter are Gaussian in nature. Simulations are also used in evaluating translation–rotation functions such as $\langle e^{i\mathbf{q}\cdot\mathbf{r}(t)}P_n(\mathbf{u}(t)\cdot\mathbf{u}(0))\rangle$, useful in the theory of neutron scattering. Here $\mathbf{u}$ is the dipole unit vector, $\mathbf{q}$ the wave vector, and $\mathbf{r}$ the position vector. The use of molecular dynamics is extended to the numerical evaluation of the planar itinerant oscillator system, with the intention of achieving numerically what is analytically intractable.

Finally, simulations are carried out in liquid nitrogen of single-particle and collective correlation functions such as those of the longitudinal and transverse spin density, and current density, which for a structured fluid contains a rotational part. The coupling between transverse spin and current density has also been simulated as the hydrodynamic counterpart of $\langle \boldsymbol{\omega}(t)\mathbf{v}(0)^T\rangle$. The nonvanishing of the matrix possibly explains the zero-frequency splitting observed recently in the depolarized spectrum of scattered light.

I. GENERAL INTRODUCTION. THEORETICAL AND EXPERIMENTAL BACKGROUND

To understand in detail the motions and interactions of about 10^{23} molecules, we would have to follow each trajectory separately over a period of time from the initial $t=0$. In a gas dilute enough so that interactions between each molecule are negligible, this is not an awesome task, since the translation of the center of mass and end-over-end molecular rotation of each molecule are for long periods of time undisturbed. The translation and rotation may safely be described separately. For certain geometries, the Schrödinger equation may be solved to yield the familiar set of rotational absorptions peaking in the far-infrared region[1,2] of the electromagnetic spectrum for some of the commoner dipolar gases. This region, from 1 cm^{-1} (30 GHz) to about 300 cm^{-1}, is often known as the submillimeter or

THz. As the gas is compressed, however, the discrete nature of these absorption lines disappears, until at liquid densities, a broad band is observed peaking with a root-mean-square angular velocity much higher than that of the $J=1$ peak of quantum theory. Figure 1, taken from a paper by Gerschel,[3] illustrates this process along the gas–liquid coexistence curve. After solidification, the broad band may split into a lattice of modes describable in terms of cooperative torsional oscillation of large domains of molecules, with the center-of-mass translation of each strongly hindered.

If we represent the absorption in terms of dielectric loss, the familiar liquid-phase Debye relaxation curve,[4] or a variant thereof, appears, peaking at a frequency much less than that of the gaseous root-mean-square angular velocity. As the density of the liquid increases, the loss peak moves to a lower frequency, while the peak in the power absorption coefficient $\alpha(\omega)$

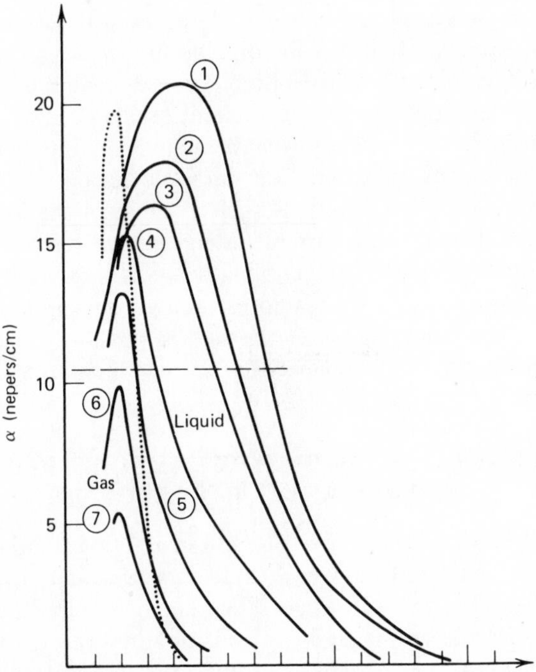

Fig. 1. Far-infrared spectra of $CHCl_3$: (1) to (5) liquid, (6) and (7) gas; *d* the density in g/cm^3. (1) is 233°K, $d=1.62$; (2) 300°K, $d=1.47$; (3) 353°K, $d=1.36$; (4) 483°K, $d=1.05$; (5) 534°K, $d=0.65$; (6) 543°K, $d=0.38$; (7) 517°K, $d=0.2$; The dotted line represents the free-rotator absorption at 534°K. The dashed line represents the "Debye plateau" at 300°K (see text). [Reproduced by permission from A. Gerschel et al., *Mol. Phys.*, **23**, 317 (1972).]

does just the opposite. Even in fluids of nondipolar molecules such as CCl_4 and benzene, whose end-over-end rotations do not cause infrared or microwave absorptions in the infinitely dilute gas, a liquid-phase broad band appears[5] (Fig. 2). Therefore, the loss and power absorption coefficient may be used as measures of molecular interaction in at least two senses.

1. The gradual inhibition of rotation and interruption of free translation, leading to the inevitable interaction and coupling between these freedoms as the liquid state is approached. Their subsequent decoupling once more in the solid, where center-of-mass translation over long distances is rare and slow.
2. The development of electrostatic and overlap interactions which give rise to temporary, induced dipole moments, and thus to rotational far-infrared absorptions in nondipolar gases[6] such as N_2 and solvents such as CCl_4.

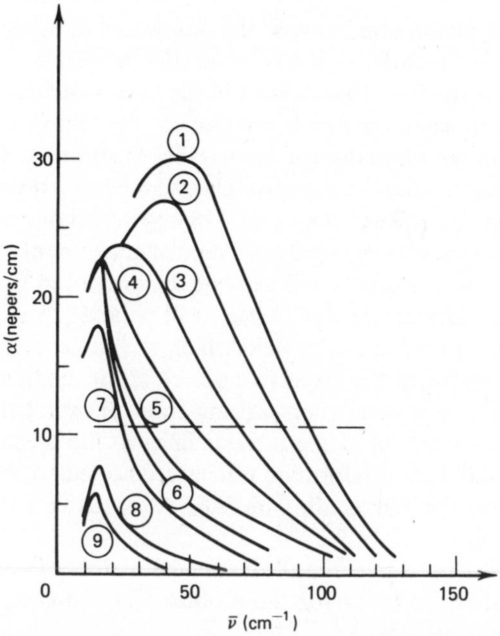

Fig. 2. Far-infrared spectra of C_6H_5F: (1) to (7) liquid, (8) and (9) gas; d the density in g/cm^3. (1) is 239°K, d=1.08; (2) 300°K, d=1.024; (3) 397°K, d=0.89; (4) 459°K, d=0.79; (5) 513°K, d=0.68; (6) 549°K, d=0.548; (7) 559.5°K, d=0.4; (8) 559°K, d=0.3; (9) 619°K, d=0.184. The dashed line is as in Fig. 3. [Reproduced by permission from *Mol. Phys.*, **23**, 317 (1972).]

To interpret these macroscopic observations in terms of molecular motion we must resort to statistical methods since, as pointed out by Hansen and McDonald,[7] the description of each individual molecular trajectory in the fluid, even if possible, is not desirable, on the grounds that a building large enough to house the billions of neatly plotted three-dimensional graphs could not be found. The major problem confronting the liquid-state dynamicist is not one of building a large enough library, but of developing an expansion of the Liouville equation free of numerous phenomenological variables (adjustable for best fit), which gives a consistent description of data from all sources. Such a process should also yield a satisfactory functional for the intermolecular potential energy. A secondary problem, hardly less formidable, is that of treating the "internal field." This means that whenever a solution describing the ensemble dynamical situation is found at the microscopic level, the relation of the displaced field to the measuring field must also be predictable before the macroscopic band shapes can be calculated. A discussion by Scaife[8] is the clearest formulation of this problem in the author's experience. We summarize it here at the outset.

The electrical interaction between the dipole and its surroundings has no influence on the dynamical behavior of the molecule in a nondipolar medium because the field fluctuations in the cavity occupied by the dipole are so rapid as to leave the dipole unaffected. The dipole and the reaction field it sets up in the surroundings are parallel at all times. Thus the experimentalist wishing to study in the zero-THz waveband a dipolar molecule in a dilute nondipolar solvent may use a theory of dipole autocorrelations. When the molecule is surrounded by a dipolar medium of the same molecular type, however, the dipole will be strongly influenced by the fluctuating cavity field. In addition, the dipole itself will polarize its surroundings and thereby produce a further cavity field which will in general not be parallel to the dipole because of the dispersive nature of the medium. A division of the cavity field into a component arising from the surroundings and one generated by the reaction of the surroundings to the preceding motion of the dipole is valid only if saturation effects are neglected. No such division is possible unless the surrounding medium responds in a linear fashion to all applied fields.

If, for the sake of argument, all dynamical memory is neglected, and a Langevin equation accepted for the motion of the dipole, in two dimensions one has the relation

$$I\ddot{\theta}+\xi\dot{\theta}-\mu\mathbf{F}(t)\sin(\psi-\theta)=\lambda(t) \tag{I.1}$$

where I and ξ are constants, $\boldsymbol{\lambda}$ a random force, $\mathbf{F}(t)$ the electric field acting on the dipole $\boldsymbol{\mu}$, ψ the angle between this field and a particular fixed direction, and θ the angle between this direction and the dipole. The field consists of a fluctuating cavity field $\mathbf{F}_c(t)$ and a reaction field $\mathbf{F}_r(t)$, whose magnitude and direction depends on the motion of the dipole up to time t. Clearly, we cannot calculate the dielectric permittivity $\varepsilon(\omega)$ from a solution of (I.1) since such a solution would require a knowledge of the fields $\mathbf{F}_c(t)$ and $\mathbf{F}_r(t)$, which themselves depend on $\varepsilon(\omega)$. In other words, the most we may hope for analytically is an approximate solution for $\varepsilon(\omega)$ based on some self-consistent technique. Some notable advances in dealing with this dipole–dipole coupling problem have been made by Zwanzig,[9] Bellemans and others[10] and Cole,[11] who consider a lattice of translationally fixed dipoles. Deutch[12] has summarized the situation up to 1976. Coffey[13] has recently extended the work of Budo to show that a distribution of macroscopic relaxation times is to be expected when dipole–dipole coupling effects are present. Scaife[8] points out, however, that there seem to be cases (such as water) where dipole–dipole coupling is very strong but only one macroscopic relaxation time appears. This suggests that, in such cases, the concept of a "microscopic relaxation time" is meaningless, and that we are in fact dealing with a cooperative process involving many molecules.

A. Liouville and Fokker–Planck Equations for the Canonical Ensemble

Setting aside for the moment the vexing question of the internal field, we proceed in this section to describe the fundamental equations of classical mechanics for the N-particle ensemble. These equations must be solved by approximate methods, the purpose of which is to extract a correlation function (or spectrum) with the maximum of realism and the minimum number of adjustable, or phenomenological variables. Our justification for the use of classical rather than quantum mechanics has been given by van Vleck and Weisskopf,[14] who outlined conditions under which rotational quantum lines in the microwave and far infrared give way to broad loss bands in the dense fluid. Some remarks by Lobo, Robinson, and Rodriguez[15] illuminate this question with particular clarity. Suffice it to say that classical theory proves adequate except in such cases as[16] HF and NH_3 dissolved in liquid SF_6, where the fluid state far infrared contours clearly show up the presence of residual, broadened $J \rightarrow J+1$ absorptions (Fig. 3). We shall refer to these specific examples later, and we use as a description some classical perturbations of the quantum delta functions.

Consider, therefore, a linearly independent set $[A_j(t)]$, $j=1,\ldots,n$, of real-valued (implicitly) time-dependent dynamical variables of the given N-particle system. Assume that an ensemble average $\langle\cdot\rangle$ can be defined for the system and that the set of all possible dynamical variables is a real

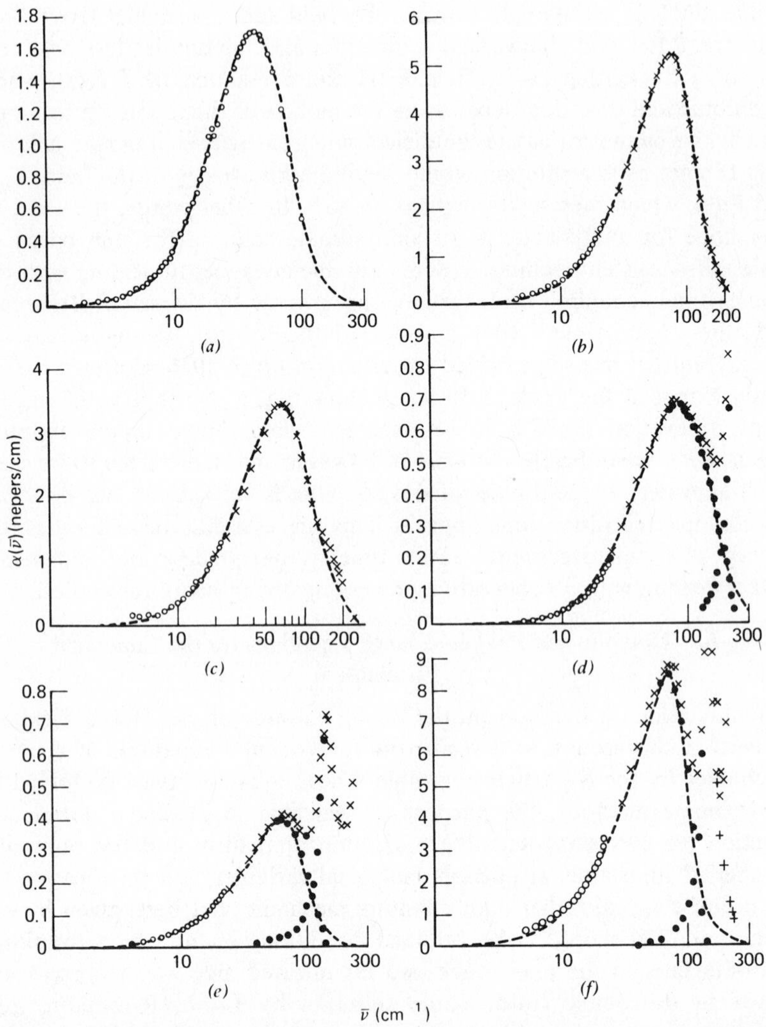

Fig. 3. (*a*) ⊙ Experimental absorption for CCl_4(l) at 296°K. (---), Mori 3 variable theory (Section V); (*b*) (⊙, ×), absorption observed for benzene(*l*) at 296°K; (---), theory; (*c*) (⊙, ×), absorption of CS_2(*l*) at 296°K; (---), theory; (*d*) (⊙, ×) absorption of cyclohexane at 296°K. The high-frequency proper mode is extrapolated using a model of collision-disturbed vibrators. This technique is also used in (*e*) and (*f*); (---), theory; (*e*) (⊙, ×), absorption of *trans*-decalin at 296°K; (----), theory; (*f*) (⊙, ×), absorption of 1,4-dioxan at 296°K; (----), theory. [Reproduced by permission from *J. Chem. Soc. Faraday Trans.* 2, **72**, 1196 (1976).]

Hilbert space in which the inner product is defined by $(A,B)\equiv\langle AB\rangle$. Without loss of generality we may assume that each particle is a molecular asymmetric top within whose rigid but polarizable framework is embedded a dipole vector $\boldsymbol{\mu}$. The set $[A_j(t)]$ spans an n-dimensional subspace of the Hilbert space. Each variable obeys the classical Liouville equation of motion:

$$\dot{A}_j(t)=\{A_j(t),H\}\equiv i\mathfrak{L}A_j(t) \tag{I.2}$$

where H is the Hamiltonian of the system and $i\mathfrak{L}$ the Liouville operator. The formal solution is

$$A_j(t)=\exp(i\mathfrak{L}t)A_j(0) \tag{I.3}$$

where the propagator exp $(i\mathfrak{L}t)$ is an orthogonal operator. The time evolution of $A_j(t)$ is therefore a rotation in Liouville space, with

$$\langle A_j(t)A_k(t)\rangle=\langle A_j(0)A_k(0)\rangle \qquad j,k=1,\ldots,n \tag{I.4}$$

From now on we confine the discussion to correlation functions of the form of (I.4), but the general theory is equally applicable to multiparticle correlation functions, and may be used to determine the relation between the two different kinds, as for example in the paper by Kivelson and Madden.[17] This problem is closely linked with that of the internal field.

If $\mathbf{A}(t)$ denotes the $n\times 1$ column vector with elements $A_j(t)$, and $\mathbf{A}^T(t)$ its corresponding row vector, we let $\langle A(t)A^T(t)\rangle$ denote the $n\times n$ matrix with elements $\langle A_j(t)A_k(t)\rangle$. From (I.4) it follows that

$$\langle A(t)A^T(t)\rangle=\langle A(0)A^T(0)\rangle \tag{I.5}$$

Mori[18] shows that $A(t)$ evolves in time according to the equation

$$\dot{\mathbf{A}}(t)=i\Omega_{\mathbf{A}}\mathbf{A}(t)-\int_0^t d\tau\,\phi_{\mathbf{A}}(t-\tau)\mathbf{A}(\tau)+\mathbf{F}_{\mathbf{A}}(t) \tag{I.6}$$

Because of its similarity to Langevin's equation of 1906 for translational Brownian motion,[19] this is sometimes known as a generalized Langevin equation. It is a form of the Liouville equation which proves particularly useful in generating approximate solutions, or models, of the required autocorrelation function. In (I.6) $\Omega_{\mathbf{A}}$ is a resonance frequency operator, the matrix kernel $\phi_{\mathbf{A}}(t)$ is a memory function (or effective Liouvillian), and $\mathbf{F}_{\mathbf{A}}(t)$ is a generalized random force or torque propagated in a special way from $\mathbf{A}(0)$. These quantities are specified in terms of the projection operators $\hat{P},\hat{Q}$ defined by

$$\hat{P}\mathbf{G} \equiv \langle GA^T(0)\rangle\langle A(0)A^T(0)\rangle^{-1}\mathbf{A}(0)$$
$$\hat{Q} \equiv \hat{1} - \hat{P}$$

where $\hat{P}$ projects an arbitrary vector $\mathbf{G}$ into the subspace spanned by $\mathbf{A}(0)$, and $\hat{Q}$ projects into the orthogonal complement of this subspace. The appropriate relations are

$$i\Omega_{\mathbf{A}} = \langle\dot{\mathbf{A}}(0)\mathbf{A}^T(0)\rangle\langle\mathbf{A}(0)\mathbf{A}^T(0)\rangle^{-1}$$
$$\mathbf{F}_{\mathbf{A}}(t) = \exp(i\dot{Q}\mathfrak{L}t)\dot{Q}\dot{\mathbf{A}}(0)$$
$$\phi_{\mathbf{A}}(t) = \langle\mathbf{F}_{\mathbf{A}}(t)\mathbf{F}_{\mathbf{A}}^T(0)\rangle\langle\mathbf{A}(0)\mathbf{A}^T(0)\rangle^{-1} \tag{I.7}$$

The meaning of the generalized force $\mathbf{F}_{\mathbf{A}}(t)$ should not be confused with that of generalized force in classical lagrangian dynamics. In the Mori formalism of (I.6) and (I.7), $\mathbf{F}_{\mathbf{A}}(0)$ is the component of $\dot{\mathbf{A}}(0)$ orthogonal to $\mathbf{A}(0)$, whereas $\mathbf{F}_{\mathbf{A}}(t)$ is propagated from $\mathbf{F}_{\mathbf{A}}(0)$ by the special propagator $\exp(i\dot{Q}\mathfrak{L}t)$. It is easily shown that

$$\langle\mathbf{F}_{\mathbf{A}}(t)\mathbf{A}^T(0)\rangle = \mathbf{0} \tag{I.8}$$

The simplest physical realizations[20] of (I.6) occur, for example, in the rotational Brownian motion of a spherical top, or a disk about a fixed axis through its center perpendicular to its plane. Equation (I.6) then becomes

$$\frac{D\mathbf{J}}{Dt} + \int_0^t d\tau\,\phi_{\mathbf{J}}(t-\tau)\mathbf{J}(\tau) = \mathbf{F}_{\mathbf{J}}(t) \tag{I.9}$$

where $\mathbf{A}=\mathbf{J}$ is the angular momentum of the rotator, the operator $\Omega_{\mathbf{J}}$ is null, $\phi_{\mathbf{J}}(t)$ is a time-dependent friction tensor, and $\mathbf{F}_{\mathbf{J}}(t)$ is a random driving torque of finite correlation time with mean $\langle\mathbf{F}_{\mathbf{J}}(t)\rangle = 0$. Equation (I.9) is such that the components of each vector are referred to principal body axes, and D/Dt denotes rate of change relative to these axes, so that $DJ_k/Dt = \dot{J}_k$. In the case of the disk there is only one component, referred to the fixed axis. For more complicated geometries nonlinear terms appear in (I.9). Asymmetric-top Brownian angular diffusion has been treated in definitive detail recently by Morita,[21] and McConnell and others[22] for the case where the memory matrix $\phi_{\mathbf{J}}$ consists of delta functions in time. The more general angular equation

$$\dot{\mathbf{J}}(t) - i\Omega_{\mathbf{J}}\mathbf{J}(t) + \int_0^t d\tau\, \phi_{\mathbf{J}}(t-\tau)\mathbf{J}(\tau) = \mathbf{F}_{\mathbf{J}}(t) \tag{I.10}$$

is formally valid for the asymmetric top, but then the quantities $\Omega_{\mathbf{J}}$, $\phi_{\mathbf{J}}$, and $\mathbf{F}_{\mathbf{J}}$ may not have obvious physical meaning.

All the approximations of (I.6) prior to this have been concerned either with pure translation or pure rotation, the latter being complicated considerably by geometrical factors. Some popular models of the last decade are molded into the Mori framework in Table I (see Section I.B). In the next section we extend the formalism to cover rototranslational movements by deriving suitable approximations for the correlation functions associated with the column vector $\begin{bmatrix} \mathbf{v} \\ \boldsymbol{\omega} \end{bmatrix}$. It turns out that the zero-THz loss in dipolar solutes is markedly dependent upon the degree of rotation–translation coupling. Models devised in terms of pure rotation thus only oversimplify the dynamics considerably.

Reverting to general A, (I.6) can be solved using elementary Laplace transforms, giving

$$\dot{\mathbf{A}}(t) = C_{\mathbf{A}}(t)\mathbf{A}(0) + \int_0^t d\tau\, C_{\mathbf{A}}(t-\tau)\mathbf{F}_{\mathbf{A}}(\tau) \tag{I.11}$$

where $C_{\mathbf{A}}(t)$ is a matrix of normalized correlation functions given by

$$\begin{aligned} C_{\mathbf{A}}(t) &= \mathfrak{L}_a^{-1}\left\{\left[s - i\Omega_{\mathbf{A}} + \tilde{\phi}_{\mathbf{A}}(s)\right]^{-1}\right\} \\ &= \langle \mathbf{A}(t)\mathbf{A}^T(0)\rangle\langle \mathbf{A}(0)\mathbf{A}^T(0)\rangle^{-1} \end{aligned} \tag{I.12}$$

Here $\tilde{\phi}_{\mathbf{A}}(s)$ denotes the Laplace transform of $\phi_{\mathbf{A}}(t)$ and $\mathfrak{L}_a^{-1}$ denotes inverse Laplace transformation.

If the vector $\mathbf{F}_{\mathbf{A}}(t)$ is a multivariate Gaussian process with mean $\langle \mathbf{F}_{\mathbf{A}}(t)\rangle$ and if the initial vector $\mathbf{A}(0)$ at time $t=0$ is given, it follows from (I.11) that $\mathbf{A}(t)$ at time $t>0$ is also a multivariate Gaussian process with mean

$$\langle \mathbf{A}(t)\rangle = C_{\mathbf{A}}(t)\mathbf{A}(0) + \int_0^t d\tau\, C_{\mathbf{A}}(t-\tau)\langle \mathbf{F}_{\mathbf{A}}(\tau)\rangle$$

(The Gaussian assumption is fundamental to further progress in this section, but later we present molecular dynamics data for N_2 which clearly shows that it is only approximately valid.)

Restricting ourselves to the special case $\langle \mathbf{F}_{\mathbf{A}}(t)\rangle = 0$ admits the treatment of rotational Brownian motion, for example, and simplifies the algebra generally. This remains true for rototranslation. Then $\mathbf{A}(t)$ is conditionally distributed with probability density function

$$p(\mathbf{A}(t);t|\mathbf{A}(0)) = (2\pi)^{-(1/2)n}(\det V(t))^{-1/2} \times \exp\left[-\tfrac{1}{2}(\mathbf{A}(t)-C_{\mathbf{A}}(t)\mathbf{A}(0))^T V^{-1}(t)(\mathbf{A}(t)-C_{\mathbf{A}}(t)\mathbf{A}(0))\right] \tag{I.13}$$

with variance–covariance matrix

$$\begin{aligned} V(t) &= \langle(\mathbf{A}(t)-C_{\mathbf{A}}(t)\mathbf{A}(0))(\mathbf{A}(t)-C_{\mathbf{A}}(t)\mathbf{A}(0))^T\rangle \\ &= \langle\mathbf{A}(0)\mathbf{A}^T(0)\rangle - C_{\mathbf{A}}(t)\langle\mathbf{A}(0)\mathbf{A}^T(0)\rangle C_{\mathbf{A}}^T(t) \end{aligned} \tag{I.14}$$

We see from (I.14) that the elements of $V(t)$ can be calculated directly from the elements of the correlation matrix $C_{\mathbf{A}}(t)$, so that if the elements of $C_{\mathbf{A}}(t)$ can be found experimentally or otherwise, the conditional probability density function $p(\mathbf{A}(t);t|\mathbf{A}(0))$ can be calculated.

It is clearly essential to know whether or not the generalized force $\mathbf{F}_{\mathbf{A}}(t)$ is a Gaussian process. This depends both on the nature of $\mathbf{A}(t)$ and on the geometry. For example, if we consider the angular velocity $\boldsymbol{\omega}$ of the sphere or disk mentioned above, the corresponding random driving torque $\mathbf{F}_{\omega}(t)$ may be assumed Gaussian (although non-Markovian), so $\boldsymbol{\omega}$ is conditionally Gaussian. If, however, we consider the total Euler angle

$$\theta(t) = \int_0^t \omega_1(\tau)\,d\tau$$

it is not obvious whether $F_\theta(t)$ is Gaussian. Here θ is the total angle turned through in time t rather than the angular orientation which is restricted to the range $-\pi \le \theta \le \pi$. In fact, the angular distribution has a wrapped normal distribution which we shall consider later.

One possible way of investigating the Gaussian or non-Gaussian nature of $\mathbf{F}_{\mathbf{A}}(t)$ for a given $\mathbf{A}(t)$ is to expand the operator $\exp(i\hat{Q}\mathfrak{L}t)$ in (I.7). Since $\hat{Q}$ and $\mathfrak{L}$ commute, and $\hat{Q}^m = \hat{Q}$ for $m \ge 1$, we have

$$\mathbf{F}_{\mathbf{A}}(t) = \hat{Q}\dot{\mathbf{A}}(0) + t\hat{Q}i\mathfrak{L}\dot{\mathbf{A}}(0) - \tfrac{1}{2}t^2\hat{Q}\mathfrak{L}^2\dot{\mathbf{A}}(0) + \cdots$$

As a simple illustration we may consider $F_\theta(t)$ for the disk rotating in its plane under a Gaussian restoring torque $\Gamma(t) = -\tau[\theta(t)-\theta(0)]$, where τ is a constant. The Liouville operator is

$$i\mathfrak{L} = I\dot{\theta}\frac{\partial}{\partial\theta} - \tau\theta\frac{\partial}{\partial\dot{\theta}}$$

where I is the moment of inertia of the disk, so that

$$(i\mathfrak{L})^{2m}\dot{\theta} = (-I\tau)^m\dot{\theta}$$
$$(i\mathfrak{L})^{2m+1}\dot{\theta} = I^m(-\tau)^{m+1}\theta$$

Since $\hat{Q}\theta(0)=0$ and $\hat{Q}\dot{\theta}(0)=\dot{\theta}(0)$, it follows that

$$F_\theta(t)=\dot{\theta}(0)\left[1-\frac{1}{2!}I\tau t^2+\frac{1}{4!}(I\tau)^2t^4-\cdots\right]=\dot{\theta}(0)\cos\left[(I\tau)^{1/2}t\right]$$

Similarly, we can show that $F_{\dot{\theta}}(t)=(\Gamma(0)/I)\cos[(I\tau)^{1/2}t]$, so that $F_{\dot{\theta}}(t)$ is Gaussian while the nature of $F_\theta(t)$ is determined by that of $\dot{\theta}(0)$.

The conditional probability density functions discussed above are the most informative functions which can be extracted from a statistical treatment of the microscopic ensemble dynamics, more so than the correlation functions, which are integrals over the density functions. We now show that these density functions satisfy a generalized diffusion equation, similar to that of Fokker–Planck.[23] This extends our earlier analogy with the classical theory of Browian motion.

To simplify the analysis, we write $C \equiv C_{\mathbf{A}}(t)$, make the change of variable $\mathbf{B}=C^{-1}\mathbf{A}$, and introduce the probability density function $q(\mathbf{B};t|\mathbf{B}(0)) = p(\mathbf{A};t|\mathbf{A}(0))$. Clearly, $\mathbf{B}(0)=\mathbf{A}(0)$ since $C(0)=1$. Differentiating q, we find that

$$\frac{\partial q}{\partial t} = -\tfrac{1}{2}\left[(\det V)^{-1}\frac{d}{dt}(\det V)\right]q-\tfrac{1}{2}(\mathbf{B}-\mathbf{B}(0))\dot{M}(\mathbf{B}-\mathbf{B}(0))q \tag{I.15}$$

and

$$\frac{\partial q}{\partial \mathbf{B}} = -M(\mathbf{B}-\mathbf{B}(0))q$$

Here $M=C^TV^{-1}C$. Using the identity

$$M^{-1}\dot{M}+\frac{d}{dt}(M^{-1})M=0$$

we have

$$\frac{\partial}{\partial \mathbf{B}}\cdot\left[\frac{d}{dt}(M^{-1})\frac{\partial q}{\partial \mathbf{B}}\right]=\left[\mathrm{tr}(M^{-1}\dot{M})-(\mathbf{B}-\mathbf{B}(0))\dot{M}(\mathbf{B}-\mathbf{B}(0))\right]q$$

From the identities

$$(\det V)^{-1}\frac{d}{dt}(\det V)=\mathrm{tr}(V^{-1}\dot{V})$$

and

$$\mathrm{tr}(V^{-1}\dot{V})+\mathrm{tr}(M^{-1}\dot{M})=2\,\mathrm{tr}(\dot{C}C^{-1})$$

(I.15) may be rewritten in the form

$$\frac{\partial q}{\partial t}=-\mathrm{tr}(\dot{C}C^{-1})q+\frac{1}{2}\frac{\partial}{\partial \mathbf{B}}\cdot\left[\frac{d}{dt}(M^{-1})\frac{\partial q}{\partial \mathbf{B}}\right] \tag{I.16}$$

Returning to the vector **A**, we find that

$$\frac{\partial q}{\partial t}=\frac{\partial p}{\partial t}+\frac{\partial p}{\partial \mathbf{A}}\cdot\dot{C}C^{-1}\mathbf{A}$$

and

$$\frac{\partial}{\partial \mathbf{B}}\cdot\left[\frac{d}{dt}(M^{-1})\frac{\partial q}{\partial \mathbf{B}}\right]=\frac{\partial}{\partial \mathbf{A}}\cdot\left[C\frac{d}{dt}(M^{-1})C^{T}\frac{\partial p}{\partial \mathbf{A}}\right]$$

which upon substitution into (I.16) finally gives the generalized Fokker–Planck equation

$$\frac{\partial p}{\partial t}=-\frac{\partial}{\partial \mathbf{A}}\cdot(\dot{C}_{\mathbf{A}}C_{\mathbf{A}}^{-1}\mathbf{A}p)+\frac{1}{2}\frac{\partial}{\partial \mathbf{A}}\cdot\left[C_{\mathbf{A}}\frac{d}{dt}(M^{-1})C_{\mathbf{A}}^{T}\frac{\partial p}{\partial \mathbf{A}}\right] \tag{I.17}$$

Equation (I.17) has the associated initial condition

$$p(\mathbf{A};0|\mathbf{A}(0))=\delta(\mathbf{A}-\mathbf{A}(0))$$

where δ denotes the Dirac delta function.

Equation (I.17) is exact for non-Markovian, Gaussian systems. Particular forms of this equation have been discussed by Adelman[24] when the elements of **A** are uncorrelated, and also for phase-space variables of the form $\mathbf{A}=\begin{bmatrix}\mathbf{a}\\ \dot{\mathbf{a}}\end{bmatrix}$. Here we shall be concerned mainly with orientation, angular velocity, and coupled angular and linear velocity. For this purpose we need probability densities of the form $p(\mathbf{A}(t);t|\mathbf{A}(0),\dot{\mathbf{A}}(0))$ as well as those of (I.13). Writing $\dot{\mathbf{A}}$ for **A** in (I.6), we obtain

$$\ddot{\mathbf{A}}(t)=i\Omega_{\dot{\mathbf{A}}}\dot{\mathbf{A}}(t)-\int_0^t d\tau\,\phi_{\dot{\mathbf{A}}}(t-\tau)\dot{\mathbf{A}}(\tau)+\mathbf{F}_{\dot{\mathbf{A}}}(t)$$

which may be solved to give

$$\mathbf{A}(t)=\mathbf{A}(0)+X_{\dot{\mathbf{A}}}(t)\dot{\mathbf{A}}(0)+\int_0^t d\tau\,X_{\dot{\mathbf{A}}}(t-\tau)\mathbf{F}_{\dot{\mathbf{A}}}(\tau)$$

where $X_{\dot{\mathbf{A}}}(t)$ is a cross-correlation matrix given by

$$X_{\dot{\mathbf{A}}}(t)=\mathfrak{L}_a^{-1}\left\{\left[s^2-i\Omega_{\dot{\mathbf{A}}}+s\tilde{\phi}_{\dot{\mathbf{A}}}(s)\right]^{-1}\right\}=\langle\mathbf{A}(t)\dot{\mathbf{A}}^T(0)\rangle\langle\dot{\mathbf{A}}(0)\dot{\mathbf{A}}^T(0)\rangle^{-1}$$

so that

$$\dot{X}_{\dot{\mathbf{A}}}(t)=C_{\dot{\mathbf{A}}}(t)$$

and

$$X_{\dot{\mathbf{A}}}(t)=\int_0^t C_{\dot{\mathbf{A}}}(\tau)\,d\tau+X_{\dot{\mathbf{A}}}(0) \tag{I.18}$$

$X_{\dot{\mathbf{A}}}(t)$ is also related to $C_A(t)$ by

$$X_{\dot{\mathbf{A}}}(t)=-\dot{C}_{\mathbf{A}}(t)\langle\mathbf{A}(0)\mathbf{A}^T(0)\rangle\langle\dot{\mathbf{A}}(0)\dot{\mathbf{A}}^T(0)\rangle^{-1}$$

so that

$$C_{\mathbf{A}}(t)=1-\int_0^t d\tau\,X_{\dot{\mathbf{A}}}(\tau)\langle\dot{\mathbf{A}}(0)\dot{\mathbf{A}}^T(0)\rangle\langle\mathbf{A}(0)\mathbf{A}^T(0)\rangle^{-1}$$

and

$$C_{\dot{\mathbf{A}}}(t)=-\ddot{C}_{\mathbf{A}}(t)\langle\mathbf{A}(0)\mathbf{A}^T(0)\rangle\langle\dot{\mathbf{A}}(0)\dot{\mathbf{A}}^T(0)\rangle^{-1}$$

The conditional probability density function

$$\begin{aligned}p(\mathbf{A}(t);t|\mathbf{A}(0),\dot{\mathbf{A}}(0))&=(2\pi)^{-(1/2)n}(\det W(t))^{-1/2}\\&\quad\times\exp\left[-\tfrac{1}{2}\left(\mathbf{A}(t)-\mathbf{A}(0)-X_{\dot{\mathbf{A}}}(t)\dot{\mathbf{A}}(0)\right)^T W^{-1}(t)\right.\\&\quad\left.\times\left(\mathbf{A}(t)-\mathbf{A}(0)-X_{\dot{\mathbf{A}}}(t)\dot{\mathbf{A}}(0)\right)\right]\end{aligned} \tag{I.19}$$

is inferred when $\langle\mathbf{F}_{\dot{\mathbf{A}}}(t)\rangle=\mathbf{0}$, with variance–covariance matrix

$$\begin{aligned}W(t)&=\langle\left(\mathbf{A}(t)-\mathbf{A}(0)-X_{\dot{\mathbf{A}}}(t)\dot{\mathbf{A}}(0)\right)\left(\mathbf{A}(t)-\mathbf{A}(0)-X_{\dot{\mathbf{A}}}(t)\dot{\mathbf{A}}(0)\right)^T\rangle\\&=2(1-C_{\mathbf{A}}(t))\langle\mathbf{A}(0)\mathbf{A}^T(0)\rangle-X_{\dot{\mathbf{A}}}(t)\langle\dot{\mathbf{A}}(0)\dot{\mathbf{A}}^T(0)\rangle X_{\dot{\mathbf{A}}}^T(t)\\&=2\int_0^t d\tau\,X_{\dot{\mathbf{A}}}(\tau)\langle\dot{\mathbf{A}}(0)\dot{\mathbf{A}}^T(0)\rangle-X_{\dot{\mathbf{A}}}(t)\langle\dot{\mathbf{A}}(0)\dot{\mathbf{A}}^T(0)\rangle X_{\dot{\mathbf{A}}}(t)\end{aligned} \tag{I.20}$$

Thus $p(\mathbf{A}(t);t|\mathbf{A}(0),\dot{\mathbf{A}}(0))$ is readily calculated from (I.18) to (I.20) from a knowledge of $C_{\dot{\mathbf{A}}}(t)$. This probabilty density also satisfies a generalized Fokker–Planck equation analogous to (I.17).

B. The Mori Continued Fraction

Mori[18] has shown that the Laplace transform, $\tilde{C}_{\mathbf{A}}(s)$, of $C_{\mathbf{A}}(t)$ has a continued-fraction representation, and that the time evolution of the vector $\mathbf{A}(t)$ is determined by the singularities of $\tilde{C}_{\mathbf{A}}(s)$. Tractable approximations to $C_{\mathbf{A}}(t)$ are thus generated by finite approximants of the continued fraction, and it is this basic idea which is the essential leaven running through this chapter. The derivation of the continued fraction has been discussed in this series by Berne and Harp,[25] so that we quote the result:

$$\tilde{C}_{\mathbf{A}}(s) = \mathfrak{L}_a[\langle \mathbf{A}(t)\mathbf{A}^T(0)\rangle\langle \mathbf{A}(0)\mathbf{A}^T(0)\rangle^{-1}]$$

$$= \cfrac{1}{s - i\Omega_0 + \cfrac{1}{s - i\Omega_1 + \cfrac{1}{\cdots \cfrac{1}{s - i\Omega_{n-1} + \tilde{C}_{\mathbf{A},n}(s)\phi_{\mathbf{A},n}(0)}\cdot\phi_{\mathbf{A},n-1}(0)}\cdot\phi_{\mathbf{A},2}(0)}\cdot\phi_{\mathbf{A},1}(0)} \tag{I.21}$$

Here

$$i\Omega_j \equiv \langle \dot{\mathbf{f}}_j\mathbf{f}_j^T\rangle\langle \mathbf{f}_j\mathbf{f}_j^T\rangle^{-1}$$

and

$$\phi_{\mathbf{A},j}(0) = \langle \mathbf{f}_j\mathbf{f}_j^T\rangle\langle \mathbf{f}_{j-1}\mathbf{f}_{j-1}^T\rangle^{-1}$$

where the **f** vectors are given by

$$\mathbf{f}_0 = \mathbf{A} \qquad \hat{P}_0 = \hat{P}$$

$$\mathbf{f}_j = \left[\hat{1} - \sum_{k=0}^{j-1}\hat{P}_k\right] i\mathfrak{L}\mathbf{f}_{j-1} \qquad j \geq 1 \tag{I.22}$$

$$\dot{\mathbf{f}}_j = \left[\hat{1} - \sum_{k=0}^{j-1}\hat{P}_k\right] i\mathfrak{L}\mathbf{f}_j \qquad j \geq 1 \tag{I.23}$$

where $\hat{P}_k$ projects into the subspace spanned by $\mathbf{f}_k(0)$. From (I.22) and (I.23) we find

$$\mathbf{f}_0 = \mathbf{A}$$

$$\mathbf{f}_1 = \dot{\mathbf{A}} - \left[\langle \dot{\mathbf{A}}\mathbf{A}^T\rangle\langle \mathbf{A}\mathbf{A}^T\rangle^{-1}\right]\mathbf{A}$$

$$\begin{aligned}\mathbf{f}_2 = \ddot{\mathbf{A}} - \big[&\langle \dot{\mathbf{A}}\mathbf{A}^T\rangle\langle \mathbf{A}\mathbf{A}^T\rangle^{-1} + \langle \ddot{\mathbf{A}}\mathbf{f}_1^T\rangle\langle \mathbf{f}_1\mathbf{f}_1^T\rangle^{-1} \\ &- \langle \dot{\mathbf{A}}\mathbf{A}^T\rangle\langle \dot{\mathbf{A}}\mathbf{f}_1^T\rangle\langle \mathbf{A}\mathbf{A}^T\rangle^{-1}\langle \mathbf{f}_1\mathbf{f}_1^T\rangle^{-1}\big]\dot{\mathbf{A}} \\ - \big[&\langle \ddot{\mathbf{A}}\mathbf{A}^T\rangle\langle \mathbf{A}\mathbf{A}^T\rangle^{-1} - \langle \dot{\mathbf{A}}\mathbf{A}^T\rangle^2\langle \mathbf{A}\mathbf{A}^T\rangle^{-2} \\ &- \langle \dot{\mathbf{A}}\mathbf{A}^T\rangle\langle \ddot{\mathbf{A}}\mathbf{f}_1^T\rangle\langle \mathbf{A}\mathbf{A}^T\rangle^{-1}\langle \mathbf{f}_1\mathbf{f}_1^T\rangle^{-1} \\ &+ \langle \dot{\mathbf{A}}\mathbf{A}^T\rangle^2\langle \dot{\mathbf{A}}\mathbf{f}_1^T\rangle\langle \mathbf{A}\mathbf{A}^T\rangle^{-2}\langle \mathbf{f}_1\mathbf{f}_1^T\rangle^{-1}\big]\mathbf{A}\end{aligned}$$

where $\dot{\mathbf{A}} = i\Omega\mathbf{A}$. Thus $\phi_{\mathbf{A},j}$ and Ω_j are related to the moments of the frequency distribution function of $C_{\mathbf{A}}(t)$. For example, when $\mathbf{A}$ is a scalar,

$$\langle\omega^n\rangle \equiv \int_{-\infty}^{\infty} \omega^n P(\omega)\,d\omega = i^{-n}\left[\frac{d^n}{dt^n} C_A(t)\right]_{t=0} \tag{I.24}$$

where $$P(\omega) \equiv \frac{1}{\pi}\mathrm{Re}\left[\tilde{C}_A(i\omega)\right] = \frac{1}{2\pi}\int_{-\infty}^{\infty} C_A(t)\exp(-i\omega t)\,dt$$

As $n \to \infty$ (I.21) leads to an infinite continued fraction, with the time evolution of $C_{\mathbf{A}}(t)$ determined by the singularities of this fraction in the complex s-plane.

Many molecular models of fluid dynamics in the literature are approximants of this expansion. Indeed, for reasons of consistency, it is difficult to see a model approximating the Liouville equation which does not fit into the expansion represented by (I.21). Table I shows a selection of these, where $\mathbf{A}$ is the angular momentum $\mathbf{J}$ or the dipole orientation $\mathbf{u}$. The real power of the continued fraction is its general validity for classifying not only molecular but also hydrodynamical and ferromagnetic phenomena within the same theoretical framework. The generalized hydrodynamics developed in the last few years[7,26,27] has proved useful in explaining the observations made by polarized and depolarized light scattering, and the results of computer simulation. The special brand of mode–mode coupling[12,28] used has been slow in influencing molecular theories, but we suggest later in this section how molecular rototranslation may be described.

Mori in his original paper discusses how, if we truncate the continued fraction at a particular memory matrix $\tilde{C}_{\mathbf{A},n}(s)$, it is often possible (Table I) to introduce one of the following approximations or any of their combinations: (1) long-time approximation, (2) a perturbation procedure, (3) a high- or low-temperature approximation, and (4) a short-time approximation.

1. *Long-Time Approximation*

Consider a scalar A and denote the first spectral moment by $\Omega_0 = \langle\omega\rangle$. When $A(t)\exp(-i\Omega_0 t)$ is a slowly varying function of time in a certain time scale Δt, the function $\tilde{C}^{(0)}_{A,n}(s) \equiv \tilde{C}_{A,n}(s + i\Omega_0)$ will be insensitive to s in a small region around the origin in the complex s-plane (i.e., at low frequencies). We can then neglect the s-dependence of $\tilde{C}_{A,n}(s)$, thus truncating the continued fraction with

$$\tilde{C}^{(0)}_{A,n}(s) \simeq \xi_{A,n} \equiv \int_0^{\infty} C_{A,n}(t)\exp(-i\Omega_0 t)\,dt \tag{I.25}$$

TABLE I
Some Dynamical Models in Mori's Formalism[a]

Memory function	Model	Comments
$\phi_1^{(\mathbf{u})}(t) = D\delta(t)$, where D is a diffusion coefficient, δ the Dirac delta function	Debye, rotational diffusion[37] of the inertia-less spherical top with embedded dipole	Used extensively to describe dielectric experiments, but leads to a plateau in $\alpha(\omega)$, the power absorption coefficient at frequencies high with respect to the Debye relaxation time. $C_{\mathbf{u}}(t) = e^{-t/\tau_0}$.
$\phi_1^{(\mathbf{u})}(t) = \phi_1^{(\mathbf{u})}(0)e^{-\gamma_1 t}$ before ensemble averaging	Gordon m diffusion[16,34] linear molecule; space reorientation	Instantaneous collisions perturb the angular momentum vector (**J**) in direction but not in magnitude. Power spectrum diverges logarithmically around $\omega = 0$. Binary collision model. Mean-square torque singular at every impact. Does not follow the observed THz frequency shifts in ω_{max}.
$\phi_1^{(\mathbf{u})}(t) = \phi_{FR}^{(\mathbf{u})}(t)e^{-\gamma\|t\|}$ where $\phi_{FR}(t)$ is the free-rotor ensemble memory function	Gordon J diffusion.[16,34] Linear molecule; space reorientation	As for M diffusion, **J** randomized both in direction and magnitude onto a Poisson distribution. A slow return to spectral transparency at high THz frequencies of the far infrared. Does not follow changes in ω_{max}.

$\phi_1^{(\mathbf{u})}(t) = \phi_1^{(\mathbf{u})}(0)\exp(-\gamma t)\cos\omega_0 t$	Damped planar libration[38,39]	$\delta(\omega)$ falls off asymptotically at high frequencies as ω^{-2}, as for J diffusion. ω_0 is the libration frequency, synonymous with $\omega_{\max}$, the far-infrared peak frequency. $C_{\mathbf{u}}(t)$ a sum of three exponentials, as for the itinerant planar librator.
$\phi_1^{(\mathbf{u})}(t) = \phi_1^{(\mathbf{u})}(0)\exp(-\gamma t)$	Maximum information entropy.[25]:space orrientation of CO	$\gamma = (\pi/4)^{1/2}\phi_1(0)\int_0^{\infty}\psi(t)\,dt; \psi(t) = \langle\omega_{\perp}(0)\cdot\omega_{\perp}(t)\rangle$ where $\omega_{\perp}$ is perpendicular to $\mathbf{u}$. THz spectrum too sharp compared with experimental data unless γ is varied for best fit.
$\phi_2^{(\mathbf{u})}(t) = {}^{FR}\phi_2^{(\mathbf{u})}(t)\psi(t)$	Desplanques	$\psi(t)$ as above. In general, the theoretical spectrum $\alpha(\omega)$ is too sharp in the far infrared.
$\phi_2^{(\mathbf{u})}(t) = \beta\phi_{2FR}^{(\mathbf{u})}(t)\exp(-\lvert t\rvert/\tau)$	Bliot and Constant[40,41]	$\phi_{2FR}^{(\mathbf{u})}(t)$ is the second memory of the free rotor. β and τ are stochastic variables, the former being proportional to the mean-square torque. Too sharp in the far infrared for broad bands. $C_{\mathbf{u}}(t)$ even up to t^4 of its Taylor series.
$\phi_2^{(\mathbf{u})}(t) = \phi_2^{(\mathbf{u})}(0)\exp(-\gamma t)$	First used by Barojas, Levesque, and Quentrec in a simulation[42] of liquid nitrogen	Used extensively[43] by Evans for semiempirical analyses of a wide range of zero-THz data. Used by Drawid and Halley[44] for Heisenberg paramagnetism (see text).

TABLE I *(Continued)*

Memory function	Model	Comments
$\phi_1^{(J)}(t)=\phi_1^{(J)}(0)\delta(t)$	Asymmetric top, Langevin equation for rotational Brownian diffusion[21,22] in space	Produces a high-frequency return to transparency but no THz resonance (or Poley absorption—see text).
$\phi_2^{(\omega)}(t)=\phi_2^{(\omega)}(0)\exp(-\gamma t)$	This approximant produces $\tilde{C}_\omega(s)$ formally identical with planar itinerant libration[29]	Here $\omega\equiv\dot{\theta}$ is the angular velocity of a disk librating in a plane, (I.30).
Eqs. (I.50)–(I.53)	A zeroth-order approximant of the system of Damle and others[29] for neutron scattering	Leads to $\tilde{C}_\omega(s)$ formally identical with that from (I.31). A physical system identical with that of (I.30). A finite friction, however, exists between ring and annulus. May be interpreted also in terms of dipole–dipole coupling, (I.54).
Modeling performed on the memory function of **J** by Lindenberg and Cukier[45]		Generalized stochastic model for molecular rotational motion in condensed media, encompassing all previous stochastic models of rotational relaxation as specialized cases. Close in spirit to the approach of Kivelson and Keyes.[46]

[a] $\phi_n(t)$ is the nth memory of the dipole unit vector (**u**) or angular momentum (**J**).

It may be shown that

$$\tilde{C}_A(s+i\Omega_0)\simeq g_{n-1}(s)/g_n(s) \tag{I.26}$$

where $g_j(s)$ is a jth-degree polynomial defined by

$$g_j(s)=\left[s+i(\Omega_0-\Omega_{n-j})\right]g_{j-1}(s)+\phi_{A,n-j+1}(0)g_{j-2}(s) \quad j\geq 2 \tag{I.27}$$

$$g_0(s)=1, \qquad g_1(s)=s+i(\Omega_0-\Omega_{n-1})+\lambda_{n-1} \tag{I.28}$$

with

$$\lambda_{j-1}\equiv\phi_{A,j}(0)\xi_j=\phi_{A,j}(0)/\left[i(\Omega_0-\Omega_j)+\lambda_j\right]$$

The singularities of (I.26) are given by the zeros of $g_n(s)$. Denoting these zeros by $s_\alpha=i(\beta_\alpha-\Omega_0)-\gamma_\alpha, \alpha=1,\ldots,n$, and taking the inverse Laplace transform of (I.26), we have

$$C_A(t)\simeq\sum_{\alpha=1}^{n}R_\alpha\exp\left[(i\beta_\alpha-\gamma_\alpha)t\right] \tag{I.29}$$

where R_α is the residue of the ratio $g_{n-1}(s)/g_n(s)$ at the pole s_α.

The approximation in (I.25) corresponds to the description of $A(t)\exp(-i\Omega_0 t)$ in the time scale Δt distinctly larger than the decay time τ_n of $C_{A,n}(t)$. Mori calls this the nth-order long-time approximation around $s=i\Omega_0$. Equation (I.25) is valid if all the n poles are located in the left-half s-plane inside a semicircle center $s=i\Omega_0$, and radius $1/\tau_n$. These poles represent slow processes, and the higher-frequency components which correspond to the remaining singularities of $C_A(t)$ are represented by the nth-order random force $f_n(t)$, whose correlation time is much smaller than the relaxation time of the slow processes.

Rotational Motions: The Planar Itinerant Librator. Let A be the rate at which an angle θ is swept out by a dipole rotating in a plane. It is easily shown that the operators $i\Omega_0$, $i\Omega_1$, and $i\Omega_2$ vanish identically. Introducing the third-order long-time approximation around $s=0$, we have

$$\tilde{C}_{\dot\theta}(s)=\frac{s^2+\lambda_2 s+\phi_{\dot\theta,2}(0)}{s^3+\lambda_2 s^2+\left[\phi_{\dot\theta,1}(0)+\phi_{\dot\theta,2}(0)\right]s+\phi_{\dot\theta,1}(0)\lambda_2} \tag{I.30}$$

This expression for $\tilde{C}_{\dot\theta}(s)$ may be identified with a simple model[29] of the angular Brownian motion of a dipole immersed in a bath of interacting particles. We shall return to the experimental evaluation of this later—it has been used extensively[30] to describe the zero-THz absorptions in fluids and glasses. The quantities λ_2, $\phi_{\dot\theta,1}(0)$, and $\phi_{\dot\theta,2}(0)$ may be given solid physical meanings in terms of the disk/annulus system illustrated in Fig. 4. It is assumed in the rotational dynamics leading to (I.30) that a librating central molecule together with its cage of neighbors may be represented by an

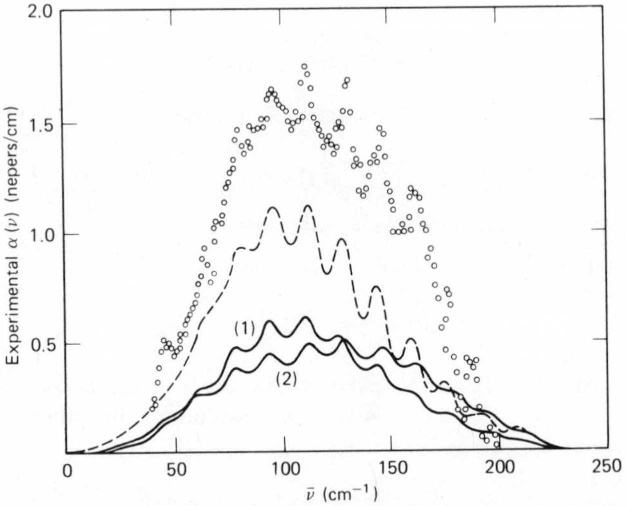

Fig. 4. Experimental absorption of HBr/SF_6 liquid mixture at 296°K. (---), Frenkel/Wegdam formalism (see text); ordinate scale unnormalized. (—), (1) truncated Mori formalism of Bliot and Constant, $\beta_2=1$; (2) $\beta_2=1.2$. Both curves (1) and (2) are unnormalized to the experimental data on the ordinate scale. [Reproduced by permission from *Chem. Phys. Lett.*, **42**, 331 (1976).]

annulus which is free to rotate about a central axis, perpendicular to its plane. Concentric and coplanar with the annulus is a disk which is free to rotate about the same central axis. (The words "disk" and "annulus" are used only as a convenient scematic description of the model; the theory will apply to a body of arbitrary shape provided that it is constrained to rotate in two dimensions.) The disk carries a dipole μ lying along one of its diameters, the orientation of the dipole is specified by an angle $\theta(t)$ relative to a fixed direction, and the position of a point on the rim of the annulus is specified by an angle $\psi(t)$ relative to the same fixed direction. The mechanical interaction between the central molecule and its neighbors is a restoring torque acting on the dipole, which is proportional to the angular displacement $\theta(t)-\psi(t)$ of the dipole in space. Furthermore, if it is stipulated that on the average each cage containing a molecule behaves in the same way, so that we may study the behavior of a single cage containing a dipole and examine how the dipole orientation changes when the measuring field is switched off, then the equations of motion are[13]

$$\begin{gathered} I_1\ddot{\psi}(t)+I_1\beta_1\dot{\psi}(t)-I_2\gamma^2[\theta(t)-\psi(t)]=I_1\dot{W}_1(t) \\ I_2\ddot{\theta}(t)+I_2\beta_2\dot{\theta}(t)+I_2\gamma^2[\theta(t)-\psi(t)]=I_2\dot{W}_2(t) \\ \beta_1=\xi_1/I_1 \qquad \beta_2=\xi_2/I_2 \end{gathered} \tag{I.31}$$

Here I_1 is the moment of inertia of the annulus; I_2 that of the disk; $\xi_1\dot{\psi}(t)$ and $\xi_2\dot{\theta}(t)$ are the frictional couples acting on the annulus and disk, respectively, arising from the thermal motion of the surroundings; $\dot{W}_1$ and $\dot{W}_2$ are Wiener processes representing random couples acting on the annulus and disk; and γ is the natural angular frequency of oscillation when the annulus is held stationary.

In the case $\beta_2 = 0$ (no friction between annulus and disk), the angular velocity correlation function $\langle\dot{\theta}(t)\dot{\theta}(0)\rangle$ is given by the inverse Laplace transform of (I.30) with

$$\lambda_2 = \beta_1 \qquad \phi_{\dot{\theta},1}(0) = \gamma^2 \qquad \phi_{\dot{\theta},2}(0) = \left(\frac{I_2}{I_1}\right)\phi_{\dot{\theta},1}(0) \tag{I.32}$$

Writing $J = I\dot{\theta}$ for the angular momentum of the disk librating and diffusing in its plane, then from (I.17) we have

$$\frac{\partial p}{\partial t} = \tilde{\beta}(t)\left[\frac{\partial}{\partial J}(Jp) + kTI_2\frac{\partial^2 p}{\partial J^2}\right] \tag{I.33}$$

where the probability density function is

$$p(J;t|J(0)) = \left[2\pi kTI_2(1 - C_J^2(t))\right]^{-1/2}\exp\left[-\frac{(J - C_J(t)J(0))^2}{2kTI_2(1 - C_J^2(t))}\right] \tag{I.34}$$

and

$$\begin{aligned} C_J(t) &= \langle J(t)J(0)\rangle/(kTI_2) \\ \tilde{\beta}(t) &= -\dot{C}_J(t)/C_J(t) \end{aligned} \tag{I.35}$$

with

$$p(J;0|J(0)) = \delta(J - J(0))$$

The probability density functions calculated from (I.33) to (I.35) exhibit widely different decays for each pair of $\phi_1(0), \phi_2(0)$ values (dropping the subcript J), and are directly interpretable in terms of the structure and dynamics of the molecular fluid they represent. We consider two cases in order of increasing ratio $\phi_2(0)/\phi_1(0)$; the parameters used to calculate $C_J(t)$ are given in Table II. In each case we adopt the normalization $(2kT/I_2) = 1$, and take $J(0) = 0$. In Figs. 5*a* and 6*a* the probability densities are displayed at various times ranging from near the origin to ~2 psec. In

TABLE II
Parameters Used to Calculate $C_J(t)$ of (I.30)

Fig.	T(°K)	$\beta_1(2kT/I_2)^{1/2}$	$\phi_1(0)(2kT/I_2)$	$\phi_2(0)(2kT/I_2)$	$10^{-12}(2kT/I_2)^{1/2}$
5,7	296	4.0	1.0	8.0	2.25
6,8	340	2.0	0.78	200.0	1.44

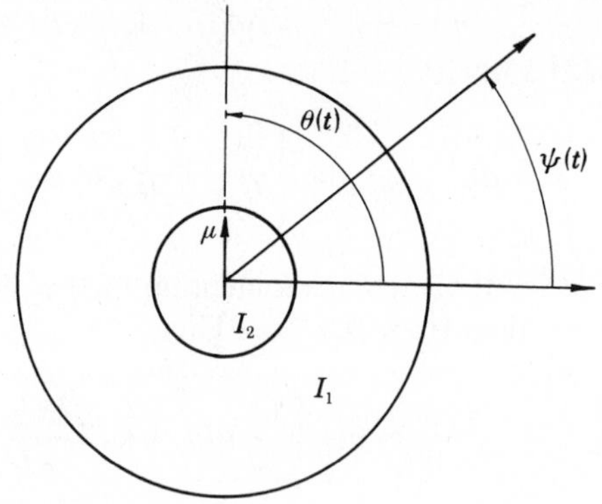

Fig. 5. Schematic of the disk–annulus representation of (I.30). [Reproduced by permission from *Proc. R. Soc. Lond.*, **356**, 269 (1977).]

Figs. 6*b* and 7*b* the decay of the peak height is shown as a function of time. The curve in Fig. 6*b* decays smoothly; the torque is not subject to abrupt changes on the average (i.e., the torque derivative is not large compared with the torque itself). The model fluid as a whole is not one where the angular momentum for individual dipoles is restricted to favored values for any significant period of time, since $p(J;t|J(0))$ broadens out rapidly to the equilibrium distribution. Collisions take place with a low mean transfer of momentum.

On the other hand, the probability density function associated with a ratio $\phi_2(0)/\phi_1(0)\gg 1$ reveals over 2 psec quite different features (Fig. 7). In this case the disk is very heavy and the torque derivative is large compared with the torque itself. This means abrupt changes in the root-mean-square torque which acts on the annulus. The oscillations in p imply that periodically it becomes increasingly probable to find a mean angular momentum which corresponds physically to a torsional motion at the bottom of an energetically favorable potential well. Here a molecule is encaged along a well-defined axis for relatively long times, since at each successive peak maximum in p the distribution width is narrow. Hard energetic collisions are needed to affect the motion of the heavy disk.

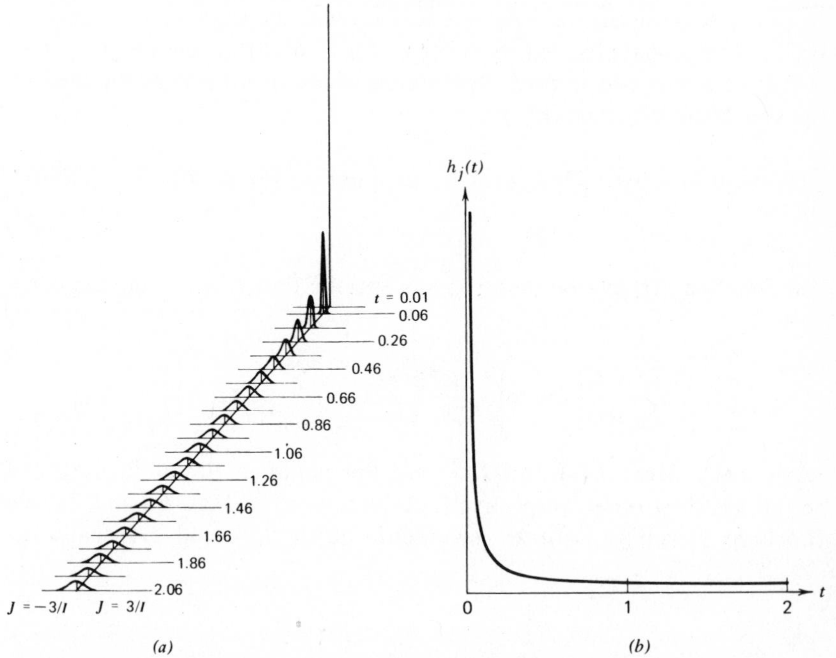

Fig. 6. (*a*) $p(J; t/J(o))$ for β_1, $\phi_1(o)$ and $\phi_2(o)$ as in Table II. p is plotted on the vertical axis, J on the horizontal axes (from $-3/I$ to $3/I$), and time t on the diagonal axis. (*b*) Plot of pdf peak height vs. time (psec). [Reproduced by permission from *Mol. Phys.*, **35**, 864 (1978).]

$C_J(t)$ may also be used to calculate $p(\theta(t); t|\theta(0), \dot{\theta}(0))$ which describes the torsional oscillation of the dipole $\boldsymbol{\mu}$ in an itinerant librator of (I.30). If $\theta(t)$ is the total angle turned through in time t, then from (I.19) and (I.20)

$$p(\theta(t); t|\theta(0), \dot{\theta}(0)) = \left[\frac{2\pi kT}{I_2}\left(2\int_0^t X_J(\tau)\,d\tau - X_J^2(t)\right)\right]^{-1/2} \times \exp\left[-\frac{(\theta(t)-\theta(0)-X_J(t)\dot{\theta}(0))^2}{\left(\frac{2kT}{I_2}\right)\left(2\int_0^t X_J(\tau)\,d\tau - X_J^2(t)\right)}\right] \tag{I.36}$$

where

$$X_J(\tau) = \int_0^t C_J(\tau)\,d\tau$$

In practice, however, we wish $\theta(t)$ to represent the angular orientation at time t. The probability density function for *restricted* to the range $(-\pi,\pi)$ is that of a wrapped normal distribution which may be approximated by the von Mises distribution:

$$p(\theta(t);t|\theta(0),\dot{\theta}(0)) \simeq [2\pi I_0(\alpha(t))]^{-1}\exp[\alpha(t)\cos(\theta(t)-\theta(0)-X_J(t)\dot{\theta}(0))] \tag{I.37}$$

The function $\alpha(t)$ can be found at any specific time t by solving the equation

$$\frac{I_1(\alpha(t))}{I_0(\alpha(t))} = \exp\left[-\frac{kT}{I}\left(\int_0^t X_J(\tau)\,d\tau - \tfrac{1}{2}X_J^2(t)\right)\right]$$

numerically. Here $I_0(\cdot)$ and $I_1(\cdot)$ are the modified Bessel functions of zeroth and first order, respectively. As illustrated in Figs. 6*a* and 7*a*, the functions $p(\theta;t|\theta(0),\dot{\theta}(0))$ are symmetric about $\theta(0)$, and eventually die

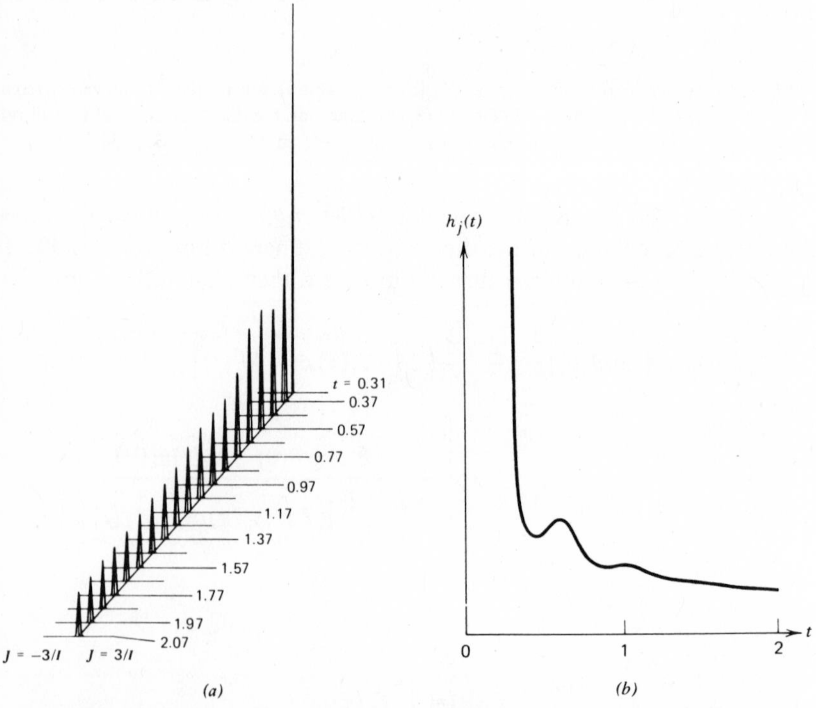

Fig. 7. Key as in Fig. 6; parameters and pdf's defined in the text and Table II.

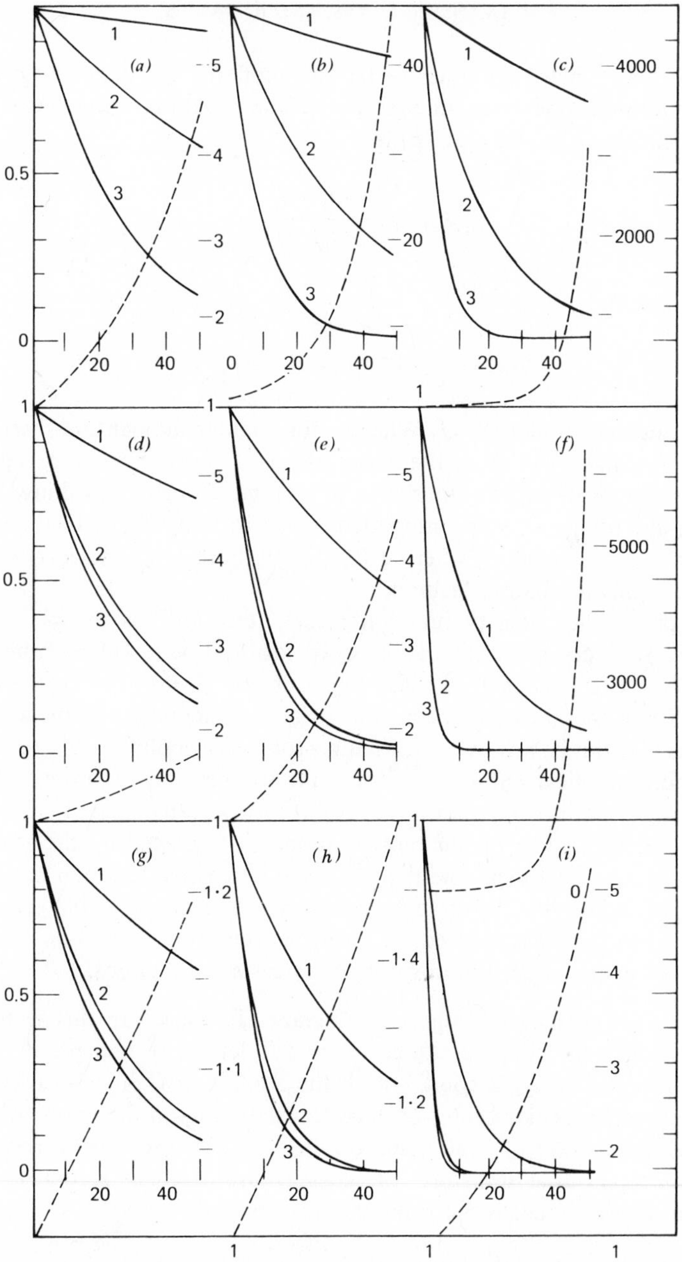

Fig. 8. (1) $C_0(\mathbf{Q},t)$; (2) $C_1(\mathbf{Q},t)$; (3) $C_1(\mathbf{o};t)$; (----), the ratio $C_1(\mathbf{Q},t)/C_1^{(u)}(\mathbf{Q},t)$ for $Q=1$, $\kappa=0.4$. (*a*) $\lambda=0.1$, $\beta_\omega=50$; (*b*) $\lambda=0.1$, $\beta_\omega=20$; (*c*) $\lambda=0.1$, $\beta_\omega=10$; (*d*) $\lambda=0.5$, $\beta_\omega=50$; (*e*) $\lambda=0.5$, $\beta_\omega=20$; (*f*) $\lambda=0.5$, $\beta_\omega=5$; (*g*) $\lambda=1.0$, $\beta_\omega=50$; (*h*) $\lambda=1.0$, $\beta_\omega=20$; (*i*) $\lambda=1.0$, $\beta_\omega=5$. λ is the roughness parameter, κ the mass-distribution parameter of Berne and Montgomery. The parameters above are in reduced units specified by these authors. [Reproduced by permission from *Faraday Discuss. Chem. Soc.*, **66** (1978).]

down to a flat distribution as $t \to \infty$. The similarity in shape to Gaussian distributions disappears as time increases. The equation of the decay curve of peak height (Figs. 6*b* and 7*b*) is

$$h_\theta(t) = \frac{\exp[\alpha(t)]}{2\pi I_0(\alpha(t))}$$

as opposed to

$$h_J(t) = \left[2\pi kTI_2(1 - C_J^2(t))\right]^{-1/2}$$

for the angular momentum J. Whereas for angular momentum there is an onset of oscillation in $h_J(t)$, the decay of $h_\theta(t)$ is free of them, except perhaps for the most rapid torque rate of change. Roughly speaking, this is because the oscillations are integrated out in forming $X_J(t)$. In the underdamped regime, however, where $\phi_2(0) < \phi_1(0)$, both $h_J(t)$ and $h_\theta(t)$ will decay in a highly oscillatory fashion.

The probability density function $p(\theta(t); t|\theta(0), \dot{\theta}(0))$ is the planar reorientational counterpart of the translational van Hove function[31] $G_s(\mathbf{r}; t|\mathbf{0})$, where r is center-of-mass displacement. G_s may be obtained from atomic fluids by scattering thermal neutrons incoherently and inelastically. In molecular fluids, however, translation-rotation coupling is of great practical and theoretical importance,[32] and the self-part of the observable van Hove function is in general the joint probability density function $G_s(\mathbf{r}, \boldsymbol{\Omega}; t|\mathbf{0}, \boldsymbol{\Omega}(0))$. (For planar reorientation $\boldsymbol{\Omega}$ is replaced by the scalar θ.) Moreover, tractable expressions for the self van Hove function have been developed in the first Born approximation, when coupling between rotation and translation is ignored. In the next section we show how this coupling may be included in joint probability density functions.

Rotation–Translation Coupling—Itinerant Libration. In this section we first deal with the coupled autocorrelation functions of immediate interest to neutron scattering. Second, within the context of (I.30), we discuss the influence of molecular center-of-mass translation upon the momentum autocorrelation $\langle J(t)J(0)\rangle$ and hence on the zero-THz loss profile, related directly to the Fourier transform of $\langle \cos\theta(t)\cos\theta(0)\rangle$. The zero-THz power absorption coefficient is the Fourier transform[30] of

$$\left\langle \frac{d}{dt}(\cos\theta(t))\left(\frac{d}{dt}(\cos\theta(t))\right)_{t=0}\right\rangle$$

when the influence of the internal field is not considered.

The first kind of autocorrelation function is, for general space rototranslation,[32]

$$C_k(\mathbf{q};t)=\langle P_k(\boldsymbol{\mu}(t)\cdot\boldsymbol{\mu}(0))\exp[i\mathbf{q}\cdot\Delta\mathbf{r}(t)]\rangle \tag{I.38}$$

where k is a positive integer, $P_k(x)$ the Legendre polynomial of degree k, $\mathbf{q}$ the scattering wave vector, and $\Delta\mathbf{r}(t)=\mathbf{r}(t)-\mathbf{r}(0)$ the displacement in time t. The $C_k(\mathbf{q};t)$ in neutron scattering theory are approximated by

$$C_k^{(u)}(\mathbf{q};t)=\langle P_k(\boldsymbol{\mu}(t)\cdot(\boldsymbol{\mu}(0))\rangle\langle\exp[i\mathbf{q}\cdot\Delta\mathbf{r}(t)]\rangle \tag{I.39}$$

Berne and Montgomery[32] have lately shown, however, that the maximum deviation between $C_k(\mathbf{q};t)$ and $C_k^{(u)}(\mathbf{q};t)$ occurs for wave numbers commonly found in thermal neutron scattering, and it was pointed out that the effect of the coupling would increase for structured molecules as opposed to the rough spheres which they considered analytically. In this work, Berne and Montgomery adopt the Chandler[33] binary collision approximation (Table I)—an earlier approximant of (I.21) than that resulting in (I.30). When translational effects are ignored, this approximant becomes the Gordon J-diffusion model for spherical tops.[34] Furthermore, they adopt (1) a second-order expansion for the Laplace transform of the free-particle rotation–translation correlation function, and (2) a partial curtailment to first order of (1) in obtaining $C_k(\mathbf{q};s)$, the Laplace transform of $C_k(\mathbf{q};t)$. If, however, we dispense with approximation (2) and keep to second order, we obtain

$$\tilde{C}_k(\mathbf{Q};s)=\frac{s^2+2\beta_k(Q)s+[\beta_k^2(Q)-k(k+1)-Q^2]}{s^3+2\beta_k(Q)s^2+\beta_k^2(Q)s+\beta_k(Q)[k(k+1)+Q^2]} \tag{I.40}$$

Here $\mathbf{Q}$ denotes the dimensionless wave vector and

$$\beta_k(Q)=\beta_w\left[\frac{k(k+1)+\left\{\dfrac{(1+\lambda)K+1}{\lambda}\right\}Q^2}{k(k+1)+Q^2}\right] \tag{I.41}$$

where λ is a slip coefficient, K a loading parameter, and the dimensionless angular velocity relaxation rate of the rough sphere fluid. Equation (I.40) is similar in structure to (I.30), and its inverse Laplace transform may be recovered analytically. In Fig. 9 we see that for $Q=1$, (1) the ratio $C_1(\mathbf{Q};t)/C_1^{(u)}(\mathbf{Q};t)$ always increases (from unity) with time; (2) for fixed λ, the ratio increases more rapidly the smaller the value of β_ω; and (3) the

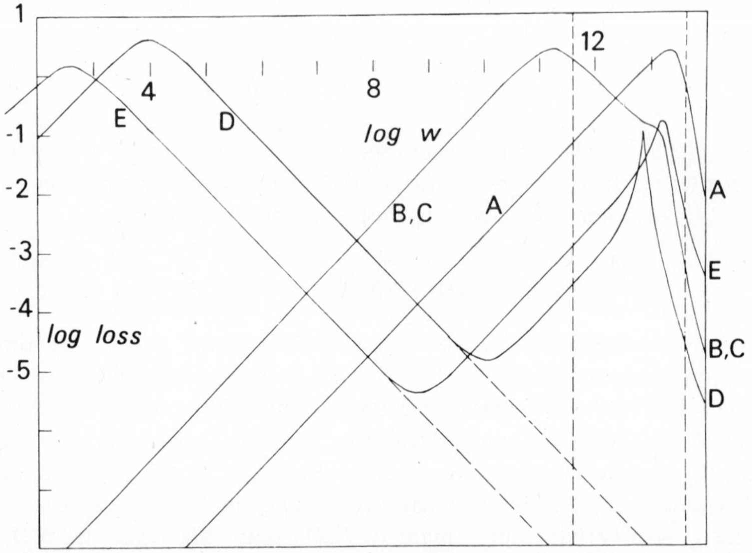

Fig. 9. Log-log plot (schematic) of loss against angular frequency expected from various classes of dipolar compounds, as simulated by (I.31). (*A*) Water free from hydrogen bonding (in dilute solution); gaslike rototranslation in the liquid state. The Debye and Poley portions of the loss are indistinguishable. (*B*), (*C*) Liquids such as benzonitrile and plastic crystals such as $(CH_3)_3Cl$, where the Poley absorption is clearly resolved on the high-frequency side of the Debye curve. (*D*) Mesophase, such as the nematic of 4-cyano-4′*n*-heptyl biphenyl. The Poley absorption is considerably sharper than in cases (*B*) and (*C*) and well separated on the frequency scale. (*E*) Glasses, such as those of CH_2Cl_2 in decalin (see text). Here the β process is shown at low frequency together with the high frequency far-infrared adjunct.

ratio is greatest for small λ. Thus for given $\mathbf{Q}$ the coupling effect is small when β_ω is large and collisions are rough. It is greatest when slipping conditions are invoked in a dilute fluid where the angular velocity relaxes fairly slowly.

For structured molecules, rather than consider $C_1(\mathbf{q};t)$, we discuss an alternative and simpler means of investigating the interplay between spin angular velocity and linear velocity. For a diatomic, for example, one may consider the total velocity autocorrelation function for an atom. If $\boldsymbol{\mu}$ is the interatomic displacement vector, then the total velocity $\mathbf{v}_a$ of an atom is

$$\mathbf{v}_a = \mathbf{v} + \tfrac{1}{2}\boldsymbol{\omega} \times \boldsymbol{\mu} \tag{I.42}$$

where $\mathbf{v}$ denotes the center-of-mass velocity and $\boldsymbol{\omega}$ the spin angular velocity about the center of mass. Thus the autocorrelation function of $\mathbf{v}_a$ contains information on both linear and angular velocities. The relation between the autocorrelations of $\mathbf{v}_a$ and $\mathbf{v}$ can easily be found if $\boldsymbol{\omega}$ is constrained to lie in a fixed direction, as in the planar itinerant librator, so that $\dot{\theta}(t)=\omega(t)$. For (I.42) we have

$$\langle \mathbf{v}_a(t)\cdot\mathbf{v}_a(0)\rangle = \langle \mathbf{v}(t)\cdot\mathbf{v}(0)\rangle + \langle \mathbf{v}(t)\cdot\boldsymbol{\omega}(0)\mathbf{x}\boldsymbol{\mu}(0)\rangle + \tfrac{1}{4}\langle \boldsymbol{\mu}(t)\cdot\boldsymbol{\mu}(0)\boldsymbol{\omega}(t)\cdot\boldsymbol{\omega}(0)\rangle \tag{I.43}$$

since

$$\langle \boldsymbol{\omega}(t)\mathbf{x}\boldsymbol{\mu}(t)\cdot\mathbf{v}(0)\rangle = \langle \mathbf{v}(t)\cdot\boldsymbol{\omega}(0)\mathbf{x}\boldsymbol{\mu}(0)\rangle$$

The central term in the right-hand side of (I.43) describes the effect of rotation–translation coupling and would vanish in a decoupled approximation; the third term describes the coupling between reorientation and spin angular velocity, which vanishes only in the limit $t\to\infty$. Although it is easily shown that

$$\langle \boldsymbol{\mu}(t)\cdot\boldsymbol{\mu}(0)\boldsymbol{\omega}(t)\cdot\boldsymbol{\omega}(0)\rangle = \frac{d^2}{dt^2}\langle \boldsymbol{\mu}(t)\cdot\boldsymbol{\mu}(0)\rangle$$

this relation hides the dependence of the third term on the joint conditional probability density function $p(\omega(t),\theta(t);t|\omega(0),\theta(0))$, which describes a cylindrical distribution of importance in its own right. We demonstrate its use by calculating $\langle \boldsymbol{\mu}(t)\cdot\boldsymbol{\mu}(0)\boldsymbol{\omega}(t)\cdot\boldsymbol{\omega}(0)\rangle$ explicitly.

Since $\boldsymbol{\mu}(t)\cdot\boldsymbol{\mu}(0)=\mu^2\cos(\theta(t)-\theta(0))$, the required correlation function is defined by

$$\langle \boldsymbol{\mu}(t)\cdot\boldsymbol{\mu}(0)\boldsymbol{\omega}(t)\cdot\boldsymbol{\omega}(0)\rangle$$
$$=\mu^2\int_{-\infty}^{\infty}\int_{-\infty}^{\infty}\int_{-\pi}^{\pi}\int_{-\pi}^{\pi}\omega\omega_0\cos(\theta-\theta_0)p(\omega,\theta;t|\omega_0,\theta_0)p(\omega_0,\theta_0)\,d\omega\,d\omega_0\,d\theta\,d\theta_0 \tag{I.44}$$

where $\omega(0)=\omega_0$, $\theta(0)=\theta_0$, and $p(\omega_0,\theta_0)$ is the joint density for the initial distribution of ω and θ at time $t=0$. Since ω and θ are statistically independent variables, and the initial distribution of ω may be assumed Maxwellian, we may write

$$p(\omega_0,\theta_0)=\left(\frac{2\pi k}{I_2}\right)^{1/2}\exp\left[-\frac{\omega_0^2}{(2kT/I_2)}\right]p(\theta_0)$$

where the initial distribution $p(\theta_0)$ satisfies

$$\int_{-\pi}^{\pi}p(\theta_0)\,d\theta_0=1$$

but otherwise need not be specified for the purpose of the present calculation. Since $\theta(t)$ is restricted to the range $-\pi \leq \theta \leq \pi$, the cylindrical distribution has a marginal distribution for ω which is normal, and a marginal distribution for θ which is wrapped normal. It may be specified by generalizing a well-known result in the theory of wrapped distributions.[35] If x is a normally distributed random variable with mean $\langle x \rangle$ and variance $\sigma_x^2 = \langle (x - \langle x \rangle)^2 \rangle$, the corresponding wrapped variate $\tilde{x} = x(\mathrm{mod}\, 2\pi)$ has a density

$$p(\tilde{x}) = \frac{1}{2\pi}\left[1 + 2\sum_{n=1}^{\alpha}\{\cos(n\langle x \rangle)\cos(n\tilde{x}) + \sin(n\langle x \rangle)\sin(n\tilde{x})\}\right]\exp\left(-\tfrac{1}{2}n^2\sigma_x^2\right)$$

Extending this result to the case of two normal variables, only one of which is wrapped, we obtain the expression

$$\begin{aligned} p(\omega, \theta; t|\omega_0, \theta_0) = \frac{1}{(2\pi)^2}\int_{-\infty}^{\infty}\Bigg[&\exp\left(is\langle \omega \rangle - \tfrac{1}{2}s^2\sigma_\omega^2 - is\omega\right) \\ &+ 2\sum_{N=1}^{\infty}\{\cos(s\langle \omega \rangle + n\langle \theta \rangle)\cos(s\omega + n\theta) \\ &+ \sin(s\langle \omega \rangle + n\langle \theta \rangle)\sin(s\omega + n\theta) \\ &\times \exp\{-\tfrac{1}{2}(s^2\sigma_\omega^2 + 2ns\,\mathrm{cov} + n^2\sigma_\theta^2)\}\Bigg]ds \end{aligned} \tag{I.45}$$

where the covariance term is defined by

$$\mathrm{cov} = \langle (\omega - \langle \omega \rangle)(\theta - \langle \theta \rangle) \rangle$$

In terms of the normalized correlation functions $C_\omega(t)$ and $X_\omega(t) = \int_0^t C_\omega(\tau)\,d\tau$, we have

$$\begin{aligned} \langle \omega \rangle &= C_\omega(t)\omega_0 \\ \langle \theta \rangle &= \theta_0 + X_\omega(t)\omega_0 \\ \sigma_\omega^2 &= \frac{kT}{I_2}(1 - C_\omega^2(t)) \\ \sigma_\theta^2 &= \frac{kT}{I_2}\left(2\int_0^t X_\omega(\tau)\,d\tau - X_\omega^2(t)\right) \\ \mathrm{cov} &= \frac{kT}{I_2}X_\omega(t)(1 - C_\omega(t)) \end{aligned} \tag{I.46}$$

Equation (I.44) may be integrated using (I.45) and (I.46) and some standard integrals to give

$$\langle \boldsymbol{\mu}(t)\cdot \boldsymbol{\mu}(0)\boldsymbol{\omega}(t)\cdot \boldsymbol{\omega}(0)\rangle = \mu^2 \frac{kT}{I_2}\left[C_\omega(t) - \frac{kT}{I_2} X_\omega^2(t)\right] \exp\left[-\frac{kT}{I_2}\int_0^t X_\omega(\tau)\,d\tau\right] \tag{I.47}$$

But from (I.36) it may be shown that

$$\begin{aligned}\langle \boldsymbol{\mu}(t)\cdot\boldsymbol{\mu}(0)\rangle &\equiv \mu^2 \int_{-\pi}^{\pi}\int_{-\pi}^{\pi}\int_{-\infty}^{\infty} \cos(\theta-\theta_0)p(\theta;t|\theta_0,\omega_0)p(\theta_0,\omega_0)\,d\theta\,d\theta_0\,d\omega_0 \\ &= \mu^2 \exp\left[-\frac{kT}{I_2}\int_0^t X_\omega(\tau)\,d\tau\right] \end{aligned} \tag{I.48}$$

and hence that

$$\langle \boldsymbol{\mu}(t)\cdot\boldsymbol{\mu}(0)\boldsymbol{\omega}(t)\cdot\boldsymbol{\omega}(0)\rangle = \langle \boldsymbol{\mu}(t)\cdot\boldsymbol{\mu}(0)\rangle \left[\langle \boldsymbol{\omega}(t)\cdot\boldsymbol{\omega}(0)\rangle - \left(\frac{kT}{I_2} X_\omega(t)\right)^2\right] \tag{I.49}$$

Thus the full decoupled approximation to (I.43) reads

$$\langle \mathbf{v}_a(t)\cdot\mathbf{v}_a(0)\rangle = \langle \mathbf{v}(t)\cdot\mathbf{v}(0)\rangle + \langle \boldsymbol{\mu}(t)\cdot\boldsymbol{\mu}(0)\rangle \left[\langle \boldsymbol{\omega}(t)\cdot\boldsymbol{\omega}(0)\rangle - \left(\frac{kT}{I_2} X_\omega(t)\right)^2\right]$$

Notice also that analogously with (I.38) and (I.39), for space reorientation of the diatomic we have

$$\langle \boldsymbol{\mu}(t)\cdot\boldsymbol{\mu}(0)\boldsymbol{\omega}(t)\cdot\boldsymbol{\omega}(0)\rangle \neq \langle \boldsymbol{\mu}(t)\cdot\boldsymbol{\mu}(0)\rangle\langle \boldsymbol{\omega}(t)\cdot\boldsymbol{\omega}(0)\rangle \qquad t>0$$

A more direct study of rotation–translation coupling may be attempted by considering the correlation matrix for the vector $\begin{bmatrix}\mathbf{v}\\ \boldsymbol{\omega}\end{bmatrix}$ under the assumption that the matrices $\langle \mathbf{v}(t)\boldsymbol{\omega}^T(0)\rangle$ and $\langle \boldsymbol{\omega}(t)\mathbf{v}^T(0)\rangle$ are not null when $t>0$. Within a memory formalism we therefore propose that past rotations influence future translations and past translations influence future rotations. Instantaneously at time t, of course, the statistical correlation between $\mathbf{v}(t), \boldsymbol{\omega}(t)$, and any of their time derivatives must vanish. As shown below, it is a simple matter to generalize the Mori continued-fraction approximant corresponding to planar itinerant libration, with the introduction of only one further phenomenological parameter specifying the off-diagonal elements of the third memory matrix of $\begin{bmatrix}\mathbf{v}\\ \boldsymbol{\omega}\end{bmatrix}$. [It should be noted that in planar motion the matrix elements of $\langle \mathbf{v}(t)\boldsymbol{\omega}^T(0)\rangle$ do not all vanish, even though the scalar correlation $\langle \mathbf{v}(t)\cdot\boldsymbol{\omega}(0)\rangle$ does.] It is seen that the new "coupling" parameter also appears in the transforms of $\langle \mathbf{v}(t)\mathbf{v}^T(0)\rangle$ and $\langle \boldsymbol{\omega}(t)\boldsymbol{\omega}^T(0)\rangle$, which emphasizes the inadequacy of a decoupled approximation, even when studying autocorrelations.

Since the components of $\mathbf{v}$ are mutually statistically uncorrelated, the elements of the correlation matrix of $\begin{bmatrix} \mathbf{v} \\ \boldsymbol{\omega} \end{bmatrix}$ are typified by those of the 2×2 matrix

$$C(t) = \begin{bmatrix} \dfrac{\langle v(t)v(0)\rangle}{\langle v^2(0)\rangle} & \dfrac{\langle v(t)\omega(0)\rangle}{\langle \omega^2(0)\rangle} \\[2ex] \dfrac{\langle \omega(t)v(0)\rangle}{\langle v^2(0)\rangle} & \dfrac{\langle \omega(t)\omega(0)\rangle}{\langle \omega^2(0)\rangle} \end{bmatrix} \tag{I.50}$$

where v and ω are scalar components. Interpreting (I.21) in matrix form, the Laplace transform of $C(t)$ is

$$\begin{aligned} \tilde{C}(s) &= \left[s + \tilde{C}_1(s)\phi_1(0)\right]^{-1} \\ &= \left[s + \{s + \tilde{C}_2(s)\phi_2(0)\}^{-1}\phi_1(0)\right]^{-1} \end{aligned} \tag{I.51}$$

where $\qquad \tilde{\phi}_1(s) = \tilde{C}_1(s)\phi_1(0) \quad \text{and} \quad \tilde{\phi}_2(s) = \tilde{C}_2(s)\phi_2(0)$

are the transforms of the first and second memory matrices, and the $\phi_j(0)$ are defined by

$$\phi_j(0) = \langle \mathbf{f}_j\mathbf{f}_j^T\rangle\langle \mathbf{f}_{j-1}\mathbf{f}_{j-1}^T\rangle^{-1} \qquad j = 1,2$$

with

$$\mathbf{f}_0 = \begin{bmatrix} v(0) \\ \omega(0) \end{bmatrix} \qquad \mathbf{f}_1 = \begin{bmatrix} \dot{v}(0) \\ \dot{\omega}(0) \end{bmatrix} \qquad \mathbf{f}_2 = \begin{bmatrix} \ddot{v}(0) + \dfrac{\langle \dot{v}^2(0)\rangle}{\langle v^2(0)\rangle} v(0) \\[2ex] \ddot{\omega}(0) + \dfrac{\langle \dot{\omega}^2(0)\rangle}{\langle \omega^2(0)\rangle} \omega(0) \end{bmatrix}$$

We find that

$$\phi_j(0) = \begin{pmatrix} \phi_{tj} & 0 \\ 0 & \phi_{rj} \end{pmatrix}$$

where

$$\phi_{t1} = \frac{\langle \dot{v}^2(0)\rangle}{\langle v^2(0)\rangle} \qquad \phi_{r1} = \frac{\langle \dot{\omega}^2(0)\rangle}{\langle \omega^2(0)\rangle}$$

and

$$\phi_{t2}=\frac{\langle \ddot{v}^2(0)\rangle}{\langle \dot{v}^2(0)\rangle}-\frac{\langle \dot{v}^2(0)\rangle}{\langle v^2(0)\rangle} \qquad \phi_{r2}=\frac{\langle \ddot{\omega}^2(0)\rangle}{\langle \dot{\omega}^2(0)\rangle}-\frac{\langle \dot{\omega}^2(0)\rangle}{\langle \omega^2(0)\rangle}$$

These terms are identical with the corresponding terms in the (decoupled) planar itinerant librator model. To attempt a description of rototranslational coupling consider, first, the first-order approximant in (I.51) defined by

$$\tilde{C}_1(s)\phi_1(0)=\begin{pmatrix}\lambda_t & \lambda_{tr}\\ \lambda_{rt} & \lambda_r\end{pmatrix}$$

where the λ's may be interpreted as frictional parameters. It follows that

$$\tilde{C}(s)=\begin{pmatrix}s+\lambda_t & \lambda_{tr}\\ \lambda_{rt} & s+\lambda_r\end{pmatrix}^{-1}$$
$$=\frac{1}{\Delta(s)}\begin{pmatrix}s+\lambda_r & -\lambda_{tr}\\ -\lambda_{rt} & s+\lambda_t\end{pmatrix}$$

where

$$\Delta(s)=(s+\lambda_t)(s+\lambda_r)-\lambda_{tr}\lambda_{rt}$$

Writing $\qquad b=2(\lambda_t+\lambda_r),\ c=\lambda_t\lambda_r-\lambda_{tr}\lambda_{rt}, \qquad$ we find

$$\langle v(t)v(0)\rangle$$
$$=\begin{cases}\langle v^2(0)\rangle\exp(-bt)\left[\cos(c-b^2)^{1/2}t+\dfrac{\lambda_r-b}{(c-b^2)^{1/2}}\sin(c-b^2)^{1/2}t\right] & c>b^2\\ \langle v^2(0)\rangle\exp(-bt)\left[\cosh(b^2-c)^{1/2}t+\dfrac{\lambda_r-b}{(b^2-c)^{1/2}}\sinh(b^2-c)^{1/2}t\right] & c<b^2\end{cases}$$

$$\langle \omega(t)v(0)\rangle$$

$$=\begin{cases}-\dfrac{\langle\omega^2(0)\rangle\lambda_{tr}}{(c-b^2)^{1/2}}\exp(-bt)\sin(c-b^2)^{1/2}t & c>b^2\\ -\dfrac{\langle\omega^2(0)\rangle\lambda_{tr}}{(b^2-c)^{1/2}}\exp(-bt)\sinh(b^2-c)^{1/2}t & b^2<c\end{cases}$$

with similar expressions for $\langle\omega(t)\omega(0)\rangle$ and $\langle\omega(t)v(0)\rangle$. Since $\langle v(t)\omega(0)\rangle = \langle\omega(t)v(0)\rangle$, it follows that

$$\lambda_{tr}\langle\omega^2(0)\rangle = \lambda_{rt}\langle v^2(0)\rangle$$

and hence only three of the four parameters $\lambda_t, \lambda_r, \lambda_{tr}, \lambda_{rt}$ are independent. It can be seen that the translational autocorrelation function $\langle v(t)v(0)\rangle$ involves not only the translational parameter λ_t, but also the rotational parameter λ_r linked via the coupling parameter λ_{tr}; and similarly for the rotational autocorrelation. When $\lambda_{tr}=0$, the dependence reduces to one parameter and the cross-correlation $\langle v(t)\omega(0)\rangle$ vanishes identically. Moreover, when $\lambda_{tr}\neq 0$, the cross-correlation $\langle v(0)\omega(0)\rangle$ vanishes, as it should.

To obtain the generalized model for planar itinerant libration we simply define the third memory matrix by

$$\tilde{C}_3(s)\phi_3(0) = \begin{pmatrix} \lambda_t & \lambda_{tr} \\ \lambda_{rt} & \lambda_r \end{pmatrix}$$

and hence find

$$\tilde{C}(s) = \frac{1}{\Delta_2(s)}\begin{pmatrix} s+\left[s+(s+\lambda_r)\dfrac{\phi_{t2}}{\Delta(s)}\right]\dfrac{\phi_{r1}}{\Delta_1(s)} & -\dfrac{\lambda_{tr}\phi_{r2}\phi_{r1}}{\Delta(s)\Delta_1(s)} \\ -\dfrac{\lambda_{rt}\phi_{t2}\phi_{t1}}{\Delta(s)\Delta_1(s)} & s+\left[s+(s+\lambda_t)\dfrac{\phi_{r2}}{\Delta(s)}\right]\dfrac{\phi_{t1}}{\Delta_1(s)} \end{pmatrix}$$

where

$$\Delta_1(s) = \left[s+(s+\lambda_t)\frac{\phi_{r2}}{\Delta(s)}\right]\left[s+(s+\lambda_r)\frac{\phi_{t2}}{\Delta(s)}\right] - \frac{\lambda_{tr}\lambda_{rt}\phi_{t2}\phi_{r2}}{\Delta^2(s)}$$

and

$$\Delta_2(s) = \left\{s+\left[s+(s+\lambda_t)\frac{\phi_{r2}}{\Delta(s)}\right]\frac{\phi_{t1}}{\Delta_1(s)}\right\}$$

$$\times\left\{s+\left[s+(s+\lambda_r)\frac{\phi_{t2}}{\Delta(s)}\right]\frac{\phi_{r1}}{\Delta_1(s)}\right\} - \frac{\lambda_{tr}\lambda_{rt}\phi_{t2}\phi_{t1}\phi_{r2}\phi_{r1}}{\Delta^2(s)\Delta_1^2(s)}.$$

We now have:

$$\lambda_{tr}\phi_{r2}\phi_{r1} = \lambda_{rt}\phi_{t2}\phi_{t1}$$

and thus all correlation functions involve the seven independent phenomenological parameters ϕ_{t1}, ϕ_{t2}, ϕ_{r1}, ϕ_{r2}, λ_t, λ_r, and λ_{tr}. When the coupling parameter $\lambda_{tr}=0$, the determinants $\Delta(s)$, $\Delta_1(s)$, and $\Delta_2(s)$ reduce to simple factors, the autocorrelation functions become dependent on only three parameters (in each case), and the cross-correlations vanish identically.

It is the authors' belief that *all* continued-fraction-approximant models should contain coupling parameters such as λ_{tr}, except of course for cases where the correlation matrices can be shown to vanish at all times for reasons of symmetry.

Further Applications of the Long-time Approximation. Consider the case in (I.31), where β_2 is finite. The dynamical system is then described by (I.6) with

$$A(t)=\begin{bmatrix}\dot{\theta}(t)\\ \dot{\psi}(t)\end{bmatrix}$$

$$\phi_A=\begin{bmatrix}\beta_2\delta(t)+\gamma^2 & -\gamma^2\\ -(I_2\gamma/I_1)^2 & \beta_1\delta(t)+(I_2\gamma/I_1)^2\end{bmatrix} \tag{I.52}$$

$$F_A=\begin{bmatrix}\dot{W}_2\\ \dot{W}_1\end{bmatrix} \qquad \text{[with a null resonance operator]}$$

Equation (I.30) becomes

$$C_{\dot{\theta}}(s)=$$

$$\frac{kT}{I_2}\,\frac{s^2+\beta_1 s+(I_2\gamma/I_1)^2}{s^3+s^2(\beta_1+\beta_2)+\left(\beta_1\beta_2+\gamma^2\left(1+(I_2/I_1)^2\right)\right)s+\gamma^2\left(\beta_1+\beta_2(I_2/I_1)^2\right)} \tag{I.53}$$

In Fig. 9 we plot some zero-THz dielectric loss profiles obtained from (I.53) for various friction coefficients β_1 and β_2 and force constant. The effect of β_2 is to broaden the γ resonance, and with $\beta_1 \gg \beta_2$ it is interesting to note the loss profile may extend over many decades of frequency. This might be typical of the situation in glasses, disordered solids, or liquid crystals, where a cage of nearest-neighboring molecules under collective reorientation would do so, if at all, only very slowly compared with the libration of the inner molecule, taking place at THz frequencies. The version of the long-time approximation embodied in (I.50) to (I.53) may then be used to link together loss peaks separated on the frequency scale by many decades. In these solids therefore, whatever the precise validity of (I.53) it is needless to emphasize that any study of rotational motions of the permanent dipole in the condensed phase should not end at frequencies lower than those of the far infrared. Conversely, data above 10 cm^{-1} should not be analyzed without taking into account the low-frequency loss. In the extreme of charge carrier hopping in semiconductors or chalcogenide glasses,[36] the peak of the loss is often at too low a frequency to measure; nevertheless, the librational movement of molecular frameworks in between the occasional hopping of charges will produce a quasiharmonic Poley type of absorption (a resonance around the frequency γ) at THz frequencies. This is to be expected on the grounds of (I.53), which is of course an approximation of the Liouville equation, neglecting all cooperative movements embodied in the resonance operator Ω_A of (I.6).

It is remarkable that to (I.53) may be ascribed another interpretation in terms of the Brownian motion of two interacting dipoles as in Fig. (I.10). The following conditions are imposed.

1. The potential between the dipoles has the form

$$V(\theta-\psi)=\tfrac{1}{2}I_2\gamma^2(\theta-\psi)^2 \tag{I.54}$$

2. The measuring field is applied in the same plane of rotation of the dipoles μ_1 and μ_2 which rotate about an axis about the common center perpendicular to the plane containing $\boldsymbol{\mu}_1$ and $\boldsymbol{\mu}_2$. β_1 and β_2 are the opposing friction coefficients arising from the surroundings. This interpretation of (I.53) is restrictive in the sense that it still applies to planar motions, but by varying the terms in the column vector **A** and memory matrix ϕ_A the model is capable of further improvement. Developments may be made to account for more realistic potentials than that of (I.54), and of space reorientation. In the present context γ is a measure of the dipole–dipole coupling strength.

The two physical processes embodied in Figs. 5 and 10 may be distinguished experimentally by noticing that in the itinerant oscillator we might expect $I_1>I_2$ by geometrical considerations, but in the dipole interaction model $I_1=I_2$ for a neat fluid. Further, in a dilute solution of dipolar molecules in nondipolar solvents, where a theory of autocorrelations is valid, dipole–dipole coupling is considerably weakened, whereas the THz resonance remains. It seems unlikely that a theory of dipole–dipole resonance can account for the persistent resonance band. The problem of kinematics vs. electrostatics in the context of (I.54) reduces to one of de-

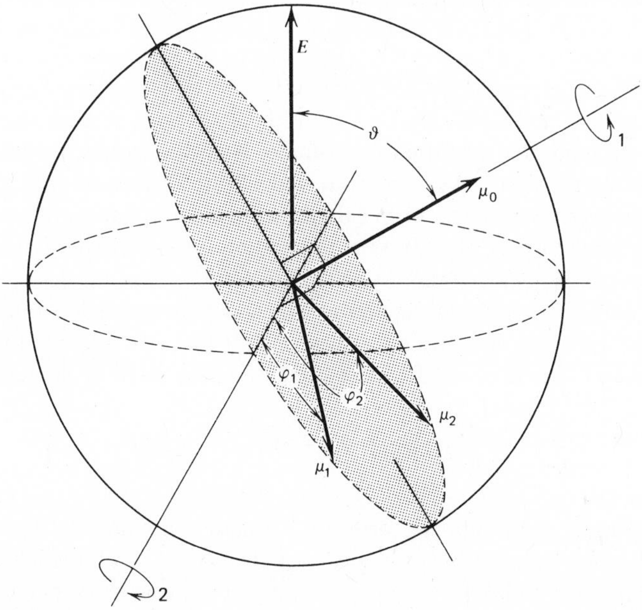

Fig. 10. Dipole–dipole interaction as an interpretation of (I.53) (where θ of Fig. I.5 is zero) and $(\phi_2-\phi_1)\equiv(\theta-\psi)$ in the notation of (I.31). [Reproduced by permission from *Mol. Phys.*, (1979), **37**, 473]

termining, experimentally, whether $I_1=I_2$ is in closer agreement with the data than $I_1>I_2$. This is resolved in Section II.C.

C. Collective Variables

The effect of a finite resonance operator on the theoretical zero-THz loss profile has removed, like rotation/translation coupling, some of the restrictions of the simple Langevin equation. The resonance operator $i\Omega_A$ remains finite in the following instances:

1. The coupling between even and odd variables is finite.
2. When time-reversal symmetry is destroyed by an external pertubation such as a magnetic field.

The role of the resonance operator $i\Omega_A$ may be described most clearly by making the separation

$$\frac{d}{dt}A(t)\equiv F(t)=F_1(A(s),t\geq s\geq t_0)+F_2(t,t_0) \tag{I.55}$$

where the Langevin equation is now

$$\frac{d}{dt}A(t)=i\Omega_A-\beta A(t)+f(t) \tag{I.56}$$

Here $A(t)$ denotes a normal coordinate, Ω is the angular frequency of a monochromatic sound wave or spin wave, and $f(t)$ represents a random force, which is a stochastic rather than mechanical quantity, as usual. Here $f(t)$ is a random part of $F(t)$, thus depending on the thermodynamic state of the system. F_1 is a functional of $A(s)$ depending also on the past history of $A(t)$, and F_2 represents the terms which depend explicitly on the other degrees of freedom. Expanding F in terms of $A(s), t\geq s\geq t_0$, the linear term obtained has a generalized form of the systematic part of the equation

$$\frac{d}{dt}A(t)=\int_{t_0}^{t}\theta(t-s)A(s)\,ds+f(t) \tag{I.57}$$

and the sum of the nonlinear terms of F uniquely defines the quantity $f(t)$. The collective description of many-particle systems by much fewer variables than the number of degrees of freedom is possible if and only if the fluctuations due to $f(t)$ are negligible.

Extracting the linear term in (I.55) is equivalent to projecting $A(t)$ into the subspace of Hilbert space spanned by A in (I.6). This may be envisaged geometrically as projecting $A(t)$ onto the A-axis. The projection of a column vector G onto this axis is then $\hat{P}G=\langle \mathbf{G}\mathbf{A}^T\rangle\langle \mathbf{A}\mathbf{A}^T\rangle^{-1}\mathbf{A}$. The operation in (I.55) is then equivalent for a column vector $\mathbf{A}$ of splitting $i\Omega\mathbf{A}$ into its projection and vertical components:

$$\dot{\mathbf{A}}=\langle\dot{\mathbf{A}}\mathbf{A}^T\rangle\langle\mathbf{A}\mathbf{A}^T\rangle^{-1}+(1-\hat{P})\dot{\mathbf{A}}\equiv i\Omega_{\mathbf{A}}+(1-\hat{P})\mathbf{A} \tag{I.58}$$

The matrix $i\Omega_A$ therefore has eigenvalues which determine the temperature-dependent eigenfrequencies of collective oscillations. Very pronounced effects on the power spectrum (e.g., the Rytov splitting of Rayleigh scattering) are describable in terms of these oscillations. The effect of an external magnetic field of time-reversal symmetry has been discussed in detail by Berne and Harp[25] in this series, so we restrict our discussion to the main points of interest.

Consider a Hilbert space of dynamical variables whose invariant parts are set to zero. Denote the inner product of two variables F and G by $\langle FG\rangle$. F and G are, for example, linear combinations of the Hermitian functions of particle coordinates $(\mathbf{r}_j)$ and momenta $(\mathbf{p}_j)$, and of spins $\mathbf{s}_j$

which are even or odd with respect to time reversal ·

$$(\mathbf{r}_j \to \mathbf{r}_j; \mathbf{p}_j \to -\mathbf{p}_j; \mathbf{s}_j \to -\mathbf{s}_j)$$

Collective, or hydrodynamic variables such as mass density or momentum density are such quantities (see Section III.D.1). Denoting the system Hamiltonian by H_0 and the external magnetic field by $\mathbf{H}$, we have

$$H_0(\mathbf{H}) \to H_0(-\mathbf{H}) \tag{I.59}$$

$$\langle F(t)G(0)\rangle_{\mathbf{H}} = \varepsilon_F \varepsilon_G \langle F(-t)G(0)\rangle_{-\mathbf{H}} \tag{I.60}$$

Where ε_F and ε_G are 1 or -1, according as whether F or G is even or odd. $-\mathbf{H}$ indicates the reversal of the external magnetic field. Now consider a column vector $\mathbf{A}$ of such variables as F and G. In the absence of the magnetic field, the linear or odd variables are orthogonal:

$$\langle FG\rangle = 0 \qquad if\ \varepsilon_F\varepsilon_G = -1 \tag{I.61}$$

This is not so in the external perturbation, and the coupling between even and odd variables gives rise to collective oscillations. Denoting the signatures of A_j by ε_j, the determinant of $\langle \mathbf{A}\mathbf{A}^T\rangle$ is invariant under time reversal and the cofactor of the (i,j) element changes its sign by $\varepsilon_i\varepsilon_j$. The (i,j) element of the inverse matrix of $\langle A_i A_j\rangle$ is transformed accordingly. Therefore, the projection operator is invariant under time reversal:

$$[\hat{P}G(t)]_{\mathbf{H}} \to \varepsilon_G[\hat{P}G(-t)]_{-\mathbf{H}} \tag{I.62}$$

Now arrange the set $\mathbf{A}$ such that the first m variables are even and the rest are odd. Denote the even and odd parts by A_e and A_o. Then

$$\Omega_{\mathbf{A}}(t)_{\mathbf{H}} = \begin{bmatrix} -\Omega_{ee} & \Omega_{eo} \\ \Omega_{oe} & -\Omega_{oo} \end{bmatrix}_{-\mathbf{H}} \tag{I.63}$$

where Ω_{ee} is the submatrix of Ω consisting of the elements of the first m rows and columns and Ω_{eo} that of the first m rows and last $(n-m)$ columns. Ω_{oe} and Ω_{oo} are defined similarly.

Another type of symmetry relation is that between the Fourier transform components. Denote those of the local densities of physical quantities by F_k and G_k. Since there is no inhomogenous field applied, we have, from the translational invariance,

$$\langle F_k(t)G_q(0)\rangle = 0 \qquad \text{if} \qquad \mathbf{k} \neq \mathbf{q}$$

This is characteristic of linear phenomena. If one assumes inversion symmetry

$$\langle F(\mathbf{r})G(\mathbf{0})\rangle = \langle F(-\mathbf{r})G(\mathbf{0})\rangle$$
$$\langle F_k(t)G_k(0)\rangle = \langle F_k(t)G_{-k}(0)\rangle \tag{I.64}$$

In the absence of a magnetic field or similar type of external application, these symmetry relations are simplified. $\langle \mathbf{A}(t)\mathbf{A}^T(0)\rangle$ is split into two disjoint submatrices and the projection takes the form

$$\hat{P}G = \langle GA_e\rangle\langle A_eA_e\rangle^{-1}A_{es} + \langle GA_o\rangle\langle A_oA_o\rangle^{-1}A_o \tag{I.65}$$

The diagonal elements of (I.63) then vanish.

1. Relation between Multi- and Single-Particle Correlation Functions

Since the collective oscillations in a fluid are significant spectroscopically a theorem is needed to explain the relation between many- and single-particle correlation functions, (i.e., to provide a macro–micro correlation). Assume therefore that the collective elements $A(p,q;t)$ of the column vector $\mathbf{A}$ may be defined as sums over monomolecular elements $\alpha(p,q;t)$:

$$A(p,q;t) = \sum_{i=1}^{N} \alpha_i(p,q;t) \tag{I.66}$$

where N is the number of molecules in the system. For example, the multimolecular or macroscopic dipole moment would be defined by

$$\mathbf{M}(p,q;t) = \sum_{i=1}^{N} \boldsymbol{\mu}_i(p,q;t) \tag{I.67}$$

and the dielectric tensor by

$$\varepsilon(p,q;t) = \sum_{i=1}^{N} \varepsilon_i(p,q;t) \tag{I.68}$$

A evolves in time according to (I.6). The continued fraction may then be employed as an approximate solution. Alternatively, we may use the language of Kivelson and Keyes, where the correlation function of the dielectric tensor is proportional to that of a primary variable which is slowly varying, and its time evolution is calculated by means of a pair of coupled transport equations in which the primary variable is coupled to a secondary variable. This is a rapidly varying quantity dependent on intermolecular forces. Equation (I.30), for instance, may be derived in this fashion by using three orientational variables interrelated by three coupled transform equations. The macro–micro correlation theorems discussed by Kivelson and co-workers[17,46,47] are useful in that they demonstrate that collective correlation matrices of the form

$$J(t)=\langle \mathbf{A}(t)\mathbf{A}^T(0)\rangle\langle \mathbf{A}(0)\mathbf{A}^T(0)\rangle^{-1} \tag{I.69}$$

must have the same general mathematical structure as the corresponding matrix of autocorrelation functions.

The macro–micro correlation theorems are most useful in those systems where superimposed correlations due to, for example, dipole–dipole interaction are not overwhelmingly important. The theorems assert that the theory of autocorrelations used in the following sections may be utilized when dealing with spectral band shapes dependent on correlations between many particles. Multimolecular correlations may always be built up from autocorrelations—an expression of the domino theory. It is interesting to note in concluding this section that hydrodynamic interactions between well-separated solute particles in a fluid medium may persist via a mechanism of translation/rotation mixing. The rotation of a dipole sets up a velocity perturbation which imposes a torque on another. Wolynes and Deutch[48] have considered a many-particle coupled translation/rotation model in which the coupled Brownian diffusion of solute particles is described using anisotropic potentials in a continuum solvent, the motion of which is described hydrodynamically. There is a long-range dynamical orientation correlation between solute particles which again (see Section I) manifests itself in the appearance of off-diagonal coupled diffusion constants in the N-particle translation/rotation Langevin equation describing the system. If we follow this paper and consider a collection of N Brownian solute molecules labeled with position coordinates $\mathbf{X}_i, i=1,\ldots,N$, and orientation coordinated $\mathbf{\Omega}_i, i=1,\ldots,N$, then for infinitesimally small step Brownian motion (delta memories), the configurational probability density function $P([X,\Omega]=[0])$ is locally conserved and therefore satisfies the Fokker–Planck equation

$$\partial P/\partial t=-\sum_i\left[\nabla_{x_i}\cdot\mathbf{J}_{x_i}+\mathbf{L}(i)\cdot\mathbf{J}_{\Omega_i}\right] \tag{I.70}$$

where $\mathbf{J}_{x_i}$ and J_{Ω_i} are the probability current densities along the positional and orientational coordinates of the ith particle. $\mathbf{L}(i)$ is the operator $\mathbf{u x}\nabla_{\mathbf{u}_i}$, where $\mathbf{u}_i$ is a unit vector fixed in the molecule. For Brownian particles the probability current densities are linearly related to the deviation of the configurational probability density from its equilibrium value

$$J_{xi} = -\sum_j \left[D_{ij}^{xx}(Q)\cdot(\nabla_{\mathbf{x}_j} P + \beta \nabla_{\mathbf{x}_j} U) + D_{ij}^{x\Omega}(Q)\cdot(\mathbf{L}_j P + \beta P \mathbf{L}_j U) \right] \tag{I.71}$$

$$J_{\Omega i} = -\sum_j \left[D_{ij}^{\Omega x}(Q)\cdot(\nabla_{\mathbf{x}j} P + \beta \nabla_{\mathbf{x}j} U) + D_{ij}^{\Omega\Omega}(Q)\cdot(\mathbf{L}_j P + \beta P \mathbf{L}_j U) \right] \tag{I.72}$$

where $U(Q)$ is the potential of mean force of the system. The equilibrium probability distribution $P_{\text{eq}}(Q)$ is related to $U(Q)$ according to

$$P_{\text{eq}}(Q) = \text{const} \exp[-\beta U(Q)] \tag{I.73}$$

The potential of mean force will be affected by any collective fluctuation. The diffusion tensors $D(3\times 3)$ are functions of the Brownian particles' configuration. If it is assumed that the the time scales of momentum and configuration change may be separated in a dense fluid, then the D tensors are integrals over the time autocorrelation matrices defined by the tensor product:

$$D_{ij}^{xx}(Q) = \int_0^\infty \langle \mathbf{v}_i(0)\mathbf{v}_j(t)\rangle_Q \, dt \tag{I.74a}$$

$$D_{ij}^{x\Omega}(Q) = \int_0^\infty \langle \mathbf{v}_i(0)\boldsymbol{\omega}_j(t)\rangle_Q \, dt \tag{I.74b}$$

$$D_{ij}^{\Omega x}(Q) = \int_0^\infty \langle \boldsymbol{\omega}_i(0)\mathbf{v}_j(t)\rangle_Q \, dt \tag{I.74c}$$

$$D_{ij}^{\Omega\Omega}(Q) = \int_0^\infty \langle \boldsymbol{\omega}_i(0)\boldsymbol{\omega}_j(t)\rangle_Q \, dt \tag{I.74d}$$

A more acceptable theory for high-frequency spectroscopy would involve memory terms and a Fokker–Planck equation of the form of (I.7). However, using delta memories the Langevin equations for rotation–translation again take the form

$$M_i \dot{\mathbf{v}}_i = -\sum_j \left[\xi_{ij}^{xx}\cdot\mathbf{v}_j + \xi_{ij}^{x\Omega}\cdot\boldsymbol{\omega}_j \right] + \mathbf{F}_i(t) \tag{I.75}$$

$$I_i \dot{\boldsymbol{\omega}}_i = -\sum_j \left[\xi_{ij}^{\Omega x}\cdot\mathbf{v}_j + \xi_{ij}^{\Omega\Omega}\cdot\boldsymbol{\omega}_j \right] + \mathbf{T}_i(t) \tag{I.76}$$

The friction tensors ξ originate hydrodynamically as follows. The translation or rotation of a Brownian particle causes a flow in the surrounding solvent. The moving solvent then exerts a flow of force and torque on the other particles. The outcome of this is that the $\underline{D}$ tensors are directly related to Oseen's tensors, which are the Green's functions for the steady state, linearized, incompressible-fluid, Navier–Stokes equations. Thus

$$D_{ij}^{xx} = kT\left[(\xi_0^{xx})^{-1}\delta_{ij} + (1-\delta_{ij})T_{ij}^{xx}\right] \tag{I.77}$$

$$D_{ij}^{\Omega x} = kT(1-\delta_{ij})T_{ij}^{x\Omega} \tag{I.78}$$

$$D_{ij}^{\Omega x} = kT(1-\delta_{ij})T_{ij}^{\Omega x} \tag{I.79}$$

$$D_{ij}^{\Omega\Omega} = kT\left[(\xi_0^{\Omega\Omega})^{-1}\delta_{ij} + (1-\delta_{ij})T_{ij}^{\Omega\Omega}\right] \tag{I.80}$$

Here ζ_{ij} is the Kronecker delta. The "rotation–rotation" Oseen tensor $T^{\Omega\Omega}$ is simple in form and takes a formally identical interparticle distance dependence as that between the interaction of dipoles. The long-range translation–rotation and rotation–translation tensors $D^{x\Omega}$ and $D^{\Omega x}$ take a more involved form.

D. Experimental Methods for Rotational Correlation Functions

In this section we shall describe the technique of Fourier transform spectroscopy which we use to obtain power absorption coefficients and refractive indices in the THz region of the electromagnetic spectrum (2 to $300\,\text{cm}^{-1}$). The subject of submillimeter spectroscopy is developing explosively and "definitive" articles and books[1,2] are rapidly outdated. Here we shall summarize our own methods and attempt to put them in perspective. The tables published by Berne and Harp and by Williams[51] are especially useful in describing the various correlation functions amenable to experimental measurement. The advantage of zero- TH_z spectroscopy (far infrared and dielectric spectroscopy considered in unison) is that the evolution of the measurable autocorrelation function (that of the rotational velocity[52]) mirrors in great detail the initial decay of the orientational autocorrelation functions that are also observed in depolarized Rayleigh scattering. Experimentally, dielectric spectroscopy in isolation produces an exponential correlation function that is easily reproducible theoretically by a great number of models which are seemingly quite different in dynamical origin. The absurd situation then arises that the classical Debye model of rotational diffusion (infinitesimally small changes of angular momentum taking place infinitely fast in an inertialess sphere) produces the same type of loss curve as a model of 180° hopping from potential wells. Both models

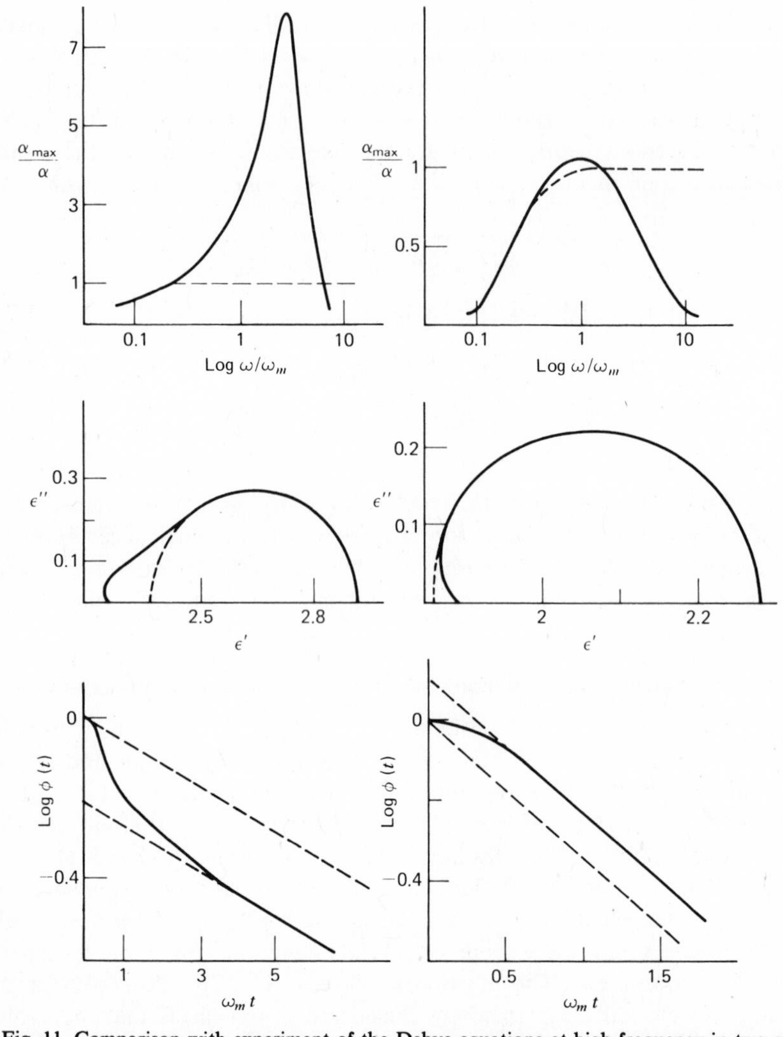

Fig. 11. Comparison with experiment of the Debye equations at high frequency in two extreme cases where the Poley absorption is pronounced (left column, a solution of CH_3CN in CCl_4 5% mole fraction), and small (right column, 20% mole fraction $(CH_3)_3CCl$ in hexane at 298°K). The dotted curve is the theoretical absorption according to the Debye model of rotational diffusion. (Reproduced by permission from P. Desplanques, Thèse d'État, Lille, 1973, p. 45.)

are clearly unacceptable when considered in relation to infrared power absorption coefficients in liquids (Fig. 11, from a thesis by Desplanques), since both produce an unrealistic plateau absorption.

The unique advantage of making both loss measurements [those on $\varepsilon''(\omega)$] and power absorption measurements on the same sample arises from the fundamental relations

$$\alpha(\omega)=\frac{2\sqrt{2}\,\pi\varepsilon''(\bar{\nu})\bar{\nu}}{\left(\left(\varepsilon'(\bar{\nu})^2+\varepsilon''(\bar{\nu})^2\right)^{1/2}+\varepsilon'(\bar{\nu})\right)^{1/2}}$$
$$=2\pi\bar{\nu}\varepsilon''(\bar{\nu})/n(\bar{\nu}) \tag{I.81}$$

Here $\bar{\nu}$ is the wave number (in cm^{-1}), related to the angular frequency (ω) by $\omega=2\pi\bar{\nu}c$. $\varepsilon'(\bar{\nu})$ is the frequency-dependent dielectric permittivity, and $n(\bar{\nu})$ the refractive index. Measurements on $\varepsilon''(\bar{\nu})$ and $\varepsilon'(\bar{\nu})$ yield accurately the long-time behavior of the multiparticle orientational correlation functions of the dipole vector $\boldsymbol{\mu}$, and measurements on $\alpha(\bar{\nu})$ and $n(\bar{\nu})$ do the same for its second derivative at short times. In contrast, measurements of depolarized Rayleigh scattering yield the equivalent of $\varepsilon''(\bar{\nu})/\bar{\nu}$, so that the high-frequency wing information is often lost as instrumental noise. Notice that $\alpha(\bar{\nu})$ is approximately $\bar{\nu}\varepsilon''(\bar{\nu})$, so that high-frequency information is "enlarged" by the frequency multiplication. To illustrate this, in Fig. 12 we present the zero-THz absorption profile for CH_2Cl_2 at 298 K in terms of $\varepsilon''(\bar{\nu})$ and $\alpha(\bar{\nu})$, along with some dynamical models of the autocorrelation functions. All of these succeed in matching $\varepsilon''(\bar{\nu})$, but none very well $\alpha(\bar{\nu})$. Alternatively, in the time domain, the Fourier transform of the Lorentzian $\varepsilon''(\omega)/\omega$ gives the multiparticle orientational correlation function (if we set aside the internal field modification), which for the $\varepsilon''(\omega)$ curve of Fig. 13 would be nearly a featureless exponential decay. In contrast, the direct Fourier transform of $\alpha(\omega)$ yields $-\ddot{C}_m(t)$, which is oscillatory, as illustrated in Figs. 14 to 19. Brot has indicated[52] that a further advantage of making $\alpha(\omega)$ measurements is that the many-particle different versions of internal field correction turn out to be almost identical numerically and have little affect on the shape of $-\ddot{C}_m(t)$ when normalized at the origin. Despite the kinematic complications pointed out by van Kanynenburg and Steele,[55] $-\ddot{C}_m(t)$ is numerically often identical[30] to the angular velocity correlation function. In the absence of memory effects this is a single exponential, so that Figs. 14 to 19 are clearly indicitive of the need for more realistic approximants of (I.6) when **A** refers to angular velocity.

By consideration of $\mathbf{A}=\begin{bmatrix}\mathbf{v}\\ \boldsymbol{\omega}\end{bmatrix}$ in the context of (I.6) we have shown that rototranslation influences $\alpha(\omega)$ through the relations between $\langle\mathbf{v}(t)\cdot\mathbf{v}(0)\rangle$,

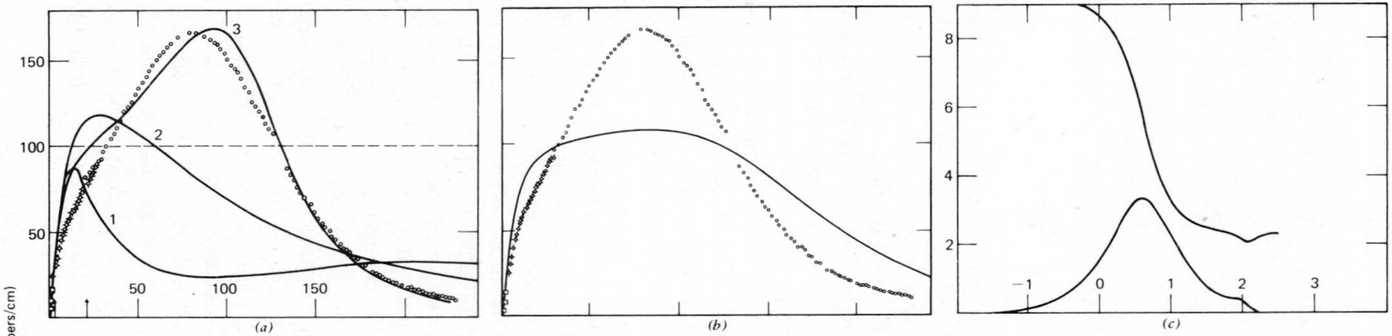

Fig. 12*a*. Absorption of CH_2Cl_2 (liquid) in the far infrared compared with some dynamical models. (*a*) (1) Asymmetric top, Langevin equation; (2) Chandler binary collision approximation; (3) Mori theory, iterating on γ and $K_1(o)$ for least-mean-squares best fit. (*b*) Mori theory, γ, $K_0(0)$ and $K_1(o)$ fixed. (*c*) Microwave loss and dispersion, *all* models give a good fit.

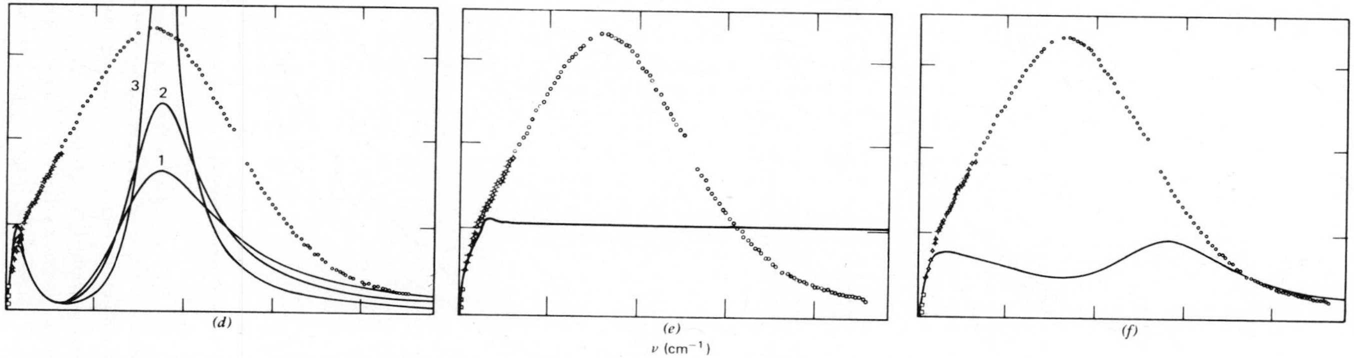

(*d*) (1), (2), (3), itinerant librator, decreasing β_2. (*e*) and (*f*) illustrate the disasters which sometimes result by allowing too much freedom of choice of parameters in a model when least-mean-squares fitting. In this case the itinerant librator (see original text for more details). Even in (*e*) and (*f*) the microwave data are fitted very well. Ordinates: $\alpha(\bar{\nu})$ (nepers/cm); abscissas: $\bar{\nu}$ (cm^{-1}). [Reproduced by permission from *J. Chem. Soc. Faraday Trans.* 2, 1978, **74**, 2143.]

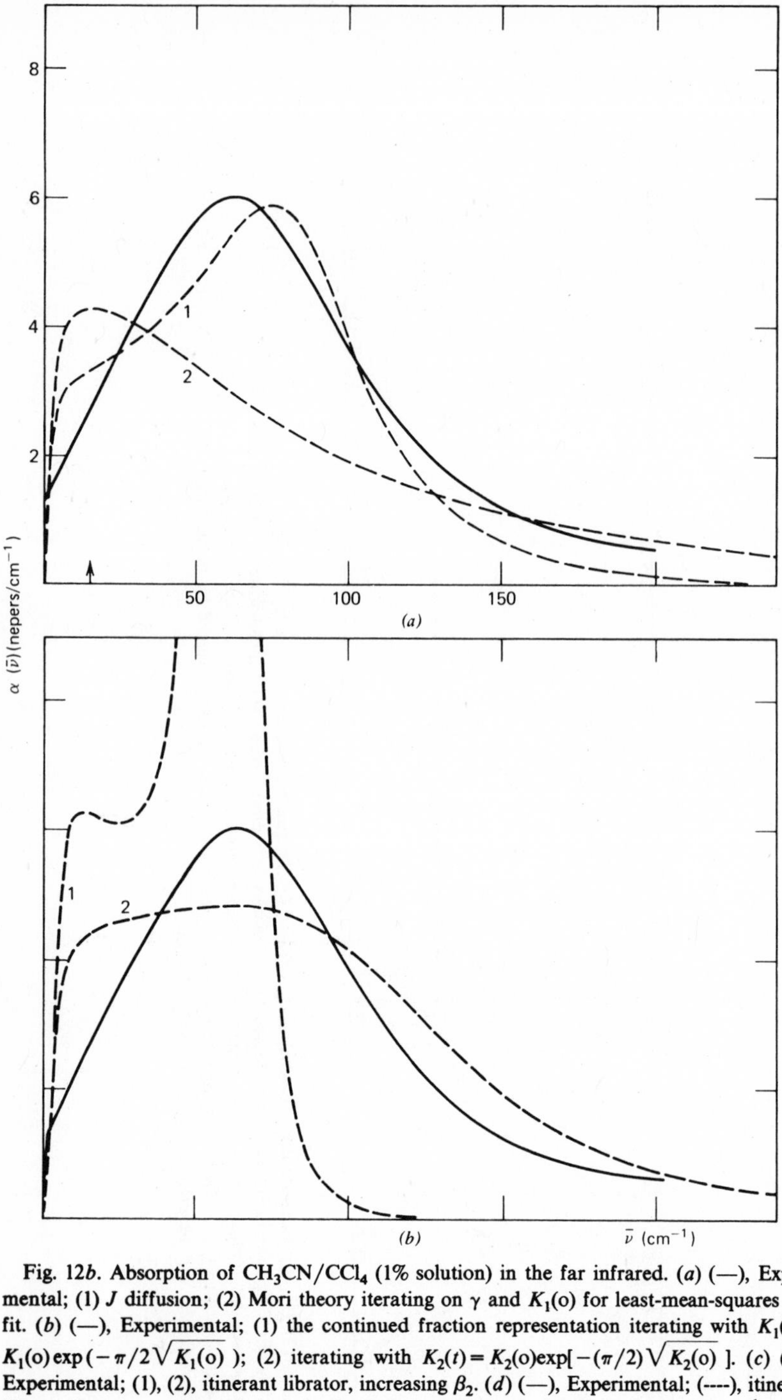

Fig. 12*b*. Absorption of CH_3CN/CCl_4 (1% solution) in the far infrared. (*a*) (—), Experimental; (1) *J* diffusion; (2) Mori theory iterating on γ and $K_1(o)$ for least-mean-squares best fit. (*b*) (—), Experimental; (1) the continued fraction representation iterating with $K_1(t) = K_1(o)\exp(-\pi/2\sqrt{K_1(o)}\ )$; (2) iterating with $K_2(t) = K_2(o)\exp[-(\pi/2)\sqrt{K_2(o)}\ ]$. (*c*) (—), Experimental; (1), (2), itinerant librator, increasing β_2. (*d*) (—), Experimental; (----), itinerant librator iterating on both β_1 and β_2. Ordinates: $\alpha(\bar{\nu})$ (nepers/cm); abscissas: $\bar{\nu}$ (cm^{-1}).

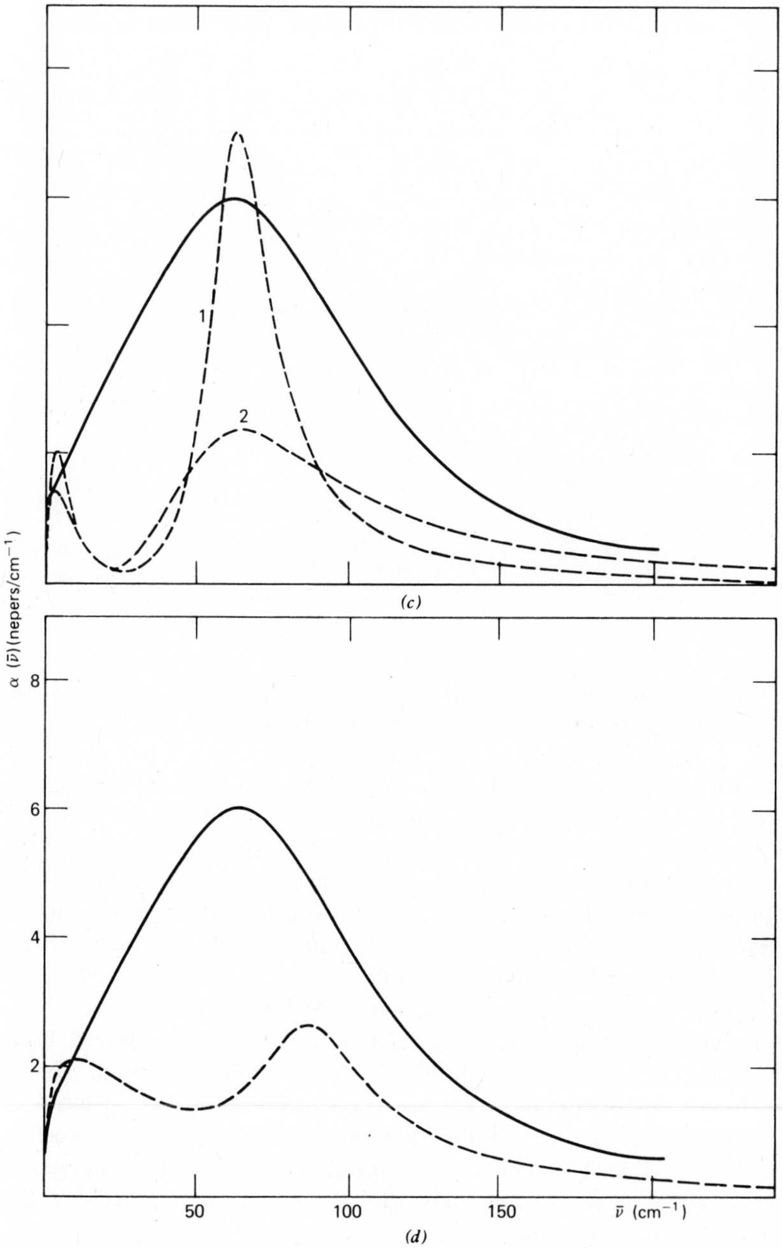

Fig. 12*b*. Continued

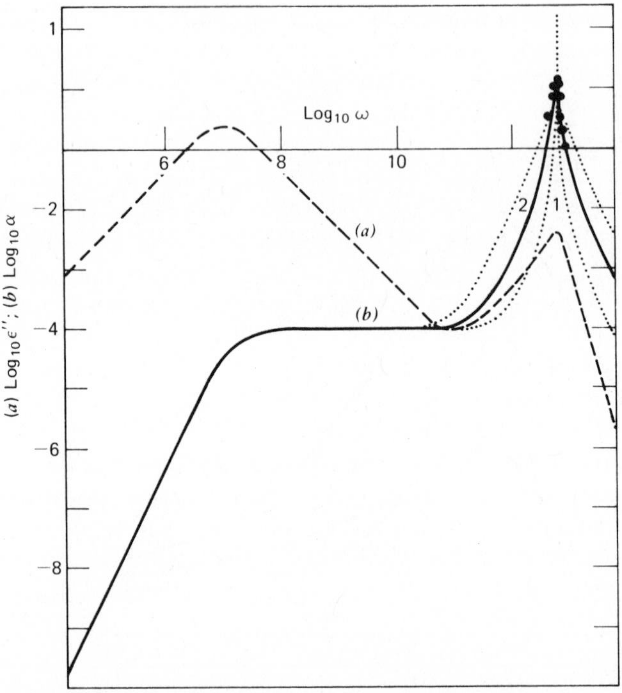

Fig. 13. Log-log plot (schematic) of (1) $\alpha(\omega)$, (2) $\epsilon''(\omega)$, and (3) $\epsilon''(\omega)/\omega$ (proportional to the Rayleigh wing). (●), Denote data in the far infrared for CH_2Cl_2/decalin glass.

$\langle\boldsymbol{\omega}(t)\cdot\boldsymbol{\omega}(0)\rangle$, $\langle\mathbf{v}(t)\cdot\boldsymbol{\omega}(0)\rangle$, and $\langle\boldsymbol{\omega}(t)\cdot\mathbf{v}(0)\rangle$. The macro-micro correlation theorem then ensures that the same effects will be present for the multiparticle correlation functions. However, the advantages of measuring $\varepsilon''(\omega)$ and $\alpha(\omega)$ in liquids as opposed to incoherent neutron scattering is that with the former we measure only resultant orientations of dipole vectors, due to rototranslation, whereas with the latter the time-of-flight spectrum reflects all aspects of the molecular rototranslations, due to the large momentum transfer, and is far too richly informative to be interpretable.

We shall, for historical reasons, refer to the broad far-infrared band rising above the Debye plateau in Fig. 11 as the Poley absorption.[56] In dipolar liquids this is essentially the zero-THz profile expressed in terms of $\alpha(\omega)$ (in nepers/cm). Although the total rotational absorption cross-section per molecule (A_0/N) should remain constant through various changes of material state, it is extraneously and intrinsically affected by the mechanisms outlined below.

1. The observed A_0/N sometimes differs by an internal field factor from that which is estimated theoretically. The correction is frequency-dependent, but the function $\alpha(\omega)$ is not distorted to any troublesome degree.

2. The available sum rules for A_0/N derived by Gordon[34] and by Brot[52] take no account of electrodynamic contributions. These may be thought of in terms of multipole–multipole effects, particularly the dipole–dipole effects described by Kirkwood. Neither is any account taken of long-range hydrodynamical effects as described already. The sum rule refers strictly to autocorrelations of single molecules. Large differences are known between the experimental and theoretical estimates of the absorption cross-section for many dipolar liquids. They cannot be attributed solely to collision induced absorptions of temporary dipole moments (Section D) which originate in the effect on any given dipole of the combined electrodynamic fields of all its neighbors at an instant t. These fluctuate in direction and magnitude with time but have a finite average value measurable by a broad submillimeter absorption[5] consisting of molecules with no permanent dipole moment, such as CCl_4. This induced absorption contains information on the intermolecular potential energy and may be used to study molecular motions and interactions in nondipolar gases, liquids, and plastic crystals.

Therefore, we adopt the criterion that to measure the extent of cross-correlations and induced absorption, it is necessary to use in dilute solution a strongly dipolar species such as CH_2Cl_2. Figure 20 then shows that a shift in the wavenumber of maximum absorption ($\bar{\nu}_{max}$) to low frequencies with dilution is a marked characteristic of the spectrum. Even at infinite dilution a large difference remains, however, between $\bar{\nu}_{max}$ and the root-mean-square angular velocity (ω_J) of the infinitely dilute gas. This is the most useful feature of the spectrum and the one which theoretical models find difficult to reproduce. It strongly discriminates against the use of gas-phase models (M and J diffusion) in the liquid. These, in their simplest form, produce profiles centered always at ω_J.

The constancy of (A_0/N) with dilution of CH_2Cl_2 in CCl_4 is evidence for the fact that the Poley absorption in this case is hardly affected by collision-induced absorption (proportional to a power of N). The extent of the induced absorption may be measured roughly from the fact that in CCl_4, $\alpha(\bar{\nu})$ peaks at 1.5 nepers/cm, whereas in CH_2Cl_2 it reaches 160 nepers/cm. In dilute solution, induced absorption will still be present because of the interaction between CH_2Cl_2 and CCl_4 molecules. The types of information available from the zero-THz profile are summarized in Table III.

1. *Some Instrumental Details*[57]

The Michelson Interferometer. A Fourier transform spectrometer (N.P.L./Grubb-Parsons "cube") (Fig. 21) is used for submillimeter spectroscopy because in such an instrument a simultaneous observation of all the elements of the spectrum (multiplex record) is kept, which has the

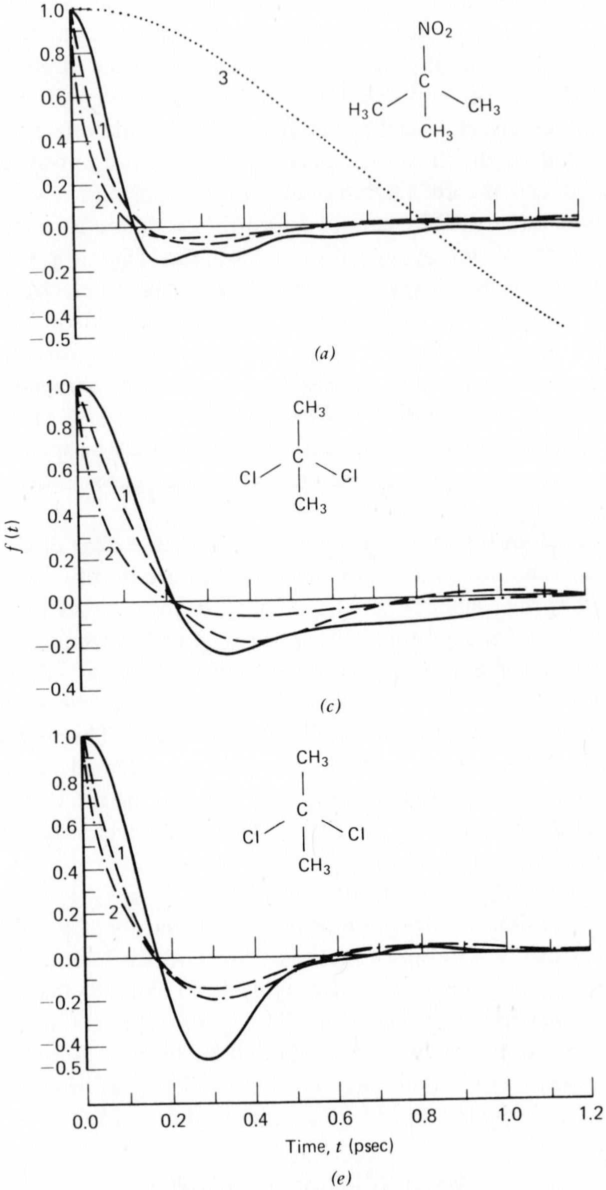

Fig. 14. (—), Experimental value of $-\ddot{C}_M(t)$ from a direct Fourier transform of the far infrared $\alpha(\omega)$, normalized to unity at the origin. (---), (1) and (2) are models (see text) typifying the need for a realistic description at short times (least-mean-squares fitting). (· · · · ·), (3) Free-rotor function equivalent to $-\ddot{C}_M(t)$. (*a*) Rotator phase, 294°K; (*b*) rotator phase, 273°K, and (·---·---·), 219°K; (*c*) liquid at 295°K; (*d*) liquid at 241°K; (*e*) rotator phase at 235°K; (*f*) rotator phase at 192°K. Abscissa: time, *t* (psec). [Reproduced by permission from *J. Chem. Soc. Faraday Trans. 2*, **71**, 2051 (1975).]

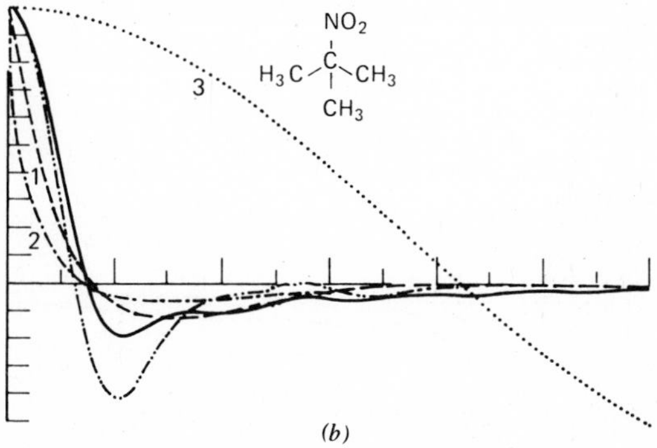

(b)

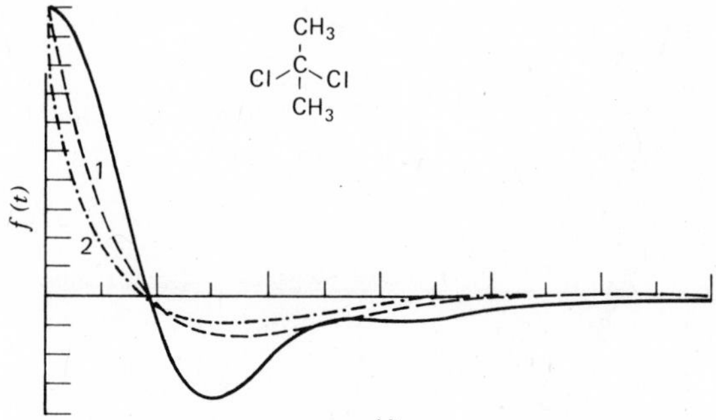

(d)

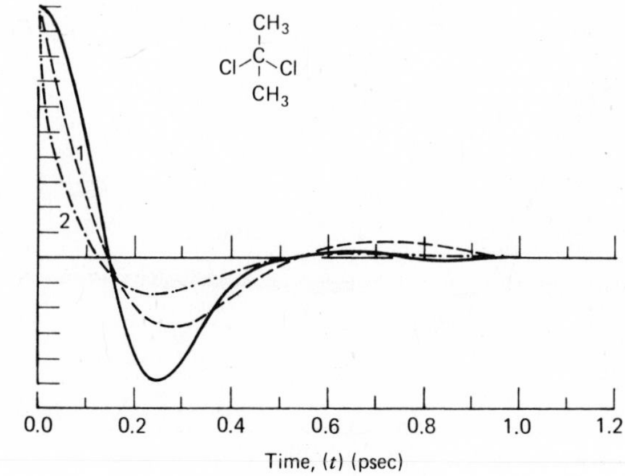

(f)

Fig. 14. Continued

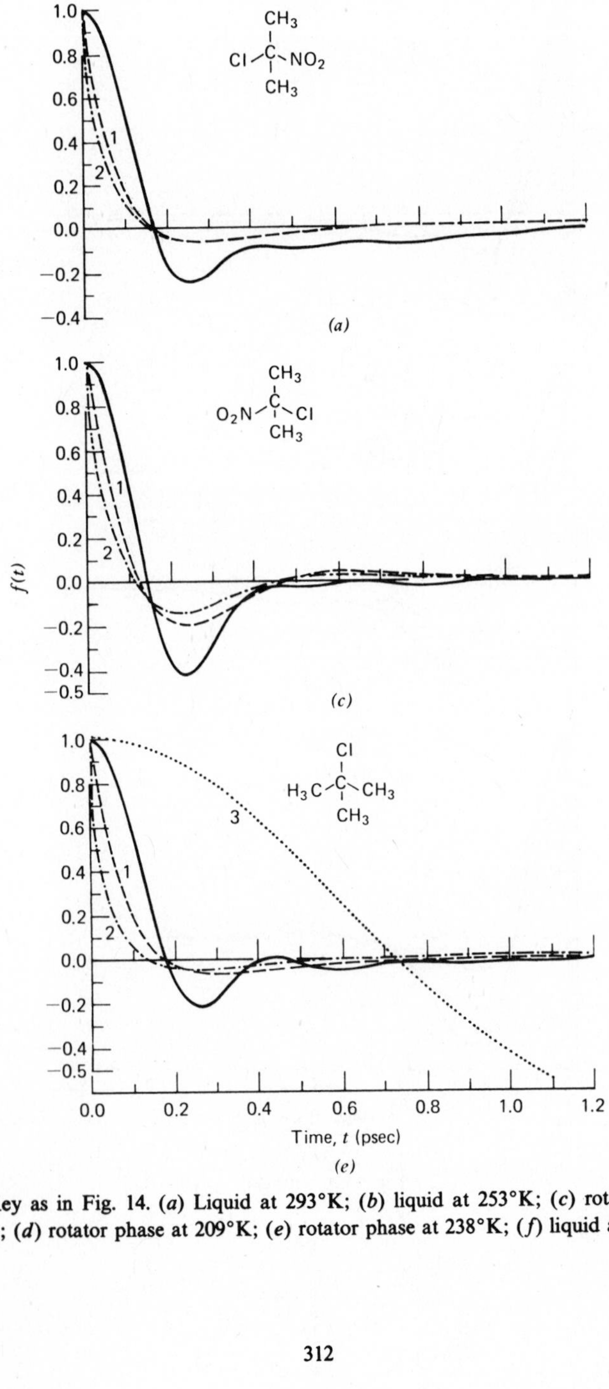

Fig. 15. Key as in Fig. 14. (*a*) Liquid at 293°K; (*b*) liquid at 253°K; (*c*) rotator phase at 233°K; (*d*) rotator phase at 209°K; (*e*) rotator phase at 238°K; (*f*) liquid at 274°K.

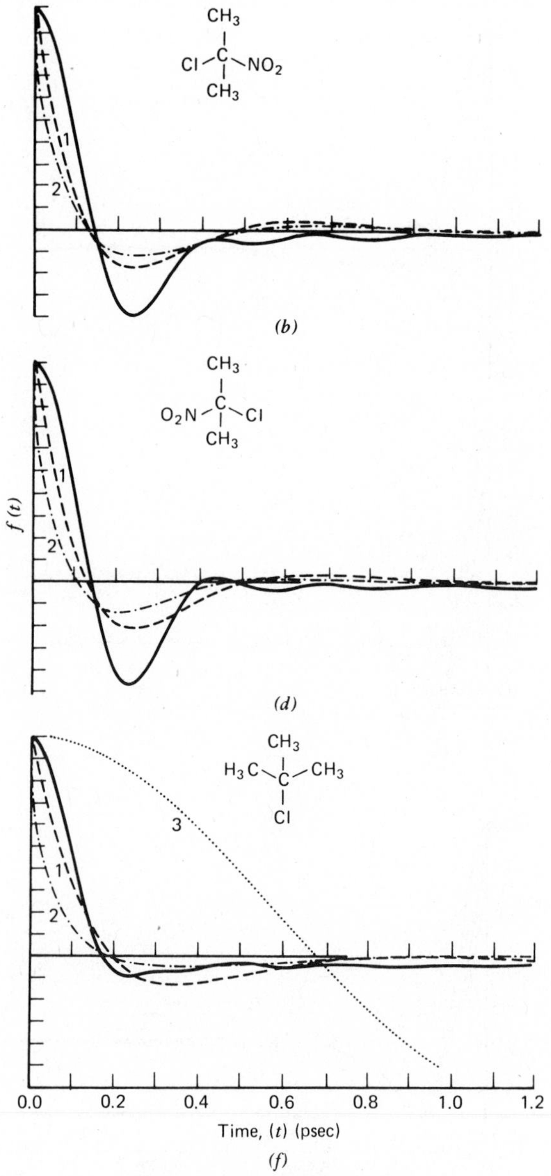

Fig. 15. Continued

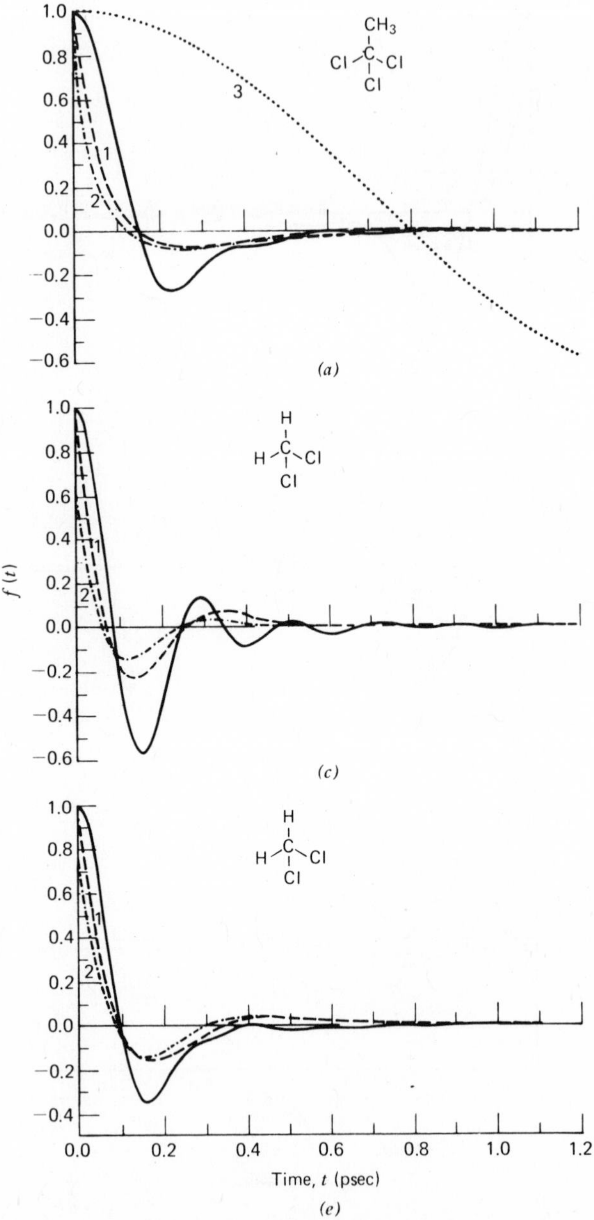

Fig. 16. Key as in Fig. 14. (*a*) Rotator phase at 233°K; (*b*) liquid at 293°K; (*c*) liquid at 188°K; (*d*) liquid at 249°K; (*e*) liquid at 298°K; (*f*) liquid at 293°K.

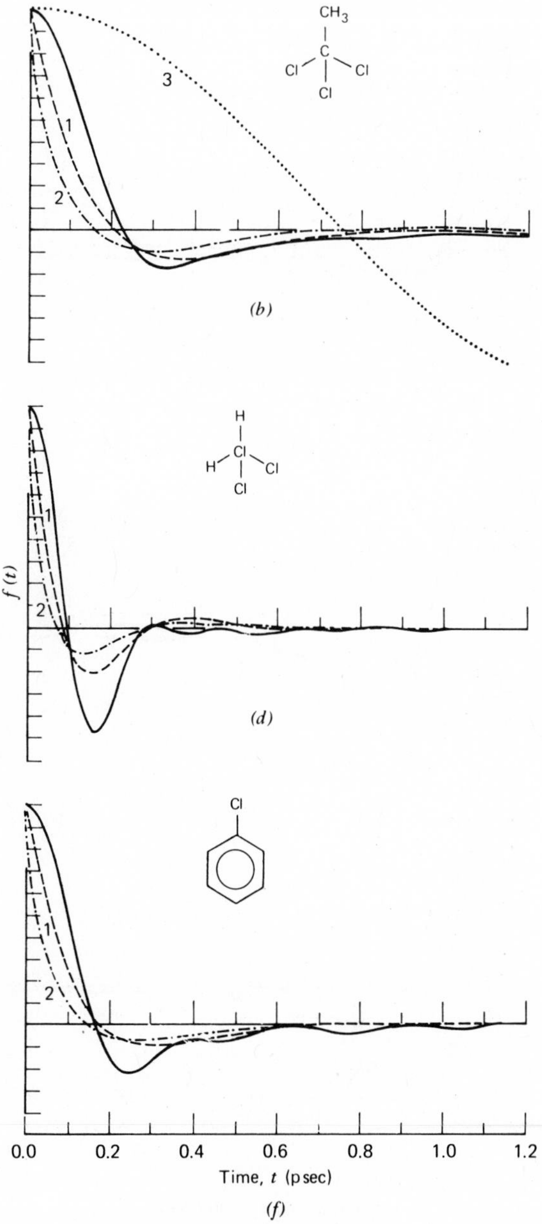

Fig. 16. Continued

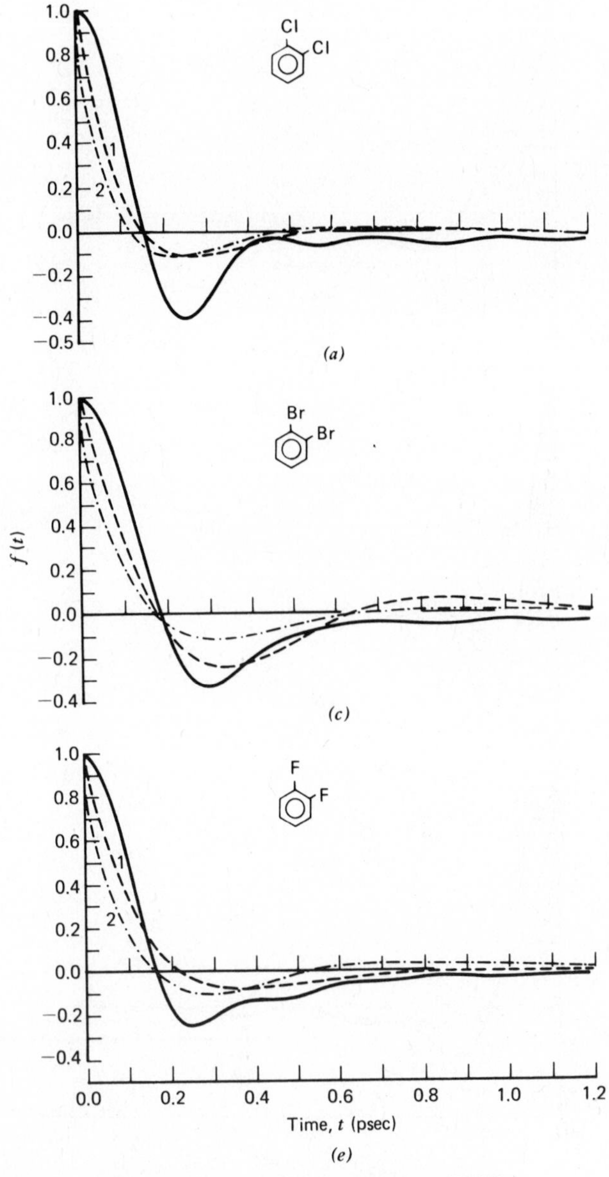

Fig. 17. Key as in Fig. 14. Liquids at 293°K.

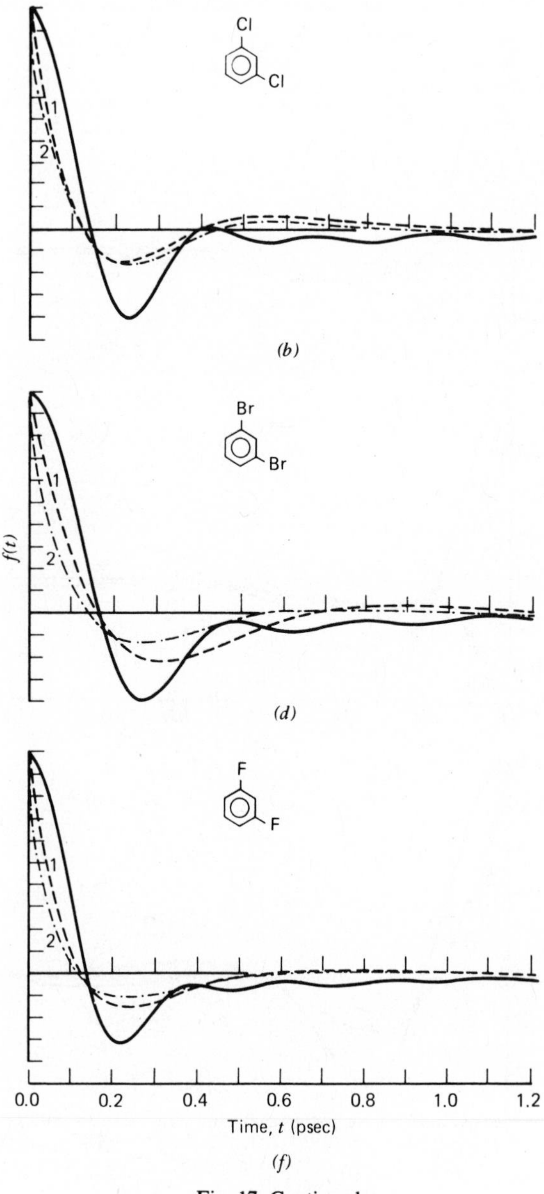

Fig. 17. Continued

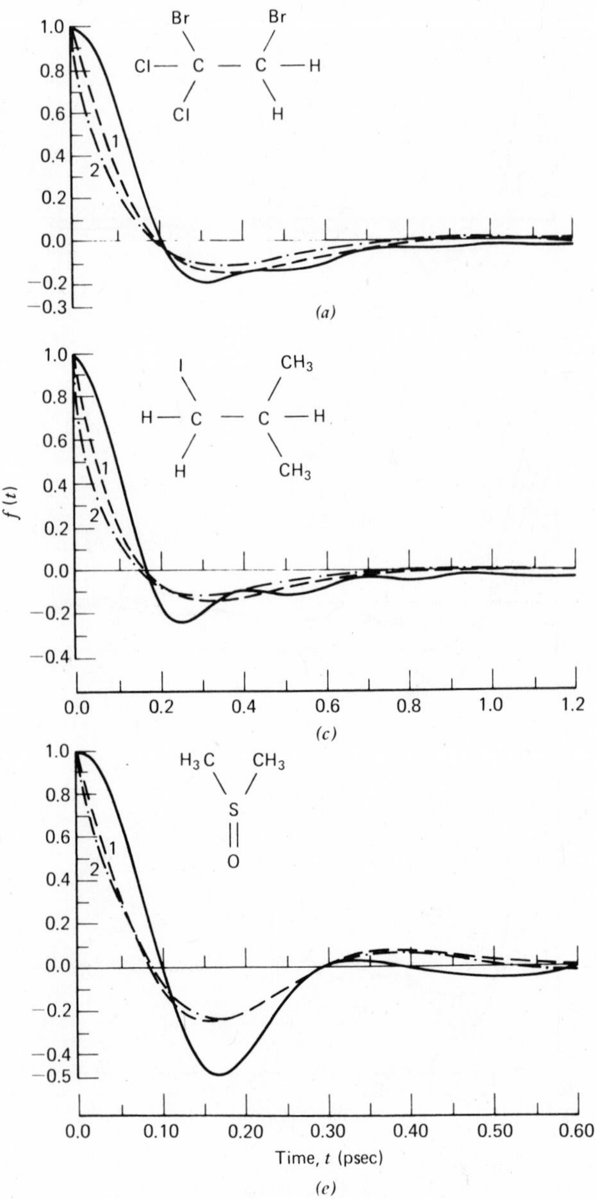

Fig. 18. Key as in Fig. 14. (*a*) to (*d*) Liquids at 295°K; (*e*) and (*f*) liquids at 293°K.

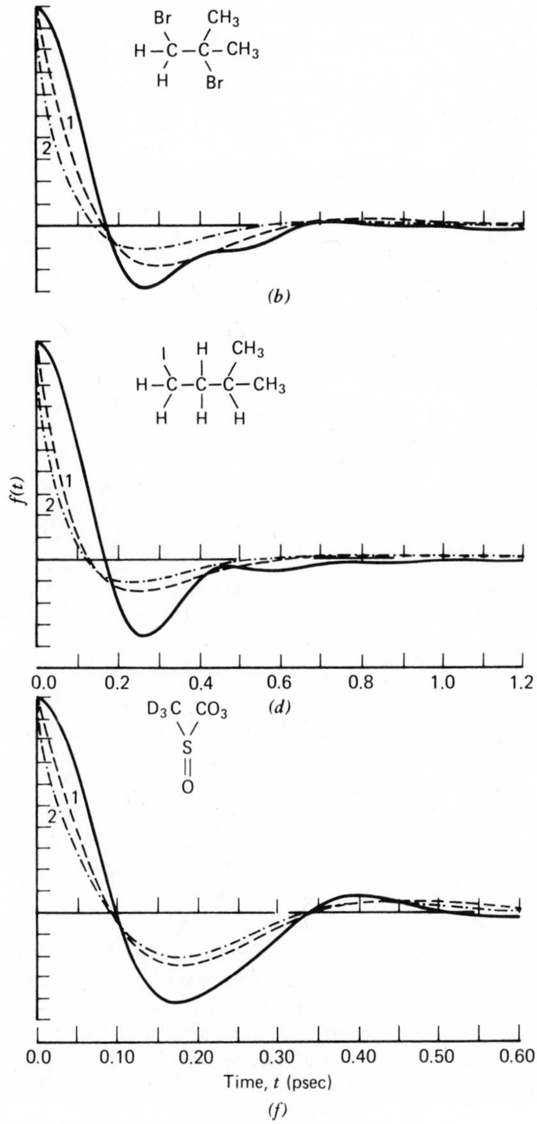

Fig. 18. Continued

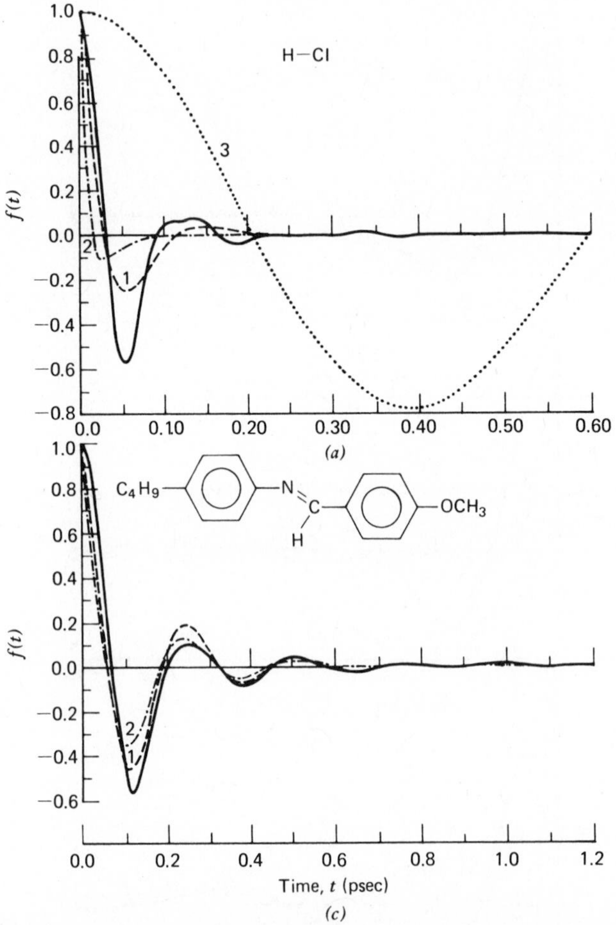

Fig. 19. (*a*) to (*c*) Key as in Fig. 14. (*a*) Rotator phase (100°K); (*b*) liquid at 293°K; (*c*) nematic phase at 296°K; (*d*) curve 1—frequency domain curve of $\alpha(\bar{\nu})$ predicted by an early form of itinerant oscillation at 238°K for the rotator phase of *t*-butyl chloride. Integration of this curve up to only 400 cm^{-1} produces the spurious oscillations of curve 2, which is the correlation function. It is necessary (because of the asymptotic ω^{-2} behavior at high frequences) to integrate up to 3000 cm^{-1} before these disappear. Abscissas: upper, time, *t* (psec); lower, $\bar{\nu}$ (cm^{-1}).

advantage of a better signal-to-noise ratio than a conventional grating instrument, whereby the spectrum is scanned element by element. A highly efficient optical system is needed because of the low emission in the far infrared of conventional sources such as the mercury discharge lamp. Although some 1500 laser lines are known in the far infrared, the need for broad-band spectroscopy ensures the continuing usefulness of the inter-

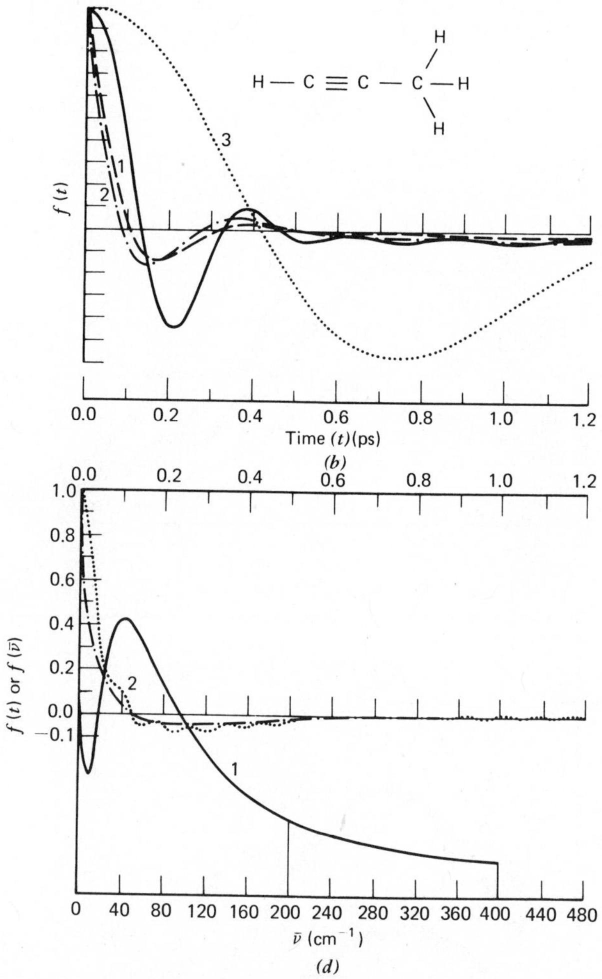

Fig. 19. Continued

ferometric system that indicated for work in compressed gases, liquids, and solids. The spectrometer consists basically of a Michelson two-beam interferometer connected, off-line, to a digital computer which selects the component frequencies from the interference record (interferogram) observed by the detector as the path difference X between the two partial beams of the interferometer is varied.

The power variation at the detector is recorded as a function of the path difference X in Fig. 22. This is twice the mirror displacement. The beam divider is a thin film of stretched dielectric, usually poly(ethylene

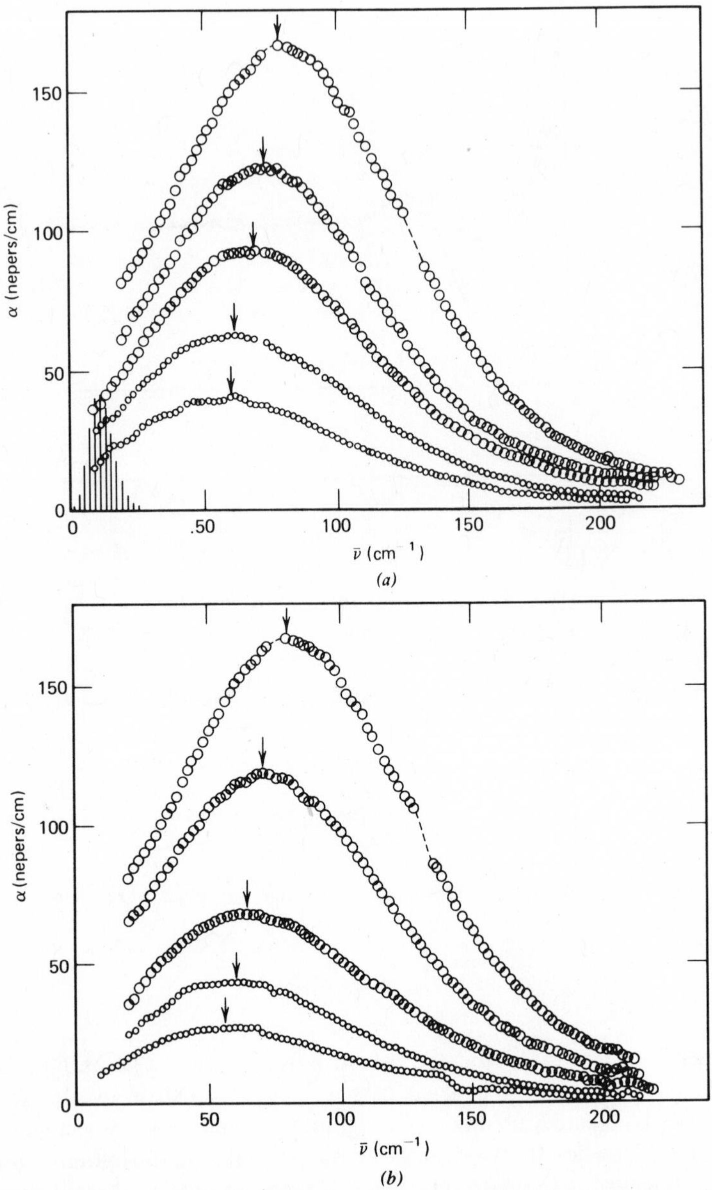

Fig. 20. (*a*) Absorption of dichloromethane in CCl_4 at 298°K, corrected for solvent absorption. From top to bottom: pure CH_2Cl_2 ($N = 9.4 \times 10^{21}$ molecules/cm^3); $N = 7.22 \times 10^{21}$ molecules/cm^3; $N = 4.93 \times 10^{21}$ molecules/cm^3; $N = 3.25 \times 10^{21}$ molecules/cm^3; $N = 1.97 \times 10^{21}$ molecules/cm^3. (*b*) The same in decalin solvent. The stick spectrum represents some $J \rightarrow J+1$ ($\Delta K = 0$) lines for quantized free rotation of CH_2Cl_2, regarded as an approximate symmetric top. [Reproduced by permission from *J. Chem. Soc. Faraday Trans. 2*, **74**, 343 (1978).]

TABLE III

Information Available from Far-Infrared Broad Bands of Rotational and Intermolecular Origin

Measurement	Physical significance
$A_0/N = \frac{1}{N}\int \alpha(\bar{\nu})\,d\bar{\nu}$ band	For a dilute dipolar gas, may be used to estimate the dipole moment $\boldsymbol{\mu}$, given an accurate sum rule. For a compressed nondipolar gas, may be used to estimate its first multipole moment subject to theoretical constraints (e.g., bimolecular collisions) and molecular symmetry. In nondipolar liquids and plastic crystals, A_0/N may be compared with its equivalent in the compressed gas and an estimate made of the cancellation of dipole inductions due to intermolecular potential symmetry. The constancy or otherwise of A_0/N with dilution of a dipolar solute in a nondipolar solvent is a useful method of investigating molecular association and complex formation and of probing liquid crystalline environments.
$\bar{\nu}_{max}$, the maximum power absorption frequency	As a dilute dipolar gas is compressed, $\bar{\nu}_{max}$ shifts gradually to higher frequency as the overall $\alpha(\bar{\nu})$ contour broadens (Fig. 1). At liquid densities $\bar{\nu}_{max}$ is removed a long way from its value in the gas phase (by as much as 100 cm^{-1}). It is therefore strongly dependent on intermolecular potential energy. Successive approximants of the Mori continued fraction may be used to model $\bar{\nu}_{max}$ and the contour $\alpha(\bar{\nu})$. Equation (I.30) is such that the mean-square torque is well defined and measurable from $\bar{\nu}_{max}$. Higher approximants would relate $\bar{\nu}_{max}$ to this and its derivatives.
$\alpha(\bar{\nu})$, the zero-THz power absorption coefficients	This is a probability distribution of wavenumber (i.e., a spectral function). Its Fourier transform is thus a correlation function, relating random reorientations of dipoles taken in a temporal sequence. The correlation function obtained by Fourier transformation of $\alpha(\omega)$ is $$-\ddot{C}_m(t) \equiv \langle \dot{\boldsymbol{\mu}}(0)\cdot \sum_{i\neq j} \dot{\boldsymbol{\mu}}_j(t)\rangle$$ where $\boldsymbol{\mu}$ is the permanent molecular dipole moment. In its ability to yield $-\ddot{C}_m(t)$ directly, accurately, and in detail, the zero-THz $\alpha(\bar{\nu})$ profile is uniquely placed as a probe into the short-time details of molecular dynamics. Any model of the fluid state must match up with $C(t)$ satisfactorily as well reproduce $C_m(t) = \langle \boldsymbol{\mu}_i(0)\cdot \sum_{i\neq j} \boldsymbol{\mu}_j(t)\rangle$, measurable from the dielectric loss.

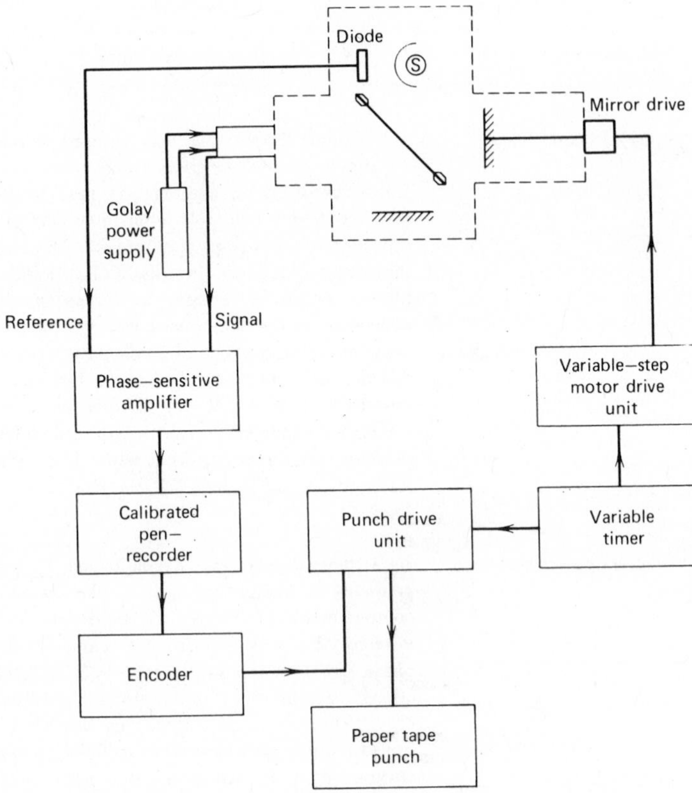

Fig. 21. Schematic of the Grubb–Parsons/N.P.L. interferometer.

terephthalate). Internal reflections in this beam divider determine the transmissivity of the interferometer and cause it to be strongly frequency-dependent, so that different beam dividers are needed for different frequency regions.

Submillimeter detector technology is rapidly advancing, but the Golay pneumatic cell is conveniently operable at ambient temperatures. The faster and more sensitive Rollin detector is helium-cooled and extends the spectral range to overlap with the microwave klystron frequencies. Two types of interferogram are usually used (Fig. 23), resulting from amplitude and phase modulation of the detected signal. Phase modulation discriminates in favor of the interferogram range, resulting in a higher signal-to-noise ratio in the detected signal. The modulation is achieved by aperiodic displacement of one of the mirrors through a distance of the

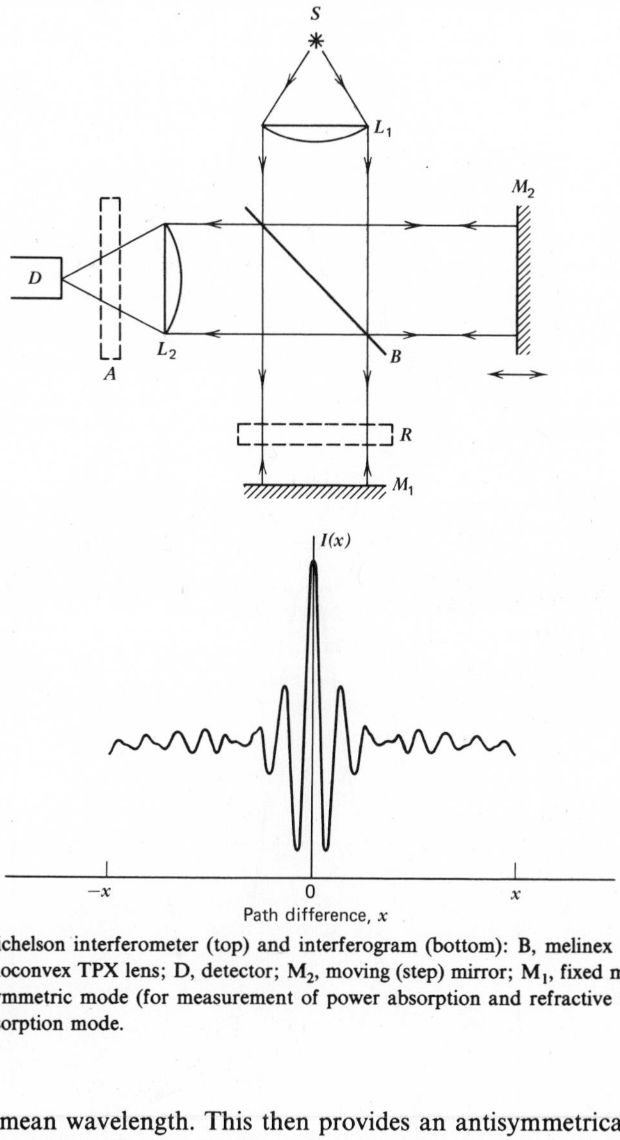

Fig. 22. Michelson interferometer (top) and interferogram (bottom): B, melinex beam divider; L, planoconvex TPX lens; D, detector; M_2, moving (step) mirror; M_1, fixed mirror; R, sample in asymmetric mode (for measurement of power absorption and refractive index); a sample in absorption mode.

order of a mean wavelength. This then provides an antisymmetrical interference record (Fig. 23). At far-infrared frequencies the magnitude of the oscillations and tolerances in the quality of the motion are such that a relatively simple vibrator may be used to sinusoidally drive the mirror on a loudspeaker coil. This development by Chamberlain[57] removes the following disadvantages of amplitude modulation of the detected signal.

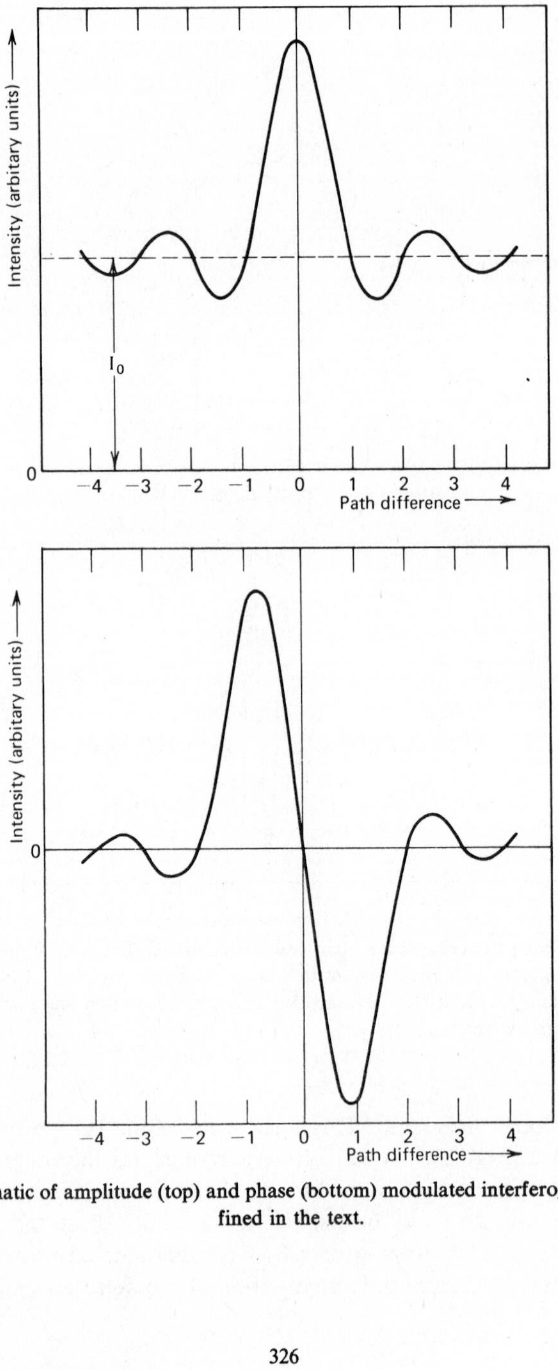

Fig. 23. Schematic of amplitude (top) and phase (bottom) modulated interferograms. I_o as defined in the text.

1. The beam dividers cause the interferometer to have a transmissivity that is strongly frequency-dependent. In the polarizing interferometer, devised by Martin and Pupplett[58] and discussed later, the transmissivity is constant up to a cutoff frequency which is inversely proportional to the grid spacings of the polarizers and beam divider through which the incident and emergent beam passes. Features of the transmissivity are improved by phase modulation in both types of interferometer.
2. Considering the grand maximum [point I(0) in Fig. 22] of a typical amplitude-modulated interferogram, the maximum of the interference fluctuations is a little more than half the total signal, but the detector is required to record all of the signal. Phase modulation allows the detector to record only that part which is varying with path difference and eliminates the background. Thus lower amplification gains are required, electronic noise is reduced, and source noise is removed. Stray rays emitted by the interferometer are discarded.
3. A rotating chopper, as used in amplitude modulation, cuts off half the available radiation. In the Martin–Pupplett instrument the signal is modulated by rotating the beam divider, and this square-function modulation gives a superior spectrum to both the sinusoidal amplitude modulation and sinusoidal phase modulation.

Computation of the Spectrum from the Interferogram. If the interferometer is irradiated with monochromatic radiation of luminance $B_0(\bar{\nu}_0)$ at a wavenumber $\bar{\nu}_0$, and if the reflectivity and transmission of the beam divider are 50% the energy detected will be $B_0(\bar{\nu}_0)/2$. However, whatever the reflectivity of the beam divider, the average intensity in the two beams is always equal, since each beam undergoes one reflection and one transmission before reaching the detector. It follows in general that

$$I_0(x)=\tfrac{1}{2}B_0(\bar{\nu}_0)+\tfrac{1}{2}B_0(\bar{\nu}_0)\cos(2\pi\bar{\nu}x) \tag{I.82}$$

where x is the path difference between the beams and equals twice the distance traveled by the movable mirror. Cosine fringes are therefore observed. For a polychromatic source of luminous density $B(\bar{\nu})$,

$$I(x)=\tfrac{1}{2}\int_0^{\infty}B(\bar{\nu})\,d\bar{\nu}+\tfrac{1}{2}\int_0^{\infty}B(\bar{\nu})\cos(2\pi\bar{\nu}x)\,d\bar{\nu} \tag{I.83}$$

We define the interferograms as

$$F(x)=\tfrac{1}{2}\int_0^{\infty}B(\bar{\nu})\cos 2\pi\bar{\nu}x\,d\bar{\nu} \tag{I.84}$$

and note in passing that $I(x)-F(x)$ is eliminated with phase modulation.

$F(0)$ is the amplituded-modulated grand maximum where all frequencies interfere constructively. The detected spectral power $B(\bar{\nu})$ and the interferogram are a Fourier transform pair whence

$$B(\bar{\nu}) = \int_{-\infty}^{\infty} \cos 2\pi\bar{\nu}x\, dx \tag{I.85}$$

By digitally recording $F(x), B(\bar{\nu})$ is calculated numerically. Equations (I.84) and (I.85) apply to the ideal case when the interferogram is perfectly symmetrical about $x=0$ [i.e., $F(tx)$ is an even function of x]. With imperfect alignment the following equations are used:

$$F(x) = \int_{-\infty}^{\infty} B(\bar{\nu}) \exp(2\pi i\bar{\nu}x)\, d\bar{\nu}$$

$$B(\bar{\nu}) = \int_{-\infty}^{\infty} F(x) \exp(-2\pi i\bar{\nu}x)\, dx \tag{I.86}$$

Assuming that $F(x)$ is real implies that $\tilde{B}(\bar{\nu})$ is Hermitian; that is, if $B(\bar{\nu}) = p(\bar{\nu}) - iq(\bar{\nu})$, then p is an even function of $\bar{\nu}$ and q is an odd function of $\bar{\nu}$. Phase modulation implies that

$$I(x) = \int_{-\infty}^{\infty} B(\bar{\nu})\, d\bar{\nu} + \int_{-\infty}^{\infty} B(\bar{\nu}) \cos[2\pi\bar{\nu}x - \phi(\bar{\nu})]\, d\bar{\nu} \tag{I.87}$$

where $\phi(\bar{\nu})$ is the phase difference due to any residual asymmetry. On placing an isotropic specimen in front of the detector, the expression for the center power becomes

$$I_0(x) = I_0 + F_0(x)$$

where

$$I_0 = \int_{-\infty}^{\infty} B_0(\bar{\nu})\, d\bar{\nu}$$

$$F_0(x) = \int_{-\infty}^{\infty} B_0(\bar{\nu}) \cos[2\pi\bar{\nu}x - \phi(\bar{\nu})]\, d\bar{\nu} \tag{I.88}$$

and

$$B_0(\bar{\nu}) = \tau(\bar{\nu}) B(\bar{\nu})$$

Here $\tau(\bar{\nu})$ is the transmissivity of the specimen. Fourier-transforming (I.88) yields

$$\tilde{S}(\bar{\nu}) = B_0(\bar{\nu}) \exp[-i\phi(\bar{\nu})]$$

$$= \int_{-\infty}^{\infty} F_0(x) \exp(-2\pi i\bar{\nu}x)\, dx \tag{I.89}$$

from which the transmission spectrum $\tau(\bar{\nu})$ may be recovered for the sample.

Truncation and Sampling of the Interferogram. For the purposes of numerical Fourier transformation, we must truncate the limits of the integrals in (I.83) to (I.89) and must also record the interferogram digitally, using a stepping motor to drive the mirror. The integration limits cannot be infinite when the independent variable is path difference derived from the displacement of the mirror. If the maximum path difference is X, the spectral resolution is then determined by $\Delta\bar{\nu}=1/X$. If we truncate the integral in the range $-X$ to X, the interferogram function $F(x)$ is effectively multiplied by a rectangle function $\pi(x/2X)$, where

$$\begin{aligned} \pi(x/2X) &= 1 \qquad |x|<X \\ &= 0 \qquad |x|>X \end{aligned} \tag{I.90}$$

The transform of $F(x)\pi(x/2X)$ is the convolutuion

$$g(\bar{\nu})*2X\sin(2\pi x\bar{\nu})/(2\pi x\bar{\nu}) \tag{I.91}$$

which results in a physically meaningless negative region in the final spectrum. To overcome this problem a weighted rectangular function is used which causes a loss of resolution. This is $\pi(x/2X)W(x)$, where $W(x) = \cos^2(\pi x/2X)$.

If a finite number ($2N$) of ordinates of the interferogram are sampled, the sampling theorem of information theory shows that the spectrum will be meaningfully divisible into N frequency elements. The maximum spacing (Δx) at which the ordinates of the interferogram need to be measured is then given by

$$\Delta x = 1/(2|\bar{\nu}_1 - \bar{\nu}_2|) \tag{I.92}$$

a relation which also defines the cutoff frequency of the instrument, past which the spectrum will be distorted by folding (periodicity in the Fourier series used to calculate the Fourier transform).

Synchronous Detection. The real interferogram $F(x)$ is aperiodically scanned at a constant speed and recorded using phase modulation followed by synchronous amplification in which the periodic detector output is amplified, filtered, and mixed with a reference signal derived from the modulation process. This leads to a dc output that is directly related to the peak-to-peak detector output. The process is outlined below. A schematic representation of the electronics is displayed in Fig. 21.

STEP 1

Suppose that the radiation of power $J(y)$ is modulated at a frequency f to give at any instant t a signal $J(y,ft)$ which can be expressed as a Fourier series:

$$J(y,ft)=\tfrac{1}{2}K^0(y)+\sum_{l=1}^{\infty}\left[K^{(l)}(y)\cos 2\pi lft+Z^{(l)}(y)\sin 2\pi lft\right] \quad \text{(I.93)}$$

where $K^{(l)}(y)$ and $Z^{(l)}(y)$ are the Fourier coefficients. In the case of phase modulation we take the time origin so that $J(y,ft)$ is odd with respect to t. The detected signal is then given by

$$J(y,ft)=\sum_{l=1}^{\infty} Z^{(l)}(y)\sin 2\pi lft \quad \text{(I.94)}$$

STEP 2
The output from the detector is $uJ(y,ft)$, where u is the responsivity (signal delivered per unit power input).

STEP 3
The detector signal is amplified with a gain of g.

STEP 4
It is passed through a filter tuned to the frequency f to give

$$D(y,ft)=guZ^{(1)}(y)\sin 2\pi ft \quad \text{(I.95)}$$

STEP 5
In a synchronous recording this is multiplied and phase-linked with a reference signal, also of frequency f. This results in a slowly varying recorded signal:

$$V(y)=(2/\pi)guZ^{(1)}(y) \quad \text{(I.96)}$$

and a series of time-dependent terms of frequency multiples of f.

STEP 6
A series of low-pass filters removes the time-dependent terms to give $V(y)$ as the output signal. $Z_1(y)$ is the coefficient of the fundamental term in the series expansion of $J(y,ft)$. In amplitude modulation the recorded signal is proportional to the total power $J(y)$ that would reach the detector in the absence of modulation. With phase modulation this is not so.

The path difference at any instant t is not y but $y+j(ft)$, where j is the jitter amplitude function. The corresponding power at the detector is then

$$J(y,ft)=J+G(y,ft) \quad \text{(I.97)}$$

where $$G(y,ft)=\int_{-\infty}^{\infty} B_0(\bar{\nu})\cos[2\pi\bar{\nu}y-\phi(\bar{\nu})+2\pi\bar{\nu}j(ft)]\,d\bar{\nu}$$

Expanding the cosine, we obtain

$$\cos(2\pi\bar{\nu}j(ft))=\tfrac{1}{2}\mu^0(p\bar{\nu})+\sum_{l=1}^{\infty}\mu^{(l)}(p\bar{\nu})\cos 2\pi lft$$

$$\sin(2\pi\bar{\nu}j(ft))=\sum_{l=1}^{\infty}\alpha^{(l)}(p\bar{\nu})\sin 2\pi lft \tag{I.98}$$

where p is the jitter amplitude. Comparison with (I.93) yields

$$\begin{aligned} Z_1(y)&=-\int_{-\infty}^{\infty}\alpha^{(1)}(p\bar{\nu})B_0(\bar{\nu})\sin[2\pi\bar{\nu}y-\phi(\bar{\nu})]\,d\bar{\nu}\\ &=G\phi(y) \end{aligned} \tag{I.99}$$

The recorded signal $V(y)=(2/\pi)guZ^{(1)}(y)$ now contains no y-independent terms and is the phase-modulated interferogram. The effect of phase modulation is to modify the transmitted power spectrum by the wave-number-dependent factor

$$|\alpha^{(1)}(p\bar{\nu})|$$

For sinusoidal modulation,

$$j(ft)=p\sin 2\pi ft \tag{I.100}$$

and $$\alpha^{(1)}(p\bar{\nu})=2J_1(2\pi\bar{\nu}p) \tag{I.101}$$

where J_1 is the first-order Bessel function of the first kind. Limitations in the final spectrum due to variations in this factor can be minimized by matching $\alpha^1(p\bar{\nu})$ with the transmission characteristic of the interferometer, governed largely by beam-divider effects. Square function modulation produces a spectrum superior to sinusoidal modulation. Such a square function is possible with a rotating chopper as used in a polarizing interferometer.

The Polarizing Interferometer.[58] *To* measure absorptions in the very far infrared range 2 to 40 cm^{-1}, we have access to the apparatus designed originally by Martin and Pupplett. This uses a wire-mesh beam divider and similar meshes as polarizer and analyzer. Figure 24 is a schematic. The polarizer (P_1) is a circular grid with reflection and transmission coefficients close to 100% from frequencies approaching zero up to $1/(2d)\,cm^{-1}$. It is

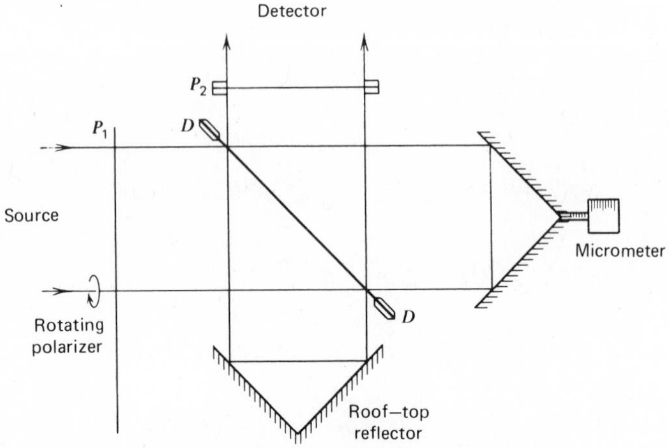

Fig. 24. Schematic of a polarizing interferometer. [Reproduced by permission from *J. Chem. Soc. Faraday Trans. 2*, **74**, 59 (1978).]

spun about its center point acting as a polarizing chopper blade and producing square function modulation. A second polarizer (P_2) is held fixed. The beam divider windings are at 45° to those of P_2. Roof-top reflectors replace the Michelson plane mirrors.

A collimated beam is plane-polarized at P_1 and then divided by the flat wire-grid polarizer D into a beam A, polarized with its *E*-vector at 90° to the second beam B. Then A and B are recombined at the wire grid D and finally pass through the polarizing analyzer P_2, which has its axis parallel to or at 90° to P. The reflectors M_1 and M_2 act as polarization rotators. For a monochromatic source, the beam is elliptically polarized after recombination at D with an ellipticity varying periodically with increasing path difference between beams A and B. For the *E*-vector leaving the beam divider,

$$\mathbf{E}_j = \frac{a}{\sqrt{2}}\mathbf{n}\cos(\omega t + \Delta_A) + \frac{a}{\sqrt{2}}\mathbf{t}\cos(\omega t + \Delta_B) \tag{I.102}$$

where $\boldsymbol{n}$ is the unit vector normal to the plane of the paper (Fig. 24), and t that in the plane and transverse to the direction of propagation. Δ_A and Δ_B are phase shifts for beams A and B. For the *E*-vector leaving the polarizer P_2,

$$\begin{aligned} |E_0| &= \mathbf{E}_j \cdot \mathbf{p} = (a/2)\left[\cos(\omega t + \Delta_A) + \cos(\omega t + \Delta_B)\right] \\ &= a\cos(\omega t + \bar{\Delta})\cos(\Delta/2) \end{aligned} \tag{I.103}$$

where $\bar{\Delta}$ is the mean of Δ_A and Δ_B and

$$\Delta = \Delta_A - \Delta_B = (2\pi/\lambda)x \tag{I.104}$$

where x is the path difference in the interferometer. Here $\boldsymbol{p}$ is a unit vector in the direction of the optical axes of P_1 and of P_2. After leaving P_2, the beam is plane-polarized with an amplitude that varies periodically with path difference in the same way as in a Michelson interferometer. From (I.103) the emergent intensity is given by

$$I_p = \langle |E_0^2| \rangle = \frac{a^2}{2}\cos^2\left(\frac{\Delta}{2}\right) = \frac{a^2}{4}(1+\cos\Delta) \tag{I.105}$$

Alternatively, if the axis of P_2 were rotated by $\pi/2$:

$$|E_0| = \mathbf{E}_j \cdot \mathbf{p}^1 \qquad \text{where} \qquad \mathbf{p}\cdot\mathbf{p}' = 0$$

and hence

$$I_t = \frac{a^2}{4}(1-\cos\Delta) \tag{I.106}$$

The transmissivity of the instrument is constant up to a cutoff frequency which is inversely proportional to the grid spacing. The resolution is not limited by departures from the flatness of the grid or by any phase retardation in the polarizing grid.

2. *Significance of the Experimental Results*[59]

Spectra are always of the absorption type, so that the sample whose absorption coefficient is to be measured is placed between the interferometer and the detector. By comparing sample thicknesses d_2 and d, we obtain

$$\alpha(\bar{\nu}) = (d_2 - d_1)^{-1}\log_e(I_0(\bar{\nu})/I(\bar{\nu})) \tag{I.107}$$

where $I_0(\bar{\nu})$ is the background power spectrum and $I(\bar{\nu})$ is that for an increase in sample thickness $(d_2 - d_1)$. In this section we discuss briefly the sources of uncertainty in $\alpha(\bar{\nu})$ arising from the experimental setup. The quality and reproducibility of the spectra are optimum when the product $(d_2 - d_1)\alpha_{max}$ lies between 0.5 and 1.5, which amounts to an overall absorption by the specimen of about 60 to 85% of the incident radiation. Under these conditions phase-modulated spectra (e.g., those of Fig. 20) are repeatable to within 2% and reproducible between different laboratories. Figure 25 shows the far-infrared absorptions of chloroform under amplitude- and phase-modulation conditions.

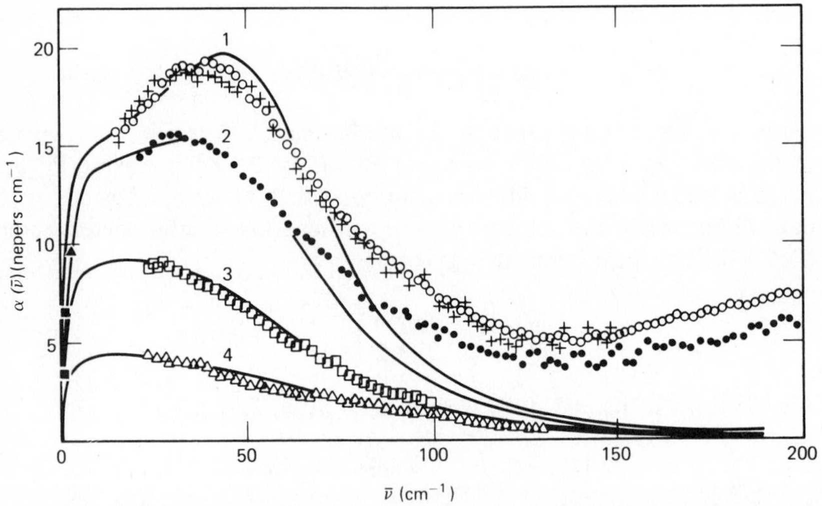

Fig. 25. Far infrared spectrum of liquid chloroform. ○, Phase modulation: +, amplitude modulation; ▲, klystron (112 GHz); ■, microwave measurements. Ordinate: $\alpha(\bar{\nu})$ (nepers/cm)^1; abscissa: $\bar{\nu}$ (cm^{-1}).

The systematic uncertainties that arise from a particular instrument and set of conditions are typically small (Fig. 25). As Fig. 12 shows, however, ±2% uncertainty is small enough to be insignificant when making comparisons with the theoretical models available. The data in Fig. 26 were taken with different window materials and different detectors (Golay and Rollin types). Further good agreement is usually obtainable, with skill, between different types of beam splitter in the Michelson interferometer.

A source of systematic uncertainty in the "cube" design is the highly convergent beam through the sample, effective collimation being difficult to reconcile with the all-important need for maintaining maximum flux throughput via sometimes highly absorbing pressure windows. These convergence effects have been discussed by Fleming[60] but are small when the difference $(d_2 - d_1)$ is small, which is always the case for intense absorbers such as dipolar liquids. With longer sample paths the data may be checked against those obtained with accurately collimated laser radiation. When $(d_2 - d_1)$ is of the order of the wavelength of the incoming radiation, spurious peaks sometimes arise as a sine wave superimposed on the computed spectrum due to multiple internal reflections occurring within the sample cell. These have been recognized in the work of Baise[114] on compressed nitrous oxide. Experience has shown that the cell window material is important in determining the extent of this effect. *Z*-cut crystalline quartz

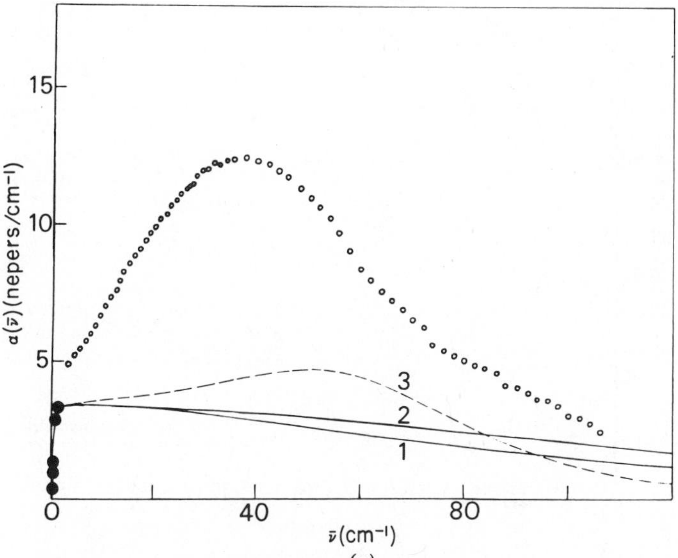

(a)

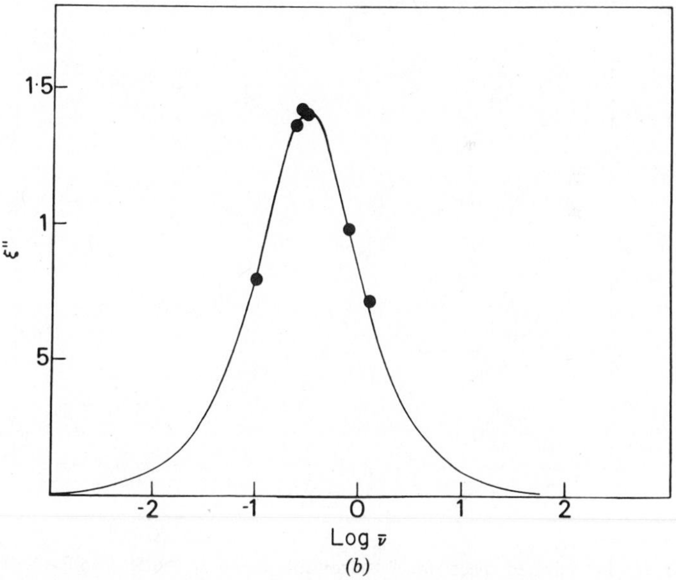

(b)

Fig. 26. (*a*) Agreement between various experimental sources for liquid bromobenzene. ●, Microwave (klystron) spot frequencies; ○, polarizing interferometer; ⊕, Grubb–Parsons/N.P.L. "cube" interferometer. (1), (2), (3) Various theoretical models (see original reference). (*b*) — Calculated loss curve for liquid bromobenzene using Langevin's equations for an asymmetric and a symmetric top molecule. The theoretical curve from Mori three-variable theory is also coincident. ● Experimental loss data. [Reproduced by permission from *J. Chem. Soc. Faraday Trans. 2*, **73**, 1074 (1977).]

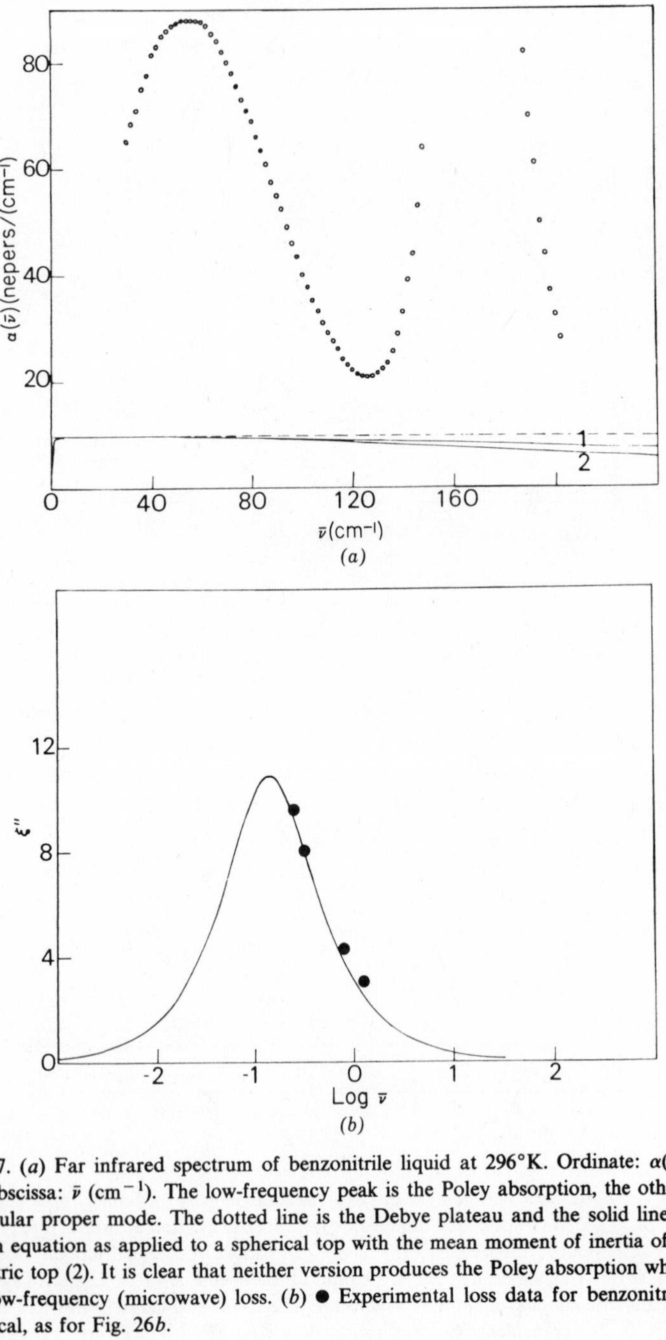

Fig. 27. (*a*) Far infrared spectrum of benzonitrile liquid at 296°K. Ordinate: $\alpha(\bar{\nu})$ neper cm^{-1}; abscissa: $\bar{\nu}$ (cm^{-1}). The low-frequency peak is the Poley absorption, the other an intramolecular proper mode. The dotted line is the Debye plateau and the solid lines (1) the Langevin equation as applied to a spherical top with the mean moment of inertia of the true asymmetric top (2). It is clear that neither version produces the Poley absorption when fitted to the low-frequency (microwave) loss. (*b*) ● Experimental loss data for benzonitrile. (—), Theoretical, as for Fig. 26*b*.

windows, when optically parallel, enhance the effect, which is eliminated by wedging very slightly. There is usually less of an effect with polymeric window material, so that these are used wherever practicable. These effects may be recognized from the relation

$$\Delta\bar{\nu} = \frac{1}{2n(d_2 - d_1)}$$

which defines the frequency spacing of the spurious peaks in terms of the sample refractive index n.

One further cause of experimental uncertainty arises from the intense higher-frequency proper modes of absorption as illustrated in Fig. 27. These are intramolecular in origin (vibrational, difference modes, etc.), but their intensity with respect to the Poley absorption is sometimes enough to block off almost the whole of the flux throughput. The interpretative difficulty is one of extrapolation and resolution of the Poley band from the proper mode. For the purposes of theoretical analysis we chose our spectra carefully to avoid these complications.

Sample-Cell Technology. The design of sample cells is beyond the scope of this chapter, but it has an essential bearing on the quality of the final spectra because of the need to maximize flux throughput. The reader is referred to original articles for examples of designs for use with dilute and compressed gases, nondipolar liquids, dipolar liquids, mesophases, and solids. Specialized designs are available[50] for studies of liquid water, an enormously intense absorber in the far infrared,[1] and for ultra-accurate standards measurements. One concern common to all designs is the preferability of evacuating the whole system (interferometer, sample cell, and detector) to eliminate water vapor, which absorbs strongly in the far infrared. For the same reason special care ought to be taken to eliminate moisture from the samples being investigated.

II. EVALUATION OF MODELS OF FLUID-STATE MOLECULAR DYNAMICS*

In Section I we have been concerned with rationalizing the growth of modeling techniques for the fluid-state molecular dynamics in terms of a continued-fraction expansion of the Liouville equation, for which all such models are approximate solutions. In this we compare the theoretical and experimental data for selected dipolar systems, ranging from the moderately compressed gas (Section II.B.2), through to the glassy condition where the zero-THz profile is a continuum loss peaking at frequencies

*In this section note that $K_n(t) \equiv \phi_{n+1}(t)$ for the memory functions.

sometimes separated by many decades on the log scale (Section II.F). Features of this progression have been illustrated in terms of one particular model, the itinerant oscillator/librator, in Fig. 9. In Section III, a further dimension is added by the use of machine simulations of a structured (diatomic) molecule based on the algorithm developed recently by Tildesley and Streett.[61]

At the outset, in Section II.A some fundamental points are discussed which remain at issue and will probably be best tackled not by the invention of more models but by the use of molecular dynamics and Monte Carlo simulations.[61] The most intractable problems—that of the internal field and the related one of electrodynamic coupling (dipole-dipole being the longest in range)—have already been introduced. However, we have taken the line that these should not distract us from our major tasks of relating, whenever possible, theories of statistical autocorrelations of dipole vectors to zero-THz absorption profiles. We have also defined the liquid state as that where rotation–translation coupling is the most pronounced characteristic. This is, however, with the hindsight of only a few months, when the results of Section I.B first suggested themselves to us. Before that no clear-cut mathematical formulation of the effect of linear velocity on $\langle \omega(t)\cdot\omega(0)\rangle$ has been included in the purely rotational models of Table I. The new theory of translation–rotation coupling being yet in its infancy, we proceed on the understanding that the use of purely reorientational formalism for the liquid-phase molecular dynamics is a widespread approximation, similar to the use of (I.39) in neutron scattering. In stating this we might be doing an injustice to Wolynes and Deutch and to others who may have considered such effects from a hydrodynamic point of view.

A. The Dipolar Autocorrelation Function and Dielectric Loss—The Fluctuation/Dissipation Theorem[62]

The mean observed value of $A_j(t)$ in (I.2) is given by

$$\langle A(t)\rangle = \int_\Gamma A(p,q)\,{}^1f_A(p,q,t)\,dp\,dq \tag{II.1}$$

where Γ denotes the phase space (of p and q) and ${}^1f_A$ is the first-order space distribution function akin to the probability density functions of Section I.A. In (II.1) and the following, we drop the subscript j for convenience. For an isolated system,

$$\frac{d\,{}^1f_A}{dt} = 0 = \frac{\partial\,{}^1f_A}{\partial t} + \Omega\,{}^1f_A \tag{II.2}$$

By considering the change H' brought about in the equilibrium Hamiltonian H_0 by an external perturbation $E(t)$, we have

$$H' = -X(p_r, q_r)E(t) \tag{II.3}$$

where X is a property of the system (canonical ensemble of r molecules) deemed responsible for the increase in the Hamiltonian. If $E(t)$ is an electric force field, then X will have the units of the electric dipole moment. The new distribution function is now

$$f = f_0 + f'(t) \tag{II.4}$$

where $f \equiv {}^1f_A$ and

$$\frac{\partial f'}{\partial t} = -\mathfrak{L}_0 f' - \mathfrak{L}' f_0 - \mathfrak{L}' f' \tag{II.5}$$

in the same notation. The term $\mathfrak{L}'f'$ is the product of operator and functional increments, and since the perturbation is weak is considered as negligible. Equation (II.5) thus becomes

$$\frac{\partial f'}{\partial t} = -\mathfrak{L}_0 f' + E(t)\mathfrak{L}_X f_0 \tag{II.6}$$

where the operator $\mathfrak{L}_X$ is defined by

$$\mathfrak{L}_X = \sum_r \left[\frac{\partial X}{\partial p_r}\frac{\partial}{\partial q_r} - \frac{\partial X}{\partial q_r}\frac{\partial}{\partial p_r} \right] \tag{II.7}$$

Integrating (II.6) yields

$$f'(t) = \int_{-\infty}^{t} \exp(-\mathfrak{L}_0(t-t'))E(t')\mathfrak{L}_X f_0 \, dt' \tag{II.8}$$

If it is assumed now that the system can be observed by a study of the property Y (related to X), then from (II.1) and (II.8) for Y we may write

$$\langle Y(t) \rangle = Y_{\text{eq}} + \int_{-\infty}^{t} E(t')\,dt' \int_\Gamma Y(t')\exp(-(t-t')\mathfrak{L}_0) \tag{II.9}$$

$$\times \mathfrak{L}_X f_0(t')\,dp\,dq$$

$$\equiv Y_{\text{eq}} + \int_{-\infty}^{t} E(t)\Psi_{YX}(t,t')\,dt' \tag{II.10}$$

In (II.10) we may deduce that

$$\Psi_{YX}(t,t') = \int_{\Gamma} Y(t-t')\mathfrak{L}_X f_0(0)\, dp\, dq \tag{II.11}$$

from the fact that $\mathfrak{L}_X f_0$ is stationary in time, since the operator is taken at equilibrium. Similarly, Y is stationary, since it depends only on coordinates and momenta. Ψ_{YX} depends only on the change $\delta t = t - t'$ and is the response of Y to a unit impulse of E. It is thereby known as the response function, or, because of the change δt, as the after-effect function. Equations (II.1) to (II.11) limit the relation of molecular statistical mechanics to macroscopic zero-THz bandshapes to the regions of linear response where the probe force field $E(t)$ is small. We shall not stray outside this region in this section. In the canonical ensemble

$$f_0 \propto \exp(-H_0/kT)$$

so that

$$\mathfrak{L}_X f_0 = f_0 \dot{X}/kT \tag{II.12}$$

Denoting means over p, q, and t by $\langle\cdot\rangle$, (II.11) becomes

$$\begin{aligned}\Psi_{YX}(t) &= \frac{1}{kT}\langle \dot{X}(0)\, Y(t)\rangle \\ &= \frac{1}{kT}\langle X(0)\, \dot{Y}(t)\rangle \\ &= -\frac{1}{kT}\frac{d}{dt}\langle X(0)\, Y(t)\rangle\end{aligned} \tag{II.13}$$

Therefore,

$$\langle Y(t)\rangle = Y_{\text{eq}} - \frac{1}{kT}\int_{-\infty}^{t} E(t')\dot{\psi}(t-t')\, dt' \tag{II.14}$$

where $\psi(t) = \langle X(0)\, Y(t)\rangle$ is a multiparticle correlation function. Equation (II.14) allows calculation of the forced response of the system from the correlation functions of the spontaneous fluctuations at equilibrium of the properties X and Y. It is of general validity for linear response, where without loss of generality $E(t)$ may be represented by a single sinusoidal component:

$$E(t) = E_0 \operatorname{Re}[\exp(i\omega t)]$$

$$Y(t) = \operatorname{Re}[\exp(i\omega t)]$$

Taking $Y_{eq}=0$, we have then the fluctuation–dissipation theorem in the form

$$\langle Y(t)\rangle = E_0 \mathrm{Re}\left[\exp(i\omega t)\int_0^\infty \exp(-i\omega t_0)\Psi_{XY}(t_0)\,dt_0\right] \qquad \text{(II.15)}$$

1. *Application to Dielectric Susceptibility*

The complex dielectric susceptibility ($X^*(\omega)$) is related to the complex permittivity ($\epsilon^*(\omega)$) by

$$X^*(\omega)=\frac{\epsilon^*(\omega)-1}{4\pi}=\frac{\langle m_z(t)\rangle}{VE_0\exp(i\omega t)} \qquad \text{(II.16)}$$

where $\langle \mathrm{m}_z(t)\rangle$ is the resultant dipole moment in the direction (z) of the probe field E. V is the volume of the system and $\langle \mathrm{m}_z(t)\rangle/V$ is the polarization.

From (II.15),

$$X^*(\omega)=\int_0^\infty \exp(-i\omega t)\Psi_z(t)\,dt \qquad \text{(II.17)}$$

where

$$\Psi_z(t)=-\frac{1}{kT}\langle \mathbf{M}_z(0)\cdot\dot{\mathbf{M}}_z(t)\rangle$$

Since in an isotropic fluid

$$\langle \mathbf{M}_z(0)\cdot\mathbf{M}_z(t)\rangle=\tfrac{1}{3}\langle \mathbf{M}(0)\cdot\mathbf{M}(t)\rangle$$

we have finally

$$\frac{\epsilon^*(\omega)-1}{\epsilon^*(\omega)+2}=-\frac{4\pi}{9kTV}\int_0^\infty \exp(-i\omega t)\dot{\psi}(t)\,dt \qquad \text{(II.18)}$$

where we have used the internal field relation

$$E_0^*(\omega)/E^*(\omega)=\frac{\epsilon^*(\omega)+2}{3}$$

to relate the external applied field E_0 to the field E which figures in the Maxwell equations for the system (regarded as a macroscopic sphere *in vacuo*). Equation (II.18) is generally valid in this case, whatever the origin of the resultant dipole moment **M** (atomic polarization, ionic, orientational, etc.), $\psi(t)$ reduces to an autocorrelation only:

1. For a fluid where the molecular fluctuations are uncorrelated.
2. Where the molecules are only slightly polarizable so that $\epsilon^*(\omega)-1$ is small.
3. Where deformation polarization is negligible.

Under these conditions:

$$\begin{aligned} C_{\mathbf{M}}(t) &= \langle \mathbf{M}(0)\cdot\mathbf{M}(t)\rangle = \sum_{ij} \langle \boldsymbol{\mu}_i(0)\cdot\boldsymbol{\mu}_j(t)\rangle \\ &= N_{\mathbf{u}}^2\langle \boldsymbol{\mu}(0)\cdot\boldsymbol{\mu}(t)\rangle \end{aligned} \tag{II.19}$$

where $\mathbf{u}=\boldsymbol{\mu}/|\boldsymbol{\mu}|$. Thus

$$\begin{aligned} \epsilon'(\omega)-1 &= \frac{4\pi N\mu^2}{3kTV}\left[1-\omega\int_0^\infty C_{\mathbf{u}}(t)\sin\omega t\,dt\right] \\ \epsilon''(\omega) &= \frac{4\pi N\mu^2}{3kTV}\int_0^\infty C_{\mathbf{u}}(t)\cos\omega t\,dt \end{aligned} \tag{II.20}$$

where $C_{\mathbf{u}}(t)$ is the dipole autocorrelation function $\langle \mathbf{u}(t)\cdot\mathbf{u}(0)\rangle$. In (II.20),

$$4\pi\mu^2/3kTV=(\epsilon_0-1)=4\pi\chi_0$$

In a pure, strongly dipolar liquid such as CH_2Cl_2, (II.19) and (II.20) are still approximately valid, but dilution has an appreciable effect on the position of $\bar{\nu}_{max}$ (Fig. 20). The micro–macro correlation theorem implies that the decay of $\langle \mathbf{M}(t)\cdot\mathbf{M}(0)\rangle/\langle \mathbf{M}(0)\cdot\mathbf{M}(0)\rangle$ should still be identical with but faster or slower than that of $\langle \mathbf{u}(t)\cdot\mathbf{u}(0)\rangle/\langle \mathbf{u}(0)\cdot\mathbf{u}(0)\rangle$. However, the relations among $\epsilon''(\omega)$, $\epsilon'(\omega)$, and $C_{\mathbf{u}}(t)$ are altered from those of (II.20). The physical origin is that each molecular dipole polarizes its neighbors, which in turn react upon it as discussed in Section I. Scaife[8] has explained why the solution to this problem can only be approximate, but Kivelson and Madden[17] have developed formal relationships and provided a systematic framework, based on the Mori continued fraction, for introducing successive approximants relating $C_{\mathbf{u}}(t)$ to $C_{\mathbf{M}}(t)$. The ratio of relaxation times of $C_{\mathbf{M}}(t)$ and $C_{\mathbf{u}}(t)$ is, in this theory,

$$\frac{\tau_{\mathbf{M}}}{\tau_{\mathbf{u}}}=\frac{1+Nf}{1+N\dot{f}} \tag{II.21}$$

Here $1+Nf$ is a factor akin to Kirkwood correlation g which measures the correlation between neighboring dipoles, and $\dot{f}$ is a dynamic orientation time correlation function for the factor $[d(\cos C_{\mathbf{m}}(t))/dt]$ of two different dipoles. The difficulty is that $\dot{f}$ is determined formally only in terms of a projected Liouville operator. Furthermore, the factor $[1+Nf]$ is sample-shape-dependent and must always be identified with a particular geometry (e.g., the sphere *in vacuo* used above). Perhaps the treatment of the internal field most in the spirit of this review is that of Sullivan and Deutch,[63] who have introduced the projection operator: $\hat{P}G=\mathbf{M}\cdot\langle\mathbf{MM}\rangle^{-1}\cdot\langle\mathbf{M}G\rangle$ independent of geometrical factors. Formally, then,

$$\frac{\epsilon^*(\omega)-1}{\epsilon(0)-1}=\mathfrak{L}_a\big(-\dot{C}_p(t)\big)$$

$$\equiv\left[1+i\omega\tau_p(\omega)\right]^{-1} \tag{II.22}$$

where

$$\tau_p^{-1}(\omega)\propto\int_0^\infty \exp(-i\omega t)k(t)\,dt$$

$$k(t)=\mu^{-2}\langle\mu_Z(1)\exp\left[i(1-\hat{P})\mathfrak{L}t\right]\dot{M}_Z\rangle$$

While $k(t)$ is a multiparticle correlation function, it is a local quantity that does not depend on sample and surroundings. The projector $\hat{P}$ removes the correlation over long ranges arising from dipole–dipole coupling or hydrodynamically.

Nevertheless, the formal nature of (II.22) remains, although a number of approximations can be incorporated within its framework (rather like Table I). We do not propose to review these numerous equations here, but proceed strategically on the basis that for moderately dipolar fluids the problem in numerical terms is considerably deflated, as demonstrated by Brot[52] and by, Greffe, Goulon, Brondeau, and Rivail.[64] Brot has considered the internal-field problem in a group of intercorrelated molecules constrained in a cavity just large enough for correlations to be negligible outside. Using the Lorentz field, he obtains

$$C_{\mathbf{u}}(t)=\frac{27kT}{4\pi^3N\mu^2}\int_0^\infty\frac{\alpha(\bar{\nu})n(\bar{\nu})\cos(2\pi\bar{\nu}ct)\,d\bar{\nu}}{\bar{\nu}^2\left(\left[\epsilon'(\bar{\nu})+2\right]^2+\epsilon''(\bar{\nu})^2\right)} \tag{II.23}$$

and with the Onsager field,

$$C_{\mathbf{u}}(t)=\frac{27kT(2\epsilon_0+\epsilon_\infty)^2}{4\pi^3N_\mu^2(\epsilon_\infty+2)^2}\int_0^\infty\frac{\alpha(\bar{\nu})n(\bar{\nu})\cos(2\pi\bar{\nu}ct)\,d\bar{\nu}}{\bar{\nu}^2\left(\left[\epsilon'(\bar{\nu})+2\epsilon_0\right]^2+\epsilon''(\bar{\nu})^2\right)} \tag{II.24}$$

Above 10 cm^{-1} (i.e., in the far infrared), $\epsilon'(\bar{\nu}) \doteqdot \epsilon_\infty$ and $\epsilon''(\bar{\nu})$ is small [although $\alpha(\bar{\nu})$ is large]. Using these approximations, both these equations reduce to

$$C_{\mathbf{u}}(t)=\frac{3kT}{4\pi^3 N_u^2}\left[\frac{9n}{(n^2+2)^2}\right]\int_0^\infty \frac{\alpha(\bar{\nu})}{\bar{\nu}^2}\cos(2\pi\bar{\nu}ct)\,d\bar{\nu} \tag{II.25}$$

where the square brackets enclose a nondispersive correction. Numerically, therefore, $C_{\mathbf{u}}(t)$ is independent of any severe internal field correction above 10 cm^{-1}. Therefore, the shift in $\bar{\nu}_{max}$ in Fig. 20, and the features of Fig. 1, may be ascribed to the molecular dynamics. $C_{\mathbf{u}}(t)$ is a real, even function of time, so that it can be expanded as a Taylor series:

$$\langle \mathbf{u}(t)\cdot\mathbf{u}(0)\rangle=\langle u^2(0)\rangle-\frac{t^2}{2!}\langle \dot{u}^2(0)\rangle+\frac{t^4}{4!}\langle \ddot{u}^2(0)\rangle-\cdots \tag{II.26}$$

with $\langle u^2(0)\rangle = 1$ by definition. For a linear molecule

$$-\ddot{C}_{\mathbf{u}}(0)=\langle \dot{u}^2(0)\rangle=2\frac{kT}{I}$$

where I is the moment of inertia, since for rotational kinetic energy $\frac{1}{2}I\omega^2$ becomes on average kT. The mean-square acceleration $\langle \ddot{u}^2(0)\rangle$ is the sum of a radial part (centripetal acceleration, owing to the fact that the vector $\mathbf{u}$ is of a fixed length), independent of interactions and having the value $8k^2T^2/I^2$; and a tangential part $\langle O(V)^2\rangle/I^2$ produced by the mean-square torque $\langle O(V)^2\rangle$ that the environment produces in the molecule. Therefore, (II.26) becomes

$$C_{\mathbf{u}}(t)=1-\frac{kT}{I}t^2+\left[\frac{k^2T^2}{3I^2}+\frac{\langle O(V)^2\rangle}{24I^2}\right]t^4-\cdots$$

Gordon[34] has calculated the first two odd moments, which appear in quantum mechanical treatments of $C_{\mathbf{u}}(t)$, as well as some quantum corrections to the even moments. Classically, we have

$$\ddot{C}_{\mathbf{u}}(t)=\frac{3kTVnc}{4\pi N^2\mu}\int_{-\infty}^{\infty}\exp(i\omega t)\alpha(\omega)\,d\omega \tag{II.27}$$

Taking (II.27) at $t=0$ produces a relation for the integrand absorption intensity per molecule for all rotational-type zero-THz absorptions:

$$\int_0^\infty \frac{V}{N}\alpha(\omega)\,d\omega=\frac{4\pi^2\mu^2}{3Inc} \tag{II.28}$$

which is Gordon's sum rule, derived classically by Brot.[52]

B. A Continued-Fraction Expansion of $C_{\mathbf{u}}(t)$

Under certain conditions we may write (I.6) with $\mathbf{A}=[\mathbf{u}]$. Therefore we have, with these restrictions,

$$\dot{\mathbf{u}}(t)=-\int_0^t K_0(t-\tau)\mathbf{u}(\tau)\,d\tau+\mathbf{f}(t)$$

where

$$K_0(t)=\langle\mathbf{f}(t)\cdot\mathbf{f}(0)\rangle/\langle\mathbf{u}(0)\cdot\mathbf{u}(0)\rangle \tag{II.29}$$

Here $\mathbf{f}(t)$ is a stochastic quantity whose correlation function defines the memory $K_0(t)$. This relation is referred to often as the second fluctuation–dissipation theorem derived initially by Kubo.[65] Equation (II.29) may be solved to give

$$\dot{C}_{\mathbf{u}}(t)=-\int_0^t K_0(t-\tau)C_{\mathbf{u}}(\tau)\,d\tau \tag{II.30}$$

which may be expanded in the Mori continued-fraction form

$$\tilde{C}_{\mathbf{u}}(s)=\frac{C_{\mathbf{u}}(0)}{s+\tilde{K}_0(s)}=\frac{C_{\mathbf{u}}(0)}{s+\cfrac{K_0(0)}{s+\tilde{K}_1(s)}}=\cdots \tag{II.31}$$

in the space (s) of Laplace transforms. Now if

$$C_{\mathbf{u}}(t)=\sum_{n=0}^{\infty}a_n\frac{t^{2n}}{(2n)!}$$

as is required by time-reversal symmetry for the classical autocorrelation function of $\mathbf{u}$, it follows that

$$K_0(t)=\sum_{n=0}^{\infty}{}^0k_n t^{2n}/(2n)!$$

$$K_1(t)=\sum_{n=0}^{\infty}{}^1k_n t^{2n}/(2n)!$$

$$K_2(t)=\sum_{n=0}^{\infty}{}^2k_n t^{2n}/(2n)! \tag{II.32}$$

Therefore, solving for ${}^0k_n$ in terms of a_N gives for $N\geq 1$:

$${}^0k_N=-a_{N+1}-\sum_{n=1}^{N}{}^0k_{N-n}a_n \tag{II.33}$$

so that the coefficients ${}^0k_N$ are known in terms of a_N and their precursors. The coefficients a_N themselves are given in terms of experimental spectral moments $(1/N_0)\int_0^\infty \omega^n \alpha(\omega)\,d\omega$, with N_0 as the molecular number density. Ideally, therefore, measurement of the experimental spectral moments is enough to define the memory functions $K_0(0),\ldots,K_N(0)$. In reality, of course, we are limited by experimental uncertainty in the high-frequency side of the absorption band for $n>1$, so that to obtain a tractable expression for $C_{\mathbf{u}}(t)$ we must minimize the number of parameters $K_0(0),\ldots,K_N(0)$ by truncating the continued fraction at a level N where the decay of $K_N(t)$ is so sharp as to be virtually a delta function compared with that of $C_{\mathbf{u}}(t)$ itself. The value of N at which this is so is not known a priori, and must therefore be arrived at by physical intuition similar to that used in deriving (I.30). We aim therefore at a compromise—avoiding too early an approximant (some of the models in Table I) and also avoiding adjustable parameters, which obscure the true quality of the fit between theory and experiment. If we can, in this way, devise a method for evaluating $K_0(0),\ldots,K_N(0)$ from the experimental zero-THz bandshape, then effectively this is providing us with information on the mean intermolecular potential energy in terms of the mean-square torque [related to $K_1(0)$ via (II.26)], the mean of the torque derivative squared [via $K_2(0)$], and so on. In the next section we discuss the recent attempts at this problem and thereafter the results in the gaslike environment. The relation of extended diffusion modes such as the m and J of Gordon to the continued fraction is explained.

1. Approximants of the Continued Fraction for $\tilde{C}_{\mathbf{u}}(s)$

Since molecular interaction is the essential element in the characteristic shift to high frequencies of $\tilde{\nu}_{\max}$ (in Fig. 1, for example) and since the intermolecular potential is basic in this context, a truncation of (II.131) is unrealistic unless the final expression for $C_{\mathbf{u}}(t)$ involves at least the equilibrium averages $K_0(0)$ and $K_1(0)$. This minimum requirement is fulfilled by the assumption that $K_2(t)$ decays as a delta function at the origin, $t=0$. This implies that

$$K_1(t)=K_1(0)\exp(-\gamma t) \tag{II.34}$$

where γ is a characteristic decay frequency to be determined. Now $\tilde{C}_{\mathbf{u}}(s)$ has the same form as (I.30):

$$\tilde{C}_{\mathbf{u}}(s)=\frac{s^2+\gamma s+K_1(0)}{s^3+\gamma s^2+(K_0(0)+K_1(0))s+\gamma K_0(0)} \tag{II.35}$$

and $C_u(t)$ is recoverable analytically. This is essentially also the form used by Kivelson and Madden in relating $C_u(t)$ to $C_M(t)$, and was first considered by Barojas, Levesque, and Quentrec[42] to explore some molecular dynamics simulations of nitrogen using the atom–atom potential of Section III. Recently, the same approximant has been considered carefully by Drawid and Halley[44] in order to calculate a time-dependent spin–spin correlation function for the classical Heisenberg model of ferromagnetism. Using the relationships devised by these authors, it may be deduced that (II.35) may also be derived from

$$K_0(t) = K_0(0)\exp(-K_1(0)t^2/2) \tag{II.36}$$

when it seems clear that

$$\gamma = \left(\frac{\pi}{2}K_1(0)\right)^{1/2} \tag{II.37}$$

Equations (II.36) and (II.37) ensure that $K_0(t)$ have the correct first two terms in Taylor's series expansion (II.32). The "mean-square-torque" term $K_1(0)$ in (II.35) may be fixed by differentiating the model absorption coefficient [essentially the Fourier–Laplace transform of (II.35)] with respect to ω, whereupon $K_1(0)$ can be obtained from the measured peak frequency $\bar{\nu}_{max}$. Since $K_0(0)$ is the mean-square angular velocity, (II.35) may be used for reproducing experimental data without recourse to least-mean-squares iteration. Alternatively, it is sometimes convenient to iterate on γ and $K_1(0)$ so that (II.37) may be tested out empirically.

It is possible also to take the next approximant represented by

$$K_2(t) = K_2(0)\exp\left[-(\pi/2K_1(0))^{1/2}t\right] \tag{II.38}$$

and to estimate both $K_1(0)$ and $K_2(0)$ without recourse to least-mean-squares iteration by differentiating both $\epsilon''(\omega)$ and $\alpha(\omega)$ with respect to ω. Since $\epsilon''(\omega)$ peaks at a much lower frequency than $\alpha(\omega)$, this leads to two simultaneous equations for $K_1(0)$ and $K_2(0)$.

The absorption coefficient may be recovered from $\tilde{C}_u(s)$ using equations such as (II.23) to (II.25). The autocorrelation function is a sum of three complex exponentials, which, when the denominator discriminant in (II.35) is negative, becomes

$$C_u(t) = \left(\frac{\cos\beta t}{1+\Gamma} + \frac{1}{\beta}\left(\frac{\alpha_1 + \Gamma\alpha_2}{1+\Gamma}\right)\sin\beta t\right)\exp(-\alpha_1 t) + \frac{\Gamma}{1+\Gamma}\exp(-\alpha_2 t) \tag{II.39}$$

where

$$\Gamma = \frac{-2\alpha_1(\alpha_1^2 + \beta^2)}{\alpha_2(3\alpha_1^2 - \beta^2 - \alpha_2^2)}$$

with

$$\alpha_2 = -\frac{1}{2}(s_1 + s_2) + \frac{\gamma}{3} \qquad \alpha_1 = \frac{1}{2}(s_1 + s_2) + \frac{\gamma}{3} \qquad \beta = \frac{\sqrt{3}}{2}(s_1 - s_2)$$

The parameters s_1 and s_2 are defined by

$$s_1 = \left[-\frac{B}{2} + \left(\frac{A^3}{27} + \frac{B^2}{4} \right)^{1/2} \right]^{1/3}$$

$$s_2 = \left[-\frac{B}{2} - \left(\frac{A^3}{27} + \frac{B^2}{4} \right)^{1/2} \right]^{1/3}$$

where

$$A = K_0(0) + K_1(0) - \frac{\gamma^2}{3} \qquad B = \frac{\gamma}{3}\left(\frac{2\gamma^2}{9} + 2K_0(0) - K_1(0) \right)$$

The Taylor series of (II.39) is even to the fourth power of time, but has a term in t^5, and all odd terms thereafter are also nonzero. The form of (II.39) although cumbersome is useful since it is derivable from (I.30) *and* also (II.53). It represents the autocorrelation functions $C_{\mathbf{u}}(t)$ and $C_{\dot{\theta}}(t)$ therefore for at least four seemingly unrelated models of the fluid state.

2. *Molecular Orientations in the Compressed Gas*

With only the mean-square torque properly defined [through $K_1(0)$] it is interesting to evaluate the usefulness of (II.35) in a compressed gas of dipolar molecules before proceeding to the liquid state as embodied in, for example, pure benzonitrile at ambient temperature. In so doing we may observe, by compressing the gas into the liquid, how changes of bandshape and intensity result from the additional constraints which may be imposed on rototranslational freedom on going from one phase to the next.

We have chosen for experimental convenience the symmetric top CH_3Cl and the asymmetric top CH_2CF_2. In the limit of free rotation, the *infinite* continued fraction reduces to the classical Kummer function for $C_{\mathbf{u}}(t)$, but the approximant embodied in (II.35) does not, as shown in Fig. 28 with

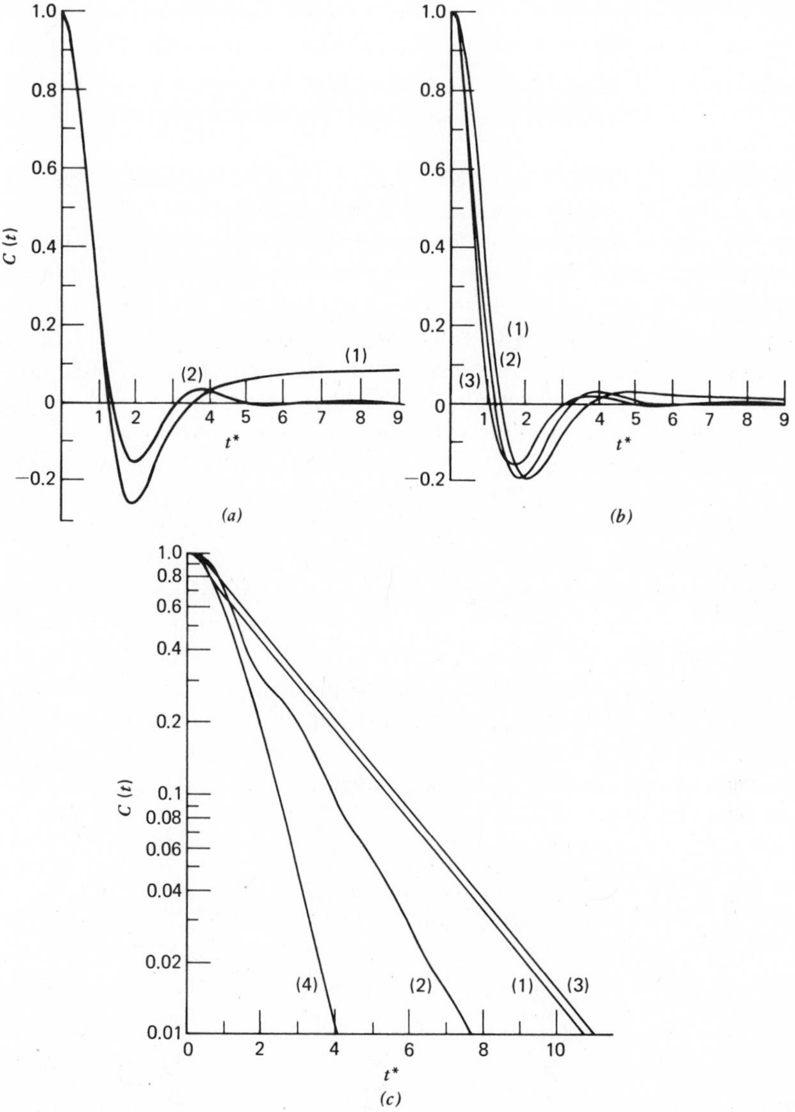

Fig. 28. (*a*) $t^* = (kT/I_B)^{1/2}t$. (1) True free-rotor orientational autocorrelation function for CH_3Cl at 296°K; (2) best fit of three variable theory to the free-rotor $J \to J+1$ contour. (*b*) (1) $C_u(t)$ for CH_3Cl(g) at 5.8 bars, 296°K; (2) *J*-diffusion model with $\tau^* = (kT/I_B)^{1/2}\tau = 4.0$ as the reduced time between collisions; (3) $C_u(t)$ for CH_3Cl + ethane(*g*) at 33.5 bars, 296°K. (*c*) (1) $C_u(t)$ for CH_3Cl (liquid), at 296°K; (2) $C_u(t)$ for CH_3Cl + ethane(*l*) at 296°K; (3) *J*-diffusion model with $\tau^* = 0.2$; (4) *J*-diffusion model with $\tau^* = 0.5$. [Reproduced by permission from *J. Chem. Soc. Faraday Trans. 2*, **72**, 1907 (1976).]

$\langle O(V)^2 \rangle = 0$ and $K_1(0) = 4kT/I_B$, where I_B is the moment of inertia about the dipole axis in CH_3Cl. In the limit of vanishing torque, therefore, (II.35) reduces to the Kummer function for classical free rotation at short times only, when the free-rotor Taylor series is approximated adequately by its first few terms.

A typical fit[66] of $\alpha(\omega)$ from (II.35) to the CH_3Cl *compressed* gas[67] is shown in Fig. 28 together with that for a gaseous mixture of CH_3Cl in ethylene (to remove dipole–dipole coupling). The torque term $K_1(0)$ in the latter is almost twice that needed to fit the free-rotor contour, as can be seen in Table IV.

The fit for pure liquid CH_3Cl (Fig. 28*c*) is good, with $K_1(0)$ considerably increased—reflecting the increase in the apparent root-mean-square torque. In a mixture with ethane in the liquid state, $K_1(0)$ needs to be surprisingly small for best fit since the observed band is surprisingly narrow, with the half-width reflected in the low value of γ needed (Table IV). Within the limitations of the early approximant used for $C_{\mathbf{u}}(t)$ we may take this as a rough indication of the less severe angular constraints in the ethane environment. This is reflected in the form of $C_{\mathbf{u}}(t)$ (Fig. 28*c*), which becomes exponential at long times only, in contrast to that of the pure liquid, which exhibits logarithmic decay very quickly, suggestive of small-step rotational diffusion. Below $t^* = (kT/I_B)^{1/2} t = 1$, $C_{\mathbf{u}}(t)$ for liquid CH_3Cl exhibits oscillatory behavior. This shows up in the far infrared as the Poley absorption (i.e., the high-frequency extension of the microwave absorption). This short-time behavior of $C_{\mathbf{u}}(t)$ reflects the torsional oscillatory, or librational, motion of **u** about an axis which is diffusing through the fluid

TABLE IV
Parameters γ and $K_1(0)$ Least Mean Squares Fitted to Compressed Gaseous Data for CH_3Cl and CH_2CF_2[a]

System	Pressure (bars)	$\left(\frac{2kT}{I_B}\right)K_1(0)$	$\left(\frac{2kT}{I_B}\right)^{1/2}\gamma$	$10^{40}I_B(\text{g}\cdot\text{cm}^2)$
CH_3Cl	5.8	3.61	3.74	63.12
CH_3Cl/ethane	39.3	4.21	3.74	63.12
CH_3Cl/ethane(l)	—	4.00	1.90	63.12
$CH_3Cl(l)$	—	14.39	4.29	63.12
CF_2CH_2	35.2	3.67	3.86	80.45
$CF_2CH_2(l)$	—	6.15	3.91	80.45
CH_3Cl ($J \to J+1$ contour)	—	2.51	2.54	63.12

[a]Reproduced by permission from *J. Chem. Soc. Faraday Trans*. 2, **72**, 1907 (1976).

(i.e., performing itinerant oscillation–libration). As Fig. 28*c* shows, the *J*-diffusion model for the symmetric top is incapable of reproducing this oscillating form for $C_u(t)$ since essentially the concept of libration is as ill-defined in this theory as is the mean-square torque. It is too early an approximant of the continued fraction.

In the asymmetric rotor CH_2CF_2 (Fig. 28), the essential difference between the compressed gas and liquid is seen more clearly in terms of $C_u(t)$. This is damped in the liquid and never becomes negative, as in the compressed gas.

If, then, we are prepared to regard γ and $K_1(0)$ for convenience as variable and to accept the limitations inherent in (II.35), the truncation procedure outlined in Section II.B.1 may be used to elucidate the compressed gas molecular dynamics, and comparison with the liquid may be made quantitatively. The stricter approach is to evaluate γ and $K_1(0)$ with no least-mean-squares fitting. This would soon bring out the deficiencies of (II.35), as is seen in the case of CH_2Cl_2 in Fig. 28. In fact, Table IV shows clearly that (II.137) is generally not obeyed, so that least-mean-squares fitting tends to mask the theoretical difficiencies in favor of empiricism.

3. *Gaslike Behavior in the Liquid State: Water Free of Hydrogen Bonding*[68]

This is discernible in systems such as hydrogen halide/SF_6 mixtures, and ammonia dissolved in the same liquid solvent: for here the broadened remnants of the $J \rightarrow J+1$ rotational contour clearly appear. These systems will be considered presently, but the motions of free, unbound water molecules in organic solvents are equally extraordinary, since $C_u(t)$ is very markedly nonexponential, with correlation times typically ca. 0.1 psec. The loss curves $\epsilon''(\omega)$ are asymmetric and peak at frequencies almost in the mid infrared, in great contrast to the "ordinary" behavior exhibited in the microwave region by the pure liquid. In producing these curves the empirical approach with (II.35) has proven useful since it is found possible, by varying γ and $K_1(0)$, to reproduce closely the far-infrared absorption data of Pardoe and Gebbie[69] for *very dilute* solutions of water in cyclohexane, carbon tetrachloride, and benzene. The results are illustrated in Fig. 29 along with $C_u(t)$, calculated therefrom. Table V shows the refined γ and $K_1(0)$ values.

The motion of the water molecules is gaslike in the sense that $C_u(t)$ is gaslike, displaying a negative region in cyclohexane, for example, at $\sim$0.11 psec. Therefore, there is a probability that a majority of molecules will have swung through greater than $\pi/2$ of the solid angle at times greater than 0.11 psec. The overall correlation time is 0.10 ps, much shorter than

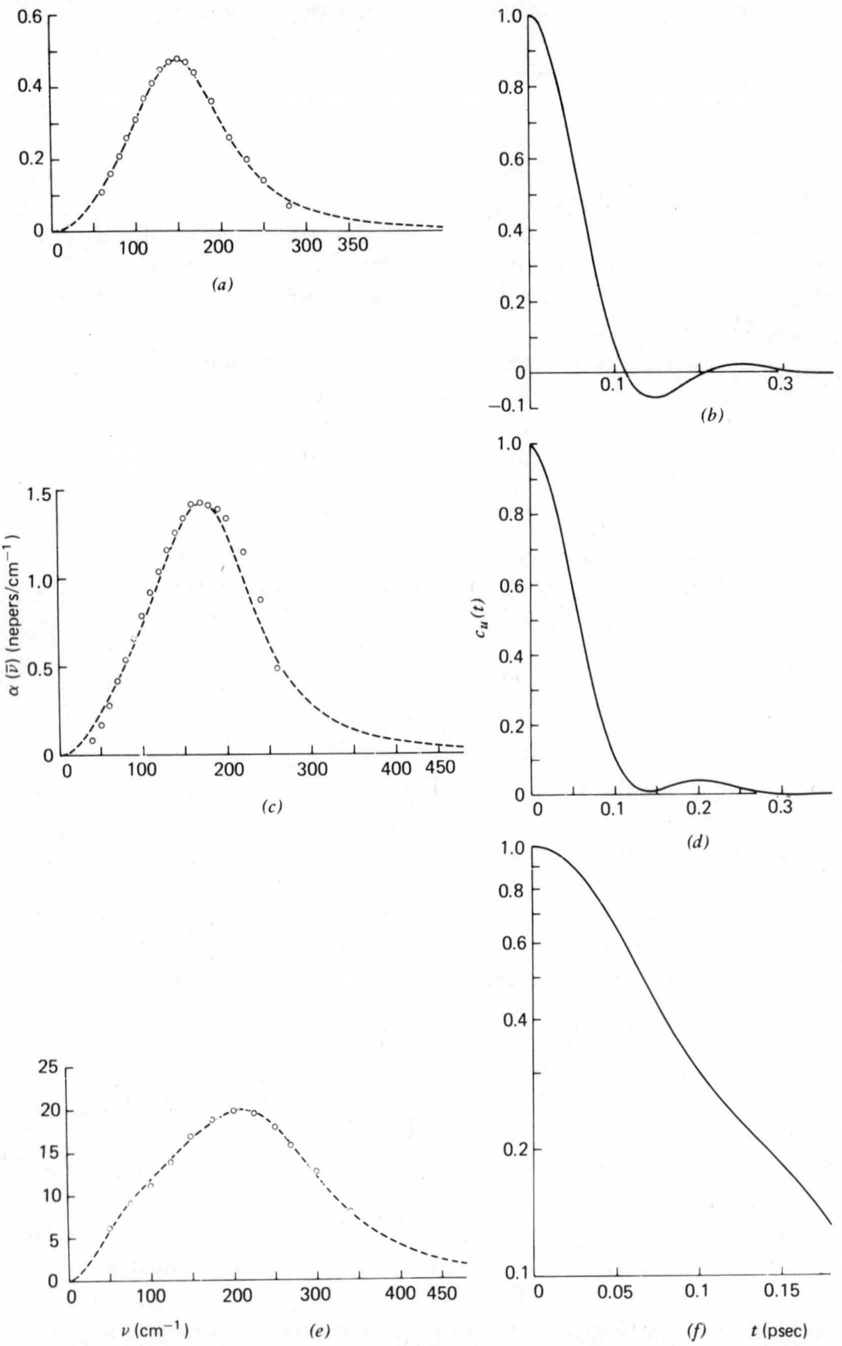

Fig. 29. (*a*) Absorption of a 0.011% w/w solution of water in cyclohexane at 296°K, corrected for solvent; (----), Mori theory best fit. (*b*) $C_u(t)$ for the absorption of Fig. 29*a*. (*c*) Absorption of a 0.01% w/w solution of water in CCl_4 at 300°K, corrected for solvent; (----), Mori theory best fit. (*d*) $C_u(t)$ for H_2O+CCl_4. (*e*) Absorption of a 0.06% w/w solution of water in benzene at 300°K, corrected for solvent; (----), Mori theory best fit. (*f*) $C_u(t)$ for H_2O+benzene. [Reproduced by permission from *J. Chem. Soc. Faraday Trans. 2*, **72**, 2143 (1976).]

TABLE V
Parameters γ and $K_1(0)$ for Solutions of Water[a]

Solution	T	$(2kT/I_B)^{-1}K_0(0)$	$(2kT/I_B)^{-1}K_1(0)$	$\left(\frac{2kT}{I_B}\right)^{-1/2}\gamma$
H_2O + cyclohexane (0.011% w/w)	296	1.11	3.00	2.58
$H_2O + CCl_4$ (0.01% w/w)	300	1.18	3.39	2.37
H_2O + benzene (0.06% w/w)	300	1.16	6.19	2.82

[a]Reproduced by permission from *J. Chem. Soc. Faraday Trans.* 2, **72** (1976).

the Debye relaxation time in pure liquid water [the inverse of $\omega_{max}(\epsilon'')$], the epitome of hydrogen bonding.[70] The loss curve in cyclohexane is *asymmetric on the log scale* and peaks at 140 cm^{-1}. The oscillations in $C_{\mathbf{u}}(t)$ are damped in benzene but not completely so (Fig. 29*f*).

At this stage it is convenient to describe the quantized rotation observable in hydrogen halides such as HBr in liquid SF_6 in terms of the extended diffusion models of gaslike molecular dynamics developed over the last 15 years in numerous articles and reviews.[53] They are, of course, discretely classical in the sense that the $J \rightarrow J+1$ lines are broadened by classical statistical mechanics. Lindenberg and Cukier have generalized the concepts involved to an extent where direct comparison with the continued fraction is possible for higher-order approximants than those used in the well-known *m*- and *J*-diffusion models.[45]

With reference to Table I, the latter is an approximant defined by the relation

$$K_0(t) = K_{FR}(t)\exp(-|t|/\tau) \tag{II.40}$$

where $K(t)$ is the overall memory function and $K_{FR}(t)$ that associated with an ensemble of free rotors. τ is the mean time between elastic collisions which randomize the molecular angular momentum (**J**) in magnitude and direction. The correlation function associated with (II.40) is then the Laplace transform of

$$\begin{aligned}\tilde{C}_{\mathbf{u}}(s) &= \frac{\tilde{C}_{FR}(s+1/\tau)}{1-\tau^{-1}\tilde{C}_{FR}(s+1/\tau)}\\ &= \sum_{n=0}^{\infty}\frac{1}{\tau^n}\tilde{C}_{FR}^{n+1}(s+1/\tau)\end{aligned} \tag{II.41}$$

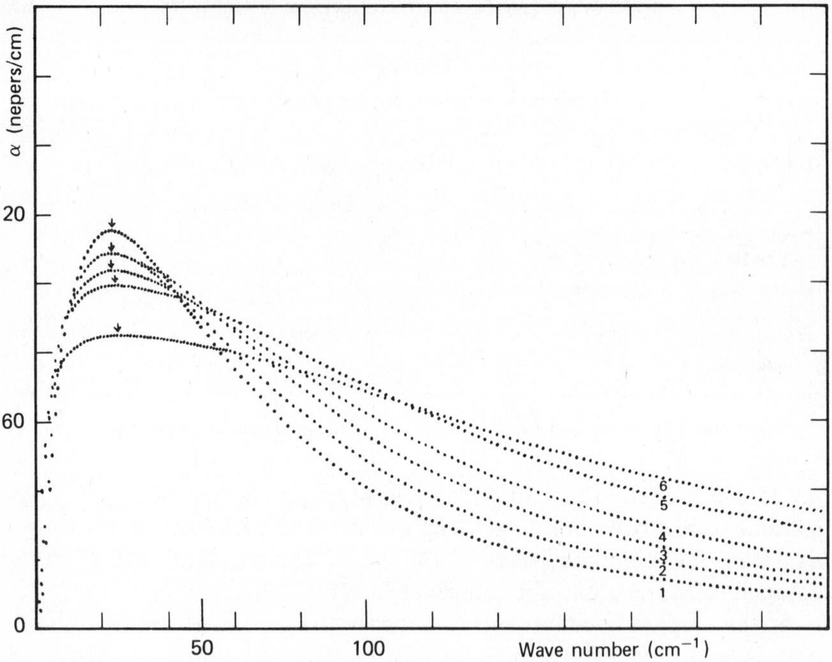

(a)

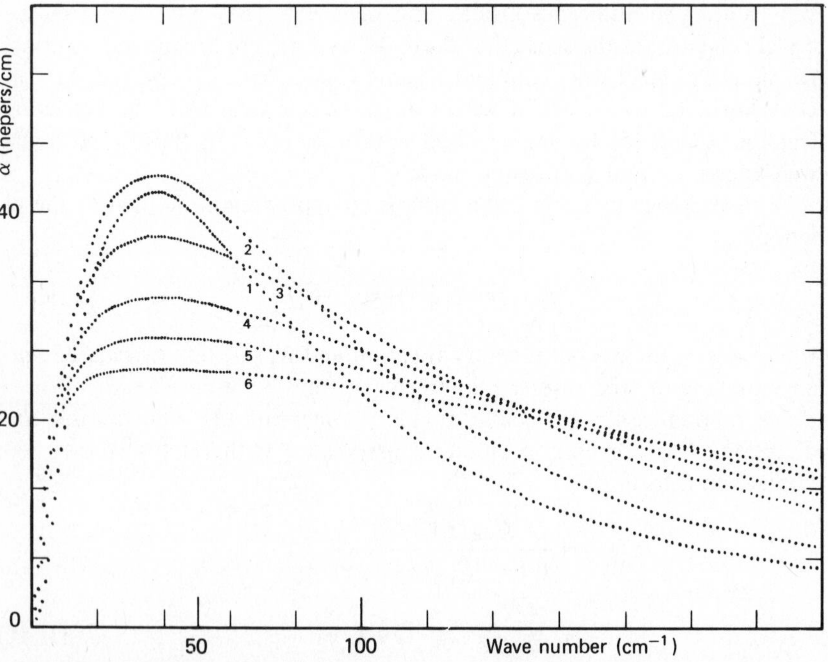

Fig. 30. Type of variation with temperature of the symmetric-top Langevin equation for (*a*) CHF_3 with (1)–(6) decreasing τ; (*b*) CH_3F with (1)–(6) decreasing τ. [Reproduced by permission from *J. Chim. Phys.*, **75**, 527 (1978).]

which is identical with the form devised by Gordon for linear molecules. It has a Taylor expansion with a term proportional to t^3, since the concept of elastic collisions implies that the torque is singular at each collision and therefore has no definable mean. The spectral consequences are illustrated in Fig. 30 for the methyl halides considered by Gerschel.[3] Experimentally, $\bar{\nu}_{max}$ moves to higher frequencies, but theoretically it remains at the frequency ω_J corresponding to the root-mean-square angular velocity of the N-particle ensemble, the high-frequency return to transparency being far too gradual. At one stage in the derivation of $C_{\mathbf{u}}(t)$ for the m diffusion model, the assumption is made that

$$K_0(t) = K_0(0)\exp(-\gamma_1 t) \tag{II.42}$$

so that again the mean-square torque is undefined. The zero-THz band can be used to evaluate this model, which is also not capable of moving $\bar{\nu}_{max}$ to higher frequencies. In view of this straightforward procedure it is surprising that the same indications have been so long in forthcoming with techniques such as Rayleigh scattering, infrared and Raman line broadening, and neutron scattering.[71]

In fact, the conspicuous experimental shift in $\bar{\nu}_{max}$ (along the gas–liquid coexistence curve of Fig. 1, for example) demands the assumption of correlated collisions. This has been the basis for some modifications to the original m and J diffusion concepts whereby it becomes possible to vary the amount of angular momentum transferred during a collision and the statistics governing the distributions of intervals in between. However, the modifications still result in the aphysical t^3 of the Taylor expansion and the models are still approximants of the degree represented by (II.40) and (II.42). Denoting by $\cos(\gamma(J))$ the average cosine of the angular momentum vector, it is found[16] that the Poley absorption is ill-matched unless collisions are assumed statistically correlated and that the angular momentum is such that $\cos[\gamma(\mathrm{J})] < 0$.

Chandler[33] has discussed the translational and rotational diffusion of rough hard spheres starting from the Liouville equation, and has found that the derivation from Poisson behavior in dense systems means that the first term in the *cumulant* expansion of the memory kernel is insufficient to describe the dynamics of the system. This happens when the rotational motion couples strongly to collective modes in the system. It may be argued that the amount of rotational energy and momentum transfer during a collision roughly corresponds, for dense systems, to the average change in these quantities during a time of the order of the interval between collisions. Correlated collisions imply an oscillatory angular momentum correlation function (see Fig. 14 to 19) which is found in Section III with molecular dynamics systems of strongly anisotropic molecules. Restricted to binary collision operators, the Chandler formalism reduces to the original J-diffusion model represented by (II.40), where the angular momentum is randomized onto a Poisson distribution. Frenkel, Wegdam, and van der Elsken[16] have shifted $\bar{\nu}_{max}$ in $CH_3CN(l)$ in the context of this model by invoking a statistical distribution which is basically Poisson but with an additional "pseudophonon" peak adjustable for best fit, with the zero-THz data. However, the intermolecular potential is still ill-defined (in terms of hard-sphere collisions). The usefulness of molecular dynamics simulations in this context has been demonstrated by O'Dell and Berne,[43] who have collected data on rough-sphere ensembles which meet almost all the criteria dictated by Gordon for m and J diffusion. However there is an enormous disparity between prediction and observation of correlation functions from the numerical solution of the equation of motion. Part of the discrepancy seems to be in fact that collisions in the molecular frame can, in reality, randomize only two of the three components of momentum. A similar numerical simulation of the planar itinerant oscillator produces results much more in accord with the analytical analysis, and is described in Section III.

However, some modifications to the amount of momentum and energy transfer introduced by Frenkel and Wegdam[72] are useful in describing the $HBr/SF_6(l)$ spectrum and correlation functions of Figs. 4 and 37, where the $J \rightarrow J+1$ lines are preferentially broadened (those on the low-frequency side more so than the others). Using this phenomenon, Frenkel, Wegdam, and van der Elsken[16] have obtained much valuable information on the compressed-gas and liquid-phase molecular dynamics of the hydrogen halides. Some computer simulations by Frenkel[73] complement these data. In Fig. 4 we have used an approximant defined by

$$K_1(t) = \beta K_{1FR}(t)\exp(-|t|/\tau) \tag{II.43}$$

and introduced by Bliot and Constant[40] who describe the parameters β and τ as purely stochastic in origin. By using a set of $J \rightarrow J+1$ lines for the free-rotor ensemble $C_{FR}(t)$, the variation of the computed $HBr/SF_6(l)$ spectrum with β is demonstrated in Fig. 4. No preferential broadening is possible with this model,[41] but $\bar{\nu}_{max}$ is shifted by very small changes in β.

C. Continued-Fraction Expansion and the Dense Liquid State

By "dense liquid" we mean liquids such as benzonitrile or *tert*-butyl chloride which may be classed as ordinary solvents whose critical temperatures lie well above ambient. These are characterized in the far infrared by the large difference observable between the Poley frequency maximum $\bar{\nu}_{max}$ and the peak of the $J \rightarrow J+1$ set of free-rotor transitions, which is classically its root-mean-square angular velocity. In this section we will describe how the absorption in this phase may be used to discriminate between some of the currently popular models of the fluid state which have been classified in Table I as approximants of the Mori continued fraction. It is therefore logical to commence with a description of the work carried out recently on the approximant where the first memory function is a delta function at $t=0$. In this case the (I.6) reduces to the Langevin equation for rotational Brownian motion:

$$\dot{\mathbf{J}}(t)+\zeta_{\mathrm{R}}\cdot\mathbf{J}(t)=\Gamma(t) \tag{II.44}$$

where **J** is the angular momentum, ζ_R is the rotational friction tensor, and $\boldsymbol{\Gamma}(t)$ is a random torque which is stationary and Gaussian, having an infinitely short correlation time, so that $\langle\boldsymbol{\Gamma}(t)\cdot\boldsymbol{\Gamma}(0)\rangle=2D_{\mathbf{J}}\delta(t)$, where $\delta(t)$ is the delta function in time, and $D_{\mathbf{J}}$ the rotational diffusion coefficient. No correlation exists between **J** and **Γ**. In a coordinate frame defined by the principal axes of inertia, we have

$$I\cdot\dot{\omega}+\omega\mathrm{x}(I\cdot\omega)+\zeta_R\cdot(I\cdot\omega)=\boldsymbol{\Gamma}(t) \tag{II.45}$$

The solution of this nonlinear stochastic differential equation is simplified by neglecting the molecular inertia. For the asymmetric top, Perrin's equations then follow, where in the absence of an internal field correction,

$$\frac{\epsilon^*(\omega)-\epsilon_\infty}{\epsilon_0-\epsilon_\infty}=\frac{1}{\mu^2}\left[\frac{\mu_x^2(D_y+D_z)}{i\omega+D_y+D_z}+\frac{\mu_y^2(D_x+D_z)}{i\omega+D_x+D_z}+\frac{\mu_z^2(D_x+D_y)}{i\omega+D_x+D_y}\right] \tag{II.46}$$

where $D_i=kT/I_i\beta_i$, and $\mu^2=\mu_x^2+\mu_y^2+\mu_z^2$. Here μ_i is the component of the permanent dipole along the principal axis denoted by i, and β_i is the friction coefficient. By substituting $D=D_i$ in (II.46) the result simplifies further to the particular case of the rotational Brownian motion of the inertialess spherical top as treated by Debye. Converting $\epsilon''(\omega)$ to power absorption gives the plateau value:

$$\alpha(\omega)\rightarrow(\epsilon_0-\epsilon_\infty)/nc\tau_0 \tag{II.47}$$

as $\omega\rightarrow\infty$, where τ_D is the Debye relaxation time, and n is defined by $\epsilon_\infty^{1/2}$. Recently, McConnell and others,[22] and Morita,[21] have independently solved (II.45) with a full consideration of inertial effects. To give an idea of the complexity of this undertaking, we present some of their final equations below and evaluate them against some of the most accurate and comprehensive zero-THz data available.

For certain geometries, such as those of the needle and sphere with embedded dipoles, the use by McConnell of Bogoliubov–Mitropolsky matrices yields the following equations, for the complex polarizability

$\alpha^*(\omega)$ [not to be confused with $\alpha(\omega)$, the power absorption coefficient]. For the sphere, for example,

$$\frac{\alpha'(\omega)}{\alpha(0)} =$$

$$1-(F(0))^{-1}\Bigg[\frac{\omega'^2}{G'^2+\omega'^2}-\frac{2\gamma\omega'^2}{(1+G')^2+\omega'^2}$$

$$+\gamma^2\left\{\frac{3/2\omega'^2}{(2+G'^2)+\omega'^2}-\frac{4(1+G')\omega'^2}{[(1+G')^2+\omega'^2]^2}-\frac{2\omega'^2}{(1+G')^2+\omega'^2}\right\}$$

$$-\gamma^3\left\{\frac{6(1+G')^2\omega'^2-2\omega'^4}{[(1+G')^2+\omega'^2]^3}+\frac{8(1+G')\omega'^2}{[(1+G')^2+\omega'^2]^3}+\frac{2\omega'^2}{(1+G')^2+\omega'^2}\right.$$

$$\left.+\frac{3(2+G')\omega'^2}{[(2+G')^2+\omega'^2]^2}+\frac{(17/9)\omega'^2}{(3+G')^2+\omega'^2}\right\}+\cdots\Bigg] \qquad \text{(II.48)}$$

$$\frac{\alpha''(\omega)}{\alpha(0)} =$$

$$(F(0))^{-1}\Bigg[\frac{G'\omega'}{G'^2+\omega'^2}-\frac{2\gamma(1+G')\omega'}{(1+G')^2+\omega'^2}$$

$$-\gamma^2\left\{\frac{2[(1+G')^2-\omega'^2]\omega'}{[(1+G')^2+\omega'^2]^2}+\frac{2(1+G')\omega'}{(1+G')^2+\omega'^2}-\frac{(3+(3/2)G')\omega'}{(2+G')^2+\omega'^2}\right\}$$

$$-\gamma^3\left\{\frac{2[(1+G')^3\omega'-3(1+G')\omega']^2}{[(1+G')^2+\omega'^2]^3}+\frac{4[(1+G')^2-\omega'^2]\omega'}{[(1+G')^2+\omega'^2]^2}\right.$$

$$+\frac{2(1+G')\omega'}{(1+G')^2+\omega'^2}+\frac{(3/2)[(2+G')^2-\omega'^2]\omega'}{[(2+G')^2+\omega'^2]^2}$$

$$\left.+\frac{\omega'\left(\frac{17}{3}+\left(\frac{17}{9}\right)G'\right)}{(3+G')^2+\omega'^2}\right\}+\cdots\Bigg] \qquad \text{(II.49)}$$

where

$$\gamma = kT/I\beta^2 \qquad \omega' = \omega/\beta$$

$$G' = 2\gamma + \gamma^2 + (7/6)\gamma^3 + (25/18)\gamma^4 + \cdots \tag{II.50}$$

$$(F(0))^{-1} = 1 + 2\gamma + (9/2)\gamma^2 + (125/9)\gamma^3 + \cdots \tag{II.51}$$

There are similar expressions for the needle (the dipole axis) reorienting in space. For $\omega \to \infty$ both expressions (for the sphere and needle) reduce to

$$\frac{\alpha^*(\omega)}{\alpha(0)} \to \frac{2\gamma}{\omega'^2} - \frac{2i\gamma}{\omega'^3} \tag{II.52}$$

in agreement with the work of Sack on the same problem. McConnell and others have solved (II.45) for the asymmetric top, and Morita the corresponding Eulerian equations of motion. We shall show presently that their results are entirely consistent numerically. However, for all geometries the memoryless (II.44) cannot reproduce the zero-THz Poley absorption, but merely causes a slow high-frequency return to transparency of the plateau value [e.g., (II.47)]. Nevertheless, the powerful theoretical methods used in solving (II.44) should be extended to the general (I.6). For the asymmetric top, Morita shows that

$$\frac{\epsilon^* - \epsilon_0}{\epsilon_0 - \epsilon_\infty} = 1 - s \sum_{i = x,y,z} \frac{m_i^2}{\mu^2} A^{(i)}(s)$$

where $s = i\omega$, and $\mu^2 = M_x^2 + M_y^2 + M_z^2$, in which M_i is the permanent dipole moment along the principal axis denoted by i, and β_i the corresponding frictional component. $A^{(i)}(s)$ is defined as follows:

$$A^{(i)}(s) = \left[s + \frac{p_2^{(i)}(q_1^{(i)} + r_1^{(i)}) - p_1^{(i)}(q_2^{(i)} + r_2^{(i)})}{q_1^{(i)} r_2^{(i)} - r_1^{(i)} q_2^{(i)}} \right]^{-1} \tag{II.53}$$

where

$$p_1^{(i)} = -4kT/I_j(s + 2\beta_j)(s + \beta_i + \beta_k)(s + \beta_i + \beta_j)$$

$$p_2^{(i)} = 4kT/I_k(s + 2\beta_k)(s + \beta_i + \beta_j)(s + \beta_i + \beta_k)$$

$$q_1^{(i)} = 4(s + \beta_j)(s + 2\beta_j)(s + \beta_i + \beta_j)(s + \beta_i + \beta_k)$$
$$+ 8kT/I_j(s + \beta_i + \beta_j)(s + \beta_i + \beta_k) + \frac{4kT}{I_i}(s + 2\beta_j)(s + \beta_i + \beta_k)$$
$$+ 4kT/(I_i I_j I_k)\left[(I_i - I_k)^2 (s + 2\beta_j)(s + \beta_i + \beta_j) \right]$$

$$q_2^{(i)} = \frac{4kT}{I_i I_k}(s+2\beta_k)\left[(I_k - I_i)(s+\beta_i+\beta_j) + (I_j - I_i)(s+\beta_i+\beta_k)\right]$$

$$r_1^{(i)} = \frac{4kT}{I_1 I_2}(s+2\beta_j)\left[(I_j - I_i)(s+\beta_i+\beta_j) + (I_k - I_i)(s+\beta_i+\beta_k)\right]$$

$$\begin{aligned} r_2^{(i)} = {} & 4(s+\beta_k)(s+2\beta_k)(s+\beta_i+\beta_j)(s+\beta_i+\beta_k) \\ & + \frac{8kT}{I_k}(s+\beta_i+\beta_j)(s+\beta_i+\beta_k) + \frac{4kT}{I_i}(s+2\beta_k)(s+\beta_i+\beta_j) \\ & + \frac{4kT}{I_i I_j I_k}(I_j - I_i)^2(s+2\beta_k)(s+\beta_i+\beta_k) \end{aligned}$$

In the foregoing, $(i=x, j=y, k=z)$, $(i=y, j=z, k=x)$, and $(i=z, j=x, k=y)$ in cyclic permutation. We denote by I_x, I_y, and I_z the principal moments of inertia. This equation reduces to Perrin's equation when inertial effects are neglected. $\epsilon^*(\omega)$ is the complex permittivity. Internal field effects are neglected.

The solution of McConnell and others is, on the other hand,

$$\alpha^*(\omega) = \frac{1}{3kT}\sum_{x,y,z}\left[\frac{D_y(D_y+D_z+\beta_y)}{(D_y+D_z+i\omega)(D_y+D_z+\beta_y+i\omega)} + \frac{D_z(D_y+D_z+\beta_z)}{(D_y+D_z+i\omega)(D_y+D_z+\beta_z+i\omega)}\right]\mu_x^2 \qquad \text{(II.54)}$$

Here $D_x = kT/I_x\beta_x$, and so on. The frictional couples with respect to the rotating principal axes of inertia are $I_x\beta_x\omega_x$, $I_y\beta_y\omega_y$, and $I_z\beta_z\omega_z$. Equations (II.53) and (II.54) produce virtually identical numerical results for all β_x, β_y, and β_z of interest.

For the symmetric top, where the components μ_x and μ_y are zero, the Fokker–Planck–Kramers equation for the probability density function $p(\omega_x, \omega_y, \omega_z, \theta, \phi, \psi; t)$ in angular velocity/Euler space yields the Laplace transform of the dipole autocorrelation function $C_{\mathbf{u}}(t) = \langle \boldsymbol{\mu}_z(t) \cdot \boldsymbol{\mu}_z(0) \rangle$ according to Morita as

$$\tilde{C}_\mu(s) = \cfrac{\mu_z^2/3}{s + \cfrac{2kT/I_x}{s+\beta_x+\cfrac{2kT/I_x}{s+2\beta_x+\cfrac{4kT/I_x}{s+3\beta_x}} + \cfrac{(kT/I_x)(I_z/I_x)}{s+\beta_x+\beta_z+\cfrac{(2kT/I_x)(I_z/I_x)}{s+\beta_x+2\beta_z+\cfrac{2kT/I_x}{s+2\beta_z}}}}} \qquad \text{(II.55)}$$

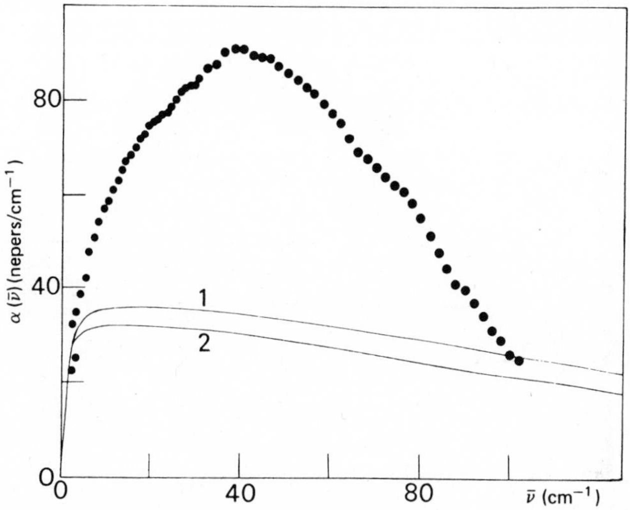

(a)

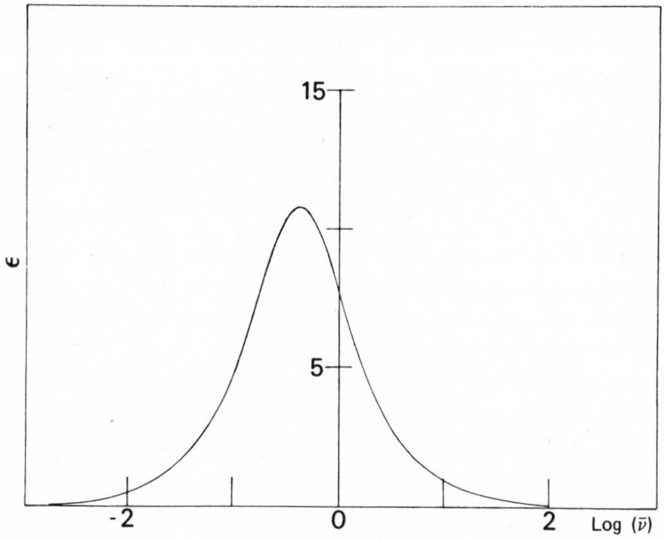

(b)

Fig. 31. (*a*) ●, Absorption of 2-chloro-2-nitropropane(*l*) at 293°K, taken with a P.O. polarizing interferometer. (—), (1) Spherical-top Langevin equation; (2) Asymmetric-top Langevin equation. (*b*) Calculated loss curve for 2-chloro-2-nitropropane, using (II.49) or (II.54). Ordinates: (*a*) α (nepers/cm), (*b*) $\epsilon''(\bar{\nu})$; abscissas: (*a*) $\bar{\nu}$ (cm^{-1}), (*b*) $\log_{10}(\bar{\nu})$ (cm^{-1}).

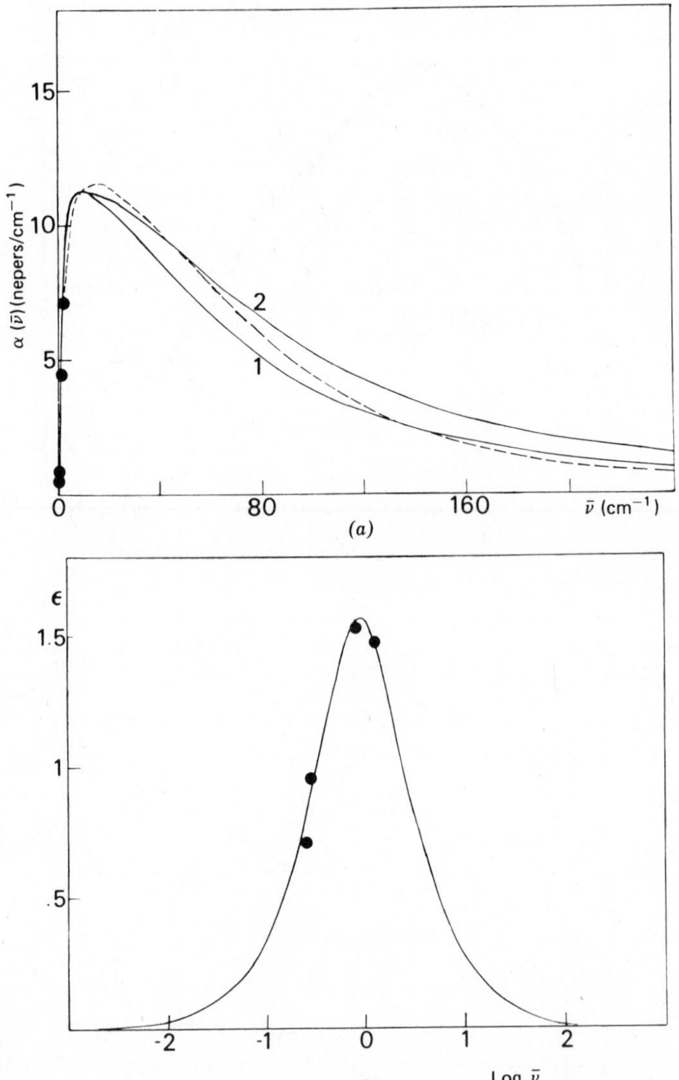

Fig. 32. Key as in Fig. 31. Liquid fluorobenzene. (----), (3) Mori theory applied to $C_u(t)$, best fit to the microwave data shown in Fig. 32b as ●.

If $\omega_i(t)$ $(i = x,y,z)$ is the angular velocity about the principal axis labelled i and I_i is the moment of inertia, then $I_i \beta_i \omega_i(t)$ is the damping torque.

Numerically (Figs. 30 to 35), (II.55) reduces, in the appropriate limits, to the sphere and needle expressions as given by McConnell [e.g., (II.48) and (II.49)]. In evaluating (II.48) to (II.55) against zero-THz data we proceed as follows with the friction parameters. The only parameter of (II.50) and

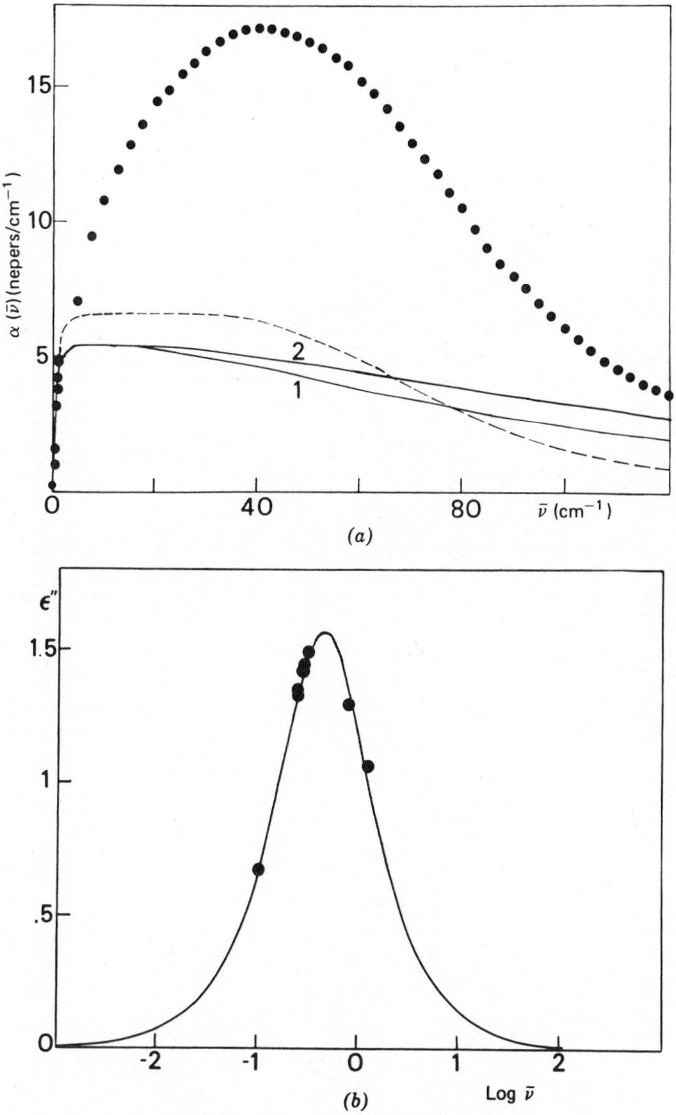

Fig. 33. Key as in Fig. 32. Chlorobenzene, iodobenzene, and nitrobenzene.

those for the needle, is β, which is directly observable from the loss peak $\epsilon''_{max}(\omega)$. When the components β_x, β_y, and β_z are all nonzero (as, e.g., in liquid 2-chloro-2-nitropropane (Fig. 31), they are estimated by shape-factor analysis. Therefore, there are no adjustable parameters for best fit. Figure 30 shows some comparisons with Gerschel's data for fluoroform at 293°K, in the liquid phases. (The internal field corrections are discussed also by

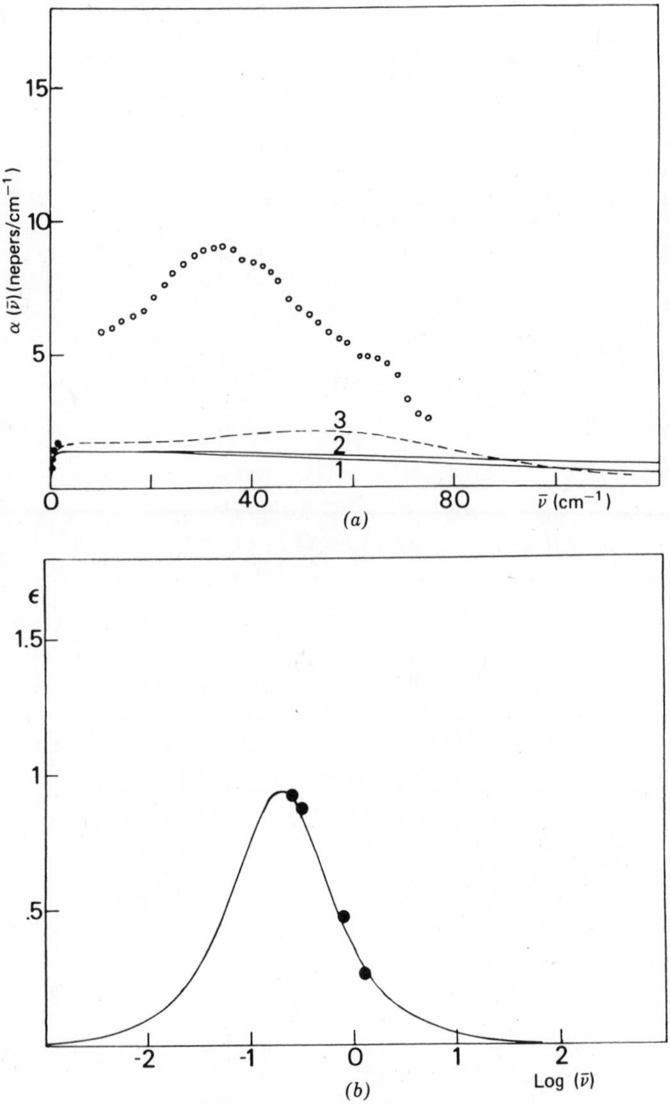

Fig. 34. Key as in Fig. 32. Iodobenzene.

Gerschel.) The needle, sphere, and symmetric-top representations yield absorption and dispersion profiles which are superimposable to a high degree of proximity, but for realistic values of the molecular parameters, β_z in (II.55) is virtually redundant (i.e., has little effect on the band contour once β_x is determined). The Poley absorption is obviously not described. Equation (II.35), on the other hand, produces a satisfactory match for both the

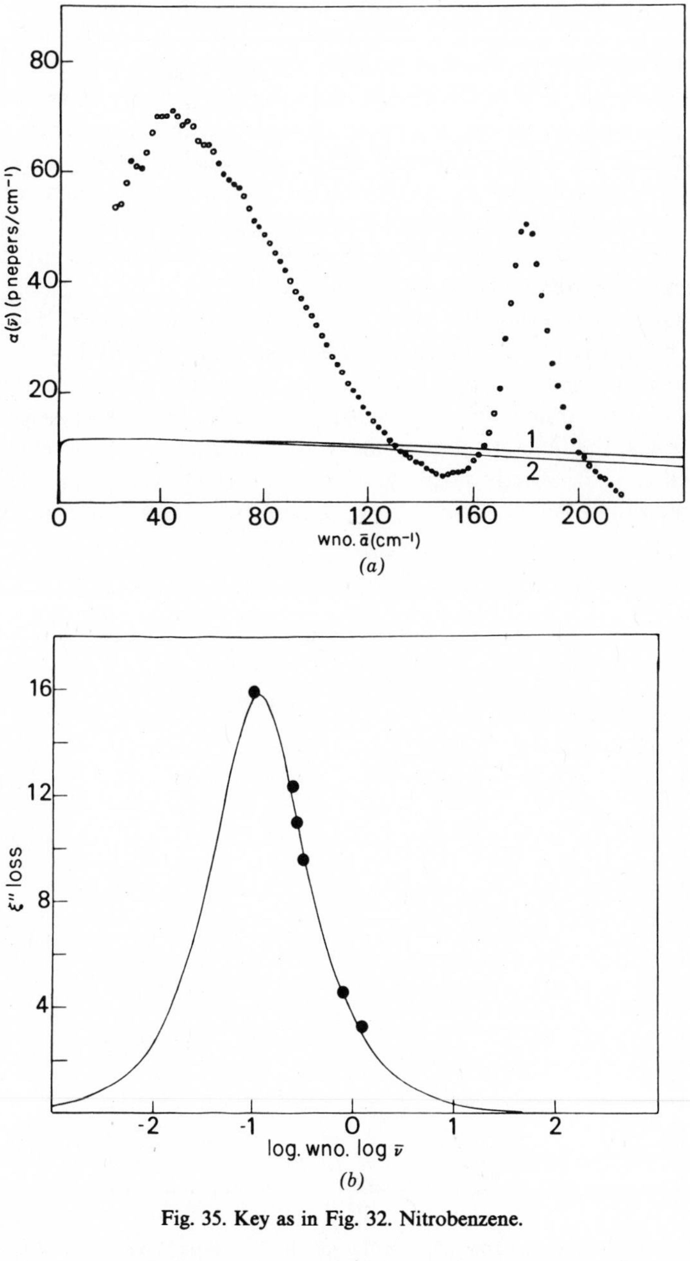

Fig. 35. Key as in Fig. 32. Nitrobenzene.

loss peaks in the zero-THz range (Fig. 44). These two loss peaks (Debye and Poley) having separated clearly in liquid MeF at 133°K offer a severe test for any approximants of (I.6), especially so when empiricism is eliminated [i.e., if γ and $K_1(0)$ are not adjusted for best fit as in Fig. 12]. An apparently good, but deceptive, agreement with the low-frequency loss (or exponential tail of $C_\mu(t)$) may be achieved easily with any one of (II.49) to (II.55), regardless of the molecule's true symmetric-top geometry. Before the development of Michelson interferometry and submillimeter spectroscopy it was common practice to use Debye's equations for any shape of molecule, regardless of the inertia tensor.

Figures 30 to 35 for the asymmetric-top halogenobenzenes demonstrate how insensitive the inertia-corrected equations are to the Poley absorption. Since the intermolecular potential is so ill defined in (II.44), the theoretical results for the sphere and asymmetric top are similar. Nevertheless, we reiterate that the theoretical methods involved are powerful enough to be extended to (I.6). Strictly speaking, of course, (II.44) was never intended by Langevin and Perrin for *molecular* dynamics, but for those of a massive

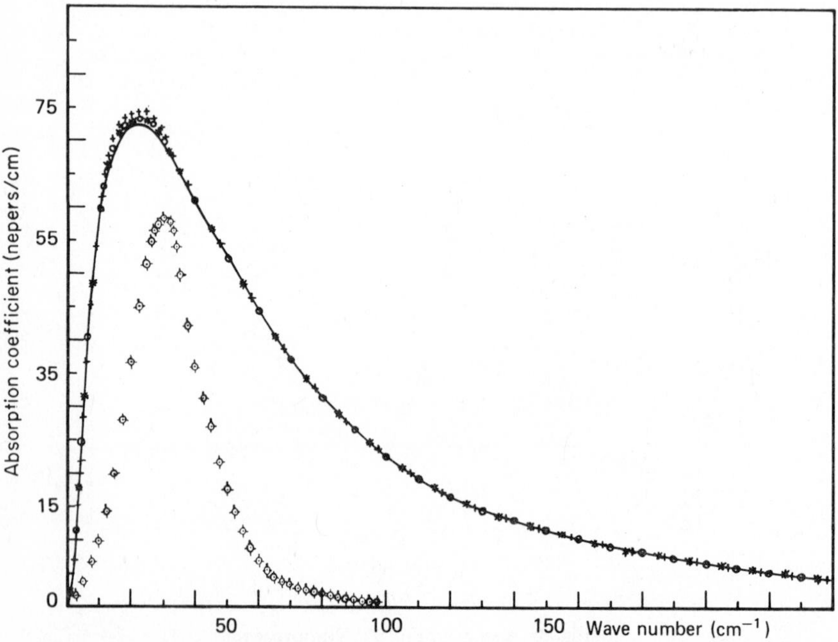

Fig. 36. ⌖ Far-infrared absorption of fluoroform (liquid) at 293°K. (—), Morita's (II.55) for the symmetric top, needle, or spherical top; (****), McConnell's (II.48) for a needle; O, McConnell's for a spherical top; +, Morita's model for an asymmetric top reduced to the symmetric top. [Reproduced by permission from *J. Chim. Phys.*, **75**, 526 (1978).]

particle immersed in a bath of lighter ones. Equations (II.49) to (II.55) should then be tested out against molecular dynamics simulation of this system. Finally, of course, a critique such as this detracts nothing from the immense labor involved in producing (II.49) to (II.55).

As an example of the usefulness[74] of the theoretical techniques developed by McConnell and others, we indicate briefly how they can be used to derive, using a simple kinematic principle, the correlation function $C_{\mathbf{u}}(t)$ from the approximant represented by (I.30). Consider, then, the motion of the dipole vector $\mathbf{u}(t)$ as it librates in a plane with angular velocity $\boldsymbol{\omega}(t)$. Kinematics yield the relation

$$\boldsymbol{\omega}(t)=\omega(t)\mathbf{k} \tag{II.56}$$

where $\mathbf{k}$ is a unit vector and $\omega(t)=|\boldsymbol{\omega}(t)|$. Let I be the moment of inertia of the disk drawn out by the rotating dipole about the axis $\mathbf{k}$. Then

$$\mathbf{k}\cdot\mathbf{u}_0=0 \qquad \mathbf{k}\times(\mathbf{k}\times\mathbf{u}_0)=-\mathbf{u}_0$$

where $\mathbf{u}_0\equiv\mathbf{u}(0)$. Using an expansion as per McConnell and others:

$$\begin{aligned}\langle\mathbf{u}(t)\rangle=\mathbf{u}_0\Bigg[1&-\int\int_{0\le t_1\le t_2\le t}\langle\omega_2\omega_1\rangle\,dt_1\,dt_2\\&+\int\cdots\int_{0\le t_1\le\cdots\le t_4\le t}\langle\omega_4\omega_3\omega_2\omega_1\rangle dt_1\cdots dt_4\\&-\int\cdots\int_{0\le t_1\le\cdots\le t_6\le t}\langle\omega_6\omega_5\omega_4\omega_3\omega_2\omega_1\rangle dt_1\cdots dt_6+\cdots\Bigg]\end{aligned} \tag{II.57}$$

Since the ω's are Gaussian random variables, we have

$$\begin{aligned}\langle\omega_{i_1}\cdots\omega_{i_{2n+1}}\rangle&=0\\\langle\omega_{i_1}\cdots\omega_{i_{2n}}\rangle&=\sum\prod_{i_r>i_s}\langle\omega_{i_r}\omega_{i_s}\rangle\end{aligned} \tag{II.58}$$

Writing (I.6) for ω,

$$I\dot{\omega}+I\int_0^t K_0^{(\omega)}(t-\tau)\omega(\tau)\,dt=I\Theta \tag{II.59}$$

Replacing $K_0^{(\omega)}$ by the truncated continued fraction leading to (I.30) or

(II.35) implies that

$$\dot{\Theta}(t)+\int_0^t K_1^{(\omega)}(t-\tau)\Theta(\tau)\,d\tau=\Theta_1$$

$$\dot{\Theta}_1(t)+\gamma^{(\omega)}\Theta_1(t)=\dot{\Theta}_2(t) \tag{II.60}$$

Solving (II.59) and (II.60) leads to the relation

$$\begin{aligned}\langle\omega_k\omega_l\rangle=c^2[&A_0\exp(-\alpha_1|t_k-t_l|)+B_0\exp(-\alpha_2|t_k-t_l|)\\&+C_0\exp(-\alpha_3|t_k-t_k|)]\end{aligned} \tag{II.61}$$

where the discriminant (Δ_0) in the denominator cubic of (II.35) is positive. (The case $\Delta_0<0$ complicates matters by making the exponentials in (II.61) complex.) In (II.61) $\alpha_1,\ldots,\alpha_3$, C and A_0, B_0, C_0 are constants related to $\gamma^{(\omega)}$, $K_0^{(\omega)}(0)$, and $K_1^{(\omega)}(0)$. The second integral in (II.57) can be written as

$$\frac{3}{4!}\left[2\int_0^t dt_2\int_0^{t_2}\langle\omega(t_2)\omega(t_1)\rangle\,dt_1\right]^2$$

and all successive integrals, as shown by McConnell et al., follow this pattern. Accordingly,

$$\langle\mathbf{u}(t)\cdot\mathbf{u}(0)\rangle=\exp\left[-\int_0^t(t-\tau)\langle\omega(\tau)\omega(0)\rangle\,d\tau\right] \tag{II.62}$$

A more general version of (II.61) is derivable via the probability density functions of Section I.A. This equation is useful in being the closed form for the orientational autocorrelation function of the planar itinerant librator. Using (II.61) in (II.62) and expanding in a Taylor series, there are no linear or t coefficients in $C_{\mathbf{u}}(t)$ and

$$\begin{aligned}C_{\mathbf{u}}(t)&=\exp(-kTt^2/2I+O(t^3))\\&\to\exp(-kTt^2/2I)\end{aligned} \tag{II.63}$$

the free-rotor limit, at very short times, when $O(t^3)$ is negligibly small. At the extreme $t\to\infty$, we have

$$C_{\mathbf{u}}(t)\to\exp(-t/\tau) \tag{II.64}$$

where τ is a constant with the dimensions of time. Note, that in 3–D (Fig.

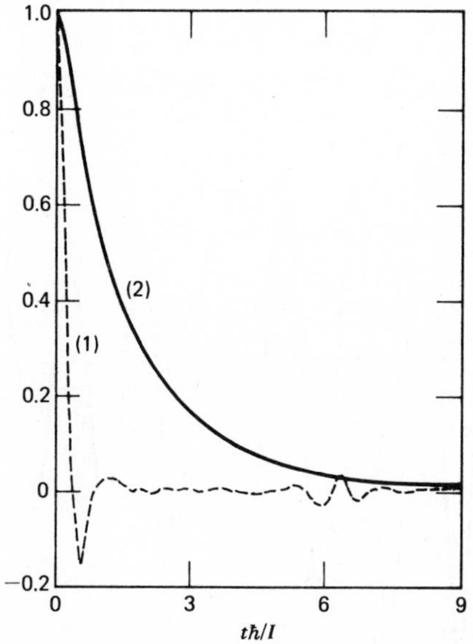

Fig. 37. (----), (1) $C_{\mathbf{u}}(t)$ from the Frenkel–Wegdam curve of Fig. 4. (—), (2) Normalized angular momentum autocorrelation function. The abscissa is in time units of $t\hbar/I$. [Reproduced by permission from *Chem. Phys. Lett.*, **42**, 331 (1976).]

37), the free-rotor and long time behaviours are in general more complicated. Writing (II.62) then as

$$C_{\mathbf{u}}(t)=\exp(-f(t))$$

it may be shown that in three dimensions, an approximate equivalent form is

$$C_{\mathbf{u}}(t)=\exp(-2f(t))$$

so that the space itinerant librator would produce a $C_{\mathbf{u}}(t)$ decaying faster than that of the planar itinerant librator. To link (II.59) and (II.60) to (I.31), we have

$$\beta_2=0 \qquad K_0^{(\omega)}(0)=\gamma^2 \qquad \left[\text{i.e., the } \gamma \text{ of (I.31)}\right]$$
$$K_1^{(\omega)}(0)=(I_2/I_1)K_0^{(\omega)}(0)$$
$$\gamma^{(\omega)}=kT\tau_D/I_1 \tag{II.65}$$

with τ_D as the inverse in the peak loss frequency produced theoretically.

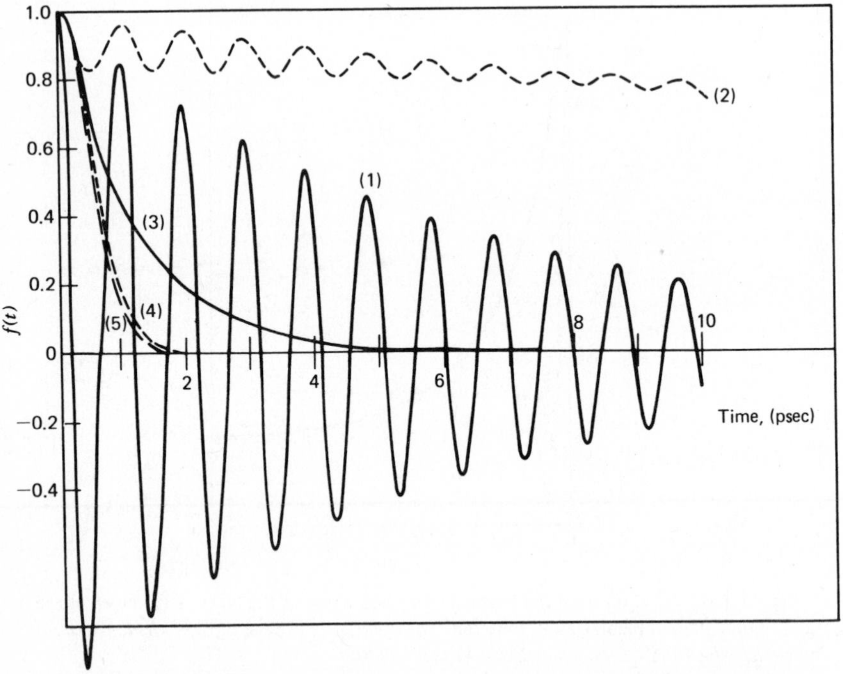

Fig. 38. Angular velocity and orientational autocorrelation functions for motion in two dimensions of the itinerant oscillator. (—) (1) Underdamped $C_{\omega}(t)$; (3) overdamped $C_{\omega}(t)$. (····) (2) Underdamped $C_{u}(t)$ corresponding to (1); (4) overdamped $C_{u}(t)$ corresponding to (3); (5) $\exp(-kTt^2/2I)$, the $C_{u}(t)$ for free rotation of a Maxwellian ensemble of dipoles in two dimensions. For (1) to (5), $T=296°K$, and $I=10^{-38}$ gm·cm^2. [Reproduced by permission from *Mol. Phys.*, **34**, 973 (1977).]

Figure 38 illustrates some curves for the planar itinerant oscillator which should be compared with those simulated (Figs. 70a to e) in Section III.E by computer molecular dynamics with rough annuli within which are constrained disks.

1. *Evaluation of Models in Liquid CH_2Cl_2: Comparison with Other Techniques*

The reasons for and criteria employed in evaluating the foregoing models for $CH_2Cl_2(l)$ are as follows:

1. We have already made the point that the induced absorptions here are likely to be negligible. These are treated further in Section II.E.
2. The complete zero-THz band shape is available for liquid CH_2Cl_2 and in solutions thereof. The data are accurate to within a few percent.

3. At a long enough time after the initial $t=0$, the molecular interactions and motions in the densely packed N-molecule ensemble evolve such that the decay in the correlation functions $\langle A_j(t)A_j(0)\rangle$ of Section I.A is exponential [(II.64) is an example]. This is to say that after suffering many interruptions in its initial trajectory, the probability of finding a molecule with, for example, an orientation $\mathbf{u}(t)$, given an initial $\mathbf{u}(0)$, decays eventually as a Gaussian in time. [This is not so in the autocorrelation function of Fig. 37, since Frenkel and Wegdam consider there an ensemble of interacting *quantized* rotors, where $C_{\mathbf{u}}(t)$ is periodic.] Therefore, we assume in what follows that the *low-frequency* loss curve should be modeled accurately and $\alpha(\omega)$ of the THz region extrapolated therefrom. Using this method it will be shown that deviations from the experimental observations are large in the far infrared, and improvements in the modeling are generally needed.
4. The motion of CH_2Cl_2 molecules in the pure liquid have been monitored by sweep frequency, microwave klystron, and interferometric techniques (both Michelson and Martin–Pupplett). This molecule is chosen for reasons of its intense absorption, especially in the THz range. A recent review[75] has emphasized the considerable extent to which parallel studies have been carried out by incoherent, inelastic neutron scattering, and other techniques. Rotational diffusion and extended rotational diffusion $(m;J)$ have been used for the interpretation of these results.
5. CH_2Cl_2 has been used as a probe into the mesophase and glassy environments. The results of these investigations are dealt with in sections (II.D) and (II.E).
6. CH_2Cl_2 is an asymmetric top and thus endowed with interesting rotational dynamic properties. The glassy results of Section II.F show quite clearly that the shape factor (van der Waals contour) is far from that of a spherical or "pseudospherical" molecule. ($I_x=26.3\times10^{-40}$ g·cm^2; $I_y=278\times10^{-40}$ g·cm^2; $I_z=253\times10^{-40}$ g·cm^2.)
7. The evaluation of (I.31) presents no conceptual difficulty if we remember that they apply, in the dynamic interpretation, to the libration *in a plane* of the dipole vector **u** embedded in the *asymmetric top*. The terms "disk" and "annulus" are conveniences of the mathematical development.

Results and Improvements. These are summarized in Fig. 13*a*.

1. Equation (II.44) is interesting for CH_2Cl_2 since two broad peaks are produced in $\alpha(\bar{\nu})$ (Fig. 12). However, the Poley resonance is, of course, missing. In physical terms the mean intermolecular potential energy is inaccurately defined (e.g., its derivative with respect to orientation, the root-mean-square torque, is meaningless in this context). Therefore, the concept

of libration within potential wells is also meaningless. There is room for further progress by using successive approximants if the geometrical problems and their effect on the Gaussian nature of (I.6) can be overcome.

2. Equation (II.41) was fitted to the experimental data by least-mean-squares optimization of τ. The fit is poor since the model $\bar{\nu}_{max}$ remains at the value

$$\bar{\nu}_{max}=\langle\omega^2\rangle^{1/2}=\left[kT\left(\frac{1}{I_x}+\frac{1}{I_y}\right)\right]^{1/2}=21.5\ \mathrm{cm}^{-1}$$

the root-mean-square angular velocity. Varying τ broadens the theoretical band asymmetrically. Progress in this type of model may be pursued analytically by using the idea of inelastic collisions, as has been demonstrated by computer simulations in rough spheres[61] and disks, and hydrodynamically using slip–stick interactions.[26] Satisfactory definition of a time of collision, and thus of a mean-square torque would enable the desired shift in $\langle\omega^2\rangle^{1/2}$ to take place. Obviously, zero-THz band shapes are sensitive measures of analytical realism in this respect, and will also be useful in evaluating density expressions such as those of Chandler[33] on the memory operator. These allow for the effect of multiple collisions.

3. Using (II.34) to (II.37) without recourse to fitting of any kind leads to rather too broad a band in comparison with the data (Fig. 12). Iterating on γ and $K_2(0)$ separately improves the fit, but in doing so obscures the fact that further approximants are needed before the mean intermolecular potential energy is satisfactorily described. The validity of some truncations at the level of (II.34) to (II.37) has also been investigated for the monohalogenobenzenes[59] using different versions of the memory function $K_1(t)$. A fairly realistic result is obtained only from an exponential $K_1(t)$ which gives an integrated absorption intensity A of about half that observed in the far infrared when the loss curve is matched correctly at lower frequencies. In Fig. 12 some results for a Gaussian and Lorentzian $K_1(t)$ are given which reveal the sensitivity of the Mori series to the form of truncation used. Naturally, all these forms reduce to the same type of loss curve at low frequency, and are indistinguishable without far-infrared data. The halogenobenzenes are less favorable cases than CH_2Cl_2 with which to fit data over the whole of the zero-THz range since the collision-induced component is relatively much stronger. Fitting the (II.34) to (II.37) to microwave data alone is not a satisfactory procedure because one is effectively fitting a simple exponential form with more than one phenomenological variable.

4. Using the notation of (I.31) (the itinerant librator) and the approximate geometrical relation $I_1 = 10I_2$ for various values of β_2, it is obvious in Fig. 12 that the resonance around $\bar{\nu}_{max}$ is too narrow theoretically. The original concept with $\beta_2 = 0$ would therefore produce an even sharper peak. However, iteration on I_1 and β_2 rectifies matters at the expense of physical realism in the disk–annulus sense, since for best fit $I_1 < I_2$ (mathematically corresponding to the overdamped case). A parallel result was found by Damle and others[29] for the translational space itinerant oscillator where β_2 and β_1 are made time-dependent memory functions. This result is reasonable, however, in the context $\beta_2 = 0$ since it implies merely that $K_1^{(\omega)}(0) > K_0^{(\omega)}(0)$ (i.e., that the rate of change of torque is large). Equation (I.31) is in one sense a zeroth-order approximant of the system devised by Damle and others,[29] and perhaps successive approximants would improve matters if the problem of too many adjustable variables could be overcome. Another possibility for improvement is the inclusion of rotation translation coupling as discussed in Section I.C. Some of the adjustable variables could be evaluated separately by molecular dynamics simulations. Using $I_1 = I_2$; $\beta_1 = \beta_2 = kT\tau_D / I_2$ in the dipole interaction representation of (I.31) produces a poor fit (Fig. 12). This suggests that a more realistic potential V is of dipole–dipole interaction is necessary. However, $K_0^{(\omega)}(0)$ and $K_1^{(\omega)}(0)$ are not wholly electrostatic in origin and an interpretation of the far-infrared absorption *solely* in these terms is obviously to be avoided.

Recently, Brier and Perry[75] have obtained time-of-flight neutron scattering data on liquid CH_2Cl_2 and have tested their results with four models of the liquid state. They have also reviewed critically the available NMR, infrared / Raman, and depolarized Rayleigh scattering work. The attempts at evaluating the anisotropy of the molecular angular motion using these techniques sometimes end in confusion and contradiction. The *m*- and *J*-diffusion models were used regarding CH_2Cl_2 as an inertial symmetric top (i.e., assuming axial symmetry of angular motion about that axis of least inertial moment). [This assumption is however contradicted by NMR results, giving $\tau_2(H—H) \neq \tau_2(C—H)$.] An Egelstaff-Schofield form was used for the translational correlation function in the usual decoupling approximation [e.g., (I.39)]. The conclusions drawn as to the efficacy of the model *within this approximation* are very similar to those of the zero-THz data. Changing the values of the time between collisions alters the magnitude of the inelastic peak, its position remaining virtually constant. The maximum of the predicted inelastic intensity distribution occurs at much too low an energy transfer (frequency) for both models; therefore, within the context of extended diffusion, treating CH_2Cl_2 as a spherical top (Fig. 12) or as a symmetric top (neutron scattering) makes little difference to the final result, which is poor. It is important to realize that this does not imply that the data are insensitive to asymmetry of orientational or rototranslational motions, but rather that unrealistic and oversimplified assumptions lead to oversimplified and unrealistic results. In common with nearly all neutron-scattering studies of molecular motion, Brier and Perry work with the assumption of complete decoupling of rotation from translation. Berne and Montgomery[32] have demonstrated the severity of this restriction by showing that the analytical rototranslational neutron scattering spectrum of a rough-sphere fluid is very different from that of a smooth-sphere fluid. Molecular structure will increase the coupling. A self-consistent approach is clearly needed to this problem, which would be indicated within the framework of the (I.6). This would aim at an appreciation of the spectrum as in Section I.C.

Studies of the depolarized Rayleigh wing complement the zero-THz band. Brier and Perry have also discussed the depolarized Rayleigh, Raman, infrared and NMR data available for liquid CH_2Cl_2. Ideally, NMR and infrared/Raman band shapes provide data on single-molecule motion. The correlation times available from these techniques are confusingly disparate. An explanation is attempted based on the asymmetric-top Langevin equation, which in the light of Fig. 12, is meaningless. It is significant that only the zero-THz $[\alpha(\bar{\nu})]$ dielectric absorption shows up clearly enough the discrepancy between rotational diffusion and observation. Another feature is that interpretation in terms of jump models gives directly contradictory results. This is hardly surprising, since without far-infrared data the 180° jump and infinitesimally small jump model both fit the available loss data exactly. A similar kind of indistinguishability is present when jump models are used in the theory of incoherent neutron scattering, as demonstrated in a review by Janik (Ref. 43, p. 45).

It is clear that the available data from all sources on liquid CH_2Cl_2 have been interpreted using many different models, with each of which are associated (usually) adjustable parameters, so that an overall viewpoint is not attainable. We propose a scheme to remedy this to a modest degree.

A computer simulation, using an empirical intermolecular potential (e.g., atom–atom[61] Lennard-Jones interactions) should be carried out on CH_2Cl_2, and desired quantities such as the mean-square torque, various autocorrelation functions, and collective correlation functions extracted. These should then be compared with the values obtained from self-consistent Mori approximants used with the zero-THz and depolarized Rayleigh data, these being free from the uncertainties of vibrational relaxation, hot bands, and so on. The effect of cross-correlations may be easily estimated from the zero-THz data by dilution, and compared with those simulated. Isotropy or otherwise of angular orientations may be simulated in detail and compared with the considerable amount of NMR data available. Rototranslational effects may be simulated and compared with the available neutron scattering data. In this way it may be possible to refine the empirical intermolecular potential by evaluation against the spectroscopic data, especially if these were available over a broad-enough range of number density and temperature. Alternatively, if the empirical potential were considered adequate, the efficacy of the continued-fraction approximation could be measured when truncated at various levels.

D. Internal Rotations: Relation with ^{13}C NMR Relaxation of the Zero-THz Absorption

We digress a little in this section to discuss the concept of *internal* libration within the framework of a nonrigid species such as *p*-dimethoxybenzene or dimethyl carbonate. This is conveniently studied[76] by means of both zero-THz data and ^{13}C NMR (T_1 and N.O.E.) relaxation. One reason is to emphasize the point that different relaxation techniques used in a cohesive study of a selected problem may be more incisive than studies undertaken separately. A second is that rotational correlation functions may apply to systems of molecules with internal freedoms, and the far infrared is a particularly suitable frequency range with which to determine the kinetics of motions such as those of the methoxy-and methyl-group internal rotations. A quantitative analysis in terms of a "chemical relaxation process" permits an estimation of both the kinetic constant $k_{cis \to trans}$ of the dielectrically "active" *cis*/*trans* isomerism of the *para*-dimethoxybenzene molecule, and of the jumping rate of the methyl group from any of its three equivalent positions. The methoxy torsional modes appear in the far infrared and it is also possible to assign to this frequency range the methyl torsions.

A number of low-frequency (microwave) studies has been made in the pure liquid phase or in dilute solution on compounds having one or more methoxy groups. In these previous investigations, the only practicable way of estimating qualitatively the contribution of the group rotation to the overall dielectric relaxation was by comparison within a homologous series of compounds having roughly the same molecular shape. Such analyses have been hazardous because of changes in dipole moment (magnitude or orientation in the molecular frame), internal field, microscopic viscosity, and very often in the barrier height to internal rotation itself (e.g., in the case of electron donating or withdrawing aromatic substituents). More complete and quantitative information may be extracted on the different internal motions by simultaneous measurement of the longitudinal relaxation time (T_1) and of the nuclear Overhauser enhancement factor (η), with the zero-THz electromagnetic absorption. In this section we describe such

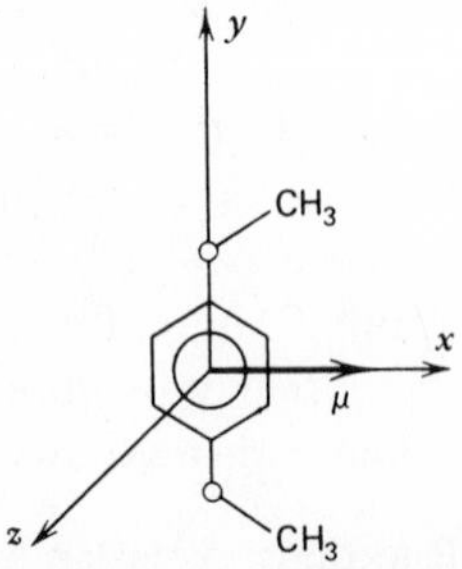

Fig. 39. *p*-Dimethoxybenzene molecule in the *cis*-position; **μ** represents the direction of the resultant dipole. [Reproduced by permission from *Mol. Phys.*, **30**, 974 (1975).]

a joint study of a solution of *p*-dimethoxybenzene in tetrachloroethylene (4.17_6 mol/*p*-DMB dm^3), chosen for experimental convenience. There is in the solute only one internal rotation axis, and along this, owing to the molecular symmetry, there is no component of the electric dipole (Fig. 39).

Plots of ϵ'' vs. ϵ' at 298°K, 323°K, and 348°K are shown in Fig. 40 and far-infrared absorption spectra obtained at 298°K for both the normal and the deuterated compounds in Figs. 41 and 42. The overlap between the microwave interferometric measurements (University of Nancy I) and the results obtained by Michelson free-space interferometry in the region 3 to 28

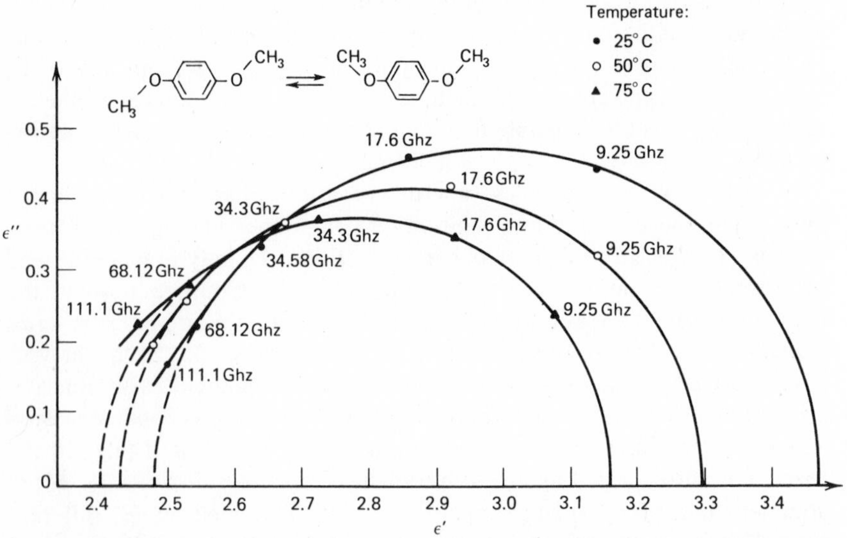

Fig. 40. Cole–Cole plots for a solution of a 4.176 mol/dm^3 *p*-dimethoxybenzene in C_2Cl_4 at 298°K, 323°K, and 348°K. The frequencies of the (ϵ'',ϵ') measurements are indicated on each plot. The dotted lines correspond to a semicircular extrapolation. [Reproduced by permission from *Mol. Phys.*, **30**, 976 (1975).]

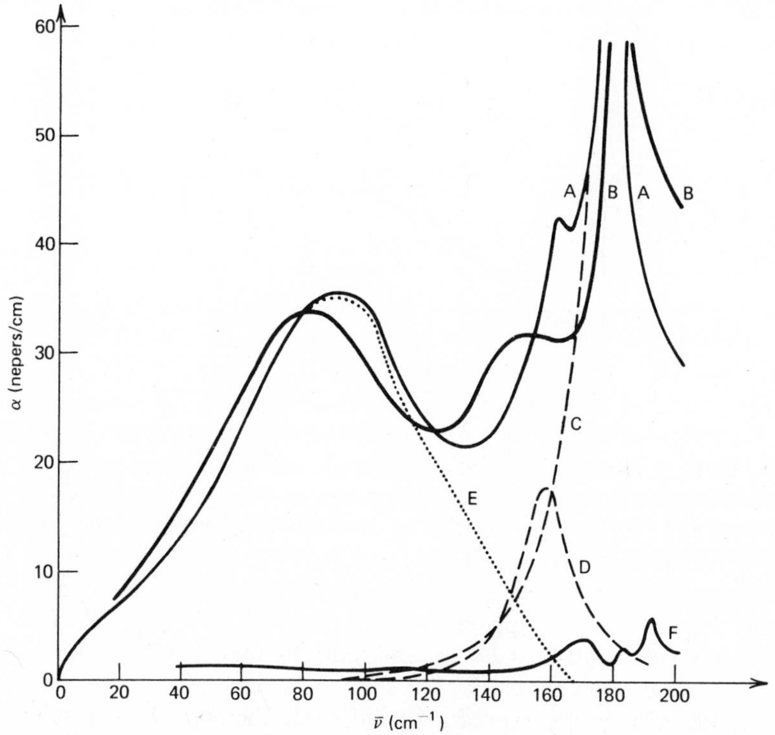

Fig. 41. (-), (*A*) Far-infrared absorption of *p*-dimethoxybenzene/C_2Cl_4 (4.176 mol/dm^3) at room temperature. (—), (*B*) Deuterated compound absorption at the same concentration. (----), (*C*), (*D*) Idealized line shapes for some of the higher-frequency absorptions. (·····), (*E*) Unresolved low-frequency band extracted from (*A*). (—), (*F*) Solvent absorption. [Reproduced by permission from *Mol. Phys.*, **30**, 977 (1975).]

cm^{-1} is satisfactory, Fig. 42. The complete zero-THz band is thus accurately defined. NMR relaxation measurements were separately made on the signal relative to the four equivalent ortho aromatic carbons, and on the line due to the methyl carbons. The solvent peak and those of the C_6D_6 reference are well separated from those of the solute.

The microwave results support the existence of only one "resultant" relaxation time accounting for the *low-frequency* part of the dielectric process. At times shorter than ca. 2 psec, however, the computed "pseudo" rotational velocity correlation function is oscillatory. These reflect both the librational motions as a whole of the *cis* conformers and the internal torsional motion of the methoxy groups, but occur at too short a time (<2 psec) to affect the exponential behavior of the vectorial dipolar correlation function (DVCF) at the time scale of the microwave measurements. The resultant value of $\mu_z = 4.5 \pm 0.4$ D calculated from the far-infrared band is

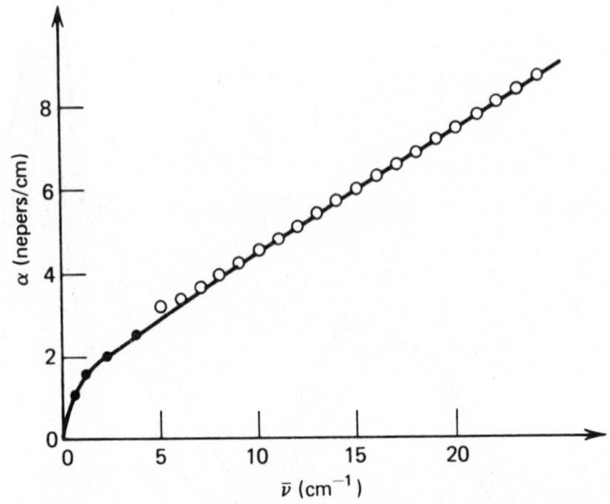

Fig. 42. Detail of the low-frequency absorption spectrum of *p-dimethoxybenzene*/C_2Cl_4 at room temperature. ●, Microwave interferometry with klystron sources; ○, Michelson interferometry with a He(*l*)-cooled Rollin detector. [Reproduced by permission from *Mol. Phys.*, **30**, 978 (1975).]

much larger than the apparent dipole moment obtained by dielectric measurements, 1.5(4) D at 298°K, and indicates that the main contribution to this band has another origin. The shift of the band center on deuteration of the methoxy groups supports quantitatively the hypothesis of torsional modes of the methoxy groups.

If, for the sake of argument, we accept the oversimplified hypothesis of isotropic molecular rotational diffusion as being adequate to reproduce the low-frequency loss data, and the aromatic ortho carbon NMR data, then the isotropic diffusion constant D_0 is related to an NMR relaxation time τ_c^{NMR} by

$$6D_0 = \left(\tau_c^{\mathrm{NMR}}\right)^{-1} \tag{II.66}$$

For the symmetric-top diffusion on the other hand, τ_c^{NMR} has to be replaced by $F(D_x, D_y, D_z)$ defined for a planar molecule by

$$F(D_x, D_y, D_z) = \frac{3}{4D_r}\left[(D_x + D_s)\cos^2\phi + (D_y + D_s)\sin^2\phi - \frac{(D_y - D_x)}{D_z + D_s}\sin^2\phi\cos^2\phi\right] \tag{II.67}$$

where ϕ (in our case, 30°) defines the orientation of the internuclear vector

r_{CH} in the molecular frame, D_r and D_s being given by

$$D_r = 3(D_x D_y + D_y D_z + D_z D_x)$$
$$D_s = \tfrac{1}{3}(D_x + D_y + D_z)$$

The consequences of neglecting this anisotropic character lead to a difference, however, of less than -10 to -15% in the ratio $p=3$ expected in the rotational diffusion limit between τ_c^{NMR} and the dielectric relaxation time τ_μ^{Diel} estimated after correction for the internal field.

Similarly, the methyl carbon relaxation time $\tau_{CH_3}^{NMR}$ is estimable after consideration of the problem of methyl group rotation about a fixed but arbitrary axis in the molecular frame. $\tau_{CH_3}^{NMR}$ can be related to the jumping rate $2R/3$ of the methyl group from any of its three equivalent positions by

$$\tau_{CH_3}^{NMR} = \frac{1}{2D_0}\left[1/(3+R/2D_0)\right] \tag{II.68}$$

where again the isotropic diffusion coefficient has been used for lack of knowledge of the anisotropic diffusion coefficients $D_\perp$ and $D_\parallel$.

This restriction to rotational diffusion is unsatisfactory in the light of the foregoing behavior of the model in the far infrared, but the complexity of the formulas is already beyond the data available. The use of NMR relaxation in isolation is therefore prone to vagueness of analysis, just in the way that a semicircular low-frequency Cole–Cole plot is about the least discriminating imaginable. It is not surprising therefore that the measured ratio

$$\rho_{app} = \tau_\mu^{Diel} / \tau_c^{NMR}$$

is considerably less than 3 (Table VI). This may be caused by:

1. Substantial contribution from internal methoxy group rotation to the dielectric relaxation phenomenon.
2. Breakdown of the Debye–Perrin rotational diffusion model (i.e., reorientation of the whole molecule with memory effects).
3. Strongly anisotropic rotational diffusion of the whole molecule.

The first explanation may be supported quantitatively using the model of dielectrically active "chemical relaxation processes," as previously proposed by Williams and Cook[77] and Goulon, Canet, Evans, and Davies.[76]

TABLE VI
Dielectric and NMR Relaxation Times for *p*-Dimethoxybenzene

T	298°K		323°K		348°K	
^{13}C NMR relaxation	C_{ortho}	C_{CH_3}	C_{ortho}	C_{CH_3}	C_{ortho}	C_{CH_3}
η	1.6(5)	1.7(5)	1.7(6)	1.8(0)	1.7(7)	1.8(7)
T_1/s	5.5(4)	5.1(2)	7.8	6.9(5)	10.1	9.6(5)
τ_C^{NMR}/psec	7.7(5)	2.7(3)	5.1(3)	2.0(7)	3.9(8)	1.5(5)
Dielectric relaxation	τ_μ^H	$\tau_\mu^{P.G.}$	τ_μ^H	$\tau_\mu^{P.G.}$	τ_μ^H	$\tau_\mu^{P.G.}$
τ_μ/psec	12.2	11.1	7.8(8)	7.1(9)	5.5(3)	5.0(8)
$\rho_{app}=\tau_\mu/\tau_{C(ortho)}^{NMR}$	1.8(2)	1.6(5)	1.5(3)	1.4(1)	1.3(8)	1.2(7)

[a] $\tau_\mu^H=\tau_c^{Diel}$; $\tau_\mu^{P.G.}=((2\epsilon_0+\epsilon_\infty)/3\epsilon_0)\tau_c^{Diel}$.

If one assumes then that the molecule takes up two planar, cis (dipolar) and trans (nondipolar) configurations, the rotation of the methoxy groups gives rise to both *cis/trans* isomerism and *cis/cis* inversion mechanisms. We can therefore summarize the internal motions using a triangular kinetic scheme (Fig. 43). The dipolar autocorrelation function is found, then, to have the following time dependence:

$$\langle \mu(t)\cdot\mu(0)\rangle = \tfrac{1}{2}\langle \mu(0)\cdot\mu(0)\rangle \times\left[\exp(-t/\tau_2)+\exp(-t/\tau_1)\right]\exp(-t/\tau_0) \quad \text{(II.69)}$$

where τ_0 characterizes the reorientational process of the whole cis conformer, considered as a rigid molecule, and τ_1 and τ_2 are given by

$$1/\tau_1=k_{21}+2k_{12}\equiv k_{21}\left[1+K_{eq}\right]$$
$$1/\tau_2=k_{21}+2k_{22}\equiv k_{21}\left[1+K'\right] \quad \text{(II.70)}$$

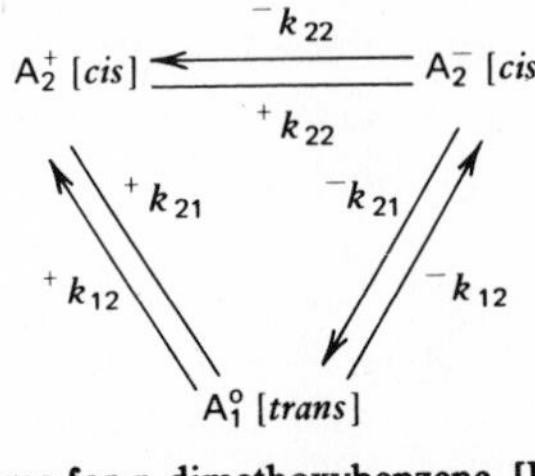

Fig. 43. Triangular kinetic scheme for *p*-dimethoxybenzene. [Reproduced by permission from *Mol. Phys.*, **30**, 984 (1975).]

According to the Curie principle stated by Prigogine and Mazur, this form of the dipolar correlation function assumes a total statistical independence between the internal (chemical) process and the external (diffusional) process. However, the analysis leading to (II.69) and (II.70) would be more convincing if associated with a model other than that of rotational diffusion, so that point 2 above ought to be considered in greater detail. Logarithmic plots of the inverse of the two correlation times of Table VI produce the apparent activation enthalpies

$$\Delta G_{\mathrm{NMR}} = 2.2(6) \text{ kcal/mol}$$

$$\Delta G_{\mathrm{Diel}} = 3.3(4) \text{ kcal/mol}$$

The difference between the two values may be an additional indication that the NMR and dielectric relaxation processes are different in nature, the former, on the basis of hypothesis 1, being insensitive to any group rotation.

A barrier height hindering the methoxy-group torsion may be evaluated from the far infrared data. The theoretical analysis of the torsion in completely asymmetric molecules remains, as usual, a very complicated problem, but it is possible, using the methods of Goulon, Canet, Evans, and Davies, to predict a band shift on deuteration from $\bar{\nu}_H = 91 \cdot 88 \text{ cm}^{-1}$ to $\bar{\nu}_D = 82 \text{ cm}^{-1}$. This is in excellent agreement with the experimental results, the barrier is then calculated as $5 \cdot 88$ kcal/mol (2058 cm^{-1}).

Therefore, there is little doubt that the methoxy-group internal rotation is fast enough to contribute significantly to the dielectric relaxation. The potential barrier hindering this motion is about $5 \cdot 3$ kcal/mol, estimated from the kinetic scheme of Fig. 43. This is sufficiently close to the $5 \cdot 88$ kcal/mol estimated above to be acceptable. Similarly, a value of $1 \cdot 8 \pm 0.3$ kcal/mol may be estimated for the apparent activation energy for the methyl rotation. It is also worth noting that if our results are acceptable in terms of the idea that the methyl rotation should be faster than the methoxy-group rotation of the methyl group, our evaluation of the height of the barrier hindering the methyl internal rotation might suggest an appreciable coupling of both the methyl and methoxy internal librations.

This first attempt to investigate the dynamics of internal motions by comparison of zero-THz and ^{13}C NMR relaxation data has run up against the problem of interpretation in terms other than rotational diffusion (isotropic at that). However, we have been able to deduce:

1. That internal methoxy-group rotation contributes significantly to the dielectric process.
2. Quantitative estimates of the kinetic parameters governing the internal librations and activation enthalpies.

E. Mesophases: Liquid and Plastic Crystals, Disordered Solids

The number of specialist articles and review series devoted to various researches into liquid crystal phenomena is steadily growing.[78] This section deals with the special insight these mesophases provide into the isotropic liquid state, in that the orientation correlation function $C_{\mathbf{u}}(t)$ and its second derivative reflect the anisotropy of the molecular rotational characteristics brought about essentially by the molecular geometry. The alignment along the director axis (Section I) is the long-range consequence of the restricted torsional oscillation starting at the level of the nearest-neighbor cage, and reflected in the far infrared by a sharp and high-frequency Poley band (or bands) whose low-frequency loss adjunct peaks at megahertz frequencies, typically, in the aligned condition (Fig. 9). In consequence, $C_{\mathbf{u}}(t)$ is virtually a pure exponential decay, whereas its second derivative, the Fourier transform of $\alpha(\omega)$, is highly oscillatory (Fig. 44). The object of this section is to demonstrate how the zero-THz profile in phases such as the nematic may be used to aid in the evaluation of the molecular dynamics. With such an objective the first far-infrared study of the nematic phase [of *p*-methoxybenzylidene-*p′ n*-butylaniline (MBBA)] was carried out independently by Bulkin and Lok[79] and by Evans, Davies, and Larkin[80] in 1973.

The difference between the aligned nematic phase and the plastic crystalline or disordered solid mesophase also considered here is that the molecular rototranslation in the former evidently prohibits crystallization, or even solidification. This degree of dynamic freedom is propagated by the asymmetric van der Waals contours (constantly fluctuating due to intramolecular motions) of molecules such as MBBA which have liquid crystalline properties. It is possible, in consequence, to supercool the aligned nematic phase, the sample remaining a viscous fluid. The main feature of the MBBA far-infrared spectrum (taken in unaligned, aligned, isotropic solution, and solid states) is a strong and broad-band peaking at 130 cm^{-1} (Fig. 44). In the pure isotropic phase this shifts slightly to lower frequencies (123 cm^{-1}). The band seems almost to disappear in very dilute solution, and broadens considerably on heating a moderately dilute solution of MBBA in cyclooctane (Fig. 45). At the same time the peak moves to a lower frequency. Thus the absorption is markedly environment-sensitive, the near-neighbor interactions involved being strong in the pure nematic phase. It is justifiable to conclude therefore that its origin is torsional oscillation of the MBBA resultant dipole vector $\boldsymbol{\mu}$ occurring at a higher frequency (given the MBBA moment of inertia effective in determining this motion) than is usual for isotropic, dipolar liquids such as the halogenobenzenes or CH_2Cl_2. On this basis Evans, Davies, and

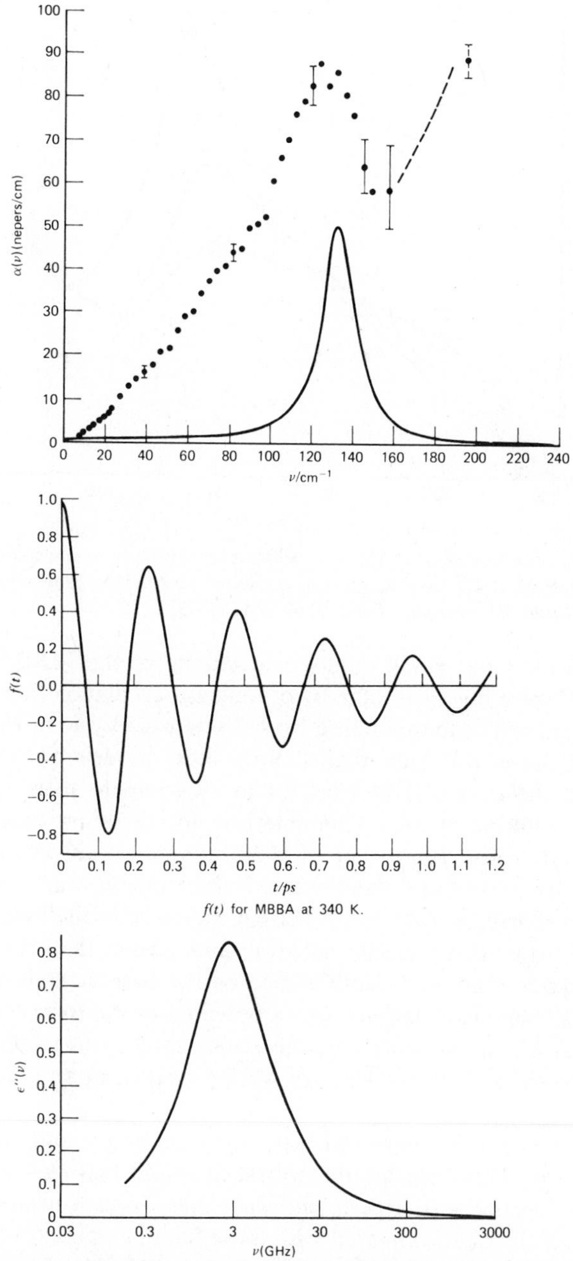

Fig. 44. (*a*) ●, Experimental absorption for *p*-methoxybenzylidene-*p*′*n*-butyl aniline (MBBA) in the nematic phase at 340°K. (—), Mori theory best fit. (*b*) $-\ddot{C}_{\mathbf{u}}(t)$ for MBBA at 340°K. (*c*) Loss curve calculated for MBBA at 340°K. Observed $\epsilon''(\bar{\nu})=0.70$; observed critical frequency = 2.6 GHz; calculated = 2.7 GHz. [Reproduced by permission from *J. Chem. Soc. Faraday Trans. 2*, **72**, 1169 (1976).]

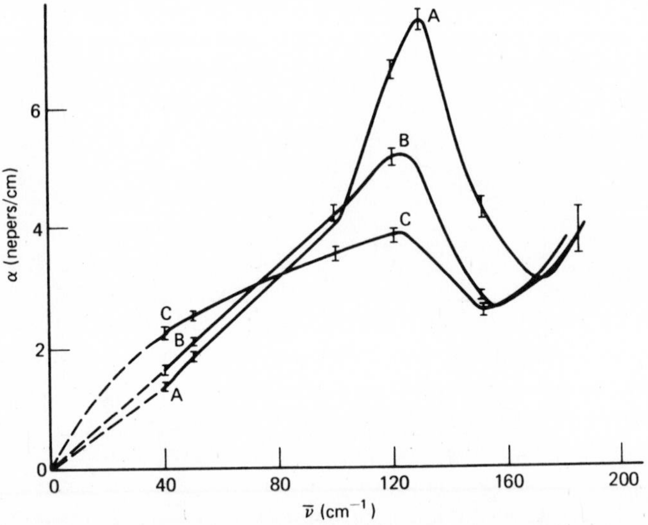

Fig. 45. (*A*) Spectrum of a 10.2% w/w solution of MBBA in cyclooctane (corrected for solvent absorption) at 297±0.5°K. (*B*) 341±1°K; (*C*) 338±1°K. [Reproduced by permission from *J. Chem. Soc. Faraday Trans. 2*, **69**, 1011 (1973).]

Larkin[80] carried out some model calculations of the MBBA absorption. The first of these involved a theory of itinerant oscillation due to Hill[81] and Wyllie,[82] and of random hopping from discrete sites due to Brot[52]—an extension of the *m* diffusion model. Both these models are early approximants and the examples of Figs. 14 to 19 compare them against direct Fourier transforms of $\alpha(\omega)$. Characteristically, the short-time behavior of approximants earlier than that of (I.30) compare badly with experiment (Fig. 19). To reproduce the observed spectrum, it was found that the potential well experienced by an MBBA molecule in the field of its neighbors needs to be considerably narrower and steeper than that of isotropic dipolar liquids. The well depth estimated for best fit agreed surprisingly well with a rough calculation using a potential of the form $V=a\exp(-br)-c/r^6$, in which the only intermolecular interactions considered were those between the benzene rings of MBBA packed in an idealized geometry.

Using (II.35) in the empirical fashion by varying γ and $K_1(0)$ gives the results of Fig. 44, where the theoretical $\alpha(\omega)$ and loss curves are in fairly satisfactory accord with experiment. The "mean-square-torque" term $K_1(0)$ takes on a high value compared with those found empirically in fluids such as CH_2Cl_2. Across the series[83] MBBA : propyne : $CBrF_3$: $CClF_3$: CHF_3, for example, the apparent mean-square torques decrease roughly in the ratio 200 : 25 : 10 : 10 : 8. This trend is the one expected on the assumption that the greater the molecular geometrical anisotropy, the greater the mean barrier to torsional oscillation. For a high-mean-square torque the microwave and far-infrared parts of the zero-THz band must be widely separated, whereas for a low-mean-square torque (as in CHF_3, for example) the two parts are virtually fused into one.

Despite the apparently good fit obtained with γ and $K_1(0)$, it must be emphasized that any modeling such as the above of the rototranslational dynamics in MBBA must of necessity be crude and approximate because mathematical tractability demands the use of rigid, whole molecule libration, using a very simple representation of the intermolecular potential. The MBBA zero-THz profile is assuredly environment-sensitive but is best described as arising from the librations of a dipole within a flexible framework, the motions of which are determined by and in turn determine the character of the nearest-neighbor and less immediate environment. In a flexible molecular framework the previous section demonstrated the degree of extra complexity engendered even within the restricted limits of rotational diffusion—it is difficult, for example, to estimate the moment of inertia dyadic, which is time-dependent. In addition, the long-range correlations are of greater import in the mesophase, so that it is likely that collective motions are favored. The continuum theory of the mesophase as reviewed, for example, by de Gennes has to be matched by molecular theories if a cohesive picture is to be built up. Computer simulations run into the difficulties of swarm sizes being larger than the grid or cube of molecules set up initially. Finally, it has been observed that in solid MBBA the 130-cm^{-1} band splits into at least four partially resolved peaks, so that there is a possibility, as pointed out by Sciesinska, Sciesenski, Twardowski, and Janik,[84] that the torsional vibrations and other low-frequency internal modes of the MBBA molecule account for all the absorption below 170 cm^{-1}. They cite the evidence of changes in the spectra which they associate with different phases of MBBA solid and in the persistence of the absorption in solution. However, the 130-cm^{-1} band is dilution-sensitive as regards its shape and peak position, so that intermolecular sensitivity is detectable.

To extend the zero-THz monitor to phases such as the cholesteric (of cholesteryl oleate and cholesteryl oleyl carbonate, for example), it is more fruitful to look indirectly at the effect of the environment on small amounts of rigid, intensely dipolar solute molecules such as CH_2Cl_2 used as dynamical probes. (This technique is extended to glasses in the next section.) The following advantages accrue:

1. Cross-correlations terms between guest molecules (dynamic and electrostatic), not amenable to ready mathematical analysis, are minimized

in dilute solution in the cholesteric, nematic, or aligned nematic solvent.

2. The probe can be chosen to be particularly suitable for model simulation of its absorption profile (i.e., to be rigid and intensely absorbing).
3. The influence of a liquid crystalline environment on molecular motion may be measured directly against the equivalent spectra in an isotropic solvent such as CCl_4.

The far-infrared spectra of the mesophases themselves are often rich in detail but consequently very difficult to model. For example, we monitor in this section the alignment of 4-cyano-4′-*n*-heptyl biphenyl (7CB) with ac and dc electric fields of up to 7 kV/cm, and with magnetic fields, and Fig. 46 shows the appearance of extra peaks underlying the structure in the unaligned condition.[85] However, an attempt has been made[86] to use (II.35) with 7CB— for best fit to the low-frequency loss data, the THz peak corresponding to rigid end-over-end torsional oscillation is very sharp, centered at over 100 cm^{-1}, but obviously (in the light of the electric and magnetic field work) one of many possible such absorptions, all markedly environment sensitive. The effect of applying an increasing dc electric field to the nematic phase of 7CB is shown in Fig. 46. The overall intensity of the absorption decreases across the whole of the far-infrared range and the spectrum is split into peaks hitherto unresolved in the unaligned condition. An intensity decrease on application of an external field has been observed in the Raman by Schwartz and Wang.[87] In the nematic phase of two compounds, striking changes in the relative intensity of several Raman bands were observed as a function of applied electric field strength. An explanation was given in terms of the collective stabilization due to the large ensemble of molecules aligned by the field. However, the appearance of so many extra peaks in the far infrared is entirely novel. If these peaks are all intermolecular in origin, partially so, then one possible explanation of their appearance is that the increased alignment under the effect of a field accentuates the underlying lattice modes [i.e., brings out single-crystal-type behavior, but with the residual translational freedom (NMR studies) associated with the liquid crystal phase still being retained]. This kind of dynamical effect is indicated also by the fact that the overall intensity drops due to polarization of the radiation reaching the detector (i.e., the aligned nematic phase is acting as a polarizer). The question of what happens to the torsional oscillating Poley absorption in these circumstances is an interesting one. It was in an attempt to describe quantitatively this process and its low-frequency counterpart, the loss peak observed by Moutran in 7CB at 6 MHz that (II.35) was used for 7CB as described already.[86]

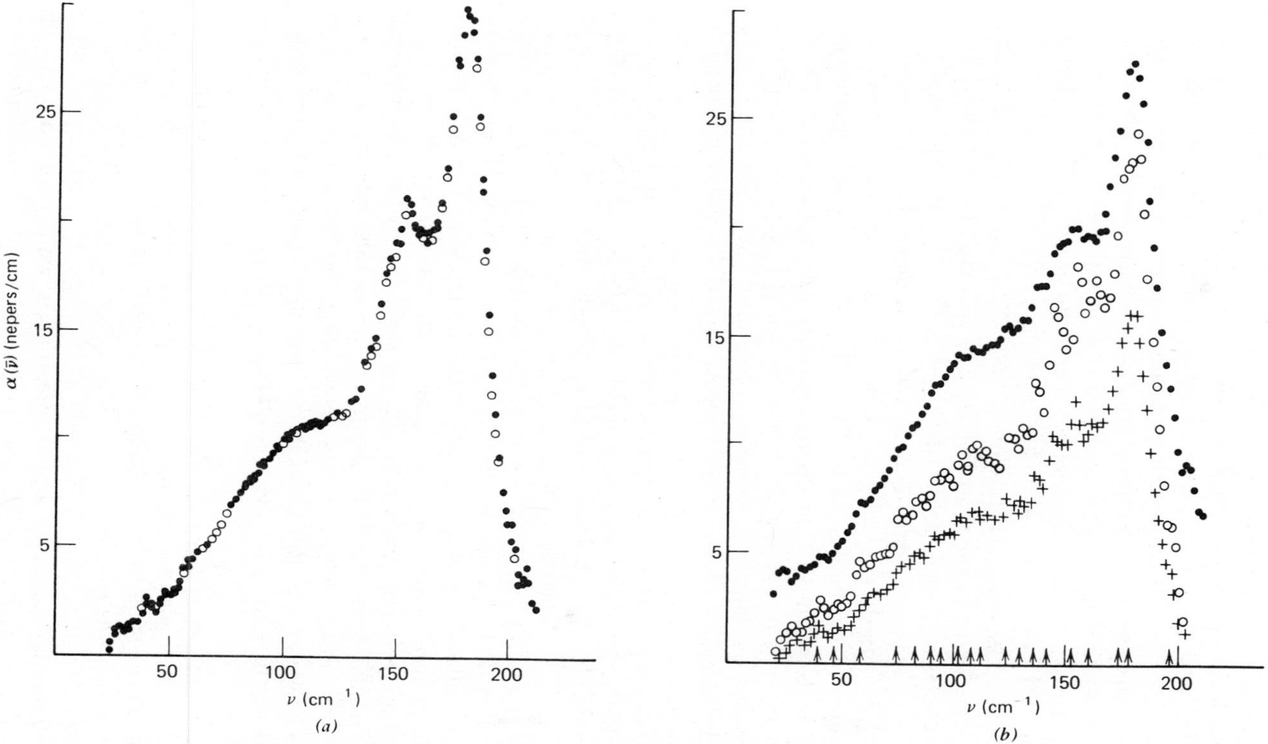

Fig. 46. (*a*) Nematic phase of 7CB at 299°K, unaligned (no applied field, no treatment of the window surfaces): ○, far-infrared absorption. (*b*) ●, Aligned nematic phase of 7CB at 299°K, applied field of 2.0 kV/cm^1; ⊙, applied field of 3.0 kV/cm^1; +, applied field of 7.1 kV/cm^1. [Reproduced by permission from *J. Chem. Soc. Chem. Commun.*, 268 (1978).]

The observation of the probe Poley absorption,[54] on the other hand, is designed to avoid for the moment the difficulties of quantitative interpretation associated with the far-infrared spectra of pure liquid crystals. We chose CH_2Cl_2 because its far-infrared absorption is intense and well defined (Fig. 20). The far-infrared broad-band absorptions in CH_2Cl_2 have been measured carefully in isotropic solutions in CCl_4, decalin, cholesteryl linoleate, cholesteryl oleyl carbonate, and 7CB in order to bring out by direct comparison unusual dynamical effects on the CH_2Cl_2 molecules themselves. Whereas the CH_2Cl_2 band maximum ($\bar{\nu}_{max}$) shifts by about 30 cm^{-1} to lower frequency on dilution in both CCl_4 and decalin, there is a smaller corresponding change when CH_2Cl_2 is dissolved in cholesteryl linoleate and cholesteryl oleyl carbonate (Fig. 47). This may be attributed to a persistence of statistical cross-correlations (time-dependent Kirkwood *g*-factor) which vanish gradually in isotropic solvents. The observed integrated intensity per molecule (A/N) of CH_2Cl_2 is decreased significantly compared with that in CCl_4 or decalin. However, the opposite effect is observed in the microwave region, where the CH_2Cl_2 apparent dipole moment increases on dilution in cholesteryl linoleate. Thus there is an inhibition of the intensity of the Poley process of CH_2Cl_2 when dissolved in molecules such as those which form a cholesteric phase. The integrated absorption intensity vs. molecular number density is plotted in Fig. 47 for the CH_2Cl_2 Poley band in various solvents at 298°K. These are carbon tetrachloride, decalin, cholesteryl oleyl carbonate, and 7CB. In CCl_4 and decalin, A/N is constant within the experimental uncertainty over the whole range of dilution, while it is clear that dilution in the solvents which have liquid-crystal-type phases reduces A/N considerably. This reflects an unusual constraint on angular movement (polarization) which persists when the concentrations of CH_2Cl_2 are such that no liquid crystalline properties are apparent on a macroscopic scale (e.g., when birefringence has disappeared from the 7CB solutions): this is substantiated by recent Kerr effect studies[88] where it was shown that the beginnings of liquid crystal behavior can be discerned in the "isotropic" phase long before the transition temperature into the mesophase, which is cloudy in visual appearance.

In the mesophase itself, this type of partial ordering was first observed using NMR methods of studying benzene in a nematic phase. A spectrum is obtained consisting of broad bands attributable to the solvent, superimposed on which was a series of sharp lines. Benzene acquires a preferential orientation due to solvent–solute interactions and its NMR spectrum is governed dipole–magnetic dipole interactions which are uniquely intramolecular in origin. The benzene molecules retain a translational freedom with respect to the nematic solvent which explains the sharpness of

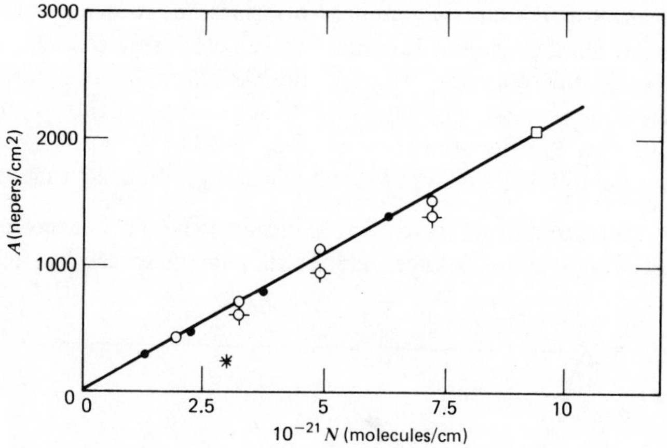

(a)

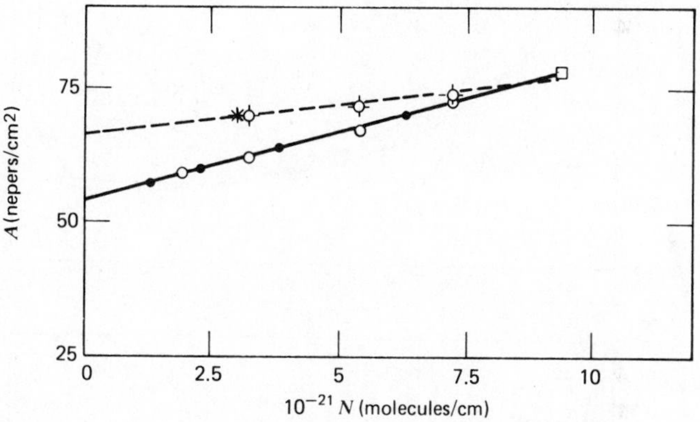

(b)

Fig. 47. (*a*) Plot of integrated absorption intensity (A) against N for all environments at 298°K: ⊙, in CCl_4; □, pure $CH_2Cl_2(l)$; ●, in decalin; -ϕ-, in cholesteryl oleyl carbonate (c.c.o.); *, in cholesteryl linoleate (C.L.). (—) Best straight line through the CCl_4 and decalin data. (*b*) Plot of $\bar{\nu}_{max}$ against number density in CCl_4 and decalin: (—) best straight line through the CCl_4 and decalin data; (---) best straight line through the cholesteric data; □, pure CH_2Cl_2; ⊙, CCl_4 solution; ●, decalin solution; -ϕ-, in cholesteryl oleyl carbonate; *, in cholesteryl linoleate. [Reproduced by permission from *J. Chem. Soc. Faraday Trans. 2*, **74**, 346 (1978).]

the NMR lines. The orientation is in the direction of the principal magnetic field. Since this discovery analogous NMR studies have shown that most molecules are preferentially oriented in a nematic phase. The effect of this on its far-infrared Poley absorption is retained in CH_2Cl_2 well into the apparently isotropic condition[54] Additional dynamical information is of course available in the zero-THz range because the band shape of the

Poley absorption contains dynamical information at short times in the orientational autocorrelation function, while NMR studies yield areas beneath a correlation function, and not the details of its analytical dependence. We illustrate this point in Fig. 48 where by roughly reproducing, using (I.30), the Fourier transform of $\alpha(\omega)$ of CH_2Cl_2 in cholesteryl oleyl carbonate, the following related functions may be produced analytically.

1. The orientational acf or dielectric decay function $\langle \cos\theta(t)\cos\theta(0)\rangle$, where θ is the angle between the dipole and the measuring field (I.30).

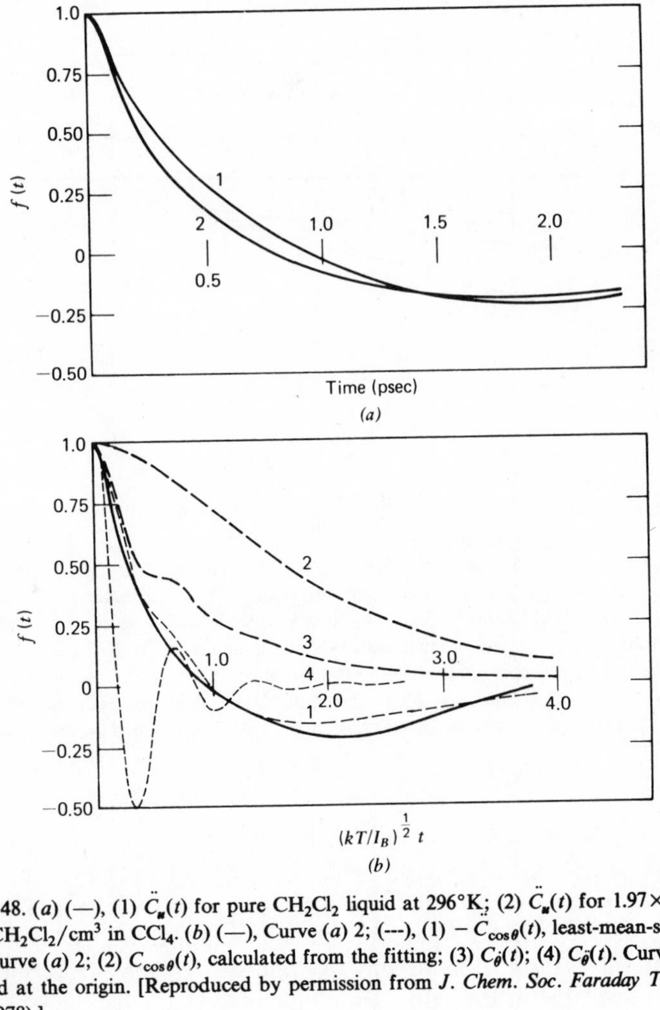

Fig. 48. (*a*) (—), (1) $\ddot{C}_u(t)$ for pure CH_2Cl_2 liquid at 296°K; (2) $\ddot{C}_u(t)$ for 1.97×10^{21} molecules CH_2Cl_2/cm^3 in CCl_4. (*b*) (—), Curve (*a*) 2; (---), (1) $-\ddot{C}_{\cos\theta}(t)$, least-mean-squares best fit to curve (*a*) 2; (2) $C_{\cos\theta}(t)$, calculated from the fitting; (3) $C_{\dot\theta}(t)$; (4) $C_{\ddot\theta}(t)$. Curves are normalized at the origin. [Reproduced by permission from *J. Chem. Soc. Faraday Trans. 2*, **74**, 350 (1978).]

2. The torque acf which mirrors the molecular librations by oscillating about the abscissa (time axis) [i.e., $\langle\ddot{\theta}(t)\ddot{\theta}(0)\rangle$].
3. The angular velocity acf the area beneath which is the NMR spin–rotation relaxation time.

1. Plastic, Disordered, or Amorphous Solids

We define these phases in terms of the continuing rotational freedom of individual molecules whose translation is very strongly hindered. In terms of the zero-THz profile different plastic and disordered crystals exhibit the full range of frequency coverage sketched in Fig. 9. Molecules with symmetric van der Waals contours, such as $(CH_3)_3CCl$, absorb with essentially the same zero-THz loss profile in the plastic crystalline phase as in the liquid just above the melting point. Any residual difference may then be attributed to the effect of rototranslation in the liquid as opposed to pure libration. Some of the rotational velocity correlation functions of these plastic phases are shown in Figs. 14 to 19 and have been discussed in greater detail by Haffmanns and Larkin[89] and by Davies.[90] In this section we are concerned more with the disordered and amorphous solids giving rise to a zero-THz profile with widely separated loss maxima on the frequency scale (Fig. 9). The rotational freedom remaining in these phases ensures that at THz frequencies a remnant of the liquid Poley-type absorption will remain as an indication of the torsional oscillation of the molecular dipole. This torsional oscillation will not be confined to one potential well over a long period of time, and a gradual movement through larger angles will give rise to an adjunct of the THz loss peaking at kHz frequencies and lower. The complete profile, sometimes covering much more than a dozen decades of frequency, must be amenable to treatment by an equation such as (I.6) represented by approximants such as (I.30), which are sophisticated enough to approximate $C_{\mathbf{u}}(t)$ adequately at short times. On these grounds alone the THz peak and low-frequency peak in the overall loss should form parts of the same continuous function of frequency. In terms of $C_{\mathbf{u}}(t)$ and its second derivative, the former decays exponentially from about 0.5 ps onwards, taking upwards of milliseconds and sometimes much larger, but the latter is oscillatory, being damped to zero in roughly the time that $C_{\mathbf{u}}(t)$ takes to become exponential. Theories of the low-frequency dielectric loss in disordered solids have usually been based on rotational diffusion (or alternatively on inertialess charge carrier hopping) which match the decay characteristics of $C_{\mathbf{u}}(t)$ but leave its second derivative undefined and produce not the required THz resonance but the Debye plateau. The contribution the far infrared can make to the molecular dynamics in these media has therefore been ignored. In this section we attempt to remedy this by showing that even the simplest form (I.30) of approximant capable of shifting $\bar{\nu}_{max}$ in the THz region may be used to reproduce the overall features of the *complete* zero-THz profile.

The itinerant librator as described in (I.30) and (I.31) is particularly well suited geometrically to describe the loss in the disordered phase of the hexasubstituted benzenes, since these are known to rotate in a plane about their hexad axes (Fig. 49). In pentachloronitrobenzene (PCNB), for example, Aihara, Kitazawa, and Nohara in 1970 detected a loss peaking between 30 Hz and 1 MHz in the temperature range 293 to 372°K with a large energy barrier to rotation.[91] An entropy difference between the stationary and transitional positions was calculated on the basis of plane reorientation between two opposite wells, ignoring the effect of molecular inertia. In the far infrared a peak at 38 cm^{-1} has been identified recently as librational in origin (Fig. 50) by invoking the harmonic approximation for reorientation of six-fold symmetry in the manner of Darmon and Brot,[92] who assume that the angular movement of the molecule occurs in a fixed crystalline potential. The libration frequency $\bar{\nu}_0$ is then defined for

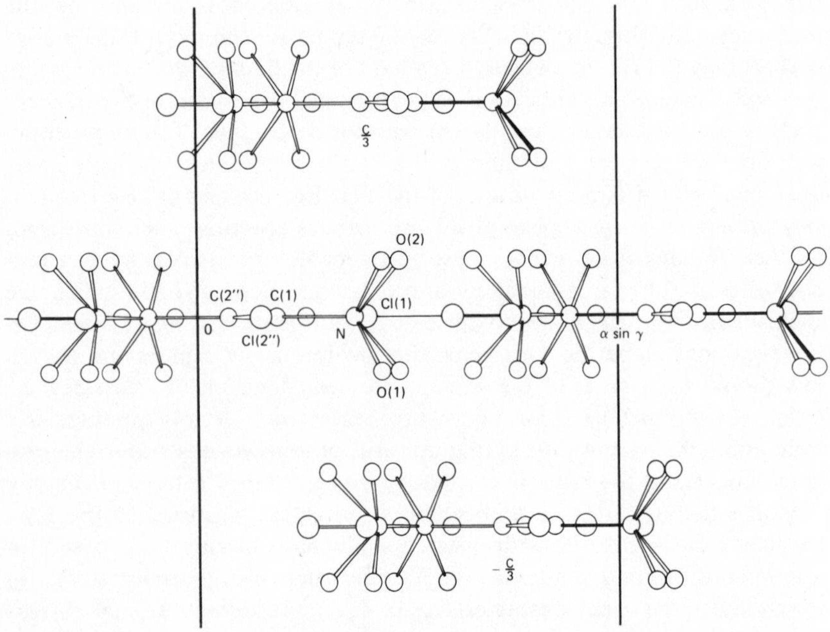

Fig. 49. Projection of the pentachloronitrobenzene crystal structure. [Reproduced by permission from *Acta Crystallogr.*, **30B**, 1546 (1974).]

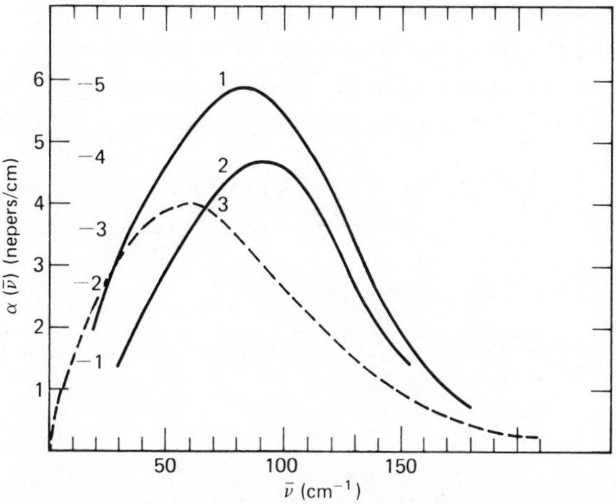

Fig. 50. Far-infrared part of the total loss profile in CH_2Cl_2 solutions. (1) Glass at 118°K (inner scale); (2) glass at 114°K (outer scale); (3) solution (liquid) at 298°K. Ordinate: $\alpha(\bar{\nu})$ (nepers/cm); abscissa: $\bar{\nu}$ (cm^{-1}). [Reproduced by permission from *Chem. Phys. Lett.*, **56**, 529 (1978).]

simple symmetries such as that of benzene by

$$\bar{\nu}_0 = \frac{3}{\pi c}\left(\frac{V}{2I}\right)^{1/2} \tag{II.71}$$

where I is the moment of inertia about the hexad axis. Naturally, (II.71) is an approximation for PCNB, especially since the large NO_2 group will set up potential barriers to rotation of different magnitude, resulting in an observed distribution of dielectric "relaxation times" which become nearly identical only at about 372°K. No account is taken by (II.71) of intermolecular coupling. This results in a distribution of librational frequencies and sets up vibrational waves throughout the lattice. Equation (II.71) has been used[92] to predict the observed Raman or far-infrared peak libration frequencies in plastic crystalline benzene, furane, and some other hexasubstituted benzenes which all lie in the range 30 to 60 cm^{-1}.

The frequency $2\pi\bar{\nu}_0 c$ from (II.71) may be identified with ω_0 of (I.31) in order to reproduce theoretically the required zero-THz profile of PCNB, which ranges experimentally from 30 Hz to 38 cm^{-1}. The factor β_1 of (I.31) is related to the low-frequency loss through τ_D, the inverse of the loss peak frequency. Therefore, the only phenomenological variable to be evaluated empirically is β_2, which has the effect of broadening the 38-cm^{-1}

resonance. No force fitting of the molecular parameters of the itinerant librator model is attempted. The moment of inertia of the annulus I_1 is estimated using the X-ray data of Tanaka, Iwasaki, and Aihara.[93] In the limit $kT/I_2\gamma^2 \langle 0.1$, the complex polarizability from (I.31) reduces to

$$\alpha_\mu^*(s)=\frac{kT}{I_2}\left[\frac{x(\Omega_0^2-\omega^2)+\omega\beta_1 y-i(\omega\beta_1 x-y(\Omega_0^2-\omega^2))}{x^2+y^2}\right]$$
$$x=\omega^2(\omega^2-x_3)+x_1(x_4-x_2\omega^2)$$
$$y=\omega(x_4+x_1(x_3-\omega^2)-x_2\omega^2) \tag{II.72}$$

with

$$x_1=\frac{kT}{I_1\beta_1}\left[\frac{\beta_1\omega_0^2}{\beta_1\omega_0^2+\beta_2\Omega_0^2}\right]$$
$$x_2=\beta_1+\beta_2$$
$$x_3=\omega_0^2+\Omega_0^2+\beta_1\beta_2$$
$$x_4=\beta_2\Omega_0^2+\beta_1\omega_0^2$$

Values of the various parameters used for PCNB are listed in Table VII.

For $\beta_2=0$, the THz resonance at the frequency $(\gamma/2\pi c)=38\ \text{cm}^{-1}$ is too sharp and Debye loss curves are produced theoretically at the low frequencies, which are too narrow (Fig. 51) in comparison with the broad experimental data. The effect of increasing β_2 is most clearly depicted in terms of the absorption coefficient $\alpha(\omega)$ of Fig. 52, where the THz resonance is depicted theoretically as rising above the intermediate Debye plateau.

TABLE VII
Parameters for PCNB Used in the
Itinerant Librator Model (I.31) ($I_1/I_2=10$)

Temperature (°K)	τ_D (sec)	$\frac{kT}{I_2\gamma^2}$	γ/β_1
293	1.1×10^{-4}	0.002	6.4×10^{-6}
313.4	1.3×10^{-5}	0.002	4.9×10^{-5}
333	3.1×10^{-6}	0.002	2.0×10^{-4}
353.7	8.2×10^{-7}	0.0025	6.9×10^{-4}
372.1	2.7×10^{-8}	0.0026	2.0×10^{-3}

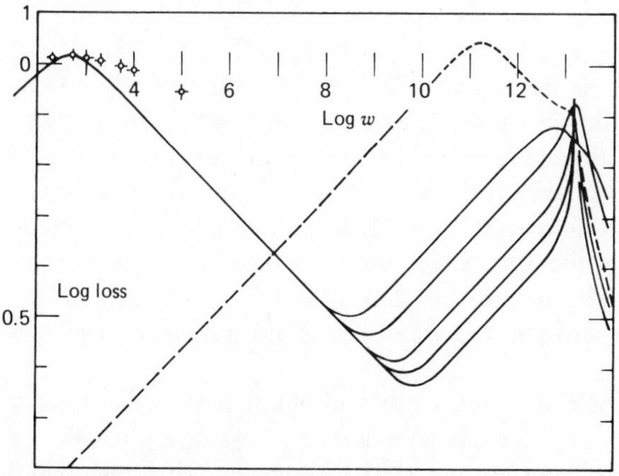

Fig. 51. $Log_{10}(\epsilon'')$ vs. $\log(\omega)$ representation of the total loss profile in glassy CH_2Cl_2/decalin at 114°K (1.97×10^{21} molecule/cm^3 of CH_2Cl_2). -⌖-, Some experimental points indicating that the observed low-frequency loss is broader than the model loss. The various high-frequency curves are for $\beta_2=0.1$, 2, 5, 10, and 20 TH_z. This illustrates the broadening effect of this friction coefficient. (----), Loss curve of a solution of CH_2Cl_2 in decalin at 296°K (room temperature). [Reproduced by permission from *Chem. Phys. Lett.*, **56**, 529 (1978).]

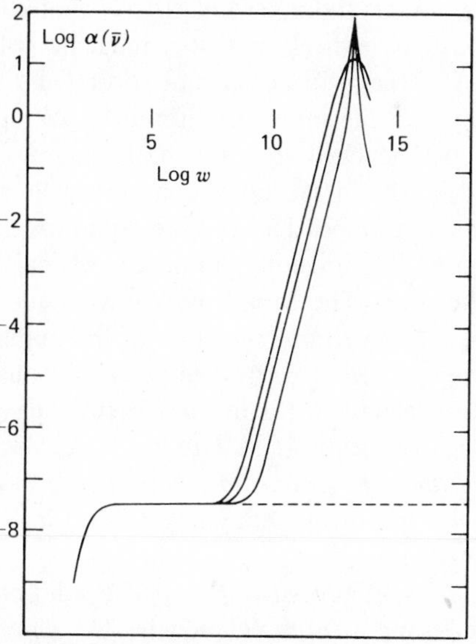

Fig. 52. $\mathrm{Log}\,\alpha(\omega)$ vs. $\log(\omega)$ representation of the absorption in glassy CH_2Cl_2/decalin. Note the horizontal Debye plateau, which continues as $\omega\rightarrow\infty$ for classical theories of Brownian motion. The far-infrared resonance rises steeply above this at 90 cm^{-1} (114°K) in itinerant oscillation. [Reproduced by permission from *Chem. Phys. Lett.*, **56**, 529 (1978).]

Taken overall, then, the approximant of (I.6) represented in (I.31) produces a fairly realistic picture of the overall zero-THz profile in TCTMB which may be extended to the losses observed generally in amorphous and disordered solids, as reviewed recently by Lewis.[36] Here it was emphasized that despite large differences in composition and structure among these solids, it is remarkable that the ac conductivity $[\sigma(\omega)=\omega\epsilon''(\omega)]$ varies according to an apparently simple law, albeit over a restricted range of low frequencies in the zero-THz range. In fact, this law is known to break down at frequencies greater than about 10^{10} Hz, and must do so in the THz range where the Poley resonance still remains on general theoretical grounds.

Examples of these solids are vanadium phosphate glass, doped silica, aluminium oxide, amorphous selenium, organic polymers, and molecular solids such as *trans*-carotene. In dielectric terms the response in these solids is coming from the high-frequency side of a loss peak [i.e., where $\epsilon''(\omega)$ decreases as ω increases], the peak itself and the low-frequency side beyond being inaccessible to conventional experimentation, owing to their very low frequencies. One of the objects of the review was to show that the classical concept of rigid dipole reorientation and that of localized hopping charge carriers can be unified and shown to be capable of describing similar phenomena in a noncrystalline solid. In view of the foregoing arguments, such models, to be entirely realistic, must be able also to produce an equation for the loss that will be realistic in the THz region (the far infrared). This is to say that the molecular libration, taking place as usual at THz frequencies, must be built into the basic equations for elementary polarization induced by donor–acceptor charge transfers. As it stands, the carrier hopping theory produces Debye-type equations for the elementary loss, which is integrated to give the total dielectrical response over all barrier heights in the solid. The broad, very low frequency loss may then be calculated theoretically without recourse to the empirical concept of a spread in Debye times. However, any Debye-type formalism is physically unrealistic when extrapolated to far-infrared frequencies, where the sharp quasiresonant Poley absorption is still to be expected *whenever there is enough molecular freedom of rotation to give rise to a low-frequency loss or ac conductivity*. This is true even though the broad loss peaks at frequencies too low to be measured.

In physical terms the spatial distribution in the solid consists of localized states which are oscillating at high frequencies. At a much lower range of frequencies, charge hopping occurs from one site to another, giving rise to the observed ac conductivity. This may be formulated by considering the elementary polarization

$$p(t)=V^{-1}\left[\sum_D e(1-f_i)r_i\cos\theta_i-\sum_A ef_ir_i\cos\theta_i\right] \tag{II.73}$$

where e is the electronic charge, f_i the probability that a state i (donor or acceptor) is occupied by an electron, r_i the distance of this state from a chosen origin, the angle between $\mathbf{r}_i$ and $\mathbf{F}(t)$, the probe field (not necessarily the internal field). If the vector $\mathbf{r}_i$ remains fixed in space, Debye-type equations for the elementary loss are obtained after certain simplifying assumptions:

$$\Delta\epsilon''(\omega)=e^2r_{ij}^2\frac{\cos^2\theta}{kT}\,\frac{\omega P_{ij}P_{ji}}{\omega_{ij}\left(\omega^2+\omega_{ij}^2\right)} \tag{II.74}$$

where P_{ij} and P_{ji} are transition rates between sites, and $\omega_{ij}=P_{ij}+P_{ji}$ is the characteristic relaxation frequency for the pair element in question. However, the dipole moment determining the polarization is $e\mathbf{r}_{ij}$, which is affected by a torsional oscillation in $\mathbf{r}_{ij}$ at far infrared frequencies. Thus in (II.73), if $\mathbf{r}_{ij}$ were given a characteristic libration frequency, $\cos\theta_i$ would become time-variant and the necessary high-frequency Poley adjunct would appear. The *overall* time correlation function for the reorientation of $e\mathbf{r}_{ij}$ would then be well behaved. It remains to be seen whether these ideas are mathematically tractable, but certainly librational-type absorptions in the far infrared should be observed for all the amorphous solids mentioned by Lewis as well as in zeolites and clathrates with dipolar guests, where whole molecule rototranslation, as opposed to libration, is the more important consideration. Pardoe and Fleming[94] have recently observed such bands in certain inorganic glasses. It may be mentioned finally that the libration of $e\mathbf{r}_{ij}$ would depend inversely as usual on an effective molecular or intermolecular moment of inertia, thus bringing the charge-carrier hopping model into line with the concept of inertial effects in the far infrared. It should be emphasized that in this type of theory we are not necessarily confined to molecular dipole autocorrelations, since hopping occurs between different librating molecular frameworks. Naturally, $\mathbf{r}_{ij}$ does not coincide with the molecular permanent dipole vector, in general.

F. Zero-THz Absorptions in Glasses and Viscous Liquids

The influence of viscosity in the medium surrounding a particle or molecule undergoing rototranslation in the fluid state is of basic interest. Viscosity is a factor which can be varied conveniently through orders of magnitude by supercooling solvents such as decalin or o terphenyl. This section purposes to demonstrate the influence of a highly viscous environment on the rotational dynamics of small dipolar molecules (in particular, CH_2Cl_2) in glassy and viscous liquid decalin and other convenient solvents. These manifest themselves in a similar fashion to those described in Section II.E [i.e., over a frequency range extending from audio frequencies (kHz) and below to the far infrared (or THz)]. This is in very marked contrast to the loss in the equivalent room-temperature solution, where it is restricted (as in CH_2Cl_2/decalin, for example) to the microwave (GHz) on the low-frequency side.

In Figs. 50 and 51 we illustrate the low- and high-frequency parts of the experimental loss in a solution containing 1.97×10^{21} molecules/cm^3 of CH_2Cl_2 in glassy decalin at 114°K as represented by the itinerant librator. A temperature difference of 4°K is enough to move the loss peak through almost two decades of frequency, so that the Arrhenius activation enthalpy for the process is high. The far-infrared part of the loss is shown over the temperature range 109 to 113°K in Fig. 50 as the absorption coefficient $\alpha(\omega)$. There is a slight movement to *higher* frequency as the temperature is lowered. There is also a slight drop in intensity. In comparison, the same concentration of CH_2Cl_2 in decalin liquid at 293°K produced a far-infrared peak at 61 cm^{-1}, almost 50 cm^{-1} below that of the glassy solution at 109°K. The low-frequency loss peak, which appears at 3.67 cm^{-1} in the high microwave at 293°K, has of course shifted downward by many decades in the glass.

For $(kT/I_2\gamma^2) < 0.1$ (as in this case) (II.72) may be employed to calculate analytically the zero-THz multidecade loss profile. The optimum value of β_2 for best fit to the far-infrared data is 10 THz, while $\beta_1 = 3.6 \times 10^{10}$ THz at 109°K. The vast difference between the two friction coefficients reflects the difference between single-molecule libration and collective reorientation of the cage of nearest neighbors—the latter, involving some translation, being much the more energetic and slower process. It is of course only by collective efforts that the encaged dipole ever manages to surmount the potential well in which it is librating at THz frequencies. Such an oversimplified model as (I.31) represents may be expected to reproduce only the gross features in the available experimental data [e.g., a distribution of inverse frequencies (or relaxation times) is seen at low frequencies]. However, the most urgent point here again is that the loss profile is not confined to the kHz frequencies but continues into the THz or far-infrared range where the Poley resonance is still clearly defined.

In the supercooled viscous liquids studied by Johari and others,[95] two absorption peaks (α and β processes) are often observed in molecules containing rigid, asymmetric, aromatic probes such as halogen and methyl-substituted benzenes and naphthalenes. Below the glass transition only the secondary relaxation is still observed. The loss curve we see in glassy CH_2Cl_2/decalin rapidly shifts up to microwave frequencies over a very narrow range of temperature at the glass-to-liquid transition temperature (T_g), in contrast to the larger molecules studied by Johari. With very fine temperature control an α and β spectrum appears in this small temperature range for CH_2Cl_2 also, so that in the viscous liquid the overall loss profile peaks *three times*, in the audio, MHz, and THz frequency regions. It is reasonable to suggest, therefore, that there exists in the viscous CH_2Cl_2/decalin liquid three loss peaks in the zero-THz profile which in general may be designated as primary, secondary, and tertiary processes of the overall dynamical evolution. The tertiary (γ) process is that of libration of the guest dipole showing up at THz frequencies, the other two being well documented by Johari and others and by Williams.[96] The β process represents the influence of nearest-neighbor cage fluctuations on the libration of the γ process, creating a diffusion of the encaged molecule from one energy well to another. It is the remnant of the liquidlike rotational process (more precisely rototranslational of course) first described by Debye in terms of inertialess spherical diffusion. The α process is one of bulk reorientation (inclusive of next nearest neighbors, etc.) and is the slowest. In the glass the cooperative motion becomes infinitely slow, and the α loss peak moves to zero frequency, leaving the β and γ processes to be described by our simplistic itinerant librator (i.e., simplistic in concept, almost intractable analytically).

The far infrared γ peak (i.e., the Poley absorption) is shifted dramatically to higher frequencies in the glassy CH_2Cl_2/decalin solution, (i.e., from 60 cm^{-1} at 298°K to 116 cm^{-1} in the glass). To put this in perspective, the root-mean-square angular velocity for a freely rotating CH_2Cl_2 ensemble is classically 21.2 cm^{-1}, so that the change in $\gamma = 2\pi\bar{\nu}_{max}c$ produced by a glassy environment is commensurate with that produced by condensing the infinitely dilute gas into the liquid at ambient temperature. At the same time, the β part of the Zero-THz profile moves from 110 GHz at 298°K to 5 kHz in the glass at 111°K. Very much *smaller* shifts to higher frequency in γ have been observed[125] by compressing under *kilobars* of external pressure liquids such as C_6H_5Cl and CS_2.

In contrast to the behavior of CH_2Cl_2/decalin glass, the enthalpy of activation of the β process of 10% v/v fluorobenzene/decalin is very low (18 kJ/mol). The predominant β process is observed to peak here at 77°K (i.e., 50°K below the glass transition). Assuming that the β process is due to rotation about the sixfold axis in this case, a simple model of harmonic libration such as that of Darmon and Brot produces a resonant delta function at 60 cm^{-1} using the enthalpy of activation measured at kHz frequencies. The observed γ peak is at 56 cm^{-1}. Across the halogenobenzene series the glassy β process shifts to lower frequencies, and is hardly detectable in bromobenzene. There is a corresponding shift to higher frequencies in the γ part of the overall loss, but not nearly as pronounced as that in CH_2Cl_2/decalin.

Tetrahydrofuran/decalin glass is also interesting and contrasts the CH_2Cl_2/decalin system in that the γ shift is much less and the enthalpy of activation again much smaller.

An oscillatory angular velocity autocorrelation function may be extracted from the best fit of (I.31) to both glassy β and γ parts. In Section III we simulate nitrogen in a high-temperature disordered lattice by computer molecular dynamics and find that this oscillatory behavior is also characteristic of the linear velocity autocorrelation function $C_{\mathbf{v}}(t)$. This increase of the oscillatory character is of course accompanied by the large γ shift. The β shift implies that $C_{\mathbf{u}}(t)$ will decay, almost exponentially, much more slowly in the glass.

III. MACHINE SIMULATIONS OF ROTATIONAL AND TRANSLATIONAL CORRELATION FUNCTIONS

There are available reviews[25,61] and some books[7,28] partially devoted to this topic, usually referenced under "molecular dynamics" and "Monte Carlo methods." Within the scope of this chapter the most incisive use of computer time may be made by pitting these techniques against our preconceptions of the fluid state. Modeling demands a degree of intuition before any equations may be set down on paper. The computer yields enough information about a small ensemble of molecules ($N \doteqdot 10^2$ or 10^3) to sound its depth. The "molecular dynamics" technique is particularly useful in our context since the Liouville (or Newton) equations are solved for a given intermolecular potential, the results being that for each molecule its trajectory is defined over a fraction of a picosecond or longer in terms of the first five derivatives of orientation and position. Therefore, it is possible to draw up a picture of our artificial droplet to a degree of detail which is itself almost as puzzling as a contemplation of the original fluid. Statistics therefore appear in terms of autocorrelation functions and multiparticle correlation functions, the latter being more difficult to compute, since a great deal more averaging is involved.

Some of the models of Table I have already been evaluated using this technique. Rahman,[97] in 1964, using 864 potentials representing argon atoms, demonstrated that the velocity autocorrelation function of the assemblage displayed a negative region out at long times (a few ps). The

Langevin equation (I.6), with $\mathbf{A}=[\mathbf{v}]$, $i\Omega_{\mathbf{A}}$ null, and $\phi_{\mathbf{A}}$ a delta function, is therefore oversimplified, since $\langle \mathbf{v}(t)\cdot\mathbf{v}(0)\rangle$ is exponentially decaying (but not when rotation is considered). Berne and Harp[25] in this series have simulated numerically the first memory function of CO, with a modified Stockmayer potential. Since then the number of simulations has grown—for example, the J-diffusion testing with rough spheres has already been mentioned. The development with which we are involved here is the extension to atom-atom Lennard-Jones interactions in diatomics of the molecular dynamics technique initiated by Barojas, Levesque, and Quentrec,[42] Cheung and Powles,[98] Streett and Tildesley,[99] and Singer and others.[100] In this section we use the algorithm developed by Streett and Tildesley to add a further dimension to the experimental evaluations of Section II. We simulate also, using disks bound within rough annuli, the analytical results of (I.31) with $\beta_2=0$. Essentially, this allows us to evaluate how well rough annulus/rough annulus interactions reproduce the Wiener statistics. If the match between the analytical results and the simulation is satisfactory in two dimensions, it will be reasonable to extend the simulation to three dimensions, and to simulate joint probability density functions which are intractable analytically. Throughout this section the following dimensionless, or reduced units are employed.

Bond length:	L^*	=	L/σ	(bond length/atom diameter)
Temperature:	T^*	=	kT/ε	
Density:	ρ^*	=	$\rho\sigma_e^3$	(where σ_e is the diameter of a sphere having a volume equal to that of the diatomic)
Pressure:	P^*	=	$P\sigma_e^3/\varepsilon$	

Here ε is defined through the fact that the potential energy of two diatomic molecules interacting via an atom–atom potential is the sum of four interactions between pairs of atoms not on the same molecules. For the Lennard-Jones model the atom–atom interactions take the form

$$U^{\mathrm{LJ}}(r)=4\varepsilon\left[\left(\frac{\sigma}{r}\right)^{12}-\left(\frac{\sigma}{r}\right)^{6}\right] \tag{III.1}$$

where r is the distance between atoms on different molecules.

For a purely repulsive (hard) diatomic the equivalent potential is

$$\begin{aligned} U^{LJR}(r)&=4\varepsilon\left[\left(\frac{\sigma}{r}\right)^{12}-\left(\frac{\sigma}{r}\right)^{6}+\frac{1}{4}\right] \qquad (r/\sigma)\le 2^{1/6} \\ &=0 \qquad (r/\sigma)>2^{1/6} \end{aligned} \tag{III.2}$$

A. The Molecular Dynamics Method

We briefly review the method involved, following Streett and Tildesley, who base their algorithm in turn on that developed by Cheung and Powles. The equations of motion are written in vector form:

$$M\ddot{\mathbf{r}}_i = \mathbf{F}_i \tag{III.3}$$

$$ML^2\boldsymbol{\omega}_i = 4\mathbf{T}_i \tag{III.4}$$

where $\mathbf{r}_i$ is the center-of-mass coordinate for molecule $i, \boldsymbol{\omega}_i$ its angular velocity, and $\mathbf{F}_i$ and $\mathbf{T}_i$ are the net force and torque exerted on particle i by all other particles. Forces and torques are computed for all molecular pairs having center-to-center separations less than $2.5\sigma + L$, where L is the interatomic separation. This ensures that all atom-atom interactions at distances of 2.5σ or less are counted. At this distance the potential energy of two atoms interacting via the L-J potential (III.1) is of the order of 1% of the well depth. The virial theorem is used (as per Cheung and Powles) to correct the computed pressure and energy for long-range interactions.

Equations (III.3) and (III.4) are integrated numerically by means of a fifth-order predictor-corrector method due to Gear.[101] Since the particles involved are, ideally, linear, all centers of force within a molecule lie on its axis. As a consequence, vectors representing the torque, angular acceleration, and higher derivatives of angular position are always perpendicular to the axial vector **L** of the molecule. This allows the use of (III.4) rather than a second-order equation for angular position. Simulations in this section are carried out with a cube of 256 diatomic molecules arranged initially on an α-nitrogen lattice (fcc). Periodic boundary conditions are used which ensure that when a molecule leaves one side of the cube during the course of the simulation, another replaces it with the coordinates (x, y, z) displaced by the cube side length. After a complicated initial step, the simulation is allowed to run for about 1600 time steps [in units of $(M\sigma^2/\varepsilon)^{1/2}$], each of 0.0016 after rejecting the first few unstable steps. These units each correspond to a real time of the order of 10^{-15} sec. The calculated pressures and configurational internal energies are in excellent agreement with those calculated by Singer and others,[100] who have used a completely different molecular dynamics algorithm based on a different method of solving the equations of motion. The unpublished Monte Carlo calculations of Streett and Tildesley are in excellent agreement with their molecular dynamics data for $L^* = 0.3292(N_2)$, as well as with those of Cheung.

1. Computation of Correlation Functions

The fifth-order predictor–corrector algorithm used means that the first five derivatives of orientation and position may be stored on magnetic tape for future statistical analysis. For any element **A** of (I.6), its autocorrelation function may be calculated using the running time average

$$C_{\mathbf{A}}(t)=\frac{1}{T}\int_0^T\sum_{j=1}^{256}\mathbf{A}_j(\tau)\cdot\mathbf{A}_j(\tau+t)\,d\tau \tag{III.5}$$

using different initial times. In (III.5) T is the total time over which the simulation runs with j molecules. We notice that in any algorithm which conserves the total linear momentum ($\Sigma_i M\dot{\mathbf{r}}_i=\mathbf{0}$), the normalized autocorrelation function and cross-correlation of velocity and force will decay identically. This is because

$$\begin{aligned}\mathbf{v}_j(0)\cdot\sum_{k\neq j}\mathbf{v}_k(t)&=\mathbf{v}_j\cdot\left(\sum_{k=1}^{N}\mathbf{v}_k(t)-\mathbf{v}_j(t)\right)\\&=-\mathbf{v}_j(0)\cdot\mathbf{v}_j(t)\end{aligned} \tag{III.6}$$

Therefore, it is possible to calculate collective correlation functions only when these take forms such as those of the longitudinal and transverse current and spin densities of hydrodynamical theory:

$$C(\mathbf{k},t)=\frac{1}{T}\int_0^T J^*(\mathbf{k},\tau)J(\mathbf{k},t+\tau)\,d\tau \tag{III.7}$$

where the wave vector $\mathbf{k}$ stands for $k_\perp$ or $k_\parallel$, and T is the total simulation time. We have, for current densities,

$$\begin{aligned}\mathbf{J}(\mathbf{k},t)=N^{-1/2}\sum_j\Big[&M\mathbf{v}_j^{(1)}(t)\exp\big(i\mathbf{k}\cdot\mathbf{r}_j^{(1)}(t)\big)\\&+M\mathbf{v}_j^{(2)}(t)\exp\big(i\mathbf{k}\cdot\mathbf{r}_j^{(2)}(t)\big)\Big]\end{aligned}$$

where $\mathbf{v}_j^{(1)}(t)$ is the velocity of the first atom of the jth molecule, $\mathbf{v}_j^{(2)}(t)$ that of the second atom. In Section IV we build these up from individual molecular vectors and thus attempt to bridge the gap between molecular and hydrodynamic theories.

We shall illustrate the use of molecular dynamics simulations in evaluating the approximation, (I.31), when the equations are applicable to space-itinerant oscillation of the molecular linear velocity $\mathbf{v}$.

B. Translational Motion—Simulations and Itinerant Oscillation

The itinerant oscillator model for motion in atomic fluids and uncoupled linear motion in molecular fluids was developed by Sears[102] in 1965 following some speculative remarks by Frenkel. Unfortunately, Sears's paper is mathematically a little flawed, as was pointed out by Damle and others.[29]

In this section we shall use the simplest version of this model consistent with the concept involved to calculate $C_{\mathbf{v}}(t)$, the linear velocity autocorrelation function. The analytical $C_{\mathbf{v}}(t)$ is then compared with that simulated using (III.1), (III.3), and (III.4). The self part of the van Hove correlation function $G_s(\mathbf{r},t)$ is evaluated analytically and compared with the experimental neutron scattering results of Dassannacharya and Rao[103] on liquid argon, and the theoretical $C_{\mathbf{v}}(t)$ is also compared with the computer simulation of this function for liquid argon carried out by Rahman.[97] This is a good check on internal and interexperimental consistency, since $G_s(\mathbf{r},t)$ can be expressed in terms of $C_{\mathbf{v}}(t)$ using the techniques of Section I. By evaluating the speed acf (that of $|\mathbf{v}|$) and that of the direction of the velocity, following Berne and Harp,[25] it is shown that a constant-speed approximation is valid in treating translational properties of fluids, confirming their results for CO.

Equation (I.6), for uncoupled, linear motion of the center of mass of an atom or molecule of mass m, reduces to

$$\dot{\mathbf{v}}(t)+\int_0^t K(t-\tau)\mathbf{v}(\tau)\,d\tau=\mathbf{f}(t)/m \tag{III.8}$$

where $\mathbf{f}(t)$ is defined by

$$K(t)=\langle \mathbf{f}(t)\cdot\mathbf{f}(0)\rangle\frac{m}{3kT}$$

Since $\langle \mathbf{f}(t)\cdot\mathbf{v}(0)\rangle=0$, we have the further relations

$$\dot{C}_{\mathbf{v}}(t)=-\int_0^t K(t-\tau)C_{\mathbf{v}}(\tau)\,d\tau \tag{III.9}$$

and the Mori series

$$\frac{\partial}{\partial t}K_{n-1}(t)=-\int_0^t K_n(t-\tau)K_{n-1}(\tau)\,d\tau \tag{III.10}$$

where $n=0,\ldots,N$ are positive integers. In this notation $K_{-1}(t)\equiv C_{\mathbf{v}}(t)$. No intermode coupling or cross-correlations [describable by $G_d(\mathbf{r},t)$, the distinct part of the van Hove function] are accounted for in these equations. Truncating (III.10) with

$$K_1(t)=K_1(0)\exp(-\gamma t)$$

produces a result for $\tilde{C}_{\mathbf{v}}(s)$ formally identical with that for the equations of motion:

$$m\ddot{\mathbf{q}}_1(t) + m\gamma\dot{\mathbf{q}}_1(t) - m_1 K_0(0)[\mathbf{q}-\mathbf{q}_1] = m_1\dot{\mathbf{W}}_1(t)$$

$$m\ddot{\mathbf{q}}(t) + mK_0(0)(\mathbf{q}-\mathbf{q}_1) = \mathbf{0} \tag{III.11}$$

$$K_1(0) = (m/m_1)K_0(0)$$

$$K_0(0) = \omega_0^2$$

Here m is the mass of the atom or molecule whose coordinate is $\mathbf{q}$ and which is surrounded by a diffusing "cage" of such particles whose center of mass is at $\mathbf{q}_1$ and whose total mass is m_1. The inner particle m is harmonically bound at a frequency ω_0 to the diffusing cage with a restoring force constant $K_0(0)$. We note that $\mathbf{v} \equiv \dot{\mathbf{q}}$. A frictional force $\gamma\dot{\mathbf{q}}_1$ acts in opposition to the diffusing cage and $\dot{\mathbf{W}}_1(t)$, which is represented by a statistical Wiener process, is the force on the cage caused by "random" collisions. The two versions of $C_{\mathbf{v}}(t)$ [from (III.8) and (III.11)] take the form of (II.39).

The van Hove function $G_s(\mathbf{r},t)$ for self correlations may be evaluated by considering (III.8) in the form

$$\ddot{\mathbf{r}}(t) + \int_0^t K(t-\tau)\dot{\mathbf{r}}(\tau)\,d\tau = \mathbf{f}(t)/m \tag{III.12}$$

and classically from (III.11) by a method to be described shortly. In (III.12) we have $\dot{\mathbf{r}} \equiv \mathbf{v} \equiv \dot{\mathbf{q}}$, the velocity of the tagged inner particle of mass m. $G_s(\mathbf{r},t)$ is the probability of finding this particle at $\mathbf{r}$ at time t given that it could be found at $\mathbf{r}=0$ when $t=0$. Using (I.33) and (I.34), the probability density function

$$p(\mathbf{r}(t),\mathbf{r}(0),\mathbf{v}(0);t) = \left[\frac{3}{2\pi B(t)}\right]^{3/2} \exp\left[-\frac{3|\mathbf{y}(t)|^2}{2B(t)}\right] \tag{III.13}$$

where

$$\mathbf{y}(t) = \mathbf{r}(t) - \mathbf{r}(0) - \Gamma_{\mathbf{v}}(t)\mathbf{v}(0)$$

$$= \frac{1}{m}\int_0^t \Gamma_{\mathbf{v}}(t)\mathbf{f}(t-\tau)\,d\tau$$

is the solution of (III.12). In (III.13)

$$\Gamma_v(t) = \mathfrak{L}_a^{-1}[s(s+\tilde{K}(s))]^{-1}$$

$$= \int_0^t \frac{\langle \mathbf{v}(t)\cdot\mathbf{v}(0)\rangle}{\langle \mathbf{v}(0)\cdot\mathbf{v}(0)\rangle} dt \tag{III.14}$$

$$B(t) = \frac{3kT}{m}\left[2\int_0^t \Gamma_v(t)\,dt - \Gamma_v^2(t)\right] \tag{III.15}$$

To obtain $G_s(\mathbf{r},t)$ we must average over all initial $\mathbf{v}(0)$ values so that

$$G_s(\mathbf{r},t) = \left[\frac{3}{2\pi B(t)}\right]^{3/2} \exp\left[-\frac{3|\mathbf{r}(t)|^2}{2B(t)}\right] \tag{III.16}$$

Equation (III.16) links $G_s(\mathbf{r},t)$ directly to $C_v(t)$. In classical Brownian translational theory, $C_v(t) = \exp(-\beta t)$, so

$$\Gamma_v(t) = (1-\exp(-\beta t))/\beta$$

and

$$B(t) = \frac{3kT}{m\beta^2}(2\beta t + 4e^{-\beta t} - e^{-2\beta t} - 3)$$

in agreement with the calculations of Uhlenbeck and Ornstein.[104] In our case the equivalent expressions are

$$\Gamma_v(t) = x_0[1 - \exp(-\alpha_1 t)(\cos\beta t + x_1 \sin\beta t) + x_2(1-\exp(-\alpha_2 t))]$$

where

$$x_0 = \frac{2\alpha_1 + \Gamma\alpha_2}{(1+\Gamma)(\alpha_1^2+\beta^2)} \qquad x_1 = \frac{\alpha_1^2 - \beta^2 + \Gamma\alpha_1\alpha_2}{\beta(2\alpha_1 + \Gamma\alpha_2)} \qquad x_2 = \frac{\Gamma(\alpha_1^2+\beta^2)}{\alpha_2(2\alpha_1+\Gamma\alpha_2)}$$

1. *Probability Density Functions from (III.11)*

Without loss of generality one may consider for purposes of computation the behavior of the ith component ($i = 1,2,3$) of (III.11), which may be written[105] in the matrix form

$$\dot{\mathsf{X}}(t) = \mathsf{A}\mathsf{X}(t) + \mathsf{B}\dot{\mathsf{W}}_i(t) \tag{III.17}$$

where

$$\mathsf{A} = \begin{bmatrix} 0 & 0 & 1 & 0 \\ 0 & 0 & 0 & 1 \\ -\omega_0^2 & \omega_0^2 & 0 & 0 \\ \Omega_0^2 & -\Omega_0^2 & 0 & -\beta \end{bmatrix} \qquad \mathsf{B} = \begin{bmatrix} 0 \\ 0 \\ 0 \\ 1 \end{bmatrix} \qquad \mathsf{X} = \begin{bmatrix} X_1 \\ X_2 \\ X_3 \\ X_4 \end{bmatrix} = \begin{bmatrix} R_i \\ r_i \\ \dot{R}_i \\ \dot{r}_i \end{bmatrix}$$

Here

$$\Omega_0^2 = (m_1/m)\omega_0^2 \qquad \mathbf{R} = \mathbf{q}$$

Equation (III.17) may be solved formally to give

$$\mathsf{X}(t) = (\exp \mathsf{A}t)\mathsf{X}_0 + \int_0^t \exp[\mathsf{A}(t-\tau)]\mathsf{B}\xi(d\tau) \tag{III.18}$$

with

$$W_i(t_2) - W_i(t_1) = \xi(t_2 - t_1)$$

The van Hove function may be calculated from (III.17) and (III.18) by virtue of the fact that it is the Gaussian probability density function of $X_1(t)$. Thus

$$G_s(X_1, t) = \left[\frac{3}{2\pi\langle Y_1^2(t)\rangle}\right]^{3/2} \exp\left[-\frac{3Y_1^2(t)}{2\langle Y_1^2(t)\rangle}\right] \tag{III.19}$$

It turns out from the formal solution, (III.18), that

$$Y_1(t) = \int_0^t \omega_0^2 \int_0^{t-\tau} g_2(u)\, du\, \xi(d\tau)$$

so that

$$\langle Y_1^2(t)\rangle = \omega_0^4 C^2 \int_0^t \left[\int_0^{t-\tau} g_2(u)\, du\right]^2 d\tau$$

with

$$g_2(t) = \left[(\alpha_1 - \alpha_2)^2 - \beta^2\right]^{-1} \left[\frac{(\cos\beta t - (\alpha_2 - \alpha_1)\sin\beta t)}{\beta} e^{-\alpha_1 t} - \exp(-\alpha_2 t)\right]$$

The constant C^2 is deduced from the limit at long times of (III.19).

2. *Comparisons with Molecular Dynamics. Simulations of N_2*

The atom-atom computed force, velocity, speed, and direction of velocity autocorrelation functions are shown in Fig. 53 along with the least-mean-squares best fits for the force, $(m\dot{\mathbf{v}})$ acf's calculated from (II.39) with ω_0^2, Ω_0^2, and γ as variables (Table VIII).

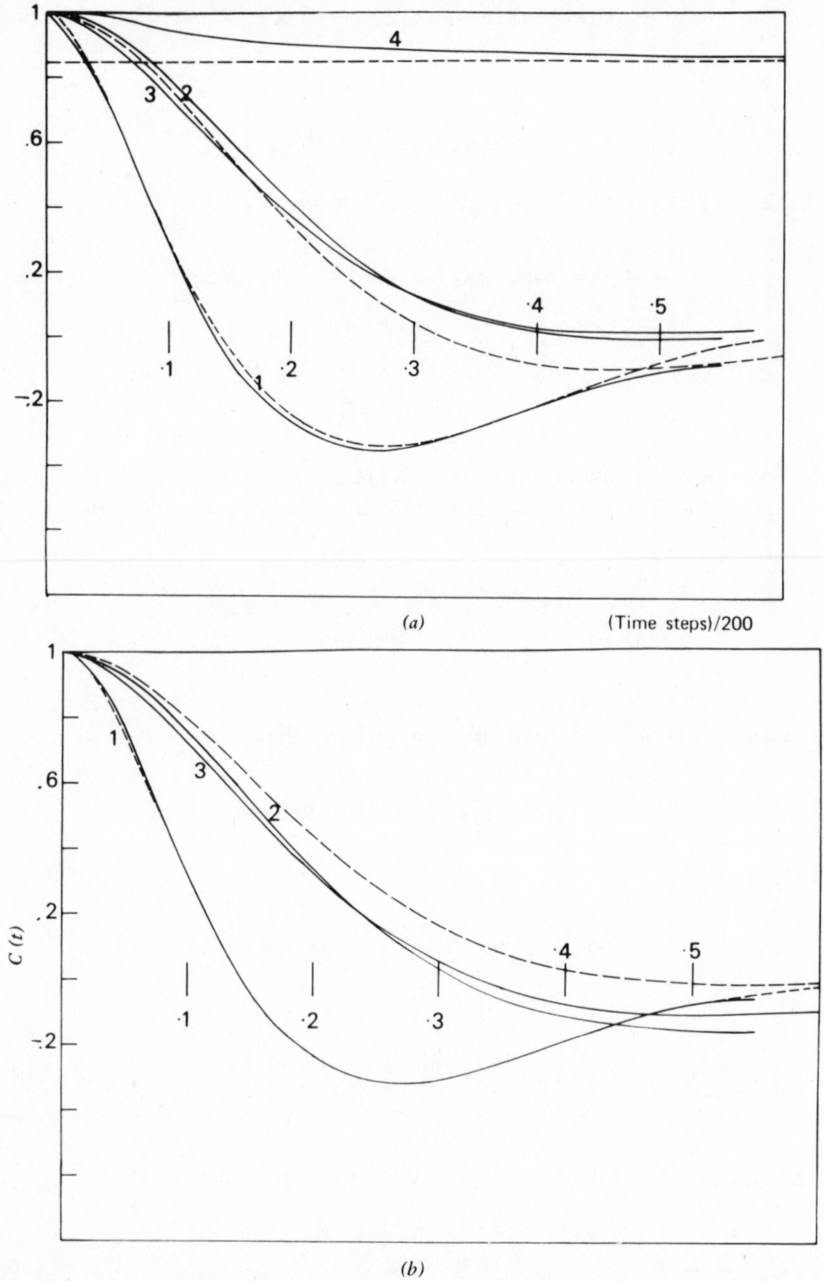

Fig. 53. (*a*) (—), (1) Atom–atom simulation of the force acf $C_F(t)$ for a reduced interatomic separation, L^* of 0.1. (—), (2) Simulated velocity acf. (—), (3) Simulated direction of velocity acf. (—), (4) Simulated speed acf, the horizontal line is at $8/(3\pi)$. (----), (1) Least-mean-squares best fit of the itinerant oscillator to the simulated $C_F(t)$. (----), (2) $C_v(t)$ (itinerant oscillator), calculated with the ω_0^{2*}, Ω_0^{2*}, and γ^* estimated by fitting $C_F(t)$. (*b*) $L^*=0.3$; (*c*) $L^*=0.5$; (*d*) $L^*=0.7$. Ordinate: $C(t)$; abscissa: (time steps)/200. [Reproduced by permission from *J. Mol. Struct.*, **46**, 395 (1978).]

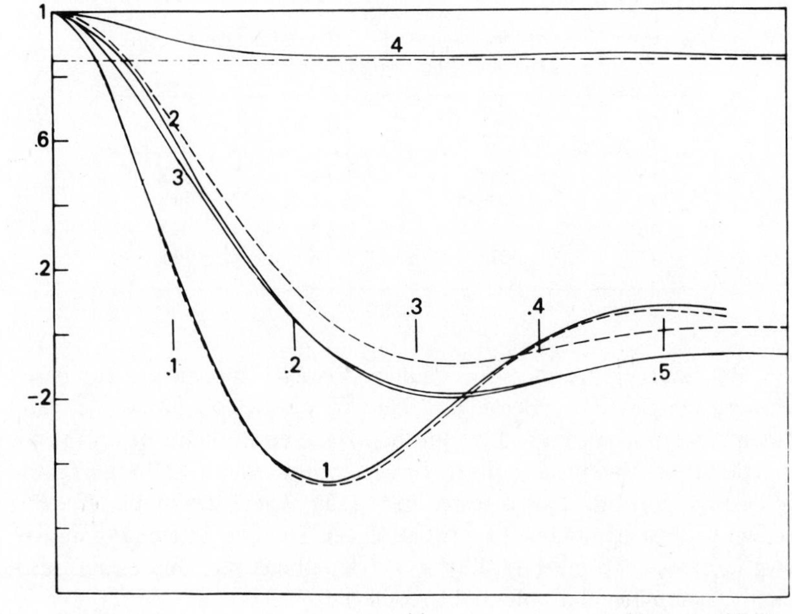

(c)

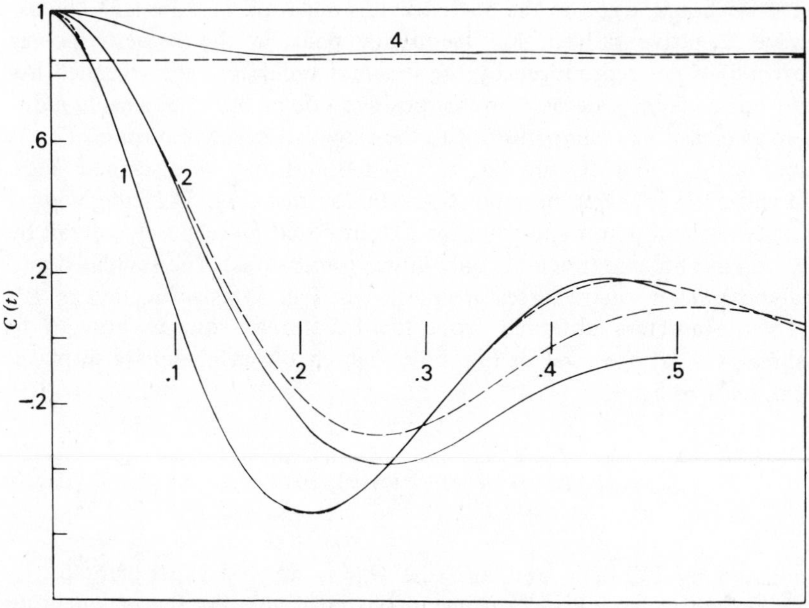

(d)

Fig. 53. Continued

TABLE VIII
Parameters for Least-Mean-Squares Best Fit to Molecular Dynamics Data of $C_F(t)$ ($\rho^* = 0.64$, $T^* = 2.3$)

L^*	ω_0^{2*}	Ω_0^{2*}	γ^*
0.1	56.5	313.9	38.6
0.3	47.8	263.3	29.1
0.5	80.6	213.3	21.4
0.7	127.9	250.6	30.8

ω_0^{2*}, Ω_0^{2*}, and γ^* are in reduced units. Velocity spectra are compared with those simulated by Berne and Harp[25] (on carbon monoxide), and Rahman (on argon) in Fig. 54. From the formal equivalence of $C_{\mathbf{v}}(s), \omega_0^2$ is proportional to the mean-square force, computed as $\langle F^2 \rangle$ using the atom–atom algorithm. This is tested in Fig. 55, where the simulated $\langle F^2 \rangle$ is plotted against ω_0^2 obtained by fitting $C_{\mathbf{F}}(t)$. The overall trend is similar, but ω_0^2 increases the more rapidly as L^* lengthens (i.e., the more anisotropic the intermolecular potential becomes).

Figure 54 shows the least-mean-square best fit to the velocity acf computed for liquid argon in the 864-particle simulation of Rahman. The extended negative tail (or low-frequency peak in the velocity power spectrum) is not reproduced by the itinerant oscillator. The so-called hydrodynamic tail is a decay from the positive side of the $C_{\mathbf{v}}(t)$ axis, and difficult to measure in comparison with the extended negative portion. This is found again in the CO simulation (Fig. 54) and may be discerned (Figs. 53*a* and 53*d*) in the atom–atom $C_{\mathbf{v}}(t)$. In contrast (Fig. 54*d*), the angular velocity acf and power spectrum for CO are fitted more closely overall by the itinerant librator (tractable only in two dimensions). The simulated and analytical mean square forces are plotted in Fig. 55 as a function of L^*.

The parameters obtained from the least-mean-squares best fit to Rahman's $C_{\mathbf{v}}(t)$ are used in Fig. 56 to match the mean-square displacement, defined by

$$\langle \Delta_r^2 \rangle = 2 \int_0^t (t-\tau) \langle \mathbf{v}(\tau) \cdot \mathbf{v}(0) \rangle \, d\tau \qquad \text{(III.20)}$$

simulated by Rahman, and also the $G_s(\mathbf{r}, t)$ derived experimentally by Dasannacharya and Rao,[103] using incoherent, inelastic, thermal neutron scattering.

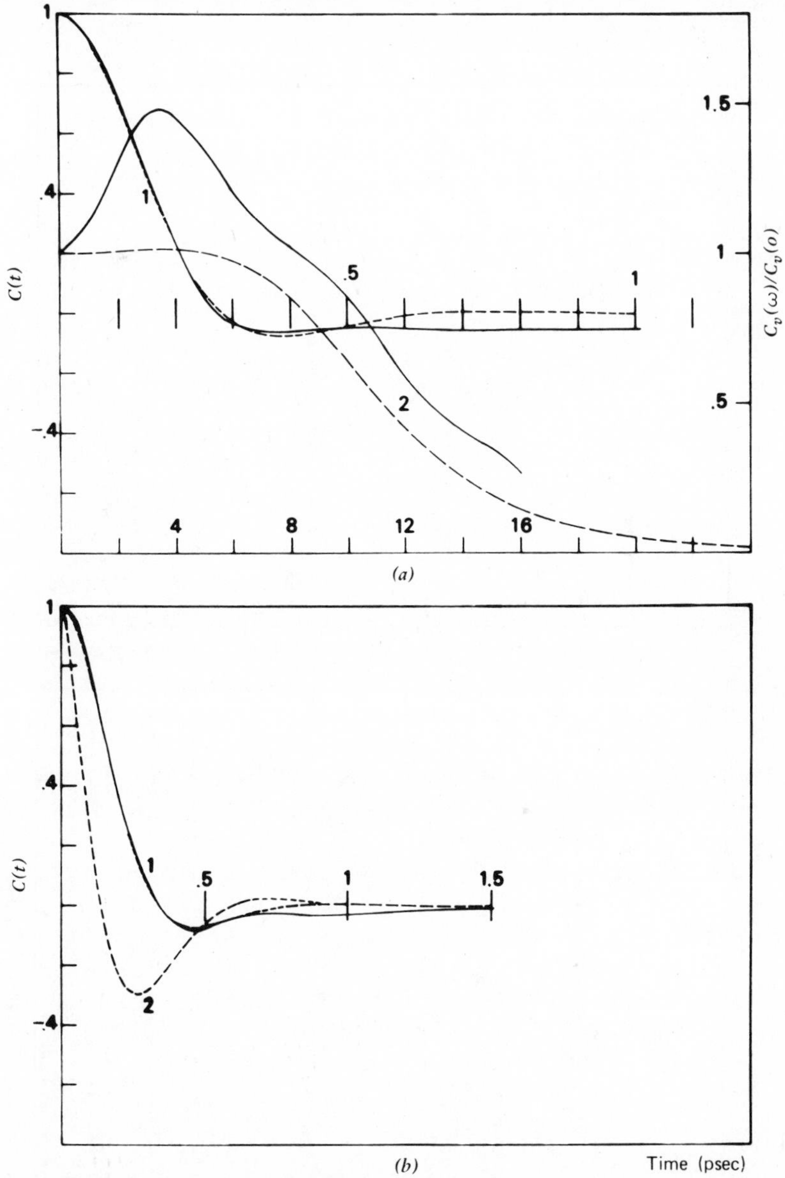

Fig. 54. (*a*) (—), (1) Berne and Harp computed velocity acf for CO. (—), (2) Computed, velocity power spectrum. (----), (1) Itinerant oscillator least-mean-squares best fit to the velocity acf. (---), (2) Corresponding normalized velocity power spectrum. Ordinates: left—$C(t)$; right—$C_v(\omega)/C_v(\mathrm{o})$. Abscissas: top, time (psec); bottom, frequency (THz). (*b*) (—), Rahman $C_v(t)$, simulated for liquid argon by Rahman. (----), (1) Itinerant oscillator, best fit. (---), (2) $C_F(t)$ estimated from the $C_v(t)$ best fit. Ordinate: $C(t)$; abscissa: time (psec). (*c*) (—), Rahman-simulated, normalized, velocity, power spectrum. (---), (1) Velocity power spectrum calculated from the itinerant oscillator best fit to $C_v(t)$ (Fig. 53). (---), (2) Itinerant oscillator normalized force spectrum. Ordinate: intensity; abscissa: frequency (THz). (*d*) (—), (1) Berne and Harp simulated angular velocity acf for liquid CO. (---), (1) Best fit to (1) of the itinerant librator in a plane. (—), (2) Simulated normalized angular velocity power spectrum. (---), (2) Itinerant oscillator normalized power spectrum calculated from fitting the acf. Ordinates: left—$C_\omega(t)$; right—$C_\omega(\omega)/C_\omega(\mathrm{o})$; abscissas: top, time (psec); bottom, ω (THz). [Reproduced by permission from *J. Mol. Struct.*, **46**, (1978).]

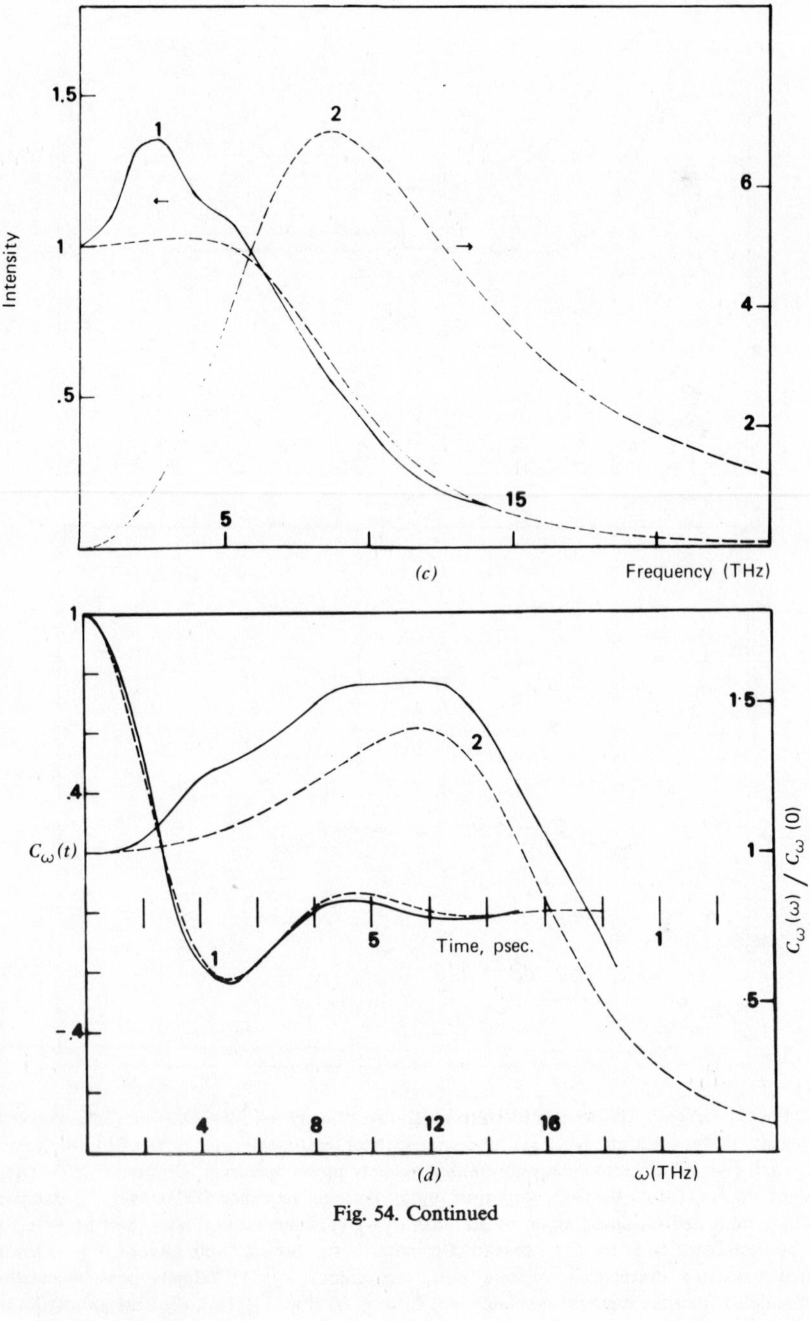

Fig. 54. Continued

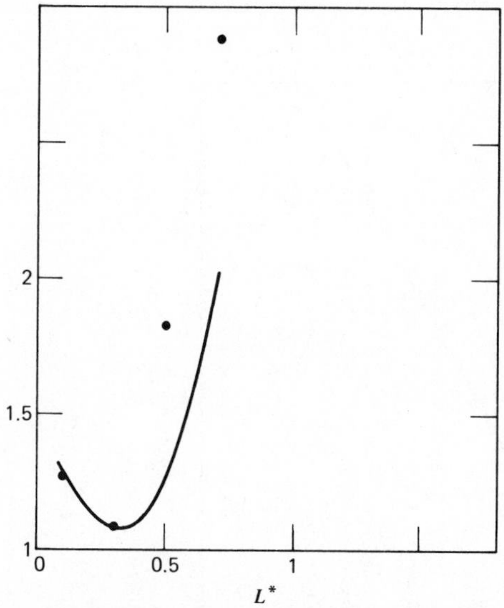

Fig. 55. Plot of (—), $\langle F^2 \rangle$ and (●), ω_0^2 vs. L^* normalized at $L^* = 0.3$. [Reproduced by permission from *J. Mol. Struct.*, **46** (1978).]

In Section I it is of interest to know whether the Wiener process is, in fact, a justifiable statistical representation of random force and velocity. We adopt the method of computing acf's of moments of velocity and force to investigate this further. For example, the second moment of velocity (or kinetic energy) acf,

$$C_{2v}(t) = \langle v^2(t) v^2(0) \rangle / \langle v^4(0) \rangle$$

should be related to $C_v(t)$ by

$$C_{2v}(t) = \tfrac{3}{5}\left[1 + \tfrac{2}{3} C_v^2(t)\right] \qquad \text{(III.21)}$$

where the probability density function of velocities is Gaussian. Similarly,

$$C_{4v}(t) = \left(225 + 600 C_v^2(t) + 120 C_v^4(t)\right)/945 \qquad \text{(III.22)}$$

and so on, as evaluated by Berne and Harp. The functions $C_{2v}(t)$ and $C_{4v}(t)$ are calculated analytically using the atom-atom $C_F(t)$ to optimize ω_0^2, and Ω_0^2, and γ. They can also be simulated independently using the atom-atom algorithm, and the two sets of functions are compared in Fig. 57.

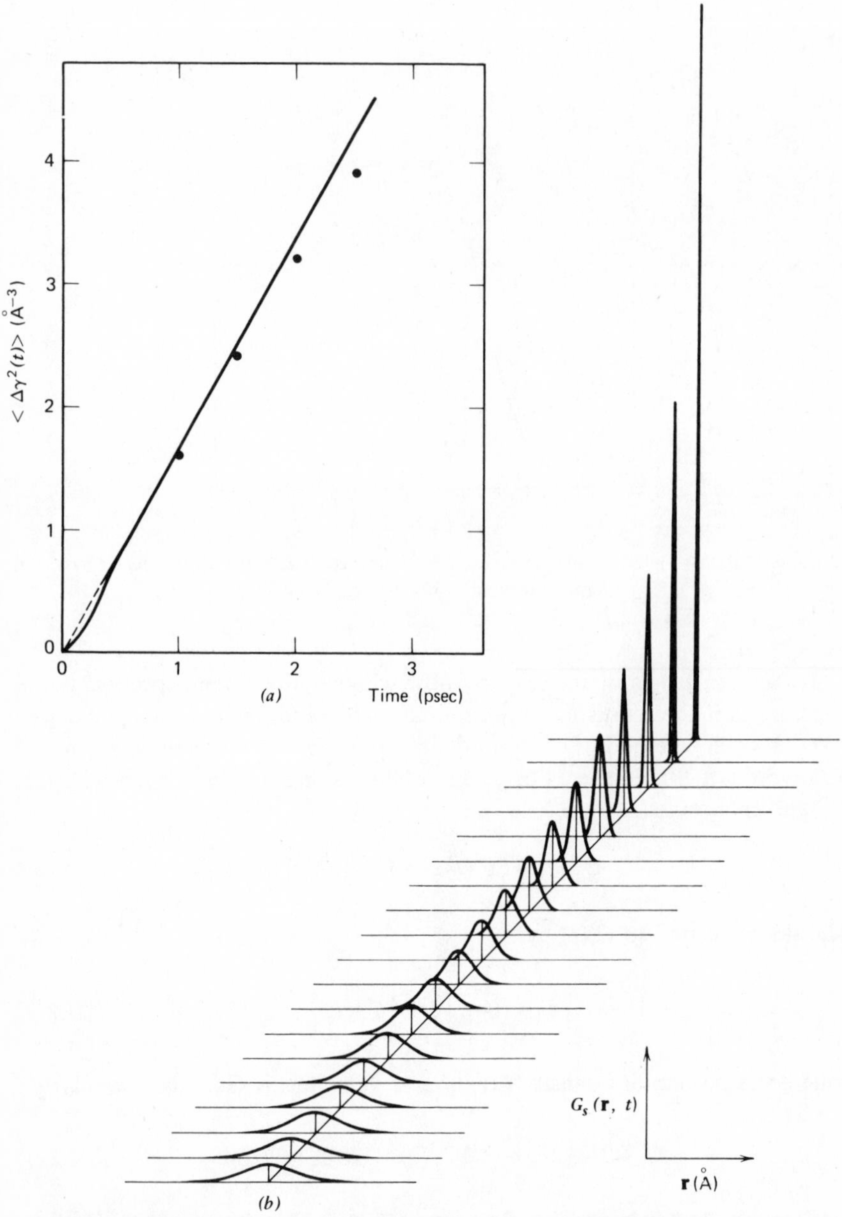

Fig. 56. (*a*) Plot of mean-square displacement. (—) $\langle \Delta r^2(t) \rangle$ calculated from the itinerant oscillator fitting to Rahman's $C_v(t)$. ●, Mean-square displacements computed independently by Rahman. Ordinate: $\langle \Delta r^2(t) \rangle$ (Å); abscissa: time (psec). (*b*) Plot of $G_s(\boldsymbol{r}, t)$ calculated for the itinerant oscillator from fitting the Rahman $C_v(t)$ function. Ordinate: $G_s(\boldsymbol{r}, t)$ (Å^{-3}); abscissa: $\boldsymbol{r}$ (Å). [Reproduced by permission from *J. Mol. Struct.*, **46** (1978).]

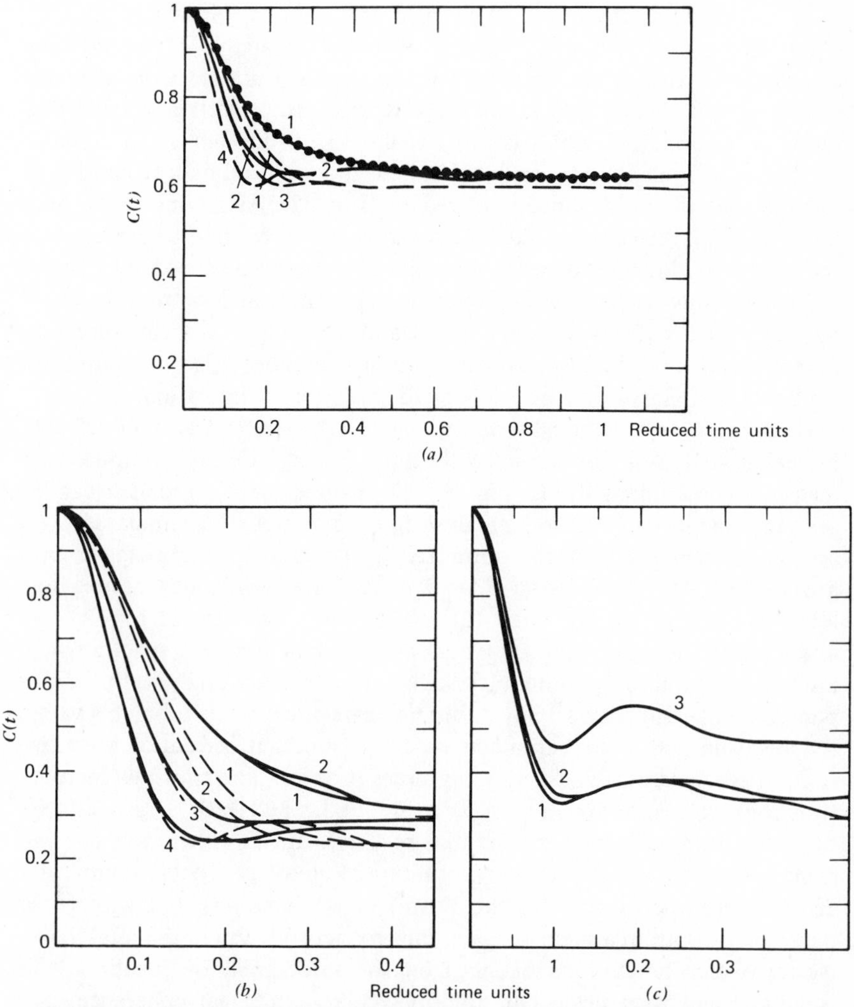

Fig. 57. (*a*) Kinetic-energy acf's. (—), (1) $L^* = 0.1$ Atom–atom simulation: (—) (2) $L^* = 0.5$; (OOO), (1) $L^* = 0.3$. Itinerant oscillator, calculated from fitting $C_F(t)$: (----), (1) $L^* = 0.1$; (2) $L^* = 0.3$; (3) $L^* = 0.5$; (4) $L^* = 0.7$. The horizontal line represents the Gaussian limit. Ordinate: $C(t)$; abscissa: reduced time units. (*b*) As for (*a*), $C_{4v}(t)$. (*c*) $\langle F^2(t)F^2(0)\rangle / \langle F^4(0)\rangle$, atom–atom potential. (1) $L^* = 0.1$; (2) $L^* = 0.3$; (3) $L^* = 0.5$. Ordinate: $C(t)$; abscissa: time steps. [Reproduced by permission from *J. Mol. Struct.*, **46**, (1978).]

In the classical theory of uncoupled translational Brownian motion, $C_{\mathbf{v}}(t)$ decays exponentially,[104] and is therefore incapable of reproducing any negative parts of the computed velocity acf's. Further, in the classical case the mean-square force is not defined, since $\exp(-\gamma|t|)$ is not differentiable at the origin. Not only is $\langle F^2 \rangle$ well defined (through ω_0^2) in itinerant oscillation, but also $C_{\mathbf{F}}(t)$ can be followed, by optimizing ω_0^2, Ω_0^2, and β as L^*, the interatomic distance is increased (Fig. 53). As L^* increases, both $C_{\mathbf{F}}(t)$ and $C_{\mathbf{v}}(t)$ become markedly oscillatory, such being the case also for the acf of the direction of velocity, while in contrast the speed acf (that of $|\mathbf{v}|$) consistently and quickly decays to its theoretical long-time value of $8/(3\pi)$. The similarity between $C_{\mathbf{v}}(t)$ and the acf of velocity direction favors theories with a constant-speed approximation, as was pointed out by Berne and Harp, who first suggested this type of simulation.

Knowing $C_{\mathbf{F}}(t)$ analytically means that $C_{\mathbf{v}}(t)$, $\langle \Delta r^2(t) \rangle$, and $G_s(\mathbf{r},t)$ may be calculated and compared with those independently computed or measured experimentally. In Fig. 53 this is done for $C_{\mathbf{v}}(t)$, and it can be seen that there is a consistent small difference between the simulated $C_{\mathbf{v}}(t)$ and that calculated from the optimized $C_{\mathbf{F}}(t)$, although the main features are similar. At $L^* = 0.3$ and $L^* = 0.5$ there are indications of negative long-time tails in the simulated $C_{\mathbf{v}}(t)$. This tail is well defined for CO and argon, and causes low-frequency peaks in the velocity power spectra which are not reproduced by itinerant oscillation as treated analytically in this paper. Damle and others[29] have obtained agreement with Rahman's velocity spectrum with a six-parameter model of itinerant oscillation with two friction coefficients, two fluctuating forces **A** and **B**, and thus two memory functions corresponding to $\langle \mathbf{A}(t)\cdot\mathbf{A}(0) \rangle$ and $\langle \mathbf{B}(t)\cdot\mathbf{B}(0) \rangle$, respectively, the latter being assumed exponential or Gaussian. In either case two parameters were needed for their definition. Equations (III.11) compose a zeroth-order approximant of the Damle et al. equations, but with fewer parameters. Both treatments neglect the cross-correlation in the total velocity correlation function as distinct from the autocorrelation function. This is tantamount to a neglect of intermolecular dynamical coherence, embodied in $G_d(\mathbf{r},t)$ the distinct van Hove correlation function, which is the probability of finding *another* particle at **r** given one at the origin initially.

Light- and neutron-scattering experiments are interpretable generally in terms of the sum

$$G_s(\mathbf{r},t) + G_d(\mathbf{r},t) = \frac{1}{n} \langle n(\mathbf{r},t) n(\mathbf{0},0) \rangle$$

where the time-dependent particle density $n(\mathbf{r},t)$ is given by

$$n(\mathbf{r},t) = \sum \delta(\mathbf{r} - \mathbf{r}_i(t))$$

and it is never straightforward to separate $G_s(\mathbf{r},t)$, usually estimated on a molecular basis, from $G_d(\mathbf{r},t)$ estimable on a hydrodynamic basis. However, this has been attempted experimentally for liquid argon at 84.5°K and therefrom found to be Gaussian within the uncertainty. The Rahman simulations of $C_v(t)$ is carried out at 94.4°K, but it is instructive to compare $G_s(\mathbf{r},t)$ calculated from the itinerant oscillator fitting to $C_v(t)$ at 94.4°K with the $G_s(\mathbf{r},t)$ estimated at 10°K lower. The results are illustrated in Fig. 56. The mean-square displacement is reproduced well, but this is in any case rather insensitive to environmental effects on molecular motion compared with van Hove's functions. The overall features of the experimental $G_s(\mathbf{r},t)$ are reproduced [e.g., the itinerant oscillator decays to zero at about the same $\mathbf{r}$ values for given t, but the experimental $G_s(\mathbf{r},t)$ is always much the larger in magnitude]. The greatest difference is at $t=0.1$ psec, where the experimental $G_s(\mathbf{r},t)$ is 27 Å^{-3} at $\mathbf{r}=0$ and the itinerant oscillator about 6.5 Å^{-3}.

Rahman[97] has demonstrated that the simulated $G_s(\mathbf{r},t)$ in argon displays an initial non-Gaussian behavior lasting until 10 psec, and in the atom–atom simulation of Fig. 57 it seems that up to 200 or more time steps (ca. 1 psec) the Gaussian limit in $C_{2v}(t)$, 0.6, is not reached. This is confirmed in Fig. 57*b*, where the simulated $C_{4v}(t)$ does not reach its equivalent limit of 0.2381. In Fig. 57*c* the acf $\langle F^2(t)F^2(0)\rangle/\langle F^4(0)\rangle$ is displayed for $L^*=0.1$, 0.3 and 0.5, and it is clear that no common, single-valued (or

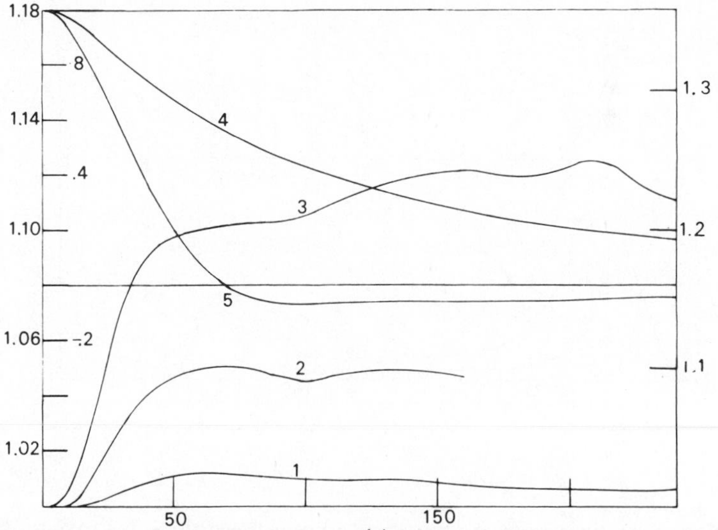

(*a*)

Fig. 58. Rotation–translation speed-correlation functions for $T^*=2.32$, $\rho^*=0.643$. (1) $\dfrac{\langle|\mathbf{v}(0)||\boldsymbol{\omega}(t)|\rangle}{\langle|\mathbf{v}(0)||\boldsymbol{\omega}(0)|\rangle}$; (2) $\dfrac{\langle v^2(0)\cdot\omega^2(t)\rangle}{\langle v^2(0)\cdot\omega^2(0)\rangle}$; (3) $\dfrac{\langle v^4(0)\cdot\omega^4(t)\rangle}{\langle v^4(0)\cdot\omega^4(0)\rangle}$; (4) $\dfrac{\langle\mathbf{v}(0)\cdot\mathbf{v}(t)\rangle}{\langle\mathbf{v}(0)\cdot\mathbf{v}(0)\rangle}$; (5) $\dfrac{\langle\boldsymbol{\omega}(0)\cdot\boldsymbol{\omega}(t)\rangle}{\langle\boldsymbol{\omega}(0)\cdot\boldsymbol{\omega}(0)\rangle}$. (*a*) $L^*=0.200$; (*b*) $L^*=0.2392$ (N_2); (*c*) $L^*=0.500$. In (*a*) to (*c*), all curves are computed averaging over 1500 times steps.

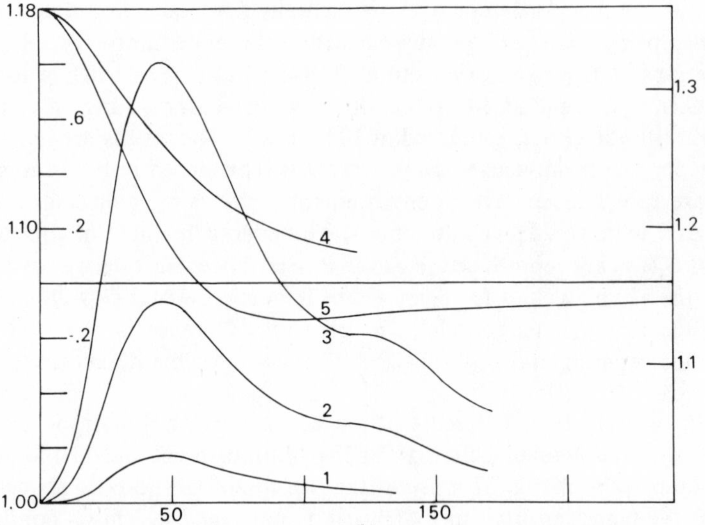

(b)

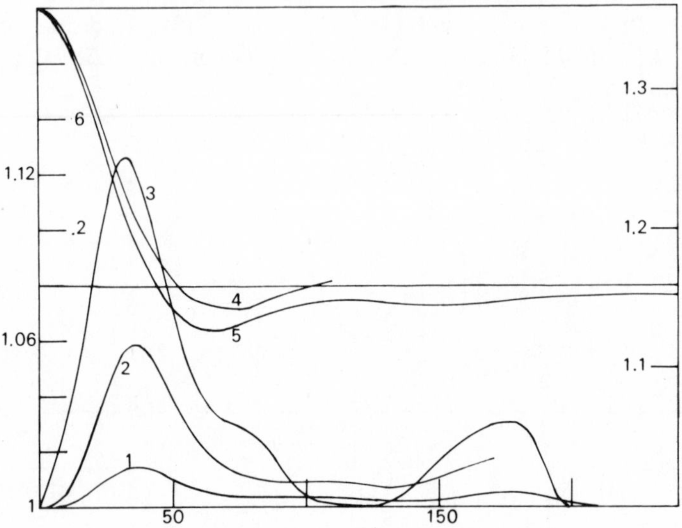

(c)

Gaussian) long-time limit is arrived at among these three potentials. Thus generally it seems that **v**, **f** (the projected force), and $G_s(\mathbf{r}, t)$ are non-Gaussian variates, as well as being non-Markovian, even in atomic fluids. The analytical treatment of such behavior is, of course, more difficult and compounded by rotation/translation coupling (Fig. 58) and short time effects (Fig. 59).

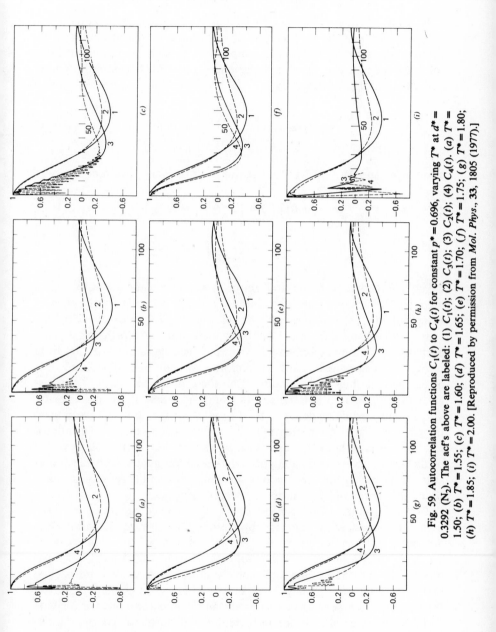

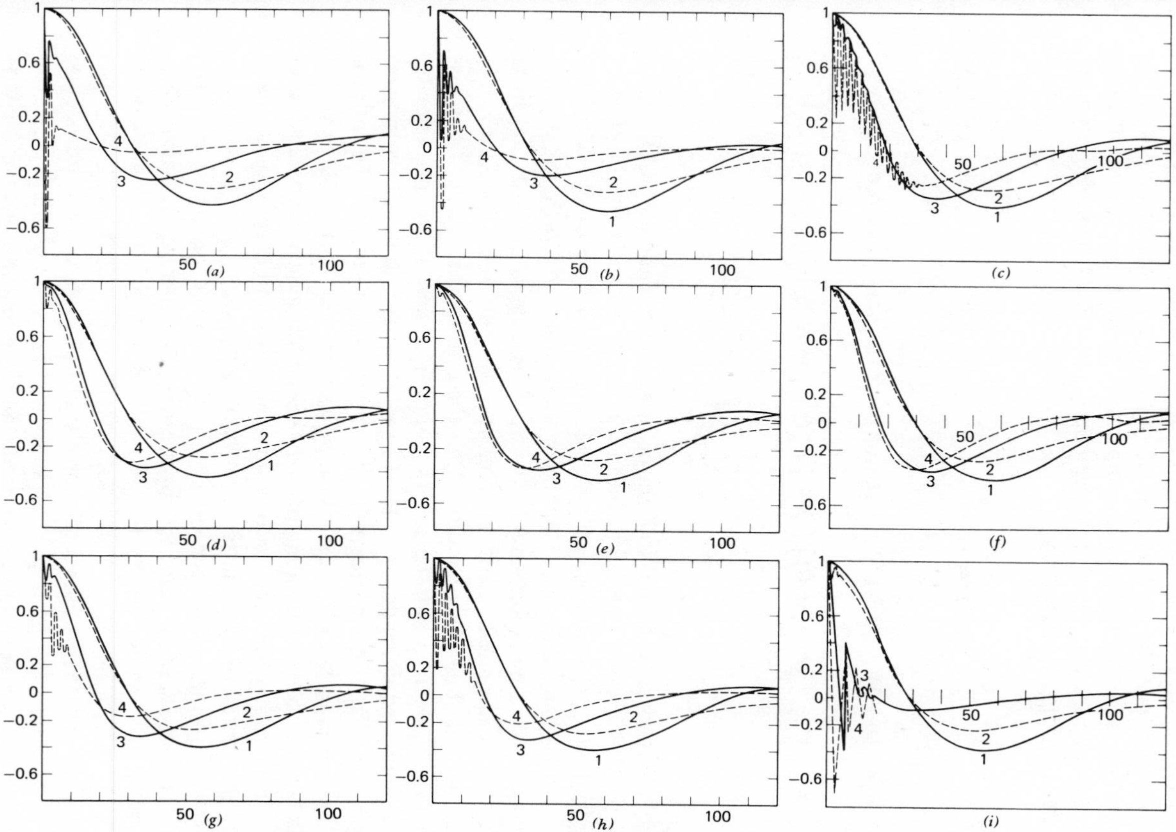

Fig. 59. Autocorrelation functions $C_1(t)$ to $C_4(t)$ for constant $p^* = 0.696$, varying T^* at $d^* = 0.3292$ (N_2). The acf's above are labeled: (1) $C_1(t)$; (2) $C_3(t)$; (3) $C_2(t)$; (4) $C_4(t)$. (*a*) $T^* = 1.50$; (*b*) $T^* = 1.55$; (*c*) $T^* = 1.60$; (*d*) $T^* = 1.65$; (*e*) $T^* = 1.70$; (*f*) $T^* = 1.75$; (*g*) $T^* = 1.80$; (*h*) $T^* = 1.85$; (*i*) $T^* = 2.00$. [Reproduced by permission from *Mol. Phys.*, **33**, 1805 (1977).]

3. *Translations in Glassy, or Amorphous, Nitrogen*

By using a reduced density $\rho^* = 0.80$ and a reduced temperature $T^* = 2.00$, nitrogen may be simulated in a metastable solid state induced by kilobars of external pressure at a temperature beyond the normal melting point. In this case $G_s(\mathbf{r}, t)$ is markedly non-Gaussian, as expressed by non-zero values of $a_n(t)$, defined by

$$a_n(t) = \frac{\langle [\mathbf{r}(t) - \mathbf{r}(0)]^{2n} \rangle}{\alpha_n \langle [\mathbf{r}(t) - \mathbf{r}(0)]^2 \rangle^n} - 1 \tag{III.23}$$

where $\alpha_n = (2n+1), \ldots, 5.3.1/3^n$. The link with the van Hove function is

$$\langle [\mathbf{r}(t) - \mathbf{r}(0)]^{2n} \rangle = \int r_1^{2n} G_s(\mathbf{r}_1, t)\, d\mathbf{r}_1$$

$$\equiv \frac{1}{N} \sum_{i=1}^{N} [\mathbf{r}_i(t) - \mathbf{r}_i(0)]^{2n}$$

where $\mathbf{r}_1 = \mathbf{r}(t) - \mathbf{r}(0)$. Berne and Harp, in their simulation of CO with a modified Stockmayer potential, found $a_n(t)$ to be moderately sensitive to variations in N, the number of molecules used. $a_n(t)$ become less significant as N is changed from 256 to 500 or thereabouts. Accordingly, it is expected that our simulation would overstress these deviations [of $a_n(t)$ from zero] to an unspecified extent. Further, it is difficult to estimate the effect of our periodic boundary conditions on these functions. Berne[106] has discussed their effect on long tails in the autocorrelation of angular velocity. The complicated dependence of $a_n(t)$ $(n = 2, 3, 4)$ upon time (from an arbitrary $t = 0$) (Fig. 60) is not correlated with statistical noise in the ratio of rotational to translational kinetic energy, as illustrated in Fig. 60. It would be difficult to follow these analytically with the techniques available at present except perhaps in the case of computer argon and other atomic fluids where the curves $a_n(t)$ are simpler in overall form.

The mean-square displacement of argon atoms as simulated by Rahman[97] is reproduced satisfactorily by a process of itinerant oscillation (Fig. 56), but Fig. 61 shows clearly that the rate of diffusion in the glass is far too high analytically. This analytical rate is again calculated by a least-mean-squares best fit to the force correlation function. Only at times close to the start are the simulated and analytical functions similar; thereafter the former flatten out and increase only slowly with a tendency to oscillate as in a clathrate solid. It seems therefore that an improved rototranslational itinerant oscillation model should be capable of taking this high-density behavior in the glass in its stride. The rotational constraints are clearly discernible in Fig. 62, where elongation of interatomic distance is

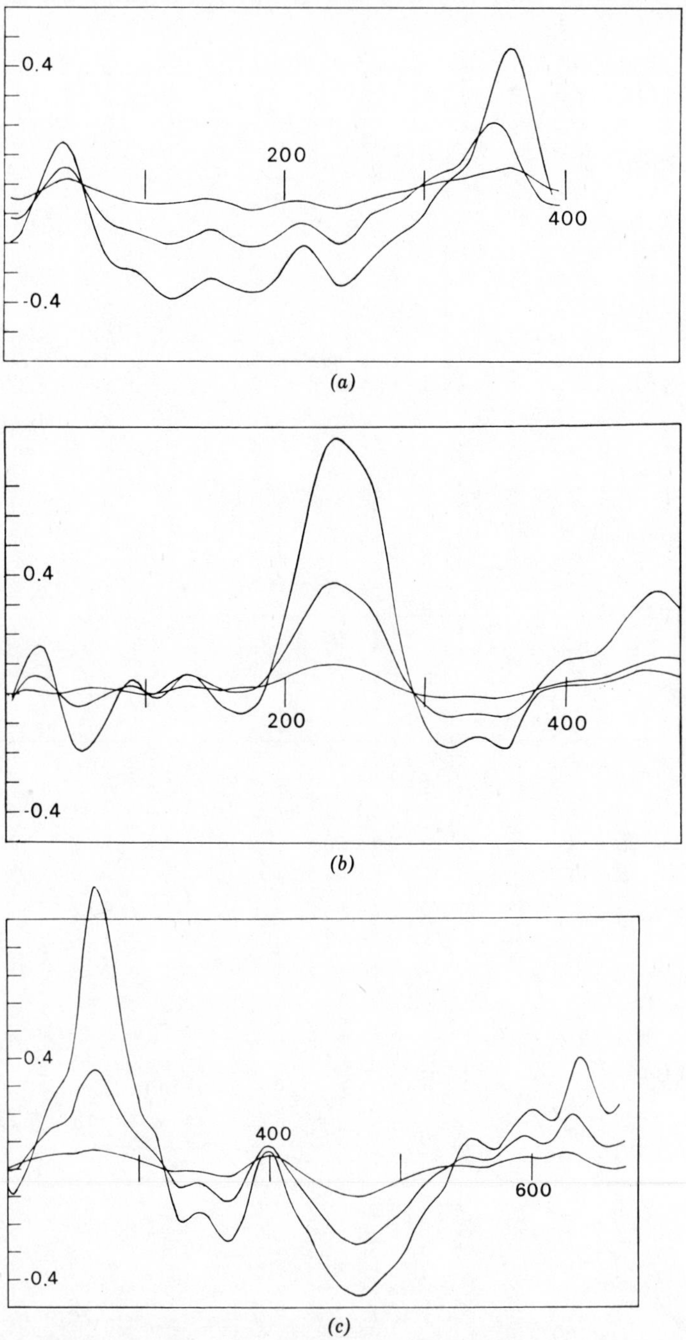

Fig. 60. (*a*) Plot of $a_n(t)$ (see text) for $n=2,3,4$; $L^*=0.200$. (*b*) As in (*a*) $L^*=0.3292$. (*c*) As in (*a*) $L^*=0.400$ (200 to 600 time steps). (*d*) As in (*a*) $L^*=0.400$ (after rejecting 800 steps). (*e*) Ratio of kinetic to rotational energy. (—), After 800 steps; (---), 200 to 600 steps. Abscissas: (*a*) to (*e*), time steps (see text).

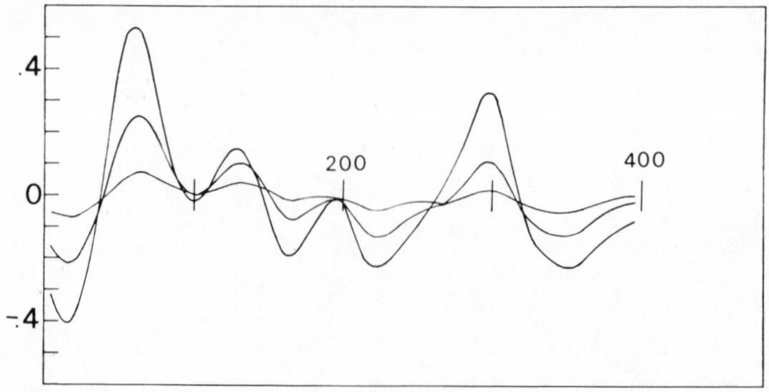

(d)

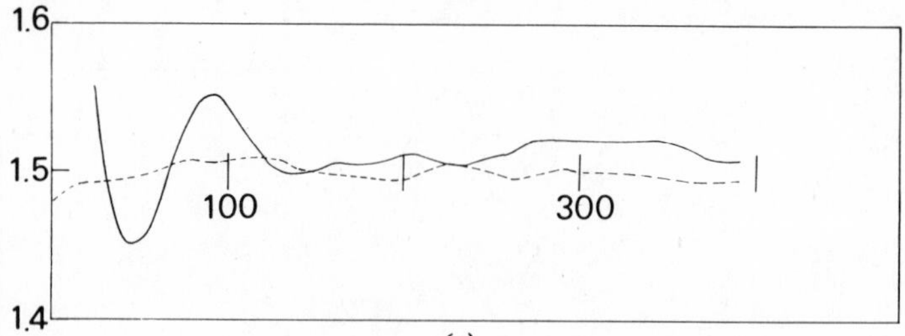

(e)

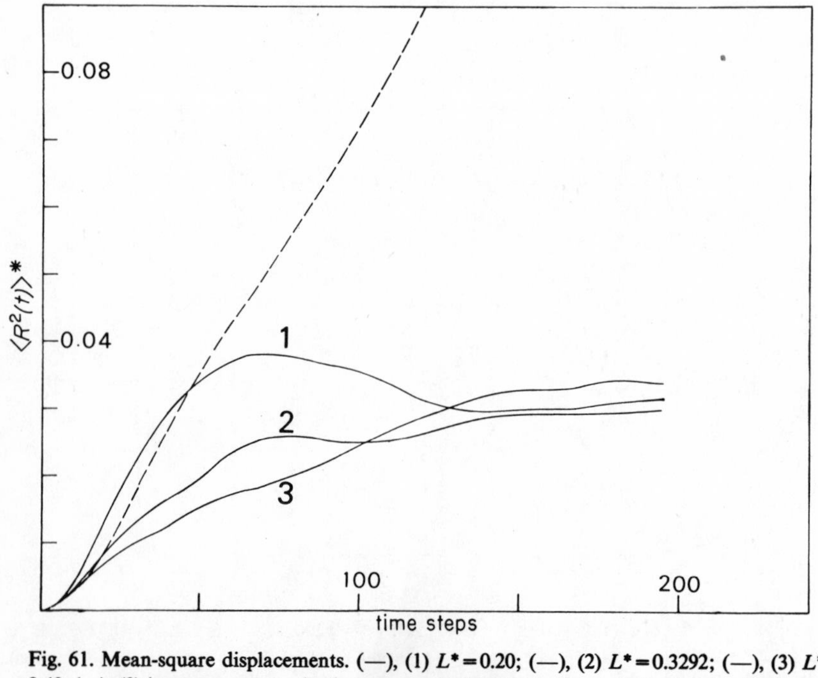

Fig. 61. Mean-square displacements. (—), (1) $L^*=0.20$; (—), (2) $L^*=0.3292$; (—), (3) $L^*=0.40$. (---), (3) i.o. mean-square displacement calculated from the optimized $C_F(t)$. Ordinate: $\langle R^2(t)\rangle^*$; abscissa: time steps.

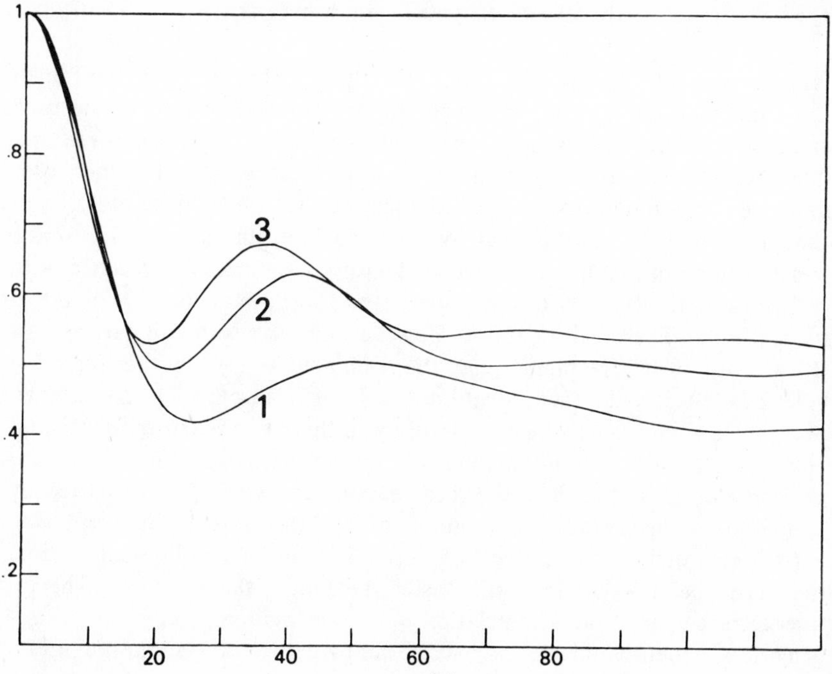

(a)

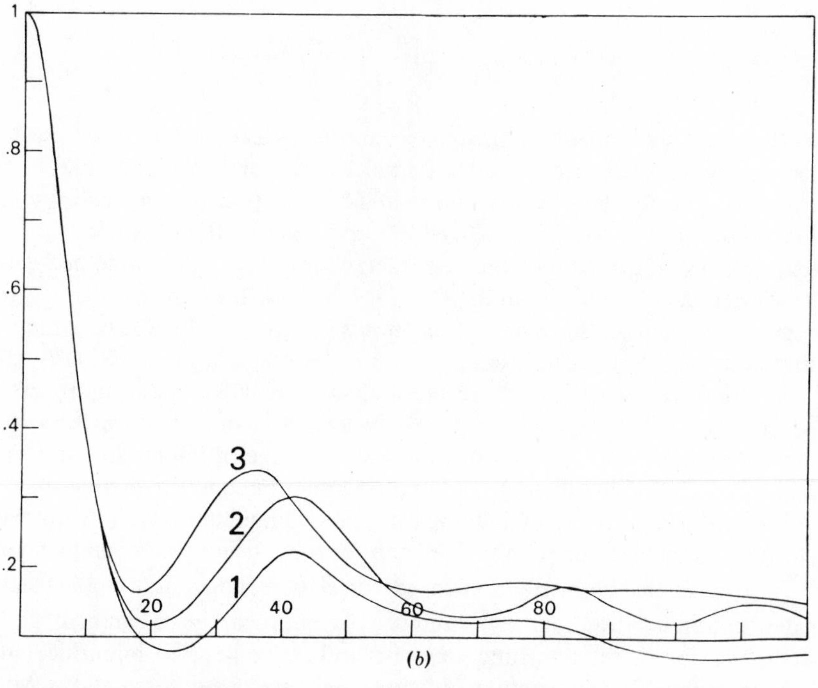

(b)

Fig. 62. (a) $\langle F^2(t)F^2(0)\rangle/\langle F^4(0)\rangle$; (b) $\langle F^4(t)F^4(0)\rangle/\langle F^8(0)\rangle$: (1) $L^* = 0.200$; (2) $L^* = 0.3292$; (3) $L^* = 0.400$. Abscissas: time steps.

the only variable. Figure 62*b* brings out the coupling through the fourth force moment acf, a function which for free translation falls at the time origin immediately to its long-time limit. The falloff is collision-free and thus the same for all elongations up to 10 time steps or so, but thereafter molecular interaction occurs and the three curves behave differently, even though ρ^* and T^* are identical for each. In fact, throughout Fig. 62 the constant long-time limits expected of Gaussian statistics are reached with difficulty (if at all) for both the second- and fourth-moment acf's of velocity and force. There is little doubt, however, that this is due in part to statistical noise in the averaging [e.g., the simulated $C_{4v}(t)$ function for $L^*=$ 0.3292 falls below the theoretical limit of 0.2381]. In other instances, however, these moment acf's behave similarly to those computed by Berne and Harp for CO, in that a limiting constant is reached at long times. Finally, it is clear that acf's of molecular speed must be conserved fairly well in any system [decaying, as they do, to only $8/(3\pi)$ of their initial value], but even here there are discernible effects of elongation on the translational motion, the decay being oscillatory and faster the longer the molecule. The influence of center-of-mass translation upon the rotational velocity in nitrogen and symmetrical diatomics may be simulated (so far only in the liquid) using the functions

$$\langle|\mathbf{v}(0)||\omega(t)|\rangle \qquad \text{and} \qquad \langle \mathbf{v}^{2n}(0)\omega^{2n}(t)\rangle$$

$n=1,2$, which are products of scalar quantities (speeds) which have no directive properties. These are illustrated in Fig. 58 for $L^*=0.2$, 0.3292 (N_2), and 0.50. Despite the statistical noise at $L^*=0.50$ (due to the similarity in the decays of $\langle\mathbf{v}(t)\cdot\mathbf{v}(0)\rangle/\langle\mathbf{v}(0)\cdot\mathbf{v}(0)\rangle$ and $\langle\omega(t)\cdot\omega(0)\rangle/\langle\omega(0)\cdot\omega(0)\rangle$ at longer times), it can be seen that the interaction is very sensitive to molecular elongation. $\langle\mathbf{v}(0)\cdot\omega(t)\rangle$ and $\langle\mathbf{v}^{2n+1}(0)\cdot\omega^{2n+1}(t)\rangle$ are zero for molecular symmetries such as those of N_2, but not necessarily so for lower symmetries (e.g., C_{2v}, the angular triatomic). As L^* shortens, the interaction functions are longer-lived and more featureless than those in the more pronounced dumbbell represented by N_2, where oscillatory features become more discernible. At $L^*=0.5$ there is a sharp surge of interaction at periodic intervals and the interaction functions are pronouncedly oscillatory.

We are in the course of investigating these simulation results with the stochastic Liouville equation in the Mori form by using the column vector $\begin{bmatrix}\mathbf{v}\\ \omega\end{bmatrix}$ to evaluate the interaction-speed correlation functions using a truncated continued fraction. By forcing a fit between the theoretical and simulated autocorrelation functions of $\mathbf{v}$ and ω, we hope to reproduce, at least in outline, the interaction functions simulated here. Simulations will then be extended to triatomic molecules of C_{2v} symmetry and the cycle repeated. In this way we hope to discern stochastic features in how the center-of-mass molecular translation affects $\langle \boldsymbol{\omega}(0)\cdot\boldsymbol{\omega}(t)\rangle$ and therefore the spectral functions

$$P_n\langle \mathbf{u}(t)\cdot\mathbf{u}(0)\rangle$$

in the case $\langle \mathbf{v}(0)\cdot\boldsymbol{\omega}(t)\rangle \neq 0$. This will enable us to check whether these latter are directly or indirectly sensitive to molecular translation, and if so, in what way. Are they, in the purist sense, rotational or strictly rototranslational? For which symmetries is it then valid to use a purely rotational stochastic Liouville equation? We hope to answer these fundamental questions aided by computer simulation.

C. Simulations of High-Derivative Autocorrelation Functions

In this section we investigate the time decay of correlation functions of high derivatives of position and orientation. This is necessary because the equilibrium (zero-time) values of successive memory functions $K_n(t)$ are defined in terms of

$$C^{(n)}(0) = \langle A^{(n)}(0)A^{(n)}(0)\rangle$$

where (n) denotes the nth derivative of A. Thus, if the $C^{(n)}(t)$ are intricate functions of time, then by implication so are the $K_n(t)\,(n=0,\ldots,N)$. Specifically, we aim to see whether the exponential approximation $K_1(t)$ is worthwhile. The decay with time of each member of the set $[K_0(t), K_1(t),\ldots,]$ is determined by that of the successive $C^{(n)}(t)$. It is difficult and time-consuming to calculate the memory functions $K_n(t)$ directly because of their definition only in terms of projection operators, but such is not the case for $C^{(n)}(t)$, since the predictor–corrector algorithms used in molecular dynamics to solve the initial equations depend for their usefulness upon the calculation and storage of up to the first five derivatives of a dynamical variable A, typically the center-of-mass velocity $\mathbf{v}$ or the total angular velocity $\boldsymbol{\omega}$. The autocorrelation functions

$$C_1(t) = \langle \dot{\mathbf{v}}(t)\cdot\dot{\mathbf{v}}(0)\rangle,$$
$$C_2(t) = \langle \ddot{\mathbf{v}}(t)\cdot\ddot{\mathbf{v}}(0)\rangle,$$
$$C_3(t) = \langle \dot{\boldsymbol{\omega}}(t)\cdot\dot{\boldsymbol{\omega}}(0)\rangle,$$
$$C_4(t) = \langle \ddot{\boldsymbol{\omega}}(t)\cdot\ddot{\boldsymbol{\omega}}(0)\rangle$$

are simulated using the atom-atom algorithm, over a range of temperature

(T^*) at constant number density (p^*), and vice versa. In the first calculation an interatomic distance (L^*) of 0.3292 was used, corresponding to N_2. In the second we used the best available Lennard-Jones parameters for nitrogen with an interatomic distance of 0.4, so that we do not deal in this case with a "real" molecule. The statistical stability of the computed acf's is judged as usual by using different numbers of time steps (ca. 5×10^{-15} sec) for the ensemble averaging.[107] No difference could be perceived between runs of 200 and 400 steps. All the results shown here were obtained with the latter. A few runs of up to 1600 steps were carried out initially to look for any drift in quantities such as the mean-square torque and force as well as thermodynamic data. The first 100 or 50 steps of each run are unstable and are rejected in forming averages of any kind, and no acf is plotted beyond about 0.5 psec. Real time was divided into batches of up to 1200 decimal seconds each of CDC 7600 (U.M.R.C.C.) time.

It is clear from Fig. 59 that $C_2(t)$ and $C_4(t)$ are intricate functions of time, being oscillatory sometimes, sometimes very rapidly decaying, but more often with a decay on the same time scale as C_1 and C_3. Therefore, were we to take the set $[\boldsymbol{\omega}, \dot{\boldsymbol{\omega}}, \ddot{\boldsymbol{\omega}}]$ or $[\mathbf{v}, \dot{\mathbf{v}}, \ddot{\mathbf{v}}]$ in a three-variable formalism, the autocorrelation functions of each would decay usually on much the same time scale and the "fast variable" hypothesis would be inapplicable.

The complicated analytical form of members of the set $[K_0(0), \ldots, K_n(0)]$ for n greater than 2 may be clearly demonstrated as below for the orientation, dipole unit vector $\mathbf{u}$ by evaluating the Maclaurin expansion and coefficients of $\langle \mathbf{u}(t)\cdot\mathbf{u}(0)\rangle$. Here we carry this out for the coefficients up to that of t^8 in a symmetric-top or linear model. The terms to be evaluated are

$$M_0 \equiv 1 \text{ (the coefficients of } t^0\text{)}$$

$$M_2 = \langle \mathbf{u}^2(0)\rangle \qquad M_4 = \langle \ddot{\mathbf{u}}^2(0)\rangle$$

$$M_6 = \langle \dddot{\mathbf{u}}^2(0)\rangle \qquad M_8 = \langle \ddddot{\mathbf{u}}^2(0)\rangle$$

The method employed is an extension of that of Desplanques,[39] which aims at an expansion of each coefficient in terms of $\boldsymbol{\omega}_\perp$, the component of the angular velocity perpendicular to the C_{3v} axis. This is useful since its time derivative is a torque component, itself a derivative of a potential with respect to orientation.

Gordon has evaluated M_2 and M_4 for a linear molecule, and Desplanques has extended this to the symmetric-top symmetry. We have

$$M_2 = \langle \dot{\mathbf{u}}(0)\cdot\dot{\mathbf{u}}(0)\rangle = 2kT/I_B \tag{III.24}$$

$$M_4 = \langle \ddot{\mathbf{u}}(0)\cdot\ddot{\mathbf{u}}(0)\rangle$$

$$= 2\left(2\frac{kT}{I_B}\right)^2\left(1+\frac{I_A}{4I_B}\right)+\frac{\langle O(V)^2\rangle}{I_B^2} \tag{III.25}$$

where I_B and I_A are the usual moments of inertia of the asymmetric top and $\langle O(V)^2\rangle$ denotes the mean-square torques, V being the mean intermolecular potential. This is in the direction perpendicular to the C_{3v} axis. Thus $K_1(0)$ for $\mathbf{u}$ is proportional to $\langle O(V)^2\rangle$. The expansion of m_6 and m_8 may be accomplished by repeated differentiation of the relation

$$\dot{\omega}=\dot{\mathbf{u}}\times\ddot{\mathbf{u}}+\mathbf{u}\times\dddot{\mathbf{u}} \tag{III.26}$$

so that

$$\begin{aligned}M_6&=\langle\dddot{\mathbf{u}}(0)\cdot\dddot{\mathbf{u}}(0)\rangle\\&=\langle\omega_\perp^6\rangle+\langle\omega_\perp^2(\dot{\omega}_\perp)^2\rangle+\langle(\ddot{\omega}_\perp)^2\rangle\\&\quad+\tfrac{5}{2}\left\langle\left[\frac{d}{dt}(\omega_\perp^2)\right]^2\right\rangle-\left\langle\omega_\perp^2\frac{d^2}{dt^2}(\omega_\perp^2)\right\rangle\end{aligned} \tag{III.27}$$

The coefficient of t^6 is related to the mean-square torque derivative term (the third) and also *four others* of the same dimension. m_6 is related to $K_2(0)$ and therefore throughout its domain of existence $(t>0)$ must reflect the time decay of all the vector terms in the right-hand side of (III.27), and is consequently an intricate function of time.

Differentiating (III.26), we have

$$\omega_\perp=2\dot{\mathbf{u}}\times\dddot{\mathbf{u}}+\mathbf{u}\times\ddddot{\mathbf{u}} \tag{III.28}$$

giving

$$\begin{aligned}\langle\ddddot{\mathbf{u}}(0)\cdot\ddddot{\mathbf{u}}(0)\rangle&=\langle\ddot{\omega}_\perp^2\rangle-2\langle\omega_\perp^8\rangle-4\langle\omega_\perp^2(\ddot{\omega})^2\rangle\\&\quad+2\langle\dot{\omega}_\perp^4\rangle+10\left\langle\omega_\perp^2\left(\frac{d}{dt}(\omega_\perp^2)\right)^2\right\rangle-9\left\langle\omega_\perp^4\frac{d^2}{dt^2}(\omega_\perp^2)\right\rangle\\&\quad-5\left\langle(\dot{\omega}_\perp)^2\frac{d^2}{dt^2}(\omega_\perp^2)\right\rangle-\frac{5}{4}\left\langle\left(\frac{d^2}{dt^2}(\omega_\perp^2)\right)^2\right\rangle\\&\quad+3\left\langle\frac{d}{dt}(\omega_\perp^2)\left[\frac{d^3}{dt^3}(\omega_\perp^2)-\frac{3}{2}\frac{d}{dt}(\dot{\omega}_\perp)^2-\frac{3}{2}(\omega_\perp^4)\right]\right\rangle\end{aligned}$$

The coefficient m_8 is related to $K_3(0)$.

The overall time dependence of functions such as these is supplied in great detail by a molecular dynamics calculation and one of the more obvious results (Fig. 59) is that both $C_2(t)$ and $C_4(t)$ seem to be sensitive to small changes in T^* or ρ^*, whereas $C_1(t)$ and $C_3(t)$ are not. C_2 and C_4, being probes into the extreme short-time dynamical properties of the molecular ensemble, are revealing details about changes in the linear and angular acceleration, those changes which must be taking place during the course of an interaction or "collision." The fact that all the acf's exhibit negative regions is not surprising in view of the fact that the velocity and angular velocity acf's themselves oscillate out to fairly long times (see previous sections) at higher values of ρ^*. It is not surprising either that C_2 and C_4 oscillate so rapidly in comparison with C_1 and C_3, the same relationship has been observed spectroscopically, of course, for $\langle \mathbf{u}(t)\cdot\mathbf{u}(0)\rangle$ and $\langle \dot{\mathbf{u}}(t)\cdot\dot{\mathbf{u}}(0)\rangle$. It is merely another expression of the increased sensitivity of derivative auto correlation functions to short-time, or high-frequency phenomena.

An interesting fact of the molecular motion is revealed when C_2 and C_4 are plotted together as in Figs. 59, since the types of decay at each different T^* and ρ^* resemble each other so closely. When one function is rapidly oscillatory, then so is the other, and the same is true when both are long-lived. This seems to be indicative of a great deal of those translation-rotation effects, typified in the extreme case by the propeller action. It is known that in the Markov limit such coupling is rigorously zero for symmetry inclusive of $C_{\infty v}$ and $D_{\infty h}$. Needless to say, the Markov limit is unrealistic (delta memories) but under inversion $\mathbf{v}\rightarrow-\mathbf{v}$ and $\omega\rightarrow\omega$, whereas the Liouville operator remains unchanged. This implies that

$$C_{v\omega}(0) = -C_{v\omega}(0) = 0$$

but $C_{v\omega}^{(2n)}(t) \equiv \langle v^{2n}(0)\omega^{2n}(t)\rangle$ is finite (Figs. 58). It needs to be emphasized that a similar decay rate for the velocity and angular velocity autocorrelation functions is not itself indicative of coupling, since in the absence of any intermolecular interaction, both normalized autocorrelation functions would remain indefinitely at unity.

To close this section we emphasize that the intricate nature of these functions does not imply that the memory function expansion of Mori is unusable, but the protagonists of two and three formalisms should note in particular the longevity of the autocorrelation functions of some of the higher derivatives. Also, we are of course always considering an artificial droplet of liquid.

D. General Dynamic Properties of Computer Nitrogen: Collective Correlation Functions

In this section we use the Tildesley–Streett algorithm to study the behavior of five autocorrelation functions: linear and angular velocity, orientation, torque, and force under the following conditions: (1) increasing number density (ρ^*) at constant temperature (T^*) and interatomic distance (L^*), (2) increasing L^* at constant T^* and ρ^*, and (3) increasing T^*

at constant ρ^* and L^*. The following indications appear: (1) The mean-square torque and mean-square force can exhibit maxima or minima as a function of ρ^* or L^*, but over a restricted range seem linear in T^* at constant ρ^* and L^*. (2) Autocorrelation functions of high derivative of the interatomic vector $\mathbf{u}$ or the angular velocity $\boldsymbol{\omega}$ decay generally on the same time scale as the vectors themselves and become more complicated, as described in the previous section. (3) The effect of elongation at constant ρ^* on dynamical properties such as the above is much more pronounced than is that of ρ^* at constant L^*, indicating that hard-core anisotropy is the important factor in the determination of, for example, nematic behavior. By increasing L^* we effectively change from a pseudospherical molecular shape to a dumbbell, and so measure the effect of increasing geometrical anisotropy on spectral properties.

The results are illustrated in Figs. 63 to 66 in terms of several different autocorrelation functions and as plots of mean-square torque and mean-square force against reduced number density ρ^* at a constant temperature, and vice versa. The features of these functions can be used to criticize the models in Table I. It is clear that purely rotational diffusion is inadequate to explain the simulations, even in its inertia-corrected form, where the angular momentum autocorrelation function is a single exponential, and where the torque autocorrelation function is not defined. The M and J diffusion models for the motion of the interatomic vector $\mathbf{L}^*$ may be derived quite easily using projectors whereby the autocorrelation functions of $\mathbf{L}^*$ and $\dot{\mathbf{L}}^*$ are slowly decaying compared with that of $\ddot{\mathbf{L}}^*$. Since the autocorrelation function $\langle \ddot{\mathbf{L}}^*(0)\cdot\ddot{\mathbf{L}}^*(t)\rangle$ has the units of angular acceleration, it is related to the torque autocorrelation function, which in Fig. 59 decays on the same time scale as $\langle \mathbf{L}^*(0)\cdot\mathbf{L}^*(t)\rangle$. It seems that $\langle \boldsymbol{\omega}(t)\cdot\boldsymbol{\omega}(0)\rangle$ sometimes decays on a much longer scale, but more often it does not. Even the autocorrelation function $\langle \ddot{\boldsymbol{\omega}}(t)\cdot\ddot{\boldsymbol{\omega}}(0)\rangle$ (previous section) sometimes takes longer to decay than $\langle \mathbf{L}^*(t)\cdot\mathbf{L}^*(0)\rangle$, and apparently is a much more complicated function of time than the latter. The Mori continued-fraction representation depends partly for its usefulness on the hope that autocorrelation functions of derivatives of $\mathbf{L}^*$ might have simpler time dependencies than that of $\mathbf{L}^*$ itself. It seems that successive kernels in Mori's series are more complicated than envisaged (on the results of Section III.C), but Singer et al.'s[100] simulation of memory functions shows that these are much less so than are high derivative acf's such as $\langle \ddot{\boldsymbol{\omega}}(t)\cdot\ddot{\boldsymbol{\omega}}(0)\rangle$. Also, of course, the memory function series is a useful way of generating the whole time dependence of an autocorrelation function knowing only the *short-time* behavior of, say, its second memory function. We discuss hereafter some specific systems.

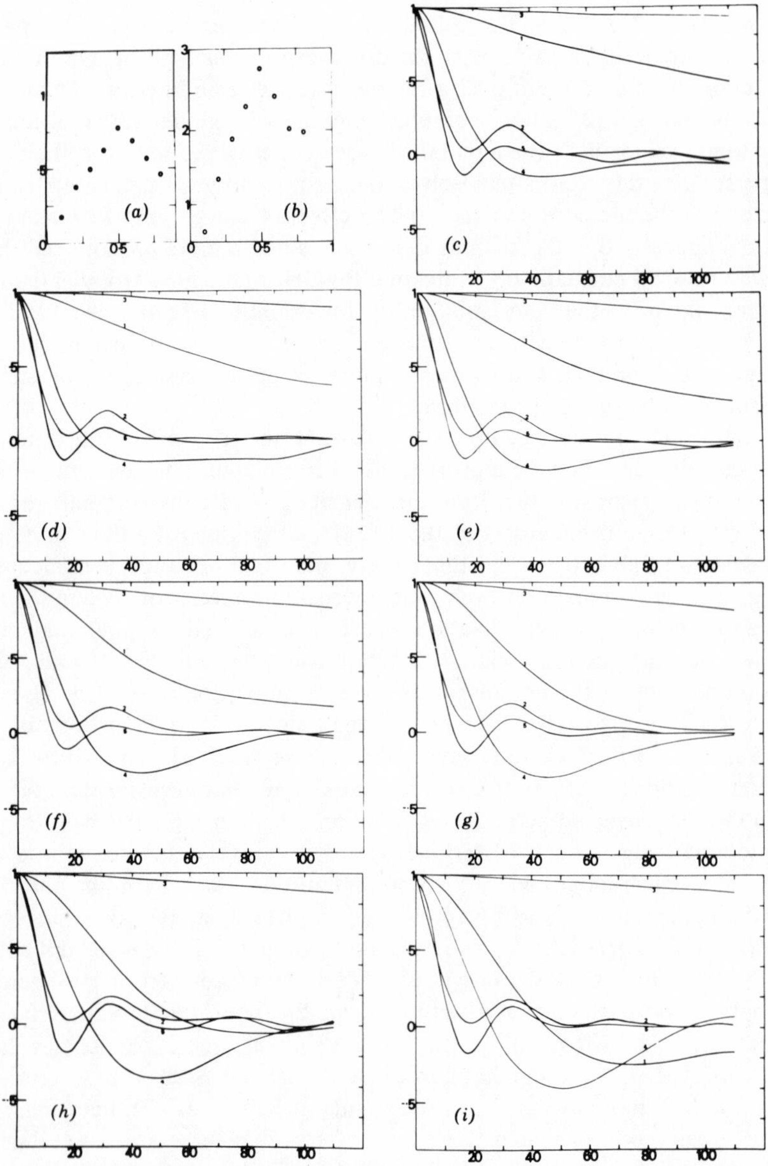

Fig. 63. Left to right, top to bottom ($T^* = 3.6$, $L^* = 0.1$): (*a*) Plot of $10^3\langle Tq^2 \rangle$ vs. ρ^*. (*b*) Plot of $\langle F^2 \rangle$ vs. ρ^*. (*c*) Autocorrelation functions for $\rho^* = 0.2$: (1) velocity; (2) orientation; (3) angular velocity; (4) force; (5) torque. (*d*) As in (*c*), $\rho^* = 0.3$. (*e*) As in (*c*), $\rho^* = 0.4$. (*f*) As in (*c*), $\rho^* = 0.5$. (*g*) As in (*c*), $\rho^* = 0.6$. (*h*) As in (*c*), $\rho^* = 0.7$. (*i*) As in (*c*), $\rho^* = 0.8$. Ordinates: (*a*) $10^3(Tq^2)$, (*b*) $\langle F^2 \rangle$; abscissas: (*a*) ρ^*, (*b*) ρ^*, (*c*) to (*i*) time steps. [Reproduced by permission from *Adv. Mol. Rel. Int. Proc.*, **11**, 295 (1977).]

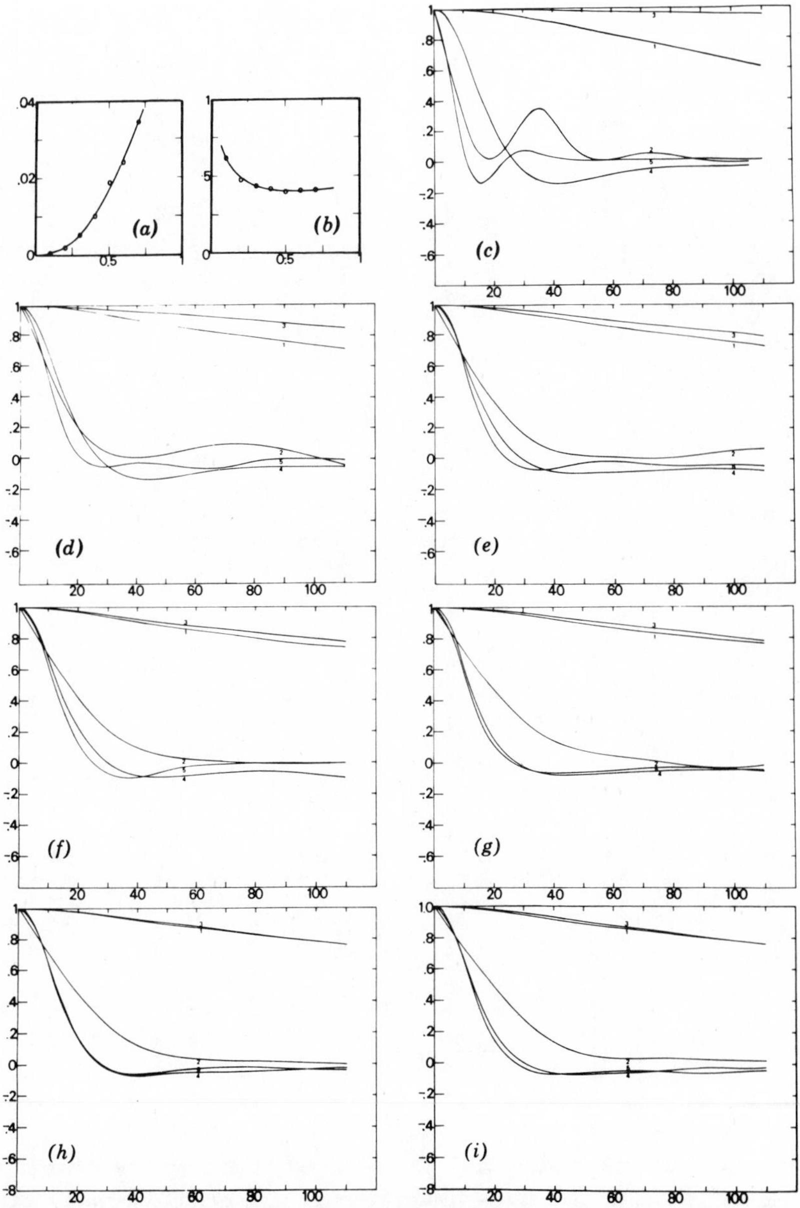

Fig. 64. Left to right, top to bottom ($\rho^*=0.1$; $T^*=3.6$): (*a*) $\langle Tq^2\rangle$ vs. L^*. (*b*) $\langle F^2\rangle$ vs. L^*. (*c*) autocorrelation functions: (1) velocity; (2) orientation; (3) angular velocity; (4) force; (5) torque, $d^*=0.1$. (*d*) as in (*c*), $d^*=0.2$. (*e*) as in (*c*), $d^*=0.3$. (*f*) as in (*c*), $d^*=0.4$. (*g*) as in (*c*) $d^*=0.5$. (*h*) as in (*c*), $d^*=0.6$. (*i*) as in (*c*), $d^*=0.7$. Ordinates: (*a*) $\langle Tq^2\rangle$, (*b*) $\langle F^2\rangle$; abscissas: (*a*) L^*, (*b*) L^*, (*c*) to (*i*) time steps. [Reproduced by permission from *Adv. Mol. Rel. Int. Proc.*, **11**, 295 (1977).]

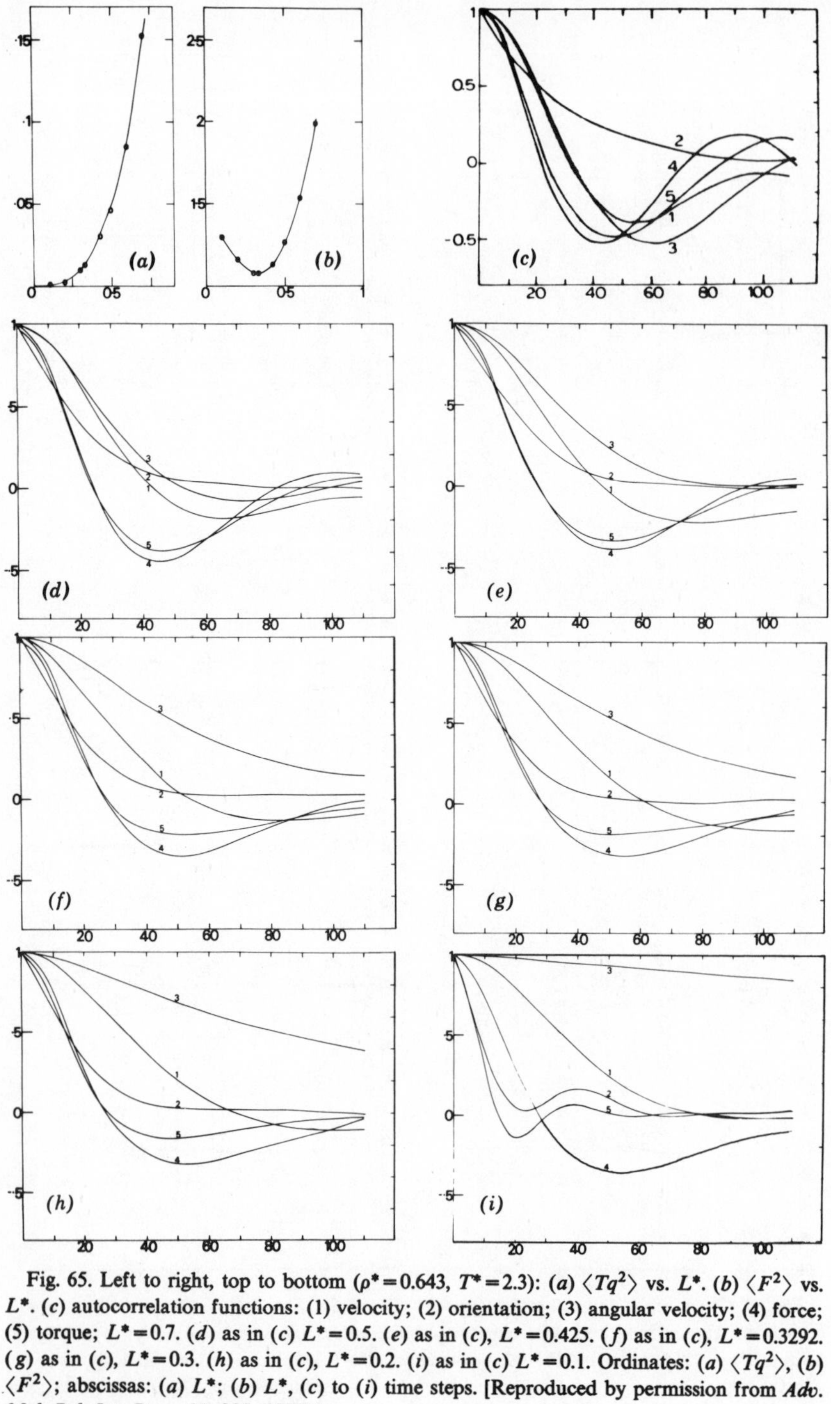

Fig. 65. Left to right, top to bottom ($\rho^*=0.643$, $T^*=2.3$): (*a*) $\langle Tq^2 \rangle$ vs. L^*. (*b*) $\langle F^2 \rangle$ vs. L^*. (*c*) autocorrelation functions: (1) velocity; (2) orientation; (3) angular velocity; (4) force; (5) torque; $L^*=0.7$. (*d*) as in (*c*) $L^*=0.5$. (*e*) as in (*c*), $L^*=0.425$. (*f*) as in (*c*), $L^*=0.3292$. (*g*) as in (*c*), $L^*=0.3$. (*h*) as in (*c*), $L^*=0.2$. (*i*) as in (*c*) $L^*=0.1$. Ordinates: (*a*) $\langle Tq^2 \rangle$, (*b*) $\langle F^2 \rangle$; abscissas: (*a*) L^*; (*b*) L^*, (*c*) to (*i*) time steps. [Reproduced by permission from *Adv. Mol. Rel. Int. Proc.*, **11**, 295 (1977).]

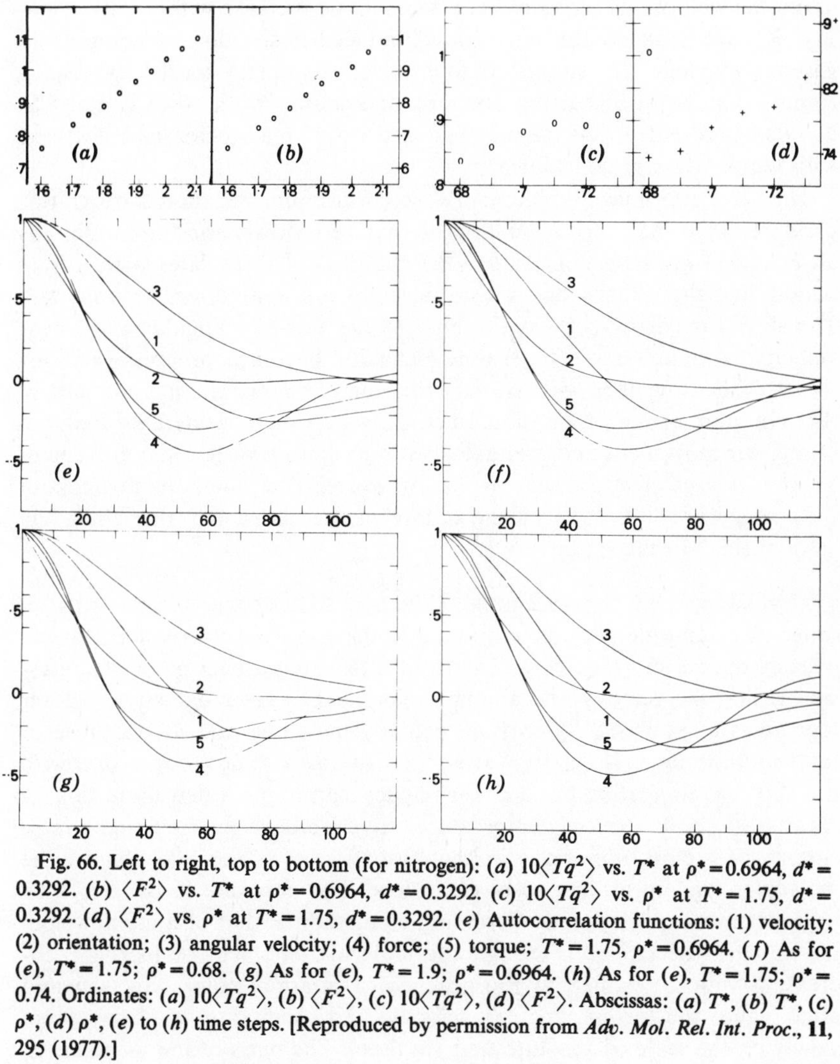

Fig. 66. Left to right, top to bottom (for nitrogen): (*a*) $10\langle Tq^2\rangle$ vs. T^* at $\rho^*=0.6964$, $d^*=0.3292$. (*b*) $\langle F^2\rangle$ vs. T^* at $\rho^*=0.6964$, $d^*=0.3292$. (*c*) $10\langle Tq^2\rangle$ vs. ρ^* at $T^*=1.75$, $d^*=0.3292$. (*d*) $\langle F^2\rangle$ vs. ρ^* at $T^*=1.75$, $d^*=0.3292$. (*e*) Autocorrelation functions: (1) velocity; (2) orientation; (3) angular velocity; (4) force; (5) torque; $T^*=1.75$, $\rho^*=0.6964$. (*f*) As for (*e*), $T^*=1.75$; $\rho^*=0.68$. (*g*) As for (*e*), $T^*=1.9$; $\rho^*=0.6964$. (*h*) As for (*e*), $T^*=1.75$; $\rho^*=0.74$. Ordinates: (*a*) $10\langle Tq^2\rangle$, (*b*) $\langle F^2\rangle$, (*c*) $10\langle Tq^2\rangle$, (*d*) $\langle F^2\rangle$. Abscissas: (*a*) T^*, (*b*) T^*, (*c*) ρ^*, (*d*) ρ^*, (*e*) to (*h*) time steps. [Reproduced by permission from *Adv. Mol. Rel. Int. Proc.*, **11**, 295 (1977).]

$T^*=3.6$, $L^*=0.1$, $\rho^*=0.2$ to 0.8 (Fig. 63). This set of data is intended to simulate the effect of increasing reduced number density on a roughly spherical molecule at constant reduced temperature. Throughout, the orientational autocorrelation function is inertia-dominated, decaying rapidly and exhibiting a secondary maximum similar to that observed by Kneubühl and Keller[38] for the exceptionally free rotations of HF and HCl in SF_6 solvents. The torque acf tends to oscillate at the higher ρ^*. It is significant that both the mean-square torque and mean-square force go through a minimum at $\rho^*=0.5$, whereas theories of hard collisions and harmonic well oscillations predict a monatomic increase with number density. Recent work on the far-infrared induced absorption of compressed gaseous ethylene has suggested that the mean-square torque exhibits a turning point with increasing pressure, and similar work of CCl_4 and CS_2 liquids has revealed that the mean-square torque may increase or decrease with temperature at constant ρ^*.

The force and linear velocity acf's have maximum variation through this pressure range. At $\rho^*=0.2$ the latter decays very slowly and the former has an extended negative tail. At $\rho^*=0.7$ the force acf oscillates with a long period, and the velocity acf in turn develops the well-known negative tail. The slowly decaying acf's at the high ρ^* are those of angular and linear velocity, although their short-time behavior becomes progressively different. These, together with the fact that there is now an apparent shift in the minimum of the orientational acf, shows up the considerable freedom of angular movement and gradual constraint upon translational movement as ρ^* increases. The changes in the autocorrelation functions throughout their range are indicative of the nature of the collision rather than any "structuring" in the fluid.

$T^*=3.6$, $\rho^*=0.1$, $L^*=0.2$ to 0.7 (Fig. 64). The torque increases by an order of magnitude with elongation, and the angular velocity acf simultaneously decays more quickly. The most interesting aspect of this progression is that the velocity and angular velocity acf's get progressively closer together and at $d^*=0.7$ decay at virtually the same rate as do those of force and torque. It is apparent from a comparison of this progression with the first that elongation has a larger impact than number density in this respect at constant reduced temperature. Here we have the first vague indications of factors important in the formation of a nematic phase—the molecular log-jam leading to birefringence. The only one of the five acf's to decay more quickly with increasing elongation is that of angular velocity, and this occurs even though the molecular moment of inertia is increasing with L^*, which in the absence of intermolecular effects would alone cause the torque, angular velocity, and orientation to decay more slowly on the scale of absolute time (in psec). The mass of the molecule is, of course, unaffected by elongation.

$\rho^*=0.643$, $T^*=2.3$, $L^*=0.1$ to 0.7 (Fig. 65). At this density there is a pronounced change from gaslike to liquidlike behavior as the elongation L^* increases at constant T^*. For example, the angular velocity acf decays slowly at $L^*=0.1$, but is oscillatory at $L^*=0.7$, where so are all the others except the orientational. Hard-core anisotropy must be an important factor in the determination of "structure" in liquids and, in this limit, in the appearance of nematic properties, where the orientational acf has been observed to be a slowly decaying exponential, and where the memory function, which is $\langle \dot{\mathbf{L}}^*(0)\cdot\dot{\mathbf{L}}^*(t)\rangle$ at $t=0$, oscillates very rapidly. In an Einstein solid the velocity acf is a pure cosine, but the nematic phase is characterized by rotational (albeit restricted) and translational freedom, and the forms of the five acf's may well be an extreme version of ours at $d^*=0.7$ for the Lennard-Jones dumbbell.

The linear velocity acf at $\rho^*=0.643$ decays initially much more quickly for the longer molecules, the torque and orientational acf's more slowly, but at $L^*=0.1$, the decay time of the orientational acf's at $\rho^*=0.643$ and $\rho^*=0.1$ are virtually identical, although the linear velocity acf decays much faster at the higher number density. On the other hand, for $L^*=0.7$, the velocity acf at the *lower* number density is slower to decay. For the longest molecule the decay time of the orientation is much longer at the higher number density.

The effect of elongation at constant reduced number density is much more pronounced than the effect of ρ^* at constant elongation, and the built-in hard-core repulsive part of our double Lennard-Jones potential seems dominant in promoting oscillations in some of our acf's.

$T^*=1.75$, $L^*=0.3292$, $\rho^*=0.68$ to 0.74 (Fig. 66). In this range the five acf's change very little, but the mean-square torque and force trend upward in value with increasing ρ^*, although there is an inflexion, or maybe a slight maximum, at $\rho^*=0.715$. The elongation corresponds to that of N_2, so our results are for a real molecule at constant reduced temperature. It is interesting to note that the mean-square torque for ethylene, calculated from far-infrared pressure-induced absorption, shows a minimum as pressure decreases at constant temperature. Ethylene is isoelectronic with nitrogen.

It is clear from the varied forms of autocorrelation function displayed here that further experiments (such as scattering of laser radiation, depolarization of fluorescence, infrared, and Raman wings) will be more fruitful when carried out simultaneously on one selected fluid, so that several different aspects of the motion of $\mathbf{L}^*$ can be discerned. It is clear that a study of $\langle \mathbf{L}^*(t)\cdot\mathbf{L}^*(0)\rangle$ done by one experimental technique in isolation disposes of a lot of information by statistical averaging; and in fact experiments on Debye relaxation at frequencies below those of the far infrared look at the long-time tail of this autocorrelation function, which is almost always exponential. Such experiments, although carried out with great experimental skill and effort, thus yield the minimum information about the trochilics of a typical molecular fluid, even when this happens to be dipolar. Now we are fortunate in having available large computers which, with admittedly rough-and-ready intermolecular potentials, can be used to yield useful, complementary, incisive data.

1. Collective Motions—Hydrodynamics and the Mori Equation

One of the clearest illustrations of the use of (I.6) in the field of hydrodynamics is that by Lallemand. A further incisive discussion on the topic is that of Kruus.[28] In this section we simulate functions such as the current and spin density from the molecular level, and attempt to fathom the gap between molecular and hydrodynamic theories of the fluid condition. This relation will continue to be intractable without the aid of computer simulation techniques, whereby information on individual molecules may be averaged to produce the correlation functions employed in the hydrodynamic equations of mass, momentum, and energy conservation leading to the linearized Navier-Stokes equation and to the Brillouin peaks of scattered light.

In this section we present some preliminary results on the simulation of current and spin density correlation functions for N_2 in the liquid and glassy states starting from the atom–atom potential and Newton's equations for individual molecular motions. The results are discussed very generally in terms of generalized hydrodynamic theory as first propounded by

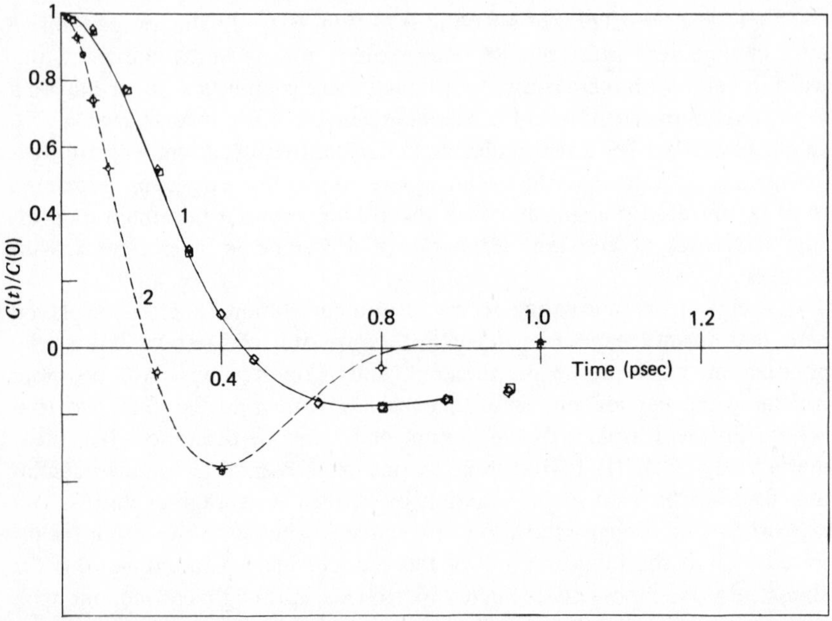

Fig. 67. (1) Autocorrelation function of velocity for N_2 in the liquid state at a reduced number density (ρ^*) of 0.643, reduced temperature (T^*) of 2.32. □, Cross-correlation of velocity, 1000 time steps; ◇, cross-correlation of velocity, 1600 time steps. (2) Autocorrelation function of force for N_2 in the same liquid state. -⊙- Cross-correlation function 1600 time steps; ⊙, cross-correlation function 1000 time steps. Ordinate: $C(t)/C(\text{o})$; abscissa: time (psec). [Reproduced by permission from *Chem. Phys. Lett.*, (1978).]

Zwanzig et al. for computer argon. This is the hydrodynamic equation formally equivalent to (I.6). Both the longitudinal and transverse current-density correlation functions are expected to be purely exponential decays on the basis of classical hydrodynamic theory, and so is the transverse spin-density correlation function. The spin-density correlation functions (longitudinal and transverse) are evaluated similarly. To check on the reliability of the statistics in the simulation of these collective correlation functions, the cross-correlation functions of velocity and force were evaluated using 1600 time steps and checked against the equivalent autocorrelation functions. Figure 67 shows that the decay of each is satisfactorily similar. All the hydrodynamic functions were thereafter evaluated using 1600 time steps of ca. 5×10^{-15} sec after rejecting the first 200.

In Fig. 68 are illustrated some longitudinal current density correlation functions $C_{\|}(\mathbf{k},t)$ for large and intermediate values of momentum transfer (wave vector) $\mathbf{k}$. In the N_2 fluid, $C_{\|}(\mathbf{k},t)$ is oscillatory with pronounced negative regions as $\mathbf{k}$ increases. Even with $|\mathbf{k}|=0.1$ (in reduced units of $1/\sigma$, where σ is the Lennard-Jones parameter for N_2), the correlation function is far from exponential, as is the assumption of classical Navier–Stokes equations. Our calculations of the transverse current correlation

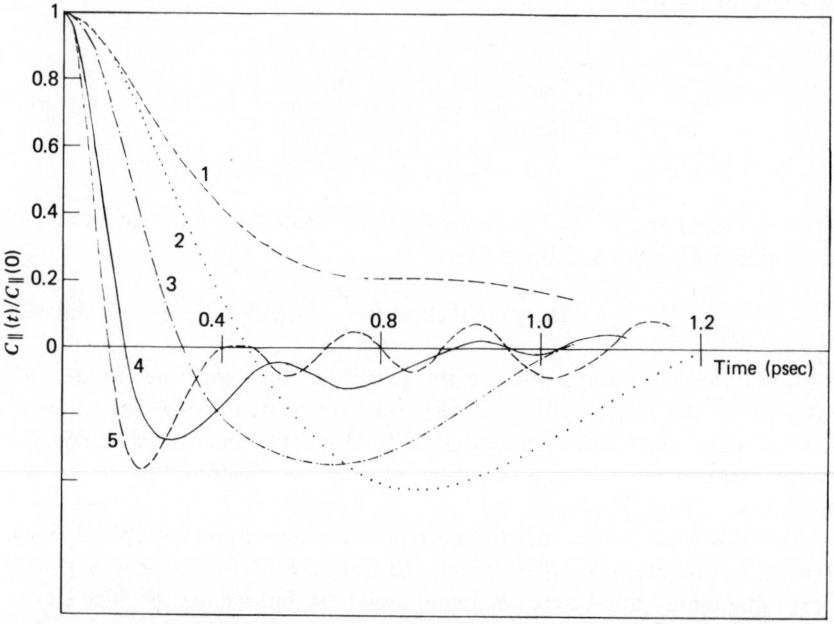

Fig. 68. Longitudinal current density correlation functions for different values of momentum transfer $\mathbf{k}$ (in reduced units of σ). (1) $|\mathbf{k}|^2=0.3$; (2) 3; (3) 15; (4) 45; (5) 60. Ordinate: $C_{\|}(t)/C_{\|}(0)$; abscissa: time (psec). [Reproduced by permission from *Chem. Phys. Lett.*, (1978).]

function indicate the same results. It is clear that a generalized hydrodynamic formalism is needed to account for the correlation functions simulated in this paper. The form suggested by Ailawadi et al. is

$$\frac{\partial}{\partial t} C_{\parallel}(\mathbf{k},t) = -\int_0^t K_l(\mathbf{k},t-t')C_{\parallel}(\mathbf{k},t')\,dt' \tag{III.29}$$

where the memory function $K_l(\mathbf{k},t)$ may be expressed as the sum

$$K_l(\mathbf{k},t) = k^2\left[\frac{kT}{m}S(\mathbf{k}) + \phi_{\parallel}(\mathbf{k},t)\right] \tag{III.30}$$

Here $S(\mathbf{k})$ is the equilibrium structure factor determined by a **k**-dependent compressibility, and $\phi_{\parallel}(t)$ is an after effect, or memory function, describing the delayed response of the longitudinal part of the stress tensor to a change in the rate of shear. We propose here, very tentatively, to develop (III.29) into a Mori continued fraction and in order to maintain compatibility between macroscopic and microscopic levels the series of equations in the Mori expansion of (III.29) may be truncated at the three-variable level already widely used in molecular theories of itinerant oscillations cited in this section.

$$\frac{\tilde{C}_{\parallel}(\mathbf{k},s)}{\tilde{C}_{\parallel}(\mathbf{k},0)} = \left[s + \cfrac{K_l^{(0)}(\mathbf{k},0)}{s + \cfrac{K_l^{(1)}(\mathbf{k},0)}{s+\gamma(\mathbf{k})}} \right] \tag{III.31}$$

Here $K_l^{(0)}(\mathbf{k},0)$ and $K_l^{(1)}(\mathbf{k},0)$ are the first and second memory functions of $\tilde{C}_{\parallel}(\mathbf{k},s)$ at $t=0$. $\gamma(\mathbf{k})$ is defined by

$$K_l^{(1)}(\mathbf{k},t) = K_l^{(1)}(\mathbf{k},0)\exp[-\gamma(\mathbf{k})t] \tag{III.32}$$

Naturally, (III.31) is empirical in the sense that three parameters are unknown: $K_l^{(0)}(\mathbf{k},0)$, $K_l^{(1)}(\mathbf{k},0)$, and $\gamma(\mathbf{k})$ and have to be fixed by least-mean-squares iteration to the simulated $C_{\parallel}(\mathbf{k},t)$. However, some physical significance and dependence on $|\mathbf{k}|$ may be extracted as illustrated in the article by Ailawadi et al.[108]

The transverse current density correlation function and the spin density correlation functions oscillate about the time axes in the same way as the foregoing representative curves, being drawn in Figs. 68 and 69. The transverse spin density correlation function may not be derived in any other way than by computer simulation, since neutron scattering is insensitive in this context. It is obvious that the coupling between transverse spin and

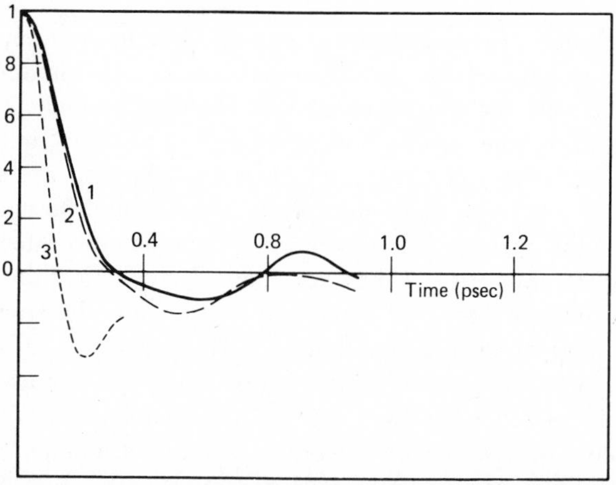

Fig. 69. (1) Longitudinal spin-density correlation function N_2, $p^* = 0.643$; $T^* = 2.32$; $|\mathbf{k}|^2 = 45$. (2) Transverse spin-density correlation function, N_2, $p^* = 0.643$; $T^* = 2.32$; $|\mathbf{k}|^2 = 30$. (3) Longitudinal current density correlation function, N_2; $|\mathbf{k}|^2 = 60$. N_2 in the glassy state at $p^* = 0.800$, $T^* = 2.30$. Abscissa: time (psec). [Reproduced by permission from *Chem. Phys. Lett.*, (1978).]

current density will be complicated in fluids of more pronounced shape anisotropy than N_2, and the limit of analytical tractability will be reached very quickly in dealing with this rotation–translation problem. This leaves only computer simulation as a practical means of investigation given a suitable intermolecular potential and improved numerical method. It would be particularly useful, therefore, in further work to solve numerically the equations of motion for a given *molecular* model, such as the itinerant oscilator, and then to determine directly whether the current and spin density correlation functions are describable by the same type of Mori approximant as are their molecular counterparts. This type of simulation would then be effectively a method of evaluating hydrodynamic functions from a specific model, of, say, a Brownian plus resonance type. In the next section we take the first step toward this goal by numerically solving (I.31) with the molecular dynamics techniques developed by Bellemans and Hermans, Kestemont, van Loon, and Finsy.[109]

E. Molecular Dynamics Simulation of the Planar Itinerant Librator

In this section we use the molecular dynamics method to simulate the system of ring/annulus itinerant libration developed by Coffey et al., (I.31), for angular planar reorientation of the asymmetric-top dipole vector. The main aim is the limited one of ascertaining to what precision the numerical solution of the equations of motion used in the numerical simulation reproduces the *stochastic* differential equations of the analytical approach. Therefrom, the simulation may be extended to fields beyond analytical tractability. The assumptions inherent in a model of molecular motion may be satisfied exactly by (for example) rough-sphere simulations, and therefrom analytical and simulated autocorrelation functions may be compared. In this way, O'Dell and Berne[108] have demonstrated clearly the limits of applicability of the J-diffusion model for spherical tops. Similarly, the itinerant librator model of molecular motion may be simulated by a two-dimensional molecular dynamics system consisting of rough rings within which are disks, bound harmonically. Exchange of linear and angular momentum occurs when two rings collide. This may be compared with the system devised analytically where the annulus of moment of inertia I_1 is subjected to Brownian motion with friction coefficient ζ. It turns out that the analytical results varying ζ may be simulated very precisely, so that the latter may be used to extend the formalism to three dimensions, or to include, for example, the effects of rotation–translation coupling.

1. *Computational Details*[109]

We use an assembly of 120 particles (i.e., disks/annuli) of total mass m and diameter D. The motion of the annulus is perturbed by collisions, between which the center of each particle moves along a straight line at constant velocity and total angular momentum. The rotational motion of the annulus and disk is governed by

$$I_1\ddot{\theta}_1(t) = -\gamma[\theta_1(t) - \theta_2(t)] \tag{III.33}$$

$$I_2\ddot{\theta}_2(t) = \gamma[\theta_1(t) - \theta_2(t)] \tag{III.34}$$

Here I_1 and I_2 are the moments of inertia of the annulus and the disk; θ_1 and θ_2 specify the position of a point on the rim of the annulus and the position of the dipole on the disk. γ is the restoring torque constant between ring and disk. The dipole on the inner disk is supposed vanishingly small so that dipole-dipole coupling is neglected. When a collision occurs, an energy transfer takes place between rotational and translational degrees of freedom, depending on the dimensionless quantity

$$\tau = \frac{4I_1}{mD^2} \tag{III.35}$$

We seek to establish how closely τ may be used to simulate the frictional torque $\zeta\theta_1(t)$ and $\lambda(t)$, the random couple of the Brownian motion assumed analytically. In the molecular dynamics the change of linear and angular velocities at a collision between particles A and B are given by the following set of equations:

$$\mathbf{v}_A' = \mathbf{v}_A + \frac{\tau}{1+\tau}\left[\mathbf{v} + \frac{1}{\tau}\mathbf{k}(\mathbf{k}\cdot\mathbf{v})\right] \quad \text{(III.36)}$$

$$\mathbf{v}_B' = \mathbf{v}_B - \frac{\tau}{1+\tau}\left[\mathbf{v} + \frac{1}{\tau}\mathbf{k}(\mathbf{k}\cdot\mathbf{v})\right] \quad \text{(III.37)}$$

$$\omega_A' = \omega_A - \frac{2}{(1+\tau)D}\mathbf{k}\times\mathbf{v} \quad \text{(III.38)}$$

$$\omega_B' = \omega_B - \frac{2}{(1+\tau)D}\mathbf{k}\times\mathbf{v} \quad \text{(III.39)}$$

Here $\mathbf{v}_A$ and $\mathbf{v}_B$ are the translational velocities and ω_A and ω_B are the angular velocities of the annulus, that of the inner disk being unaffected by the collision. The primed variables correspond to the situation just after the collision. $\mathbf{k}$ is the unit vector directed from the center of the B particle to that of A at the time of collision. $\mathbf{v}$ is the relative velocity of the points in contact:

$$\mathbf{v} = \mathbf{v}_B - \mathbf{v}_A - \tfrac{1}{2}D\mathbf{k}\times(\omega_A + \omega_B) \quad \text{(III.40)}$$

Initially, the 120 particles are arranged in 12 rows of 10 at the nodes of a triangular lattice whose dimensions are chosen to obtain the desired density d expressed as the number of particles per unit surface. Periodic boundary conditions are used, and initially the translational and angular velocities and the orientation of the two parts of each particle are randomly distributed. Reduced units of $I_1/kT=1$, $D=1$, $\alpha=\gamma/kT$, and $R=I_1/I_2$ are used. After the system has reached equilibrium it is followed for up to 1024 time intervals of $t=0.05(kT/I_1)^{-1/2}$. The two components of the unit vector parallel to the dipole, the two components of the derivative of this vector, the angular velocity of the disk, and the torque on it are recorded for each particle for subsequent calculation of correlation functions.

In Fig. 70 some autocorrelation functions for this system are illustrated. The analytical curves for least-mean-squares best fit behave similarly and are inseparable by eye for various values of ζ. This justifies the use of simulation in two dimensions, and therefore future computation in three dimensions for itinerant libration or itinerant libration–oscillation may be contemplated. These simulations will be of great utility in, for example, the evaluation of the approximation to (I.6) of (I.31). It would be useful to define the difference between these equations in the itinerant librator–oscillator system, and future work will concentrate on this. It will also be possible

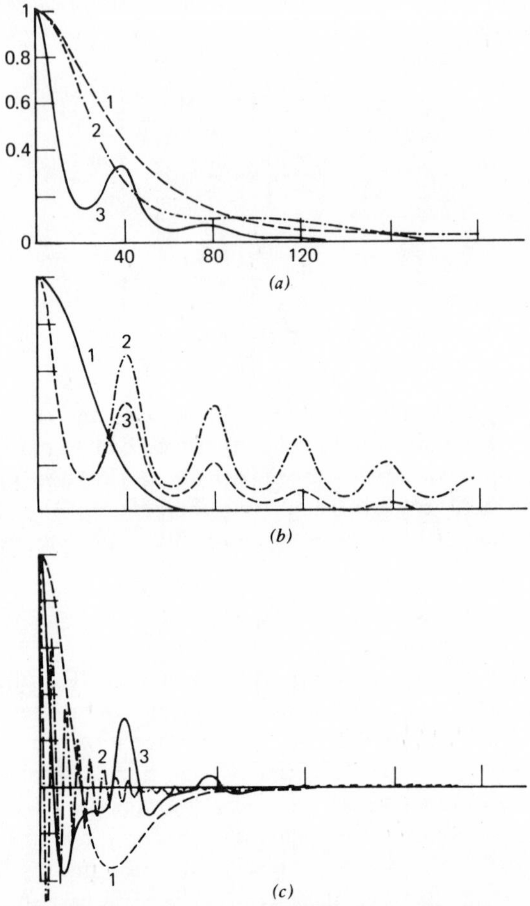

Fig. 70. Autocorrelation functions of orientation of the disk dipole vector $\boldsymbol{u}$. (*a*) (1) Reduced density $(d^*)=0.6$; $\alpha=10^3$; $R=10$, $\tau=0.50$; (2) $d^*=0.8$; $\alpha=1.0$; $R=1$, $\tau=0.50$; (3) $d^*=0.6$; $\alpha=1.0$; $R=10$, $\tau=0.50$. (*b*) (1) $d^*=0.6$; $\alpha=0.1$; $R=1$, $\tau=0.50$; (2) $d^*=1.0$; $\alpha=1.0$; $R=10$, $\tau=0.50$; (3) $d^*=0.8$; $\alpha=1.0$; $R=10$, $\tau=0.50$. (*c*), (*d*) Autocorrelation function $\langle\dot{\boldsymbol{u}}(t)\cdot\dot{\boldsymbol{u}}(o)\rangle$. (*c*) (1) Reduced density $(d^*)=0.6$; $\alpha=1.00$; $R=1.00$; $\tau=0.50$; (2) $d^*=0.6$; $\alpha=10^3$; $R=10.00$; $\tau=0.50$; (3) $d^*=0.6$, $\alpha=1.00$; $R=10.00$; $\tau=0.50$. (*d*) (1) Reduced density $(d^*)=0.6$; $\alpha=0.10$; $R=1.00$; $\tau=0.50$; (2) $d^*=1.0$; $\alpha=1.00$; $R=10.00$; $\tau=0.50$; (3) $d^*=0.6$; $\alpha=10.00$; $R=1.00$; $\tau=0.50$. (*e*), (*f*) Autocorrelation function of the disk angular velocity $\langle\omega(t)\cdot\omega(o)\rangle$. (*e*) (1) Reduced density $(d^*)=0.6$; $\alpha=1.00$; $R=1.00$; $\tau=0.50$; (2) $d^*=0.6$; $\alpha=10^3$; $R=10.00$; $\tau=0.50$; (3) $d^*=0.6$; $\alpha=1.0$; $R=10.00$; $\tau=0.50$; (*f*) (1) Reduced density $(d^*)=0.6$; $\alpha=0.10$; $R=1.00$; $\tau=0.50$; (2) $d^*=0.6$; $\alpha=10.00$; $R=1.00$; $\tau=0.50$; (3) $d^*=0.8$; $\alpha=1.00$; $R=1.00$; $\tau=0.50$. (*g*), (*h*) Autocorrelation function of the disk torque $\langle\dot{\omega}(t)\cdot\dot{\omega}(o)\rangle$. (*g*) (1) Reduced density $(d^*)=0.6$; $\alpha=0.10$; $R=1.00$; $\tau=0.5$; (2) $d^*=0.6$; $\alpha=10^3$; $R=10.00$; $\tau=0.5$; (3) $d^*=0.6$; $\alpha=1.0$; $R=10.00$; $\tau=0.5$. (*h*) (1) $d^*=0.8$; $\alpha=1.0$; $R=1.00$; $\tau=0.5$; (2) $d^*=0.6$; $\alpha=1.0$; $R=1.00$; $\tau=0.5$; (3) $d^*=0.6$; $\alpha=10.00$; $R=1.00$; $\tau=0.5$. Abscissas: time steps. [Reproduced by permission from *Chem. Phys. Lett.*, **58**, 521 (1978).]

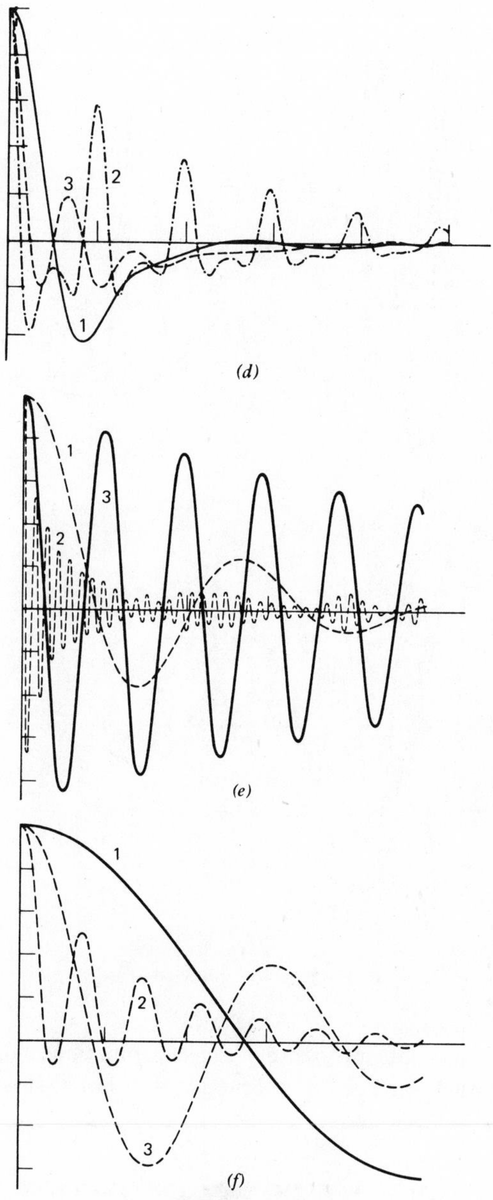

Fig. 70. Continued

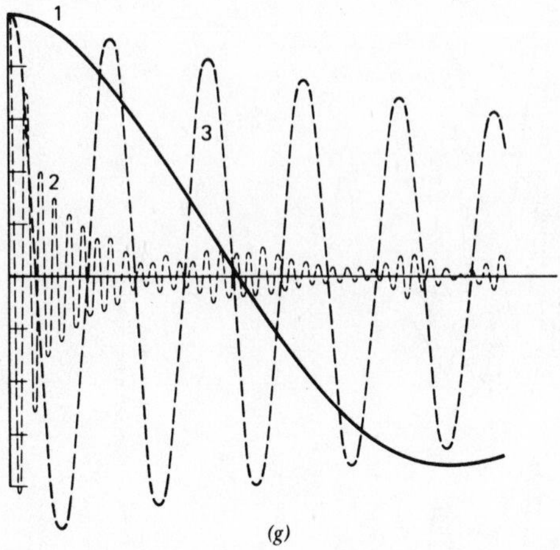

(g)

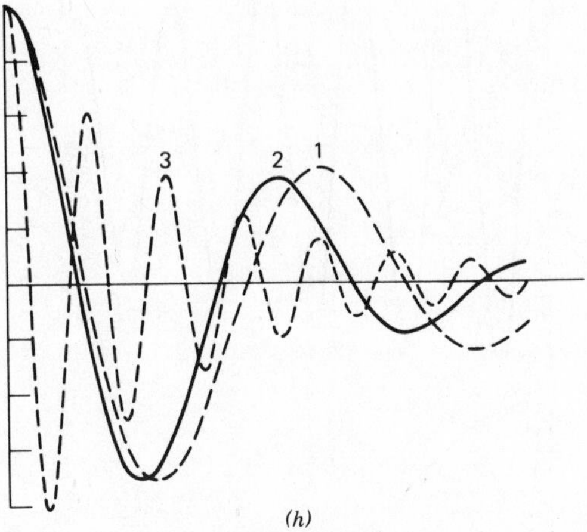

(h)

Fig. 70. Continued

numerically to simulate probability density functions which are analytically almost intractable by classical methods, although less so with those of Section I.

IV. INDUCED ZERO-THZ ABSORPTIONS AND ROTATION/TRANSLATION OF MOLECULES

We have mentioned in Section I that nondipolar liquids and compressed gases absorb in the zero-THz frequency region due to a fluctuating (time-dependent) dipole induced by the other molecules in the ensemble. Here we consider how the phenomenon may be measured in terms of the molecular rototranslations and intermolecular potential energies. Although only broad bands are observable, whose breadth corresponds roughly to the lifetime of the induced dipole, a quantum theory using a multipole expansion of the electrodynamic field has often been used as an approximate representation, using radial averaging of the Lennard-Jones potential. We review this briefly and adopt also a classical description in terms of the Mori expansion which is empirical in approach, but affords a very close fit to the available data and is useful for their intercomparison.

Early evidence of an absorption at high microwave frequencies in highly purified nondipolar liquids was presented by Whiffen.[110] The first indication that the absorptions are of a rototranslational rather than purely translational origin came via Savoie and Fournier,[111] who obtained the far-infrared spectra of CH_4 and CD_4 as the liquid and plastic crystal down to 12°K. The liquid exhibits a broad maximum at about 200 cm^{-1} for CH_4 and 150 cm^{-1} for CD_4, which the authors interpreted as $I^{1/2}$ (rotational) rather than $m^{1/2}$ (translational) dependence. Davies, Chamberlain, and Davies[112] made the first attempt to interpret these bands in terms of the torsional oscillation of a molecule within the cage formed by its neighbors. They carried out refractive-index measurements on nondipolar liquids, and discovered that these indices had shallow minima in the far-infrared region. Evidence of a substantial intermolecular mean-square torque which hinders the molecular rotational-type motions in liquid CO_2 as compared with the compressed gas put forward by Birnbaum and Rosenberg[113] and by Baise.[114] The absorption of the liquid at 273°K has an integrated intensity an order of magnitude less than that in the gas phase, where rotational type $J \rightarrow J+2$ (quadrupole-induced) band shapes predominate. Significant intermolecular forces shift the peak by about 25 cm^{-1} to higher frequencies in the liquid. The authors argue then that the large quadrupole moment of CO_2, which has a dominant effect on the gas-phase absorption, is apparently much reduced in value in the liquid. This is the result of the "local order" and the symmetry arising, for example, in three-body collisions, the effective induced dipole being smaller. Carrying this argument to the static limit, collision-induced absorption would disappear if each molecule occupied a site of inversion symmetry. Such behavior was verified by Evans[115] in the more strongly quadrupolar cyanogen, and by Baise[114] in the weakly dipolar nitrous oxide.

Thus the evidence is in favor of a torsional cum rotational diffusion type of motion as the principal source of the far-infrared absorptions in these nondipolar liquids, closely analogous to that in their dipolar counterparts, so that (I.6) may be used to predict the spectral function $\tilde{C}(s)$. In this case the relevant element of $\mathbf{A}$ is the net induced dipole moment at the instant t.

The simplest theories may be constructed classically with this one element alone. First, however, we summarize the quantum theory as developed in symmetric tops by Frost.[116]

A. Multipole-Induced Absorption in Symmetric-Top Molecules

Nondipolar molecules absorb in the zero-THz region because intermolecular electrostatic fields distort the overall symmetry of a given molecule's electron cloud, producing upon "collision" a small dipole moment that changes in magnitude and direction rapidly with time. Thus compressed gaseous mixtures of rare gas atoms absorb, whereas the components when separated and moderately pressurized do not. A pair of colliding helium atoms, for example, will not possess a resultant electronic cloud of dipolar asymmetry, whereas a helium–neon pair will modulate the electromagnetic field over a broad band of far-infrared frequencies commensurate with the most probable frequencies at which interatomic collisions occur. Atomic-induced absorption is of a purely translational origin, a mechanism that persists in molecular fluids such as hydrogen and nitrogen as the absorption $\Delta J=0$, where J is the rotational quantum number. A dipole moment set up between a pair of colliding molecules will in addition absorb by rotational means since even without relative translation of molecular centers, the effect on each other of their rotatory electrostatic fields will not cancel. A practical means of dealing with these intermolecular absorption mechanisms is to treat them separately. The rotational absorption is dealt with by expanding the field in terms of multipole tensors, which all vanish only in the case of spherical symmetry such as that of atoms.

Pseudospherical molecules such as SF_6 retain the higher multipoles (those above and including the hexadecapole for O_h symmetry), and thus display a weak, rotational-induced band at moderately higher number densities. The first nonvanishing multipole in a homogeneous diatomic such as N_2 is the quadrupole, which produces a dipole on molecule A which is modulated by the rotational motion of the inducing molecule B. The symmetry of the quadrupole moment is such that it rotates twice as fast as the molecule itself and thus produces quantum absorptions with the selection rule $\Delta J=2$, in contrast with the $\Delta J=1$ rule for the rotation of a permanent dipole. Similarly, the first nonzero multipole moment for T_d symmetry—the octopole—produces $\Delta J=3$, and the hexadecapole $\Delta J=4$. The formal quantum-mechanical equation for multipole-induced dipole absorption has been developed by Colpa and Ketelaar[117] for linear molecules, and extended as follows by Frost.[116]

The absorption is attributed to the mutually induced dipole moment $\boldsymbol{\mu}(R)$ on orientations of the molecules is expressed by expanding $\boldsymbol{\mu}(R)$ in terms of quantities $D^{\lambda_1}_{\mu_1\nu_1}(\phi_1,\theta_1,\chi_1)D^{\lambda_2}_{\mu_2\nu_2}(\phi_2,\theta_2,\chi_2)$, where a $D^{\lambda}_{\mu\nu}$ is a matrix element of an irreducible representation of the rotation group. This permits evaluation of matrix elements of $\boldsymbol{\mu}(R)$ between eigenstates of a molecular pair. The expansion coefficients are then evaluated in terms of the polarizabilities and multipole moments (of any order) of the molecules. The theory performs summations over "uninteresting" magnetic quantum numbers and produces expressions for the intensity of pressure-induced in terms of Clebsch–Gordan coefficients. The center-of-mass motion of each molecule is treated classically, with each molecule at rest, so translational absorption is ignored. The pressure-induced intensity of an absorption band is calculated in terms of quantities:

$$\sum_{m_i m_f} |\langle im_i|\boldsymbol{\mu}(R)|fm_f\rangle|^2 \tag{IV.1}$$

where $|im_i\rangle, |fm_f\rangle$ denote rotation–vibration energy eigenstates for the pair, with m_i, m_f as degenerate magnetic quantum numbers. The dipole moment $\boldsymbol{\mu}(R)$ which occurs in (IV.1) is a sum of two parts:

$$\boldsymbol{\mu}(R) = \boldsymbol{\mu}(1;R) + \boldsymbol{\mu}(2;R) \tag{IV.2}$$

where $\boldsymbol{\mu}(1;R)$ is the moment induced in molecule 1 by the electric field of molecule 2, and vice versa. The electric field at 1 due to 2 depends on the orientation ϕ_2; θ_2, χ_2 and the vibration coordinate s_2 of 2, while the polarizability of 1 depends on its orientation ϕ_1, θ_1, χ_1 and vibration coordinate s_1. Therefore, $\boldsymbol{\mu}(1;R)$ depends on the orientation and vibration coordinates of both molecules. The same applies to $\boldsymbol{\mu}(2;R)$, so

$$\boldsymbol{\mu}(R) = \boldsymbol{\mu}(R;\phi_1,\theta_1,\chi_1,s_1;\phi_2,\theta_2,\chi_2,s_2)$$

Although the molecules are interacting, it is assumed that an eigenstate of the pair is simply a product of eigenstates of the isolated molecules. This result is exact only if the intermolecular potential is independent of the Euler angles θ, ϕ, χ, and therefore the theory is restricted to the range where R is determined primarily by the central part $U(R)$ of the interaction potential. Frost now takes the intensity (I) of the absorption band as defined by Colpa and Ketelaar for a gas consisting of two types of molecule, species A of number density n_A and species B.

$$I = I_{AA} + I_{BB} + I_{AB} \tag{IV.3}$$

where

$$I_{AB}=\frac{8\pi^3\nu}{3hc}\frac{(2-\delta_{AB})}{2}n_An_B\sum_{\substack{i,f\\E_f-E_i\simeq h\nu}}\left[\frac{F_i^{AB}}{d_i^{AB}}-\frac{F_f^{AB}}{d_f^{AB}}\right]$$

$$\times\int 4\pi R^2\exp(-U_{AB}(R)/kT)\sum_{m_im_f}\left|\langle im_i|\boldsymbol{\mu}^{AB}(R)|fm_f\rangle\right|^2 dR$$

The summations over the quantum numbers i,f are restricted to those transitions $i\to f$ for which the absorbed frequency $h^{-1}(E_f-E_i)$ lies in the absorption band of approximate frequency ν. The quantities d_i, d_f are the degeneracies of the quantum numbers i, f $[d_i=(2J_1+1)(2J_2+1)]$. The quantities F_i, F_f are the fractional populations of pair states with quantum numbers i, f, respectively: for example,

$$F_i=\frac{d_i\exp(-E_i/kT)}{\sum_j d_j\exp(-E_j/kT)}$$

We emphasize that $\boldsymbol{\mu}(R)$ is the *induced* dipole moment of a pair of molecules and does not include the permanent dipole moment $\mathbf{p}^{(1)}+\mathbf{p}^{(2)}$ of the pair. The latter is independent of R and leads to a divergence of (IV.3), which gives absorption which may be attributed, *with the neglect of pressure broadening effects*, to transitions of isolated molecules. Later we shall show how pressure broadening may be accounted for classically using approximants of (I.6). Evaluation of the dipole moment matrix elements yields; for pure rotational transitions

$$\sum_{m_im_f}\left|\langle im_i|\boldsymbol{\mu}(R)|fm_f\rangle\right|^2$$

$$=\sum_{\lambda_1\lambda_2}\sum_{\mu_1\mu_2}\sum_m|F_m(R,\lambda_1,\mu_1,K_1'-K_1;\lambda_2,\mu_2,K_2'-K_2)|$$

$$\times\frac{2J_1+1}{2\lambda_1+1}C(J_1,\lambda_1,J_1';K_1,K_1'-K_1,K_1')^2$$

$$\times\frac{2J_2+1}{2\lambda_2+1}C(J_2,\lambda_2,J_2';K_2,K_2'-K_2,K_2')^2 \qquad \text{(IV.4)}$$

in terms of Clebsch–Gordan coefficients C. Here the dipole moment $\boldsymbol{\mu}(R)$ is regarded with rectangular Cartesian components $\boldsymbol{\mu}_j(R)(j=1,2,3)$ relative to the space fixed frame. These are Hermitian operators. Each of μ_1 in

(II.5) then ranges from $-\lambda_1$ to $+\lambda_1$. J and K are defined by the rotational part of the molecular eigenstate of a rigid symmetric top:

$$|JKM\rangle = \psi_{JKM}(\phi,\theta,\chi) = \left(\frac{2J+1}{8\pi^2}\right)^{1/2} D^{J}_{-M,-K}(\phi,\theta,\chi)$$

where the D values are the irreducible representations of the rotation group as defined by Rose. The expansion coefficient F_m depends on molecular parameters such as the polarizability and is in practice negligible for all but a few values of λ_1 and λ_2. To evaluate F_m we want the Cartesian components in the space fixed frame of the induced dipole moment $\mu(R)$ in the form

$$\mu_j(R) = \sum_k \alpha^{(1)}_{jk}(s_1,\phi_1,\theta_1,\chi_1) E^{(2)}_k(R; s_2,\phi_2,\theta_2,\chi_2) + \sum_k \alpha^{(2)}_{jk}(s_2,\phi_2,\theta_2,\chi_2) E^{(1)}_k(R; s_1,\phi_1,\theta_1,\chi_1) \tag{IV.5}$$

where, for example, the $\alpha^{(1)}_{jk}$ are the components of the polarizability tensor $\alpha^{(1)}$ of molecule 1, while the $E^{(2)}_k$ are the components of the electric field at molecule 1 due to molecule 2. The polarizability tensor is then cast into a spherical form relative to the space fixed frame. Let $\bar{x}_1$, $\bar{x}_2$, $\bar{x}_3$ denote the body fixed principal axes of inertia of a symmetric-top molecule, chosen so that the x_3 axis is the symmetry axis. Then there are also the principal axes of polarizability of the molecule; that is, if it is subjected to an external electric field, with Cartesian coordinates $\bar{E}_j$ relative to the $\bar{x}$-frame, the components of the induced moment are

$$\bar{\mu}_j = \sum_k \bar{\alpha}_{jk}(s)\bar{E}_k \qquad \bar{\alpha} = \begin{bmatrix} \alpha_\perp & 0 & 0 \\ 0 & \alpha_\perp & 0 \\ 0 & 0 & \alpha_\parallel \end{bmatrix} \tag{IV.6}$$

The spherical component version of (IV.6) is then

$$\bar{\mu}_m = \sum_m \bar{A}_{m'm}(s)\bar{E}_m \qquad A = \begin{bmatrix} \alpha_\perp & 0 & 0 \\ 0 & \alpha_\parallel & 0 \\ 0 & 0 & \alpha_\perp \end{bmatrix} \tag{IV.7}$$

The space fixed x-frame is then rotated into the $\bar{x}$-frame along with the components of the field and polarizability. It turns out that as far as polarizability is concerned, symmetric tops behave like linear molecules since, for both sorts, (IV.6) has two equal components.

The electric field is now expanded in terms of multipole moments. The electrostatic potential caused by a molecule at a point R, with polar coordinates R, Θ, Φ in the x-frame is then given by

$$V(R)=\sum_{\lambda=0}^{\infty}\sum_{\mu=-\lambda}^{\lambda}\left(\frac{4\pi}{2\lambda+1}\right)^{1/2}\frac{Q_\mu^\lambda Y_{\lambda\mu}^*}{R^{\lambda+1}}(\Theta,\Phi) \tag{IV.8}$$

Here the μth component of the λth multipole moment of the molecule in the x-frame is

$$Q_\mu^\lambda=\left(\frac{4\pi}{2\lambda+1}\right)^{1/2}\int\int\int \xi^\lambda Y_{\lambda\mu}(\alpha\beta)_p(s,\xi\alpha\beta)\xi^2 d\xi \sin\alpha\, d\alpha\, d\beta$$

where p is the charge density of the molecule at a point ξ with polar coordinates ξ, α, β in the x-frame. We then have for the required fields:

$$E_m(R^\pm)=\sum_\lambda E_m^\lambda(R^\pm)$$

with

$$E_m^\lambda(R^\pm)=(-1)^{m+1}(\pm 1)^{\lambda+1}\left(\sqrt{(\lambda+1)(2\lambda+3)}\,/R^{\lambda+2}\right)$$
$$\times C(\lambda+1,1,\lambda;0,-m,-m)\sum_\nu D^\lambda_{-m,\nu}(\phi\theta\chi)\bar{Q}_\nu^{\lambda *}$$

Here R^+ denotes the case $\Theta=0$ while R^- denotes $\Theta=\pi$. The multipole moment components of $\bar{Q}_\mu^\lambda$ of a symmetric top are severely limited by the fact that the x_3-axis is an n-fold axis of rotational symmetry, with $n\geq 3$. Therefore:

1. For $\lambda=1$, only $\bar{Q}_0^1$ is nonzero. Since $Q_\mu^{\lambda *}=(-1)^\mu Q^\lambda$, this quantity is real and is the usual dipole moment of the molecule.
2. For $\lambda=2$, only $\bar{Q}_0^2$ is nonzero. This real quantity is the usual quadrupole moment defined by $Q_0^2=q=q_{33}$, where

$$q_{ij}=\tfrac{1}{2}\int\int\int\rho(s,\bar{\xi}_1\bar{\xi}_2\bar{\xi}_3)\left[3\bar{\xi}_i\bar{\xi}_j-\bar{\xi}^2\delta_{ij}\right]d\bar{\xi}_1 d\bar{\xi}_2 d\bar{\xi}_3$$

   is the Cartesian quadrupole moment tensor in the principal frame.
3. For $\lambda=3$, only $\bar{Q}_0^3$ and $\bar{Q}_{\pm 3}^3$ are nonzero if $n=3$, while only $\bar{Q}_0^3$ is nonzero for $n>3$.
4. For $\lambda>3$ (hexadecapole moment and higher), the number of nonzero components $\bar{Q}_\mu^\lambda$ depends in an obvious way on n. For a linear molecule $n=\infty$, and each multipole moment has only one nonzero component $\bar{Q}_0^\lambda$.

The expansion coefficient F_m of (IV.4) may now be evaluated in terms of polarizability and multipole moments. The dipole moment induced in molecule 1 is then

$$\mu_m(1;R) = \alpha_0^{(1)} \sum_{\lambda_2} (-1)^{\lambda_2+1} \sqrt{(\lambda_2+1)(2\lambda_2+3)}\; R^{-(\lambda_2+2)}$$

$$\times \sum_{\mu_2\nu_2} (-1)^{1-\mu_2} \delta_{-\mu_2,m} \bar{Q}_{\nu_2}^{(2)\lambda_2^*} C(\lambda_2+1,1,\lambda_2;0,\mu_2,\mu_2) D_{0,0}^{0}(1) D_{\mu_2\nu_2}^{\lambda_2}(2)$$

$$+(-1)^m \sqrt{\frac{2}{3}}\; \delta^{(1)} \sum_{\lambda_2} (-1)^{\lambda_2+1} \sqrt{(\lambda_2+1)(2\lambda_2+3)}\; R^{-(\lambda_2+2)}$$

$$\times \sum_{\mu_1} \sum_{\mu_2\nu_2} (-1)^{1-\mu_2} \bar{Q}_{\nu_2}^{(2)\lambda_2^*} C(1,1,2;-m,-\mu_2,\mu_1)$$

$$\times C(\lambda_2+1,1,\lambda_2;0,\mu_2,\mu_2) D_{\mu,0}^{2}(1) D_{\mu_2\nu_2}^{\lambda_2}(2) \qquad \text{(IV.9)}$$

A similar expression for $\mu_m(2;R)$ is obtained from (IV.9) by first omitting the factor $(-1)^{\lambda_2+1}$ and then interchanging all superscripts and subscripts 1 and 2. The sum of these yields an expansion of $\mu_m(R)$ in terms of the orientation functions $D_{\mu_1\nu_1}^{\lambda_1}(1) D_{\nu_2\mu_2}^{\lambda_2}(2)$. The expansion coefficients contain the polarizabilities α_0, anisotropies δ, and the multiple moments $\bar{Q}_\mu^\lambda$ of the molecules, quantities which are functions of the vibration coordinates s. Consequently, from each coefficient in this expansion of $\mu_m(R)$, we obtain the corresponding F_m by setting all vibration coordinates to their equilibrium values, (i.e., by using the equilibrium polarizabilities and multipole moments). It follows from (IV.9) and the corresponding equation for $\mu_m(2;R)$ that an $F_m(R;\lambda,\mu_1\nu_1;\lambda_2\mu_2\nu_2)$ is nonzero only if one of λ_1 or λ_2 is 0 or 2. F_m $(\lambda_1=0,\lambda_2=0)$ is zero, since we presume that each molecule has no net charge (i.e., $\bar{Q}_0^0=0$). An F_m $(\lambda_1=0$ or $2,\lambda_2\neq 2)$ is obtained entirely from the expansion of $\mu_m(1;R)$ (i.e., such an F_m depends only on the dipole moment induced in molecule 1 by molecule 2). An F_m $(\lambda_1\neq 2,\lambda_2=0$ or $2)$ is obtained entirely from the expansion of $\mu_m(2;R)$. The coefficient F_m $(\lambda_1=2,\lambda_2=2)$ is the only one which must be calculated from the sum. The quantities

$$\sum_{m\mu_1\mu_2} |F_m(R,\lambda_1\mu_1\nu_1,\lambda_2\mu_2\nu_2)|^2$$

of (IV.4) may now be evaluated since they reduce to sums of Clebsch–Gordan coefficients. They are tabulated in Table IX.

TABLE IX
Summation of F_m in (IV.4)

λ_1	λ_2	$\Sigma\vert F_m^2\vert$
0	1	$\delta_{0\nu_1}\delta_{0\nu_2}6(\alpha_0^{(1)}\rho^{(2)})^2R^{-6}$
0	2	$\delta_{0\nu_1}\delta_{0\nu_2}15(\alpha_0^{(1)}q^{(2)})^2R^{-8}$
0	Any λ_2	$\delta_{0\nu_1}(\lambda_2+1)(2\lambda_2+1)(\alpha_0^{(1)}\vert\overline{Q}_{\nu_2}^{(2)\lambda_2}\vert)^2R^{-2(\lambda_2+2)}$
2	1	$\delta_{0\nu_1}\delta_{0\nu_2}(20/3)(\delta^{(1)}\rho^{(2)})^2R^{-6}$
2	Any $\lambda_2\neq 0,2$	$\delta_{0\nu_1}(10/9)(\lambda_2+1)(2\lambda_2+1)(\delta^{(1)}\vert\overline{Q}_{\nu_2}^{(2)\lambda_2}\vert)^2R^{-2(\lambda_2+2)}$
2	2	$\delta_{0\nu_1}\delta_{0\nu_2}(10/3)[5(\delta^{(1)}q^{(2)})^2+5(\delta^{(2)}q^{(1)})^2-6\delta^{(1)}\delta^{(2)}]R^{-8}$

Selection rules in (IV.3) arise from two sources:

1. General limitations on Clebsch–Gordan coefficients.
2. The nature of symmetric-top molecules is such that $\Sigma|F_m^2|$ is nonzero only for restricted values of ν_1 and ν_2. A term in (IV.4) is zero unless:

$$\Delta J_1 = J_1' - J_1 = 0, \pm 1, \ldots, \pm\lambda_1$$

$$\Delta K_1 = K_1' - K_1 = 0, \pm n_1, \ldots, \pm m_1 n_1 \qquad (m_1 n_1 \le \lambda_1)$$

$$\Delta J_2 = J_2' - J_2 = 0, \pm 1, \ldots, \pm\lambda_2$$

$$\Delta K_2 = K_2' - K_2 = 0, \pm n_2, \ldots, \pm m_2 n_2 \qquad (m_2 n_2 \le \lambda_2)$$

Here m_1 or m_2 is a positive integer or zero, while n_1 or n_2 denotes the rotational symmetry class of molecule 2, respectively. Since one of λ_1 or λ_2 in (IV.4) must be 0 or 2, the allowed transitions are any ΔJ_1, ΔJ_2 provided that one of $|\Delta J_1|, |\Delta J_2| \le 2$; $\Delta K_1 = \pm m_1 n_1$, $\Delta K_2 = \pm m_2 n_2$ provided that one of $m_1, m_2 = 0$; $m_1 = 0$ if $|\Delta J_2| > 2$, $m_2 = 0$ if $|\Delta J_1| > 2$. An allowed transition will contribute to (IV.4) only through terms with λ_1, λ_2 such that $\lambda_1 \ge \max[|\Delta J_1|, |\Delta K_2|]$, $\lambda_2 \ge \max[|\Delta J_2|, |\Delta K_2|]$. For example, if the dipole and quadrupole moments are the only important ones, the summation in (IV.4) is over $\lambda_1 \le 2$; $\lambda_2 \le 2$, and the selection rules reduce to $\Delta J_1 = 0, \pm 1, \pm 2$; $\Delta K_1 = 0$; $\Delta J_2 = 0, \pm 1, \pm 2$; $\Delta K_2 = 0$. If the octopole moments are also important, and if both molecules have threefold rotational symmetry, the summation in (IV.4) is over $\lambda_1 \le 3$, $\lambda_2 \le 3$, and the selection rules are

$\pm\Delta J_1 = 0, 1, 2$	0, 1, 2	3	0, 1, 2
$\pm\Delta J_2 = 0, 1, 2$	0, 1, 2	0, 1, 2	3
$\pm\Delta K_1 = 0$	3	0, 3	0
$\pm\Delta K_2 = 0, 3$	0	0	0, 3

where any combination may be chosen within a given block. For a given allowed transition in a pair of true symmetric tops, there will in general be contributions to (IV.4) from all (λ_1,λ_2) terms with λ_1 and λ_2 bounded below and above by the selection rules, by whatever are taken as the important multiple moments and by the demand that one of λ_1 or λ_2 is 0 or 2. For example, if the transition is $\Delta J_1=1$, $\Delta J_2=1$, $\Delta K_1=\Delta K_2=0$, then in general all the terms with $(\lambda_1,\lambda_2)=(1,2)$, $(2,1)$, $(2,2)$, $(2,3)$, $(3,2)\ldots$ will contribute to (IV.4). Finally, only those terms contribute in (IV.4) for which $\lambda_1+\Delta J_1$ is even.

For a pair of linear molecules the selection rules become any ΔJ_1, ΔJ_2 provided one of ΔJ_1 or ΔJ_2 is 0 or ± 2. An allowed transition will contribute only through (λ_1,λ_2) terms for which

$$\lambda_1 \geq |\Delta J_1| \qquad \lambda_1+\Delta J_1 \text{ is even}$$
$$\lambda_2 \geq |\Delta J_2| \qquad \lambda_2+\Delta J_2 \text{ is even}$$

B. Linear Molecules—The Intermolecular Potential in O_2 Gas[118]

In this section we illustrate the intricate general theory by demonstrating how the rotational absorptions of the O_2-induced dipole may be explained in terms of a modified potential consisting of a quadrupole and hexadecapole moment. The theory is applied to the case of compressed O_2 gas, whose absorption, observed by Bosomworth and Gush[119] in the region 20 to 400 cm^{-1}, could not be explained satisfactorily on the basis of quadrupole-induced dipole absorption alone. Values of $|Q|$ and $|\Phi|$, the quadrupole and hexadecapole moments, are obtained from the best fit to the experimental intensity and band shape. A justification for the use of the very short range (R^{-12}-dependent) hexadecapole field is based on the evaluation of the approximate range of the induced dipole moment. Equation (IV.4) is conveniently presented in terms of the dipole, quadrupole, octopole, and hexadecapole terms. We include the anisotropy terms (δ), but these are small.

Dipole / Induced Dipole Absorption

$$
\begin{aligned}
{}_{I}A^{\mu}_{J\to J+1} = {} & \frac{4\pi^3\mu^2N^2}{3hcZ}\int_0^{\infty} 4\pi R^{-4}\exp(-U_{AA}(R)/kT)\,dR \\
& \times(1-\exp(-hc\bar{\nu}_1(J)/kT))\exp(-E_J hc/kT)\bar{\nu}_1(J) \\
& \times\left(4\alpha_0^2(J+1)+\frac{8}{3}\delta^2\frac{(J+1)^2(J+2)}{(2J+3)}\right)
\end{aligned}
\qquad \text{(IV.10)}
$$

where

$$\bar{\nu}_1(J)=2B(J+1) \qquad \text{and} \qquad E_J=BJ(J+1)$$

Quadrupole / Induced Dipole Intensity

$$\begin{aligned}{}_lA^{Q}_{J\to J+2}=&\frac{4\pi^3Q^2N^2}{3hcZ}\int_0^\infty 4\pi R^{-6}\exp(-U_{AA}(R)/kT)dR\\ &\times(1-\exp(-hc\bar{\nu}_2(J)/kT))\exp(-E_Jhc/kT)\bar{\nu}_2(J)\\ &\times\left[9\alpha_0^2\frac{(J+1)(J+2)}{(2J+3)}+\frac{18}{5}\delta^2\left(\frac{(J+1)(J+2)}{(2J+3)}\right)^2\right]\end{aligned} \tag{IV.11}$$

where $\bar{\nu}_2(J)=2B(2J+3)$.

Octopole / Induced Dipole Intensity

$$\begin{aligned}{}_lA^{\Omega}_{J\to J+3}=&\frac{4\pi^3\Omega^2N^2}{3hcZ}\int_0^\infty 4\pi R^{-8}\exp(-U_{AA}(R)/kT)dR\\ &\times(1-\exp(-hc\bar{\nu}_3(J)/kT))\exp(-E_Jhc/kT)\bar{\nu}_3(J)\\ &\times\left[40\alpha_0^2\frac{(J+1)(J+2)(J+3)}{(2J+3)(2J+5)}+\frac{80}{3}\delta^2\left(\frac{(J+1)(J+2)}{(2J+3)}\right)^2\frac{(J+3)}{(2J+5)}\right]\end{aligned} \tag{IV.12}$$

where $\bar{\nu}_3(J)=6B(J+2)$.

Hexadecapole / Induced Dipole Intensity

$$\begin{aligned}{}_lA^{\Phi}_{J\to J+4}=&\frac{4\pi^3\Phi^2N^2}{3hcZ}\int_0^\infty 4\pi R^{-10}\exp(-U_{AA}(R)/kT)dR\\ &\times(1-\exp(-hc\bar{\nu}_4(J)/kT))\exp(-E_Jhc/kT)\bar{\nu}_4(J)\\ &\times\left[\frac{175(J+1)(J+2)(J+3)(J+4)}{2(2J+3)(2J+5)(2J+7)}\alpha_0^2\right.\\ &\left.+\frac{875}{12}\delta^2\left(\frac{(J+1)(J+2)}{(2J+3)}\right)^2\frac{(J+3)(J+4)}{(2J+5)(2J+7)}\right]\end{aligned} \tag{IV.13}$$

where $\bar{\nu}_4=4B(2J+5)$.

By comparison with oxygen, the induced zero-THz absorption in N_2 is much narrower and is fairly well simulated by the frequencies and relative intensities of the unbroadened $\Delta J=2$ rotational transitions calculated[119] with an equation similar to (IV.11). The *induced* absorption in N_2 has also been simulated by computer molecular dynamics by Jacucci, Buontempo, and Cunsolo.[120] Bosomworth and Gush[119] attributed the high-frequency part of the oxygen spectrum to a short-range overlap contribution to the dipole moment, but made no quantitative analysis of the phenomenon. However, with (IV.10) to (IV.13) it is possible to simulate the oxygen band with two contributions to the bimolecular collision-induced dipole moment, assumed to arise from the quadrupole and hexadecapole moments of the field of the second oxygen molecule, and vice versa. Oxygen has no dipole or octopole moment by symmetry. The hexadecapole field, being R^{-12}-dependent, is important only at very short separations R. Justification for its employment comes from a simple analysis given by Bosomworth and Gush involving a rough measurement of p, the range of the induced dipole moment, which may be obtained from the width of the spectrum. Classically, the spectrum is proportional to the Fourier transform of the correlation function of the dipole moment; the width function of the induced dipole is roughly equal to the duration of the collision. Thus

$$\tau=\frac{1}{2\pi\bar{\nu}_{1/2}c} \tag{IV.14}$$

where $\bar{\nu}_{1/2}$ is the width of the spectrum at half peak height. For oxygen at 300°K, $\bar{\nu}_{1/2}=160\ \text{cm}^{-1}$; thus $t=0.1$ psec. Then p can be estimated by multiplying the duration of collision (τ) by the average rate of change of the intermolecular distance ($\dot{R}_{av}$). Now $\frac{1}{2}m\dot{R}_{av}^2=\frac{1}{2}kT$, where m is the reduced mass of the colliding molecules. Thus $p=\dot{R}_{av}\tau=0.055$ nm at 300°K. The Lennard-Jones diameter (σ) of an O_2–O_2 pair is 0.792 nm; thus the induced dipole moment is practically zero until the colliding O_2 molecules enter the repulsion part of the intermolecular potential, and rises rapidly as the van der Waals contours interpenetrate. In other words, the high-frequency wing arises from the absorption of the dipole moment induced in the temporary O_2–O_2 pairs.

Values of $|Q|$ and $|\Phi|$ can be estimated (Fig. 71) by resolving the O_2 profile into a quadrupole-induced and hexadecapole-induced dipole absorption band. These are based on line spectra calculated from the even J values in (IV.11) and (IV.13) since oxygen has no odd J due to nuclear spin statistics. Of course, the considerable broadening of each line expected in practice might lead to a different overall profile than that suggested by the

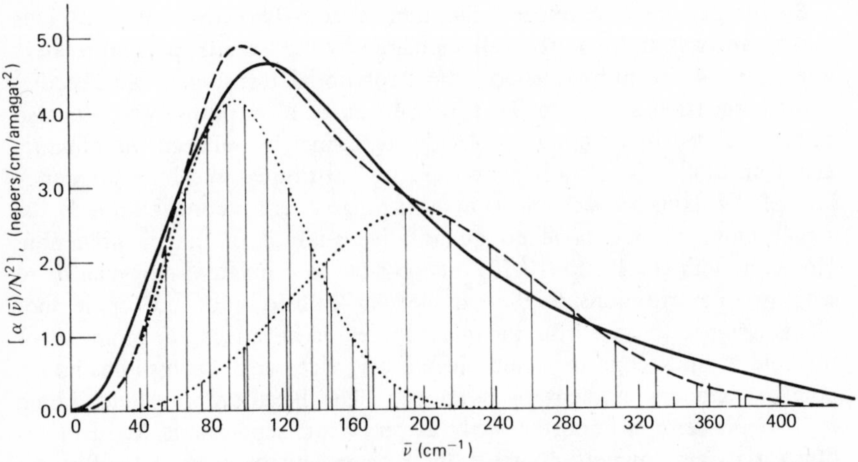

Fig. 71. (—), Experimental (Bosomworth and Gush) absorption of compressed oxygen. (····), Profiles of the $J \rightarrow J+2$ (quadrupole-induced) and $J \rightarrow J+4$ (hexadecapole-induced) dipole transitions. (---), Overall theoretical profile. Ordinate: $[\alpha(\bar{\nu})/N^2]$ (nepers/cm/amagat^2); abscissa: $\bar{\nu}$ (cm^{-1}). [Reproduced by permission from *Mol. Phys.*, **29**, 1345 (1975).]

line spectrum even if this ever exists in practice; nevertheless, a forced agreement can be obtained[118] with simple profile joining. Using $\epsilon/k = 118°\text{K}$, $\sigma = 0.346$ nm, $B = 1.45$ cm^{-1}, $\alpha_0 = 1.6 \times 10^{-24}$ cm^3, $\delta = 1.14 \times 10^{-24}$ cm^3 gives

$$\sum_{2nJ} {}_lA^{Q}_{J\rightarrow J+2} = 3.61 \times 10^{-5} N^2 Q^2 \quad \text{nepers/cm}^2/\text{amagat}^2$$

$$\sum_{2nJ} {}_lA^{\Phi}_{J\rightarrow J+4} = 4.51 \times 10^{-5} N^2 \Phi^2 \quad \text{nepers/cm}^2/\text{amagat}^2$$

$$\left(\int_0^\infty \frac{\alpha(\bar{\nu})}{N^2}\, d\bar{\nu} \right) = 8.65 \times 10^{-5} \quad \text{nepers/cm}^2/\text{amagat}^2$$

Summing (IV.11) and (IV.13) over 2 nJ, we find, for empirical (forced) agreement,

$$|Q| = 1.0 \times 10^{-40}\ \text{cm}^2 \qquad |\Phi| = 3.7 \times 10^{-60}\ \text{cm}^4$$

The value of $|Q|$ compares well with that of -1.34×10^{-40} cm^2 found by induced birefringence, and $|\Phi|$ is the "expected" order of magnitude.

The hexadecapole moment is one facet of transient O_4 formation which may be interpreted perhaps in other ways which are theoretically less convenient. A complete treatment of the O_2 absorption would have to include in addition such complicated factors as:

1. The overlap dipole contribution.
2. Translational and rotational contributions of the interference between the quadrupolar and overlap induction, and of the hexadecapole and overlap induction.
3. The pure translational ($\Delta J_1 = \Delta J_2 = 0$) contribution observable clearly in the H_2-induced absorption.[119]
4. Classical broadening[121] of the quantum stick spectra.

In the next section we use the J-diffusion model to investigate the effect factor 4 for cyanogen and its zero-THz absorption.

C. Quadrupole Induced Absorption in $(CN)_2$: Classical Broadening

In order to develop this theory we need to extract the Fourier transforms of (IV.10) to (IV.13) as follows. These equations may be manipulated into a continuous form, conveniently with time as a variable. The usual expression for a rotational absorption band shape, in terms of transitions between quantum states, is

$$I(\omega) = \frac{3\hbar c\sigma(\omega)}{4\pi^2\omega[1-\exp(-\hbar\omega/kT)]} \tag{IV.15}$$

where σ is the absorption cross-section per molecule. The Fourier transform

$$F(t) = \int_{-\infty}^{\infty} \omega^2 I(\omega)\exp(i\omega t)\,d\omega \tag{IV.16}$$

is recommended by Gordon and used here to weight the intensity toward the higher frequencies ($10 \lesssim \bar{\nu} \lesssim 450$ cm^{-1}), where accurate data are available. Here $\sigma(\omega) = \alpha(\omega)/N$, where α is the power absorption coefficient and N the number density in molecules/cm^3. The theoretical functions of time are obtained by substituting into (IV.16) using the continuous expression for the contours passing through the points of $I(\bar{\nu})$ obtained by eliminating J. For a linear molecule in the absence of broadening, one has then the theoretical contour

$$I_{\text{th}}(\bar{\nu}) \propto \left(\frac{\bar{\nu}}{2B} - \frac{2B}{\bar{\nu}}\right) Q^2 \exp\left[-\frac{hcB}{4kT}\left(\frac{\bar{\nu}}{2B} - 3\right)\left(\frac{\bar{\nu}}{2B} - 1\right)\right] A_8$$
$$\times \left[12\alpha_0^2 + \frac{24}{5}\left(\frac{\bar{\nu}}{2B} - \frac{2B}{\bar{\nu}}\right)\delta^2\right] + \frac{(\bar{\nu}+4B)(\bar{\nu}+12B)}{\bar{\nu}(\bar{\nu}+8B)}\Phi^2$$
$$\times \exp\left[-\frac{hcB}{4kT}\left(\frac{\bar{\nu}}{4B} - 5\right)\left(\frac{\bar{\nu}}{4B} - 3\right)\right] A_{12}\left[\frac{175}{96}\alpha_0^2\left(\frac{(\bar{\nu}-12B)(\bar{\nu}-4B)}{B(\bar{\nu}-8B)}\right)\right.$$
$$\left. + \frac{875}{36}\delta^2\left(\frac{\bar{\nu}}{2B} - \frac{2B}{\bar{\nu}}\right)^2\right] \tag{IV.17}$$

where

$$A_n = \frac{\pi^3 N^2}{hcZ}\int_0^\infty 4\pi^2 R^{-n} \exp(-U_{AA}(R)/kT) R^2 \, dR \tag{IV.18}$$

In the time domain, therefore,

$$C_{\text{th}}(t) = \int_0^\infty I_{\text{th}}(\bar{\nu})\bar{\nu}^2 \exp(2\pi i\bar{\nu}ct)\, d\bar{\nu} \Big/ \int_0^\infty I_{\text{th}}(\bar{\nu})\bar{\nu}^2 \, d\bar{\nu}$$

The direct Fourier transform of the experimental data can be made in the same way, giving

$$C_{\text{exp}}(t) = \int_0^\infty \frac{\bar{\nu}\alpha(\bar{\nu})\exp(2\pi i\bar{\nu}ct)\, d\bar{\nu}}{(1-\exp(-hc\bar{\nu}/kT))} \Big/ C_{\text{exp}}(0) \tag{IV.19}$$

The functions $C_{\text{th}}(t)$ and $C_{\text{exp}}(t)$ are compared for various values of $|Q|$ and $|\Phi|$ in Figs. 72 to 74 in the compressed gaseous states of oxygen, carbon dioxide, and cyanogen. Fourier transforms are carried out directly, by Simpson's rule, and via the fast Fourier transform algorithm of Cooley and Tukey, implemented in *Algol* by Singleton and developed by Baise.[122] The time functions seem to be extraordinarily sensitive to small changes in $|Q|$ and $|\Phi|$ used. This is illustrated in Fig. 72 for oxygen where the previous section's frequency-domain curve fittings, using values of $|Q| = 0.30 \times 10^{-26}$ esu, yields a $C_{\text{th}}(t)$ function which is quite severely underdamped compared with $C_{\text{exp}}(t)$. However, with slight changes, $|Q| = 0.36 \times 10^{-26}$ esu, and $|\Phi| = 0.4 \times 10^{-42}$ esu a much better fit is obtained, $C_{\text{th}}(t)$ now showing very short time "oscillations," although they are slightly displaced along the time axis from those of $C_{\text{exp}}(t)$. Analysis in the time domain,

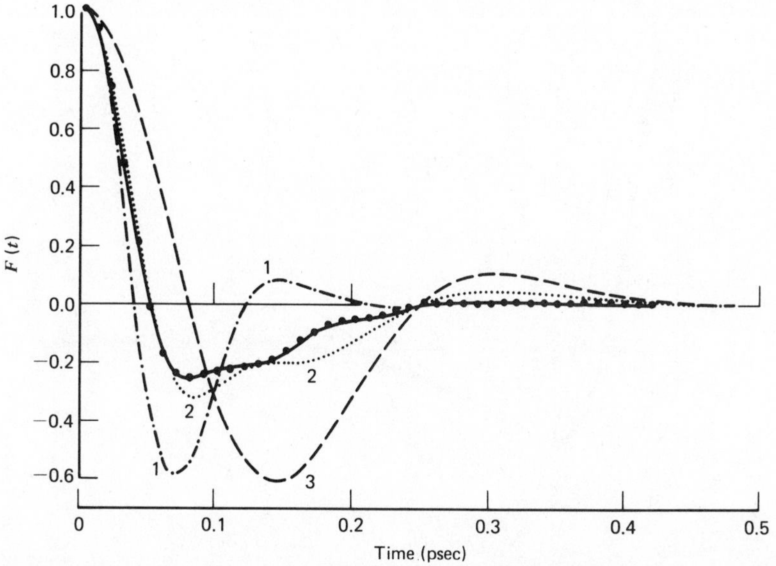

Fig. 72. Fourier transforms for O_2 gas at 300°K, 35 to 75 amagat. (●–●–●), Exp(t) as derived from two different algorithms. (----), (1) $C_{th}(t)$ calculated with $|Q|=0.30\times10^{-26}$ esu, $|\Phi|=1.1\times10^{-42}$ esu; (2) $F_{th}(t)$ with $|Q|=0.36\times10^{-26}$ esu, $|\Phi|=0.4\times10^{-42}$ esu; (3) $F_{th}(t)$ with $|Q|=0.38\times10^{-26}$ esu, $|\Phi|=0$. [Reproduced by permission from *J. Chem. Soc. Faraday Trans. 2*, **71**, 1257 (1975).]

then, is quite pronouncedly more sensitive than that in the frequency domain.

However, for cyanogen, no satisfactory fit can be obtained. The extraction of a continuous-time-domain function $C_{th}(t)$ by transforming the sum of the profiles of the $\delta(\bar{\nu})$ functions may be affected by neglect of the classical broadening of each line observed in practice (i.e., the experimental absorption is a broad band and not an assembly of lines). A broadening mechanism based on the Gordon J-diffusion will be considered presently, but a few remarks on the cyanogen spectrum are needed first. The only satisfactory feature of Fig. 74 is that of the $|Q|$ and $|\Phi|$ used confirm an intuitive expectation of certainly a large molecular quadrupole moment, and possibly a large hexadecapole moment as well. The $C_{th}(t)$ curves are not underdamped compared with the $C_{exp}(t)$ curves, which suggests that triple collisions are not important at 33.5 bars. Attempts to modify the Frost theory[116] with angle-dependent intermolecular potentials such as $U_{AA}(R)+U_{QQ}$ will have no effect on the normalized line shape because U_{QQ} is independent of the rotational state of a molecule provided that the rotational wave functions are assumed to be unperturbed. While the ab-

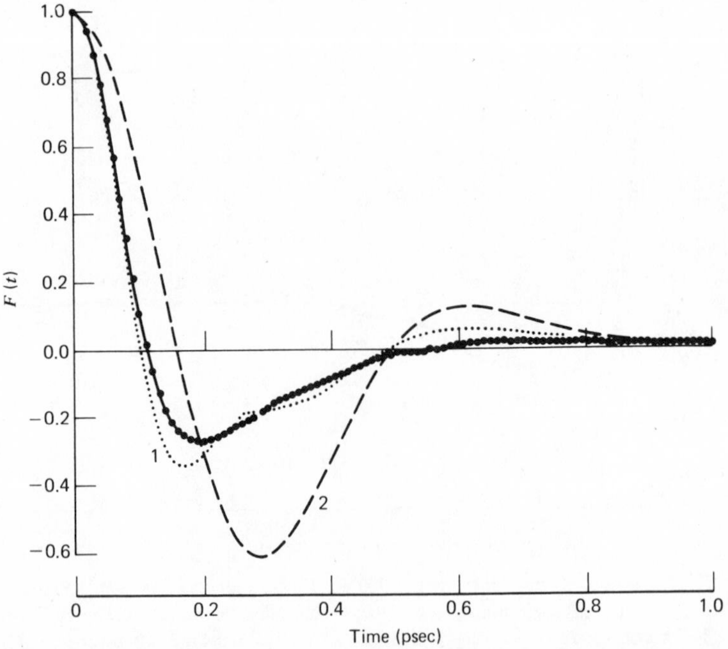

Fig. 73. Fourier transforms for CO_2 gas at 273°K, 85 amagat. (●–●–●) $C_{ept}(t)$ as derived from two separate algorithms. (----), (1) $C_{th}(t)$ with $|Q|=5.0\times10^{-26}$ esu, $|\Phi|=6.1\times10^{-42}$ esu. (2) $C_{th}(t)$ with $|Q|=5.2\times10^{-26}$ esu, $|\Phi|=0$. [Reproduced by permission from *J. Chem. Soc. Faraday Trans. 2*, **71**, 1257 (1975).]

solute values of $|Q|$ and $|\Phi|$ are very sensitive to the Lennard-Jones parameters ϵ/k and σ, the relative values of A_n will not be changed much. Therefore, ϵ/k and σ have little effect on the normalized line shape represented by $C_{th}(t)$. Quantum mechanically, therefore, a theory of pressure-induced absorption is needed which either disposes with point multipole expansions of the electrostatic field, or retains this approximation and then proceeds (albeit discordantly) to take into account the effect of molecular anisotropy on the eigenstate of a pair of molecules.

To broaden the set of $J \rightarrow J+2$ absorptions in compressed cyanogen, oxygen, or carbon dioxide, we assume that the broadened contour $C_b(t)$ has the general property of being an even function in time, and is also a solution of the integrodifferential equation

$$\dot{C}_b(t) = -\int_0^t K_b(t-\tau)C_b(\tau)\,d\tau \tag{IV.20}$$

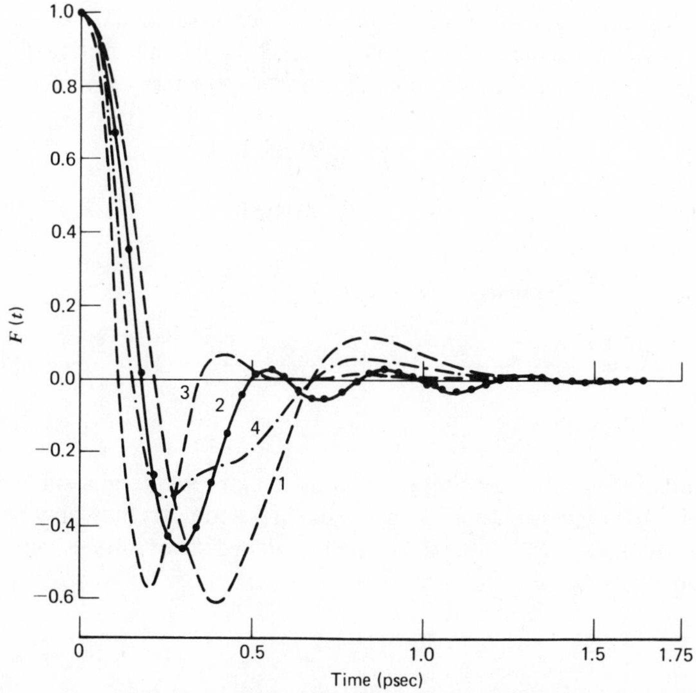

Fig. 74. Fourier transforms for $(CN)_2$ gas at 383°K, 33.5 bars. (●-●-●), $C_{\text{exp}}(t)$ as derived from two separate algorithms. (1) $C_{\text{th}}(t)$ with $|Q|=15.5\times10^{-26}$ esu, $|\Phi|=0$; (2) $C_{\text{th}}(t)$ with $|Q|=12.0\times10^{-26}$ esu, $|\Phi|=44\times10^{-42}$ esu; (3) $C_{\text{th}}(t)$ with $|Q|=9\times10^{-26}$ esu, $|\Phi|=56\times10^{-42}$ esu; (4) $C_{\text{th}}(t)$ with $|Q|=14.5\times10^{-26}$ esu, $|\Phi|=24\times10^{-42}$ esu. [Reproduced by permission from *J. Chem. Soc. Faraday Trans. 2*, **71**, 1257 (1975).]

where the memory kernel may be expanded in a set of coupled integro-differential equations analogous to the Mori expansion.

$$\frac{\partial}{\partial t}K_{n-1}^{(b)}(t)=-\int_0^t K_n^{(b)}(\tau)K_{n-1}^{(b)}(t-\tau)\,d\tau \qquad \text{(IV.21)}$$

To effect broadening we now forge a link analogous to that leading to *J*-diffusion in the case of permanent dipolar absorption:

$$K_0^{(b)}(t)=K_\delta(t)\exp(-|t|/\tau) \qquad \text{(IV.22)}$$

where $K_\delta(t)$ is associated through an equation identical with (IV.20), with

$C_\delta(t)$ the correlation function of the set of unbroadened $J \to J+2$ transitions. For bimolecular, quadrupole-induced dipolar absorption in linear, nondipolar molecules, neglecting the hexadecapole term, we have

$$C_\delta(t) = \frac{\int_0^\infty f_0(\Omega)\cos\Omega t\, d\Omega}{\int_0^\infty f_0(\Omega)\, d\Omega} \tag{IV.23}$$

for each $J \to J+2$ transition, where

$$f_0(\Omega) = \left(\frac{\Omega}{4\pi Bc} - \frac{4\pi Bc}{\Omega}\right)\exp\left[-\frac{hcB}{4kT}\left(\frac{\Omega}{4\pi Bc} - 3\right)\left(\frac{\Omega}{4\pi Bc} - 1\right)\right] \tag{IV.24}$$

For nondipolar symmetric tops, up to the quadrupole term a similar, more complicated expression may be derived. $C_b(t)$ and $C_\delta(t)$ may now be linked by equations identical with (II.41) the broadened set of $J \to J+2$ lines is extracted from

$$\begin{aligned} C_b(\omega) &= \mathrm{Re}[C_b(i\omega)] \\ &= \mathrm{Re}\left[\frac{C_\delta(i\omega+\tau^{-1})}{1-\tau^{-1}C_\delta(i\omega+\tau^{-1})}\right] \end{aligned} \tag{IV.25}$$

where

$$C_\delta(i\omega+\tau^{-1}) = \Gamma + i\Lambda$$

with

$$\Gamma(\omega) = \int_0^\infty f_0(\Omega)\left[\frac{\tau^{-1}(\Omega^2+\omega^2+\tau^{-2})}{(\Omega^2-\omega^2+\tau^{-2})^2+4\omega^2\tau^{-2}}\right]d\Omega \Big/ \int_0^\infty f_0(\Omega)\, d\Omega$$

$$\Lambda(\omega) = \int_0^\infty f_0(\Omega)\left[\frac{\omega(\Omega^2-\omega^2-\tau^{-2})}{(\Omega^2-\omega^2+\tau^{-2})^2+4\omega^2\tau^{-2}}\right]d\Omega \Big/ \int_0^\infty f_0(\Omega)\, d\Omega$$

The absorption coefficient $\alpha(\omega)$ is then given by

$$\alpha(\omega) = \frac{(\epsilon_0 - \epsilon_\infty)\omega^2}{n(\omega)c}\, C_b(\omega) \tag{IV.26}$$

where $n(\omega)$ is the frequency-dependent refractive index, and c the velocity of light. For quadrupole-induced dipole absorption in pair collisions,

$$(\epsilon_0-\epsilon_\infty)=\frac{16\pi^2}{kT}N^2\bar{\alpha}_pQ\int_0^\infty R^{-6}\exp\left[-\frac{U_{AA}(R)}{kT}\right]dR$$

with $\bar{\alpha}_p$ as the mean molecular polarizability and Q the quadrupole moment.

Figure 75 shows how (IV.26) produces an absorption which simulates the broadening and eventually fuses the $J\rightarrow J+2$ lines in cyanogen ($B=0.1570\ \text{cm}^{-1}$). A broad continuum is reached at $\tau=10$ psec, which, according to kinetic theory, corresponds to a mean free path of about 38 Å. Therefore, a continuum is reached well before triple collisions become sta tistically significant. Equation (IV.26) is matched with nitrogen data[121] in Figs. 76 and 77 (gas and liquid). An effective quadrupole moment of $|Q|=5\times10^{-40}\ \text{cm}^2$ was extracted from this curve-fitting procedure for the total dispersion $(\epsilon_0-\epsilon_\infty)$. Despite the neglect of many factors, such as translational and electronic overlap absorption, this estimate of $|Q|$ compares

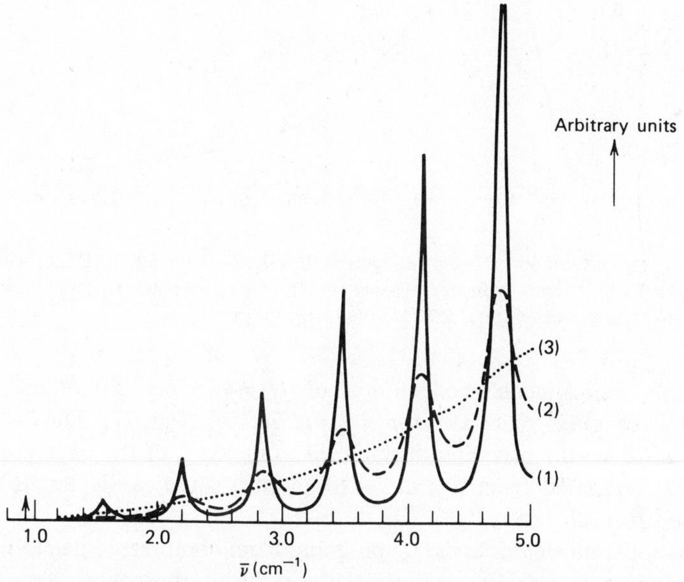

Fig. 75. First few $J\rightarrow J+2$ transitions for cyanogen broadened by (IV.26) at 350°K. (Rotational constant $B=0.1571\ \text{cm}^{-1}$.) (1) $\tau=100$ psec, (2) $\tau=35$ psec, (3) $\tau=10$ psec. Abscissa: $\bar{\nu}$ (cm^{-1}). [Reproduced by permission from *Spectrochim. Acta*, **32A**, 1253 (1976).]

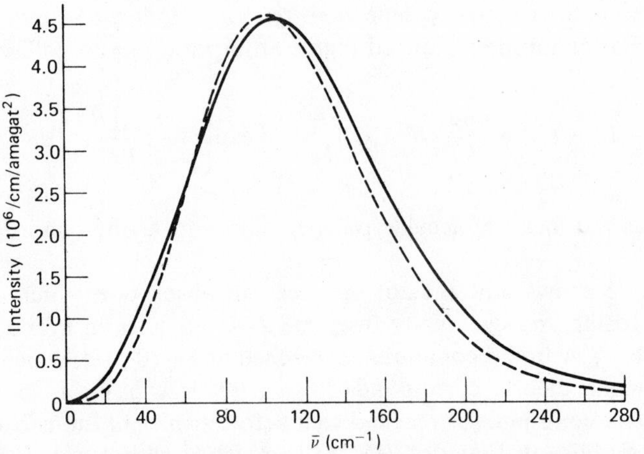

Fig. 76. (—), Experimental absorption of compressed nitrogen at 300°K. (----), Equation (IV.26) with $\tau=0.4$ psec. $|Q|=5\times10^{-40}$ cm^2. Ordinate: intensity (10^6 cm^{-1}/amagat^2); Abscissa: $\bar{\nu}$ (cm^{-1}). [Reproduced by permission from *Spectrochim. Acta*, **32A**, 1253 (1976).]

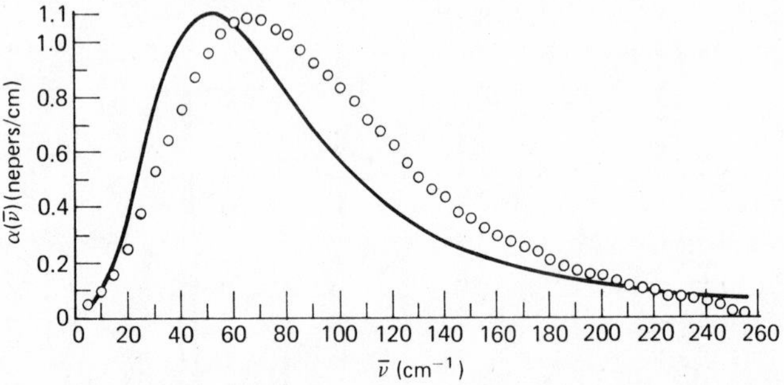

Fig. 77. O, Absorption of liquid nitrogen at 76.4°K. (—), Equation (IV.26) with $\tau=0.1$ psec, $|Q|=3\times10^{-40}$ cm^2. Ordinate: α (neper cm^{-1}); abscissa: $\bar{\nu}$ (cm^{-1}). [Reproduced by permission from *Spectrochim. Acta*, **32A**, 1253 (1976).]

favorably with Kielich's collection[123] of $|Q|=4.5-6.9\times10^{-40}$ cm^2. Equation (IV.26) is less successful for $N_2(1)$ at 76.4°K (Fig. 77). The calculated curve is for $\tau=0.1$ psec and normalized to the α_{max} of the observed band. The $|Q|$ estimated from this is 3×10^{-40} cm^2, significantly less than that deduced from the gas.

This apparent decrease in $|Q|$ on going from compressed gas to liquid is characteristic of a model of bimolecular-induced absorption. An explanation is the reduced effectiveness of multimolecular collisions in generating induced dipoles, as discussed earlier. In liquid CO_2, for example,[113] that

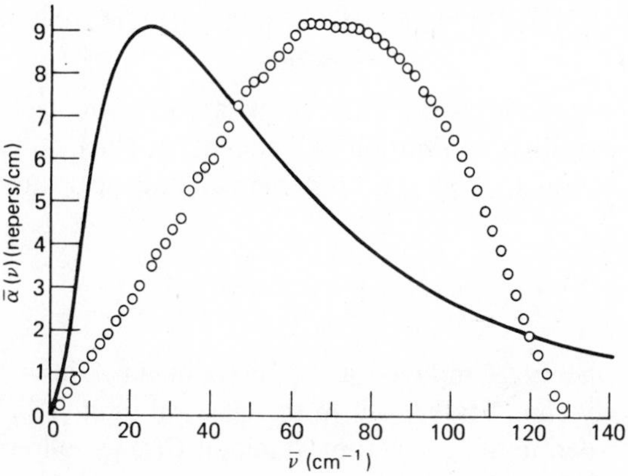

Fig. 78. Absorption of liquid cyanogen at 301°K. (—), Equation (IV.26) with $\tau = 0.1$ psec. $|Q| = 9.7 \times 10^{-26}$ esu. Ordinate: $\alpha(\bar{\nu})$ (nepers/cm); abscissa: $\bar{\nu}$ (cm^{-1}). [Reproduced by permission from *Spectrochim. Acta*, **32A**, 1253 (1976).]

part of the induced dipole moment due to quadrupolar induction is effectively canceled in the liquid, leaving essentially the contribution from shorter-range interactions.

Equation (IV.26) is not successful at all for cyanogen liquid[115] at 301°K, where analogously to the case of permanent dipole absorption, there is a large discrepancy between the observed and calculated $\bar{\nu}_{max}$ (Fig. 78). This shift in the observed $\bar{\nu}_{max}$ can be interpreted reasonably in terms of the increased amount of shorter-range interactions in the liquid phase where the torques in the rod-like molecules $(CN)_2$ will be greater than those N_2. This is saying merely that collision-interrupted free rotation is not the case in the liquid phase of the great majority of both nondipolar and dipolar liquids. The present J-diffusion type of model cannot move the position of $\bar{\nu}_{max}$ because the reorientation of the molecular angular momentum is assumed to take place during an infinitely short time, and it is thus impossible to get any information about, or take account of, the intermolecular mean-square torque.

However, a different continued-fraction representation might yield the correct absorption contour, by involving implicitly averages such as $K_0(0)$ and $K_1(0)$ which are both torque-dependent. An expression for $C_{th}(t)$ might then in principle be attainable which takes account of the contour in terms of the mean-square torque and its derivative. This method has been employed for compressed ethylene gas,[124] where the quantum theory fails even at low pressures. This empirical approach has been used successfully in nondipolar liquids, as described in the next section.

D. Absorptions in Nondipolar Liquids—Use of the Continued Fraction

Any absorption band in the infrared, whatever its molecular dynamical origin, is a probability distribution of frequencies, $C(\omega)$, and is related to a correlation function $C(t)$ by the fundamental statistical theorem, which is classically

$$C(t)=\int_0^\infty \cos\omega t\, dC(\omega) \tag{IV.27}$$

The quantum theory of induced absorptions in an N-body interaction is obviously hugely complicated, but in the classical limit (IV.27) holds quite generally if we define the correlation function $C(t)$ as follows:

$$C(t)=\sum_{i,j}\langle \boldsymbol{\mu}_j(t)\cdot\boldsymbol{\mu}_i(0)\rangle \tag{IV.28}$$

where $\boldsymbol{\mu}_i$ is the induced dipole on molecule i at time t. $C(t)$ is an orientation/interaction correlation function dependent simultaneously at time t on the orientation of a molecule with respect to all the others. We now expand $C(t)$ in a continued fraction formally identical to that of Mori, and truncate with

$$\tilde{K}_1(s)=K_1(0)/(s+\gamma) \tag{IV.29}$$

Using an equation such as (IV.26) then allows us to fit directly the frequency-domain data[5] iterating on $K_0(0)$, $K_1(0)$, and γ (Table X).

TABLE X
Parameters $K_1(0)$, $K_0(0)$, and γ for Nondipolar Liquids[a]

Liquid	Temp. (°K)	$10^{40}I_B$ (g·cm²)	$\gamma(I_B/2kT)^{1/2}$	$K_0(I_B/2kT)$	$K_1(I_B/2kT)$	$(\epsilon_0-\epsilon_\infty)$
Nitrogen	76.4	12.2	10.6	5.9	37.8	0.005
Carbon dioxide	273	71.2	11.5	8.6	51.9	0.007
CCL_4	296	484	14.2	10.9	80.6	0.002
CH_4 (rot. phase I)	76	5.34	10.6	14.7	47.9	0.009
Cyanogen	301	155	10.9	14.9	66.5	0.050
Methane	98	5.34	14.5	16.8	75.7	0.007
Benzene	296	198	12.8	20.8	100.6	0.023
CS_2	296	259	20.3	26.2	170.2	0.026
Cyclohexane	296	178	21.1	28.4	194.3	0.040
Trans-Decalin	296	1020	22.7	70.7	335.3	0.003
1,4-Dioxan	296	160	7.8	10.4	46.5	0.06

[a]Reproduced by permission from *J. Chem. Soc. Faraday Trans.* 2, **72**, 1194 (1976).

The three-variable fit is very close in the far infrared (Figs. 3 and 79) and there also is a tendency for $K_0(0)$ and $K_1(0)$ to increase as the geometrical anisotropy of each molecule. The absolute magnitude of the absorption can be related via $(\epsilon_0 - \epsilon_\infty)$ to an "effective induced dipole" or higher multipole given some simplifying assumption about the molecular dynamical and electrostatic origin of these very broad bands. The increase in $K_0(0)$ and $K_1(0)$ is illustrated in Table X in units which take account of inertial factors. The satisfactory fits to experimental data over almost three decades of frequency show that the analytical dependence of α upon $\bar{\nu}$ is that of (IV.26) in this frequency range, but the physical interpretation of $K_0(0)$, $K_1(0)$, and γ remains obscure, apart from the obvious interaction dependence. The curves for $C(t)$ then can be calculated and are illustrated in Fig. 80, where are illustrated also some predictions of the "cell" model of Litovitz and co-workers and (where applicable) the model of multipole-induced absorption in a two-molecule collision. The latter is usually inadequate in describing the more complex interactions of the condensed phase.

We note finally that in deriving (IV.26) we are assuming that the molecular ensemble obeys classical equations of motion ($\hbar \to 0$). This is consistent with our basic assumption that Mori formalism is applicable to the classical correlation fraction defined in (IV.27). This assumption rests on the broad and related generalizations which lie at the root of our present understanding of transport properties (i.e., *linear* response and fluctuation–dissipation). Classically, the latter can be derived for a canonical ensemble using the Liouville equation:

$$\frac{dB}{dt} = \sum_i \left(\frac{\partial H}{\partial p_i} \frac{\partial B}{\partial q_i} - \frac{\partial H}{\partial q_i} \frac{\partial B}{\partial p_i} \right)$$

describing the motion of B which depends on time t by the intermediacy of coordinates q_i and their conjugate momenta p_i. Using quantized mechanics the Poisson brackets are replaced by the commutator $\hbar^{-1}[H, B]$, and the relation between a classical correlation function $\langle \boldsymbol{\mu}_i(0) \cdot \boldsymbol{\mu}_j(t) \rangle$ and the quantized analog $\langle [\boldsymbol{\mu}_i(0) \cdot \boldsymbol{\mu}_j(t)] \rangle$ is

$$\int e^{i\omega t} \langle [\boldsymbol{\mu}_i(0) \cdot \boldsymbol{\mu}_j(t)] \rangle \, d\omega = (1 - e^{-\hbar\omega/kT}) \int e^{i\omega t} \langle \boldsymbol{\mu}_i(0) \cdot \boldsymbol{\mu}_j(t) \rangle \, d\omega$$

The quantum-mechanical correlation function is real and contains odd powers of t in its Maclaurin expansion. The classical correlation function contains only even powers of t, in accord with the Onsager principle of time reversibility. Mori has shown that the equation of motion of an arbitrary dynamical variable of an arbitrary system can be transformed rigorously to a linear generalized Langevin equation form; and Kubo

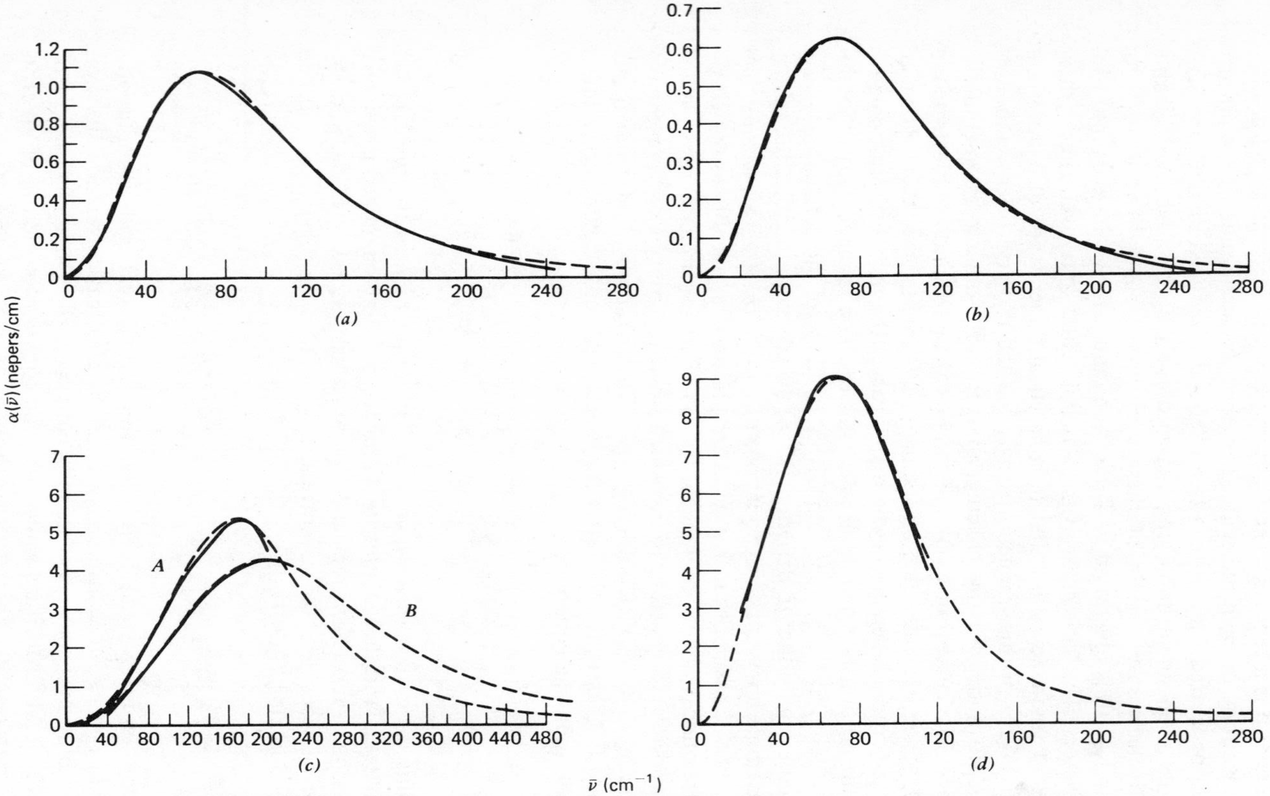

Fig. 79. (*a*) Experimental absorption for liquid nitrogen at 76.4*K. (----), Mori three-variable fit (Table X). (*b*) Liquid carbon dioxide at 273°K. (*c*) (*A*) Methane (rotator phase) at 76°K; (*B*) methane (liquid) at 98°K. (*d*) Liquid cyanogen at 301°K. [Reproduced by permission from *J. Chem. Soc. Faraday Trans.* 2, **72**, 1195 (1976).]

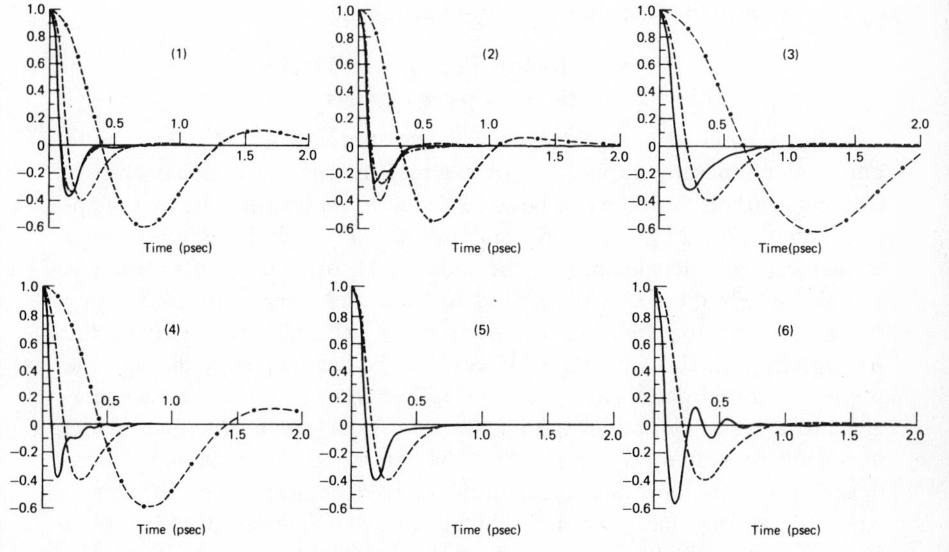

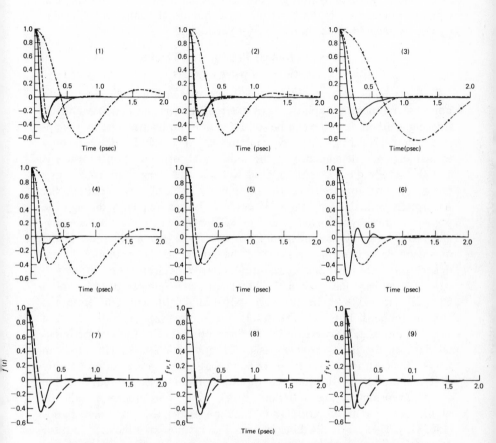

Fig. 80. Fourier transforms of $\alpha(\bar{\nu})$ for some nondipolar liquids. (1) Benzene; (2) carbon disulfide; (3) carbon tetrachloride; (4) cyclohexane; (5) *p*-difluorobenzene; (6) bicyclohexyl; (7) *trans*-decalin; (8) 1,4-dioxan; (9) trans-1,2-dichloroethylene. (—), Experimental; (---·---·---), free rotor; (---), Litovitz "cell" model; (-··-), function derived from three-variable theory. [Reproduced by permission from *J. Chem. Soc. Faraday Trans. 2*, **72**, 1195 (1976).]

shows that a subsystem of an ensemble when perturbed will relax to thermal equilibrium via the same generalized Langevin equation. Neither the arbitrary subsystem nor the variable need be quantized. That an ensemble of molecules as small as nitrogen or methane can be treated with classical equations of motion is the basis of the technique of computer molecular dynamics, including the simulations of Section III.

E. Effect of Kilobar Pressures on Liquids—Plastic Nondipolar Crystals

In this section we are concerned with the approach to the solid state in nondipolar fluids by application of external hydrostatic pressure and by freezing. Carbon disulfide can be solidified at room temperature by the application of 12 to 13 kbars of external pressure. The only far-infrared study available of the dependence of the induced absorption upon pressure is still that of Bradley, Gebbie, Gilby, Kechin, and King[125] in 1965, using prototype apparatus and pressures of up to 11.6 kbars. They found that the absorption peak shifts by about 35 cm^{-1} to higher frequency through the pressure-induced phase change at 293°K. We show here that this is equivalent to a large increase in both $K_0(0)$ and $K_1(0)$ [i.e., an increase in the mean-square torque (in the slope of the intermolecular potential dependence on orientation)]. Such pressure data are technically very difficult to come by, and we have taken the other approach (temperature) in liquid CCl_4 (298 to 343°K). The change in $K_0(0)$ and $K_1(0)$ with temperature is less pronounced, but real. The results are summarized in Table XI and Figs. 81 and 82. They were obtained[126] by fitting (IV.26) to the experimental data on the three-variable basis of Section (IV.D). At 11.6 kbars in $CS_2(l)$ it is clear from the zero-THz band that there is a greater probability that the motion of the induced dipole moment is associated with the central frequency (ω_0) of ~100 cm^{-1} (0.33 psec). This process is reflected in the behavior of the correlation function associated with these bands (Fig. 81b). The less-damped behavior at 11.6 kbars is an indication that the orientational correlation is greater.

The angular forces resulting from mechanical anisotropy seem to be enhanced at the greater pressure, an effect that can be seen reflected in the very large increase in the torque-dependent parameters $K_0(0)$ and $K_1(0)$ (Table XI). The equivalent correlation function of depolarized Rayleigh scattering has recently been observed with applied pressure by Dill, Livovitz, and Bucaro,[127] van Konynenburg and Steele,[128] and Perrot, Devaure, and Lascombe.[129] The effect of temperature is not as pronounced on $K_0(0)$ and $K_1(0)$ as that of pressure, but for $CS_2(l)$ both increase with T. This dependence of mean-square torque upon temperature is predicted both by harmonic well dynamics and hard-core collisions. Therefore, no discernible

TABLE XI
Parameters for $CS_2(l)$ and $CCl_4(l)$[a]

Liquid	Temp. (°K)	P (bars)	$10^{40}I$ ($g\cdot cm^2$)	$xK_0(0)$	$xK_1(0)$	$x^{1/2}\gamma$	$(\epsilon_0-\epsilon_\infty)$
CS_2	296	1	258.6	26.2	170.2	20.3	0.026
	293	11,600	"	79.9	247.6	32.7	0.018
	232	1	"	21.9	114.3	12.0	0.034
	315	1	"	29.9	212.0	27.0	0.020
CCl_4	296	1	242	10.9	80.6	14.2	0.019
	313	1	"	10.2	65.1	12.2	0.017
	328	1	"	9.1	51.2	10.1	0.016
	343	1	"	10.3	77.7	14.6	0.016

[a]Reproduced by permission from *J. Chem. Soc. Faraday Trans.* 2, **72**, 1206 (1976).

$$x=\frac{I}{2kT}$$

"loosening up" of internal structure can be observed in $CS_2(l)$ over the range 232 to 315°K, the intermolecular torque being determined by the thermal energy available to each molecule (kT).

The situation is different in the spherical top CCl_4 (Fig. 82) over the range 296 to 343°K, the latter being a few degrees below the boiling point at 1 bar. Both $K_0(0)$ and $K_1(0)$ *decrease* as T increases, although the values at 343°K are slightly anomalous (Table XI). Therefore, there must be a considerable increase in free volume, and thereby translational freedom, as the boiling point is approached in order to overcome the purely thermal increase ($\propto kT$) in the mean-square torque. An important indication is that the mean-square torque is always much smaller for CCl_4 than for CS_2, implying that molecular geometry plays an important part in rotational freedom of motion. This is particularly so in the plastic crystalline phase of CBr_4 dealt with below.

It is relevant to note that the nuclear magnetic resonance spin-rotation relaxation time T_1 of liquids is observed to decrease as temperature is raised, following an Arrhenius law. T_1 is inversely proportional to τ_J, the angular momentum correlation time, which thus increases with temperature. τ_J is a measure of the mean time during which a molecule seems to retain its angular momentum, and in spherical-top molecules in the liquid state it is known that spin-rotation interaction is the dominant relaxational mechanism, τ_J becoming long even at temperatures well below the critical point. In contrast, for asymmetric tops and sticklike molecules, spin-rotation interaction becomes appreciable only at high temperatures, the ratio

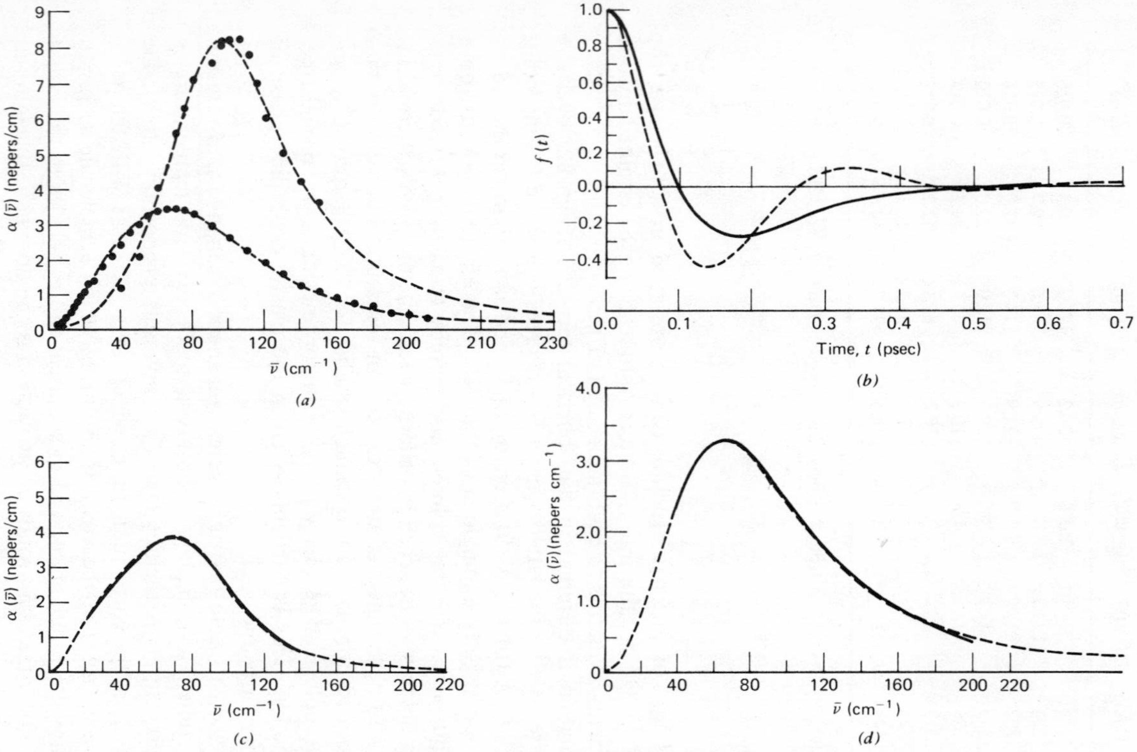

Fig. 81. (*a*) ●, Some experimental observations of the $CS_2(l)$ absorption at 11.6 kbars, 293°K. (---), Mori three-variable theory, best fit (Table X); ○, the absorption of $CS_2(l)$ at 296°K, 1 bar; (---), three-variable theory. (*b*) Fourier transforms of (*a*) (—), $CS_2(l)$ at 296°K, (---), $CS_2(l)$ at 293°K, 11.3 kbars. (*c*) (—), Experimental absorption of $CS_2(l)$ at 232°K; (---), three-variable theory (*d*) 315°K. [Reproduced by permission from *J. Chem. Soc. Faraday Trans.* 2, **72**, 1206 (1976).]

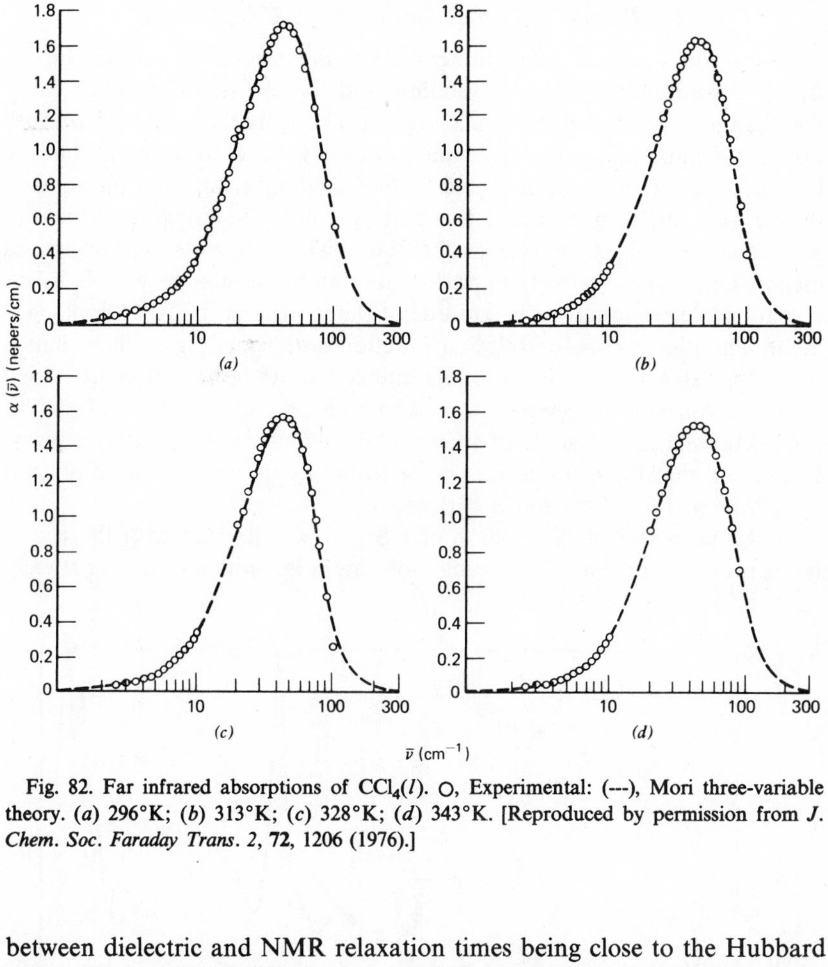

Fig. 82. Far infrared absorptions of $CCl_4(l)$. ○, Experimental: (---), Mori three-variable theory. (*a*) 296°K; (*b*) 313°K; (*c*) 328°K; (*d*) 343°K. [Reproduced by permission from *J. Chem. Soc. Faraday Trans. 2*, **72**, 1206 (1976).]

between dielectric and NMR relaxation times being close to the Hubbard ratio of 3 (rotational diffusion). In sticklike molecules, the motions about different symmetry axes is often observed to be highly anisotropic (e.g., in liquid propane near the critical point) the rotation about the C_{3v} axis is eight times faster than that about the perpendicular,[130] the contribution of spin-rotation relaxation being about equal to that of the spin-spin magnetic dipolar correlation time, and not dominant, as in spherical tops.

The apparent dispersion $(\epsilon_0 - \epsilon_\infty)$ is lower at 11.6 kbars than at 1 bar on $CS_2(l)$ and also decreases with temperature for both liquids. There remains a considerable amount of theoretical work to do, probably with the aid of computer simulations, before the observables can be interpreted in terms of molecular constants.

1. Rotational Correlations in Plastic CBr_4 Crystals

In this final section we comment upon the zero-THz induced far-infrared absorptions in plastic crystalline and liquid CBr_4 in order to study the change in rotational dynamics brought about by the increased translational constraint and packing symmetry of the solid. It turns out[131] that the barrier to rotational motion is slightly *increased* on going from the solid to the liquid a few degrees above the melting point. The integrated intensity per molecule is slightly greater in the liquid, which suggests that the spatial disposition of the electrostatic part of the intermolecular potentials is important in determining the magnitude of the molecular-induced dipole moment. The dipole cross-correlation function corresponding to these bands is compared with that of octopole-induced dipole absorption of a two molecule collision of spherical tops, and with the autocorrelation function for a Maxwellian ensemble of freely rotating molecules of this symmetry. It is found that the mean torque greater initially in the condensed phases; thereafter rotational motion is correlated.

In the plastic crystalline phase of CBr_4, in comparison with the room-temperature monoclinic form, each molecule is left with a characteristically

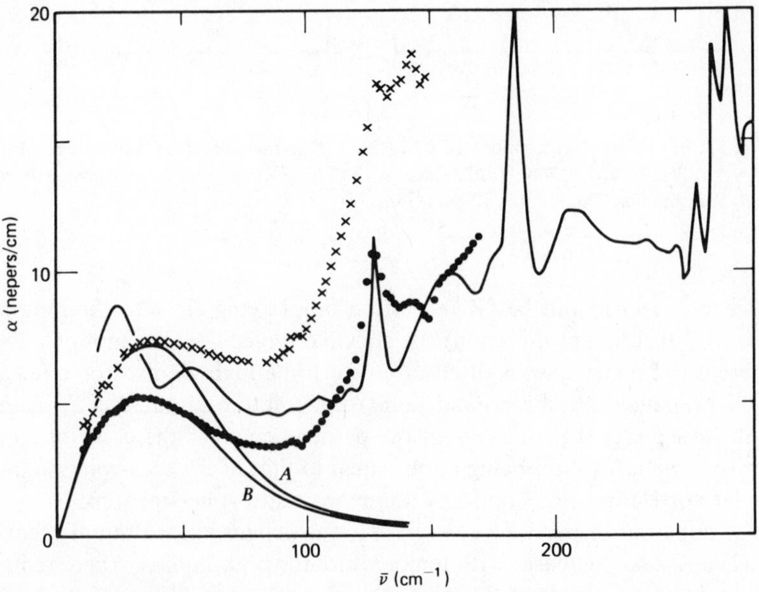

Fig. 83. Absorption of the $Th^6(Pa^3)$ monoclinic phase at 298°K. ●, Absorption of the plastic (simple cubic) single crystal at 358°K; *, absorption of the liquid at 376°K; (—), (*A*), (*B*) Mori three-variable theory. [Reproduced by permission from *J. Chem. Soc. Faraday Trans. 2*, **72**, 2147 (1976).]

generous amount of rotational freedom, but translational freedom is limited, although not entirely absent. In contrast, both types of dynamics are available and strongly coupled in the liquid. The broad bands centered near 32 cm^{-1} (Fig. 83) in both liquid and rotator phase are interpreted as intermolecular in origin, since they do not correspond to any known difference modes or overtones or fundamental, and occur at frequencies where such bands are prevalent for nondipolar molecules. In the $Th^6(Pa^3)$ monoclinic phase at 298°K this band is replaced by a doublet, both components of which are considerably broader than the fundamentals at 125 cm^{-1} and 270 cm^{-1}. In the plastic crystalline phase at 376°K, the fundamental at 124 cm^{-1} is broadened compared with the monoclinic phase, and considerably so in the liquid.

If a CBr_4 molecule whose center of mass is at the point $\mathbf{R}(\mathbf{r}, \theta, \phi, \chi)$ is assumed to develop a temporary dipole moment under the influence of the electrostatic fields of its neighbors, then the vector sum of these fields at $\mathbf{R}$ at any instant will be determined by the relative positions of all other molecules in the ensemble at that time, and consequently will be a measure of the disorder or ordering in the lattice of molecules near enough for their fields to be sufficiently influential. In order then to estimate absolutely the degree of rotational freedom retained in these condensed phases, it is profitable to compare the correlation function derived experimentally (Fig. 84)

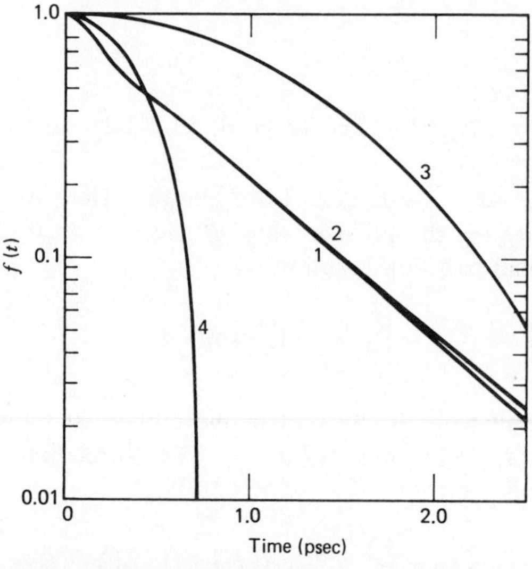

Fig. 84. Correlation functions for CBr_4 from induced far-infrared bands. (1) $C(t)$ (liquid); (2) plastic crystal; (3) $C_{FR}(t)$; (4) $C_{\Omega}(t)$. [Reproduced by permission from *J. Chem. Soc. Faraday Trans. 2*, **72**, 2147 (1976).]

with those estimated using models of rotational motion in the gas phase. The correlation function for the free rotation of a Maxwellian distribution of spherical tops is given by the classical expression

$$C_{\mathrm{FR}}(t)=\frac{2}{3}\left(1-t^2\frac{kT}{I}\right)\exp\left(-t^2\frac{kT}{2I}\right)+\frac{1}{3}$$

This is illustrated in Fig. 84 for CBr_4 at 376°K and decays much more slowly than the functions $C(t)$ of the condensed phase. Thus the effect of intermolecular forces is clearly seen in the time domain. The bimolecular octopole-induced band produces in the classical limit a correlation function $C_{\Omega}(t)$ as follows, and is compared with $C_{\mathrm{FR}}(t)$ and $C(t)$ in Fig. 84.

The total integrated intensity of an octopole-induced dipole absorption band spherical tops is given by Ozier and Fox[132] as

$$A=\sum_{J\neq J} A(J,J')$$

where

$$A(J,J')=\frac{(8\pi)^4N^2\Omega^2\alpha_0^2}{280hcZ}\int_0^{\infty}R^{-8}\exp(-U(R)/kT)dR$$
$$\times\bar{\nu}(J,J')(2J+1)(2J'+1)\left[\exp(-aJ(J+1))-\exp(-aJ'(J'+1))\right]$$

with

$$a=Bhc/kT \qquad \bar{\nu}(J,J')=B\left[J'(J'+1)-J(J+1)\right]$$

and where $\Delta J=J'-J=0$, 1, 2, 3 are allowed. Here N is the molecular number density, α_0 the polarizability, B the rotational constant (cm^{-1}), and Z rotational partition function.

$$Z=\sum_J(2J+1)^2\exp(-aJ(J+1))$$

Thus the overall band is the sum of the individual transition intensities $A(J,J+1)$, $A(J,J+2)$, and $A(J,J+3)$. The correlation function is thus given by

$$C_{\Omega}(t)=\sum_{i=1,3}^{3}\int_0^{\infty}\frac{A(J,J+i)\cos(2\pi\bar{\nu}ct)d\bar{\nu}}{{}^{(i)}\bar{\nu}\left[1-\exp(-h^{(i)}\bar{\nu}c/kT)\right]}$$

with

$$^{(1)}\bar{\nu}=2B(J+1) \qquad ^{(2)}\bar{\nu}=2B(2J+3) \qquad ^{(3)}\bar{\nu}=6B(J+2)$$

It may be shown that

$$A(J,J+1)\propto {}^{(1)}\bar{\nu}(2J+1)(2J+3)\exp\left[-aJ(J+1)\right]\left[1-\exp(-hc\,{}^{(1)}\bar{\nu}/kT)\right]$$
$$A(J,J+2)\propto {}^{(2)}\bar{\nu}(2J+1)(2J+5)\exp\left[-aJ(J+1)\right]\left[1-\exp(-hc\,{}^{(2)}\bar{\nu}/kT)\right]$$
$$A(J,J+3)\propto {}^{(3)}\bar{\nu}(2J+1)(2J+7)\exp\left[-aJ(J+1)\right]\left[1-\exp(-hc\,{}^{(3)}\bar{\nu}/kT)\right]$$

the proportionality constant being in each case the J-independent part of (I.7). We have, finally, in the classical sense

$$\begin{aligned}C_\Omega(t)\propto \int_0^\infty \Bigg[&\left(\frac{\bar{\nu}}{B}-1\right)\left(\frac{\bar{\nu}}{B}+1\right)\exp\left[-a\left(\frac{\bar{\nu}}{2B}-1\right)\frac{\bar{\nu}}{2B}\right]\\ &+\left(\frac{\bar{\nu}}{2B}-2\right)\left(\frac{\bar{\nu}}{2B}+2\right)\exp\left[-a\left(\frac{\bar{\nu}}{4B}-\frac{3}{2}\right)\left(\frac{\bar{\nu}}{4B}-\frac{1}{2}\right)\right]\\ &+\left(\frac{\bar{\nu}}{3B}\quad\right)\left(\frac{\bar{\nu}}{3B}+3\right)\exp\left[-a\left(\frac{\bar{\nu}}{6B}-2\right)\left(\frac{\bar{\nu}}{6B}-1\right)\right]\Bigg]\cos 2\pi\bar{\nu}ct\,d\bar{\nu}\end{aligned}$$

This function $(C_\Omega(t)/C_\Omega(0))$ displayed in Fig. 84 looks very little like the correlation functions of the condensed phase bands, which fall off initially faster and thereafter exponentially and more slowly. This octopolar function becomes negative after 0.7 psec, exhibits a minimum at 1.2 psec, and is damped to zero after 5.0 psec. The fact that $C_\Omega(t)/C_\Omega(0)$ falls off faster than $C_{FR}(t)$ means that a_2 is affected in some way by molecular interaction in the classical expansion

$$C(t)=1-a_2\frac{t_2}{2!}+\cdots$$

From Fig. 83 it is clear that CBr_4 molecules in both the liquid and plastic crystalline experience a torque almost immediately after the arbitrary $t=0$. This is greater in magnitude than that in a bimolecular encounter of octopole fields, since $C(t)$ falls off faster initially than $C_\Omega(t)/C_\Omega(0)$. Rotation motions are then correlated in the condensed phases, and after ~0.7 psec both $C(t)$ become exponential and decay relatively slowly compared with $C_\Omega(t)/C_\Omega(0)$. O'Dell and Berne[108] have discovered recently that rotational motion is freer in the *solid* just below the melting point in rough-hard-sphere ensembles. This seems to be the case here since $C(t)$ for the liquid falls off initially a little faster. The greater initial torque in the condensed phases can be explained superficially in terms of the greater packing density since the intermolecular potential would be much greater on average with van der Waals radii overlapping (repulsive domain) for a greater percentage of the time. However, this is, as always, too simple a view, since the packing density in the plastic solid is the greater while the mean torque is smaller. This can only mean that symmetry of packing and the resultant restriction on molecular diffusion is an important factor in determining the ease of rotational movement in the plastic phase.

ACKNOWLEDGMENT

S.R.C. is thanked for an equipment grant and a fellowship grant (to G.J.E.). M.W.E. thanks the Ramsay Memorial Trust for the 1976–1978 Fellowship. Mrs. Vincent of the Applied Mathematics Department and Mrs. Evans of the Chemistry Department are thanked for their usual high standard of typing.

Finally, M.W.E. wishes to thank publishing, technical and computer staffs for their help and advice.

REFERENCES

1. K. D. Muller and W. G. Rothschild, *Far-Infrared Spectroscopy*, Wiley-Interscience, New York, 1971. This book contains a list of all 1512 papers on the far-infrared up to 1970.
2. G. W. Chantry, *Submillimetre Spectroscopy*, Academic, London, 1971.
3. A. Gerschel, *Mol. Phys.*, **32**, 679, (1976).
4. N. E. Hill, A. H. Price, W. E. Vaughan, and M. Davies, *Dielectric Properties and Molecular Behaviour*, Van Nostrand-Reinhold, London, 1969.
5. G. J. Davies and M. Evans, *J. Chem. Soc. Faraday Trans.* 2, **72**, 1194 (1976).
6. D. R. Bosomworth and H. P. Gush, *Can. J. Phys.*, **37**, 362 (1959); **43**, 751 (1965).
7. J. -P. Hansen and I. R. McDonald, *Theory of Simple Liquids*, Academic, London, 1976.
8. B. K. P. Scaife, *Complex Permittivity*, English University Press, London, 1971, p. 35.
9. R. Zwanzig, *J. Chem. Phys.*, **38**, 2766 (1963).
10. J. Orban and A. Bellemans, *J. Chem. Phys.*, **49**, 363 (1968); J. Orban, J. van Craen, and A. Bellemans, *J. Chem. Phys.*, **49**, 1778 (1968).
11. R. H. Cole, *Mol. Phys.*, **27**, 1 (1974).
12. J. M. Deutch, *Faraday Symp. Chem. Soc.*, **11**, 1977.
13. W. T. Coffey, *Mol. Phys.*, in press.
14. J. H. van Vleck and V. Weisskopff, *Rev. Mod. Phys.*, **17**, 227 (1945).
15. R. Lobo, J. E. Robinson, and S. Rodriguez, *J. Chem. Phys.*, **59**, 5992 (1973).
16. D. Frenkel, G. H. Wegdam, and J. van der Elsken, *J. Chem. Phys.*, **57**, 2691 (1972).
17. D. Kivelson and P. Madden, *Mol. Phys.*, **30**, 1749 (1975).
18. H. Mori, *Prog. Theor. Phys.*, **33**, 423 (1965).
19. P. Langevin, *J. Phys.*, **4**, 678 (1905).
20. A. R. Davies and M. W. Evans, *Mol. Phys.*, **35**, 857 (1978).
21. A. Morita, *J. Phys. D.*, in press.
22. J. T. Lewis, J. McConnell, and B. K. P. Scaife, *Proc. R. Irish Acad. A*, **76**, 43 (1976); G. W. Ford, J. T. Lewis, and J. McConnell, *Proc. R. Irish Acad. A*, **76**, 117 (1976).

23. A. D. Fokker, *Ann. Phys.*, **43,** 812 (1914); M. Planck, *Sitzungsber. Preuss. Akad.*, 324 (1917).
24. S. A. Adelman, *J. Chem. Phys.*, **64,** 124 (1976); S. A. Adelman and B. J. Garrison, *Mol. Phys.*, **33,** 1671 (1977).
25. B. J. Berne and G. D. Harp, *Adv. Chem. Phys.*, **17,** 63 (1970).
26. B. J. Berne and R. Pecora, *Dynamic Light Scattering with Applications to Chemistry, Biology and Physics*, Wiley, New York, 1976.
27. D. A. McQuarrie, *Statistical Mechanics*, Harper & Row, 1975.
28. P. Kruus, *Liquids and Solutions, Structure and Dynamics*, Dekker, New York, 1977.
29. J. H. Calderwood and W. T. Coffey, *Proc. R. Soc. Ser. A*, **356,** 269 (1977); P. S. Damle, A. Sjölander, and K. S. Singwi, *Phys. Rev.*, **165,** 277 (1968).
30. W. T. Coffey, G. J. Evans, M. W. Evans, and G. H. Wegdam, *J. Chem. Soc. Faraday Trans. 2*, **74,** 310 (1978).
31. L. van Hove, *Phys. Rev.*, **95,** 249 (1954).
32. B. J. Berne and J. A. Montgomery, *Mol. Phys.*, **32,** 363 (1976).
33. D. Chandler, *J. Chem. Phys.*, **60,** 3508 (1974).
34. R. G. Gordon, *Adv. Magn. Reson.*, **3,** 1 (1968); R. E. D. McClung, *J. Chem. Phys.*, **57,** 5478 (1972).
35. K. V. Mardia, *Statistics of Directional Data*, Academic, New York, 1972.
36. T. J. Lewis, in M. Davies, Ed., *Dielectric and Related Molecular Processes*, Vol. 3, The Chemical Society, London, 1977, p. 186.
37. P. Debye, *Polar Molecules*, Dover, New York, 1954.
38. B. Keller and F. Kneubühl, *Helv. Phys. Acta*, **45,** 1127 (1972).
39. P. Desplanques, "Absorption dipolaire et dynamique moleculaire en phase liquide," thesis, University of Lille, 1974.
40. F. Bliot and E. Constant, *Chem. Phys. Lett.*, **29,** 618 (1974).
41. G. J. Evans, G. H. Wegdam, and M. W. Evans, *Chem. Phys. Lett.*, **42,** 331 (1976).
42. J. Barojas, D. Levesque, and B. Quentrec, *Phys. Rev. A*, **7,** 1092 (1973).
43. M. W. Evans, in Ref. 36, p. 1.
44. M. Drawid and J. W. Halley, *Magnetism and Magnetic Materials—1976*, Publ. 34, American Institute of Physics, New York, 1976, p. 211.
45. K. Lindenberg and R. I. Cukier, *J. Chem. Phys.*, **62,** 3271 (1975).
46. D. Kivelson and T. Keyes, *J. Chem. Phys.*, **57,** 4599 (1972); D. Kivelson, *Mol. Phys.*, **28,** 321 (1974).
47. M. Guillot and S. Bratos, *Mol. Phys.*, in press (1979).
48. P. G. Wolynes and J. M. Deutch, *J. Chem. Phys.*, **67,** 733 (1977).
49. P. G. de Gennes, *The Physics of Liquid Crystals*, Oxford University Press, New York, 1974, Chap. 3.
50. Eighty papers will be published in the Proceedings of the 3rd International Conference on Submillimetre Waves, Infrared Physics, held at Guildford in 1978.
51. G. Williams, *Chem. Soc. Rev.*, **7,** 89 (1978).
52. C. Brot, in Ref. 36, Vol. 2., p. 1.
53. W. A. Steele, *Adv. Chem. Phys.*, **34,** (1976).
54. G. J. Evans, C. J. Reid, and M. W. Evans, *J. Chem. Soc. Faraday Trans. 2*, **74,** 343 (1978).
55. P. van Konynenburg and W. A. Steele, *J. Chem. Phys.*, **62,** 2301, (1975).
56. J. P. Poley, *J. Appl. Sci.*, **4B,** 337 (1955).
57. J. Chamberlain, *Infrared Phys.*, **11,** 22, (1971).
58. D. H. Martin and E. Puplett, *Infrared Phys.*, **10,** 105 (1970).
59. G. J. Evans, thesis, University of Wales, 1977.

60. J. W. Flemming, *Infrared Phys.*, **10**, 57 (1970).
61. W. B. Streett and K. E. Gubbins, *Annu. Rev. Phys. Chem.*, **28**, 373 (1977).
62. J. S. Rowlinson and M. Evans, *Annu. Rep. Chem. Soc.*, **72A**, 5 (1975). This contains references to six separate and different derivations of this theorem.
63. D. E. Sullivan and J. M. Deutch, *J. Chem. Phys.*, **62**, 2130 (1975).
64. J. L. Greffe, J. Goulon, J. Brondeau, and J. L. Rivail, in J. Lascombe, Ed., *Molecular Motion in Liquids*, Dordrecht, 1974, p. 151.
65. R. Kubo, *Lectures in Theoretical Physics*, Wiley-Interscience, New York, 1959.
66. M. W. Evans, *Adv. Mol. Rel. Int. Proc.*, **10**, 203 (1977).
67. G. J. Davies, G. J. Evans, and M. W. Evans, *J. Chem. Soc. Faraday Trans. 2*, **72**, 1904 (1976).
68. M. W. Evans, *J. Chem. Soc. Faraday Trans. A*, **72**, 2138 (1976).
69. G. W. F. Pardoe and H. A. Gebbie, Symposium on Submillimetre Waves, Polytechnic Inst., Brooklyn, 1970, p. 643.
70. J. Hasted, in Ref. 36, Vol. 1.
71. J. A. Janik, in Ref. 36, p. 45.
72. D. Frenkel and G. H. Wegdam, *J. Chem. Phys.*, **61**, 4671 (1974).
73. D. Frenkel, *Faraday Symp. Chem. Soc.*, 11, (1976).
74. M. W. Evans, *Mol. Phys.*, **34**, 963 (1977).
75. P. N. Brier and A. Perry, *Adv. Mol. Int. Rel. Proc.*, in press (1978).
76. J. Goulon, D. Canet, M. Evans, and G. J. Davies, *Mol. Phys.*, **30**, 973 (1975).
77. G. Williams and M. Cook, *J. Chem. Soc. Faraday Trans. 2*, **67**, 990 (1971).
78. T. E. Faber and G. R. Luckhurst, *Annu. Rep. Chem. Soc., Lond.*, **75A**, 31 (1975).
79. B. J. Bulkin and W. B. Lok, *J. Phys. Chem.*, **77**, 326 (1973).
80. M. W. Evans, M. Davies, and I. Larkin, *J. Chem. Soc. Faraday Trans. 2*, **69**, 1011 (1973).
81. N. E. Hill, *Proc. Phys. Soc.*, **82**, 723 (1963).
82. G. A. P. Wyllie, *J. Phys. C.*, **4**, 564 (1971).
83. G. J. Evans and M. W. Evans, *J. Chem. Soc. Faraday Trans.* 2, **72**, 1169 (1976).
84. E. Sciesinska, J. Sciesinski, J. Twardowski, and J. D. Janik, *Inst. Nucl. Phys.*, 847/PS (1973).
85. G. J. Evans and M. W. Evans, *J. Chem. Soc. Chem. Commun.*, 267 (1978).
86. G. J. Evans and M. Evans, *J. Chem. Soc. Faraday Trans 2*, **73**, 203 (1977).
87. M. Schwartz and L. H. Wang, private communication to B. J. Bulkin, Ref. 79.
88. M. S. Beevers, J. Crossley, D. C. Garrington, and G. Williams, *J. Chem. Soc. Faraday Trans. 2*, **72**, 1482 (1976).
89. R. Haffmanns and I. W. Larkin, *J. Chem. Soc., Faraday Trans.* 2, **68**, 1729 (1972).
90. M. Davies, *Annu. Rep. Chem. Soc.*, **67A**, 65 (1970).
91. A. Aihara, C. Kitazawa, and A. Nohara, *Bull. Chem. Soc. Jap.*, **43**, 3750 (1970).
92. I. Darmon and C. Brot, *Mol. Cryst.* **2**, 301 (1967).
93. I. Tanaka, F. Iwasaki, and A. Aihara, *Acta Crystallogr.*, **30B**, 1546 (1974).
94. G. W. F. Pardoe and J. Flemming, to be published.
95. G. P. Johari and C. P. Smyth, *J. Chem. Phys.*, **56**, 4411 (1972); G. P. Johari and M. Goldstein, *J. Chem. Phys.*, **53**, 2372 (1970); G. P. Johari, *J. Chem. Phys.*, **58**, 1766 (1973).
96. G. Williams, in Ref. 36, Vol. 2.
97. A. Rahman, *Phys. Rev. A*, **136**, 405 (1964).
98. P. S. Y. Cheung and J. G. Powles, *Mol. Phys.*, **30**, 921 (1975).
99. W. B. Streett and D. J. Tildesley, *Proc. R. Soc. Lond.*, **348A**, 485 (1976); **355A**, 239 (1977).

100. J. W. E. Lewis and K. Singer, *J. Chem. Soc. Faraday Trans. 2*, **71**, 301 (1975); K. Singer, A. Taylor, and J. V. L. Singer, *Mol. Phys.*, **33**, 1757 (1977).
101. C. W. Gear, *Numerical Initial Value Problems in Ordering Differential Equations*, Prentice-Hall, Englewood Cliffs, N. J., 1971, pp. 148–150.
102. V. F. Sears, *Proc. Phys. Soc.*, **86**, 953 (1965).
103. B. A. Dasannacharya and K. R. Rao, *Phys. Rev. A*, **137**, 417 (1965).
104. G. E. Uhlenbeck and L. S. Ornstein, *Phys. Rev.*, **36**, 823 (1930).
105. W. T. Coffey and M. W. Evans, *Mol. Phys.*, **35**, 975 (1978).
106. B. J. Berne, *Faraday Symp. Chem. Soc.*, **11**, (1977).
107. G. H. Wegdam, G. J. Evans, and M. Evans, *Mol. Phys.*, **33**, 1805 (1977).
108. J. O'Dell and B. J. Berne, *J. Chem. Phys.*, **63**, 2376 (1975).
109. F. Hermans, E. Kestemont, R. van Loon, and R. Finsy, Proceeding of the 3rd International Conference on Submillimetre Waves, Guildford, 1978, Infrared Physics, to be published.
110. D. H. Whiffen, *Trans. Faraday Soc.*, **46**, 124 (1950).
111. R. Savoie and R. P. Fournier, *Chem. Phys. Lett.*, **7**, 1 (1970).
112. G. J. Davies, J. Chamberlain, and M. Davies, *J. Chem. Soc. Faraday Trans. 2*, **69**, 1223 (1973); G. J. Davies and J. Chamberlain, *J. Chem. Soc. Faraday Trans. 2*, **68**, 1739 (1973).
113. W. Ho, G. Birnbaum and A. Rosenberg, *J. Chem. Phys.*, **55**, 1028 (1971).
114. A. I. Baise, *J. Chem. Soc. Faraday Trans. 2*, **68**, 1904 (1972).
115. M. Evans, *J. Chem. Soc. Faraday Trans. 2*, **69**, 763 (1973).
116. B. S. Frost, *J. Chem. Soc. Faraday Trans. 2*, **69**, 1142 (1973).
117. J. P. Colpa and J. A. A. Ketelaar, *Mol. Phys.*, **1**, 343 (1958).
118. M. Evans, *Mol. Phys.*, **29**, 1345 (1975).
119. D. R. Bosomworth and H. P. Gush, *Can. J. Phys.*, **43**, 751 (1965).
120. G. Jacucci, U. Buontempo, and S. Cunsolo, *J. Chem. Phys.* **59**, 3750 (1973).
121. M. Evans, *Spectrochim. Acta*, **32A**, 1253 (1976).
122. A. I. Baise, *J. Chem. Phys.*, **60**, 2936 (1974).
123. S. Kielich, in Ref. 36, Vol. 1, pp. 192–387.
124. G. J. Evans and M. W. Evans, *Adv. Mol. Rel. Int. Proc.*, **9**, 1 (1976).
125. C. C. Bradley, H. A. Gebbie, A. C. Gilby, V. V. Kechin, and J. H. King, *Nature*, **211**, 839 (1966).
126. M. W. Evans and G. J. Davies, *J. Chem. Soc. Faraday Trans. 2*, **72**, 1206 (1976).
127. J. F. Dill, T. A. Litovitz, and J. A. Bucaro, *J. Chem. Phys.*, **62**, 3839 (1975).
128. P. van Konynenburg and W. A. Steele, *J. Chem. Phys.*, **62**, 2301 (1975).
129. M. Perrot, J. Devaure, and J. Lascombe, *Mol. Phys.*, **30**, 97 (1975).
130. J. Jonas and T. M. Di Gennaro, *J. Chem. Phys.*, **50**, 2392 (1969).
131. G. J. Davies, G. J. Evans, and M. W. Evans, *J. Chem. Soc. Faraday Trans. 2*, **72**, 2147 (1976).
132. I. Ozier and K. Fox, *J. Chem. Phys.*, **52**, 1416 (1970).

ON A THEORETICAL DESCRIPTION OF SOLVATED ELECTRONS

A. M. BRODSKY AND A. V. TSAREVSKY

Institute of Electrochemistry
Academy of the Sciences of the U.S.S.R.
Leninsky prospekt, 31,
Moscow, V-71, U.S.S.R.

CONTENTS

I. INTRODUCTION

This review is devoted to questions pertaining to a theoretical description of the behavior of thermalized excess electrons in disordered condensed media which do not exhibit intrinsic electron conductivity (i.e., electrolytes, polar and nonpolar liquids, and glasses). The phenomena that occur on injecting excess electrons into these media have in recent years become intensively studied in several areas of the physics of condensed media, chemical physics, radiation chemistry, and in biology. These phenomena are described in a series of books and reviews.[1–12]

Initially, attention was focused on localized electron states in polar liquids which were termed solvated electrons and denoted by the symbol e_s^-. The concept of solvated electron historically arose in an attempt to interpret the unique phenomena, which occur when alkai metals are dissolved in ammonia. These phenomena were discovered as early as 1863 by Weyl. Interest in solvated electrons was catalyzed especially after the discovery in the early 1960s of short-living solvated (hydrated) electrons, e_{aq}, in water and in aqueous electrolyte solutions.[13,14] It was found that solvated electrons play an important part in radiation-induced chemical conversions and in the changes in the physical properties of irradiated aqueous systems.[15,16] Subsequently, solvated electrons were found to exist in many other polar liquids, such as alcohols, ketones, amines, and so on.

States similar to electrons in polar media and sharply distinct from the excited states of electrons in crystals have been observed and studied in recent years in nonpolar liquids as well: liquified inert gases and hydrocarbons.[7,11] Excess electrons trapped in glasslike matrices at low temperatures have also been found to exhibit properties similar to those of solvated electrons.[4]

In recent years, in investigating the properties of excess electrons, use has been made of a variety of physical and chemical methods based on complicated optical flash equipment with a resolving capacity up to 10^{-12} sec, radiospectroscopy, the Hall effect, and conductivity and electrochemical measurements. The vast experimental data thus obtained have in many cases turned out to be somewhat unexpected and could not be fitted into the existing theories. One of the main results here, for instance, is the existence of different electron states. In addition to the localized states, the usually short-living "dry" electrons, which in many respects resemble the delocalized conduction electrons in metals, have been detected and studied.

A detailed investigation was made of the optical absorption curves of localized electrons that exhibit several common properties in various substances, their mixtures, and solutions. In most cases, these curves represent a solitary wide structureless peak in the visible range or in the near-ultraviolet or near-infrared regions, depending on the microscopic structure of

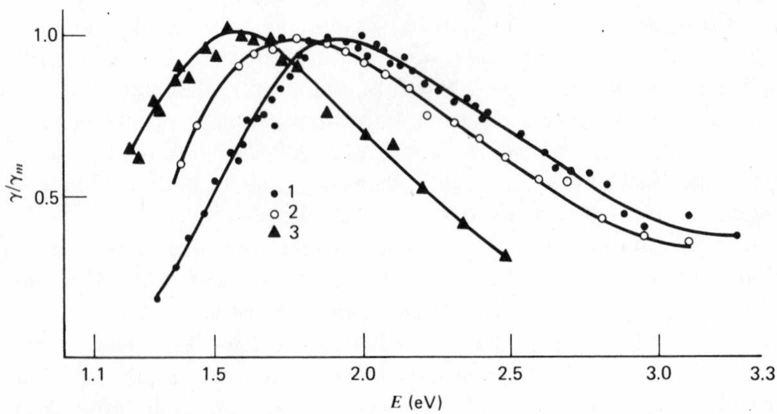

Fig. 1. Optical absorption spectrum of electrons in methanol (From Ref. 17.): 1 – $T=4°C$; 2 – $T=20°C$; 3 – $T=90°C$.

the solvent. The typical common features of the absorption curve shape which we shall repeatedly refer to in subsequent sections are shown in Fig. 1, drawn from Ref. 17. As is seen in this figure, the absorption peak is asymmetric: the increase on the low-energy side is steep, while the decrease on the high-energy side is slow. Such behavior is unusual in atomic and molecular spectroscopy, if we disregard the absorption spectra of a number of negative ions; but it has an analogy in nuclear physics. It reminds us, in particular, of the frequency dependence (under appropriate changes in the energy scale) of photofission of deuteron. The shape of the absorption peak was found to be extremely stable (e.g., in water it hardly changes from $-4°C$ to the supercritical temperature). The specific behavior of the absorption peak, essential from a theoretical viewpoint, is its shift toward the red side with increasing temperature.

Below, attention is focused on the theory of solvated electrons at low concentrations in disordered polar media. The behavior of electrons in nonpolar media and inert gases will only be touched upon in those places where it aids the understanding of electron behavior in polar media. The specific phenomena occurring at higher concentrations of solvated electrons, say in ammonia, are described in detail in several physics textbooks,[6,9,12] and we shall not, therefore, deal with them here.

In Section II we shall make a critical survey of the theoretical models proposed since the 1940s for the absorption spectra of solvated electrons. In view of the complexity of the system, a considerable number of approximations have to be used in these models, with a vaguely defined field of application. Despite the use of fitting parameters, it has not yet been possible to obtain a consistent quantitative description for all the optical characteristics observed within the framework of these theories. This is also true of the transport properties of excess electrons examined in Section III.

As the present state of the model theory is so indefinite, special importance is acquired by an analysis of spectra of solvated electrons in which model assumptions are not used at all, or are restricted to a minimum. As will be shown in Section IV, by means of these methods, including the sum rules, the virial theorem, and analysis of the threshold and high-energy behavior of optical absorption, we can draw some important qualitative and quantitative conclusions about the structure of solvated electrons. This structure corresponds to the effective short-range interaction of electrons with the medium. If a minimum number of fitting parameters are used, this interaction is in many cases well represented by the zero-radius potential. The consequences of introducing this potential into the theory of solvated electrons are examined in the last two paragraphs of Section IV. Determination of the thermodynamic characteristics of solvated electrons is dealt with in Section V.

The results that can be drawn from an analysis of the solvated electron's physical structure are summed up in Section VI. The references cited cover publications that appeared before 1978.

II. THEORETICAL MODELS OF THE STRUCTURE AND OPTICAL SPECTRA OF SOLVATED ELECTRONS

A. Initial Theoretical Models

From 1940 to 1946, Ogg[19] developed the first quantitative theory to explain the states of an electron solvated in ammonia. In this theory it is assumed that a spherical cavity of radius R is formed around the electron in ammonia, and the ammonia molecules on the cavity boundary create an effective spherical potential well with an infinitely high barrier to the electron. The electron energy is considered to be sum of the positive kinetic energy T_e, which was estimated simply from the uncertainity relation by the formula $T_e = (2\pi\hbar)^2/2mR^2$, and Born's energy of interaction with the polarization field equal to

$$V = -\frac{e^2}{R}\left(1-\varepsilon_0^{-1}\right) \tag{II.1}$$

where ε_0 is the permittivity of the medium.

The equilibrium value R_{eq} of the radius is derived from the condition of minimum total energy. Repulsion of the molecules by the electron in the Ogg's cavity was compared with the observed expansion of metal–ammonia solutions. On being excited by light, the electron leaves the cavity. The threshold frequency of the light absorbed by the solvated electron was determined from the equation

$$\hbar\Omega = T_e + V$$

The frequency Ω, and especially the large radius $R_e = 20\,\text{Å}$, obtained by Ogg are qualitatively contrary to the experiment. It is now obvious that the reason for this is the overestimated kinetic energy in this model, where the potential well has an infinitely high barrier which does not satisfactorily describe the relatively weakly bound states of e_s^-. The idea of a potential cavity, nonetheless, considerably stimulated further investigation into solvated electrons. Indeed, an analogous calculation procedure (with only minimization of the free energy of the electron-medium system, rather than of the energy) underlies the modern theory of localized states of excess electrons in inert gases.[7,9,20–25]

B. Continuous Media Models

A. S. Davydov[26] was the first to describe solvated electrons in ammonia, using the Landau–Pekar theory of large polarons that was previously developed for the electron states in ionic crystals.[27] Later, M. F. Deigen[28,29] worked out a similar approach. Within the framework of the Davidov–Deigen theory, which the authors have developed in contrast to the Ogg's cavity theory, it has been possible to obtain a number of characteristics of solvated electrons in ammonia consistent with the experiment: in the first instance, the position of the maximum in the optical absorption band which was attributed to the Franck–Condon $1s \rightarrow 2p$ transition. This theory has played a historically important part in predicting the existence and the frequency range for the optical absorption of short-living hydrated electrons, e_{aq}. On the basis of this theory many attempts have been made to refine the calculations of energy parameters and other characteristics of the ground and excited states of solvated electrons (see, e.g., Refs. 30 and 31). These results, however, on being compared with the latest data, encounter serious difficulties.

1. Modern Polaron Theory of Electrons in Disordered Media

In recent years various different versions of the polaron theory have been applied in numerous papers to determine the characteristics of excess electrons in disordered media. Today this theory is understood in a wider sense as a theory accounting for the interaction of excess electrons with various collective excitations of the medium. In view of the importance of these works for understanding the physical nature of excess electron, we shall examine the basic principles of the modern polaron theory in somewhat greater detail.

To describe the interactions with the collective excitations, use is generally made of the techniques of quantum field theory. The initial step here is that of writing the Hamiltonian in the form:[32]

$$H = H_0^{\mathrm{el}} + H_0^{\mathrm{ph}} + H_{\mathrm{int}}$$

$$H_0^{\mathrm{el}} = \frac{\mathbf{p}^2}{2m} + V_i(\mathbf{r})$$

$$H_0^{\mathrm{ph}} = \sum_{\mathbf{q},\alpha} \left(a_{\mathbf{q}}^{\alpha +} a_{\mathbf{q}}^{\alpha} + \tfrac{1}{2}\right)\hbar\omega_{\mathbf{q}}^{\alpha}$$

$$H_{\mathrm{int}} = \sum_{\mathbf{q},\alpha} \left[Q_{\mathbf{q}}^{\alpha} a_{\mathbf{q}}^{\alpha} e^{i\mathbf{qr}} + Q_{\mathbf{q}}^{\alpha *} a_{\mathbf{q}}^{\alpha +} e^{-i\mathbf{qr}} \right] \qquad \text{(II.2)}$$

where H_0^{el} is the Hamiltonian of the band electron with the coordinate operator $\mathbf{r}$, effective mass m, and momentum $\mathbf{p}$. In addition to the kinetic energy, taken for the simplicity of the quadratic law of dispersion, the potential energy V_i, corresponding either to the additional available interactions at distances of atomic order, or to interactions with large-scale fluctuations in a disordered medium, has also been included in H_0^{el}. Here H_0^{ph} stands for the energy of free collective vibrations in oscillator approximation expressed by means of secondary quantization techniques; $a_{\mathbf{q}}^{\alpha+}$ and $a_{\mathbf{q}}^{\alpha}$ denote the creation and annihilation operators of α-type phonons corresponding to collective excitations, and the function $\omega_{\mathbf{q}}^{\alpha}=\omega(\mathbf{q},\alpha)$ defines the phonon dispersion law.

The coefficients $Q_{\mathbf{q}}^{\alpha}$ in (II.2) are the interaction form factors. Summation is carried out over all the normal lattice modes α and quasimomentum $\mathbf{q}$. Below we shall omit the index α when collective excitations of one type are considered.

In the theory of solvated electrons recourse is often taken to a calculation procedure in which longitudinal optical phonons are used for collective excitations. Such collective excitations arise due to the motion of a system of dipoles which change their orientation upon the injection of a charge. In this case the so-called Fröhlich Hamiltonian is generally used, in which [33]

$$Q_{\mathbf{q}}=-i\left(\frac{\sqrt{2}\,\pi\alpha_F}{\mathfrak{V}}\right)^{1/2}\left(\frac{\hbar^5\omega^3}{m}\right)^{1/4}q^{-1}\qquad \omega_{\mathbf{q}}=\omega \tag{II.3}$$

where ω denotes the phonon frequency,* which here does not depend on q and represents the characteristic of the medium, $\mathfrak{V}$ is the volume of the system, and α_F is a dimensionless coupling constant. In the continuous-field model without spatial dispersion, it is equal to

$$\alpha_F=\tfrac{1}{2}\left(\varepsilon_\infty^{-1}-\varepsilon_0^{-1}\right)\frac{e^2}{\hbar\omega}\left(\frac{2m\omega}{\hbar}\right)^{1/2} \tag{II.4}$$

If acoustic phonons are used as the collective excitations, then we have

$$\hbar\omega_{\mathbf{q}}=|q|C_s$$

$$Q_{\mathbf{q}}=-i\sigma\left(\frac{\hbar|q|}{2Mc_s\mathfrak{V}}\right)^{1/2} \tag{II.5}$$

*The influence of disorderedness may considerably reduce the sum to much lower values of q than in the case of a regular lattice.

where C_s is the velocity of sound, M is the mass of a unit cell, and σ is the constant of deformation potential.

By analogy with (II.2) to (II.5), we can express the interaction with other types of collective interactions, say vibrons—a system of intramolecular vibrations which was introduced into the solvated electron theory in Refs. 5 and 34.

The direct effective application of the Hamiltonian (II.2) is only possible for a weak interaction H_{int} with the help of perturbation theory techniques. The other limiting case (i.e., strong interaction) is of great interest in the theory of localized electrons. A special approach has been developed for this case, in which an analog of the method of a self-consistent Hartree field is used as the main element. A highly consistent and demonstrative scheme of this approach is as follows: the total vector of state of the system $|\Psi\rangle$ is searched approximately in the form of a direct product of the state vector $|\varphi\rangle$, which only depends on the electron variables, and the state vector $|\chi\rangle$, which depends solely on the phonon field variables:*

$$|\Psi\rangle = |\varphi\rangle|\chi\rangle \tag{II.6}$$

At the same time the Hamiltonian (II.2) is replaced by an approximate Hamiltonian $\tilde{H}$ of the following type, which is evidently time-dependent:

$$\tilde{H} = H_0^{\text{el}} + H_0^{\text{ph}} + \sum_{\mathbf{q}} \left[Q_{\mathbf{q}} a_{\mathbf{q}} \rho_{\mathbf{q}}(t) + Q_{\mathbf{q}}^* a_{\mathbf{q}}^+ \rho_{\mathbf{q}}^*(t) \right]$$
$$+ \sum_{\mathbf{q}} \left[Q_{\mathbf{q}} A_{\mathbf{q}}(t) e^{i\mathbf{qr}} + Q_{\mathbf{q}}^* A_{\mathbf{q}}^*(t) e^{-i\mathbf{qr}} \right] + F(t) \tag{II.7}$$

where $\rho_{\mathbf{q}}(t)$, $A_{\mathbf{q}}(t)$, and $F_{\mathbf{q}}(t)$ are no longer operators, but are the usual functions of $\mathbf{q}$ and t. From the condition of minimum deviation of $\tilde{H}$ from H, which is formulated in the form of minimum of the functional

$$\langle\Psi|(\tilde{H} - H)^2|\Psi\rangle$$

* In the absence of local interactions of the type V_i which disturb the translation invariance, before starting the procedure described below, the motion of the center of mass has to be eliminated from the Hamiltonian (II.2), say, by means of the Bogolyubov transformation.[35] If H retains its symmetry relative to rotations, then, strictly speaking, we have at first to construct the states with fixed total angular momentum. Unless such procedures are done first, expression (II.6) will not be satisfactory.[36] In order to simplify the presentation, we shall not carry out these procedures.

it is not difficult to obtain equalities that should hold true:

$$\rho_{\mathbf{q}}(t)=\int \varphi^*(\mathbf{r},t)\varphi(\mathbf{r},t)e^{i\mathbf{qr}}d^3\mathbf{r}$$
$$\varphi(\mathbf{r},t)=\langle \mathbf{r}|\varphi(t)\rangle$$
$$A_{\mathbf{q}}(t)=\langle \chi|a_{\mathbf{q}}|\chi\rangle$$
$$F(t)=-\sum_{\mathbf{q}}\left[Q_{\mathbf{q}}A_{\mathbf{q}}(t)\rho_{\mathbf{q}}(t)+Q_{\mathbf{q}}^*A_{\mathbf{q}}^*(t)\rho_{\mathbf{q}}^*(t)\right] \tag{II.8}$$

In considering the ground states, we can discard c-number $F(t)$, not dependent on the coordinate and momentum of the electron and phonon, from the Hamiltonian (II.7) because it only introduces changes in the phase of the wave function that are of no importance in considering the stationary states. From (II.7) and (II.8) we obtain the dynamic equations which for the wave function $\varphi(\mathbf{r},t)$ of electrons can conveniently be expressed in the Schrödinger representation, and for phonon operators $a_{\mathbf{q}}$ in the Heisenberg representation:

$$i\hbar\frac{\partial\varphi}{\partial t}(\mathbf{r},t)=\left\{H_0^{\text{el}}+\sum_{\mathbf{q}}\left[Q_{\mathbf{q}}A_{\mathbf{q}}(t)e^{i\mathbf{qr}}+Q_{\mathbf{q}}^*A_{\mathbf{q}}^*(t)e^{-i\mathbf{qr}}\right]\right\}\varphi(\mathbf{r},t)$$

$$i\frac{\partial a_{\mathbf{q}}(t)}{\partial t}=\hbar^{-1}\left[\tilde{H},a_{\mathbf{q}}\right]=\omega_{\mathbf{q}}a_{\mathbf{q}}+\hbar^{-1}Q_{\mathbf{q}}^*\rho_{\mathbf{q}}^*(t) \tag{II.9}$$

In the system of (II.9) the electron (phonon) field depends on the phonon (electron) field only through the mean value of the corresponding operators, in the same way as in the Hartree method. The second equation in (II.9) is the equation of an oscillator under the influence of external force. In such cases it is more convenient to use the basis of coherent states,[37] which remain coherent during the development of the process in time; in other words, each coherent state passes into another definite coherent state corresponding to it. The coherent states not only form a complete, but even an overcomplete set, in terms of which any other state can be expanded. The coherent states contain an indefinite number of phonons and they are the eigenfunctions of the operators $a_{\mathbf{q}}$. Selecting a coherent state with an eigenvalue of the operator $a_{\mathbf{q}}$ equal to $\alpha_{\mathbf{q}}$ at $t=0$, in the Heisenberg representation we get

$$a_{\mathbf{q}}(t)|\alpha_{\mathbf{q}}\rangle=\alpha_{\mathbf{q}}(t)|\alpha_{\mathbf{q}}\rangle \qquad |\alpha(t)\rangle=\sum_{n_q=0}^{\infty}\frac{\alpha_{\mathbf{q}}^n}{n_{\mathbf{q}}!}|n_{\mathbf{q}}(t)\rangle e^{-1/2|\alpha_{\mathbf{q}}|^2} \tag{II.10}$$

where $|n_{\mathbf{q}}(t)\rangle$ is a state with a given number of phonons with a wave vector $\mathbf{q}$. Since the phonon excitations with different $\mathbf{q}$ are independent in this approximation, instead of the second operator equation in (II.9) it is sufficient to consider the ordinary differential equation for the means over coherent states:

$$A_{\mathbf{q}}(t) \equiv \langle \alpha_{\mathbf{q}} | a_{\mathbf{q}} | \alpha_{\mathbf{q}} \rangle \tag{II.11}$$

The equation for $A_{\mathbf{q}}(t)$, which follows from (II.9) and (II.10) under normalization of the coherent states, is of the form

$$i\frac{\partial A_{\mathbf{q}}(t)}{\partial t} = \omega A_{\mathbf{q}}(t) + \hbar^{-1} Q_{\mathbf{q}}^{*} \rho_{\mathbf{q}}^{*}(t) \tag{II.12}$$

Thus the second equation in (II.9) can be replaced by the equation of the classical field $A_{\mathbf{q}}(t)$. Its solution is

$$A_{\mathbf{q}}(t) = A_{\mathbf{q}}^{0} e^{-i\omega t} - i\hbar^{-1} Q_{\mathbf{q}}^{*} \int_{-\infty}^{t} e^{i\omega(t'-t)+\varepsilon(t'-t)} \rho_{\mathbf{q}}^{*}(t')\, dt' \tag{II.13}$$

The choice of solution (II.13) corresponds to the adiabatic switching on of the interaction; hence the factor

$$\exp[\varepsilon(t'-t)]$$

has been included in the integrand.

In most of the calculations carried out until recently, two additional assumptions are made either explicitly or implictly. First, the first term on the right-hand side of (II.13) is discarded, and this may be interpreted exclusively as the elimination of the influence on e_s^- of free real phonons generated in the medium even in equilibrium case due to thermal excitation (nonetheless, see Section IV.D). Second, an adiabatic approximation is made which lies in substituting $\rho_{\mathbf{q}}^{*}(t')$ in the integrand of (II.13) by $\rho_{\mathbf{q}}^{*}$ the mean of the ground electron state over a finite trajectory.* The later approximation is called for by different time scales of the motion of inert phonon and fast electron subsystems.

Under these assumptions, on substituting (II.13) into the first equation of (II.9), we obtain the following equation for the electron wave function $\varphi(\mathbf{r}, t)$:

$$i\hbar\frac{\partial \varphi}{\partial t}(\mathbf{r}, t) = \left[\frac{\mathbf{p}^2}{2m} + V_i(r) - \frac{1}{\hbar\omega}\sum_{\mathbf{q}} |Q_{\mathbf{q}}|^2 \left(\rho_{\mathbf{q}}^{*} e^{i\mathbf{q}\mathbf{r}} + \rho_{\mathbf{q}} e^{-i\mathbf{q}\mathbf{r}}\right)\right] \varphi(\mathbf{r}, t) \tag{II.14}$$

*The corrections to adiabatic approximations made in Refs. 38 and 39 may give rise to new local modes which can be interpreted as the excited states of the type "polaron + bound phonon."

In (II.14), after substituting integration for summation by the ordinary rule, the search for stationary electron function is reduced to solving the eigenvalue problem for the nonlinear integrodifferential operator:

$$E^e\varphi(\mathbf{r})=\left[\frac{\mathbf{p}^2}{2m}+V_i(r)-V_{\mathrm{sc}}(\mathbf{r},\varphi)\right]\varphi(\mathbf{r})$$

$$V_{\mathrm{sc}}=\int d^3\mathbf{r}'|\varphi(\mathbf{r})|^2 z(\mathbf{r},\mathbf{r}') \tag{II.15}$$

$$z(\mathbf{r},\mathbf{r}')=\frac{\mathfrak{V}}{(2\pi)^3}\int d^3\mathbf{q}\frac{|Q_{\mathbf{q}}|^2}{\hbar\omega}\left[e^{i\mathbf{q}(\mathbf{r}-\mathbf{r}')}+e^{i\mathbf{q}(\mathbf{r}'-\mathbf{r})}\right]$$

For the interaction with longitudinal optical phonons, from (II.3) and (II.15), we obtain

$$z(\mathbf{r},\mathbf{r}')=\frac{\sqrt{2}\,\pi\alpha_F}{\hbar\omega}\left(\frac{\hbar^5\omega^3}{m}\right)^{1/2}\frac{1}{2\pi}\frac{1}{|\mathbf{r}-\mathbf{r}'|} \tag{II.16}$$

For the interaction with acoustic phonons, the kernal $z(\mathbf{r},\mathbf{r}')$, by virtue of (II.5), proves to be local:

$$z(\mathbf{r},\mathbf{r}')=\frac{\sigma^2\hbar}{Mc_s^2}\delta(\mathbf{r}-\mathbf{r}') \tag{II.17}$$

After solving the foregoing eigenvalue problem, the total energy of the system can be expressed as

$$E=E^e+E^m \tag{II.18}$$

where E^m is the reorganization energy

$$E^m\equiv\langle\alpha|\sum_{\mathbf{q}}\hbar\omega_{\mathbf{q}}a_{\mathbf{q}}^+a_{\mathbf{q}}|\alpha\rangle=\sum_{\mathbf{q}}\frac{|Q_{\mathbf{q}}|^2|\rho_{\mathbf{q}}|^2}{\hbar\omega_{\mathbf{q}}}$$

$$=\tfrac{1}{2}\int\int d^3\mathbf{r}d^3\mathbf{r}'|\varphi(\mathbf{r})\varphi(\mathbf{r}')|^2 z(\mathbf{r},\mathbf{r}') \tag{II.19}$$

$$\equiv\tfrac{1}{2}\int d^3\mathbf{r}\varphi^*(\mathbf{r})V_{\mathrm{sc}}\varphi(\mathbf{r})$$

The expression (II.13) for $A_{\mathbf{q}}$ without free phonons has been used in calculating E^m.

The direct solution of (II.15), with any specific expression for $z(\mathbf{r},\mathbf{r}')$, is a rather complicated problem. Nonetheless, important results, relating to the ground state, can be obtained without this solution by the following technique presented in Ref. 40. Write the energy E of the ground state, assuming that there is no local interaction in (II.15) ($V_i=0$):

$$E = T_e + E^m - \langle V_{sc} \rangle \tag{II.20}$$

where

$$E^m = \tfrac{1}{2}\langle V_{sc} \rangle$$

$$T_e = \frac{\hbar^2}{2m}\int d^3\mathbf{r}\,\frac{\partial \varphi^*}{\partial \mathbf{r}}\cdot\frac{\partial \varphi}{\partial \mathbf{r}} \tag{II.21}$$

$$\langle V_{sc} \rangle = \int d^3\mathbf{r}\,\varphi^*(\mathbf{r})\,V_{sc}\,\varphi(\mathbf{r})$$

Since the energy E of the ground state is, by definition, minimal, any deviation of the normed function $\varphi(\mathbf{r})$ from its true value should lead to an increase in E. In particular, if $R^{-3/2}\varphi(\mathbf{r}R^{-1})$ is substituted for $\varphi(\mathbf{r})$, then the energy E, being a function of R, should have a minimum at $R=1$ if the bound state is realized in reality. For the classical Pecar polaron $z(\mathbf{r},\mathbf{r}')$, given by the expression (II.16), from (II.20) and (II.21) we obtain

$$E(R) = \frac{T_e}{R^2} - \frac{\langle V_{sc} \rangle}{2R} \tag{II.22}$$

$E(R)$ has only one minimum at

$$R = \frac{4T_e}{\langle V_{sc} \rangle} = 1 \tag{II.23}$$

Hence it follows that there exists a stable polaron state with the energy

$$E = E^e + E^m = -\tfrac{1}{4}\langle V_{sc} \rangle < 0 \tag{II.24}$$

Moreover, the ratio of excitation (reorganization) energy of the medium E^m to the total energy E is

$$\frac{E^m}{E} = -2 \tag{II.25}$$

For the contact electron-medium interaction of the type (II.17), the situation is qualitatively different. Here we have

$$E(R) = \frac{T_e}{R^2} - \frac{\langle V_{sc} \rangle}{2R^3} \tag{II.26}$$

For finite $R \geq 0$, a minimum is attained only at $R=0$, and this corresponds to localization at small distances, for which the initial assumptions of the theory are not valid. Therefore, we have to consider it microscopically, taking into account dispersion, exchange, and correlation effects.

The general conclusion so drawn essentially depends, as is evident from the calculations, on the dimensionality of the space. Thus, in particular, an acoustic polaron of large radius is possible[41] in a one-dimensional system in contrast to a three-dimensional case. This difference is of considerable interest in considering the electrons captured by a medium containing molecules with long hydrocarbon chains.[42]

The applicability of various approximations introduced in describing strong electron–phonon interactions is a complicated question. First, the Hartree-type self-consistent-field approximation, as is well known, cannot be characterized in terms of some expansion of a small dimensionless parameter. It is only justified in certain particular cases as a consequence of some numerical contractions. It may be said that these approximations are only satisfactory when there are large number of phonons, or large T, which are equivalent to it, in the system, since a true classical limit for the phonon equations is obtained in the basis of coherent states. The additional adiabatic approximation used certainly cannot be justified for describing the electron excited states close to the boundary of the continuous spectrum. The technique of finding the excited states with the help of variational methods is still an open question. In order to realize this method, expressions for the wave functions of lower-lying states are essential, which, unfortunately, are not known yet. Of course, in principle, it is possible to estimate indirectly the difference in the energies of the excited and the ground states, but these estimates can hardly be accurate. Often, assuming adiabaticity, the excited states in optical transitions are calculated, using the effective Hamiltonian of the ground state. The validity of this approach is quite justifiably disputed in Refs. 43 and 44. Besides the arguments given therein, we may add that in considering the optical transitions we cannot afford to discard, as is frequently the case, the time-dependent c-number factors of $F(t)$ in (II.7) which vary for different states. It is significant that the theory should qualitatively take into account those effects which arise due to the disordered state of the medium. A consecutive account of these effects and/or of the microscopic interactions, symbolically introduced in (II.2), is a matter for the future in the framework of polaron models.

Since the initial theoretical assumptions are vague, extreme importance is acquired by comparison with the experiment. As already mentioned elsewhere, several attempts have been made to apply directly the large-radius polaron theory, in which the interaction with longitudinal optical phonons is defined by the Fröhlich Hamiltonian, and microscopic interactions are discarded, in describing the solvated electrons. This theory, however, cannot explain several essential properties of solvated electrons. In terms of this theory, it has not been possible to interpret the shape of the optical absorption curve, the temperature dependence of the position of maximum on this curve, the influence of the phase state, and the pressure of the medium.* The main hurdle lies in the absence of correlation between the characteristics of solvated electrons and the permittivities of the corresponding media—the only phenomenological parameter in the polaron theory.[45–50] Nevertheless, attempts were certainly made to modify the polaron model for solvated electrons. Such trends were characterized by attempts to introduce the dependence of the effective mass of polaron on temperature and the density of the medium in order to explain the phenomena observed.[30,31] The reason for the choice of one specific type of dependence or another was left almost unexplained; today these papers can therefore be looked upon as a simple attempt to fit the parameters.

As the experimental data point out that microscopic interactions with characteristic distances of the order of the length of one or more molecules do exert an influence on the characteristics of solvated electrons, considerable effort was expended to introduce these interactions into the formal scheme of the polaron theory. One such path lies in taking into account the spatial dispersion of permittivity (i.e., the dependence of ε and ω on $\mathbf{q}$ in the expression of the phonon Hamiltonian). This naturally causes a change in the type of the form factor $Q_{\mathbf{q}}$ for large q.[51–53] But not all the existing microscopic effects (e.g., the correlation and exchange effects) can be accounted for in this manner. Besides, when the medium contains excess electrons, its function $\varepsilon(\omega, \mathbf{q})$ is unknown, and if this function is defined a priori, the theory has hardly any significance as a predictive tool. Such a definition of $\varepsilon(\omega, \mathbf{q})$ is, in fact, no way different from the choice of the effective one-particle potential for e_s^- made in the semicontinuous models to be described below.

Until recently, serious consideration was not given to the possibility of modifying the polaron theory by introducing nonlinear terms proportional to fourth or higher powers of the phonon field into the Hamiltonian. In introducing these terms, which may be regarded as an attempt to account for the interaction when the dipoles rotate through a large angle from their

* The influence of temperature and pressure on the optical spectra has not been understood fully within the framework of this theory, even for the classical problem of large-radius polaron in ionic crystals.[45] At nonzero temperatures, we cannot generally neglect free phonons [i.e., the first term on the right-hand side of (II.12)]. We shall return to this question once again at the end of Section IV.

equilibrium position, we are faced with a formal problem similar to those presently being investigated in quantum field theory. There is, in this case, an opportunity to localize polarons with spontaneous breaking of translation invariance in the same way as is done in the theory of gauge fields.[54] This localization mechanism may be looked upon as the creation of an essential homogeneity disturbance in polar medium called the "soliton." And it can only vanish by very slow tunneling through the high potential barrier on the relaxation path. Such a defect, if it corresponds to the rotation of several dipoles to mutually repelling positions, will be additionally stabilized by the electron compensating the excess positive charge and localized near it.

C. Semicontinuous Models

The introduction of microscopic interactions into the polaron theory, called the semicontinuous model, is gaining ground in the literature. It has been developed in fuller detail in the works of Jortner, Kestner, and others,[55–60] as well as in the papers similar in approach, published by Kevan and others.[61–65] Since these papers are surveyed in detail in the reviews,[1,2,4] we shall merely confine ourselves to a brief examination of their basic principles.

In order to take account of the microscopic interactions of solvated electrons, in these works the effective potential of polaron model has been modified phenomenologically to small distances. Jortner in a fundamental paper[55] made use of a one-particle Schrödinger equation with a simple centrally symmetric potential well of the type

$$V(r)=\left\{\begin{array}{ll}\dfrac{-e^2}{\varepsilon_{\text{eff}}R} & r<R \\ \dfrac{-e^2}{\varepsilon_{\text{eff}}r} & r\geq R\end{array}\right\} \qquad \varepsilon_{\text{eff}}^{-1}=\varepsilon_{\infty}^{-1}-\varepsilon_0^{-1} \tag{II.27}$$

to describe the solvated electron. In a certain sense, his choice combines the idea of Ogg's cavity with polarization interaction at large distances.

In the course of development, Jortner's model has been subjected to considerable refinement, and the microscopic parameters contained in it have been computed. Such calculations for electrons in ammonia were first reported by Copeland, Kestner, and Jortner.[56] The model was thereafter extended by Fueki, Feng, Kevan, and Christoffersen[61] for electrons in water. In these papers the interaction of the electrons with the molecules of the medium, as in the initial version of theory, is regarded to be adiabatic, centrally symmetric, and different in nature, depending on the distance from the localization center. Moreover, in Ref. 56 particularly, the following are assumed:

1. A fixed number N_s of solvent molecules are supposed to exist in the first coordination sphere forming the dipole "jacket" around the electrons. The case $N_s = 4$ or 6 is regarded to be most significant, and less importance is attached to the case $N_s = 8$ or 12.
2. The solvent molecules in the first coordination sphere interact with the electrons electrostatically just as dipoles do with a constant and induced moments, as well as with other solvent molecules via hydrogen–hydrogen repulsion and dipole–dipole interaction. At short distances the molecules are repulsed like rigid spheres of finite radius.
3. The interaction of the electron with the solvent molecules outside the coordination sphere is regarded to be of the potential type, and is divided into two parts. The first component corresponds to the interaction with the Coulomb-type long-range polarization forces. The second component is a constant potential U, which accounts for the averaged interaction of delocalized electron with the medium, and it includes not only the attraction due to polarization of the electronic shell, but also the repulsion due to the molecular rigid cores. The magnitude of U, which determines the difference in the bottom levels of continuous spectra in the medium and in vacuum, is a fitting parameter.
4. The total energy of the ground state of a system consisting of a medium and a solvated electron also contain other terms which correspond to the formation of the cavity in the solvent and the polarization of the medium. In Ref. 56 these terms are attributed to the following effects: the work done by surface tension E_{ST} needed for the formation of excess surface, dipole repulsion E_{DD} of solvent molecules in the first coordination sphere of the electron, work done against the forces of bulk pressure E_{PV}, and the changes in medium polarization E_{PM}.

Under these assumptions, a stationary Schrödinger equation is derived for the electron with a potential V, having different shapes at three portions corresponding to the cavity proper, the first coordination sphere, and the continuous interaction range, respectively:

$$V(r)=\begin{cases} \dfrac{-N_s\mu e^2}{r_d^2}-\dfrac{e^2}{\varepsilon_{\mathrm{eff}}R} & r<R \\[2ex] -\dfrac{N_s\mu e^2}{r_d^2}-\dfrac{e^2}{\varepsilon_{\mathrm{eff}}R}+U & R<r<r_d \\[2ex] -\dfrac{e^2}{e_{\mathrm{eff}}r}+U & r>r_d \end{cases} \qquad \text{(II.28)}$$

The magnitude of the dipole moment μ in (II.28) is taken in the simplest version of the theory in the form of $\mu = d\langle\cos\theta\rangle$. The mean value of cosine of the angle between the radius vector and the direction of the dipole moment $\langle\cos\theta\rangle$ is determined with the help of the Langevin formula from the theory of paramagnetism:

$$\langle\cos\theta\rangle = \mathrm{ctgh}\chi - \chi^{-1}$$
$$\chi = \frac{de_{\text{bubble}}}{\mathrm{kTr^2d}} \tag{II.29}$$

where e_{bubble} is the electron charge inside the cavity. Thus the potential $V(r)$ varies with temperature not only because of such temperature-dependent parameters as the dielectric constant contained in it, but also due to the temperature-dependent parameter $\langle\cos\theta\rangle$. In the other versions of the theory, the function $\cos\theta$, through which μ is expressed, is not replaced by its mean value directly in the expression for the potential. It is supposed to be an additional dynamic quantum coordinate which fluctuates according to a certain definite law, and averaging is carried out in the final expressions for the optical transition probability. The eigenvalues of electron energy E^e for the potential (II.28) are determined by variational techniques, using simple Coulomb functions with a varying exponent. The total energy E of the system corresponding to a solvated electron is derived by adding the reorganization energy E^m of the medium,

$$E = E^e + E^m \qquad E^m = E_{\mathrm{ST}} + E_{\mathrm{DD}} + E_{\mathrm{PV}} + E_{\mathrm{PM}} \tag{II.30}$$

to the electron energy, where E^m consists of the terms enumerated in item 4 of the initial assumptions. Variations of these terms with the variations in external conditions, say pressure, provide an opportunity to interpret the corresponding experimental dependences. The cavity dimension is determined from the condition

$$\left.\frac{dE}{dR}\right|_{R=R_{\mathrm{eq}}} = 0 \tag{II.31}$$

The validity of the theoretical scheme described above seems to be highly doubtful. In particular, the division of the total influence of the solvent on the state of the solvated electron into the influence of the molecules inside and that of the molecules outside the first coordination sphere is rather enigmatic. It is not yet clear how repeated inclusion of the same interactions is avoided when microscopic potentials and macroscopic effects, such as the surface tension in the cavity, are simultaneously introduced.

Many more general critical remarks to dispute the validity of the semicontinuous model can be made. But the phenomenon under consideration is so complicated that it is, of course, not possible to expect much of the various theoretical approaches that have been simplified due to dire necessity. But an essential question is whether this model can explain the experimental facts at all. Here we are faced with serious handicaps, the most important of which is described in Ref. 59 by the authors themselves, who are actively engaged in studying this model. The theory should, above all, explain the rather large width and the specific shape of the optical absorption curve. In models similar to the one under examination, the large width is explained as follows. The potential surface is assumed to be dependent on n fluctuating characteristics of the medium which are denoted by an n-dimensional vector $\boldsymbol{\lambda}$. It may, for example, contain the cavity radius R and the angle θ introduced above. When the vector $\boldsymbol{\lambda}$ changes, so do the matrix elements of the optical transitions, as well as the initial and the final energy levels. The probability of optical transition for a fixed $\boldsymbol{\lambda}$ can be expressed as

$$w_{fi} \sim |M|^2 \delta\left(E_i(\boldsymbol{\lambda}) - E_f(\boldsymbol{\lambda}) + \hbar\Omega\right) \tag{II.32}$$

Here $|M|^2$ is equal to the square of the modulus of the matrix element summed up over all possible finite states with energy $E_f(\boldsymbol{\lambda})$. It depends on Ω and $\boldsymbol{\lambda}$, but this dependence is neglected as compared with a much stronger dependence contained in the delta function of energy conservation. The excited state f in (II.32) is characterized by the same set of parameters $\boldsymbol{\lambda}$ as for the initial state, because the transition is supposed to take place via the Franck–Condon mechanism without changing the "slow" coordinates of the medium. On averaging over i in the initial state, we obtain the following expression for the probability of optical transition between the states of solvated electron:

$$w(\Omega) \sim \frac{|M|^2}{Z} \int d^n\boldsymbol{\lambda} \exp\left[-\frac{E_i(\boldsymbol{\lambda})}{kT}\right] \delta\left(E_i(\boldsymbol{\lambda}) - E_f(\boldsymbol{\lambda}) + \hbar\Omega\right) \tag{II.33}$$

where Z denotes the statistical sum. In the simplest case, where $\boldsymbol{\lambda}$ is a one-dimensional vector, for instance, the fluctuating cavity radius R, we have

$$w(\Omega) \sim \frac{|M|^2}{Z} \frac{\exp(-E_i(\tilde{R})/kT)}{\left|\frac{d}{dR}\left[E_i(R) - E_f(R)\right]_{R=\tilde{R}}\right|} \tag{II.34}$$

Here $\tilde{R}$ is the root of the equation

$$\hbar\Omega + E_i(R) - E_f(R) = 0 \tag{II.35}$$

which is supposed to be unique. In deriving (II.34), we made use of the well-known relation for the δ-functions:

$$\delta(f(x)) = \sum_m \frac{\delta(x-x_m)}{|f'(x_m)|} \tag{II.36}$$

where summation is carried out over all the m roots of the equation $f(x)=0$. Unduly excessive deviations from the equilibrium value of the radius R_{eq} are "cut out" by means of the exponential factor in the integrand in (II.33). On expanding this integrand in a series near the equilibrium value $R = R_{eq}$, we obtain

$$w(\Omega) \sim \exp\left[-\frac{(\Omega-\Omega_m)^2}{\Delta^2}\right] \tag{II.37}$$

where

$$\hbar\Omega_m = E_f(R_{eq}) - E_i(Req)$$

$$\Delta^2 = \hbar^2 kT \frac{dE_f}{dR}\left(\frac{d^2E_i}{dR^2}\right)^{-1}\Bigg|_{R=R_{eq}} \tag{II.38}$$

Here we have made use of the condition for the existence of minimum energy of the ground state in the equilibrium position. Contrary to the experiment, the expression (II.37) is symmetric with respect to the deviations from the maximum. Of course, this symmetry can be disturbed, say, by introducing complicated potential surfaces, and/or more than one fluctuating coordinate for the solvent. The choice of the potential surfaces is indeed rather arbitrary; nevertheless, some estimates can be made.[58,59] These estimates show that the main cause of the braodening of the absorption line within the framework of this theory should be the fluctuations in the cavity radius. But, for the maximum possible values of the parameters, the line width calculated proves to be at least half the experimental value. The line symmetry calculated is also low and shows that the transitions from the low-energy side are stronger, while the opposite is observed in experiments.

The following situation is also of great significance. Indeed, with the increasing temperature, the half-width of the e_s^- absorption line does not vary [as follows from (II.38)] proportional to $T^{1/2}$. The temperature-dependent increase in the half-width is usually rather weak, and in some cases[66] it even decreases with the temperature growth.

Besides these problems, we can mention one more difficulty. The magnitude of U (i.e., the position of the boundary of the continuous spectrum of electron in solution relative to vacuum) is regarded to be a fitting parameter. In comparing with the experimental results, it is generally taken to be quite high ($U > -0.5\,\text{eV}$). This magnitude, however, can be directly determined by measuring the photoemission at the metal-electrolyte interface.[67,68] It has been determined most reliably for water: $U = -1.3\,\text{eV}$. The negativeness of U in polar media is evidently associated with the nature of the sources of energy gain during cavity formation,* which is a fundamental question.

These difficulties have given rise to the publication of innumerable papers, in which attempts are made to overcome them without changing the semicontinuous model qualitatively. We shall list a few of these trends. Side by side with the $1s \rightarrow 2p$ transition, an attempt[57] was made to take account of the transitions to the higher p-states that cluster near the threshold of the continuous spectrum due to contribution of the Coulomb interaction to the potential. Likewise, some proposals have been made to include even the forbidden transitions of the type $1s \rightarrow 2s$. The authors of these hypotheses themselves, however, note that agreement with the experimental optical absorption curve cannot be achieved in this way.[57] The interpretation that the undue line width results from disruption of the spherical symmetry of the interaction also seems rather improbable.

D. Cluster Models

The difficulties encountered in the theories which interpret the electron–solvent interaction as the interaction of an electron with continual medium have stimulated efforts to search for alternative models in which consideration is given to microscopic interactions of solvated electrons with clusters (i.e., a small number of molecules enveloping the electron). These clusters are investigated, using quantum chemistry techniques. Such

*Cavity formation and electron localization in noble gases are essentially associated with the positiveness of U. The cause of a decrease in the density when electrons are injected into polar media with positive affinity of molecules for electrons so far remains a debatable question. Mott[18] believes that the strong polarization of molecules in water near the solvated electron leads to the formation of structures similar to Bjerrum defects in ice. An energy of the order of 1 eV under normal ice density is required for the formation of such a defect. Decreased density in this case may lead to a gain in energy. Generalizing this argument, we can say that rearrangement of the solvent is possible during electron localization, thereby leading to the formation of a region with an ordered and relatively loose structure. The overlap of such structures under high e_s^- concentrations will cause high conductivity (see Section III).

calculations of e_s^- are sometimes called *ab initio* solvated electron computations. In the language of solid-state physics, these cluster models are close to small-radius polaron models. Because to the disordered structure of the medium, the tunnel transport of electrons is rather difficult here because the equivalent position may be located quite far away. If the characteristic time of any process involving e_s^-, for instance, optical transitions, is less than the characteristic time of site-to-site jumps, the later may be disregarded in studying this process. In other words, an electron may be regarded as fixed on some definite molecular cluster.

One of the first publications along these lines was a report by Raff and Poll.[69] Within the framework of a simplified method of molecular orbitals, they calculated the states of excess electrons in dimers (water and ammonia) for some fixed molecular geometry in dimers. The excitation energy, which they equated to the difference in the energies of the lowest and higher orbitals, was found to be about 2.5 eV for $H_2 0$ and 1.63 eV for NH_3. More sophisticated computation techniques drawn from the modern quantum chemistry theory were applied in subsequent works. The number of solvent molecules, constituting the cluster, was increased and a specific steric symmetry was assigned to the clusters.

Natori and Watanabe[70] studied a model for the hydrated electron localized in a tetrahedral cavity created by removing the central water molecule from the tetrahedron, which is the basic structural element of liquid water and ice crystal.[71] Here two water molecules are rotated by OH= bonds inside the tetrahedron. Consequently, all four water molecules at the apexes of the tetrahedron become equivalent, and the center has the T_d-symmetry. A disputable point in this model is the assumption that the defect is created by the excess electron itself by removing the central molecule from the cell. According to Refs. 72 and 73, where the energy and distribution of electron density of e_s^- over the clusters in water and ammonia have been calculated, electrons are distributed over water molecules and are localized between the ammonia molecules. The optimal H–H distance in clusters was found to be 1.5 and 3.52 Å for water and ammonia, respectively. This result[72] qualitatively explains the cause for the greater expansion of liquid ammonia than water during solvation of electrons.

Byakov, Ovchinnikov and Klyachko,[74] have made a comprehensive study of the quantum chemical model for hydrated electrons, (called the configurational model of e_s^-), in which localization is supposed to take place on the structural defects already present in liquid water. We shall dwell at greater length on these calculations. According to Ref. 74, a defect, on which the solvated electron is localized, is represented by the tetrahedral elementary cell of water itself in which a few hydrogen bonds between the central molecule and the neighboring molecules are ruptured. This type of defect is commonly found in liquid water. A thermalized electron is captured, according to their assumption, at the central molecule of the defect, thereby leading to strong deformation accompanied by changes in the length of OH= bonds and the angle between them. Neighboring molecules in the H_20^- ion field also suffer rearrangement. As a result, the electron energy level is deepened and its localization range is shortened. Thus the $(H_20)_7^-$ cluster consisting of the central ion H_20^- and six water molecules forming its solvated shell is formed. The cluster is supposed to have C_{2v} symmetry.

Just as in other cluster models, in the configuration model of the solvated electron, too, the short-range interactions within the region of defect are regarded to be dominant, and the long-range polarization interactions as insignificant. In Ref. 74, four different possible configurations are assigned to the central ion; their parameters are listed in Table I. The energy of the ground state is calculated in a multipole approximation for the interaction of electrons with the molecules of the first coordination sphere and for the interactions of these molecules with one another. The multiplet expansion is justified on the grounds that the dipole is about one order less in size than the intermolecular distance. The wave function of the electron ground state, containing three variable parameters k, γ, and w, is expressed as

$$\Psi_{\rm O}(r)=\left[\gamma^2+2(1+S_1+2\gamma S_2)\right]^{-1/2}(\gamma\Psi_{{\rm O},2s}(|\mathbf{r}-\mathbf{R}_{\rm O}|)+\Psi_{{\rm H},2s}(|\mathbf{r}-\mathbf{R}_{\rm H1}|)$$
$$+\Psi_{{\rm H},2s}(|\mathbf{r}-\mathbf{R}_{\rm H2}|)) \tag{II.39}$$

where $\mathbf{R}_{\rm O}$, $\mathbf{R}_{\rm H1}$, and $\mathbf{R}_{\rm H2}$ are the radius vectors of the atoms of oxygen and hydrogen, respectively. The variational function was taken to be

$$\Psi_{{\rm O},2s}(r)=\left(\frac{k^5}{3\pi}\right)^{1/2} re^{-kr}$$
$$\Psi_{{\rm H},2s}(r)=\left(\frac{w^3}{\pi}\right)^{1/2}(1-wr)e^{-wr} \tag{II.40}$$
$$S_1=\langle\Psi_{{\rm H},2s}(\bar{\mathbf{r}}-\mathbf{R}_{\rm H1})|\Psi_{{\rm H},2s}(\mathbf{r}-\mathbf{R}_{\rm H2})\rangle$$
$$S_2=\langle\Psi_{{\rm H},2s}(\mathbf{r}-\mathbf{R}_{\rm H1})|\Psi_{{\rm O},2s}(\mathbf{r}-\mathbf{R}_{\rm O})\rangle$$

TABLE I
Configurations of the Central Ion in $(H_2O_7)^-$ Cluster

Configuration type	OH bond length (Å)	OH angle (deg)	Number of energy minima of the ground state
I	0.96	105	21
II	1.01	109.5	22
III	1.08	115	22
IV	1.10	116	21

TABLE II
Energy of Optical Transition for Equilibrium Configurations of $(H_2O_7)^-$ Cluster

Configuration	Transition energy range (eV)
I	1.52–1.95
II	1.24–2.02
III	1.68–1.99
IV	1.81–1.98

The minima of the ground state were found by means of variational techniques. The number of these minima for each configuration is presented in Table I. The value of γ for all cluster configurations is 10 to 20, and it indicates that the spherically symmetric function $\Psi_{O,2s}$ of the central ion centered on oxygen atom makes a major contribution to Ψ_O. The electron localization range, equal to $\sim k^{-1}$, according to calculations, was found to be $\sim$1.3 Å (i.e., it approximately coincides with the size of the water molecule). The typical cluster size is about 3.18 Å. On comparing it with the mean intermolecular distances in liquid water and ice crystal (2.76 Å and 2.9 Å, respectively), it was concluded that the molecular packing in clusters is looser than it is in water.

The energy of the lowest-lying excited state,* calculated by the same procedure, suggests that this state is either weakly bonded or generally delocalized for all the cluster configurations. The difference between this energy and the position of the lowest level in the continuous spectrum does not exceed 0.014 eV. This conclusion is in agreement with the results of analysis presented in Section IV, according to which the optical absorption of a hydrated electron pertains to a transition from the localized state to a continuous spectrum. The energies of optical transitions, calculated in Ref. 74 for the equilibrium configurations of the cluster, are listed in Table II. The experimental absorption maximum (1.72 eV in water) lies within the range found for the transition energies.

The next step followed in Ref. 74 for calculating the shape of the absorption band of a hydrated electron consists in statistical averaging over a set of equilibrium configurations of the cluster. The theory developed earlier for electron transitions in a polar liquid[75,76] underlies the calculations. Since for a continuous spectrum of dielectric absorption in a disordered medium, division of the system into a fast (electron) component and a slow

*Note that the validity of the calculation of excited states in this particular model and in other cluster models is far more difficult to justify than is calculation of the ground state, because of high delocalization of the electron wave function in these states and, consequently, the attendant need to include the interaction of electron with a larger number of solvent molecules.

component which generates inertia polarization is not unique, the probability of transition in this theory was expressed directly through the observed spectrum $\varepsilon(\Omega)$. An essential assumption in the calculation is that the macroscopic (without spatial dispersion) Maxwell equation is supposed to be applicable for a medium. Thus the cross-section of absorption for an l-configuration cluster is expressed in the form

$$\sigma_l(\Omega) = \frac{2\pi d_l^2 \Omega}{3n\hbar c} \int_{-\infty}^{\infty} dt \exp\{i(\Omega - \Omega_l)t + \hbar^{-1} F_l(t)\} \tag{II.41}$$

where

$$F_l(t) = B_l \int_{-\infty}^{\infty} \frac{\varepsilon''(\omega)}{|\varepsilon(\omega)|^2} \frac{\mathrm{ch}(\hbar\omega/2_K T - \omega t) - \mathrm{ch}(\hbar\omega/2kT)}{\mathrm{sh}(\hbar\omega/2kT)} \frac{d\omega}{\omega^2};$$

$$\varepsilon(\omega) = \varepsilon'(\omega) + i\varepsilon''(\omega) \tag{II.42}$$

$\hbar\Omega_l$ and d_l are the energy and dipole moment of the transition for the lth configuration, and

$$B_l = \frac{E_{p,l}}{\pi}(\varepsilon_\infty^{-1} - \varepsilon_0^{-1}) \tag{II.43}$$

Here $E_{p,l}$ is the energy of reorganization of the medium under optical transition. The resultant absorption cross-section is

$$\sigma(\Omega) = \sum_{l=1}^{N} \sigma_l(\Omega) w_l g_l \tag{II.44}$$

$$w_l = \frac{\exp(-E_{0,l}/kT)}{\sum\limits_{l-1}^{N} \exp(-E_{0,l}/kT)} \tag{II.45}$$

where $E_{0,l}$ is the energy of the ground state of the lth configuration and g_l is its degeneration factor.

The dielectric loss spectrum used in the calculation is of the form

$$\varepsilon(\omega) = \varepsilon_\infty + \frac{\varepsilon_0 - \varepsilon_1}{1 + i\omega\tau_d} + \frac{\varepsilon_1 - \varepsilon_\infty}{2}\left(\frac{1 - i\omega_r\tau_r}{1 - i(\omega + \omega_r)\tau_r} + \frac{1 + i\omega_r\tau_r}{1 - i(\omega - \omega_r)\tau_r}\right) \tag{II.46}$$

with parameters derived from a comparison with experiment.[77] The second term in (II.46) determines the Debye losses (at 300°K, $\varepsilon_0=78$; $\varepsilon_1=4.9$; $\tau_d=8.5\times10^{-12}$ sec), while the third term expresses the resonance effects ($\omega_r=6.67\times10^{-12}\,\text{sec}^{-1}$, $\tau_r=3.84\times10^{-14}$ sec). The temperature-dependent variation of τ_d was described by the Stokes–Einstein equation:

$$\tau_d=\frac{4\pi\eta}{kT}a^3 \tag{II.47}$$

where η is the viscosity of water.

The general shape of the envelope of absorption bands of the hydrated electron calculated in this model is in satisfactory agreement with the experimental curve plotted in the temperature range −50 to +100°C. At higher temperatures the agreement worsens, although the calculations give a general picture of the spectrum pattern, in particular the shift of the maximum with temperature. The physical reason for the temperature shift of maximum is that the contribution from the initial higher-energy states having high degeneracy increases with the increasing temperature. In the experiments, however, no resolutions were observed in the absorption band corresponding to the discrete structures of the initial states participating in optical transitions.

The ESR spectrum of hydrated electrons has also been calculated within the framework of the configurational model.[74] The main results are almost the same for all the 14 protons and are approximately equal to 4 gauss; that is, despite the spatial nonequivalence of the molecules, all the protons in the cluster are equivalent in ESR spectrum. The electron density distribution is almost symmetrical relative to the protons of the central ion. The anisotropic constant of dipole–dipole superfine interaction for the protons of this ion is close to zero and is about 3 gauss for the protons of solvated shell. A quantitative comparison of these results with experiment is not possible because of the high anisotropy of the superfine interactions that determines the significant part played by the forbidden transitions. We can only say that the width of the envelope does not conflict with the experiment, and the experimental results obtained for double electron–nuclear resonance suggest a cluster size close to the value calculated in the configurational model.[78]

Cluster calculations of solvated electron, despite the advances made, call into question their effectiveness because of the complexity, the use of innumerable approximations, and serious hardships in describing the excited states. Simple models of potential wells with short-range interaction[79,80] are therefore being examined of late to describe the spectra. In this manner one may attempt to account phenomenologically for the influence of the interactions of solvated electrons with molecular clusters, as well as with the collective excitations of the medium. This approach will evidently give satisfactory results only in those cases where the influence of complicated chemical interactions can be expressed in terms of a small number of parameters with simple physical meaning.

III. TRANSPORT PROPERTIES

A. Nonpolar Dielectrics

At present a highly comprehensive microscopic theory of mobility is only available for electrons in liquid and dense gaseous helium.[7,9,20–25,81–83] The results of this theory are generally used in the qualitative analysis of the transport properties of electrons in other disordered media. In helium various mobility mechanisms are realized depending on the density of the medium. For a gas density $n<n_c$, where n_c is the density corresponding to possible localization* of electrons in cavities, and if the inequality

$$n^{-1/3} \gg \lambda = \left(\frac{\hbar^2}{2mkT} \right)^{1/2} \tag{III.1}$$

is satisfied for the de Broglie wave λ of electrons, the mobility of excess electrons is determined by the scatter of the electron wave at separate molecules in the medium. In this case,[81] the mobility can, as is known, be described by the expression

$$\mu = \frac{3e}{32nf_0^2(2\pi mT)^{1/2}} \tag{III.2}$$

where f_0 is the amplitude of scattering of electrons with zero energy at separate gas molecules.

The inequality (III.1) begins to be violated with increasing gas density. And the concept of scattering at separate gas molecules loses its sense when $\lambda^{-3}<n<n_c$. Nonetheless, use may be made of the so-called "optical potential" method[84] to calculate the mobility of electrons in helium in this range. The electron-medium interaction is expressed as follows:[81]

$$u = \frac{2\pi\hbar^2}{m} a_o n \tag{III.3}$$

where a_0 is the length of scattering of electrons at a gas molecule, and scattering is supposed to occur during potential fluctuations δu:

$$\delta u = \frac{2\pi\hbar^2}{m} a_0 \delta n \tag{III.4}$$

*The theory of electron localization at cavities in helium is described, e.g., in Refs. 20–25.

corresponding to the density fluctuations δn. The interaction (III.4) is formally similar to the deformation potential used in the theory of electrons in semiconductors,[85] the only difference being in the physical meaning of the coefficients. We can therefore directly apply the results of the theory of electron–phonon interaction in semiconductors, and thus obtain the following expression for the mobility:

$$\mu = \frac{g}{8\pi}\left(\frac{C_p}{C_v}\right)\frac{e}{na_0^2(mT)^{1/2}} \tag{III.5}$$

Surprisingly, the dependence of the mobility on density and temperature in (III.2) and (III.5) are the same, although the physics of the phenomena are totally different. This coincidence of the functional dependences is probably the cause of the astonishing fact that (III.2) satisfactorily describes the experimental results over a wide density range, right up to $n \sim 5 \times 10^{20}\,\mathrm{cm}^{-3}$.[86] At such densities condition (III.1) is evidently disrupted, and in reality the mobility is determined by the interaction of electrons with the collective vibrations of the medium (i.e., acoustic phonons).

At $n > n_c$, almost all the electrons in the medium are localized in cavities. The mobility of these localized electrons depends, in the first approximation, on the Stokes diffusion of cavities:

$$\mu = \frac{e}{4\pi\eta R_e} \tag{III.6}$$

where η is the viscosity of gas, and R_e is the equilibrium radius of the cavity. The magnitude of R_e is estimated with the help of the theory of localization of electrons in helium mentioned above.

Certain difficulties have been encountered in extending this theory to other gases, such as hydrogen, neon, krypton, and gaseous hydrocarbons. These difficulties arise because of the need to take account of the relatively strong electron-medium interaction and because the short-range interaction with the medium cannot be described here by (III.4). This results from the fact that the low-energy cross-section of scattering of electrons at hydrocarbon molecules (of even low molecular weight) even at about 1 eV is strongly dependent on the electron momentum. In particular, the Ramsaver–Townsend effect[87] takes place in this energy interval. The simple optical potential theory has not yet been generalized to this case, and it is not possible to study the electron properties using the general pseudopotential theory[88,89] because of the need to account for the disorderedness of the medium.

Several workers[20–25,81–83] have already attempted to account for the polarization interaction. Apparently, for hydrogen it is sufficient to elucidate the influence of polarization on the effective magnitude of a_0 in (III.3) as has been done, say, in Ref. 25. Further calculations are to be made just as in the case of helium. For hydrocarbons, however, there is an additional need to take into consideration the nonsphericity and other physical details of the molecular geometry, because some experimental evidence is available to show that the electron mobilities in isomers with different degrees of branching of carbon chains are different.[90]

Because of the lack of a mathematically based microscopic theory of mobility of excess electrons in hydrocarbons, as well as in noble gases at densities $n \sim n_c$, a phenomenological partial localization theory was proposed to explain the experimental data. This model is based on the concepts of the "percolation theory" in inhomogeneous media.[7,91-93] It is assumed that the whole volume can be divided into "allowed" and "forbidden" regions. In the allowed region electrons are delocalized and possess high band-type mobility, whereas in forbidden regions only those states are possible in which electrons are localized on the molecules or clusters of molecules of the medium and have far lesser mobility. Tunneling of electrons from one allowed region to another is neglected. If the potential energy of the electron $V(\mathbf{r})$ corresponds to some region, then for electrons with energy E this region will be allowed one if $E > V(\mathbf{r})$, and forbidden if $E < V(\mathbf{r})$. The total allowed volume for electrons with energy E is determined as

$$C(E) = \int_{-\infty}^{E} p(V)\,dV \tag{III.7}$$

where $p(V)$ is the probability that the potential energy $V(r)$ in the volume lies between V and $V+dV$. When $C(E) > C^*$, where C^* is a certain critical value, the allowed regions are assumed to form a connected system with nonzero total conductivity. The threshold of conductivity energy E_c is determined from the relationship

$$C(E_c) = C^* \tag{III.8}$$

in an obvious way.

The observed mobility is supposed to be equal to

$$\mu = \mu_f h(E) \tag{III.9}$$

where μ_f is the mobility in allowed regions, and the function $h(E)$ is zero when $E < E_c$ and is equal to $0 < h < 1$ when $E > E_c$. Kirkpatrick[94] had experimentally found that $C^* = 0.1$ to 0.3 and $h(E) = (C - C^*)^\gamma$, where $\gamma = 1.5$ to 1.6. For continuous media C^* is often taken to be equal to 0.18. In this model, self-localization of electrons via the interaction with the medium is disregarded, and the electron states are supposed to be functions of the state of the medium which fluctuates in an independent manner.

On the basis of the concepts of the "percolation theory," Bruggeman[95] and Londaner[96] have proposed a theory of conductivity for alloys composed of components with different electron mobilities μ_0 and μ_1. For the effective electron mobility, they assumed the following expression:

$$\mu=\mu_0\left\{A+\left[A^2+\tfrac{1}{2}X\right]^{1/2}\right\} \tag{III.10}$$

where
$$X=\frac{\mu_0}{\mu_1}$$
$$A=\tfrac{1}{2}\left[\left(\tfrac{3}{2}C-\tfrac{1}{2}\right)(1-X)+\tfrac{1}{2}X\right]$$

Here C is the volume occupied by the medium with the conductivity μ_0. With the help of (III.10), Kestner and Jortner[97] succeeded in satisfactorily describing the metal-insulator transition in sodium-ammonia solutions, assuming that μ_0 and μ_1 correspond to the conductivities of "good" and "bad" conducting regions in concentrated $Na-NH_3$ solutions. They also applied (III.10) to the case of electrons in liquid hydrocarbons, assuming that μ_0 corresponds to the mobility in allowed regions, and μ_1 to that of forbidden regions where electron localization is possible. Moreover, they assumed, that, first, all the electrons are monoenergetic with an energy E, and second, that the distribution $p(V)$ is Gaussian; hence

$$C(E)=(2\pi\Gamma^2)^{-1/2}\int_{-\infty}^{E} e^{-(V-\bar{V})^2/2\Gamma^2}\,dV \tag{III.11}$$

where Γ is the distribution width. The mean potential was taken in the form

$$\bar{V}=\bar{V}_0+\bar{V}_p+\bar{V}_d \tag{III.12}$$

where $\bar{V}_0$, V_p, and V_d are the mean (over the volume) interactions of electrons with the medium under short-range repulsion of electrons from molecules, under polarization attraction, and under dipole repulsion between the molecules of the medium, respectively. The work function was determined as

$$U_0=\bar{T}_0+\bar{V}_p+\bar{V}_d \tag{III.13}$$

where $\bar{T}_0$ is the mean kinetic energy of the electron's motion in a field of uniformly distributed repulsive centers. This energy was estimated in the same way as for noble gases. In the simplest case of "optical potential," it was found that $\bar{V}_0 = \bar{T}_0$. Under these assumptions the fraction of the admissible volume is given by the expression

$$C(E_t) = (2\pi)^{-1/2} \int_{-\infty}^{(E_t - U_0)/\Gamma} e^{-z^2/2} dz \tag{III.14}$$

The fitting parameters of the theory are μ_0, μ_1, Γ, and E_t, which are assumed to be the same for all hydrocarbons. A comparison of the theory with experiment[98] has shown that the predicted U_0-dependence of μ is satisfactory for a number of hydrocarbons when the electron conductivity varies by four orders of magnitude within reasonable values of the fitting parameters.[7] A study of mobility in mixtures containing *n*-hexane and neopentane has, however, shown that the optimal values of U_0 are not in agreement with its experimental values[99] if the theoretical curve is fitted to the experimental curve by varying the work function U_0. This discrepancy suggests that either the model is unsatisfactory or the parameters μ_0, μ_1, γ, and E_t are not in fact the same for all the hydrocarbons, which means the phenomenological theory has no longer any advantages.

It was assumed in Ref. 100, which is similar in its approach to Ref. 98, that $\mu_0 \gg \mu_1$, $\Gamma^2 = kT^2C_V$, where C_V is the specific heat capacity of the medium. The probability of delocalization was determined by the same expression (III.11). The value of E_t was compared with the energy of the localized state. Thus the number of fitting parameters is reduced to only one (μ_0). Of course, agreement with experiment is somewhat worsened; nevertheless, it is quite acceptable, except for the same experiments with *n*-hexane–neopentane mixtures. Note that the data obtained for the conductivity of methane ($\mu = 100\,\text{cm}^2/v\cdot\text{sec}$, $U_0 \approx 0$) are not in agreement with any of the expressions proposed for the U_0-dependence of μ.

In addition to the models based on the ideas of the percolation theory, the two-level model is a simple scheme generally used to interpret the kinetic characteristics of excess electrons in a disordered condensed medium. In this model two electron states are supposed to exist with equilibrium number of the population. The upper-level state is supposed to be of band nature and the observed mobility of electron μ is defined by the expression

$$\mu = \mu_f \frac{c_f}{c_l + c_f} \tag{III.15}$$

where μ_f is the mobility in delocalized state, and c_f and c_l are the equilibrium concentrations of delocalized and localized electrons, respectively. Sometimes (III.15) is expressed as

$$\mu = \mu_f \frac{\tau_f}{\tau_l + \tau_f}$$

where τ_f and τ_l are the lifetimes of an electron in delocalized and localized states, respectively. The fitting parameters in this model are μ_f and the "activation energy" E_a, which defines the ratio between c_f and c_l, equated to the difference of free energies of delocalized and localized states. This simple model is quite adequate to interpret the data on the kinetics of electron capture by acceptors.[101]

B. Mobility of Electrons in Disordered Polar Dielectrics

While definite advances have been made in the microscopic theory to explain the transport properties of electrons in nonpolar liquids, the mechanism of conductivity in polar liquids is still obscure. Few experimental data are available on the temperature dependence of conductivity and on the influence of disorderedness. It is not possible to draw a final conclusion solely on the basis of this experimental evidence. We can only speculate that the transport of excess electrons may take place either via excitation of electrons from a localized state into "conduction band," by means of jumps over clusters (in accordance with the small-radius polaron theory),[102,103] or through diffusion of the whole clusters. Recently, a paper was published[104] whose main aim was to develop a program for investigation of a possible scheme to calculate the transport of electrons in polar media from the microscopic point of view. Spreading of an electron packet initially localized near a center with time is investigated in this paper. Some vacancy, trap, individual molecule, or cluster can act as this center. The expression derived for the current contains two terms, of which one corresponds to diffusion of the center of initial electron localization, the second to electron transfer in a "frozen medium." The second type of effect is estimated by means of the techniques of the small-radius polaron theory for a disordered medium. The formula thus obtained allows us, in principle, to calculate the conductivity if the nature of the medium, electron characteristics of the clusters, vacancies, and other parameters are known beforehand.

At present, it is rather problematic to conduct this program, and various types of phenomenological theories have to be used in treating and interpreting the experimental results.

C. Theory of Reactivity of Solvated Electrons

With the development of the theory of electron mobility in liquid dielectrics, the concepts of the microscopic explanation of electron capture by an acceptor have also undergone changes. Fuller experimental information on this process was obtained in the studies on the formation of O_2^- in the reactions with oxygen dissolved in hydrocarbons,[11,105,106] liquid hydrogen,[107] and in noble gases.[81,83,108] The initial theory of electron capture by an acceptor was developed on the assumption that only one electron state exists, characterized by a certain mobility and a rate constant for the reaction with the acceptor. These kinetic curves were constructed in the usual manner by solving the diffusion equation under appropriate boundary conditions. Although the time and acceptor concentration dependences of the relaxation current curves calculated in this approach are in many cases in satisfactory agreement with experiment, the values of optimal parameters (mobility, activation energy, etc.) often conflict with their experimentally determined values.[7,11] This situation has stimulated interest, on the one hand, to work out a consistent theory to explain the chemical reactions of localized electrons with due regard for tunneling effects, and, on the other hand, to develop a hypothesis to interpret the reactions of electrons with two types of states, of which one is delocalized. A solid foundation for the chemical reactions of delocalized electrons in polar media was laid after it was possible to make picosecond and nanosecond measurements of flash pulses which permit measurement of these reaction rate constants. Another tool for measuring these rates is to apply the photoelectrochemical techniques.

Frankevich and Yakovlev[11] were the first to investigate the two-state model to explain the reaction of electrons with an acceptor. They assumed that the electron reacts both in localized and in delocalized states. Later, other papers were published in which the formal diffusion description of electron capture and neutralization are generalized with due regard for the two possible electron states. In these works (see Ref. 109), instead of one diffusion equation, a system of two differential equations is written corresponding to different electron states, and the equations are interrelated by means of some equilibrium constant. It is rather difficult to solve such equations exactly. The usual simplification is to assume that the mobility (diffusion coefficients) of electrons in a localized state is far less than the mobility (diffusion coefficient) of a quasifree electron. Under this assumption it is possible to preserve the functional shapes of the relaxation curves, and the magnitudes of fitting parameters become more natural. This approach has been applied in Ref. 106, where the generalized Noyes formula[110] has been used for the effective reaction constant:

$$K_{\text{eff}} = \frac{pk_l + (1-p)k_f}{1 + \dfrac{pk_l + (1-p)k_f}{R_r D_{\text{eff}}}}$$

where p is the probability of localization; k_l and k_f are the rate constants of reactions with the acceptor in localized and in quasifree states, respectively; D_{eff} is the effective electron diffusion coefficient; and R_r is the reaction radius. They succeeded in explaining the observed monotonicity of

$K_{\rm eff}$ vs. μ (electron mobility) curves for various hydrocarbons. Qualitatively, it is explained as follows. For hydrocarbons having low μ, the reaction proceeds in a diffusion regime and the rate constant is proportional to the mobility. With increasing mobility, electron delocalization becomes greater and greater, thereby leading to a decrease in the rate constant, provided that the reaction with the acceptor in the localized state takes place much faster than in the delocalized state. Using this model, the dependence of the effective activation energy of the reaction on the effective electron mobility in hydrocarbons has also been explained.[100]

IV. ANALYSIS OF OPTICAL SPECTRA OF SOLVATED ELECTRONS WITH THE HELP OF NONMODEL RELATIONSHIPS

A. The Sum Rules and Virial Theorem

Optical dispersion relationships and the sum rules that follow from them have since long been in use in nuclear and atomic spectroscopy, in the study of systems with a large number of electrons included.[111] In recent years new sum rules for the optical constants of condensed media have been derived by combining the dispersion relations and the "superconvergence" rules. Some of them have already been verified experimentally.[112b] The sum rules are especially effective in studying solvated electrons, for the following two reasons. First, the whole absorption spectrum of excess electron is experimentally known for many cases, and it ends in the near-ultraviolet or a much lower frequency range. This spectrum is hardly overlapped by the intrinsic absorption of the solvent and has a distinct threshold.*

This situation is a direct consequence of the fact that the bond energy of solvated electron (and consequently the characteristic absorption frequencies) is much less than the bond energy (absorption frequency) of the molecular electron of the medium. Hence there arises the possibility of applying the sum rules in deriving the relationships for variations of optical constants of the medium when an excess electron is injected. This is the system for which, probably, the new optical sum rules and their generalizations may be most effectively applied. Second, the sum rules contain only the characteristics of the ground state. This fact is quite important, because, as follows from the previous sections, hardly any reliable information about the excited states is so far available.

* Additional changes are, in principle, possible in the absorption of the medium in infrared and ultraviolet regions. These absorption changes may be associated with the shift in the frequency of intrinsic vibrations of the medium due to the interaction of charged particles with the molecules of the medium, or may be caused by new local vibrations near e_s^- (similar to the polaron–phonon coupled states in ionic crystals[38,39,115]). So far, however, there is no experimental evidence to prove the existence of perceptible effects of this type for the system under consideration (see, however, Refs. 117–119).

We shall first take up the sum rules which are usually termed "f-sums" in literature. These rules follow from the Kramers–Kronig relations for the real $n(\Omega)$ and imaginary $k(\Omega)$ parts of the complex refractive index of a medium without electron conductivity:

$$n(\Omega)-1=\frac{2}{\pi}P\int_0^{\infty}\frac{\Omega' k(\Omega')}{\Omega'^2-\Omega^2}d\Omega'$$

$$k(\Omega)=-\frac{2\Omega}{\pi}p\int_0^{\infty}\frac{n(\Omega')-1}{\Omega'^2-\Omega^2}d\Omega' \tag{IV.1}$$

Besides (IV.1), use is also made of a general condition: namely, in the high-frequency region a medium behaves like a free electron gas:

$$n(\Omega)-1 \underset{\Omega\to\infty}{\to} -\frac{1}{2}\frac{\Omega_p^2}{\Omega^2} \tag{IV.2}$$

$$\Omega_p^2=\frac{4\pi N^2}{m}$$

where N is the number of electrons per unit volume and m is their vacuum mass. As regards $k(\Omega)$, it is assumed that when $\Omega\to\infty$, the magnitude of $k(\Omega)$ tends to zero faster than Ω^{-2}:

$$k(\Omega)=o(\Omega^{-2}) \qquad \Omega\to\infty \tag{IV.3}$$

Equations (IV.3) and (IV.2) underlie the so-called conditions of superconvergence. They are applied in studying the limits of dispersion relationships of the type (IV.1) when Ω tends to infinity or zero. In this case, in addition,[112–114] we can derive the following relations:

$$\int_0^{\infty}\Omega k(\Omega)\,d\Omega=\frac{\pi}{4}\Omega_p^2 \tag{IV.4a}$$

$$\int_0^{\infty}\Omega k(\Omega)[n(\Omega)-1]\,d\Omega=0 \tag{IV.4b}$$

$$\int_0^{\infty}[n(\Omega)-1]\,d\Omega=0 \tag{IV.4c}$$

$$\int_0^{\infty}\Omega n(\Omega)k(\Omega)\,d\Omega=\frac{\pi}{4}\Omega_p^2 \tag{IV.5}$$

The first two relationships are valid even for a medium with electron conductivity. Since $[n(\Omega)-1]$ is positive at low frequencies and negative at high frequencies, from a combination of these two relations, due to the positiveness of $k(\Omega)$, it follows that any change in $k(\Omega)$ at high frequencies should entail a change in $k(\Omega)$ at low frequencies, and vice versa. The relationship (IV.6c) is modified in the presence of electron conductivity due to the singularity of $\varepsilon=(n+ik)^2$ when $\Omega=0$. Thus we can obtain

$$\int_0^\infty \{[n(\Omega)-1]^2-k^2(\Omega)\}\,d\Omega=-2\pi^2\sigma \tag{IV.6}$$

where σ is the electron conductivity at zero frequency.

When excess electrons are injected into the system, $n(\Omega)+\Delta n(\Omega)$ and $k(\Omega)+\Delta k(\Omega)$ are to be substituted for $n(\Omega)$ and $k(\Omega)$, respectively. On combining the relationships (IV.4) to (IV.6) for optical constants, before and after injecting excess electrons, we obtain the following equations:

$$\begin{gathered}\int_0^\infty \Omega\Delta k(\Omega)\,d\Omega=\frac{\pi e^2N_e}{m}\\ \int_0^\infty \Delta n(\Omega)\,d\Omega=0\\ \int_0^\infty \{\Delta n(\Omega)[n(\Omega)-1]-\Delta k(\Omega)k(\Omega)+\tfrac{1}{2}[\Delta n^2(\Omega)-\Delta k^2(\Omega)]\}\,d\Omega=-\pi^2\sigma\end{gathered} \tag{IV.7}$$

Of special use is the first equation, which contains the concentration of excess electrons N_e, because, owing to the rapid decrease of $\Delta k(\Omega)$, integration is actually limited by the frequency range accessible to observation. The second relationship in (IV.7) is valid in the absence of electron conductivity, both before and after the injection of excess electrons. The third of these relationships accounts for the possibility of electron conductivity when excess electrons are injected. The last two of these relationships can be used to check the consistency of the results, and also to verify the possibility of generation, in addition to localized electron states, of conduction electrons in systems where the lifetime of excess electrons is very short and direct measurement of conductivity is difficult.* In such cases, however, it is necessary to measure $n(\Omega), k(\Omega)$ in a wide frequency range. This is, incidentally, possible today, especially if use is made of synchrotron radiation in ellipsometric measurements.†

*The conductivity due to the injection of solvated electrons into a dielectric medium can be measured not only with the help of (II.7), but also by the behavior of $\Delta k(\Omega)$ under low Ω. In this case the analytical function $\sigma(\Omega)$ has a pole at $\Omega=0$, and it can be distinguished from the experimental absorption spectrum. Pole-type behavior of $\sigma(\Omega)$ with a pole at $\Omega=i\tau$ situated on the imaginary axis and determined by decays should also be true for electrons moving in a field with large-scale fluctuations of density or other characteristics of the medium. Attempts have been made in Ref. 120 to apply the formulas following from the pole-type behavior in investigating the behavior of electrons in picosecond and nanosecond radiolysis.

†Note that the derivation of (IV.4) to (IV.7) is rather general. In fact, use is made of only dispersion relationships associated with the casuality principle and the asymptotic behavior of optical constants at high energies. This high-energy behavior is uniquely associated with the behavior of an electron system within a short time of the switching on of perturbations.[112b]

If $\Delta k(\Omega)$ is nonzero only in the transmission band of the medium, then we may approximately take that

$$\Delta k(\Omega) \approx \frac{1}{2n} \Delta \operatorname{Im} \varepsilon(\Omega) \tag{IV.8}$$

where n is the refractive index of the medium taken to be approximately constant in the frequency range which is significant, and $\Delta \operatorname{Im} \varepsilon(\Omega)$ is the change in the imaginary part of the complex permittivity after the injection of solvated electrons. For $\Delta k(\Omega)$, additional sum rules are derived with the help of (IV.8) from the sum rules already known for $\varepsilon(\Omega)$.[112–114]

We shall express the sum rules to be used subsequently in terms of the experimentally measurable extinction coefficient γ of the solvated electron. It is expressed as

$$\gamma = \ln 10 \frac{2\Omega \Delta k(\Omega)}{cN_e} \tag{IV.9}$$

where N_e is the number of solvated electrons (in moles) per liter and c is the velocity of light in vacuum.

From the dispersion relationships for optical constants and the sum rules for $\varepsilon(\Omega)$, we obtain, in particular:

$$\begin{aligned}
S_{-2} &= \int_0^\infty \gamma(\Omega)\Omega^{-2}\,d\Omega = A\frac{\alpha}{c} \\
S_{-1} &= \int_0^\infty \gamma(\Omega)\Omega^{-1}\,d\Omega = A\frac{2e^2}{3n\hbar c}\langle i|\mathbf{\hat{r}}_s^2|i\rangle \\
S_0 &= \int_0^\infty \gamma(\Omega)\,d\Omega = A\frac{e^2}{mc} \\
S_1 &= \int_0^\infty \gamma(\Omega)\Omega\,d\Omega = A\frac{2e^2}{3mn\hbar c}\langle i|\hat{T}|i\rangle \\
A &= 6.02\times 10^{20}\times 2\pi^2 \ln 10
\end{aligned} \tag{IV.10}$$

The expression on the right-hand side of (IV.10) contains the statical polarizability denoted through α and the mean quadratic radius

$$\langle i|\mathbf{\hat{r}}_s^2|i\rangle = \frac{\Delta\langle i|\hat{P}^2|i\rangle}{2N_e e^2} \tag{IV.11}$$

where $\mathbf{\hat{P}}$ is the polarization operator, and the mean kinetic energy,

$$\langle i|\hat{T}|i\rangle = \frac{\Delta\langle i|\hat{\mathfrak{T}}|i\rangle m}{2N_e e^2} \tag{IV.12}$$

where $\hat{T}$ is the kinetic-energy operator and $\hat{\mathfrak{J}}$ is the current operator. The means (IV.11) and (IV.12) are interrelated not only with the motion of intrinsic excess electron, but also with changes in the behavior of the medium on the injection of excess electrons. In the approximation, in which the motion of the solvated electron is described by the effective one-particle Hamiltonian, the right-hand side of (IV.11) is equal to the simple mean of the square of the solvated electron coordinate **r**, while the right-hand part of (IV.12) is equal to the mean of $\mathbf{p}^2/2m_e$, where **p** is the momentum and m_e is the effective mass of the solvated electron. Here note should be taken of the correction which accounts for the difference in the external wave field $\mathfrak{E}$ and the microscopic field acting on the electron. The relationships (IV.1) and (IV.4) were used to derive the formulas for S_{-2} and S_0. These relationships are exact and are not associated with the assumption that n is approximately constant in the frequency interval in which $\Delta k(\Omega)$ is nonzero. By virtue of the constancy of $n(\Omega)$, we can take n out of the sign of integration in the expressions of S_{-1} and S_1.

Finally, we shall make use of the fact that the functional type of the high-frequency behavior of the extinction coeffiecient γ is known. This coefficient coincides, up to the numerical factor, with the total cross-section of dipole transition from the localized s-state to continuous spectrum; and its asymptotic behavior in the high-energy limit is known.[121] Using this behavior, we obtain

$$\gamma \to \beta\Omega^{-7/2} \tag{IV.13}$$

where β is weakly dependent of Ω and is determined by the type of the matrix element of transition to continuous spectrum. By means of (IV.13), we can discriminate the main range of change in (IV.1) by choosing some such sufficiently large frequency Ω so that (IV.13) can be applied approximately. In particular, from (IV.1) we find that

$$\int_0^{\Omega} \gamma(\Omega')\,d\Omega' = A\frac{e^2}{mc} - \int_{\Omega}^{\infty} \gamma(\Omega')\,d\Omega' = A\frac{e^2}{mc} - \frac{2}{5}B\Omega^{-5/2} \tag{IV.14}$$

Note that this formula (IV.14) cannot be obtained in a general form for $k(\Omega)$ in an arbitrary system, because, in contrast to $n(\Omega)$, the asymptotic behavior $k(\Omega)$ can be different depending on the characteristics of the initial state. Relationships similar to (IV.14) provide an opportunity to verify the consistency of optical measurement results.

We shall illustrate the direct application of these formulas in the analysis of spectra of solvated electrons, by determining the extinction coefficients, as done in Ref. 121, at the maximum of optical absorption of solvated electrons for several substances with the help of (IV.10). Table III lists the results derived with the help of e_s^- optical absorption curves that follow from the relationship (IV.10) for S_0:

$$\gamma_m = A\frac{e^2}{mc}\left(\int_0^\infty \frac{D(\Omega)}{D_m}\,d\Omega\right)^{-1} \tag{IV.15}$$

where γ_m and D_m are the extinction coefficient and the optical density of e_s^- at the absorption maximum, respectively. For the sake of comparison, this table also lists the results of the usual measurements of γ_m, for which the concentration of solvated electrons is necessarily to be determined in an independent chemical experiment.

Table III shows that the sum rules can be applied to many cases with sufficient reliability, limiting ourselves only to the data relating to the optical frequencies. The expression (IV.15) can also be applied to determine the extinction coefficients and, consequently, the concentration of e_s^- in many particular systems only from the optical absorption data with an accuracy sufficient for several applications. Incidentally, the expression (IV.15) was used, for example in Ref. 122 to verify the correctness of the extinction coefficient determined for several solvents.

The agreement of the expressions for the oscillator strength corresponding to (IV.10) within the limits of experimental error allows us to draw the important conclusion that the absorption band of solvated electrons includes all the basic transitions, including the transition into continuous spectrum.

An analysis of the high-energy tail in the absorption spectrum may serve as a criterion for verifying the presence of the transition to continuous spectrum. Figure 2 shows the results obtained from the curves of optical absorption of solvated electrons in water and ammonia with the help of

TABLE III
Comparison of Experimental and Calculated Absorption Coefficients

Substance	$\gamma_m \times 10^{-4}$ liter/mol·cm		Mean distribution
	Experiment	Theory	radius (Å)
Water at 25°C	1.84±0.36	2.3±0.2	2.2
Ammonia	4.92±0.14	5.4±0.3	3.3
Tetrahydrofurane	4.0	4.0	–
Methyltetrahydrofurane	3.9	3.9	–
Diethyl ether	3.5	3.5	–

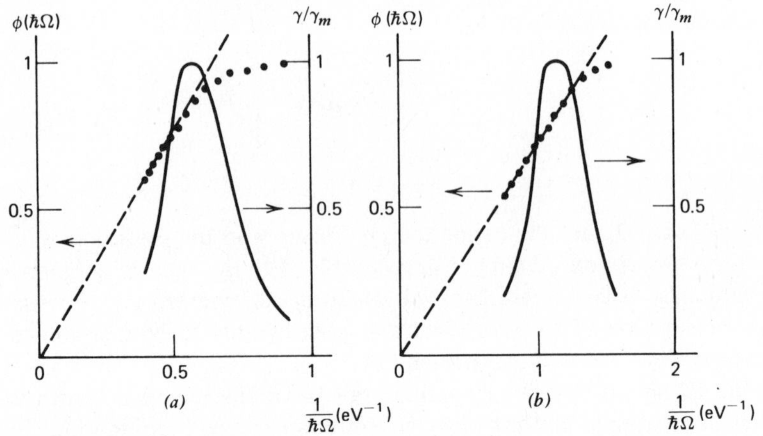

Fig. 2. Spectral characteristics of solvated electrons in water (*a*) and ammonia (*b*). (From Ref. 6.) Thick curves are the absorption spectra of solvated electrons; dots denote the function $\varphi = \left[1-(mc/Ae^2)\int_0^{\Omega}\gamma(\Omega')\,d\Omega'\right]^{2/5}$. According to (IV.14), the points should lie on a straight line in (φ, Ω^{-1}) coordinates as $\Omega^{-1}\to 0$.

(IV.14). It follows from this figure that the high-energy asymptote is indeed in satisfactory agreement with (IV.14).

The values of different physical characteristics of solvated electron can be found from (IV.10); in particular, the mean radius of charge density distribution in the initial state can be derived from the expression for S_{-1}. The results thus obtained are also presented in Table III. Note that the values of the radius of e_s^- in water and ammonia are consistent with the kinetic collison radius[8] estimated from independent experiments.

It is interesting to compare the expression for S_1 with the model theories in which the electron is supposed to be described by a one-particle Hamiltonian with a certain centrally symmetric potential well. Consider a potential of the type (II.27). From the generalized virial theorem for the initial s-state in such a potential well, we obtain

$$0=\langle i\left|\left[-i\hbar\left(\frac{\partial}{\partial r}+\frac{1}{r}\right)r, H\right]\right|i\rangle$$

$$=2i\hbar\langle i\left|\frac{p^2}{2m_e}-\frac{e^2}{\varepsilon_{\text{eff}}r}\theta(r-R)\right|i\rangle \tag{IV.16}$$

in which the mean value of the commutator of the operator $-i\hbar(\partial/\partial r+r)r$ with a Hamiltonian in the stationary state has obviously been taken to be zero. From (IV.16) and (II.27) it follows that the initial energy E_i should satisfy the equality

$$E_i = \langle i|H|i\rangle = -\left\langle i\left|\frac{p^2}{2m_e}\right|i\right\rangle - \frac{e^2}{\varepsilon_{\text{eff}}R}\langle i|\theta(R-r)|i\rangle \qquad \text{(IV.17)}$$

When $R=0$, the expression (IV.17) is transformed into the usual virial theorem for a Coulomb field. The last term in (IV.17) characterizes the fraction of electron density concentrated in the "cavity" at $r<R$. By virtue of (IV.10), at 25°C for water we obtain $\langle i|p^2/2m_e|i\rangle \approx 2.6\,\text{eV}$. Accordingly, $-E_i \geq 2.6$ eV, but this would, however, have been contrary to the experiment and to the fact that the sum rule of S_0 is satisfied without the need for introducing absorption, beginning from a frequency of $\Omega \approx -\hbar^{-1}E$.*

Furthermore, it can be seen from the sum rule (IV.10) that the observed asymmetry and the spectrum width are mainly determined by the characteristics of the initial state. If there is a distribution for the initial states, defined by the normalized distribution function $\rho_i(T)$, and if this function is dependent on temperature, then for S_1, instead of (IV.10), we can write the following expression:

$$S_1 = \int_0^\infty \gamma(\Omega)\Omega\, d\Omega = A\frac{2e^2}{3mn\hbar c}\int \langle i|\hat{T}|i\rangle \rho_i(\hat{\text{T}})\, di \qquad \text{(IV.18)}$$

Generally, ρ_i is taken to be proportional to $\exp[-E_i/kT]$, where E_i is the energy parameter dependent on the microscopic properties of the medium. The expression (IV.10) for S_0 remains even if a distribution over the initial states exists, evidently unchanged.

B. Analysis of Threshold Regions in Optical Absorption Curves

The general analysis, discussed in the previous section, shows that the photoabsorption spectrum of solvated electrons in many cases corresponds to the transition of electron from the localized initial state to delocalized state. This hypothesis is in agreement with the experimental evidence on the existence of a "conduction band" for excess delocalized "dry" electrons observed in flash[8] and photoemission studies.[68]

If we assume that the absorption of light by electrons corresponds to transition from the localized state to continuum, then without the need for any additional model hypothesis, we can determine the important characteristics of solvated electrons from the shape of the threshold region of

*Analysis of the square-well-potential model has been made in Ref. 123.

the photoabsorption line. According to the general theory of photogeneration of particles,[124] the dependence of the threshold region of the cross-section on the final energy of particles is not sensitive to the finer details of the characteristics of the system. Thus, in the formation of particles with a quantum number of angular moment l_f, the dependence of the cross-section on the final momentum in the case of short-range interaction is of the form

$$\sigma \sim k_f^{(2l_f+1)/2} \qquad k_f = \left[\frac{(\hbar\Omega + E_i)2m}{\hbar^2} \right]^{1/2} \to 0 \tag{IV.19}$$

At the same time, if Coulomb interaction is present in the final state, the cross-section at the threshold does not depend on the final energy and behaves like

$$\sigma \sim \text{const} \qquad k_f \to 0 \tag{IV.20}$$

When there is some virtual level in the continuum, the cross-section near the threshold behaves like

$$\begin{aligned} &\sigma = \sigma_1 + \sigma_2 + \sigma_3 \\ &\sigma_1 = \frac{Ck_f^3}{(k_i^2 + k_f^2)^3} \qquad \sigma_2 = \frac{Ck_f^3}{k_f^6 + (b_1 k_f^2 - a_1)^2} \\ &\sigma_3 = \frac{C \cdot 4k_f^4 (k_i k_f^2 + b_1 k_f^2 - a_1)}{(k_i^2 + k_f^2)\left[k_f^6 + (b_1 k_f^2 - a_1)^2\right]} \\ &k_i = (2m|E_i|/\hbar^2)^{1/2} \end{aligned} \tag{IV.21}$$

The constant C depends only on the characteristics of the initial state:

$$C = \frac{32\pi e^2 k_i}{3\hbar c n} \tag{IV.22}$$

This expression suggests that an additional maximum can appear in the long-wave part of the spectrum. Such a splitting in the spectrum of solvated electrons was probably observed in Ref. 125. This additional maximum is associated with the virtual $2p$-level (p-resonance) and is described by the expression for σ_2 in (IV.21); moreover, σ_3 corresponds to the inter-

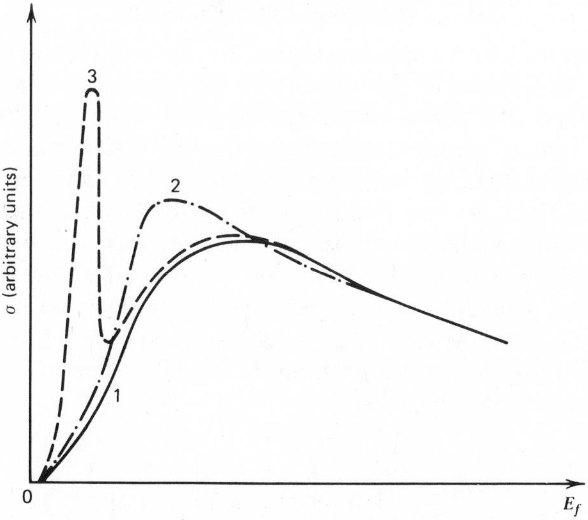

Fig. 3. General pattern of photoabsorption curves in small-radius potential model (II.33). The parameter a decreases in passing from curve 1 to curve 3.

ference effect due to various contributions to σ. The contributions of σ_2 and σ_3 may turn out to be quite significant at the initial portion of the optical absorption curve, even when the additional peak is not well pronounced. The presence of σ_2 and σ_3 results in a sharp growth in the absorption curve on the long-wave side. Various alternatives that are possible here are illustrated in Fig. 3. A general description of the spectrum can be made by matching (IV.21) with the asymptotic dependence $\sigma \sim \Omega^{-7/2}$ in the high-frequency limit.

Direct optical experiments most convincingly demonstrate the transition of solvated electrons from a discrete level to continuum. Here it is apt to recall the work[126] that reports on the recent experiment conducted to verify the suitability of formula (IV.19) for describing the long-wave part of the electron absorption spectrum in frozen CH_3OH. The threshold frequency of the electron photoconductivity in this matrix was found from the long-wave part of the solvated electron spectrum by extrapolating with the help of (IV.19). The direct independent measurements of photoconductivity made with the help of laser irradiation gave a coinciding value for the threshold of the photoconductivity of solvated electrons. A satisfactory description of the threshold regions of the e_s^- spectra by (IV.19) corresponding to the presence of only short-range component of the potential and the absence of perceptible scatter of the energy of the initial state acquires special importance in understanding the physical nature of solvated electrons.

C. Zero-Radius Potential Model

In previous sections we have found that the absorption band of solvated electrons is in satisfactory agreement with the concept on the transition from a localized state into a continuous spectrum; moreover, the effective interactions responsible for the localization are of a short-range type. It is therefore natural, first, to examine the suitability of short-range effective potentials, of which the zero-radius potential is the limiting case,[127] for describing the initial state. This potential needs the least number of fitting parameters characterizing the interaction of solvated electrons as compared to other potentials, say the rectangular-type potential well used in Refs. 79 and 123. In the case of zero-radius potential, the cross-section of photoabsorption with a transition from the initial 1s-state into continuum is given by the following expression:

$$\sigma(\Omega, E_i) = \frac{16\sigma\pi^2 e^2}{3c\hbar}\left(\frac{2k_i}{1-k_i\rho_0}\right)\left(\frac{k_f}{k_f^2+k_i^2}\right)^3 \tag{IV.23}$$

where

$$\hbar^2 k_i^2 = 2m|E_i|; \qquad \hbar^2 k_f^2 = 2m(E_i + \hbar\Omega)$$

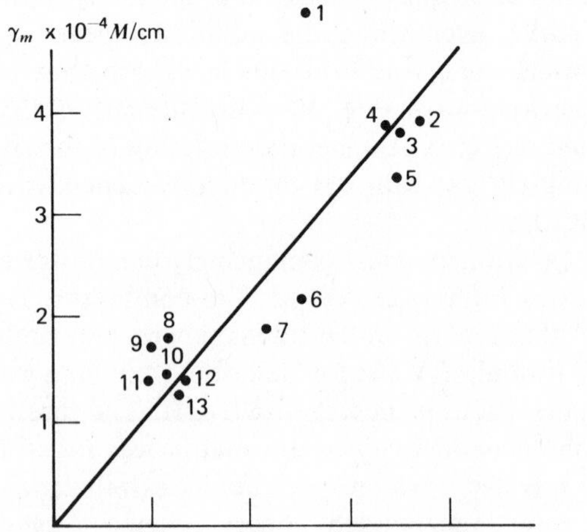

Fig. 4. Extinction coefficient γ_m of solvated electrons in various liquids vs. the parameter $(\hbar\Omega_m)^{-1}$. 1, ammonia; 2, diethyl ether; 3, methyltetrahydrofurane; 4, tetrahydrofurane; 5, dimethyloxyethane; 6, 1,3-propylene diamine; 7, ethylene diamine; 8, water; 9, methanol; 10, ethanol; 11, ethylene glycol; 12, isopropylene alcohol; 13, n-propanol.

In (IV.23) a correction to account for the effective radius (ρ_0) in the initial state has been included.

Figure 4, drawn from Ref. 128, shows the experimental data, and the straight line corresponds to the maximum of (IV.23) for $\rho_0 = 0$. For all substances, except ammonia, the extinction coefficient at the maximum is in satisfactory agreement with the dependence (IV.23) within the limits of experimental error. The discrepancy in the case of ammonia, which exceeds the experimental error, can be attributed, in accordance with the results of independent experiments on the expansion of ammonia solutions, to the relatively large size of the inital state of electrons in ammonia. In this case we have to introduce a finite value of ρ_0, which for ammonia should be equal to $\rho_0 \sim 1$ to 2 Å, so that agreement with the experiment is achieved. For other substances shown in Fig. 4, obviously we have $\rho_0 < 1$ Å. According to (IV.23) the frequency of light at maximum absorption is

$$\hbar\Omega_m \approx 2|E_i| \tag{IV.24}$$

The thresholds of several substances determined with the help of (IV.24) and from the threshold regions of curves of photoabsorption of solvated electrons by means of (IV.19) are listed in Table IV. In some cases slight differences in the threshold values obtained by these two methods may be due (if we exclude the experimental errors) both to the need to take account of the interaction type in a more exact manner (Coulomb interaction in the final state, in particular) and to the fact that the initial region of the photoabsorption curve may, unlike the position of Ω_m, suffer considerable deformation owing to the interactions with the vibrations of the medium. The simple relationship (IV.24) does, nevertheless, hold satisfactorily true for most of the systems investigated.

Now we shall consider the polarizability α. In general,[127] the formula

$$\alpha = K\frac{e^2\hbar^2}{mE_i^2} \tag{IV.25}$$

holds valid, where K is a numerical factor equal to $\frac{1}{16}$ in the case of zero-radius potential, and which, if corrections are introduced to the effective radius, takes the form

$$K = \tfrac{1}{16}(1 - k_i\rho_0)^{-1} \tag{IV.26}$$

Figure 5 shows the values of α obtained by means of (IV.10) from the experimental data on the photoabsorption of solvated electrons. The line

TABLE IV
Ratio between the Positions of Maxima in Absorption Spectra and the Threshold Energy $-E_i$[a]

Substance	$\hbar\Omega_m/2$(eV)	$-E_i$(eV)
C_2H_5OH	1.2	1.6
C_2H_5OH[b]	1.25	1.45
n-C_3H_7OH	1.2	1.5
n-C_3H_7OH[c]	1.2	1.2
i-C_4H_9OH	1.2	1.3
n-C_4H_9OH	1.15	1.25
n-$C_5H_{11}OH$	1.15	1.2
i-C_3H_7OH	1.2	1.0
$NH_2—CH_2—CH_2—NH—CH_2—CH_2—NH_2$	0.63	0.5
$(CH_3)_2—CH—CH_2—H_2$	0.6	0.6
$C_5H_{11}NH_2$	0.6	0.6
$CH_3—CH_2—CH(CH_3)(NH_2)$	0.5	0.6
$(CH_3)_2—CH—H—CH—(CH_3)_2$	0.4	0.4
$(CH_3—CH_2)_2—N—CH_2—CH_3$	0.35	0.3
$(CH_3)_2—N—CH_2—CH_2—CH_2—NH_2$	0.4	0.35
$NH_2—(CH_2)_2—OH$	0.85	0.85
$NH_2—(CH_2)_3—OH$	0.8	0.7
$NH_2—CH_2—CH(OH)(CH_3)$	0.75	0.6
$NH_2—(CH_2)_2—NH—(CH_2)_2—OH$	0.75	0.5
$NH_2—(CH_2)_2—NH—(CH_2)(OH)(CH_3)$	0.65	0.5
$(CH_3)_2—CH—NH—CH—(CH_3)_2+C_2H_5OH$	1.0	0.8
H_2O, 25°C[d]	0.9	0.95
H_20, −4°C[d]	0.9	1.0

[a]From Ref. 131.
[b]From Ref. 132.
[c]From Ref. 126.
[d]From Ref. 8.
except as noted.

corresponding to (IV.25) for $\rho_0=0$, is also shown in this figure. It can be seen that the results are in satisfactory agreement with (IV.25), and the discrepancies can be attributed both to the experimental inaccuracies and to the influence of the effective radius, which may be positive or negative.[127] In case of monotonic potential the numerical factor contained in (IV.25), which is equal to $\frac{1}{16}$ for the zero-radius potential, is the least of all the possible values.[127] In particular, for the Coulomb potential it is more than 70

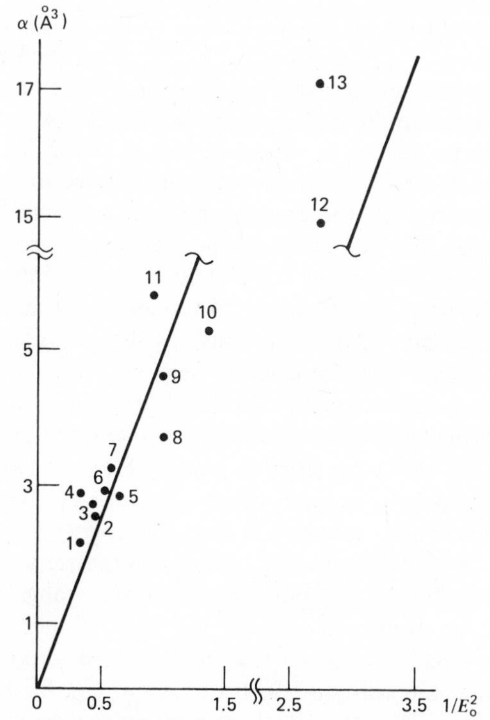

Fig. 5. Polarizability of solvated electron vs. the parameter $(\hbar\Omega_m)^{-2}$. 1, CH_3OH; 2, n-C_3H_7OH; 3, C_2H_5OH; 4, 10 *M* KOH; 5, HO–C–C–O–C–C–OH; 6, n-C_4H_9OH, 7, i-C_4H_9OH; 8, i-C_3H_7OH; 9, H_2O, 25°C; 10, NH_2–C–C–OH; 11, H_2O, −4°C; 12, C–C–C(C)(NH_2); 13, NH_2–C–C–NH–C–C–NH_2.

times greater, and is equal to $\frac{9}{2}$. This additionally corroborates the fact that the Coulomb-type long-range interactions in this case are of no significance. The relationship (IV.25) with $K=\frac{1}{16}$ can also be used in the determination of E_i.

D. Interactions with the Vibrations of the Medium in the *Zero-Radius Potential Model*

In the framework of the zero-radius potential model the description of e_s^- optical spectra is incomplete, as no explanation is found of the influence on the spectra of the interactions of solvated electrons with the collective vibrations of the medium. These interactions evidently determine the temperature dependence of the absorption spectra. So far there is no general explanation of this dependence. A surprising feature here is the considerable and almost linear red shift of the maximum on the absorption curve, its shape remaining almost unchanged as the temperature increases. Data on the influence of pressure on the spectra of solvated electrons make it possible to determine the derivative $(\partial\hbar\Omega_m/\partial T)_\rho$ under constant density, and thereby to eliminate the influence of such effects as the change in the effective mass. In particular, as shown in Ref. 6, $(\partial\hbar\Omega_m/\partial kT)_\rho \approx -16$ for solvated electrons in ammonia. It should be mentioned that in any version of the perturbation theory, for electron–phonon interactions in the adiabatic approximation, the sign of the temperature effect should be opposite to the sign observed. The reason for this, as is known from the perturbation theory, is the drop in the energy of the ground localized level when account is taken of only the first nonvanishing second-order terms with respect to the interaction with the external field.

In this section we shall examine the effects resulting from strong interaction with real collective vibrations at finite temperatures [133]. Previously, such an interaction was disregarded in passing from (II.13) to (II.14). We shall discuss this problem for the Fröhlich interaction (II.3) in a long-wave approximation under the assumption that the energy of phonons $\hbar\omega$ is much less than that of all the other energy parameters involved in the problem. These conditions are quite realistic for the problem under consideration. Further on, we shall consider the field of thermal phonons in long-wave approximate and in (II.15) substitute the zero-radius potential for the sum V_i and the self-consistent interaction V_{sc}. Under the assumptions made for the analysis of optical absorption spectra, we have to consider a Schrödinger equation of the type

$$i\hbar\frac{\partial\varphi}{\partial t}+\left[\frac{\hbar^2}{2m}\nabla^2+e\mathfrak{E}\,\mathbf{r}\cos\Omega t+\mathbf{r}(\mathbf{B}_1\cos\omega t+\beta_2\sin\omega t)\right]\varphi$$

$$=-\frac{2\pi}{\mathcal{K}}\delta^3(\mathbf{r})\frac{\partial}{\partial r}(r\varphi(r,t)) \qquad \text{(IV.27)}$$

Here, according to (II.3) and (II.4)

$$\mathbf{B}_1=2\left(\frac{\sqrt{2}\,\pi\alpha_F}{\mathfrak{B}}\right)^{1/2}\left(\frac{\hbar^5\omega^3}{m}\right)^{1/4}\sum_{\eta}\left(|b^0_{\mathbf{q}}|\cos\delta_q\right)\frac{\mathbf{q}}{q}$$

$$\mathbf{B}_2=2\left(\frac{\sqrt{2}\,\pi\alpha_F}{\mathfrak{B}}\right)^{1/2}\left(\frac{\hbar^5\omega^3}{m}\right)^{1/4}\sum_{\mathbf{q}}\left(|b^0_{\mathbf{q}}|\sin\delta_q\right)\frac{\mathbf{q}}{q} \qquad \text{(IV.28)}$$

where $|b^0_{\mathbf{q}}|$ is the amplitude and $\delta_{\mathbf{q}}$ is the phase of the vibration with momentum $\mathbf{q}$. In (IV.27) $\mathfrak{E}$ denotes the amplitude of the electromagnetic field with a frequency Ω. Hereafter this field is taken in the first nonvanishing approximation of the perturbation theory. In (IV.27) $\mathcal{K}$ stands for the inverse scattering length.[127]

Equation (IV.27) is the equation for a particle moving in a zero-radius potential field and experiencing the action of an external time-dependent fields. The procedure for solving such problems is given in Ref. 127. Here one of the possible realizations of a random phonon field $\{|\beta_{\mathbf{q}}^0|, \delta_{\mathbf{q}}\}$, over which statitistical averaging is taken, is used as the time-dependent external force.

We shall solve (IV.27) in the limit of small ω. Thus, we shall neglect the exponentially small terms in the observed quantities, in particular, in the probability of transition to continuous spectrum without absorption of a quantum $\hbar\Omega$. Since this problem, as far as we know, has not been dealt with in the literature, we shall take it up in somewhat greater detail. First, in (IV.27) make the unitary transformation

$$\varphi \to \exp\left\{-\frac{i}{\hbar}\left[-\frac{i\hbar}{m}\nabla\mathbf{u}+\mathbf{r}\mathbf{u}'+\frac{1}{2m}\int^{t}(\mathbf{u}')^2\,dt'+\frac{\mathbf{u}\mathbf{u}'}{2m}\right]\right\}\varphi; \qquad \mathbf{u}'=\frac{\partial\mathbf{u}}{\partial t} \tag{IV.29}$$

where the Hertz vector

$$\mathbf{u}=\frac{e\mathcal{E}}{\Omega^2}\cos\Omega t+\frac{1}{\omega^2}(\mathbf{B}\cos\omega t+\mathbf{B}_2\sin\omega t) \tag{IV.30}$$

By virtue of the formula

$$e^{i(\mathbf{A}_1\mathbf{p}+\mathbf{A}_2\mathbf{r})}=e^{i\mathbf{A}_1\mathbf{p}}e^{i\mathbf{a}_2\mathbf{p}}e^{(1/2)[\mathbf{A}_1,A_2]} \tag{IV.31}$$

it is not difficult to show that the expression (IV.29) is the generalization of the Kramers–Henneberger transformation used in electrodynamics.[129] With its help, (IV.27) is transformed as

$$i\hbar\frac{\partial\varphi}{\partial t}+\frac{\hbar^2}{2m}\nabla^2\varphi=-\frac{2\pi}{\mathcal{K}}\delta^3\left(\mathbf{r}-\frac{\mathbf{u}(t)}{m}\right)\frac{\mathbf{r}-\mathbf{u}(t)/m}{|\mathbf{r}-\mathbf{u}(t)/m|}$$
$$\times\left(\frac{\partial}{\partial\mathbf{r}}+\frac{i}{\hbar}\mathbf{u}'\right)\left(\left|\mathbf{r}-\frac{\mathbf{u}(t)}{m}\right|\varphi(\mathbf{r},t)\right) \tag{IV.32}$$

Besides (IV.32), by using the expression for the free Green function we can write the following integral equation:

$$\varphi(\mathbf{r},t)=\frac{2\pi}{\mathcal{K}}\frac{i}{\hbar}\left(\frac{m}{2\pi\hbar i}\right)^{3/2}\int_{-\infty}^{t}dt'\int d^3\mathbf{r}'(t-t')^{-3/2}$$

$$\times\exp\left[i\frac{m}{\hbar}\frac{|\mathbf{r}-\mathbf{r}'|^2}{2(t-t')}\right]\delta^3\left(\mathbf{r}'-\frac{u(t')}{m}\right)\frac{\mathbf{r}'-\mathbf{u}(t')/m}{|\mathbf{r}'-\mathbf{u}(t')/m|}$$

$$\times\left(\frac{\partial}{\partial\mathbf{r}'}+\frac{i}{\hbar}\mathbf{u}'\right)\left(|\mathbf{r}'-\frac{\mathbf{u}(t)}{m}|\varphi(\mathbf{r}',t')\right) \tag{IV.33}$$

From (IV.33) it follows that the function $\varphi(r,t)$ is singular at $\bar{\mathbf{r}}=\mathbf{u}(t)/m$, and

$$\varphi(\mathbf{r},t)\underset{\left|\mathbf{r}-\frac{\mathbf{u}(t)}{m}\right|\to 0}{\to} c(t)\left[\left|\mathbf{r}-\frac{\mathbf{u}(t)}{m}\right|^{-1}-\mathcal{K}+i\frac{\mathbf{u}'}{\hbar}\frac{\mathbf{r}-\mathbf{u}(t)/m}{|\mathbf{r}-\mathbf{u}(t)/m|}\right. \tag{IV.34}$$

Note that the limit (IV.34) depends on the direction in which $(\mathbf{r}-\mathbf{u}(t)/m)$ tends to zero. In order to find the function $c(t)$, we shall equate the nonsingular terms on the right-hand and left-hand sides of (IV.33), under the condition that $(\mathbf{r}-\mathbf{u}(t)/m)\to 0$.

$$\left(-\mathcal{K}+i\frac{\mathbf{u}'(t)}{\hbar}\frac{\mathbf{r}-\mathbf{u}(t)/m}{|\mathbf{r}-\mathbf{u}(t)/m|}\right)c(t)$$

$$=\sqrt{\frac{m}{\hbar}}\,(2\pi i)^{-1/2}\int_{-\infty}^{t}\left\{-2\frac{\partial c(t')}{\partial t'}(t-t')^{-1/2}-ic(t')\frac{|\mathbf{u}(t)-\mathbf{u}(t')|}{m\hbar(t-t')^{5/2}}\right.$$

$$\left.-2ic(t')\left[\frac{(\mathbf{u}(t)-\mathbf{u}(t'))\mathbf{u}'(t')+\hbar^{-1}\mathbf{u}'(t')(\mathbf{r}-m^{-1}\mathbf{u}(t'))}{m\hbar(t-t')^{3/2}}\right]\right\}$$

$$\times\exp\left[i\frac{|\mathbf{r}-m^{-1}\mathbf{u}(t')|^2m}{2\hbar(t-t')}\right]dt' \tag{IV.35}$$

Integration by parts has to be carried out in (IV.33) in order to derive (IV.35). After dividing the integral in (IV.35) into the sum of integrals over the intervals $(-\infty,t-\varepsilon)$ and $(t-\varepsilon,t)$, it can be shown by scaling the argument in the integrand that the terms, depending on the direction in which the vector $(\mathbf{r}-\mathbf{u}(t)/m)$ tends to zero, on the right-hand and left-hand sides of (IV.35), are equal to one another. The value of $c(t)$ is accordingly determined by the simplified equality

$$\mathcal{H}c(t)=\frac{2}{\sqrt{2\pi i}}\sqrt{\frac{m}{\hbar}}\int_{-\infty}^{t}\left\{\frac{1}{(t-t')^{1/2}}\frac{\partial c(t')}{\partial t'}+i\frac{c(t')}{m\hbar(t-t')^{3/2}}\right.$$
$$\left.\times\left[\frac{|\mathbf{u}(t)-\mathbf{u}(t')|^2}{2(t-t')}+\mathbf{u}'(t')(\mathbf{u}(t)-\mathbf{u}(t'))\right]\right\}$$
$$\times\exp\left[i\frac{|\mathbf{u}(t)-\mathbf{u}(t')|^2}{2m\hbar(t-t')}\right]dt' \qquad \text{(IV.36)}$$

in which only the terms of the first nonvanishing order in $\mathfrak{E}$ have to be retained. By virtue of the type of the transformation (IV.29), we shall search $c(t)$ in the form

$$c(t)=[c_0(t)+c_1(t)]\exp\left[\frac{i}{2m\hbar}\int^{t}[\mathbf{u}_B'(t')]^2dt'\right]\times e^{-(i/\hbar)Et}$$
$$\mathbf{u}_B=\frac{\mathbf{B}_1\cos\omega t+\mathbf{B}_2\sin\omega t}{\omega^2} \qquad \text{(IV.37)}$$

where $c_0(t)$ is of zero order and $c_1(t)$ is of first order in $\mathfrak{E}$. From the structure of (IV.36) it follows that C_0 is a periodic function of t with a period $2\pi\omega^{-1}$. The unknown constant introduced in (IV.37) is the quasienergy in the sense defined in Ref. 123. For $c_0(t)$ and $c_1(t)$ we get

$$\mathcal{H}c_0(t)=\frac{2}{\sqrt{2\pi i}}\sqrt{\frac{m}{\hbar}}$$
$$\times\int_{-\infty}^{t}\left(\frac{c_0'(t')}{(t-t')^{1/2}}+\left\{\frac{i}{2m\hbar}(u_B')^2-\frac{i}{\hbar}E+i\frac{1}{m\hbar(t-t')^{3/2}}\right.\right.$$
$$\left.\left.\times\left[\frac{|\mathbf{u}_B(t)-\mathbf{u}_B(t')|^2}{2(t-t')}+\mathbf{u}_B'(t')(\mathbf{u}_B(t)-\mathbf{u}_B(t'))\right]\right\}c_0(t')\right)$$
$$\times\exp\left[-\frac{i}{2m\hbar}\int_{t'}^{t}(\mathbf{u}_B'(t''))^2dt''+\frac{i}{2m\hbar}\frac{|\mathbf{u}_B(t)-\mathbf{u}_B(t')|^2}{(t-t')}\right]dt'$$

$$\mathcal{K}c_1(t) = \frac{2}{\sqrt{2\pi i}}\sqrt{\frac{m}{\hbar}}$$

$$\times \int_{-\infty}^{t} \Bigg\{ c_1'(t') + ic_s(t')\left(\frac{(\mathbf{u}_B')^2}{2m\hbar} - \frac{E}{\hbar}\right) - 2ic_0(t')\frac{e\mathcal{E}\sin\Omega t}{\Omega\cdot 2m\hbar}\mathbf{u}_B'(t') + \frac{ic_0(t')}{m\hbar}$$

$$\times \left[\frac{(\mathbf{u}_B(t) - \mathbf{u}_B(t'))}{(t-t')^2}\frac{e\mathcal{E}}{\Omega^2}(\sin\Omega t - \sin\Omega t') \right.$$

$$- \frac{\mathbf{u}_B(t) - \mathbf{u}_B(t')}{(t-t')}\frac{e\mathcal{E}}{\Omega}(\cos\Omega t - \cos\Omega t')$$

$$\left. - \frac{e\mathcal{E}}{\Omega^2}(\sin\Omega t - \sin\Omega t')\frac{\mathbf{u}_B'(t) - \mathbf{u}_B'(t')}{(t-t')} \right]$$

$$+ \frac{ic_1(t')}{m\hbar}\left[\frac{(\mathbf{u}_B(t) - \mathbf{u}_B(t'))^2}{2(t-t')^2} + \frac{\mathbf{u}_B(t) - \mathbf{u}_B(t')}{(t-t')}\mathbf{u}_B'(t') \right]$$

$$- \frac{c_0(t')}{(m\hbar)^2}\left[\frac{(\mathbf{u}_B(t) - \mathbf{u}_B(t'))^2}{2(t-t')^2} - \frac{\mathbf{u}_B(t) - \mathbf{u}_B(t')}{(t-t')}\mathbf{u}_B'(t') \right]$$

$$\times \left[\frac{\mathbf{u}_B(t) - \mathbf{u}_B(t')}{(t-t')^2}\cdot\frac{e\mathcal{E}}{\Omega^2}(\sin\Omega t - \sin\Omega t') - \int_{t'}^{t}\mathbf{u}_B'(t'')\cdot\frac{e\mathcal{E}}{\Omega}\cos\Omega t''\,dt'' \right] \Bigg\}$$

$$\times (t-t')^{-1/2}\exp\left[-\frac{i}{2m\hbar}\int_{t'}^{t}(\mathbf{u}_B'(t''))^2\,dt'' + \frac{i}{2m\hbar}\frac{|\mathbf{u}_B(t) - \mathbf{u}_B(t')|^2}{(t-t')} \right.$$

$$\left. + \frac{iE}{\hbar}(t-t') \right] dt' \qquad \text{(IV.38)}$$

In the asymptotic limit $\omega \to 0$, integration in (IV.38) can be carried out by the steepest-descent method after changing over to a new variable τ^2 by means of the substitution $t - t' = \tau^2$. Thus, for $c_0(t)$ we obtain

$$i\sqrt{E}\,\mathcal{K}c_0(t) = \sqrt{2m}\left(c_0'(t) - \frac{i}{\hbar}Ec_0(t)\right) \qquad \text{(IV.39)}$$

From (IV.39), by virtue of the periodicity of $c(t)$ and (IV.39), for E, c_0, c_1 we obtain

$$c_0 = \frac{\sqrt{\mathcal{K}}}{2\pi} \qquad c_1 \equiv 0 \qquad E = -\frac{\hbar^2\mathcal{K}^2}{2m} \qquad \text{(IV.40)}$$

The constant c_0 is determined by the normalization condition. It is of no significance in further derivations. The equality $E=-\hbar^2\mathcal{K}^2/2m$ thus obtained is consistent with the boundary condition assumed in Ref. 130 for a particle moving in a zero-radius potential and interacting with the outer low-frequency field.

After substituting (IV.40) into (IV.33), we obtain an expression for the wave function $\varphi(r,t)$ in the form of an integral. From this expression we can construct the density of radial current j_r. Eliminating the terms that are exponentially small as $\omega\to 0$, and averaging over the perion $2\pi w^{-1}$, we obtain (see Appendix 1):

$$j_r=\frac{\hbar e^3}{32m^4\omega\Omega^4k^2}Re\int_{-\infty}^{\infty}rdk\int_{-\infty}^{\infty}d\tau_s\int_0^{2\pi}d\tau_2$$
$$\times\left\{\frac{(\mathcal{E}\mathbf{r})^2}{r^2}\hbar^2k^2-\frac{\hbar k}{\hbar\omega}\frac{(\mathcal{E}r)}{r}\left[(\mathbf{B}_1\mathcal{E})\sin\tau_2-(\mathbf{B}_2\mathcal{E})\cos\tau_2\right]\cos\frac{\tau_1}{2}+\frac{1}{8(\hbar\omega)^2}\right.$$
$$\left.\times\left[(\mathbf{B}_1\mathcal{E})^2(\cos\tau_1-\cos2\tau_2)+(\mathbf{B}_2\mathcal{E})^2(\cos\tau_1+\cos2\tau_2)-(\mathbf{B}_1\mathcal{E})(\mathbf{B}_2\mathcal{E})\sin\tau_2\right]\right\}$$
$$\times\exp\left\{\frac{i}{\omega}\left[\left(\frac{\hbar k^2}{2m}+\frac{\hbar\mathcal{K}^2}{2m}+\frac{\dot{\mathbf{B}}_1^2+\mathbf{B}_2^2}{4m\hbar\omega^2}-\Omega\right)\tau_1+\frac{2k}{m\omega}\sin\frac{\tau_1}{2}\left(\frac{(\mathbf{B}_1\mathbf{r})}{r}\sin\tau_2-\frac{(\mathbf{B}_2\mathbf{r})}{r}\right.\right.\right.$$
$$\left.\left.\left.\times\cos\tau_2\right)-\frac{\sin\tau_1}{2m\hbar\omega^2}\left((\mathbf{B}_1\mathbf{B}_2)\sin2\tau_2+\frac{\mathbf{B}_1^2-\mathbf{B}_2^2}{2}\cos2\tau_2\right)\right]\right\}\qquad\text{(IV.41)}$$

The expression for the current, which determines the probability of the absorption of light, is rather cumbersome. It is, however, greatly simplified after averaging over the statistical set. It is possible to arrive at the result simply by replacing the sum of the functions of independent random variables in (IV.41) by their limiting values, i.e., by replacing

$$B_1^2=B_2^2=B^2=\frac{\pi\alpha_F}{v_{wz}}\cdot\frac{\hbar^{3/2}\omega^{1/2}}{m^{1/2}}kT\qquad\text{(IV.42)}$$

where v_{wz} is the volume of Wigner-Seitz cell. This substitution corresponds to taking a high-temperature limit. Thus for the total current I we obtain

$$I=r^2\int j_r d\cos\theta_r d\varphi_r=\frac{\sqrt{2}}{12\sqrt{\pi}}\cdot\frac{e^3\mathcal{E}^3\mathcal{K}}{m^{5/2}\hbar^{3/2}\omega^{3/2}\Omega^4}$$
$$\times\mathrm{Re}\,i^{3/2}\int_0^{\infty}\frac{dr}{\tau^{3/2}}\left[B^2(\cos\tau-1)-2m\omega^2\hbar\left(\frac{\hbar\mathcal{K}^2}{2m}-\Omega\right)\right]$$
$$\times\exp\left\{\frac{i}{\omega}\left[\left(\frac{\hbar\mathcal{K}^2}{2m}-\Omega\right)\tau_1+\frac{B^2}{2m\hbar\omega^2}\left[\tau_1-4\frac{\sin^2\frac{\tau_1}{2}}{\tau_1}\right]\right]\right\}\qquad\text{(IV.43)}$$

The photoabsorption cross-section is related to the total photocurrent by the formula

$$\sigma = \frac{I\hbar\Omega}{(c\mathcal{E}^2)/8\pi} \tag{IV.44}$$

From (IV.44) and (IV.43) in the limit

$$\left(\frac{B^2}{24m\hbar}\right)^{1/3} \ll \left|\frac{\hbar\mathcal{K}^2}{2m} - \Omega\right| \tag{IV.45}$$

we obtain: if

$$\Omega - \frac{\hbar\mathcal{K}^2}{2m} > 0$$

then

$$\sigma = \frac{8\sqrt{2}\,\pi e^2\mathcal{K}\hbar^{1/2}}{3m^{3/2}c}\,\frac{\left[\left(\Omega - \frac{\hbar\mathcal{K}^2}{2m}\right)^{3/2} + \frac{\beta^2}{8m\hbar\left(\Omega - \frac{\hbar\mathcal{K}^2}{2m}\right)^{3/2}}\right]}{\Omega^3} \tag{IV.46a}$$

and if

$$\Omega - \frac{\hbar\mathcal{K}^2}{2m} < 0$$

then

$$\sigma = \frac{\pi e^2\mathcal{K}}{3\sqrt{2}\,m^2c}\,\frac{B^2}{\Omega^3}\exp\left\{-\frac{2}{3\sqrt{3}}\,\frac{\left(\frac{\hbar\mathcal{K}^2}{2m} - \Omega\right)^{3/2}}{(B^2/24m\hbar)^{1/2}}\right\} \tag{IV.46b}$$

When (IV.45) holds true, from (IV.46a) we obtain the linear negative temperature shift of the Ω_m:

$$\left(\frac{\partial\Omega_m(\rho, T)}{\partial kT}\right)_\rho =$$

$$\frac{d\left[\Omega_m(\rho_1 T) - 2\cdot\frac{\hbar\mathcal{K}^2}{2m}(\rho)\right]}{dkT} = -\frac{\pi\sqrt{2}}{4}(\mathcal{E}_\infty^{-1} - \mathcal{E}_0^{-1})\frac{\hbar^2e^2}{mv_{wz}\left(\frac{\hbar^2\mathcal{K}^2}{2m}\right)^2} \tag{IV.47}$$

Evidently this linear temperature shift of the absorption maximum has to change at temperatures when the inequality (IV.45) violates. Such effects were apparently seen in Refs. 134 and 135.
The physical cause of the shift is the difference in the period-mean kinetic energy of pulsation in the phonon field $\{\mathbf{B}_1, \mathbf{B}_2\}$ of free and bound electrons. The pulsation energy of an electron in bound state is greater than the corresponding energy in free state, which is known to be equal to $\langle \mathbf{B}_1^2 + \mathbf{B}_2^2 \rangle / 4m\hbar\omega^2$. The later has to be discarded as there is no interaction with phonons in the final state.

In the end we may note that strong interactions with free phonons may explain the absence of a long-range Coulomb component in the effective polaron potential. After phase transformation (IV.29) and averaging over the period and angles, the potential $-r^{-1}$ is transformed into

$$-r^{-1}\theta\left(r-\frac{\langle|\mathbf{B}|\rangle}{m\omega^2}\right)+\theta\left(\frac{\langle|\mathbf{B}|\rangle}{m\omega^2}-r\right)\pi^{-1}\left[r^{-1}\arcsin\frac{m\omega^2 r}{\langle|\mathbf{B}|\rangle}\right.$$

$$\left.+\frac{m\omega^2}{\langle|\mathbf{B}|\rangle}\ln\frac{(m\omega^2 r/\langle|\mathbf{B}|\rangle)}{1+\sqrt{1-(m\omega^2 r/\langle|\mathbf{B}|\rangle)^2}}\right] \qquad \text{(IV.48)}$$

It can be seen from this expression that as $T \to \infty$, a potential of the type r^{-1} is effectively smoothened out.

V. DETERMINATION OF THE THERMODYNAMIC CHARACTERISTICS OF SOLVATED ELECTRONS

The magnitude of $|E_i|$ obtained from optical data by the methods mentioned above is indeed the difference between the energy of the bottom of the conduction band for a delocalized electron and the energy of the ground state of a solvated electron (without an account of the medium reorganization energy E^m). In order to find the difference between the chemical potentials of delocalized and localized electrons $\mu_d - \mu_l$, we have to additionally take into account the entropy factor and the medium reorganization energy. By means of the Boltzmann equation we obtain the following expression for the difference between the chemical potentials of the delocalized and localized states of solvated electrons:

$$\mu_d - \mu_l = -E_i - E^m kT \ln\left[\frac{(mkT/2\pi\hbar)^{3/2} a^3 c_l}{c_d g}\right] \qquad \text{(V.1)}$$

For delocalized electrons in (V.1) we have made use of the formula for the chemical potential of a perfect gas and have introduced the degeneration g-factor, which accounts for the changes occurring in the number of states of the phonon subsystem during electron localization: the value of a is approximately equal to the mean radius of e_s^- (i.e., of the order of intermolecular distance); $c_{l(d)}$ is the concentration of localized (delocalized) electrons. Note that here by solvated electron we mean a many-body system containing even the medium excitation. In equilibrium state $\mu_d = \mu_l$ and

$$c_l / c_d = c(g) e^{-(E_i + E^m)/kT} \tag{V.2}$$

where $c(g)$ takes a value of 10^3 at room temperature even when $g = 1$.

Interestingly, our previous discussion shows that it is possible for systems to exist in which electron localization is thermodynamically advantageous at an energy of localized state greater than the energy of the delocalized electron at the bottom of the conduction band. When interaction with electromagnetic irradiation is switched on, this becomes a system with an inverse population, and the laser effect may be possible in this system due to the energy of phonon subsystem.

The energy parameters derived from the optical data can be compared with the results obtained in independent electrochemical and photoelectrochemical measurements. Table V presents the energy and thermodynamic characteristics obtained from several such measurements. The table also lists the quantity ΔS, the difference in the specific entropies of the delocalized and localized states. This quantity can be, as was done in (V.1), equated to

$$\Delta S = kT\left(c_l \ln \frac{a^3 c_l}{g} - c_d \ln\left(\left(\frac{mkT}{2\pi\hbar^2} \right)^{-3/2c_d} \right) \right) \tag{V.3}$$

Note that nowhere did we make use of the thermodynamic characteristics of electrons in vacuum, as they cannot form a uniform equilibrium system. The quantities relating to processes (II.4) and (II.5) in Table V correspond to the Richardson exponent in the appropriate data. Moreover, the equilibrium inside the medium is considered according to the Boltzmann law, without any regard for the disruption of this equilibrium due to emission into vacuum.

Unfortunately, it is not at present possible to choose exact data that allow us to calculate and compare the quantities listed in Table V as a whole. For a number of substances, for instance, H_2O and HMPA, the quantities relating to processes (II.3) and (II.5) in Table V are approximately equal to each other (1.4 eV and 0.9 eV for H_2O and HMPA, respectively). Thus, $E^m + T\Delta S$ is close to zero.

In the conclusion it should be emphasized that the simple thermodynamic relationships between the concentrations of localized and delocalized electrons can indeed prove to be different, owing to the interactions with thermal phonons.

TABLE V
Thermodynamic and Energy Characteristics of Solvated Electrons Obtained in Different Measurements

Process	Quantity determined[a]
1. Photoconductivity threshold (including spectral data)	$-E_i = E_d^e - E_s^e - [i{:}]$
2. Difference in the threshold of photoemission from metal into vacumm and electrolyte	$E_d^e + e({}_0\varphi_s - {}_m\varphi_0)$
3. Photoemission from solvated electron solutions	$E_s^e + e_0\varphi_s$
4. Thermal conductivity	$E_d^e - E_s^e - E^m - T\Delta S$
5. Thermoemission from solutions	$E_s^e + e_0\varphi_s - E^m - T\Delta S$

[a] E_s^e and E_d^e are the energy of localized solvated electrons and the position of the conduction band for a delocalized electron reckoned from vacuum level; E^m is the energy of "reorganization" of the medium; ${}_0\varphi_s$ and ${}_s\varphi_m$ (m stands for a metal, s for the solution, and 0 for vacuum) are the potential jumps between the respective systems reckoned from the zero-charge potential; ${}_0\varphi_m$ is the possible deviation of potential jump at the metal–vacuum interface from the equilibrium value in the trinary system metal–vacuum–solution; ΔS is the difference in the entropies of delocalized and localized states calculated from a formula of type (V.3).

VI. CONCLUSIONS

The development of the theory of solvated electrons, as it follows from the facts stated in the previous pages, is rather dramatic. The results of recent experiments and comprehensive analysis of data known for a comparatively long period have compelled us in many cases either to reject the highly complicated and well-developed theories, or to modify them in a radical manner. Today, with a sufficient degree of confidence we can say that the localized states of solvated electrons are formed in disordered media and that this localization corresponds to the effective interactions with a finite radius of the order of atomic dimensions. Excited delocalized states play an essential part in many phenomena. There is every reason to believe that in many cases the optical absorption of solvated electrons corresponds to the transition into such a delocalized state.

The detailed physical nature of the interaction responsible for localization and an adequate method of explaining them are still obscure questions. As reasonable alternatives, we may consider one version which localization is supposed to take place near special structural defects, and another in which localization is assumed to be of a dynamic nature and is associated with the soliton solutions in nonlinear generalization of the polaron theory. It is possible that both versions explain the same physical situation, but in different mathematical language.

In any case, there is no doubt that both local chemical interactions with clusters surrounding a solvated electron and the interactions with collective vibrations of the medium make contributions to the structure of the solvated electron. What is significant is that these complicated interactions can be used in describing many physical phenomena involving solvated electrons only through a finite number of parameters. This is precisely the key to the success achieved with such simple models as the zero-radius potential discussed in the preceding section.

Several points still remain unexplained as regards the transport and other kinetic properties of solvated electrons. These questions are closely interlinked with the problems of conductivity in disordered media being intensively studied at present in solid-state physics. The reactivity of solvated electrons, which is also still an unexplored area, is connected with transport properties. To solve several problems in the physics and chemistry of disordered condensed media, it is absolutely essential to elucidate the dynamics of electron localization. Such regularities are closely associated with the hypothesis that some precursors to solvated electrons do exist and have been observed in a number of recent experiments.[131]

Continued interest in the problems of solvated electrons and the intensive efforts of experimentalists and theoreticians in different countries hold out hope that rapid advances shall be made in the very near future in this area of science.

APPENDIX

To construct the radial current $j_r(t)$, it is sufficient to know only the asymptotic behavior of $\varphi(r,t)$ as $r\to\infty$. Substituting (IV.40) into (IV.33), and then integrating with respect to the angles, we obtain the following expression for $\varphi(r,t)$ as $r\to\infty$:

$$\varphi = -\frac{c}{\pi\hbar}\frac{r^{-1}}{(2\pi i)}\int_{-\infty}^{\infty} k\,dk\int_{-\infty}^{\infty} d\alpha\int_{-\infty}^{\infty} dt_1$$

$$\times\exp\left\{i\left[-\hbar^{-1}r\mathbf{u}'(t)-(2m\hbar)^{-1}\int_{t_1}^{t}(\mathbf{u}')^2\,dt''+\left(\frac{\hbar\mathcal{K}^2}{2m}\right)t_1\right.\right.$$

$$\left.\left.-\left(\frac{\hbar k^2}{2m}\right)(t-t_1)+k\left(r+(\mathbf{u}(t)-\mathbf{u}(t'))\cdot\frac{\mathbf{r}}{rm}\right)\right]\right\} \qquad \text{(A.1)}$$

Since from now on we require only those terms that correspond to absorption of one photon, we shall expand the integrand as a series in $\mathfrak{S}$, retain only the first-order term proportional to $\mathcal{E}e^{-i\Omega t_1}$.

The radial current $j_r(t)$ is calculated by the usual formula:

$$j_r(t)=\frac{i\hbar}{2m}\left(\varphi\frac{\partial}{\partial r}\varphi^*-\varphi^*\frac{\partial}{\partial r}\varphi\right) \tag{A.2}$$

The mean of this current over the period $2\pi\omega^{-1}$ is

$$\begin{aligned} j_r=&\frac{\mathcal{K}}{(2\pi)^3}\left(\frac{e}{m\hbar}\right)^3\frac{\hbar^4}{m^2r^2\Omega^4}\operatorname{Re}\int\int_{-\infty}^{\infty}kk'\,dk\,dk'\cdot\left(\frac{\omega}{2\pi}\right)\cdot\\ &\cdot\int_{t_0}^{t_0+2\pi/\omega}dt\int\int_{-\infty}^{\infty}\theta(t-t_2)\cdot\theta(t_2-t_1)f(k,t_1)f^*(k',t_2)\\ &\times\left\{\left[k+k'-2\frac{\mathbf{u}'(t)\cdot\mathbf{r}}{\hbar r}\right]\exp\left[i(k-k')\frac{\hbar}{2m}\right.\right.\\ &\left.\left.\times\left(\frac{2\mathbf{u}(t)\cdot\mathbf{r}}{\hbar r}-(k+k')t\right)\right]\right\}dt_1\,dt_2 \end{aligned} \tag{A.3}$$

where

$$\begin{aligned} f(k,t)=&\left[\hbar k\frac{\mathcal{E}\mathbf{r}}{r}+\tfrac{1}{2}\mathcal{E}\mathbf{u}'(t)\right]\\ &\times\exp\left\{i\left[kr+\left(\frac{\hbar\mathcal{K}^2}{2m}-\Omega+\frac{B_1^2+B_2^2}{4m\hbar\omega^2}\right)t+\frac{1}{2m\hbar}\int^t(\mathbf{u}')^2\,d\tilde{t}\right.\right.\\ &\left.\left.-k\frac{\mathbf{u}(t)\cdot\mathbf{r}}{mr}+\frac{\hbar k^2}{2m}t\right]\right\} \end{aligned}$$

and the asterisk over the function means complex conjugation. In writing (A.3) we, making use of the symmetry of the integrand, went over from integration over the "rectangular" $[-\infty,t;-\infty,t]$ to the integration over the "triangle" $[-\infty,t_2;-\infty,t]$ in the integrals with respect to t_1,t_2. The integrand in the integral with respect to t is a complete differential and the integration limits depend on t_2. Integrating with respect to t and noticing that the shift of the variables t_1,t_2 by $-2\pi/\omega$ leads only to the appearance of the phase

$$\exp\left[\frac{i\hbar(k^2-k'^2)\pi}{m\omega}\right]$$

we obtain

$$j_r = \frac{\mathcal{H}}{(2\pi)^3}\left(\frac{e}{m\hbar}\right)^3 \frac{2\omega\hbar^3}{mr^2\Omega^4}\,\mathrm{Re}\int_{-\infty}^{\infty} dt_1 \int_0^{2\pi/\omega} dt_2\,\theta(t_2-t_1)$$

$$\cdot \int\int_{-\infty}^{\infty} kk'\,dk\,dk' f(k,t_1) f^*(k',t_2)\frac{1}{k-k'} \tag{A.4}$$

The choice of the vanishing imaginary addition to the polar expression is determined by the condition of finiteness of the integrand when $t_1 \to \infty$. Integrating with respect to k' by residue theory, we find

$$j_r = \frac{\mathcal{H}e^3\omega}{(2\pi)^3 m^4 r^2 \Omega^4} 2\mathrm{Re}\int\int_{-\infty}^{\infty}\int_0^{2\pi/\omega} dk\,dt_1\,dt_2\,\theta(t_2-t_1)$$

$$\times\, k^2 f(k,t_1) f^*(k,t_2) \tag{A.5}$$

Introducing the variables $\omega(t_2-t_1)=\tau_1$ and $\omega(t_1+t_2)=2\tau_2$ we arrive at the expression (IV.41).

REFERENCES

1. J. Jortner and N. R. Kestner, Eds., *Electrons in Fluids*, Springer-Verlag, New York, 1973.
2. L. Kevan and B. C. Webster, Eds., *Electron-Solvent and Anion-Solvent Interactions*, Elsevier, Amsterdam, 1976.
3. *J. Phys. Chem.*, **79** (26) (1975).
4. L. Kevan, in M. Burton and J. Magee, Eds., *Advances in Radiation Chemistry*, Vol. 4, Wiley, New York, 1974.
5. K. Funabashi, in M. Burton and J. Magee, Eds., *Advances in Radiation Chemistry*, Vol. 4, Wiley, New York, 1974.
6. J. C. Thompson, *Electrons in Liquid Ammonia*, Clarendon Press, Oxford, 1976.
7. H. T. Davis and R. G. Brown, *Adv. Chem. Phys.*, **31**, 329 (1975).
8. E. Hart and M. Anbar, *The Hydrated Electron*, Wiley, New York, 1970.
9. M. A. Krivoglaz and A. I. Karasevsky, *Usp. Fiz. Nauk*, **111**, 617 (1973).
10. A. M. Brodsky and A. V. Tsarevsky, in *Fizika Molekul*, Naukova Dumka, Kiev, 1975, p. 77.
11. E. L. Frankevich and V. S. Yakovlev, *Int. J. Rad. Phys. Chem.*, **6**, 281 (1974).
12. *J. Phys. Chem.*, **79**(26) (1975).
13. G. W. Boag and E. Hart, *Nature*, **197**, 45 (1963).
14. E. Hart and G. W. Boag, *J. Am. Chem. Soc.*, **84**, 4090 (1963).
15. A. K. Pikaev, *Solvated Electron in Radiation Chemistry*, Nauka, Moscow, 1966.
16. V. N. Shubin and S. Kabakchi, *Theory and Methods of Radiation Chemistry of Water*, Moscow, 1969.
17. T. A. Chubakova and N. A. Bakh, *Khim. Vys. Energ.*, **4**, 373 (1978).
18. R. Catteral, N. Mott, *Adv. Phys.*, **18**, 665 (1969).
19. R. O. Ogg, *Phys. Rev.*, **69**, 668 (1964).
20. B. Springett, M. H. Cohen, and J. Jortner, *Phys. Rev.*, **159**, 183 (1967).
21. J. Hernandez and M. Silver, *Phys. Rev. A*, **2**, 1949 (1975).

22. J. Hernandez, *Phys. Rev. A*, **7**, 1755 (1973).
23. J. Hernandez, *Phys. Rev. B*, **11**, 1289 (1975).
24. L. S. Kukushkin and V. B. Shikin, *Zh. Eksp. Teor. Fiz.*, **63**, 1830 (1972).
25. A. G. Khrapak and I. T. Yakubov, *Zh. Eksp. Teor. Fiz.*, **69**, 2042, (1975).
26. A. S. Davydov, *Zh. Eksp. Teor. Fiz.*, **18**, 913 (1948).
27. S. I. Pekar, *Investigations into the Electron Theory of Crystals*, Moscow, GITTL, 1951.
28. M. F. Deigen, *Zh. Eksp. Teor. Fiz.*, **26**, 293 (1954).
29. M. F. Deigen, *Zh. Eksp. Teor. Fiz.*, **26**, 300 (1954).
30. V. M. Biakov, Ya. I. Sharanin, and V. N. Shubin, *Ber. Bunsenges. Phys. Chem.*, **75**, 628 (1971).
31. Yu. T. Mazurenko and V. K. Mukhomorov, *Opt. Spektrosk.*, **41**, 51 (1976).
32. See, for example, M. J. Lipkin, *Quantum Mechanics*, North Holland, Amsterdam, 1973.
33. R. Evrard, in *Polarons in Ionic Crystals and Polar Semiconductors*, North-Holland, Amsterdam, 1973, p. 29.
34. R. R. Dogonadze, M. Vorotyntsev and E. M. Itskovich, *J. Phys. Chem.*, **79** 2827 (1975).
35. S. V. Tyablikov, *Zh. Eksp. Teor. Fiz.*, **21**, 377 (1951).
36. S. T. Epstein, *The Variation Method in Quantum Chemistry*, Academic, New York, 1974.
37. P. Carruthers and M. Nieto, *Rev. Mod. Phys.*, **40**, 411 (1968).
38. V. I. Melnikov, *Zh. Eksp. Teor. Fiz.*, **72**, 2345 (1977); **74**, 772 (1978).
39. P. R. Shaw and G. Whitfield, *Phys. Rev. B.*, **17**, 1495 (1978).
40. D. Emin and T. Holstein, *Phys. Rev. Lett.*, **36**, 323 (1976).
41. G. Whitfield and P. Shaw, *Phys. Rev. B*, **14**, 3346 (1976).
42. I.-Y. Cheng and K. Funábashi, *Int. J. Rad. Phys. Chem.*, **6**, 497 (1974).
43. J. Devreese, in *Polarons in Ionic Crystals and Polar Semiconductors*, North Holland, Amsterdam, 1972.
44. J. Devreese, J. De Sitter, and M. Goovaerts, *Phys. Rev. B*, **5**, 2367 (1972).
45. G. Whitfield, M. Engineer, *Phys. Rev. B*, **12**, 5472 (1975).
46. U. Schindewolf, H. Kohrmann, and G. Lang, *Angew. Chem. Int. Ed. Engl.*, **8**, 512 (1969).
47. A. V. Vannikov, "Electronic Properties of Organic Semiconductors and Dielectrics," doctoral thesis, Moscow, 1975.
48. L. Magnutson, J. Richards, and J. Thomas, *Int. J. Rad. Chem.*, **3**, 295 (1971).
49. A. Mozumder, *J. Phys. Chem.*, **76**, 3824 (1972).
50. V. V. Shornikov, Abstracts, *Modern Problems of Physical Chemistry*, Moscow, 1978.
51. R. R. Dogonadze, A. A. Kornyshev, and A. M. Kuznetsov, *Teor. Mat. Fiz.*, **15**, 127 (1973).
52. R. R. Dogonadze and A. A. Kornyshev, *Phys. Status Solidi B*, **53**, 439 (1972).
53. R. R. Dogonadze and A. A. Kornyshev, *Phys. Status Solidi B*, **55**, 843 (1973).
54. R. Jackew, *Rev. Mod. Phys.*, **49**, 681 (1977).
55. J. Jortner, *J. Chem. Phys.*, **30**, 839 (1959).
56. D. A. Copeland, N. R. Kestner, and J. Jortner, *J. Chem. Phys.*, **53**, 1189 (1972).
57. J. Segal and N. R. Kestner, *J. Phys. Chem.*, **76**, 2738 (1972).
58. N. R. Kestner and J. Jortner, *J. Phys. Chem.*, **77**, 1040 (1973).
59. N. R. Kestner, J. Jortner, and A. Gaathon, *Chem. Phys. Lett.*, **19**, 328 (1973).
60. N. R. Kestner, in J. Jortner and N. R. Kestner, Eds., *Electrons in Fluids*, Springer-Verlag, New York, 1973.
61. K. Fueki, D.-F. Feng, L. Kevan, and R. Christoffersen, *J. Phys. Chem.*, **75**, 2991 (1971).
62. K. Fueki, D.-F. Feng, and L. Kevan, *J. Chem Phys.*, **59**, 6201 (1973).
63. K. Fueki, D.-F. Feng, and L. Kevan, *J. Chem. Phys.*, **61**, 3281 (1974).
64. D. F. Feng, D. Ebbing, and L. Kevan, *J. Chem. Phys.*, **61**, 249 (1974).
65. D.-F. Feng and L. Kevan, *J. Chem. Phys.*, **61**, 4440 (1974).

66. B. Michael, E. Hart, and K. Schmidt, *J. Phys. Chem.*, **75**, 2798 (1971).
67. A. M. Brodsky, Yu. Ya. Gurevich, Yu. V. Pleskov, and Z. A. Rottenberg, *Modern Electrochemistry: Photoemission Phenomena*, Nauka, Moscow, 1974.
68. V. A. Bendersky and A. M. Brodsky, *Photoemission from Metals in Electrolyte Solutions*, Nauka, Moscow, 1977.
69. L. Raff and H. Poll, *Adv. Chem. Ser.*, **50**, 173 (1965).
70. M. Natory and T. Watanabe, *J. Phys. Soc. Jap.*, **21**, 1573 (1966).
71. P. Eisenberg and W. Kauzmann, *The Structure and Properties of Water*, Oxford University Press, Oxford, 1969.
72. B. Mc Allon and B. Webster, *Theor. Chim. Acta*, **15**, 385 (1969).
73. J. M. Moskovitz, M. Boring, and J. H. Wood, *J. Chem. Phys.*, **62**, 2254 (1975).
74. V. M. Byakov, A. A. Ovchinnikov, and B. S. Klyachko, *Configuration Model of Hydrated Electrons*, Preprints 1, 2, 3, Institute of Theoretical and Experimental Physics, Moscow, 1973, 1974, 1975 (in Russian).
75. A. A. Ovchinnikov and M. Ya Ovchinnikova, *Zh. Eksp. Teor. Fiz.*, **56**, 1278 (1969).
76. R. R. Dogonadze and A. M. Kuznetsov, *Prog. Surf. Sci.*, **6**, 1 (1975).
77. H. Fröhlich, *Theory of Dielectrics*, 2nd ed., Clarendon Press, Oxford, 1958.
78. Yu. D. Tsvetkov, K. M. Salikhov, and A. G. Semenov, *Electron Spin Echo and Its Applications*, Novosibirsk University Press, Novosibirsk, 1976 (in Russian).
79. T. Kajiwara, K. Funabshi, and C. Naleway, *Phys. Rev. A*, **6**, 808 (1972).
80. A. M. Brodsky and V. A. Tsarevsky, *Elektrokhimiya*, **9**, 1971 (1973).
81. V. B. Shikin, *Usp. Fiz. Nauk*, **121**, 427 (1977).
82. T. P. Eggarter, *Phys. Rev. A*, **5**, 2496 (1972).
83. J. A. Jahnke, M. Silver, and J. P. Hernandes, *Phys. Rev. B*, **12**, 3420 (1975).
84. E. Fermi, *Ric. Sci.*, **7**, 13 (1936).
85. J. M. Ziman, *Principles of the Theory of Solids*, 2nd ed., Cambridge University Press, Cambridge, England, 1972, p. 239.
86. J. L. Levin and T. M. Sanders, *Phys. Rev.*, **154**, 138 (1967).
87. T. F. O'Malley, *Phys. Rev.*, **130**, 1020 (1963).
88. H. Ehrenzich, F. Sietz, and D. Turnball, Eds., *Solid State Physics*, Vol. 24, Academic, New York, 1970.
89. W. A. Harrison, *Solid State Theory*, McGraw-Hill, New York, 1970.
90. R. Minday, L. Schmidt, and H. Davis, *J. Chem. Phys.*, **54**, 3112 (1971).
91. N. F. Mott and E. A. Davis, *Electronic Processes in Non-crystalline Materials*, Clarendon Press, Oxford, 1971.
92. M. H. Cohen, in J. Jortner and N. R. Kestner, Eds., *Electrons in Fluids*, Springer-Verlag, New York, 1973, p. 257.
93. R. Schiller, *J. Chem. Phys.*, **57**, 2222 (1972).
94. S. Kirkpatrick, *Rev. Mod. Phys.*, **45**, 574 (1973).
95. D. A. G. Bruggeman, *Ann. Phys.*, **24**, 639 (1935).
96. R. Londaner, *J. Appl. Phys.*, **23**, 779 (1952).
97. N. R. Kestner and J. Jortner, *J. Chem. Phys.*, **59**, 26 (1973).
98. A. Mozumder and J. L. Magee, *J. Chem. Phys.*, **45**, 3332 (1966).
99. R. M. Minday, L. D. Schmidt, and H. T. Davis, *J. Phys. Chem.*, **76**, 442 (1972).
100. Yu. A. Berlyn, L. Nyikos, and R. Schiller, "Mobility of Localized and Quasi-free Excess Electrons in Liquid Hydrocarbons," Rep. KFKI-1978-16, Hungarian Academy of Sciences, Budapest, 1978.
101. H. T. Davis, L. D. Schmidt, and R. G. Brown, in J. Jortner and N. R. Kestner, Eds., *Electrons in Fluids*, Springer-Verlag, New York, 1973, p. 393.
102. Yu. A. Firsov, Ed., *Polarons*, Nauka, Moscow, 1975.
103. P. S. Zyryanov and M. I. Klinger, *Quantum Theory of Electron Transport in Crystalline Semiconductors*, Nauka, Moscow, 1976 (in Russian).

104. J. McHale and J. Symons, *J. Chem. Phys.*, **67**, 389 (1977).
105. (a) B. S. Yakovlev, Y. A. Boriev, and A. A. Balakin, *Int. J. Rad. Phys. Chem.*, **6**, 23 (1974); (b) B. S. Yakovlev and Y. A. Boriev, *Int. J. Rad. Phys. Chem.*, **7**, 15 (1975).
106. R. Schiller and L. Nyikos, *J. Phys. Chem.*, **81**, 267 (1977).
107. A. Bartels, *Phys. Lett.*, **A45**, 491 (1973).
108. A. Bartels, thesis, University of Hamberg, 1974.
109. M. Tachio and A. Mozumder, *J. Chem. Phys.*, **62**, 2125 (1975).
110. R. M. Noyes, in G. Porter, Ed., *Progress in Reaction Kinetics*, Pergamon, London, 1971.
111. D. N. Zubarev, *Nonequilibrium Statistical Thermodynamics*, Nauka, Moscow, 1971 (in Russian).
112. (a) M. Altarelli, D. L. Dexter, H. M. Nussenzveig, and D. Y. Smith, *Phys. Rev. B*, **6**, 4502 (1972); (b) M. Altarelli and D. Y. Smith, *Phys. Rev. B*, **9**, 1290 (1974).
113. A. Villani and A. H. Zimerman, *Phys. Rev. B*, **8**, 3914 (1973).
114. F. W. King, *J. Math. Phys.*, **17**, 1509 (1976).
115. I. B. Levinson and E. I. Rashba, *Usp. Fiz. Nauk*, **111**, 683 (1973).
116. T. A. Beckman and K. S. Pitzer, *J. Phys. Chem.*, **65**, 1527 (1961).
117. P. T. Rusch and J. J. Lagovski, *J. Phys. Chem.*, **77**, 210 (1973).
118. E. J. Hart, in J. J. Lagovski and M. J. Sienko, Eds., *Metal-Ammonia Solutions*, Butterworth, London, 1970, p. 413.
119. M. G. Debacker, J. N. Decpigny, and M. Lanno, *J. Phys. Chem.*, **81**, 159 (1977).
120. A. M. Brodsky, A. V. Tsarevsky, and A. K. Kostin, in press.
121. A. M. Brodsky and A. V. Tsarevsky, *Zh. Eksp. Teor. Fiz.*, **70**, 216 (1976).
122. F. You and G. Freeman, *J. Phys. Chem.*, **81**, 909 (1977).
123. A. E. Baz, Ya. B. Zeldovich, and A. M. Perelomov, *Scattering, Reaction, and Decay in Nonrelativistic Quantum Mechanics*, Nauka, Moscow, 1971 (in Russian).
124. Ta-You Wu and T. Ohmura, *Quantum Theory of Scattering*, Prentice-Hall, Englewood Cliffs, N.J., 1962.
125. V. S. Marevtsev and A. V. Vannikov, *Khim. Vys. Energ.*, **7**, 119 (1973).
126. A. K. Kostin, V. I. Zolotorevskii, V. V. Golovanov, and A. V. Vannikov, *Khim. Vys. Energ.*, **11**, 252 (1977).
127. Yu. M. Demkov and V. N. Ostrovskii, *Zero-Radius Potential Method in Atomic Physics*, Leningrad University Press, Leningrad, 1975.
128. A. M. Brodsky and A. V. Tsarevsky, *Int. J. Rad. Phys. Chem.*, **8**, 455 (1976).
129. V. Henneberger, *Phys. Rev. Lett.*, **21**, 8383 (1968).
130. L. D. Landau and E. M. Livshits, *Quantum Mechanics, Nonrelativistic Theory*, Nauka, Moscow, 1976, p. 344 (in Russian).
131. T. Shida, S. Guata, and T. Watanabe, *J. Phys. Chem.*, **76**, 3683 (1972).
132. T. Ujikawa and K. Fueki, *J. Phys. Chem.*, **79**, 2479 (1975).
133. A. M. Brodsky and A. V. Tsarevsky, in press.
134. A. M. Brodsky, A. V. Vannikov, A. V. Tsarevsky, and T. A. Chubakova, *Khim. Vys. Energ.*, in press.
135. R. S. Dixon and V. J. Lopata, *Radiat. Phys. Chem.*, **11**, 135 (1978).

NATURAL CHIROPTICAL SPECTROSCOPY: THEORY AND COMPUTATIONS

AAGE E. HANSEN

Department of Physical Chemistry
H. C. Ørsted Institute
University of Copenhagen
Copenhagen, Denmark

THOMAS D. BOUMAN

Department of Chemistry
Southern Illinois University
Edwardsville, Illinois

CONTENTS

I. INTRODUCTION

Natural optical rotatory power is the ability of a medium to rotate plane-polarized light and to exhibit different absorption and emission intensities for left and right circularly polarized light, in the absence of other external disturbances (such as mechanical stress or static magnetic fields), and both effects of course are frequency-dependent. For isotropic media the phenomenon of natural optical rotatory power requires the presence of chiral[1] molecules, and the interaction responsible for the phenomenon is discussed most readily in terms of the differential response of a chiral molecule to left and right circularly polarized light. The frequency-dependent rotation of plane-polarized light, which then results from the difference in the refractive indices of the medium toward left- and right-circularly polarized light, is referred to as optical rotatory dispersion (ORD), while the difference in the absorption of the two circularly polarized beams is called circular dichroism (CD).

Natural optical rotatory power has evolved by now into a mature field of study, both experimentally and theoretically. The quantum foundation for an understanding of the relation between natural optical rotatory power and molecular structure was laid by Hund,[2] who clarified the nonstationary character of a chiral molecule, and by Rosenfeld[3] and Condon, Altar, and Eyring,[4,5] who derived explicit expressions for ORD[3,4] and for CD.[5] These derivations, which were based on semiclassical radiation theory, provided ORD expressions valid for nonabsorbing spectral regions only. This gap was closed when Moffitt and Moscowitz[6–8] demonstrated that ORD and CD are related by a Kronig–Kramers transformation,[9,10] and used this to extend the ORD expressions into absorbing regions. Finally, the statistical averaging required to relate the microscopic quantum molecular quantities to bulk optical rotatory power was studied in detail in the work of Maaskant and Oosterhoff.[11,12] The Rosenfeld–Condon approach has been reformulated and extended by a number of authors, as outlined in Section II. Here we shall mention that the fundamental question of the nature of a chiral system has been reopened in recent years with a view toward the possible influences of the parity-violating weak interactions,[13,14] and of the quantum beats that can be expected (in principle) for nonstationary systems.[15]

Despite the availability of theoretical ORD and CD relations, expressed in terms of the molecular wave functions, the experimental and computational conditions up through the beginning of the 1960s restricted experimental efforts almost exclusively to ORD measurements,[16,17] and theoretical efforts were equally exclusively concentrated on the development of models[4,8,18–21] that avoided the actual calculation of molecular wave functions. For organic and biological systems, this point of view is clearly represented in the Moscowitz and Tinoco articles on optical activity in the 1962 volume of this series.[8,21] At that time Tinoco expressed the view that no uniformly successful theoretical approach to optical activity had yet been published, and it is perhaps characteristic of the computational problem involved that Buckingham and Stiles, more than a decade later,[22] expressed a similarly pessimistic view. In Section VIII we present a more optimistic appraisal of the present status.

The experimental picture changed in the mid-1960s when CD measurements in the visible and near ultraviolet became practicable, and by now CD measurements in the vacuum ultraviolet are being reported from a number of laboratories (see the recent review by Johnson[23] and references therein). A beginning of a shift in computational emphasis can be discerned in Moscowitz's 1957 calculation on hexahelicene, which used a delocalized π-electron molecular orbital model and treated the entire molecule as an inherently chiral entity.[8] The first variational, all-valence-electron calculation of molecular optical rotatory power, however, did not appear until the 1966 Pao and Santry calculation[24] on methylcyclohexanones. Since then, the increasing access to large computers and sophisticated program systems has provided the opportunity to perform ever more detailed computations of chiroptical properties. However, the model approaches still play a very important role, and it should become apparent from this article that there is a very fruitful interaction between model approaches and the attempts to do full quantum molecular calculations.

The changes in experimental and theoretical emphasis within the field of natural optical rotatory power are reflected in the title of this chapter. This implies that our viewpoint is spectroscopic, so that we treat only the ordinary absorption and the CD spectra and their relation to the structure of chiral molecules (we refer to Deutsche, Lightner, Woody, and Moscowitz,[25] and Schellman[26] for recent reviews of ORD and the use of Kronig–Kramers relations). In addition, we limit the scope to organic and biological molecules (see, e.g., Mason[27] and Bosnich[28] for recent reviews of optical rotatory power of inorganic complexes), and except for Section VI we shall consider only electronic absorption chiroptical spectroscopy.

II. ORDINARY AND ROTATORY ELECTRONIC INTENSITIES

The Rosenfeld–Condon expressions for ORD and CD have been reformulated and extended in a variety of ways, including quantum-field-theoretical derivations of ORD[29–35] and of CD,[35–37] and semiclassical derivations of ORD[38,39] and of CD.[26,40–42] Most of the treatments assume that the molecules are small compared to the wavelength of the impinging light beam; expressions that are valid for molecules of arbitrary size are given in Refs. 34 and 41. In addition, nonlinear ORD has been considered by Atkins and others,[30,43] and by Desorbry and Kabir[44] within classical optics, while expressions for two-photon CD have been derived by Tinoco[45] and Power.[46] Since our main concern lies in spectroscopic quantities for absorption processes, we shall use a simple semiclassical approach to sketch the derivation of a number of equivalent expressions for the ordinary and rotatory intensities.[26,41] These sets of equivalent expressions yield the same results for, respectively, ordinary and rotatory intensities when exact wave functions are employed, whereas the introduction of approximations ordinarily generates discrepancies among the formally equivalent intensities. The use of the degree of agreement among the results obtained from the equivalent expressions as a measure of the quality of the results will be a recurring theme in our discussion.

In semiclassical radiation theory, the Hamiltonian for a molecule in a harmonic time-dependent field of angular frequency ω is written as:

$$\hat{\mathcal{H}}(t) = \hat{\mathcal{H}} + \hat{\mathcal{H}}' e^{-i\omega t}. \tag{II.1}$$

$\hat{\mathcal{H}}$ is the Hamiltonian in the absence of the field, and we have retained only the frequency term appropriate for an excitation process.[47] The transition probability per unit time for an excitation from state g to state u, both of which are eigenstates of $\hat{\mathcal{H}}$, is then[47]

$$W_{gu}(\omega) = \frac{2\pi}{\hbar^2} |\langle u|\hat{\mathcal{H}}'|g\rangle|^2 \rho_{gu}(\omega) \tag{II.2}$$

where $\rho_{gu}(\omega)$ is the normalized density of final states.

Except for Section VI.B, we shall be concerned solely with electronic transitions. $\hat{\mathcal{H}}'$ hence contains only the interaction between the electromagnetic field and the molecular electrons, and we shall assume the librational model of Moffitt and Moscowitz,[6] according to which the molecular degrees of freedom are separated into vibronic (i.e., electronic and internal nuclear) modes and librational modes. The latter term is used indiscriminately for the restricted translational and rotational motion in dense media, or the rotational motion in dilute gas phase at low spectral resolution. The subscripts g and u in (II.2) then denote vibronic states, and $\rho_{gu}(\omega)$, which is accordingly interpreted as the line shape of a single vibronic line, results from the continuous librational degrees of freedom. In practice this interpretation can be used even at finite temperature, where $\rho_{gu}(\omega)$ is then slightly temperature-dependent due to a Boltzmann factor.[6]

From the transition probability (II.2) for a given vibronic excitation $g \to u$, the corresponding partial absorption cross-section is

$$\sigma_{gu}(\omega) = \frac{\hbar\omega}{I(\omega)} W_{gu}(\omega) \tag{II.3}$$

where $I(\omega)$ is the light intensity, and for a medium with N molecules per unit volume the partial vibronic absorption coefficient[48] then becomes

$$\kappa_{gu}(\omega) = N\sigma_{gu}(\omega) \tag{II.4}$$

Circular dichroism can be reported in terms of the quantity

$$\Delta\kappa(\omega) = \kappa^L(\omega) - \kappa^R(\omega) \tag{II.5}$$

where $\kappa^L(\omega)$ and $\kappa^R(\omega)$ are the absorption coefficients for left- and right-circularly polarized light, respectively. Alternatively, one can use the ellipticity $\theta'(\omega)$ defined by[4,8]

$$\tan\theta'(\omega) = \tanh[\Delta\kappa(\omega) l/4] \tag{II.6}$$

where l is the length of the light path in the medium. For small values of $\Delta\kappa(\omega)$, (II.5) and (II.6) give the following partial vibronic ellipticity per unit length:

$$\theta_{gu}(\omega) = \Delta\kappa_{gu}(\omega)/4 \tag{II.7}$$

The distinguishing feature of the circular dichroism is of course that it is a signed quantity, whereas the ordinary absorption is inherently positive.

A. Momentum (Velocity) Formulation

The most common textbook version of the operator representing the interaction between a collection of electrons (charge $-e$) and a harmonic electromagnetic field is the following momentum formulation:[49]

$$\hat{\mathcal{H}}' = \frac{e}{mc} A^\circ(\omega) \sum_s \exp\left(\frac{i\omega}{c}\boldsymbol{\varepsilon}_3 \cdot \mathbf{r}_s\right)(\boldsymbol{\varepsilon} \cdot \hat{\mathbf{p}}_s)$$
$$= -\frac{ie\hbar}{mc} A^\circ(\omega)(\boldsymbol{\varepsilon} \cdot \hat{\boldsymbol{\eta}}), \tag{II.8}$$

where the fully retarded transition moment operator is

$$\hat{\boldsymbol{\eta}} \equiv \sum_s \hat{\boldsymbol{\eta}}_s = \sum_s \exp\left(\frac{i\omega}{c}\boldsymbol{\varepsilon}_3 \cdot \mathbf{r}_s\right)\hat{\nabla}_s \tag{II.9}$$

Here $\mathbf{r}_s$ and $\hat{\mathbf{p}}_s$ are the position vector and the linear momentum operator for electron s, referred to a center in the molecule, and $A^\circ(\omega)$ is the amplitude of the electromagnetic vector potential for the light beam with propagation vector $\boldsymbol{\varepsilon}_3$ and polarization vector $\boldsymbol{\varepsilon}$. We assume in (II.8) that there are no externally applied magnetostatic fields present (see Section VI.C). The polarization vectors for left and right circularly polarized light are given by[4]

$$\boldsymbol{\varepsilon}^L = (\boldsymbol{\varepsilon}_1 + i\boldsymbol{\varepsilon}_2)/\sqrt{2}, \qquad \boldsymbol{\varepsilon}^R = (\boldsymbol{\varepsilon}_1 - i\boldsymbol{\varepsilon}_2)/\sqrt{2} \tag{II.10}$$

where the unit vectors $\boldsymbol{\varepsilon}_1, \boldsymbol{\varepsilon}_2$ together with the propagation vector $\boldsymbol{\varepsilon}_3$ span a right-handed Cartesian system, and the intensity of the field is

$$I(\omega) = \frac{\omega^2}{2\pi c}|A^\circ(\omega)|^2 \tag{II.11}$$

Equations (II.2), and (II.8) to (II.11) then combine to yield the following transition probability per unit time:

$$W_{gu}^{L,R}(\omega) = \frac{2\pi^2 e^2}{m^2\omega^2 c} I(\omega)\rho_{gu}(\omega)\Big\{|\boldsymbol{\varepsilon}_1 \cdot \langle u|\hat{\boldsymbol{\eta}}|g\rangle|^2$$
$$+ |\boldsymbol{\varepsilon}_2 \cdot \langle u|\hat{\boldsymbol{\eta}}|g\rangle|^2$$
$$\pm i\boldsymbol{\varepsilon}_3 \cdot \left[\langle u|\hat{\boldsymbol{\eta}}|g\rangle^* \times \langle u|\hat{\boldsymbol{\eta}}|g\rangle\right]\Big\} \tag{II.12}$$

where the + and − signs correspond, respectively, to left and right circular polarization.

From (II.3) to (II.5), and (II.12) the partial absorption coefficient and the partial circular dichroism due to the vibronic excitation $g \to u$ become

$$\kappa_{gu}^{(3)}(\omega) = \left\{ \kappa_{gu}^{L}(\omega) + \kappa_{gu}^{R}(\omega) \right\}/2$$

$$= \frac{2\pi^2 e^2 \hbar}{m^2 \omega c} N \left\{ |\langle u|(\boldsymbol{\varepsilon}_1 \cdot \hat{\boldsymbol{\eta}})|g\rangle|^2 + |\langle u|(\boldsymbol{\varepsilon}_2 \cdot \hat{\boldsymbol{\eta}})|g\rangle|^2 \right\} \rho_{gu}(\omega) \qquad \text{(II.13)}$$

and

$$\Delta\kappa_{gu}^{(3)}(\omega) = \left\{ \kappa_{gu}^{L}(\omega) - \kappa_{gu}^{R}(\omega) \right\}$$

$$= \frac{4\pi^2 e^2 \hbar}{m^2 \omega c} N i \boldsymbol{\varepsilon}_3 \cdot \left[\langle u|\hat{\boldsymbol{\eta}}|g\rangle^* \times \langle u|\hat{\boldsymbol{\eta}}|g\rangle \right] \rho_{gu}(\omega) \qquad \text{(II.14)}$$

The superscript 3 indicates that these expressions are derived for a collection of molecules with a fixed orientation relative to the light beam. Equations (II.13) and (II.14) are quite general, since the presence of the full retardation factor $\exp[(i\omega/c)\boldsymbol{\varepsilon}_3 \cdot \mathbf{r}_s]$ in $\hat{\boldsymbol{\eta}}_s$ (II.9) ensures that these expressions are applicable to molecules of all sizes, hence also to molecules with dimensions comparable to the wavelength $\lambda (= 2\pi c/\omega)$ of the light beam. Here the fact that the circular dichroism is a signed quantity is apparent because the vector product of the retarded transition moment with its own complex conjugate can have a positive or a negative projection onto the light propagation vector $\boldsymbol{\varepsilon}_3$.

Most molecules are much smaller than the wavelength of the light, so that the exponential in (II.9) can be expanded to first order,

$$\boldsymbol{\varepsilon} \cdot \hat{\boldsymbol{\eta}}_s = \boldsymbol{\varepsilon} \cdot \boldsymbol{\nabla}_s + \frac{i\omega}{c} (\boldsymbol{\varepsilon}_3 \cdot \mathbf{r}_s)(\boldsymbol{\varepsilon} \cdot \hat{\boldsymbol{\nabla}}_s) \qquad \text{(II.15)}$$

The partial absorption coefficient (II.13) then becomes

$$\kappa_{gu}^{(3)}(\omega) = \frac{2\pi^2 e^2 \hbar}{m^2 \omega c} N \rho_{gu}(\omega) \times \left\{ |\langle u|(\boldsymbol{\varepsilon}_1 \cdot \hat{\boldsymbol{\nabla}})|g\rangle|^2 + |\langle u|(\boldsymbol{\varepsilon}_2 \cdot \hat{\boldsymbol{\nabla}})|g\rangle|^2 \right\} \qquad \text{(II.16)}$$

while the circular dichroism (II.14) becomes

$$\Delta\kappa_{gu}^{(3)}(\omega)=\frac{8\pi^2e^2\hbar}{m^2c^2}N\rho_{gu}(\omega)$$
$$\times\{\langle g|(\boldsymbol{\varepsilon}_1\cdot\hat{\nabla})|u\rangle\langle u|(\boldsymbol{\varepsilon}_3\cdot\mathbf{r})(\boldsymbol{\varepsilon}_2\cdot\hat{\nabla})|g\rangle$$
$$-\langle g|(\boldsymbol{\varepsilon}_2\cdot\hat{\nabla})|u\rangle\langle u|(\boldsymbol{\varepsilon}_3\cdot\mathbf{r})(\boldsymbol{\varepsilon}_1\cdot\hat{\nabla})|g\rangle\} \tag{II.17}$$

since the assumed absence of a static external magnetic field implies that the wave functions can be taken to be real with impunity (see Section VI.C). We have also let $\hat{\nabla}=\Sigma_s\hat{\nabla}_s$ and $\mathbf{r}=\Sigma_s\mathbf{r}_s$. Notice that the reality of the wave function forces us to retain the lowest complex contribution in the expansion in (II.15), since a purely real transition moment matrix element would make the circular dichroism vanish according to (II.14).

For a collection of molecules with random orientations, (II.16) and (II.17) must be rotationally averaged. This is easily done recalling that $\boldsymbol{\varepsilon}_1\cdot\hat{\nabla}$ transforms as a component of a vector operator, while the operator $(\boldsymbol{\varepsilon}_3\cdot\mathbf{r})(\boldsymbol{\varepsilon}_2\cdot\hat{\nabla})$ contains a symmetric second rank tensor component, namely $[(\boldsymbol{\varepsilon}_3\cdot\mathbf{r})(\boldsymbol{\varepsilon}_2\cdot\hat{\nabla})+(\boldsymbol{\varepsilon}_3\cdot\hat{\nabla})(\boldsymbol{\varepsilon}_2\cdot\mathbf{r})]/2$, and a component of a vector operator, namely $\boldsymbol{\varepsilon}_1\cdot(\mathbf{r}\times\hat{\nabla})/2$. From the fact that the quantity $(\boldsymbol{\varepsilon}_i\cdot\mathbf{a})(\boldsymbol{\varepsilon}_i\cdot\mathbf{b})$ averages to $\mathbf{a}\cdot\mathbf{b}/3$ for any two vectors $\mathbf{a}$ and $\mathbf{b}$, whereas the product of a vector component and a second rank tensor component averages to zero, (II.16) and (II.17) provide the following rotationally averaged partial absorption coefficient:

$$\kappa_{gu}^{\nabla}(\omega)=\frac{4\pi^2e^2\hbar}{3m^2\omega c}N\rho_{gu}(\omega)|\langle g|\hat{\nabla}|u\rangle|^2 \tag{II.18}$$

and partial circular dichroism:

$$\Delta\kappa_{gu}^{\nabla}(\omega)=-\frac{8\pi^2e^2\hbar}{3m^2c^2}N\rho_{gu}(\omega)\{\langle g|\hat{\nabla}|u\rangle\cdot\langle u|\mathbf{r}\times\hat{\nabla}|g\rangle\} \tag{II.19}$$

where $\mathbf{r}\times\hat{\nabla}$ is to be interpreted as $\Sigma_s(\mathbf{r}_s\times\hat{\nabla}_s)$.

The matrix element $\langle g|\hat{\nabla}|u\rangle$ is one version of the electric dipole transition moment, while $\langle g|\mathbf{r}\times\hat{\nabla}|u\rangle$ is proportional to the angular momentum matrix element and hence to the magnetic dipole transition moment [see (II.20) and (II.23) below]. The symmetric second rank tensor, whose contribution vanishes in the rotational averaging going from (II.17) to (II.19), represents the electric quadrupole moment operator. For oriented or partially oriented ensembles of molecules, the electric quadrupole contribution can be as important as the magnetic dipole contribution,[50–53] and (II.17) [or (II.14) for very large molecules] is then the correct point of departure.

Equations (II.18) and (II.19) are called the (electric dipole) velocity expressions, as indicated by the superscripted ∇. In (II.19), as in (II.22) and (II.23) below, the fact that the circular dichroism is a signed quantity follows because it depends upon the scalar product of an electric and a magnetic dipole transition moment matrix element. The well-known symmetry rules for natural optical rotatory power follow from the same feature, as discussed further in Section IV.A.

B. Multipole (Length) Formulation

An alternative to (II.18) and (II.19) follows from the relation[49]

$$\frac{\hbar^2}{m}\langle g|\hat{\nabla}|u\rangle=\langle g|[\mathbf{r},\mathcal{H}]|u\rangle=\hbar\omega_{ug}\langle g|\mathbf{r}|u\rangle \tag{II.20}$$

which holds because the electronic position vector $\mathbf{r}_s$ commutes with everything in the molecular Hamiltonian except the kinetic-energy operator for electron s; ω_{ug} is the angular Bohr frequency for the excitation. Insertion of (II.20) into (II.18) and (II.19) yields [see also (II.29)]

$$\kappa^{r}_{gu}(\omega)=\frac{4\pi^2e^2}{3\hbar c}N\omega_{ug}\rho_{gu}(\omega)|\langle g|\mathbf{r}|u\rangle|^2 \tag{II.21}$$

and

$$\Delta\kappa^{r}_{gu}(\omega)=-\frac{8\pi^2e^2}{3mc^2}N\omega_{ug}\rho_{gu}(\omega)\langle g|\mathbf{r}|u\rangle\cdot\langle u|\mathbf{r}\times\hat{\nabla}|g\rangle \tag{II.22}$$

$$=\frac{16\pi^2N}{3\hbar c}\omega_{ug}\rho_{gu}(\omega)Im\{\langle g|\hat{\mu}_e|u\rangle\cdot\langle u|\hat{\mu}_m|g\rangle\}, \tag{II.23}$$

where $\hat{\mu}_e\equiv-e\mathbf{r}$, and $\hat{\mu}_m\equiv(ie\hbar/2mc)(\mathbf{r}\times\hat{\nabla})$ are the electric and magnetic dipole transition moment operators. Equations (II.21–II.23) are therefore called the (electric dipole) length expressions, as indicated by the superscript r.

C. Alternative Formulations

A number of other versions of the intensity expressions are possible. We shall make use of the so-called mixed expression,[54]

$$\kappa^{r\nabla}_{gu}(\omega)=\frac{4\pi^2e^2}{3mc}N\rho_{gu}(\omega)\langle g|\hat{\nabla}|u\rangle\cdot\langle u|\mathbf{r}|g\rangle \tag{II.24}$$

which is obtained by inserting (II.20) only once into (II.18). Equivalently, it can be considered the geometric mean of (II.18) and (II.21). We have used again the approximation expressed in (II.29) below.

Additional intensity expressions containing operators of the form $\hat{\nabla} V$, $\mathbf{r}\times(\hat{\nabla} V)$, and higher derivatives of the potential can be generated by use of (II.25) below. However, the few calculations that have been attempted with such transition moments have shown that they require wave functions of an accuracy beyond what is practicable for molecular many-electron systems.[55–57] On the other hand, we shall show in Section IV.B that these higher versions of the transition moments are useful in the expression of sum rules for the intensities.

D. Comments

Equation (II.20) is a special case of the off-diagonal hypervirial relation

$$\langle g|[\hat{F},\mathcal{H}]|u\rangle=(E_u-E_g)\langle g|\hat{F}|u\rangle \tag{II.25}$$

for an arbitrary operator $\hat{F}$. This equation, which provides a useful relation between off-diagonal quantities, holds if the wave functions for states g and u are exact eigenfunctions of the molecular Hamiltonian. This has apparently left the impression that (II.18) and (II.19) are more fundamental than the alternative versions of the intensity expressions, and that they should therefore be preferred in actual calculations (which necessarily involve approximate wave functions). However, it should be realized that there is nothing fundamental about the perturbation operator given in (II.8). As first shown by Goeppert-Mayer,[58] and later reformulated by a number of authors within semiclassical[59–63] as well as fully quantized[64–69] radiation theory, the Hamiltonian for a collection of charged particles in an electromagnetic field can be written in a variety of ways that differ by canonical transformations, and none of those Hamiltonians (and hence none of the intensity expressions) can claim any formal priority.

To illustrate this point we notice that the perturbation operator appropriate for the partial absorption coefficient in the velocity form given in (II.18) is

$$\hat{\mathcal{H}}'=-\frac{ie\hbar}{mc}A^{\circ}(\omega)(\boldsymbol{\varepsilon}\cdot\hat{\nabla}) \tag{II.26}$$

The canonically transformed operator obtained by Goeppert–Mayer[58] is the electric dipole interaction form

$$\hat{\mathcal{H}}''=eE^{\circ}(\omega)(\boldsymbol{\varepsilon}\cdot\mathbf{r}) \tag{II.27}$$

(for simple derivations of this operator see, e.g., Refs. 61 and 70), where

$E°(\omega)$ is the amplitude of the electric field of the light beam. Using

$$I(\omega)=\frac{c}{2\pi}|E°(\omega)|^2 \tag{II.28}$$

(II.2)–(II.4) combine to recover the length expression for the absorption coefficient given in (II.21) *provided* that we make the approximation

$$\omega\rho_{gu}(\omega)=\omega_{ug}\rho_{gu}(\omega) \tag{II.29}$$

The same consideration holds for the circular dichroism. On the one hand, this shows that the librational approximation implies that the various intensity expressions are equivalent only when the librational line width is narrow enough to justify (II.29). On the other hand, it shows also that we are free to use either the light frequency ω or the Bohr frequency ω_{ug} in (II.18), (II.19), (II.21), and (II.22), when this assumption is fulfilled.

III. BAND SHAPES AND INTEGRATED INTENSITIES

For the individual vibronic lines, (II.29) together with either of the pairs of equations (II.18), (II.19) or (II.21), (II.22) shows that the circular dichroism and the absorption coefficient differ in magnitude (and of course often in sign), but their line shape is the same, within the librational model. This implies that the so-called anisotropy ratio[20]

$$g(\omega)=\frac{\Delta\kappa(\omega)}{\kappa(\omega)} \tag{III.1}$$

is a constant for a single vibronic line. On the other hand, the distribution of the ordinary and rotatory intensities over the vibronic lines that combine into an electronic band is not necessarily the same. In general therefore, the anisotropy ratio (III.1) is a frequency-dependent quantity, and the actual variation of $g(\omega)$ over an electronic band provides valuable information about the nature of the transition. We consider here the resulting band shapes of $\Delta\kappa(\omega)$, $\kappa(\omega)$, and $g(\omega)$ for some characteristic types of molecular transitions, and discuss the extraction of quantities that are amenable to theoretical computations, namely the electronic oscillator strength (III.15) and rotatory strength (III.16).

A. Nondegenerate Transitions

In the Born-Oppenheimer adiabatic approximation[71,72,73] the vibronic wave functions are written as product functions

$$|b,\beta\rangle=\psi_b(q,Q)\chi^b_\beta(Q), \tag{III.2}$$

where q and Q stand for the electronic coordinates and the internal nuclear coordinates, respectively. The electronic wave function, $\psi_b(q,Q)$, is an eigenfunction of the electronic part of the molecular Hamiltonian, that is,

$$[\hat{T}_q + V(q,Q)]\psi_b(q,Q) \equiv \hat{H}_e\psi_b(q,Q) = E_b(Q)\psi_b(q,Q) \qquad \text{(III.3a)}$$

and the electronic eigenvalues serve as potential energy functions for the determination of vibrational wave functions according to the eigenvalue relation

$$[\hat{T}_Q + E_b(Q)]\chi_\beta^b(Q) = w_\beta^b\chi_\beta^b(Q). \qquad \text{(III.3b)}$$

Here $\hat{T}_q$ and $\hat{T}_Q$ are the kinetic energy operators for the electronic and the internal nuclear motion, respectively, and $V(q,Q)$ is the total (internal) electrostatic potential energy. This adiabatic approximation is valid in the nondegenerate limit where the electronic energy of a given state is separated from the electronic energy of all other states by an amount much larger than the characteristic vibrational quanta. In a harmonic approximation the electronic energy is expanded to second order in the nuclear coordinates, Q_ξ, to yield

$$E_b(Q) = E_b(Q^\circ) + \sum_\xi [\partial E_b/\partial Q_\xi]^\circ (Q_\xi - Q_\xi^\circ)$$
$$+ \tfrac{1}{2}\sum_\xi\sum_{\xi'} [\partial^2 E_b/\partial Q_\xi \partial Q_{\xi'}]^\circ (Q_\xi - Q_\xi^\circ)(Q_{\xi'} - Q_{\xi'}^\circ), \qquad \text{(III.4a)}$$

where Q° is a reference nuclear configuration. For the conventional normal coordinate description it is assumed that Q° is chosen to make $[\partial E_0/\partial Q_\xi]^\circ$ zero for all nuclear coordinates, where E_0 is the ground-state electronic energy, and the quadratic term is reduced to the diagonal contributions. For the ground state we therefore have

$$E_0(Q) = E_0(Q^\circ) + \tfrac{1}{2}\sum_\xi k_\xi (Q_\xi - Q_\xi^\circ)^2 \qquad \text{(III.4b)}$$

where k_ξ is the ground-state force constant for the ξth mode. In terms of the ground-state normal coordinates, an excited electronic state has an energy given by the general form (III.4a), where linear contributions arise for all displacements Q_ξ for which the excited equilibrium geometry is different from that of the ground state Q_ξ°. If one or more pairs of normal coordinates transform the same way under the symmetry elements common to the electronic ground state and the particular excited state, then the quadratic contribution in the upper-state electronic energy may contain cross terms in those pairs of normal coordinates (the so-called Duschinsky effect).[72,74]

The adiabatic wave functions in (III.2) are not exact molecular eigenfunctions, and the various equivalent intensity expressions in Section II therefore do not lead to identical results in the Born-Oppenheimer approximation. For pure vibrational transitions this poses a nontrivial problem (see Section VI.B). On the other hand, for electronic transitions the difference between the resulting intensities is characteristically of the order of typical Born-Oppenheimer correction terms, namely proportional to the ratio of vibrational to electronic quanta. By the same token, for a nondegenerate electronic transition from the ground state $|0,o\rangle$ into a state $|b,\beta\rangle$ it is consistent with the Born-Oppenheimer approximation to approximate the various factors or divisors of the frequency ω in (II.13), (II.14), (II.16), (II.18), (II.21–II.23) by the Franck-Condon frequency

$$\omega_{b0}=[E_b(Q^\circ)-E_0(Q^\circ)]/\hbar, \tag{III.5}$$

evaluated at the ground-state equilibrium position. Utilizing this observation, the particular expressions in (II.22, II.24) provide a set of useful relations for the electronic intensity distributions. After approximating the frequency factor ω in (II.22) by ω_{b0}, both of these intensities are proportional to an intensity quantity which can be written in the general form

$$I_{0,o;b,\beta}(\omega)=\rho_{0,o;b,\beta}(\omega)$$
$$\times\langle\chi_o^0(Q)|\mathbf{F}_{0b}^a(Q)|\chi_\beta^b(Q)\rangle_Q\cdot\langle\chi_\beta^b(Q)|\mathbf{F}_{b0}^h(Q)|\chi_o^0(Q)\rangle_Q \tag{III.6a}$$

using the wave functions of (III.2). Here

$$\mathbf{F}_{0b}(Q)=\langle\psi_0(q,Q)|\hat{\mathbf{F}}|\psi_b(q,Q)\rangle_q \tag{III.6b}$$

is the nuclear-coordinate-dependent electronic transition moment vector, the subscripts Q and q on the brackets in (III.6a, b) indicating integrations over nuclear and electronic coordinates respectively. The superscripts a and h on these transition moments indicate that the electronic vector operators are, respectively, anti-Hermitian (namely $\hat{\nabla}$ or $\mathbf{r}\times\hat{\nabla}$) and Hermitian (namely $\mathbf{r}$). For semirigid molecules where average displacements are small,[75] the electronic transition moments can be expanded in the nuclear displacements to yield

$$
\begin{aligned}
I_{0,o;b,\beta}(\omega) &= \rho_{0,o;b,\beta}(\omega) \\
&\times \Bigg[\mathbf{F}^a_{0b}(Q^\circ)\mathbf{F}^h_{b0}(Q^\circ)\langle\chi^0_o|\chi^b_\beta\rangle_Q\langle\chi^b_\beta|\chi^0_o\rangle_Q \\
&+ \sum_\xi \{\mathbf{F}^a_{0b,\xi}\cdot\mathbf{F}^h_{b0}(Q^\circ)+\mathbf{F}^a_{0b}(Q^\circ)\cdot\mathbf{F}^h_{b0,\xi}\}\langle\chi^0_o|\chi^b_\beta\rangle_Q\langle\chi^b_\beta|(Q_\xi-Q^0_\xi)|\chi^0_o\rangle_Q \\
&+ \sum_\xi\sum_\eta \mathbf{F}^a_{0b,\xi}\mathbf{F}^h_{b0,\eta}\langle\chi^0_o|(Q_\xi-Q^\circ_\xi)|\chi^b_\beta\rangle_Q\langle\chi^b_\beta|(Q_\eta-Q^\circ_\eta)|\chi^0_o\rangle_Q \\
&+ \tfrac{1}{2}\sum_\xi\sum_\eta \{\mathbf{F}^a_{0b,\xi\eta}\cdot\mathbf{F}^h_{b0}(Q^\circ)+\mathbf{F}^a_{0b}(Q^\circ)\cdot\mathbf{F}^h_{b0,\xi\eta}\} \\
&\times\langle\chi^0_o|\chi^b_\beta\rangle_Q\langle\chi^b_\beta|(Q_\xi-Q^\circ_\xi)(Q_\eta-Q^\circ_\eta)|\chi^0_o\rangle_Q \Bigg]
\end{aligned} \tag{III.7}
$$

to second order. Here $\mathbf{F}_{0b,\xi}\equiv[\partial\mathbf{F}_{0b}/\partial Q_\xi]^\circ$ and $\mathbf{F}_{0b,\xi\eta}\equiv[\partial^2\mathbf{F}_{0b}/\partial Q_\xi\partial Q_\eta]^\circ$. The selection rules for the matrix element $\mathbf{F}_{0b}(Q^\circ)$ follow from the symmetry elements of Q°, and if this expansion center is taken to be the ground-state equilibrium configuration, then the first term in (III.7) is referred to as the allowed contribution, whereas the contributions that depend on the nuclear displacements are called forbidden. Other choices of reference configurations for the expansion in (III.7) have been proposed.[76,77] Dekkers and Closs[77] use a different reference configuration for each vibronic line, namely the so-called r-centroid[78]

$$Q^b_\xi=\langle\chi^b_\beta|Q_\xi|\chi^0_o\rangle_Q/\langle\chi^b_\beta|\chi^0_o\rangle_Q,$$

which collapses the entire first-order contribution into the zeroth-order term of (III.7). This technique is useful for the study of details of the intensity distribution, but does obscure the division of the total intensity into allowed and forbidden components.

The partial intensity for the electronic transition $0\rightarrow b$ is defined[6] as the entire intensity distribution over the manifold of vibronic transitions corresponding to this electronic excitation, that is,

$$I_{0b}(\omega)=\sum_\beta I_{0,o;b,\beta}(\omega). \tag{III.8}$$

The distribution of the allowed part of the partial intensity over the vibronic lines is governed by the Franck-Condon factors $|\langle\chi_o^0|\chi_\beta^b\rangle|^2$, and for ordinary intensities, the presence of a forbidden part may reveal itself in a change of polarization properties through the band.[79–81] For circular dichroism, where the various contributions are signed quantities, a contribution due to a forbidden component may lead to CD curves that change sign within an electronic absorption band,[76,77,82] such CD curves are referred to as bisignate[83] (see Section III.A.2).

The intensity distribution defined by (III.8) can be characterized by its moments,[6,82,84,85] where the nth moment is given by

$$\begin{aligned} I_{0b}^{(n)} &\equiv \int d\omega\, \omega^n I_{0b}(\omega) \\ &= \hbar^{-n}\sum_\beta \langle\chi_o^0(Q)|\mathbf{F}_{0b}^a(Q)|\chi_\beta^b(Q)\rangle_Q\left[w_\beta^b - w_o^0\right]^n\langle\chi^b(Q)|\mathbf{F}_{b0}^h(Q)|\chi_o^0(Q)\rangle_Q \\ &= \langle\chi_o^0(Q)|\mathbf{F}_{0b}^a\left\{\omega_{b0} + \left[\bar{\varepsilon}_b(Q) - \bar{\varepsilon}_o(Q)\right]\right. \\ &\quad \left. + \left[\hat{t}_Q + \bar{\varepsilon}_o(Q) - \bar{\varepsilon}_o^0\right]\right\}^n \mathbf{F}_{b0}^h|\chi_o^0(Q)\rangle_{Q'} \end{aligned} \quad \text{(III.9)}$$

from (II.29, III.6a, III.8) and the normalization of the line shape function $\rho_{0,o;b,\beta}(\omega)$. Here we have introduced the reduced quantities $\hat{t}_Q = \hat{T}_Q/\hbar$, $\bar{\varepsilon}_i(Q) = [E_i(Q) - E_i(Q^\circ)]/\hbar$ and $\bar{\varepsilon}_o^0 = [w_o^0 - E_o(Q^\circ)]/\hbar$, and the last step in (III.9) follows from (III.3b) and the closure properties of the vibrational functions $\chi_\beta^b(Q)$. The lowest moment, $I_{0b}^{(0)}$, is a measure of the total intensity of this electronic excitation. Using (III.7) and (III.9) and the harmonic relation

$$\langle\chi_o^0(Q)|(Q_\xi - Q_\xi^\circ)|\chi_o^0(Q)\rangle_Q = 0 \quad \text{(for all } \xi\text{)},$$

we find

$$\begin{aligned} I_{0b}^{(0)} &= \langle\chi_o^0(Q)|\mathbf{F}_{0b}^a(Q)\cdot\mathbf{F}_{b0}^h(Q)|\chi_o^0(Q)\rangle_Q \\ &= \mathbf{F}_{0b}^a(Q^\circ)\cdot\mathbf{F}_{b0}^h(Q^\circ) + \sum_\xi C_\xi\langle\chi_o^0|(Q_\xi - Q_\xi^\circ)^2|\chi_o^0\rangle_{Q'} \end{aligned} \quad \text{(III.10)}$$

where C_ξ is a quantity that depends on the first and second derivatives of the electronic transition moments with respect to the nuclear coordinate Q_ξ. The first-order terms in (III.7) do not contribute at all to the total intensity,[6,82] and $I_{0b}^{(0)}$ differs from its fully allowed component only in terms proportional to the mean square nuclear ground-state vibrations.[6,82] The first moment provides the center of gravity of the particular intensity distribution[6,84] according to

$$\begin{aligned}\tilde{\omega}_{b0} &= I_{0b}^{(1)}/I_{0b}^{(0)}\\ &= \omega_{b0} + \frac{1}{I_{0b}^{(0)}}\langle \chi_o^0(Q)|\mathbf{F}_{0b}^a(Q)\cdot\mathbf{F}_{b0}^h(Q)\{\bar{\varepsilon}_b(Q)-\bar{\varepsilon}_o(Q)\}|\chi_o^0(Q)\rangle_Q\\ &\quad + \frac{1}{2I_{0b}^{(0)}}\langle \chi_o^0(Q)|\mathbf{F}_{0b}^a(Q)\cdot\left[\hat{t}_Q,\mathbf{F}_{b0}^h(Q)\right]-\mathbf{F}_{0b}^h(Q)\cdot\left[\hat{t}_Q,\mathbf{F}_{b0}^a(Q)\right]\}|\chi_o^0(Q)\rangle_Q\end{aligned} \tag{III.11}$$

where we have used the Hermitian and anti-Hermitian character of the electronic operators to make the contributions from the two types of transition moments appear in a symmetric fashion in the last term. Recalling that $\hat{t}_Q=\hbar/2(\Sigma_\xi\partial^2/\partial Q_\xi^2)$, (III.11) shows that for a fully allowed electronic transition between states with the same force constants, the center of gravity coincides with the Franck-Condon frequency (III.5).[84] For transitions with sizeable forbidden components the contributions from the last two terms in (III.11) in general contribute differently in ordinary and rotatory intensities (see Section III.A.2) leading to different centers of gravity for these two types of intensity.[6,82] The second moment provides a measure of the width of the absorption band according to the relation[84]

$$\tilde{\Delta}_{0b} = \left[I_{0b}^{(2)}/I_{0b}^{(0)} - \tilde{\omega}_{b0}^2\right]^{\frac{1}{2}} \tag{III.12}$$

which can be rewritten in terms of simple and repeated commutators of $\hat{t}_Q$ and the electronic transition moments, in analogy to the terms in (III.11). For a transition with negligible forbidden components, the width reduces to

$$\begin{aligned}\tilde{\Delta}_{0b} = \Big[&\langle \chi_o^0(Q)|\left[\bar{\varepsilon}_b(Q)-\bar{\varepsilon}_o(Q)\right]^2|\chi_o^0(Q)\rangle_Q\\ &-|\langle \chi_o^0(Q)|\left[\bar{\varepsilon}_b(Q)-\bar{\varepsilon}_0(Q)\right]|\chi_o^0(Q)\rangle_Q|^2\Big]^{1/2}\end{aligned} \tag{III.13}$$

in which the leading term is seen to be proportional to the difference in equilibrium nuclear geometry of the electronic states 0 and b.[84] Again forbidden contributions tend to change the width by different amounts for ordinary and rotatory absorption bands.[6,82] The third moment $I_{0b}^{(3)}$ leads to a measure of the skewness of the absorption band. From the general relations in (III.10–III.12) a number of specific expressions for the absorption band parameters of ordinary and rotatory intensities can be obtained by inserting particular models for the electronic energy differences $\bar{\varepsilon}_b(Q)-\bar{\varepsilon}_0(Q)$ and for the electronic transition moments. A number of such expressions are given by Moffitt and Moscowitz[6] and by Weigang et al.[82]

1. *Cases I, II, and III*

For ordinary and rotatory intensities the general relations given in the preceding section lead to a convenient classification advocated by Moffitt and Moscowitz[6] and Weigang et al.[82] In the notation of Moffitt and Moscowitz[6] an electronic transition in a chiral molecule belongs to *case I* when the absorption coefficient and the circular dichroism for this transition are both dominated by the allowed components of the pertinent transition moments, that is, when both types of intensity are adequately represented by the zeroth-order term in (III.7). In this case the ordinary and rotatory intensities will then exhibit progressions, where all vibronic lines have the same polarization properties and the same sign in the case of circular dichroism, and where the intensity distribution over the vibronic lines is governed by the Franck-Condon factors $|\langle\chi_0^0|\chi_\beta^b\rangle|^2$. Equations (II.22), (II.24) and (III.1), (III.7) and (III.8) then show that the anisotropy ratio becomes

$$g_{0b}(\omega)=\frac{2\omega_{b0}}{c}\,\frac{\langle\psi_0(q,Q^\circ)|\mathbf{r}\times\nabla|\psi_b(q,Q^\circ)\rangle_q\cdot\langle\psi_b(q,Q^\circ)|\mathbf{r}|\psi_0(q,Q^\circ)\rangle_q}{\langle\psi_0(q,Q^\circ)|\nabla|\psi_b(q,Q^\circ)\rangle_q\cdot\langle\psi_b(q,Q^\circ)|\mathbf{r}|\psi_0(q,Q^\circ)\rangle_q} \tag{III.14}$$

where the electronic transition moments are evaluated at the nuclear ground-state equilibrium configuration Q°. This shows that the anisotropy ratio is constant over the frequency range covered by a case I-electronic absorption band.[6] Of course this implies in turn that the band parameters of (III.11) and (III.12) are the same for the ordinary and rotatory intensity of a pure case I transition. *Case II* transitions[6] arise when the forbidden part of either the electric or the magnetic dipole transition is comparable to or larger than the corresponding allowed contribution. In terms of (III.7), this implies typically that the first-order terms dominate either the ordinary or the rotatory intensity. As noted above, a forbidden contribution in the ordinary intensity may reveal itself in a change of polarization properties through the band,[79–81] whereas for circular dichroism a forbidden contribution may lead to the so-called bisignate CD curves.[76,77,82,83] The total ordinary or rotatory intensity of a case II transition is determined by the allowed components only,[6] because first-order terms do not contribute to $I^{(0)}$ (III.10). On the other hand, the center of gravity (III.11) and the band width (III.12) will be strongly influenced by the forbidden components, so that these band parameters will be different for the ordinary and rotatory intensities of a case II transition, and the anisotropy ratio (III.1) is hence frequency-dependent for such transitions. These features of a prototypical case II transition[6] are clearly exhibited in Fig. 6 of Ref. 20.

The logical extension of the two cases considered above are the *case III* transitions, in the nomenclature of Weigang et al.[82] where the forbidden contributions are significant in both the electric and magnetic transition dipole moments. For case III transitions all the terms in (III.7) now come into play,[76,82] and all the moments, including the total intensity $I^{(0)}$, will be affected by the forbidden contributions, so that the anisotropy ratio becomes strongly frequency-dependent. Examples of bisignate CD curves analyzed in terms of a case III model are given in Ref. 82.

The discussion in this section has dealt exclusively with nondegenerate electronic transitions in molecules where the nuclei exhibit only small-amplitude vibrations[75] about an equilibrium geometry corresponding to a single well-defined molecular conformation. However it is worth adding that a strongly frequency-dependent anisotropy ratio and bisignate CD curves can result also from a solvational equilibrium involving differently solvated species or from a conformational equilibrium.[86]

2. *Integrated Intensities*

Calculations of the magnitude and distribution of ordinary and rotatory intensity over an electronic absorption band are reported for a few molecules, for example in Refs. 87a and 87b. However, the most common procedure for extracting chiroptical quantities that are amenable to theoretical evaluations is to define integrated intensity quantities, namely the dimensionless oscillator strength[88]

$$f_{0b} = \frac{mc}{2\pi^2 e^2 N} \int d\omega \kappa_{0b}(\omega) = \frac{mc}{2\pi^2 e^2 N} \sum_{\beta} \int d\omega \kappa_{0,o;b,\beta}(\omega) \qquad \text{(III.15a)}$$

$$= 4.32 \cdot 10^{-9} \int d\nu \varepsilon_{0b}(\nu) \qquad \text{(III.15b)}$$

and the rotatory strength[6,8]

$$R_{0b} = \frac{3\hbar c}{16\pi^2 N} \int \frac{d\omega}{\omega} \Delta\kappa_{0b}(\omega) = \frac{3\hbar c}{16\pi^2 N} \sum_{\beta} \int \frac{d\omega}{\omega} \Delta\kappa_{0,o;b,\beta}(\omega) \qquad \text{(III.16a)}$$

$$\approx 2.30 \cdot 10^{-39} \nu_{b0}^{-1} \int d\nu \Delta\varepsilon_{0b}(\nu). \qquad \text{(III.16b)}$$

In equations (III.15b) and (III.16b) ν is the frequency expressed in wave numbers (cm^{-1}), ν_{b0} is the Franck-Condon frequency of equation (III.5) (again in cm^{-1}), ε_{0b} is the molar extinction coefficient,[48] and the fundamental constants are inserted such that the rotatory strength is expressed in cgs units (namely erg·esu·cm/Gauss, see below). Equations (III.15b) and (III.16b) therefore provide an immediate link to the actual experimental data.

For case I and II transitions, the corresponding theoretical expressions can be obtained by combining equations (II.13), (II.18), (II.21), (II.24), (III.10), and (III.15a) to yield the following expressions for the oscillator strength:

$$f_{0b}^{(3)} = \frac{\hbar}{m\omega_{b0}} \left\{ |\langle 0|\boldsymbol{\varepsilon}_1 \cdot \hat{\boldsymbol{\eta}}|b\rangle|^2 + |\langle 0|\boldsymbol{\varepsilon}_2 \cdot \hat{\boldsymbol{\eta}}|b\rangle|^2 \right\} \tag{III.17}$$

$$f_{0b}^{\nabla} = \frac{2\hbar}{3m\omega_{b0}} |\langle 0|\hat{\nabla}|b\rangle|^2, \tag{III.18}$$

$$f_{0b}^{r} = \frac{2m\omega_{b0}}{3\hbar} |\langle 0|\mathbf{r}|b\rangle|^2 = \frac{2m\omega_{b0}}{3e^2\hbar} |\langle 0|\boldsymbol{\mu}_e|b\rangle|^2 \tag{III.19}$$

$$f_{0b}^{r\nabla} = \frac{2}{3} \langle 0|\hat{\nabla}|b\rangle \cdot \langle b|\hat{\mathbf{r}}|0\rangle \tag{III.20}$$

and (II.14), (II.19), (II.22), and (III.16a) yield the following expressions for the rotatory strengths:

$$R_{0b}^{(3)} = \frac{3e^2\hbar^2}{4m^2\omega_{b0}^2} i\boldsymbol{\varepsilon}_3 \cdot \left[\langle b|\hat{\boldsymbol{\eta}}|0\rangle^* \times \langle b|\hat{\boldsymbol{\eta}}|0\rangle \right], \tag{III.21}$$

$$R_{0b}^{\nabla} = -\frac{e^2\hbar^2}{2m^2c\omega_{b0}} \langle 0|\hat{\nabla}|b\rangle \cdot \langle b|r \times \hat{\nabla}|0\rangle, \tag{III.22}$$

$$R_{0b}^{r} = -\frac{e^2\hbar}{2mc} \langle 0|\mathbf{r}|b\rangle \cdot \langle b|\mathbf{r} \times \hat{\nabla}|0\rangle \tag{III.23a}$$

$$= Im\{ \langle 0|\hat{\boldsymbol{\mu}}_e|b\rangle \cdot \langle b|\hat{\boldsymbol{\mu}}_m|0\rangle \} \tag{III.23b}$$

where

$$\langle 0|\hat{\mathbf{F}}|b\rangle = \langle \psi_0(q, Q^\circ)|\hat{\mathbf{F}}|\psi_b(q, Q^\circ)\rangle_q.$$

The fully retarded (III.17) and (III.21) correspond to oriented molecules of all sizes, and ω_{b0} should also be inserted here in the operator $\boldsymbol{\eta}$ (II.9). Equations (III.18–III.20), (III.22), and (III.23) are appropriate for randomly oriented molecules that are small relative to the wavelength of the exciting light beam, and (III.23b) shows explicitly that the rotatory strength has the dimensions of an electric dipole times a magnetic dipole. For a transition with a fully allowed electric dipole transition moment (i.e., ~1 Debye) and a fully allowed magnetic dipole transition moment (i.e., ~1 Bohr magneton), the rotatory strength is of the order of 10^{-38} cgs units. In practice rotatory strengths are often quoted in units of 10^{-40} cgs.

For case III transitions, the second-order terms in (III.10) contribute significantly to the integrated intensities, and a theoretical calculation must accordingly include a realistic treatment of the vibronic structure. This implies in turn that the relation of molecular chirality to the sign and magnitude of the circular dichroism will be much less direct for case III than for case I or II transitions.[82] A calculation of the band parameters in (III.11) and (III.12) similarly requires a realistic model for the nuclear coordinate dependence of the electronic transition moments and of the relative position and shape of the potential energy curves $E_b(Q)$ and $E_0(Q)$.

3. *Isotopic Substitution*

In the preceding section, we have made the customary assumption that the molecular chirality leading to nonvanishing rotatory intensity is already evident in the nuclear geometry corresponding to the minimum of the electronic ground-state potential surface. Recent experiments have shown, however, that CD spectra can be measured for molecules whose chirality stems entirely from unsymmetrical isotopic substitution.[89–91] Since the Born–Oppenheimer adiabatic electronic potential-energy surfaces are unaffected by isotopic substitution,[71,72] the effects of the nuclear vibrations are crucial to an understanding of the effect.

Qualitatively, if a given vibrational mode is anharmonic, the difference in zero-point vibrational energy for two isotopes will lead to a small difference in the *average* values of the corresponding nuclear coordinates; for example, the *average* bond length of a C—D bond would be slightly shorter than the C—H bond average bond length, and we return to a chiral geometry.[92] Dezentje and Dekkers[93] formalized an analysis of this sort by expanding the vibrationally averaged rotatory strength $\langle\chi_o^0|R_{0b}(Q)|\chi_o^0\rangle$ in a Taylor expansion in the nuclear coordinates. If the expansion center is taken to be the equilibrium geometry $Q°$, then $R_{0b}(Q°)$ vanishes by assumption, and the first nonvanishing term is in second order for harmonic vibrations, but is in first order for anharmonic modes. If, instead, one expands about the *average* geometry, defined by $\bar{Q}_\xi = \langle\chi_o^0|Q_\xi|\chi_o^0\rangle_Q$, $\xi = 1,\ldots,3N-6$, then the first-order term is made to collapse into the zeroth-order term, and $R_{0b}(\bar{Q})$ is nonzero for an anharmonic potential. Approximate calculations utilizing this effective bond-length difference show the proper qualitative behavior.[92,93]

The mechanism just presented is the only one that has been explored. Other schemes that remain to be assessed are (i) a model including second order terms, (ii) construction of a potential energy surface with explicit dependence on the nuclear masses, as in the Born-Huang adiabatic approximation,[72,94] and finally (iii) a break-down of the adiabatic approximation entirely.

B. Degenerate Transitions

The term "degenerate" is used here to cover all cases where the electronic energy difference between some of the pertinent electronic states is not large enough to justify an adiabatic approximation. Three distinct molecular cases fall into this category, namely (1) due degenerate electronic states, (2) accidentally degenerate or near-degenerate electronic states, and (3) molecular exciton systems. Due degenerate electronic states arise when the symmetry of the electronic Hamiltonian at, or in the close vicinity of, the reference nuclear configuration is high enough to support degenerate representations. The treatment of the vibronic states of such molecules of course requires the machinery of the Jahn–Teller formalism.[95,96] High-symmetry chiral organic molecules are currently being synthesized[97]; however, the majority of high-symmetry chiral molecules are inorganic coordination compounds, and we shall refer to Refs. 27 and 28 for further discussion.

1. *Accidental Degeneracy and Interference*

Accidental (or near) degeneracies of related electronic states (e.g., electronic states originating from the same electronic configuration) are often referred to as pseudo-Jahn–Teller cases, and they are treated most naturally along with the genuine Jahn–Teller case.[95,98] There is, however, a different type of near-degeneracy that we shall outline in some detail, namely the interference that arises when an essentially discrete-type excitation such as a Rydberg transition is embedded in a continuous or highly dense vibrational manifold of an electronic excitation of quite different origin, such as a broad $\pi\rightarrow\pi^*$ valence electron absorption band (Fig. 1).

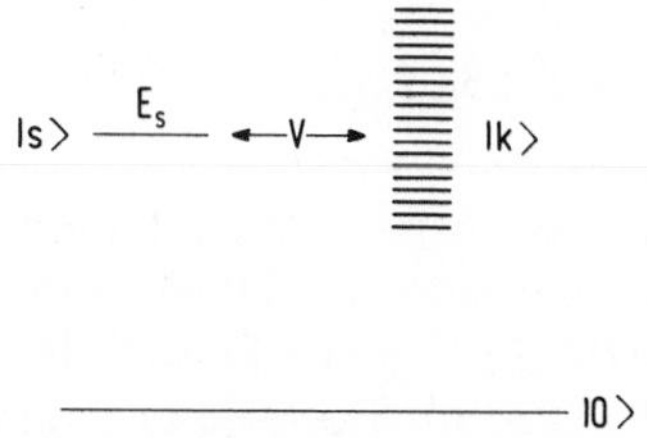

Fig. 1. Schematic illustration of a discrete state $|s\rangle$ embedded in a dense or continuous manifold $|k\rangle$.

An interference effect between discrete and continuous transitions was first discussed by Fano[99] for autoionizing atomic transitions. A number of authors have subsequently studied related interference effects in the ordinary (polarized and unpolarized) absorption spectra of molecules[100–104] and solids,[105] and recently also the circular dichroism of chiral molecules.[106] In the molecular and solid-state systems, the coupling is usually assumed to be vibronic, and the degenerate coupling is treated either by the Bixon–Jortner[100] stepladder model or by use of resolvent operators[105, 107, 47] (Green's operators). We refer to the quoted literature for these techniques.

Let $|s\rangle$ and $|k\rangle$ denote the zeroth-order discrete state and the continuous vibrational manifold, respectively; we further define the transition moments

$$\nabla_{0b} \equiv \langle 0|\hat{\nabla}|b\rangle \qquad \mathbf{r}_{0b} \equiv \langle 0|\mathbf{r}|b\rangle \qquad \mathbf{l}_{0b} \equiv \langle 0|\mathbf{r}\times\hat{\nabla}|b\rangle \tag{III.24}$$

and the dimensionless energy quantities

$$\gamma = 2\pi V \rho(E) \tag{III.25}$$

$$\varepsilon = (E - E_s)/V \tag{III.26}$$

where V is the vibronic coupling between $|s\rangle$ and $|k\rangle$ and $\rho(E)$ is the density of states in the manifold. The resulting band shapes of the ordinary (κ) and rotatory ($\Delta\kappa$) intensities in the energy range around E_s can then be written in the form[102]

$$\mathfrak{L}^{\kappa}(\varepsilon) = \frac{\rho(E_s)/E_s}{\varepsilon^2 + \gamma^2/4}\left\{|\nabla_{s0}|^2 + 2\varepsilon \nabla_{0s}\cdot\nabla_{0k} + \varepsilon^2|\nabla_{0k}|^2\right\} \tag{III.27}$$

and[106]

$$\mathfrak{L}^{\Delta\kappa}(\varepsilon) = \frac{\rho(E_s)}{\varepsilon^2 + \gamma^2/4}\left\{\nabla_{0s}\cdot\mathbf{l}_{0s} + \varepsilon\left[\nabla_{0s}\cdot\mathbf{l}_{0k} + \nabla_{0k}\cdot\mathbf{l}_{0s}\right] + \varepsilon^2\nabla_{0k}\cdot\mathbf{l}_{0k}\right\} \tag{III.28}$$

where numerical factors have been omitted. Equation (III.27) is essentially equation 21 of Ref. 102, rewritten to conform with the present (II.18) and (II.24)–(III.26), while (III.28) is given in Ref. 106 [except that the anti-Hermitian character of $\hat{\nabla}$ and $\mathbf{r}\times\hat{\nabla}$ is used to introduce a sign change and a misprint is corrected]. In the derivation of (III.27) and (III.28), we have assumed that the density $\rho(E)$, the coupling V, and the transition moments into the continuous state can be treated as constants in the energy range in the vicinity of the discrete state E_s.

Equations (III.27) and (III.28) are of the same general form, containing a Lorentzian contribution, $(\varepsilon^2+\gamma^2/4)^{-1}$, arising from the transition into the discrete state; a dispersion type contribution, $\varepsilon/(\varepsilon^2+\gamma^2/4)$, arising from cross terms of the transition moment, and an antiresonance, $\varepsilon^2/(\varepsilon^2+\gamma^2/4)$, associated with the continuous transition. All three contributions are centered at E_s [the line shift term[105] vanishes because $\rho(E)$ is assumed constant], and the width of the effect is

$$\Gamma = V\gamma = 2\pi\rho(E_s)V^2 \tag{III.29}$$

In the case of a general chiral molecule the symmetry is so low that none of the involved electric and magnetic transition moment vectors are strictly parallel or perpendicular, and in addition the various products of the transition moments can be both positive and negative. The band shapes represented by (III.27) and (II.28) can therefore become quite complex, especially in the CD, where the three contributions are signed quantities. We shall use the special case of a chiral molecule belonging to the C_2 point group to illustrate some of the important features of the resulting band shapes.

For a C_2 molecule the transition moments are either polarized along the C_2 axis, or perpendicular to this axis. When one of the transitions is polarized along the axis, and the other perpendicular to it, the cross terms vanish in (III.27) and (III.28), and we find that

$$\mathfrak{L}_{\perp}^{\kappa}(\varepsilon) = |\nabla_{0k}|^2 \frac{\rho(E_s)}{E_s} \frac{\varepsilon^2+f}{\varepsilon^2+\gamma^2/4} \tag{III.30}$$

$$\mathfrak{L}_{\perp}^{\Delta\kappa}(\varepsilon) = (\nabla_{0k}\cdot\mathbf{l}_{0k})\rho(E_s) \frac{\varepsilon^2+R}{\varepsilon^2+\gamma^2/4} \tag{III.31}$$

The anisotropy ratio (III.1) becomes (apart from some numerical factors)

$$g_{\perp}(\varepsilon) = E_s \frac{(\nabla_{0k}\cdot\mathbf{l}_{0k})}{|\nabla_{0k}|^2} \frac{\varepsilon^2+R}{\varepsilon^2+f} \tag{III.32}$$

and the notation is

$$f = \frac{|\nabla_{0s}|^2}{|\nabla_{0k}|^2} \qquad R = \frac{\nabla_{0s}\cdot\mathbf{l}_{0s}}{\nabla_{0k}\cdot\mathbf{l}_{0k}} \tag{III.33}$$

The intensity ratio f is inherently positive, and the ordinary absorption for this polarization case will hence display a symmetric resonance at $\varepsilon=0$ (i.e., $E=E_s$) when $f>\gamma^2/4$, and a symmetric antiresonance (i.e., a dip) when $f<\gamma^2/4$. For positive values of the rotatory intensity ratio R, we obtain symmetric resonance or antiresonance when R is, respectively, larger than or smaller than $\gamma^2/4$. For negative R, (III.31) always leads to an antiresonance at $\varepsilon=0$. Finally, the anisotropy ratio (III.32) is a smooth function of ε, and shows resonance or antiresonance at $\varepsilon=0$ depending upon the sign and relative magnitudes of the intensity ratios f and R. Notice that simple superposition of the intensities of the two transitions (i.e., total neglect of interference) always leads to a resonance in the ordinary absorption and in the rotatory intensity for $R>0$. On the other hand, when the $0\rightarrow s$ and the $0\rightarrow k$ transitions are both polarized along the C_2 axis, we find

$$\mathfrak{L}_{\parallel}^{\kappa}(\varepsilon)=|\nabla_{0k}|^2\frac{\rho(E_s)}{E_s}\frac{(\varepsilon+\nabla)^2}{\varepsilon^2+\gamma^2/4} \tag{III.34}$$

$$\mathfrak{L}_{\parallel}^{\Delta\kappa}(\varepsilon)=(\nabla_{0k}\cdot\mathbf{l}_{0k})\rho(E_s)\frac{(\varepsilon+\nabla)(\varepsilon+l)}{\varepsilon^2+\gamma^2/4} \tag{III.35}$$

$$g_{\parallel}(\varepsilon)=E_s\frac{(\nabla_{0k}\cdot\mathbf{l}_{0k})}{|\nabla_{0k}|^2}\frac{(\varepsilon+l)}{(\varepsilon+\nabla)} \tag{III.36}$$

where

$$l=\frac{\mathbf{u}\cdot\mathbf{l}_{0s}}{\mathbf{u}\cdot\mathbf{l}_{0k}} \qquad \nabla=\frac{\mathbf{u}\cdot\nabla_{0s}}{\mathbf{u}\cdot\nabla_{0k}} \tag{III.37}$$

and where $\mathbf{u}$ is a unit vector along the C_2 axis. The band shapes resulting from (III.34) to (III.36) are illustrated in Fig. 2*a* and *b*, which show that the ordinary and rotatory intensities for this polarization case differ markedly from simple superposition, and that the anisotropy ratio actually exhibits a strong anomaly. This anomaly lies outside the peak regions of the ordinary and rotatory intensities and is determined by the electric dipole transition moment ratio ∇ [see (III.37)]. In practice, other intramolecular couplings will prevent the ordinary absorption from vanishing exactly at $\varepsilon=-\nabla$, so that the anomaly in the anisotropy ratio will be damped. However, the presence of such a damped anomaly may still serve not only to indicate the presence of an interference, but also to provide an estimate of the ratio of the two electric dipole transition moments.

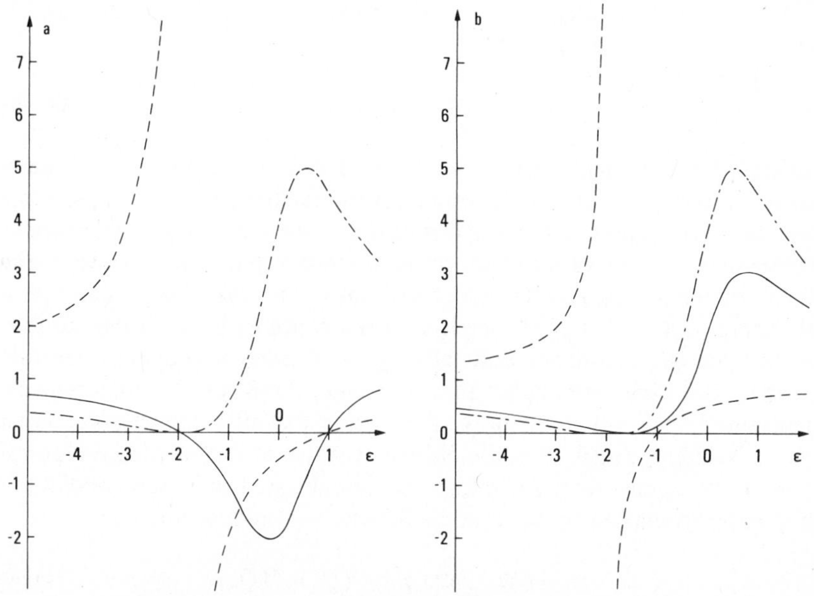

Fig. 2. (*a*) Band shapes relative to $\rho(E_s)$ resulting from (III.34) to (III.36) for $\nabla=2$, $l=-1$, and $\gamma=2$. (–·–·–), the factor $(\varepsilon+\nabla)^2/(\varepsilon^2+\gamma^2/4)$ from (III.34); (—), $(\varepsilon+\nabla)(\varepsilon+l)/(\varepsilon^2+\gamma^2/4)$ from (III.35); (–––), $(\varepsilon+l)/(\varepsilon+\nabla)$ from (III.36). (*b*) Band shapes as in (*a*), but for $\nabla=2$, $l=+1$, $\gamma=2$.

2. *Molecular Exciton Systems*

A molecular exciton system is characterized by the presence of a number of identical chromophoric groups that are so far apart that electronic overlap and exchange are assumed to be negligible.[108–112] Full equivalence of the chromophores, which implies a high degree of degeneracy in the excited electronic states in the absence of interchromophoric coupling, generally requires the presence of some element of symmetry that can carry one chromophore into the next; the sequence of amide groups in an α-helical polypeptide chain is a well-studied molecular exciton system of this nature.[21,111] A molecular crystal in which each molecule is a chromophore is another example.[108] However, even in the absence of such a symmetry element, a molecule containing formally equivalent chromophores, such as the two ethylenic chromophores in 5-methylenebicyclo[2, 2.1]hept-2-ene (Fig. 13, Section VIII) is often treated as a molecular exciton system. We shall use a simple dimeric model to outline the special vibronic problems arising in exciton systems.[84,113–119] The chiroptical properties of such systems are discussed further in Sections V.C and VII.C, and the validity of the exciton model is considered in Section VIII.

The total Hamiltonian for a system consisting of chromophores A and B can be written in the form

$$\hat{\mathcal{H}} = \hat{T}_Q^{\mathrm{A}} + \hat{T}_Q^{\mathrm{B}} + \hat{\mathcal{H}}_{\mathrm{el}}^{\mathrm{A}} + \hat{\mathcal{H}}_{\mathrm{el}}^{\mathrm{B}} + V_{\mathrm{AB}} \tag{III.38}$$

where $\hat{T}_Q^{\mathrm{A}}$ is the kinetic energy operator for the nuclear modes of chromophore A, and $\hat{\mathcal{H}}_{el}^{\mathrm{A}}$ is the Hamiltonian for the electrons in this chromophore [cf. Eqs. (III.3) and (III.4)]. It is assumed that the dimer is kept together by forces that are irrelevant to the optical properties under consideration, and the operator for the kinetic energy associated with their relative motion is therefore neglected. V_{AB} is the electrostatic coupling between the particles in the two chromophores, and the neglect of interchromophoric electron overlap and exchange implies that V_{AB} takes the form of a multipole expansion,[22,112,147] which is most often truncated after the electric dipole term. Assuming finally that the electronic states of the individual chromophores are nondegenerate, the dimeric ground state can be represented to a good approximation by the crude adiabatic[95,72] product function

$$\Phi_0 = \psi_{\mathrm{A}}^0(q_{\mathrm{A}})\psi_{\mathrm{B}}^0(q_{\mathrm{B}})\chi_o^{\mathrm{A}}(Q_{\mathrm{A}})\chi_o^{\mathrm{B}}(Q_{\mathrm{B}}) \tag{III.39}$$

where $\psi_{\mathrm{A}}^0(q_{\mathrm{A}})$ is the ground state of the electrons (q_{A}) in chromophore A, evaluated at the equilibrium geometry of that chromophore.

The electronic states $\psi_{\mathrm{A}}^1(q_{\mathrm{A}})\psi_{\mathrm{B}}^0(q_{\mathrm{B}})$ and $\psi_{\mathrm{A}}^0(q_{\mathrm{A}})\psi_{\mathrm{B}}^1(q_{\mathrm{B}})$, where chromophores A and B, respectively, are in their first excited state, are degenerate with respect to the operator $\hat{\mathcal{H}}_{\mathrm{A}} + \hat{\mathcal{H}}_{\mathrm{B}}$. If we restrict the considerations to the vibronic states generated by just these two electronic states, we can write the resulting excited states in either of the following two equivalent versions of a Longuet-Higgins representation:[95,72]

$$\Phi_\nu = \psi_{\mathrm{A}}^1(q_{\mathrm{A}})\psi_{\mathrm{B}}^0(q_{\mathrm{B}})\chi_\nu^{\mathrm{A}}(Q_{\mathrm{A}}, Q_{\mathrm{B}}) + \psi_{\mathrm{A}}^0(q_{\mathrm{A}})\psi_{\mathrm{B}}^1(q_{\mathrm{B}})\chi_\nu^{\mathrm{B}}(Q_{\mathrm{A}}, Q_{\mathrm{B}}) \tag{III.40}$$

or

$$\Phi_\nu = \Psi^+(q_{\mathrm{A}}, q_{\mathrm{B}})\chi_\nu^+(Q_{\mathrm{A}}, Q_{\mathrm{B}}) + \Psi^-(q_{\mathrm{A}}, q_{\mathrm{B}})\chi_\nu^-(Q_{\mathrm{A}}, Q_{\mathrm{B}}) \tag{III.41}$$

where the functions

$$\Psi^\pm(q_{\mathrm{A}}, q_{\mathrm{B}}) = \frac{1}{\sqrt{2}}\{\psi_{\mathrm{A}}^1(q_{\mathrm{A}})\psi_{\mathrm{B}}^0(q_{\mathrm{B}}) \pm \psi_{\mathrm{A}}^0(q_{\mathrm{A}})\psi_{\mathrm{B}}^1(q_{\mathrm{B}})\} \tag{III.42}$$

diagonalize the coupling V_{AB}. Equation (III.40) is most directly appropriate for the situations where the coupling generated by V_{AB} is weak [see (III.51)]. Here we shall use the representation in (III.41) and (III.42), which is most directly appropriate in the limit where V_{AB} generates a strong coupling between the chromophores. Equations (III.40) and (III.41) are related by a canonical transformation.[113] For real wave functions, we may let

$$E_{AA}=E_{AA}(Q)=\langle\psi_A^1|\hat{\mathcal{H}}_{el}^A|\psi_A^1\rangle+\langle\psi_A^1\psi_B^0|V_{AB}|\psi_A^1\psi_B^0\rangle + \langle\psi_B^0|\hat{\mathcal{H}}_{el}^B|\psi_B^0\rangle \qquad \text{(III.43)}$$

and

$$\gamma=\langle\psi_A^1\psi_B^0|V_{AB}|\psi_A^0\psi_B^1\rangle=\langle\psi_A^0\psi_B^1|V_{AB}|\psi_A^1\psi_B^0\rangle \qquad \text{(III.44)}$$

where the brackets indicate electronic integrations only, and where $E_{AA}(Q)$ and the analogous quantity $E_{BB}(Q)$, respectively, represent the (nuclear coordinate dependent) energy of the local electronically excited states in chromophores A and B. The nuclear functions in (III.41) are determined by a set of coupled equations that we write in the compact form

$$\left[\left\{\hat{T}_Q^A+\hat{T}_Q^B+\tfrac{1}{2}(E_{AA}+E_{BB})-W_\nu\right\}\mathbf{1}+\gamma\boldsymbol{\sigma}_3 + \tfrac{1}{2}(E_{AA}-E_{BB})\boldsymbol{\sigma}_1\right]\begin{Bmatrix}\chi_\nu^+\\ \chi_\nu^-\end{Bmatrix}=0. \qquad \text{(III.45)}$$

Here

$$\mathbf{1}=\begin{pmatrix}1&0\\0&1\end{pmatrix}\qquad \boldsymbol{\sigma}_1=\begin{pmatrix}0&1\\1&0\end{pmatrix}\qquad \boldsymbol{\sigma}_3=\begin{pmatrix}1&0\\0&-1\end{pmatrix} \qquad \text{(III.46)}$$

$\boldsymbol{\sigma}_j$ is the Pauli spin matrix, and the two equations contained in (III.45) are obtained from (III.41) by multiplying by Ψ^+ and Ψ^-, respectively, and integrating over all electronic coordinates. In what follows, we shall assume that the coupling parameter γ can be considered to be independent of the nuclear modes.

If the terms multiplying $\boldsymbol{\sigma}_3$ and $\boldsymbol{\sigma}_1$ in (III.45) vanish, then the equations are completely decoupled, and they will yield inherently degenerate sets of excitations within the chromophores. The vibronic structure of excitations in real exciton systems will hence depend upon the relative magnitudes of $(E_{AA}-E_{BB})/2$ and γ, where

$$\tfrac{1}{2}(E_{AA}-E_{BB})=\tfrac{1}{2}\left\{\langle\psi_A^1|\hat{\mathcal{H}}_{el}^A|\psi_A^1\rangle-\langle\psi_A^0|\hat{\mathcal{H}}_{el}^A|\psi_A^0\rangle\right\} - \tfrac{1}{2}\left\{\langle\psi_B^1|\hat{\mathcal{H}}_{el}^B|\psi_B^1\rangle-\langle\psi_B^0|\hat{\mathcal{H}}_{el}^B|\psi_B^0\rangle\right\} \qquad \text{(III.47)}$$

If the force constants are the same for the ground and excited chromophore states, (III.47) reduces to

$$\tfrac{1}{2}(E_{AA}-E_{BB})=\tfrac{1}{2}\left\{\sum_{\xi} l_{A,\xi}(Q_{A,\xi}-Q^{\circ}_{A,\xi})+\sum_{\xi} l_{B,\xi}(Q_{B,\xi}-Q^{\circ}_{B,\xi})\right\} \tag{III.48}$$

where $l_{A,\xi}$ is a measure of the displacement of the excited chromophore state along the $Q_{A,\xi}$ normal mode, and $Q^{\circ}_{A,\xi}$ is the ground-state equilibrium value for this mode. To compare the magnitude of the operator in (III.48) to the electrostatic coupling γ (III.44), we can evaluate the root mean square of (III.48) over the vibrational ground state to yield

$$\begin{aligned}\Delta &= \tfrac{1}{2}\left[\langle\chi_o^A\chi_o^B|(E_{AA}-E_{BB})^2|\chi_o^A\chi_o^B\rangle\right]^{\frac{1}{2}}\\ &= \tfrac{1}{2}\left[\sum_{\xi} l^2_{A,\xi}\langle\chi_o^A|(Q_{A,\xi}-Q^{\circ}_{A,\xi})^2|\chi_o^A\rangle\right.\\ &\quad\left.+\sum_{\xi} l^2_{B,\xi}\langle\chi_o^B|(Q_{B,\xi}-Q^{\circ}_{B,\xi})^2|\chi_o^B\rangle\right]=\tfrac{1}{2}(\Delta_A+\Delta_B)\end{aligned} \tag{III.49}$$

where, for identical symmetry-related chromophores, $\Delta_A=\Delta_B$ is the width of the local electronic $0\rightarrow 1$ absorption band in the chromophore. The two coupling cases[84, 115] can hence be classified by

$$\gamma \gg \Delta \qquad \text{strong coupling} \tag{III.50}$$

$$\gamma \ll \Delta \qquad \text{weak coupling} \tag{III.51}$$

In the strong coupling limit, the electronic states are so well separated that they can be used to form bases for separate adiabatic approximations for the resulting states and the band shapes and integrated intensities for this limit therefore fall into the category discussed in Section III.A. On the other hand, in the weak coupling limit the individual vibronic chromophore transitions couple via V_{AB}, and the band structure becomes rather complex.[113]

Despite the complexity of the band shapes in the intermediate and weak coupling limit, some general features have been found to persist for all coupling strengths. In analogy to (III.6a) consider the intensity quantity

$$
\begin{aligned}
I_{0;\nu}(\omega) &= \rho_{0;\nu}(\omega)\langle\Phi_0|\hat{\mathbf{F}}^a|\Phi_\nu\rangle\cdot\langle\Phi_\nu|\hat{\mathbf{F}}^h|\Phi_0\rangle \\
&= \rho_{0;\nu}(\omega)\{\mathbf{F}^a_{0+}\cdot\mathbf{F}^h_{+0}\langle\chi_o^A\chi_o^B|\chi_\nu^+\rangle_Q\langle\chi_\nu^+|\chi_o^A\chi_o^B\rangle_Q \\
&\quad+\mathbf{F}^a_{0+}\cdot\mathbf{F}^h_{-0}\langle\chi_o^A\chi_o^B|\chi_\nu^+\rangle_Q\langle\chi_\nu^-|\chi_o^A\chi_o^B\rangle_Q \\
&\quad+\mathbf{F}^a_{0-}\cdot\mathbf{F}^h_{+0}\langle\chi_o^A\chi_o^B|\chi_\nu^-\rangle_Q\langle\chi_\nu^+|\chi_o^A\chi_o^B\rangle_Q \\
&\quad+\mathbf{F}^a_{0-}\cdot\mathbf{F}^h_{-0}\langle\chi_o^A\chi_o^B|\chi_\nu^-\rangle_Q\langle\chi_\nu^-|\chi_o^A\chi_o^B\rangle_Q\},
\end{aligned}
\tag{III.52a}
$$

resulting from (III.39) and (III.41). Here

$$
\begin{aligned}
\mathbf{F}^a_{0+} &= \langle\psi_A^0\psi_B^0|\hat{\mathbf{F}}^a|\Psi^+\rangle_q \\
&= \frac{1}{\sqrt{2}}\{\langle\psi_A^0|\hat{\mathbf{F}}^a_A|\psi_A^1\rangle_q+\langle\psi_B^0|\hat{\mathbf{F}}^a_B|\psi_B^1\rangle_q\},
\end{aligned}
\tag{III.52b}
$$

with analogous expressions for the other electronic transition moments, $\hat{\mathbf{F}}_A$ being the part of the operator $\hat{\mathbf{F}}$ pertaining to the electrons in chromophore A. The relations

$$\sum_\nu|\chi_\nu^+\rangle\langle\chi_\nu^+|=\sum_\nu|\chi_\nu^-\rangle\langle\chi_\nu^-|=1 \tag{III.53a}$$

$$\sum_\nu|\chi_\nu^+\rangle\langle\chi_\nu^-|=\sum_\nu|\chi_\nu^-\rangle\langle\chi_\nu^+|=0 \tag{III.53b}$$

$$
\begin{aligned}
&\sum_\nu|\chi_\nu^\pm\rangle(w_\nu-w_o^0)\langle\chi_\nu^\pm| \\
&\qquad= \hat{T}_Q^A+\hat{T}_Q^B+\tfrac{1}{2}[E_{AA}(Q)+E_{BB}(Q)]\pm\gamma-w_o^0
\end{aligned}
\tag{III.53c}
$$

and

$$
\begin{aligned}
&\sum_\nu|\chi_\nu^+\rangle(w_\nu-w_o^0)\langle\chi_\nu^-|=\sum_\nu|\chi_\nu^-\rangle(w_\nu-w_o^0)\langle\chi_\nu^+| \\
&\qquad=\tfrac{1}{2}[E_{AA}(Q)-E_{BB}(Q)]
\end{aligned}
\tag{III.53d}
$$

where w_o^0 is the Born-Oppenheimer ground-state energy of (III.39), follow from the formal completeness of (III.41) within the space spanned by Ψ^+

and Ψ^- and from (III.45). We then obtain the following expressions for the zeroth and first moment of the intensity distribution in (III.52a)

$$I^{(0)} = \int d\omega \sum_\nu I_{0;\nu}(\omega)$$
$$= \langle \psi_A^0 | \hat{\mathbf{F}}_A^a | \psi_A^1 \rangle_q \cdot \langle \psi_A^1 | \hat{\mathbf{F}}_A^h | \psi_A^0 \rangle_q + \langle \psi_B^0 | \hat{\mathbf{F}}_B^a | \psi_B^1 \rangle_q \cdot \langle \psi_B^1 | \hat{\mathbf{F}}_B^h | \psi_B^0 \rangle_q \quad \text{(III.54)}$$

and

$$I^{(1)} = \int d\omega\, \omega \sum_\nu I_{0;\nu}(\omega) = \hbar^{-1} \sum_\nu (w_\nu - w_o^0) \int d\omega I_{0;\nu}(\omega)$$
$$= \mathbf{F}_{0+}^a \cdot \mathbf{F}_{+0}^h (\omega_A + \gamma/\hbar) + \mathbf{F}_{0-}^a \cdot \mathbf{F}_{-0}^h (\omega_A - \gamma/\hbar)$$
$$+ I^{(0)} \langle \chi_o^A \chi_o^B | \{ \tfrac{1}{2} [\bar{\varepsilon}_{AA}(Q) + \bar{\varepsilon}_{BB}(Q)] - \bar{\varepsilon}_o(Q) \} | \chi_o^A \chi_o^B \rangle_Q$$
$$+ \{ \mathbf{F}_{0+}^a \cdot \mathbf{F}_{-0}^h + \mathbf{F}_{0-}^a \cdot \mathbf{F}_{+0}^h \} \langle \chi_o^A \chi_o^B | \tfrac{1}{2} [\bar{\varepsilon}_{AA}(Q) - \bar{\varepsilon}_{BB}(Q)] | \chi_o^A \chi_o^B \rangle_Q \quad \text{(III.55)}$$

where $\omega_A = \omega_B$ is the Franck-Condon frequency for the unperturbed chromorphore, and where the reduced energy quantities $\bar{\varepsilon}_i(Q)$ are defined in analogy to the terms in (III.9). Equation (III.54) shows that the total intensity is equal to the sum of the corresponding intensities in the unperturbed chromophores. Therefore this simple vibronic excitation coupling model conserves the total intensity, independent of the strength of the coupling. On the other hand, the first moment and the center of gravity will depend on the geometry and vibronic structure of the dimeric system.

Equations (III.52a) and (III.55) simplify in a number of situations. Firstly, in the strong coupling limit the nuclear functions χ_ν^+ and χ_ν^- are never large simultaneously. In that case (III.52a) and (III.55) represent two separate progressions, one with center of gravity at approximately $\omega_A + \gamma/\hbar$ scaled by the electronic intensity $\mathbf{F}_{0+}^a \cdot \mathbf{F}_{+0}^h$, and the other with center of gravity at $\omega_A - \gamma/\hbar$ scaled by $\mathbf{F}_{0-}^a \cdot \mathbf{F}_{-0}^h$. Secondly, if the dimeric system has an element of symmetry that can carry one chromophore into the other, then for symmetry reasons the terms $\mathbf{F}_{0+}^a \cdot \mathbf{F}_{-0}^h$ and $\mathbf{F}_{0-}^a \cdot \mathbf{F}_{+0}^h$ vanish identically, and in this case the intensity distribution given by (III.52a) can again be decomposed into two separate (but in the intermediate and weak coupling limit overlapping) progressions. If the force constants are the same in the ground and excited states, (III.55) shows that these two progressions will have their centers of gravity at $\omega_A \pm \gamma/\hbar$. These features have been observed in the treatment of a specific dimer coupling case by Fulton and Gouterman,[113] and they find in addition that the widths of the two progressions are the same.

The dimeric system considered here is useful for providing a qualitative picture and defining the limiting coupling case. More detailed analyses and calculations on the exciton band shapes are given in Refs. 113 and 116–119 for dimeric systems and in Refs. 120 and 121 for polymers. It follows from the above that, unless a detailed study of the vibronic structure of a coupled chromophore system is undertaken, meaningful correlations between the computed ordinary and rotatory electronic intensities and the molecular structure of such systems are practicable only when the individual progressions discussed above can be clearly distinguished.

C. Comments

In the overwhelming majority of the literature on the calculation of rotatory intensities and the correlation of rotatory intensities with molecular structure, oscillator and rotatory strengths are evaluated or computed from (III.17)-(III.23) and are subsequently compared with the magnitude and sign of experimental intensities obtained by integrating the observed band shapes according to (III.15b) and (III.16b). We shall indeed adhere to that tradition in the following. However, as is apparent from the preceding sections, a straightforward correlation of computed and measured intensities is only practicable for case I and II nondegenerate transitions and for certain molecular exciton systems. For degenerate and near-degenerate transitions or for nondegenerate case III transitions, a careful analysis of the vibronic structure is required. The frequency dependent anisotropy ratio $g(\omega)$ (III.1) can be of very significant assistance in ascertaining the nature of the vibronic coupling in a given transition. We therefore highly recommend that $g(\omega)$ be recorded and published along with ordinary and rotatory intensities (as is certainly within the capabilities of present-day microprocessor-based instrumentation), and at the same time theoreticians should make sure that their assignment of a transition agrees with the observed anisotropy ratio.

IV. GENERAL PROPERTIES OF THE INTEGRATED INTENSITIES

A. Symmetry Considerations

The realization that the phenomenon of natural molecular optical activity in isotropic media requires that the molecules be nonsuperimposable on their mirror image (i.e., chiral) dates back to Pasteur (1848) and Lord Kelvin (1884). Within the context of the expressions developed in Section III, namely for electronic transitions arising in a semirigid molecular framework, the states of the system may be classified according to the point group of the electronic Hamiltonian corresponding to the ground-state equilibrium nuclear conformation,[75] and chirality then demands that this point group does not contain an improper axis of rotation (i.e., a rotation-inversion axis; see below). The argument may be cast in terms of the types of excitation that can be induced by the position, linear momentum, and angular momentum operators **r**, $\hat{\mathbf{p}}$, and $\mathbf{r}\times\hat{\mathbf{p}}$ that give rise to the transition moments in (III.17) to (III.23).

If the inversion operator $\hat{S}_0$ is defined by its effect on the position and linear momentum operators,

$$\hat{S}_0\mathbf{r}\hat{S}_0^\dagger = -\mathbf{r} \qquad \hat{S}_0\hat{\mathbf{p}}\hat{S}_0^\dagger = -\mathbf{p} \tag{IV.1}$$

the transformation of the angular momentum becomes

$$\hat{S}_0\mathbf{r}\times\hat{\mathbf{p}}\hat{S}_0^\dagger = \mathbf{r}\times\hat{\mathbf{p}} \tag{IV.2}$$

The operator representing a proper n-fold rotation about an axis **u** is defined as

$$\hat{R}_n^{\mathbf{u}} = \exp\left[-\frac{2\pi i}{n}(\hat{\mathbf{L}}\cdot\mathbf{u})\right] \tag{IV.3}$$

where $\hat{\mathbf{L}}$ is the angular momentum operator and **r** and $\hat{\mathbf{p}}$ transform according to

$$\hat{R}_n^{\mathbf{u}}\mathbf{r}\hat{R}_n^{\mathbf{u}\dagger} \qquad \text{and} \qquad \hat{R}_n^{\mathbf{u}}\hat{\mathbf{p}}\hat{R}_n^{\mathbf{u}\dagger} \tag{IV.4}$$

from which we obtain

$$(\hat{R}_n^{\mathbf{u}}\mathbf{r}\hat{R}_n^{\mathbf{u}\dagger})\times(\hat{R}_n^{\mathbf{u}}\hat{\mathbf{p}}\hat{R}_n^{\mathbf{u}\dagger}) = \hat{R}_n^{\mathbf{u}}(\mathbf{r}\times\hat{\mathbf{p}})\hat{R}_n^{\mathbf{u}\dagger} \tag{IV.5}$$

Equations (IV.4) and (IV.5) together with (IV.1) and (IV.2) display the polar vector character of **r** and $\hat{\mathbf{p}}$, and the axial (or pseudo-) vector character of $\mathbf{r}\times\hat{\mathbf{p}}$. [4]An improper n-fold rotation is now defined by

$$\hat{S}_n^{\mathbf{u}} = \hat{S}_0\hat{R}_n^{\mathbf{u}} = \hat{R}_n^{\mathbf{u}}\hat{S}_0 \tag{IV.6}$$

where the last of these relations follows because $\hat{S}_0$ commutes with $\hat{\mathbf{L}}$ (IV.2), and the three transition moment operators accordingly transform as follows:

$$\hat{S}_n^{\mathbf{u}}\mathbf{r}\hat{S}_n^{\mathbf{u}\dagger} = -(\hat{R}_n^{\mathbf{u}}\mathbf{r}\hat{R}_n^{\mathbf{u}\dagger}) \qquad \hat{S}_n^{\mathbf{u}}\hat{\mathbf{p}}\hat{S}_n^{\mathbf{u}\dagger} = -(\hat{R}_n^{\mathbf{u}}\hat{\mathbf{p}}\hat{R}_n^{\mathbf{u}\dagger}) \tag{IV.7}$$

$$(\hat{S}_n^{\mathbf{u}}\mathbf{r}\hat{S}_n^{\mathbf{u}\dagger})\times(\hat{S}_n^{\mathbf{u}}\hat{\mathbf{p}}\hat{S}_n^{\mathbf{u}\dagger}) = \hat{R}_n^{\mathbf{u}}(\mathbf{r}\times\hat{\mathbf{p}})\hat{R}_n^{\mathbf{u}\dagger} \tag{IV.8}$$

If the electronic Hamiltonian commutes with $\hat{R}_n^{\mathbf{u}}$, (IV.4) and (IV.5) show that **r**, $\hat{\mathbf{p}}$, and $\mathbf{r}\times\hat{\mathbf{p}}$ obey the same selection rules, and the presence of a proper axis of rotation therefore does not preclude chirality. On the other hand, if the electronic Hamiltonian commutes with $\hat{S}_n^{\mathbf{u}}$, (IV.7) and (IV.8) show that **r** and $\hat{\mathbf{p}}$ induce excitations only of odd improper rotation parity, whereas $\mathbf{r}\times\hat{\mathbf{p}}$ induces only even excitations. These two selection rules are of course mutually exclusive, and a molecule for which the point group of the ground state nuclear geometry contains an improper rotation axis therefore cannot be chiral. Since $\hat{S}_n^{\mathbf{u}}$ for n odd implies the presence of a reflection plane and for $n/2$ odd a center of inversion [see Eq. (IV.6)], these two cases cover the standard textbook criteria for optical inactivity. However, (IV.7) and (IV.8) also preclude optical activity of molecules with an $\hat{S}_4^{\mathbf{u}}$ axis, which contains no plane or center of symmetry, and it is well established[122] that molecules with a fourfold improper rotation axis are optically inactive.

In summary, for randomly oriented semirigid molecules only the pure rotation point groups C_n, D_n, T, O, and I support a chiral response to an applied radiation field. For nonrigid molecules, a similar analysis leads to the restriction that the appropriate Longuet-Higgins permutation group may not contain the "inversion" operator E^*, either alone or in combination with permutations.[75]

B. Sum Rules

The electronic oscillator and rotatory strengths (III.17) to (III.23) fulfill a number of sum rules, that is, relations of the form

$$\sum_b (E_b - E_0)^n f_{0b} = f^{(n)} \tag{IV.9}$$

and

$$\sum_b (E_b - E_0)^n R_{0b} = R^{(n)} \tag{IV.10}$$

These quantities can be considered the nth moments of the (entire) spectral distributions of ordinary and rotatory electronic intensities. The derivation of these sum rules is based on various special versions of the off-diagonal hypervirial relation (II.25) [such as (II.20)], and for the oscillator strength, one finds, for example,[123]

$$f^{(-1)} = \sum_b (E_b - E_0)^{-1} f_{0b}$$

$$= \frac{2m}{3\hbar^2}\left\{\langle 0|\sum_s \mathbf{r}_s^2|0\rangle + \langle 0|\sum_{s,t}{}' \mathbf{r}_s\cdot\mathbf{r}_t|0\rangle\right\} \tag{IV.11}$$

$$f^{(0)} = \sum_b f_{0b} = N_e \tag{IV.12}$$

$$f^{(1)} = \sum_b (E_b - E_0) f_{0b}$$

$$= -\frac{2\hbar^2}{3m}\left\{\langle 0|\sum_s \hat{\nabla}_s^2|0\rangle + \langle 0|\sum_{s,t}{}' \hat{\nabla}_s\cdot\hat{\nabla}_t|0\rangle\right\}, \tag{IV.13}$$

where the primed summations exclude the terms $s=t$. In (IV.12), which is the Thomas–Reiche–Kuhn sum rule,[123] N_e is the total number of electrons in the system. In (IV.11) the first term is related to the ground-state diamagnetic susceptibility, whereas the first term in (IV.13) is proportional to the electronic kinetic energy, which in turn is simply related to the total energy through the ordinary virial theorem[123] (at least for atoms[124]). The second terms in (IV.11) and (IV.13) depend markedly upon the degree of electronic correlation in the ground state; in fact, the second term in (IV.13) is strictly zero for a pure product (i.e., Hartree) electronic ground-state wave function. In addition, it is found[123] that the sum $f^{(-2)}$ is proportional to the static limit of the electric polarizability, while $f^{(2)}$ is proportional to the ground-state electronic density at the nucleus.

For the rotatory strength, the sum rule

$$R^{(0)}=\sum_b R_{0b}=0 \tag{IV.14}$$

was established by Condon,[4] and only recently were a number of additional sum rules obtained.[125,126] These sum rules show that

$$R^{(2)}=R^{(4)}=0 \tag{IV.15}$$

$$R^{(1)}=-\frac{e^2\hbar^3}{4m^2c}\langle 0\left|\sum_{s,t}(\mathbf{r}_s-\mathbf{r}_t)\cdot(\hat{\nabla}_s\times\hat{\nabla}_t)\right|0\rangle \tag{IV.16}$$

$$R^{(3)}=\frac{e^2\hbar^3}{4m^2c}\langle 0\left|\sum_{s,t}(\mathbf{r}_s-\mathbf{r}_t)\cdot\left[(\hat{\nabla}_s V_s)\times(\hat{\nabla}_t V_t)\right]\right|0\rangle \tag{IV.17}$$

where V_s is the (local) electrostatic potential energy of electron s. For n equal to or larger than five, $R^{(n)}$ is a complicated many-electron ground-state matrix element.[125,126] The two conspicuous features of these sum rules for the rotatory strength are that the lower even moments (i.e., $R^{(0)}$, $R^{(2)}$, and $R^{(4)}$) are strictly zero ($R^{(6)}$ has been found not to vanish[126]), and that the odd moments $R^{(1)}$ and $R^{(3)}$ are pure electron correlation effects, since (IV.16) and (IV.17) vanish identically for one-electron systems or for pure product electronic wave functions. However, $R^{(5)}$ is different from zero for one-electron as well as many-electron chiral systems. These sum rules have been tested[125] against the two classic models of optical activity,[4] namely the perturbed anisotropic harmonic oscillator model for one-electron optical activity and the coupled oscillator model. Similar sum rules have been derived for the rotatory strength of oriented systems[125] and for the high-frequency contribution to optical rotatory dispersion.[127]

Compared to ordinary intensities, the spectral distribution of the rotatory intensity is therefore quite peculiar in that the lower even moments vanish identically for all chiral molecules, and that all moments lower than the fifth vanish for one-electron systems. This difference between spectral distribution of one- and many-electron rotatory intensities is interesting since classical optics yields vanishing linear optical rotatory power for one-electron systems,[128,44] and it was therefore generally believed for a long time that only strongly coupled many-electron systems could exhibit optical activity (the coupled oscillator model[128,20] was the preferred version of such a strongly coupled system). This credo was dispelled when Condon, Altar, and Eyring[4,5] demonstrated the nonvanishing optical activity of a quantum-mechanical chiral one-electron system; however, a difference apparently remains because the nonvanishing lower moments of the rotatory intensity are totally dependent upon the coupling of electronic motion [whereas correlation plays only a minor role in the lower moments of the ordinary intensity, (IV.11) to (IV.13)].

C. Origin Dependence

The electric dipole transition moments $\langle 0|\mathbf{r}|b\rangle$ and $\langle 0|\hat{\nabla}|b\rangle$ are both invariant to a translation of the coordinate system, whereas the magnetic dipole transition moment for an excitation that is electric-dipole-allowed is not invariant, because the translation $\mathbf{r}'=\mathbf{r}-\mathbf{R}$ leads to

$$\langle 0|\mathbf{r}'\times\hat{\nabla}|b\rangle=\langle 0|\mathbf{r}\times\hat{\nabla}|b\rangle-\mathbf{R}\times\langle 0|\hat{\nabla}|b\rangle \tag{IV.18}$$

However, defining the component of $\langle 0|\mathbf{r}\times\hat{\nabla}|b\rangle$ along $\langle 0|\hat{\nabla}|b\rangle$ by

$$L'_{0b}=\langle 0|\mathbf{r}\times\hat{\nabla}|b\rangle\cdot\langle 0|\hat{\nabla}|b\rangle/|\langle 0|\hat{\nabla}|b\rangle| \tag{IV.19}$$

we observe that this component is actually invariant to a translation, so that for an electric-dipole-allowed transition, $|L'_{0b}|$ represents the smallest possible magnitude of the magnetic dipole transition moment. In the nomenclature of Moffitt,[111] (IV.19) represents the magnetic dipole transition moment relative to a point on the partial optic axis for this particular excitation. For an achiral molecule, comparison of (III.22) and (IV.19) shows that L'_{0b} vanishes identically. In other words, for an electric-dipole-allowed transition in an achiral molecule, one can always choose an origin such that the associated magnetic dipole transition moment is zero.

From the general properties of vectors, we see easily that the velocity form of the rotatory strength (III.22) is manifestly origin-invariant,[129] whereas the length form (III.23) is formally modified by a term of magnitude

$$\Delta R_{0b}^{r} = \frac{e^2\hbar}{2mc}\mathbf{R}\cdot[\langle 0|\hat{\nabla}|b\rangle \times \langle 0|\mathbf{r}|b\rangle] \tag{IV.20}$$

If the two transition moment vectors are colinear, either for symmetry reasons or because the hypervirial relation (II.25) is fulfilled for the pertinent states $|0\rangle$ and $|b\rangle$, then ΔR_{0b}^{r} vanishes, and the length form of the rotatory strength is origin-independent as well. However, actual calculations can lead to violations of the hypervirial relation, and therefore to a departure from colinearity of the two electric dipole transition moments. The angle between them, β_{0b}, may be expressed as

$$\begin{aligned}\beta_{0b} &= \cos^{-1}\left\{\frac{\langle 0|\mathbf{r}|b\rangle\cdot\langle 0|\hat{\nabla}|b\rangle}{|\langle 0|\mathbf{r}|b\rangle||\langle 0|\hat{\nabla}|b\rangle|}\right\} \\ &= \cos^{-1}\left\{\frac{f_{0b}^{r}}{(f_{0b}^{r}f_{0b}^{\nabla})^{1/2}}\right\}\end{aligned} \tag{IV.21}$$

using the oscillator strength expressions (III.18) to (III.20). Computation of both types of electric dipole transition moment thus affords the advantage of an assessment of the degree to which the hypervirial relation is violated for a particular excitation. In terms of the angle β_{0b}, the origin-dependence term ΔR_{0b}^{r} becomes

$$\Delta R_{0b}^{r} = \frac{3e^2\hbar}{4mc}|\mathbf{R}\cdot\mathbf{n}||f_{0b}^{r\nabla}\tan\beta_{0b}| \tag{IV.22}$$

where $\mathbf{n}$ is a unit vector along the direction of the vector product in (IV.20). The quantity $(3e^2\hbar/4mc)|f_{0b}^{r\nabla}\tan\beta_{0b}|$ represents the largest possible change in R_{0b}^{r} for a unit translation of the coordinate origin within the molecule. Alternatively, (IV.19) can be used to define an origin-invariant dipole-length expression for the rotatory strength, namely

$$R_{0b}^{[r]} = -\frac{e^2\hbar}{2mc}(\mathbf{r}_{0b}\cdot\mathbf{L}'_{b0}) \tag{IV.23}$$

or[130]

$$R_{0b}^{[r]} = R_{0b}^{\nabla}\left(\frac{f_{0b}^{r\nabla}}{f_{0b}^{\nabla}}\right) \tag{IV.24}$$

Equations (IV.22) and (IV.24) thus define the separation of R_{0b}^{r} into origin-independent and origin-dependent parts. $R_{0b}^{[r]}$ (IV.24) is the quantity to compare with R_{0b}^{∇} as an index of the quality of the computational model.

Finally, we note that (IV.18) suggests that the magnetic dipole transition moment may be made arbitrarily large by an appropriate translation of the origin. However, from the way in which this matrix element arises in the truncated multipole expansion of Section II, it is meaningful to consider only translations within the molecule in the context of discussions of chiroptical properties. If the excitation is only weakly electric-dipole-allowed, therefore, a translation within the molecule may still leave the magnetic dipole transition moment relatively well defined.

V. CORRELATIONS WITH MOLECULAR STRUCTURE

As discussed in the introduction, theoretical–computational approaches to chiroptical intensities fall into two (not sharply delineated) categories: essentially full-molecule all-electron calculations, and model approaches that avoid the calculation of total wave functions and concentrate instead on the properties of the chromophores and their interaction with the surrounding molecular framework. In this section we consider some aspects of the model approaches; more complete reviews are given in Refs. 22, 25, and 147.

Moscowitz[7] distinguished two types of chiral molecules, according to the local symmetry of the chromophoric group responsible for the electronic transition. The first type contains *inherently chiral* chromophores, often consisting of nonplanar π-electron systems such as in hexahelicene[7] and *trans*-cyclooctene,[132] although in a rigorous sense, all chiral molecules are inherently chiral chromophores. In such molecules, both the electric and the magnetic transition moments are nonzero, even in simple models of optical activity. The second type comprises molecules with *achiral* chromophores coupled with *chiral* surroundings to yield nonzero optical activity, for example the saturated ketones.[133,134] In this group, either the electric or the magnetic dipole transition moment of the pertinent chromophoric excitation is formally forbidden owing to the local chromophoric symmetry. The goal of approximate models of optical activity for this class of compounds is hence to relate the induced chiral response of a locally achiral chromophore to the structure of the surroundings.

The two classes just mentioned may be recognized as limiting cases of the strength of the interaction between various parts of the molecule. A sufficiently strong interaction leads simply to a redefinition of the chromophore. The molecular exciton systems discussed in Section III.B.2 are an important example of such a redefinition of the chromophore, because the interaction between the identical (nonoverlapping) chromophoric groups necessitates an approach in which this collection of groups is treated as one extended chromophore. The chiroptical intensities obtained from the simple molecular exciton model are accordingly considered separately in Section V.C.

Apart from exciton systems, the most frequent examples of chiral chromophores are unsaturated systems that are twisted out of a planar conformation, as noted above. Correlations between the observed spectra and the molecular geometry most often depend on the sense of helicity of the twisted chromophore. Examples of this are the helicity rules for dienes[135–137] and diones,[138] although Burgstahler and co-workers[139–141] have suggested that the diene chromophore be extended to include allylic axial sigma bonds. (See Section VIII for computational indications.) Since even simple model calculations on chiral chromophores involve polycenter molecular orbitals, we shall defer further discussion of this class of molecules until the general computational treatment in Section VII.

A. Achiral Chromophores

In the case where the interaction between chromophore and surroundings is sufficiently weak so that the system would not fall under the rubric of the previous paragraph, a framework put forward by Höhn and Weigang[142] provides useful models. In their treatment one formally partitions the molecule into a "chromophore," A, and one or more "perturbers," B. The zeroth-order states of the system, $|A_rB_s)$, are taken to be simple products of the states $|A_r\rangle$ and $|B_s\rangle$ of the two parts, assuming that charge transfer and electronic overlap and exchange between A and B can be neglected. The justification for the neglect of charge transfer in such a model has been considered[143] within a semiempirical (i.e., Pariser–Parr type[144]) computational scheme; we shall return to this problem in Section VIII. It is assumed further that none of the low-lying states of B are degenerate with the states of A. The chromophoric excitation $0 \rightarrow m$ may then be expressed in terms of the first-order perturbed electronic states

$$|A_iB_0\rangle = |A_iB_0) + \sum_r \sum_s \frac{(A_rB_s|V|A_iB_0)}{\hbar(\omega_{i0} - \omega_{r0} - \omega_{s0})} |A_rB_s) \qquad \text{(V.1)}$$

where $i = 0$ or m, and ω_{k0} is the transition frequency of the kth state of A or B. Here V is an electrostatic interaction potential coupling systems A and B. We shall not need to specify a particular form for V; different choices of V lead to different models for optical activity. If the chromophoric transition is formally electric-dipole-forbidden, but magnetic-dipole-allowed, then to first order,[21,143]

$$\langle A_0B_0|\mathbf{r}\times\hat{\nabla}|A_mB_0\rangle = (A_0B_0|\mathbf{r}\times\hat{\nabla}|A_mB_0) = (A_0|(\mathbf{r}\times\hat{\nabla})_A|A_m) \quad \text{(V.2)}$$

and

$$\begin{aligned}\langle A_0B_0|\boldsymbol{\mu}|A_mB_0\rangle \\ &= -\sum_{i\neq 0}\frac{(A_iB_0|V|A_0B_0)}{\hbar\omega_{i0}}(A_i|\boldsymbol{\mu}_A|A_m) \\ &\quad+\sum_{k\neq m}\frac{(A_kB_0|V|A_mB_0)}{\hbar(\omega_{m0}-\omega_{k0})}(A_0|\boldsymbol{\mu}_A|A_k) \\ &\quad-\sum_{j}\frac{(A_mB_j|V|A_0B_0)}{\hbar(\omega_{m0}+\omega_{j0})}(B_j|\boldsymbol{\mu}_B|B_0) \\ &\quad+\sum_{l}\frac{(A_0B_l|V|A_mB_0)}{\hbar(\omega_{m0}-\omega_{l0})}(B_0|\boldsymbol{\mu}_B|B_l) \end{aligned} \quad \text{(V.3)}$$

where we assume that the coordinate system is centered in chromophore A, and that $(\mathbf{r}\times\nabla)_A$ is $\Sigma_s(\mathbf{r}_s\times\hat{\nabla}_s)$ and $\boldsymbol{\mu}_A$ is $\Sigma_s\mathbf{r}_s$ or $\Sigma_s\hat{\nabla}_s$, including only the electrons in chromophore A in the summations. $\boldsymbol{\mu}_B$ is similarly summed only over the electrons in the perturber. A formally electric-dipole-allowed, magnetically forbidden transition leads to[21,143]

$$\langle A_0B_0|\boldsymbol{\mu}|A_mB_0\rangle = (A_0B_0|\boldsymbol{\mu}|A_mB_0) = (A_0|\boldsymbol{\mu}_A|A_m) \quad \text{(V.4)}$$

and

$$\begin{aligned}\langle A_0B_0|\mathbf{r}\times\hat{\nabla}|A_mB_0\rangle \\ &= \sum_{i\neq 0}\frac{(A_iB_0|V|A_0B_0)}{\hbar\omega_{i0}}(A_i|(\mathbf{r}\times\hat{\nabla})_A|A_m) \\ &\quad+\sum_{k\neq m}\frac{(A_kB_0|V|A_mB_0)}{\hbar(\omega_{m0}-\omega_{k0})}(A_k|(\mathbf{r}\times\hat{\nabla})_A|A_0) \\ +\sum_{j}\frac{(A_mB_j|V|A_0B_0)}{\hbar(\omega_{m0}+\omega_{j0})}&\left[\frac{i\omega_{j0}}{2c}\mathbf{R}_B\times(B_j|\hat{\nabla}_B|B_0)+(B_j|(\mathbf{r}\times\hat{\nabla})_B|B_0)\right] \\ +\sum_{l}\frac{(A_0B_l|V|A_mB_0)}{\hbar(\omega_{m0}-\omega_{l0})}&\left[\frac{i\omega_{j0}}{2c}\mathbf{R}_B\times(B_l|\hat{\nabla}_B|B_0)+(B_l|(\mathbf{r}\times\hat{\nabla})_B|B_0)\right] \end{aligned} \quad \text{(V.5)}$$

where $\mathbf{R}_B$ is the position vector of a center in the perturber, and $(\mathbf{r}\times\hat{\nabla})_B$ is $\Sigma_s(\mathbf{r}_s-\mathbf{R}_B)\times\hat{\nabla}_s$ summed over the electrons in the perturber. Notice, however, that because of the origin dependence of the magnetic transition moment, only the component of $(\mathbf{r}\times\hat{\nabla})_{0m}$ along $\hat{\nabla}_{0m}$ is physically meaningful in a discussion of "allowed" or "forbidden" magnetic dipole transition moments in this context (see Section IV.C). These relations must be summed over all perturbing groups B.

The first two terms in (V.3) and (V.5) induce a nonzero transition moment through mixing of the chromophore states by means of the electrostatic perturbation of the B fragment in its ground state. These are the so-called *static coupling* terms, and give rise to the "one-electron" model of Condon, Altar, and Eyring,[5, 19, 8] and to static "sector rules," the best known of which is the octant rule,[145] which provides the sign of the rotatory intensity of the $n\rightarrow\pi^*$ transition of a carbonyl chromophore in saturated chiral molecules. In general, a sector rule is a protocol for the division of the molecular framework surrounding a given type of chromophore into regions, such that the sign of the rotatory intensity induced by a perturber, for example a methyl group, can be assessed from a knowledge of the region in which the perturber is situated. It should be realized of course that such a division is characteristic not only of a specific type of chromophore, but also of a specific excitation in that chromophore. The construction of sector rules for the static coupling terms was formalized by Schellman,[131] who derived forms for the simplest pseudoscalar functions $f(X,Y,Z)$ of Cartesian perturber coordinates for which the sign of R_{0m} equals the sign of $f(X,Y,Z)$, making use of the fact that the perturbation potential V must contain a part transforming as a pseudoscalar in the point group of the chromophore. However, such symmetry arguments can give only the minimum partitioning into sign-determining regions. As an example, the C_{2v} point group of an isolated carbonyl group gives rise to a pseudoscalar of the form XY for the $^1A_2(n\rightarrow\pi^*)$ transition; that is, the space around the chromophore is divided by the symmetry planes of the carbonyl group into four quadrants, and a perturbing group placed in one of these regions gives a contribution to R_{0m} of sign characteristic of the region. On the other hand, the nodal properties of the wave functions involved leads to an additional surface[133] (bisecting the C=O bond), as anticipated in the original octant rule (Fig. 3).[145] Experimentally, evidence exists that a division into octants rather than quadrants is more appropriate for chiral ketones.[146]

The third and fourth terms in (V.3) and (V.5) couple the states of B to the chromophore states. They are the *dynamic coupling* terms, and for saturated chromophore surroundings these terms lead to the polarizability model proposed by Kirkwood[18] (see also Ref. 22). If the perturber B is a chromophore with a low-lying electronic transition, the dynamic coupling

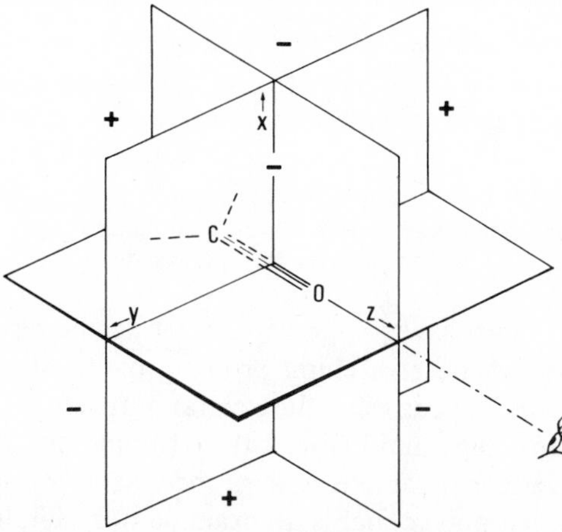

Fig. 3. Sign-determining regions of the original ketone octant rule. The carbonyl group lies in the Y-Z plane, and is normally viewed from $Z = +\infty$ to assess the positions of perturbing groups. (From Ref. 145.)

terms become a nondegenerate excitonlike interaction that can lead to large induced intensities. An example of this is provided by rigid β, γ-unsaturated ketones,[143] where the allowed electric dipole transition moment of the ethylenic chromophore couples with the magnetic dipole moment of the $n \rightarrow \pi^*$ transition of the carbonyl chromophore (through (V.2) and (V.3)) to provide this transition with a very large rotatory strength. This particular coupling is sometimes referred to as the μ-m interaction.[147]

Schellman[147] showed that the dynamic coupling terms also give rise to sector rules, and symmetry rules involving both static and dynamic coupling terms have been studied by Weigang[142,148,149] and by Stiles and Buckingham.[22,150] These studies have indicated that the dynamic coupling terms also can lead to nodal surfaces more complicated than the static coupling terms require. Second-order perturbation treatments, which introduce pairwise coupling terms between different perturbers, have been given by Yeh and Richardson,[151] and by Schipper.[152] The concept of a nodal surface loses its utility to the extent that second- or higher-order terms dominate the description of the induced transition moments. The most recent empirical modifications of the octant rule[153,154] suggest the importance of terms beyond first order.

B. Chirality Functions

The sector rules for the rotatory strength discussed in the paragraphs above are examples of the properties of a general class of pseudoscalar algebraic functions called chirality functions. These functions are polynomial expressions involving properties of a set of "ligands" attached to an

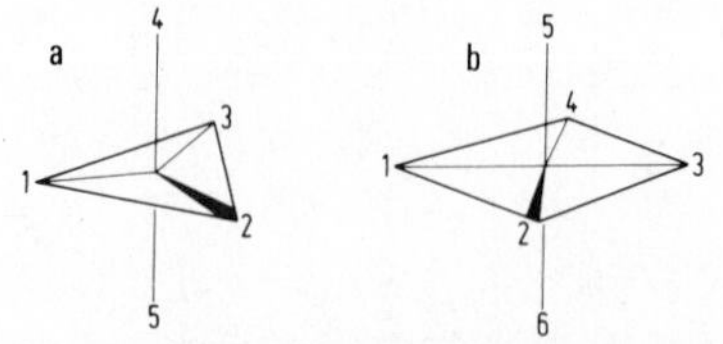

Fig. 4. (*a*) Trigonal bipyramid skeleton with five different ligands. (*b*) Tetragonal bipyramid skeleton with six different ligands.

achiral "skeleton" (such as, e.g., the tri- and tetragonal bipyramids in Fig. 4), defined so that whenever a chiral property of the system vanishes or changes sign, so does the corresponding chirality function. As an example, consider the trigonal bipyramid (Fig. 4*a*). Assuming that each ligand can be described by a scalar parameter λ_i (e.g., the mass or polarizability of the ligand), the following polynomial is an example of a chirality function for the trigonal bipyramid:

$$\chi=(\lambda_1-\lambda_2)(\lambda_2-\lambda_3)(\lambda_3-\lambda_1)(\lambda_4-\lambda_5) \tag{V.6}$$

Since any permutation of ligands on equivalent corners (e.g., interchange of ligands 1 and 3 or 4 and 5) will take one enantiomer of the molecule into the other, we notice that χ has the desired property that it takes on numerically equal but oppositely signed values for two mirror-image enantiomers. The formal algebraic and topological properties of chirality functions have been extensively studied by Ruch and co-workers.[155–158] The full implications of their formal treatment have yet to be incorporated into the common lore about chirality, but there are two concepts introduced by Ruch that we shall call attention to here.

The first novel concept, that of *homochirality*, refers to the problem of establishing a physically meaningful way of assigning a chiral molecule to one of two groups, in other words, a left–right classification based on some physical property of each ligand. Ruch[155–157] has shown that certain skeletal symmetries allow such a classification (e.g., based on the sign of a chirality function whose *form* is independent of the particular property), while other symmetries do not. Following Ruch,[156] consider again the trigonal bipyramid (Fig. 4*a*) and the chirality function χ (V.6). If the λ_i's are considered as continuous variables, we notice that χ has only "achiral zeros" (i.e., it becomes zero only for $\lambda_1=\lambda_3$ or $\lambda_4=\lambda_5$). This feature is not restricted to this particular chirality function for the trigonal bipyramid; inspection of the figure shows that no continuous variation of the ligand parameters λ_i can take one enantiomer of the system into its mirror image without passing through an achiral situation. Molecules with this property are called *homochiral*, and such molecules can be classified unambiguously as left or right. In fact, once it has been decided which ligand parameter to use as a basis, the sign of the chirality function χ can serve as an unambiguous classification of molecules built upon the trigonal bipyramid skeleton. The Cahn–Ingold–Prelog[159] R and S nomenclature for chiral substituents on a carbon atom is an example of a classification scheme of this nature, where in this case the ligand parameter is the mass of the ligand. As an example of a system that is not homochiral, consider the tetragonal bipyramid in Fig. 4*b*. If all the ligands are different initially, and ligands 1 and 2 are continuously interchanged, the molecule remains chiral during this interchange. Subsequently, ligands 3 and 4 can be interchanged continuously, and again the molecule remains chiral during the interchange. However, these two processes have now taken the initial configuration of the ligands into its mirror image without passing through an achiral arrangement, and an unambiguous left–right classification of such a nonhomochiral molecule is therefore not possible.

The second novel concept is that of *qualitative completeness*, which refers to those features that a theoretical model must possess if no chiral molecules or nonracemic mixtures of chiral molecules are predicted by the model to be optically inactive. The qualitative completeness criterion is particularly germane here, since the commonly applied, first-order sector-rule treatment in the previous section does not in fact meet this criterion in general. The violation is a consequence of the assumed additivity of perturber contributions, where pairs of identical perturbers related by reflection in one of the nodal planes give a vanishing net contribution, as do perturbers lying in the nodal surfaces. As a result, chiral systems can be envisaged for which the sector-rule model would predict zero optical activity. Examples given by Ruch[156] and by Yeh and Richardson[151] include the adamantanone derivatives shown in Fig. 5. Assuming additivity, the molecules in Fig. 5*a* and *b* become inactive because the chiral perturbers lie in

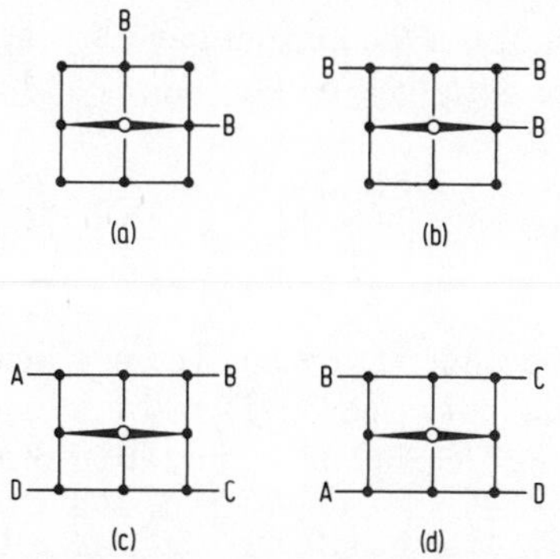

Fig. 5. Octant projections (viewed from $Z = +\infty$) of adamantanone derivatives illustrating problems of qualitative completeness in sector-rule models.

the surfaces defining the sector rules, and again assuming additivity, an equimolar mixture of 5*c* and *d* also becomes optically inactive, despite the fact that this is not a racemic mixture. Yeh and Richardson's second-order static coupling scheme provides the pairwise coupling terms among the perturbers of the form needed for qualitative completeness. Haase and Ruch[160] have devised a more general quantum mechanical model based on chirality functions, and other applications of the concept have appeared.[161]

C. Molecular Excitons

In the context of Section V.A, the molecular exciton model results when the perturber B is identical to the chromophore A, so that the nondegenerate expansions in (V.3) and (V.5) break down. Here we shall use the simple Frenkel exciton model[108, 109, 162] to obtain relations that have been used for qualitative correlations of chiroptical intensities and molecular structure. More quantitative approaches are discussed in Sections VII.C and VIII.C.

Consider a rigid molecular system consisting of N_0 identical chromophoric groups. Neglecting electronic overlap and exchange between the chromophores, the total electronic Hamiltonian is then [see (III.38) and Section VII.C]

$$\hat{\mathcal{H}} = \sum_{\mu} \hat{\mathcal{H}}_{\mu} + \tfrac{1}{2} \sum_{\mu\nu} V_{\mu\nu} \tag{V.7}$$

where $\hat{\mathcal{H}}_{\mu}$ is the Hamiltonian for the μth chromophore, and $V_{\mu\nu}$ is the multipolar interaction between chromophores μ and ν. If we use ψ_{μ}^{a} to denote the ath excited state of the μth chromophore, the electronic ground-state wave function is (See Section VII.C for a more complete treatment)

$$|0\rangle = \prod_{\mu} \psi_{\mu}^{0} \tag{V.8a}$$

while the individual states in the exciton band arising from the ath excited chromophore state are described by the linear combination

$$|q\rangle = \sum_{\mu} \left(\frac{\psi_{\mu}^{a}}{\psi_{\mu}^{0}} \right) |0\rangle C_{\mu q} \equiv \sum_{\mu} |\mu\rangle C_{\mu q} \tag{V.8b}$$

of which (III.42) is the dimeric version. The coefficients in (V.8b) are determined either by symmetry [see (V.15)] or by solving the N_0-dimensional secular equation for this degenerate perturbation problem,

$$\sum_{\mu} \left\{ \hbar(\omega_a - \omega_q)\delta_{\mu\nu} + \langle \mu | V_{\mu\nu} | \nu \rangle \right\} C_{\mu q} = 0 \tag{V.9}$$

where ω_q is the angular Bohr frequency for the excitation into the qth exciton state, and where ω_a is the unperturbed chromophoric resonance frequency. In either case we have that

$$\hbar\omega_q = \hbar\omega_a + \sum_{\mu\nu} C^*_{\mu q}\langle \mu | V_{\mu\nu} | \nu \rangle C_{\nu q} \tag{V.10}$$

Allowing for the possibility that the molecule can be so large that the fully retarded expressions must be used, the wave functions in (V.8) combine with (II.9) to yield the transition moment

$$\langle q | \hat{\boldsymbol{\eta}} | 0 \rangle = \sum_{\mu} C^*_{\mu q} \exp\left(\frac{i\omega}{c} \boldsymbol{\varepsilon}_3 \cdot \mathbf{R}_\mu \right) \langle \mu | \hat{\boldsymbol{\eta}}_\mu | 0 \rangle \tag{V.11a}$$

where $\mathbf{R}_\mu$ is the position vector to a center in chromophore μ (see below), and the local matrix element is given as

$$\langle \mu | \hat{\boldsymbol{\eta}}_\mu | 0 \rangle = \langle \psi^a_\mu | \nabla_\mu | \psi^0_\mu \rangle + \frac{i\omega}{c} \langle \psi^a_\mu | \sum_s \left[\boldsymbol{\varepsilon}_3 \cdot (\mathbf{r}_s - \mathbf{R}_\mu) \right] \hat{\nabla}_s | \psi^0_\mu \rangle \tag{V.11b}$$

where $\hat{\nabla}_\mu = \Sigma_s \hat{\nabla}_s$, and this summation and the summation in the operator in the last term in (V.11b) run over the electrons in chromophore μ only. Equations (V.11a) and (V.11b) represent a local multipole approximation because the correct phase variation between the chromophores is retained in the exponential factor in (V.11a), whereas the individual chromophores are small enough to justify the truncated expansion in (V.11b) [compare (II.15)]. From (II.14) and (V.11a), the partial circular dichroism per molecule becomes

$$\Delta\kappa^{(3)}_q(\omega) = \frac{4\pi^2 e^2 \hbar}{m^2 c} \frac{\rho_q(\omega)}{\omega} \sum_{\mu,\nu} C_{\mu q} C^*_{\nu q} \exp\left[-\frac{i\omega}{c} \boldsymbol{\varepsilon}_3 \cdot (R_\mu - R_\nu) \right] \times i\boldsymbol{\varepsilon}_3 \cdot \left[\langle \psi^a_\mu | \hat{\boldsymbol{\eta}}_\mu | \psi^0_\mu \rangle^* \times \langle \psi^a_\nu | \hat{\boldsymbol{\eta}}_\nu | \psi^0_\nu \rangle \right] \tag{V.12}$$

where $\rho_q(\omega)$ is now the band shape over the vibronic manifold of the qth electronic excitation.

Equation (V.12) is valid, strictly speaking, only in the strong exciton coupling limit [see (III.50) and (III.51)], where the line-shape functions $\rho_q(\omega)$ are well separated for the individual exciton lines. However, in order to obtain a qualitative picture of the overall CD band shape over the entire exciton band, let us assume[163–166] that $\rho_q(\omega)$ is a function of ω_q only, [i.e., $\rho_q(\omega)$

$=\rho(\omega-\omega_q)]$ and that this function can be expanded to first order, $\rho(\omega-\omega_q) =\rho(\omega-\omega_a)-(\omega_q-\omega_a)\rho'(\omega-\omega_a)$, where ω_a is the unperturbed chromophore resonance frequency, and $\rho'(\omega-\omega_a)=\partial\rho(\omega-\omega_a)/\partial\omega$. Using this expansion in conjunction with the properties of the exciton coefficients, namely

$$\sum_q C^*_{\mu q} C_{\nu q} = \delta_{\mu\nu}$$

$$\sum_q C^*_{\mu q}(\omega_q-\omega_a) C_{\nu q} = \frac{\langle \mu | V_{\mu\nu} | \nu \rangle}{\hbar} = \gamma_{\mu\nu}$$

(V.12) can be summed over all exciton states to yield

$$\Delta\kappa_q^{(3)}(\omega) = \rho(\omega-\omega_a)\sum_\mu \Delta\kappa_\mu^{(3)} - \rho'(\omega-\omega_a)\sum_{\mu,\nu} \Delta_{\mu\nu}^{(3)} \tag{V.13}$$

where

$$\Delta\kappa_\mu^{(3)} = \frac{4\pi^2 e^2 \hbar}{\omega m^2 c} i\varepsilon_3 \cdot \left[\langle \psi_\mu^a | \hat{\eta}_\mu | \psi_\mu^0 \rangle^* \times \langle \psi_\mu^a | \hat{\eta}_\mu | \psi_\mu^0 \rangle \right] \tag{V.14a}$$

is the CD contribution from the oriented μth chromophore [cf. (II.14) and (II.15)], and where

$$\Delta_{\mu\nu}^{(3)} = \frac{4\pi^2 e^2 \hbar}{\omega m^2 c} \gamma_{\mu\nu} \left\{ i\varepsilon_3 \cdot \left[\langle \psi_\mu^a | \hat{\eta}_\mu | \psi_\mu^0 \rangle^* \times \langle \psi_\nu^a | \hat{\eta}_\nu | \psi_\nu^0 \rangle \right] \right\} \times \exp\left[-\frac{i\omega}{c} \varepsilon_3 \cdot (\mathbf{R}_\mu - \mathbf{R}_\nu) \right] \tag{V.14b}$$

arises from the coupling. At this "level of resolution" the resulting exciton band shape can hence be decomposed into two contributions. The first term is the unperturbed chromophore line shape (e.g., a Gaussian) with magnitude and sign given by the sum of contributions from helically oriented but noninteracting chromophores, and the second term is the derivative of the chromophore line shape scaled by the sum of the coupling terms. If the chromophores can be considered inherently achiral, then only the second term survives, and the entire band shape takes the form of the derivative of a Gaussian. Notice also that the coupling terms in (V.14b) are scaled by the multipolar interaction $\gamma_{\mu\nu}$. This interaction falls off rapidly with the interchromophoric separation, so that $\Delta_{\mu\nu}^{(3)}$ is small for large $|\mathbf{R}_\mu - \mathbf{R}_\nu|$; the full retardation factor in (V.14b) hence does not play an important role (again, at this level of resolution), and the effective part of (V.13) and (V.14b) could have been derived as well from (II.17) [or after rotational averaging from (II.19)].

1. Helical Polymers

A prime example of the application of (V.13) is provided by the helical polymers, where a rotation Θ around the helical axis followed by a translation d parallel to the axis takes chromophore μ into chromophore $\mu+1$. Assuming cyclic boundary conditions[162] for the exciton states (see below), the coefficients become

$$C_{\mu q}=\exp\left(\frac{2\pi i}{N_0}\mu q\right) \qquad \frac{-N_0}{2}\leq q\leq\frac{N_0}{2} \tag{V.15}$$

for even N_0. The energies (V.10) therefore become doubly degenerate for $q=\pm|q|$, and singly degenerate for $q=0$. We shall also assume that the chromophoric transition is strongly electric-dipole-allowed, and that the chromophores are not inherently optically active (see Section V.C.2). In this case the chromophore center $\mathbf{R}_\mu$ can be chosen to make the second term in (V.11b) vanish (Section IV.C), and we have

$$\langle\mu|\hat{\boldsymbol{\eta}}_\mu|0\rangle=-\frac{m}{\hbar}\omega_a\langle\mu\left|\sum_s(\mathbf{r}_s-\mathbf{R}_\mu)\right|0\rangle\equiv-\frac{m}{\hbar}\omega_a D_\mu \tag{V.16}$$

where we have used (II.20) to relate the local matrix element to the electric-dipole-length chromophore transition moment. Using a molecule-fixed coordinate system with a z-axis coinciding with the helical axis, we have

$$\mathbf{R}_\mu=\begin{Bmatrix} r\cos(\mu\Theta)\\ r\sin(\mu\Theta)\\ \mu d\end{Bmatrix} \tag{V.17a}$$

and

$$\mathbf{D}_\mu=\begin{Bmatrix} D_x\cos(\mu\Theta)-D_y\sin(\mu\Theta)\\ D_x\sin(\mu\Theta)+D_y\cos(\mu\Theta)\\ D_z\end{Bmatrix} \tag{V.17b}$$

so that r is the effective radius of the helix, and D_x, D_y, and D_z are the Cartesian components of the zeroth chromophore transition moment.

We shall discuss two special cases of helical orientation relative to the light beam. For light propagation perpendicular to the helical axis, the propagation vector is given as

$$\boldsymbol{\varepsilon}_3=\mathbf{x}\cos\Phi-\mathbf{y}\sin\Phi \tag{V.18}$$

where **x** and **y** are the x and y unit vectors of the helix-fixed coordinate system, and Φ is the angle between the direction of light propagation and the helical **x**-axis. The effective size of the helix for this propagation direction is $2r$, which is much smaller than the wavelength. The exponential in (V.12) can therefore be expanded to first order, and (V.12) and (V.15) to (V.18) then combine to yield

$$\Delta\kappa_q^{\perp}(\omega)=\frac{2\pi^2e^2}{\hbar c^2}\omega_a^2 rD_yD_z\rho_q(\omega)\frac{1}{N_0}$$

$$\times\sum_{\mu,\nu}\left\{\exp[2\pi i(\mu-\nu)q/N_0]-\frac{1}{2}\exp\left[\frac{2\pi i(\mu-\nu)(q+N_0\Theta/2\pi)}{N_0}\right]\right.$$

$$\left.-\frac{1}{2}\exp\left[\frac{2\pi i(\mu-\nu)(q-N_0\Theta/2\pi)}{N_0}\right]\right\} \qquad \text{(V.19)}$$

after averaging over all values of the orientational angle Φ (V.18) because the experiments usually do not allow specification of this rotational degree of freedom. Equation (V.19) shows how the rotatory intensity for this excitation depends upon the helical parameters r and Θ and upon the tangential and axial components of the chromophoric electric dipole transition moment, D_y and D_z, and furthermore that this orientation leads to only two allowed and oppositely signed rotatory intensity bands, namely one at $q=0$ (from the first term in the summation), and one at the doubly degenerate $q=\pm N_0\Theta/2\pi$ (from the second and third terms). The $q=0$ selection rule is standard for systems with a translational symmetry,[162] whereas the $q=\pm N_0\Theta/2\pi$ is specific for the helical arrangement. The excitation energies corresponding to $q=0$ and $\pm N_0\Theta/2\pi$, respectively, are given by (V.10) and (V.14) and will depend upon the signs and magnitudes of the chromophoric interactions $\langle\mu|V_{\mu\nu}|\nu\rangle$.

For light propagating along the helical axis (i.e., $\boldsymbol{\varepsilon}_3=\mathbf{z}$), the full retardation contained in the exponential in (V.13) must be retained, and (V.13) to (V.17) yield

$$\Delta\kappa_q^{\|}(\omega)=\frac{2\pi^2e^2}{\hbar c}\frac{\rho_q(\omega)}{\omega}\omega_a^2|D_x^2+D_y^2|\frac{1}{N_0}$$

$$\times\sum_{\mu,\nu}\{\exp[2\pi i(\mu-\nu)(q-N_0d/\lambda+N_0\Theta/2\pi)]$$

$$-\exp[2\pi i(\mu-\nu)(q-N_0d/\lambda-N_0\Theta/2\pi)]\} \qquad \text{(V.20)}$$

For parallel propagation the rotatory intensity hence depends upon the radial and tangential components of the chromophoric transition moment, and there are again two allowed and oppositely signed bands, namely at $q = N_0\Theta/2\pi - N_0 d/\lambda$ and $q = N_0\Theta/2\pi + N_0 d/\lambda$. Since q must be an integer, this shows that the use of cyclic boundary conditions for the exciton states (V.15) requires that boundary conditions on the light beam be carefully matched to the exciton states, and the correct selection rules follow only when the fully retarded intensity expressions (II.14) and (V.12) are employed.

Notice that the integrated rotatory intensities sum to zero for both propagation cases,

$$\sum_q \int \Delta\kappa_q^{\perp}(\omega)\,d\omega = \sum_q \int \Delta\kappa_q^{\parallel}(\omega)\,d\omega = 0 \qquad \text{(V.21)}$$

provided we make the approximation $\omega_a^2/\omega = \omega_a$ in (V.20). This sum rule is obtained here from only the electronic wave functions, [i.e., for the strong coupling limit (III.50)]. However, by the argument leading to (III.54), (V.21) can be shown to hold also in the weak exciton coupling limit, because the individual chromophores are assumed optically inactive.

The use of the Frenkel exciton model for the optical properties of polymers was initiated by Moffitt[111] who utilized the cyclic exciton functions (V.15) in conjunction with the rotatory strength expressed in the form of the electric dipole–magnetic dipole expression in (III.22). His treatment therefore only provided the $q = 0$ and $q = \pm N_0\Theta/2\pi$ selection rules, whereas the retarded selection rule contribution $N_0 d/\lambda$ did not appear, and hence the Moffitt treatment predicted vanishing rotatory power for light propagating along the helical axis. In a subsequent communication, Moffitt, Fitts, and Kirkwood[167] presented a synthesis of the Kirkwood polarizability model[18] (see Section V.A) and the Moffitt exciton approach. For this purpose they retained the intensity expression (III.22) but relaxed the restriction of cyclic boundary conditions, and they found then that the original Moffitt treatment had left out a term that could not be considered negligible. The explicit form of this additional term was found by Tinoco[168] within the Moffitt–Fitts–Kirkwood approach, and (in retrospect) it corresponded closely to the $q = N_0\Theta/2\pi \pm N_0 d/\lambda$ selection rule of (V.20). The importance of the correct treatment of retardation when the cyclic boundary conditions are employed was anticipated by Rhodes;[169] however, the correct selection rules using cyclic boundary conditions and fully retarded interactions were first derived independently by Loxsom[170] and Deutsche[171] (see also Refs. 41, 63, 164–166), and it is now clear that the error in the original Moffitt approach lies in the use of incompatible approximations, namely (V.15), which is strictly valid only for infinitely large systems, and (II.15), which is valid only in the limit of small molecules.

The Moffitt–Fitts–Kirkwood theory was subsequently expanded into a highly useful method for the calculation of polymer chiroptical properties.[21,163,168,169,172,173] As displayed in (V.19) and (V.20), these methods are attractive because the chiroptical properties are determined by chromophoric properties and the helical geometry. In a very careful analysis of the experimental chiroptical intensities of an α-helical polymer[174] it was possible to extract exactly the contributions predicted by (V.19) and (V.20).

2. *Dimeric Systems*

In the preceding section the chromophores were assumed optically inactive, and the second term in (V.11b) could therefore be removed by an appropriate choice of the chromophore position vector $\mathbf{R}_\mu$. We shall illustrate here the importance of local optical activity contributions for the case of a coupling between two identical chromophores. The wave functions (V.8) then reduce to (III.42), and the molecule is now small enough to expand the exponential in (V.12) for all orientations, so that (III.42), (V.11b), and (V.12) combine to yield

$$\begin{aligned}\Delta\kappa_\pm(\omega) = \frac{4\pi^2e^2\hbar}{3m^2c^2}\rho_q(\omega)\{&\langle A_a|\hat{\nabla}_A|A_0\rangle\cdot\langle A_a|(\mathbf{r}_A-\mathbf{R}_A)\times\hat{\nabla}_A|A_0\rangle\\ &+\langle B_a|\hat{\nabla}_B|B_0\rangle\cdot\langle B_a|(\mathbf{r}_B-\mathbf{R}_B)\times\hat{\nabla}_B|B_0\rangle\\ &\pm\langle A_a|\hat{\nabla}_A|A_0\rangle\cdot\langle B_a|(\mathbf{r}_B-\mathbf{R}_B)\times\hat{\nabla}_B|B_0\rangle\\ &\pm\langle B_a|\hat{\nabla}_B|B_0\rangle\cdot\langle A_a|(\mathbf{r}_A-\mathbf{R}_A)\times\hat{\nabla}_A|A_0\rangle\\ &\pm(\mathbf{R}_A-\mathbf{R}_B)\cdot[\langle A_a|\hat{\nabla}_A|A_0\rangle\times\langle B_a|\hat{\nabla}_B|B_0\rangle]\}\end{aligned} \qquad \text{(V.22)}$$

in the notation of Section V.A, after rotational averaging [the same expression follows from (II.19)]. Here $\hat{\nabla}_\mu = \Sigma_s\hat{\nabla}_s$ and $(\mathbf{r}_\mu - \mathbf{R}_\mu)\times\hat{\nabla}_\mu = \Sigma_s(\mathbf{r}_s - \mathbf{R}_\mu)\times\hat{\nabla}_s$, including only the electrons in chromophore μ in the summations. The integrated rotatory intensities for the two bands given by (V.22) now sum to the total rotatory intensity of the unperturbed chromophores [compare (III.54)]. Any inherent chromophoric optical activity will hence tend to make the two dimeric circular dichroism bands differ in magnitude [in contrast to (V.19) to (V.21)].

The last term in (V.22) may be written

$$\Delta\kappa_\pm = \frac{\pm 4\pi^2e^2\omega^2}{3\hbar c^2}\rho_q(\omega)(\mathbf{R}_A-\mathbf{R}_B)\cdot(\mathbf{D}_A\times\mathbf{D}_B) \qquad \text{(V.23)}$$

[see (V.16)], and in this form it is seen to be the dimeric analog of the effect represented by (V.19) and (V.20). From (III.16) the corresponding rotatory strengths become

$$R_{\pm} = \pm \frac{1}{4} \frac{\hbar\omega_a}{137} (\mathbf{R}_A - \mathbf{R}_B)\cdot(\mathbf{D}_A \times \mathbf{D}_B) \tag{V.24}$$

utilizing the fine-structure relation $e^2/\hbar c = \frac{1}{137}$, and the transition energies for the excitations into the two dimeric states are

$$\hbar\omega_{\pm} = \hbar\omega_a \pm e^2\{\mathbf{D}_A\cdot\mathbf{D}_B - 3[(\mathbf{R}_A - \mathbf{R}_B)\cdot\mathbf{D}_A][(\mathbf{R}_A - \mathbf{B}_B)\cdot\mathbf{D}_B]/|\mathbf{R}_A - \mathbf{R}_B|^2\}/|\mathbf{R}_A - \mathbf{R}_B|^3 \tag{V.25}$$

in the electric dipole approximation.[22,147] Equations (V.23) to (V.25) have been used extensively to relate the observed CD of a dimeric molecule to the geometry and the local electric dipole transition moments, in the case of optically inactive chromophores, so that the first two terms in (V.22) vanish. However, the third and fourth terms can be removed only if the local magnetic dipole moment is known, so that the arguments of Section IV.C can be employed to locate the center of the chromophore $\mathbf{R}_\mu$. If $\mathbf{R}_\mu$ is not chosen on the basis of such an analysis, the third and fourth terms in (V.22) must be retained.[175] The importance of this observation can be seen from the fact that dimeric exciton theory using (V.23) to (V.25) was employed recently[176] as a basis for the claim that a number of absolute molecular configurations (determined by Bijvoet's method[177]) were erroneous. These particular exciton results were subsequently shown to be wrong because of incorrect assumptions about the location of the chromophore centers,[178] and proper inclusion of the third and fourth terms in (V.22) restored an absolute configurational assignment in agreement with the Bijvoet result.[178] We shall return to the problem of the correct assessment of local chromophoric transition moments in Section VIII.C.

VI. OTHER CHIROPTICAL PHENOMENA

The principal focus of this article is, as stated previously, on electronic absorption phenomena in the absence of static external fields. However, the general field of chiroptical spectroscopy includes a number of other phenomena, three of which we review briefly in this section. In addition, chiroptical scattering in the form of Rayleigh and Raman optical activity has been reviewed recently by Barron and Buckingham.[179]

A. Circularly Polarized Luminescence (CPL)

The first observation of CPL was made in 1948,[180] but it was not until the work of Emeis and Oosterhoff[181] in 1971 that the phenomenon was recognized as a structural probe of the excited electronic states of organic molecules. A theory of CPL was put forward by Snir and Schellman,[182] and extended by Steinberg and Ehrenberg,[183] and by Riehl and Richardson,[184,185] who also treated the case of an applied magnetostatic field (magnetic circularly polarized luminescence, MCPL).[185] The subject has recently been reviewed by Steinberg,[186] and by Richardson and Riehl.[187]

Experimentally, one excites a chiral system with unpolarized light, and then observes the differential emission of left- and right-circularly polarized light; alternatively, one can differentially excite a racemic mixture with circularly polarized light and observe CPL.[188] Assuming that the excited-state lifetime is sufficient to allow for thermal equilibration, much of the theoretical treatment developed in Sections II and III can be carried over to CPL; however, the appropriate Hamiltonian in the latter case is that of the nuclear equilibrium configuration of the electronic excited state. Thus CPL is to the determination of excited-state structure what CD is to ground-state structure.

The differences in the experimental and theoretical treatment of CPL and CD are largely those common to all polarized emission experiments. Thus, for example, anisotropies are induced in general in the excited-state population by a process of photoselection, even if the ground state is isotropic. The excited-state orientational distribution may be computed from a knowledge of the experimental light-beam geometry and the orientations of the transition moments within the molecule, but the relative magnitudes of the excited-state lifetime and the rotational relaxation time will govern the details of the polarization of the emitted light. These problems have been treated in Refs. 182–187.

A further experimental problem occurs because the absolute emission intensity I and CPL intensity ΔI are normally not obtainable, since that would require determination of the quantum yield as a function of frequency, but instead I and ΔI are reported in relative units. For this reason, the so-called luminescence dissymmetry,

$$g_{\text{lum}}(\omega)=\frac{\Delta I(\omega)}{2I(\omega)} \tag{VI.1}$$

takes on greater importance, and comparisons of CD and CPL spectra then involve a study of $g_{\text{lum}}(\omega)/g_{\text{abs}}(\omega)$ for a given electronic band. Studies of the structure of organic and biological systems have been reported from the laboratories of Richardson,[189] Dekkers,[190] and Steinberg.[191,192] When commercial CPL instruments become available, this technique has the potential of leading to significant advances in our understanding of upper-state geometries.

B. Vibrational Circular Dichroism

The general quantum mechanical treatment of circular dichroism is not limited to electronic transitions; a chiral response can arise equally well from a vibrational excitation. However, it was recognized at an early date[19] that the rotatory intensity for a purely vibrational transition would be expected to be lower than that for an electronic transition by 3 to 4 orders of magnitude, owing to the explicit appearance of the particle mass in the magnetic moment operator [cf. (II.23)]. In fact, CD in the infrared region of the spectrum eluded experimental detection for many years.

The ease of experimental measurements of CD depends upon the magnitude of the anisotropy ratio (III.1), and a value of 10^{-4} is well within the range of detectability of current instrumentation, at least in the near infrared. Deutsche and Moscowitz[193] developed the theory for vibrational rotatory intensities of an infinite helical polymer in anticipation that such a system would exhibit a desirably large anisotropy ratio. Both the electric and the magnetic dipole transition moments are assumed to arise from oscillations of a collection of atoms, each with a fixed partial charge (fpc). The fpc model was subsequently applied to a coupled oscillator model by Holzwarth and Chabay,[194] and Schellman[195] presented the fpc model in a form appropriate for small molecules and discussed characteristics of vibrations likely to have large anisotropy ratios. The first experimental observation of infrared CD for a liquid was reported by Holzwarth, Hsu, Mosher, Faulkner, and Moscowitz,[196] in the following year, and further experimental results followed shortly.[197–199]

Calculations carried out within the fpc model, while yielding reasonable values for the ordinary infrared intensities, appear to underestimate the rotatory intensity by an order of magnitude.[195–200] The extension of the fpc model to anharmonic vibrations by Faulkner, Marcott, Moscowitz, and Overend[200] does not alleviate this difficulty, but does allow the qualitative estimation of the rotatory strengths of overtone and combination bands, which for technical reasons may actually be easier to observe than those of fundamentals.[201,202] Basically the problem with the calculation of vibrational rotatory intensities appears to lie in the evaluation of the vibrational magnetic dipole transition moment. In a Born–Oppenheimer adiabatic approximation, the magnetic dipole transition moment for an excitation between the ground-state vibrational states α and β is given by

$$\begin{aligned}\langle\psi_0(q,Q)\chi^0_\alpha(Q)|\{(\mathbf{r}\times\hat{\nabla})_Q+(\mathbf{r}\times\hat{\nabla})_q\}|\psi_0(q,Q)\chi^0_\beta(Q)\rangle_{Q,q}\\ =\langle\chi^0_\alpha(Q)|(\mathbf{r}\times\hat{\nabla})_Q|\chi^0_\beta(Q)\rangle_Q\\ +\langle\chi^0_\alpha(Q)|\{\langle\psi_0(q,Q)|(\mathbf{r}\times\hat{\nabla})_Q+(\mathbf{r}\times\hat{\nabla})_q|\psi_0(q,Q)\rangle_q\}|\chi^0_\beta(Q)\rangle_Q\end{aligned} \qquad \text{(VI.2)}$$

[see (III.2), (III.6a), and (III.6b) for notation], where the operator now contains a nuclear as well as an electronic part. For a nondegenerate electronic ground state, the electronic integration in the second term of (VI.2) vanishes identically and the vibrational magnetic transition moment reduces to the pure nuclear contribution in the first term of (VI.2).[200] Using this representation of the magnetic dipole moment operator, the Born–Oppenheimer adiabatic approximation is hence unable to reproduce the fpc model for the magnetic transition moment (because the fpc model implies the motion of partially screened nuclei). In a similar vein, Mead and Moscowitz[203] showed that the electronic contribution to the ordinary vibrational intensity vanishes in the Born–Oppenheimer adiabatic approximation when the dipole velocity operator (i.e., $\hat{\nabla}_q + \hat{\nabla}_Q$) is employed, whereas the dipole length operator (i.e., $\mathbf{r}_q + \mathbf{r}_Q$) leads to correct results. Expressions that explicitly contain terms beyond the Born–Oppenheimer level have been presented recently for ordinary[204a] and rotatory[204b,205] vibrational intensities. Alternatively, one may follow the Mead and Moscowitz idea, and search for a representation of the magnetic dipole operator that retains the electronic contribution at the Born–Oppenheimer level.

C. Magnetic Circular Dichroism (MCD)

Because of the solenoidal character of a static external magnetic field, beams of left- and right-circularly polarized light propagating along the direction of the field are absorbed to different extents by any isotropic molecular medium. Thus the phenomenon of MCD is exhibited by all matter, not just by systems that are chiral in zero field. The static magnetic field modifies the energy levels of the molecule (e.g., by the Zeeman splitting of zero-field degenerate levels), and the information to be obtained from analysis of MCD spectra is hence somewhat different from that provided by natural CD.

In the presence of a static magnetic field, (II.9) for the electronic transition moment operator must be modified to read[206,63]

$$\hat{\boldsymbol{\eta}} = \sum_s \exp\left(\frac{i\omega}{c}\boldsymbol{\varepsilon}_3\cdot\mathbf{r}_s\right)\left\{\hat{\nabla}_s + \frac{ie}{\hbar c}\mathbf{a}(\mathbf{r}_s)\right\}$$

$$= \sum_s \exp\left(\frac{i\omega}{c}\boldsymbol{\varepsilon}_3\cdot\mathbf{r}_s\right)\left\{\hat{\nabla}_s + \frac{ie}{2\hbar c}\mathbf{b}\times\mathbf{r}_s\right\} \qquad \text{(VI.3)}$$

where $\mathbf{a}(\mathbf{r}_s)$ is the electromagnetic vector potential at the position of electron s, corresponding to the static magnetic field $\mathbf{b}$. With this modification of the operator, (II.14) holds also for the rotatory intensity of an oriented molecule in the presence of the magnetic field, provided that the eigenfunctions and energies are interpreted as those *in the presence of the static field* (i.e., the wave functions are now inherently complex, in contrast to the case of natural CD). For Frenkel exciton systems, the Hamiltonian resulting from (II.8) and (VI.3) can be subjected to a canonical transformation that leads to a simple local electric dipole approximation.[63] From this transformed Hamiltonian, general expressions have been derived for the natural and magnetic rotatory intensities of polymers.[63] One important feature is that the simple Frenkel model of Section V.D is unable to account for polymer MCD.[63,207] It is imperative to use excited-state wave functions of the form given in (VII.38), which includes more than one excited state per chromophore, because the sole effect of the magnetic field within an exciton model is to mix local excitations on the individual chromophores. The MCD of polymers was first treated by Harris[207] within time-dependent Hartree theory[40] (see Section VII.C).

For small molecules the perturbation operator that results in the electric dipole limit of (II.8) and (VI.3) still contains a static magnetic contribution (often called a bilinear term because it is linear both in time-dependent part and in the static part of the vector potential). However, a canonical transformation of this operator leads to a dipole-length perturbation operator identical to (II.27) (i.e., without bilinear terms[62]). Starting from (II.27) a development parallel to that leading to (II.13) and (II.14) provides the following expression for the magnetically induced rotatory intensity for an oriented molecule:

$$\Delta\kappa_{gu}^{\mathrm{MCD}}(\omega)=\frac{4\pi^2e^2\omega}{\hbar c}\,Ni\varepsilon_3\cdot[\langle u|\mathbf{r}|g\rangle^*\times\langle u|\mathbf{r}|g\rangle]\rho_{gu}(\omega) \qquad \text{(VI.4)}$$

This expression reemphasizes the comments made after (II.17), because in the absence of a magnetic field (i.e., for real wave functions) $\langle u|\mathbf{r}|g\rangle^*$ and $\langle u|\mathbf{r}|g\rangle$ are parallel and (VI.4) vanishes identically. On the other hand, for complex wave functions (VI.4) will yield a nonvanishing result in general. Equation (VI.4) also indicates that strong MCD intensities require electric-dipole-allowed transitions; in fact, it has been shown[208] on symmetry grounds that the MCD of an electric-dipole-forbidden transition is at least second order in any vibronic or structural perturbation. Higher-order multipolar contributions to MCD intensities are given by Stephens[206] and by Hansen and Svendsen.[62] Reference 206 also contains the magnetic modification of the spin terms.

In his 1970 contribution[206] (see also Refs. 209 and 88), Stephens presents a very complete discussion of MCD of small molecules, based on his more general version of (VI.4). By expanding the states u and g in terms of the complete set of field-free states, Stephens arrives at an expression for the total MCD ellipticity for randomly oriented molecules of the form

$$\Delta\kappa_{gu}^{\mathrm{MCD}}(\omega) = Af_1 + (B + C/kT)f_2 \qquad \text{(VI.5)}$$

where f_1 and f_2 are line-shape functions, of which f_2 is essentially the unperturbed (Gaussian) line shape, whereas f_1 is the derivative of f_2. A, B, and C are functions of electric and magnetic dipole transition moments that involve a summation over all excited states. The A and C terms arise only if the zero-field molecular symmetry supports degeneracies for the pertinent states, whereas the B terms are nonzero in general.

The interpretation of MCD spectra in terms of molecular structure is more difficult than in the case of natural CD, in part because the Zeeman splitting of energy levels complicates the vibronic analysis of an electronic MCD absorption band. [For a review of the experimental and theoretical literature through 1973, see Ref. 209.] Since MCD follows the same selection rules as does ordinary absorption, correlations have been developed for electric-dipole-forbidden chromophoric transitions subject to vibronic and structural perturbations from outside the chromophore. As mentioned above, such MCD intensities are of second or higher order,[208] and in addition the contributions to the MCD arising from different symmetries are additive.[208] Based on this analysis, Seamans and others[210–212] have been able to correlate the structures of many ketones with their MCD spectra by means of approximate models. Pi-electron systems have been studied by Rosenfield and others,[213–214] and by Eyring and others,[215] and very extensively by Michl and co-workers.[216–217]

While it is a difficult task to compute accurate rotatory intensities in the absence of static external fields (See Sections VII and VIII below), it is *a fortiori* nontrivial to compute B terms. Their calculation involves not only a sum over products of transition moments between ground and all excited states and between the excited state of interest and all others, but under the usual truncated basis set conditions, problems of origin dependence arise as well (cf. Section IV.C). Warnick and Michl[216] showed that certain model Hamiltonians can lead to origin-independent B terms, for example a semiempirical π-electron calculation including all singly excited configurations where the parameters are chosen to fulfill the hypervirial relation (II.25) (cf. Section VII.B below). However, *ab initio* calculations generally are expected to lead to origin dependence.

Seamans and Linderberg[218] adopted a rather different approach to calculating MCD intensities. Rather than employing the usual perturbation expansion in terms of the field-free eigenfunctions,[206–219] these authors incorporated the static field directly into the molecular (π-electron) Hamiltonian, and used a field-dependent atomic orbital basis set as suggested by London.[220] Gauge invariance of the results was ensured by using the random-phase approximation (RPA, see Section VII.B below). By these measures Seamans and Linderberg were able to avoid the sum over excited states, and the results obtained for several π-electron systems were in close agreement with experiment. A gauge transformation of the type invoked in generating London orbitals from field-free orbitals is similar to the above-mentioned canonical transformation of the Hamiltonian, including the static field.[63]

Because of the relatively large anisotropy ratio induced by the static magnetic field, MCD measurements are obtainable in the infrared,[221] and, as reported recently,[222–224] are of comparable magnitude for singlet-triplet and singlet-singlet ultraviolet-visible electronic transitions.[225] MCD can thus be seen to be a technique capable of yielding detailed structural information of a number of kinds; the extraction of this information, however, has not yet reached the development currently feasible in natural CD.

VII. COMPUTATION OF ELECTRONIC TRANSITION MOMENTS

The integrated ordinary and rotatory intensities (III.17) to (III.23) require the evaluation of transition energies and of transition moments $\langle 0|\hat{F}|b\rangle$ for various one-electron operators $\hat{F}$. The calculation of these quantities can proceed either in a state approach (i.e., the conventional quantum chemistry procedure where energies and wave functions are optimized individually for each pertinent state or obtained as a by-product of optimizing the ground state), or in an excitation approach, where the aim is to compute the excitation amplitudes and energies directly. In this section we shall outline a number of variants of these two approaches.

Let us assume for the moment that we are provided a complete set of molecular spin orbitals (MO's) ϕ_i, from which we construct the exact single Slater Determinant Hartree–Fock ground state wave function for an N-electron system:

$$|\Delta_0\rangle = |\phi_\alpha \phi_\beta \cdots \phi_\nu \cdots| \qquad \text{(VII.1)}$$

using Greek subscripts to denote occupied orbitals. Electron configurations that are excited with reference to $|\Delta_0\rangle$ are obtained as single excitations

$$|\alpha m\rangle = |\phi_m \phi_\beta \cdots \phi_\nu \cdots| \qquad \text{(VII.2)}$$

double excitations

$$|\alpha m,\beta n\rangle = |\phi_m\phi_n\cdots\phi_\nu\cdots|, \tag{VII.3}$$

and so forth, and the self-consistent determination of $|\Delta_0\rangle$ ensures that the Brillouin relation,[226]

$$\langle\Delta_0|\hat{\mathcal{H}}|\alpha m\rangle = 0 \tag{VII.4}$$

holds for all excitations $|\alpha m\rangle$. (We use Latin subscripts to denote orbitals unoccupied in $|\Delta_0\rangle$.) The completeness of the orbital basis, together with the one-electron nature of the operator $\hat{F}$, implies that the most general expression for a transition moment takes the form

$$\begin{aligned}\langle 0|\hat{F}|b\rangle = &\sum_{\alpha m}\{X^b_{\alpha m}\langle\alpha|\hat{f}|m\rangle + Y^b_{\alpha m}\langle m|\hat{f}|\alpha\rangle\}\\ &+\sum_{\alpha\beta} U^b_{\alpha\beta}\langle\alpha|\hat{f}|\beta\rangle\\ &+\sum_{mn} W^b_{mn}\langle m|\hat{f}|n\rangle\end{aligned} \tag{VII.5}$$

In the language of many-body theory, the occupied orbitals ϕ_α are called "hole" states, and the virtual orbitals ϕ_m are called "particle" states. The three sums in (VII.5) can accordingly be called "particle–hole," "hole–hole," and "particle–particle" contributions.

The various approaches alluded to above represent different ways of evaluating the coefficients $\{X\}$, $\{Y\}$, $\{U\}$, and $\{W\}$ and the transition energy for a given excitation $0\rightarrow b$. The transition moments are clearly off-diagonal properties, and thus ordinary variational criteria for the quality of a computed result are not applicable. Methods for estimating the error bounds on calculated ordinary electronic transition probabilities have been proposed,[227] but they do not seem useful for rotatory intensities.[228] In practice, therefore, one is left with two criteria for the reliability of the sign and magnitude of computed chiroptical intensities: first, the stability of the computed intensities with respect to variations in the computational scheme (e.g., variations in the type and size of basis set), and second, the degree of agreement among the intensities obtained from the formally equivalent expressions (III.18) to (III.20) and (III.22) to (III.23). The latter criterion will play a special role in the formulation of what we call excitation approaches (see Section VII.B).

A. State Approach

The exact stationary states $|0\rangle$ and $|b\rangle$ may be expressed as linear combinations of the configuration basis $|\Delta_0\rangle$, $\{|\alpha m\rangle\}$, $\{|\alpha m, \beta n\rangle\}$,.... Use of the Slater–Condon rules for evaluation of matrix elements of one-electron operators $\hat{F}$ between Slater determinants leads to an expression for $\langle 0|\hat{F}|b\rangle$ that may be put into the form of (VII.5). In the state approach, one obtains approximate forms for the initial and final states by optimizing their linear expansion coefficients by variational and/or perturbational calculations, and the resulting coefficients are used in (VII.5) to yield approximate transition moments.

Let us now consider the following approximate forms for the ground and low-lying excited states:

$$|0\rangle = D_0|\Delta_0\rangle + \tfrac{1}{4}\sum_{\alpha m}\sum_{\beta n} D_{\alpha m,\beta n}|\alpha m, \beta n\rangle \qquad \text{(VII.6)}$$

$$|b\rangle = \sum_{\alpha m} C^b_{\alpha m}|\alpha m\rangle \qquad \text{(VII.7)}$$

The ground-state expression (VII.6) is, in fact, correct to first order in electron correlation, and the absence of single-excitation terms in first order is a consequence of the Brillouin relation (VII.4). An equivalent perturbation expansion for the excited state is formally less well defined, but we may expect that a low-lying excited state will be adequately described by a superposition of single excitations, as in (VII.7), and small contributions from double and triple excitations. These latter terms have been left out of (VII.7) because they do not contribute to the transition moment except through the (presumably small) doubly excited component in the ground state and are hence higher-order corrections. Equations (VII.6) and (VII.7) then give the following expression for the transition moment:

$$\begin{aligned}\langle 0|\hat{F}|b\rangle &= D_0^* \sum_{\alpha m} C^b_{\alpha m}\langle \Delta_0|\hat{F}|\alpha m\rangle \\ &\quad + \sum_{\alpha m}\sum_{\beta n} D^*_{\alpha m,\beta n} C^b_{\beta n}\langle \alpha m, \beta n|\hat{F}|\beta n\rangle \\ &= \sum_{\alpha m}\left\{ D_0^* C^b_{\alpha m}\langle \alpha|\hat{f}|m\rangle + \left(\sum_{\beta n} D^*_{\alpha m,\beta n} C^b_{\beta n}\right)\langle m|\hat{f}|\alpha\rangle \right\} \end{aligned} \qquad \text{(VII.8)}$$

where we have utilized the fact that the correlation coefficients $D_{\alpha m,\beta n}$ of (VII.6) must be antisymmetric with respect to an interchange of indices α and β or m and n. Equation (VII.8) is of the form of the "particle–hole" part of the general relation (VII.5). Within this framework, the "particle–particle" and "hole–hole" intensity contributions appear only in second- and higher-order expansions of the wave functions such as were mentioned above. Equation (VII.8) shows clearly that ground-state correlation, as represented by the coefficients $D_{\alpha m, \beta n}$, is essential for a correct first-order description of electronic intensities.[54,229–231]

Various approximate schemes for computing transition moments arise from further truncations in (VII.6) and (VII.7) and hence in (VII.8). Retention of only the terms in $\langle \Delta_0 | \hat{F} | \alpha m \rangle$ (i.e., total neglect of ground-state correlation) yields the so-called Tamm–Dancoff approximation (TDA), in the notation of Dunning and McKoy[232] (also called monoexcited CI), while restriction of this summation to a single term $\alpha \rightarrow m$ gives the single transition approximation (STA). We shall refer to schemes that retain at least some terms beyond TDA as "CI," although use of the latter term to refer to any level of truncated TDA is common in the optical activity literature. (For a general survey of CI methods, see Refs. 233–235.) It is inherent in the state approach that once the size of the orbital basis and of the configurational set has been decided, the linear coefficients in the wave functions are then determined by variational and/or perturbational methods with no further constraints.

Examples of STA and TDA calculations of optical properties abound, at least in the sense that the terms retained in (VII.8) are of the STA or TDA form, but either the orbital or the configuration basis may be severely truncated. The results will be discussed in Section VIII; at this point we shall only describe briefly some variants of STA and TDA methods that have also been employed in rotatory strength calculations.

In the STA approach, the virtual orbital receiving the excited electron is generated in a potential due to N rather than $N-1$ electrons, as is well known. The IVO (improved virtual orbital) method, due to Huzinaga and Arnau,[236] invokes a unitary transformation of the virtual orbital set to generate new virtual orbitals appropriate for excitation from a particular occupied orbital ϕ_α. These new virtual orbitals are conceived as resulting from diagonalization of a modified Fock matrix, but may also be obtained from a partial TDA scheme in which the set of configurations is restricted to only those single excitations out of orbital ϕ_α. Similarly, the set of occupied orbitals may be improved with respect to an excitation into a particular virtual orbital ϕ_m by a calculation involving all excitations *into* ϕ_m. This is called the IGO (improved ground orbital) method.[237]

Another modified STA scheme is the so-called Hartree–Fock–Slater (HFS) or X_α method,[238,239] in which the Hamiltonian operator is a sum of only one-electron operators, and the potential is local. The virtual orbitals as well as the occupied ones in this scheme are generated in an $N-1$ electron potential.[240] As we shall see presently, this method has other features that may make it a desirable approximate scheme for calculating rotatory strengths. Finally, we mention the so-called Δ(SCF) method,[240] in which separate Hartree–Fock functions are obtained for the ground and excited states. The resulting sets of MO's are not mutually orthogonal, but the transition moment may still be computed as

$$\langle 0|\hat{F}|b\rangle = \sum_i \sum_j \langle \phi_i^0|\hat{f}|\phi_j^b\rangle D_{ij}$$

where D_{ij} is the signed minor of the MO overlap matrix $\langle \phi_i^0|\phi_j^b\rangle$.

All these various state approaches are based upon obtaining energies and wave functions for individual stationary states, and this can mean that the overall balance required for the fulfillment of the hypervirial relations (II.20) and (II.25) is not necessarily attained; and indeed nearly all the methods we have just described fail to provide agreement between the results obtained from the formally equivalent intensity expressions. For the simple STA method, the violation of (II.20) is a consequence of the nonlocal character of the Hartree–Fock potential.[241] If the coefficients in the states $|0\rangle$ and $|b\rangle$ are determined variationally, the sole exceptions to this violation of (II.20) and (II.25) are for (1) a full CI treatment involving all possible levels of excitation—which in fact would yield the exact result in the limit of a complete basis, and (2) the HFS method. The latter method yields transition moments obeying a hypervirial relation because of the use of a local potential in this method.[242]

B. Excitation Approach

Equation (VII.8) can be rewritten in the form

$$\langle 0|\hat{F}|b\rangle = \sum_{\alpha m} \{X_{\alpha m}^b \langle \Delta_0|\hat{F}|\alpha m\rangle + Y_{\alpha m}^b \langle \alpha m|\hat{F}|\Delta_0\rangle\} \tag{VII.9}$$

where we have introduced the quantities

$$X_{\alpha m}^b \equiv D_0^* C_{\alpha m}^b \tag{VII.10a}$$

$$Y_{\alpha m}^b \equiv \sum_{\beta n} D_{\alpha m,\beta n}^* C_{\beta n}^b = \sum_{\beta n} (D_{\alpha m,\beta n}^*/D_0^*) X_{\beta n}^b \tag{VII.10b}$$

and utilized the Slater–Condon rules to rewrite the one-electron matrix elements in terms of single excitations out of, and deexcitations into, the

Hartree–Fock ground state. Inspection of (VII.6) to (VII.8) and (VII.9) shows that there are fewer independent parameters in the transition moment expression than there are in the configuration expansions of the wave functions for the individual states, and the various excitation approaches that have evolved in the last decade can be viewed as attempts to evaluate directly the amplitudes $X^b_{\alpha m}$ and $Y^b_{\alpha m}$ of (VII.9). In this section we shall discuss these approaches in the wave-function language of (VII.6) to (VII.10),[243] whereas the conventional derivations and presentations of these approaches are phrased in the form of second quantization.[244,245]

The central idea is now to attempt to determine the amplitudes $X^b_{\alpha m}$ and $Y^b_{\alpha m}$ in (VII.9) such that the transition moments induced by a one-electron operator $\hat{F}$ and by its commutator with the electronic Hamiltonian $[\hat{F}, \hat{\mathcal{H}}]$ are related by the hypervirial equation (II.25). In order to justify the use of (VII.9) for the transition moments of $[\hat{F}, \hat{\mathcal{H}}]$ it is necessary to require that this commutator also be a one-electron operator;[246] apart from this restriction the operator $\hat{F}$ can be assumed completely general. Inserting (VII.9) into both sides of (II.25), we find

$$\sum_{\alpha m} \{\langle\Delta_0|[\hat{F}, \hat{\mathcal{H}}]|\alpha m\rangle X^b_{\alpha m} + \langle\alpha m|[\hat{F}, \hat{\mathcal{H}}]|\Delta_0\rangle Y^b_{\alpha m}\}$$
$$= \hbar\omega_{b0} \sum_{\alpha m} \{\langle\Delta_0|\hat{F}|\alpha m\rangle X^b_{\alpha m} + \langle\alpha m|\hat{F}|\Delta_0\rangle Y^b_{\alpha m}\} \tag{VII.11}$$

The commutator matrix elements are evaluated in our complete Hartree–Fock basis by inserting the following exact resolution of the identity:

$$1 = |\Delta_0\rangle\langle\Delta_0| + \sum_{\alpha m} |\alpha m\rangle\langle\alpha m| + \tfrac{1}{4}\sum_{\alpha m}\sum_{\beta n} |\alpha m, \beta n\rangle\langle\alpha m, \beta n| + \cdots \tag{VII.12}$$

and (VII.4), (VII.11), and the one-electron nature of $\hat{F}$ then lead to

$$\langle\Delta_0|\hat{F}\hat{\mathcal{H}}|\alpha m\rangle = \sum_{\beta n} \langle\Delta_0|\hat{F}|\beta n\rangle\langle\beta n|\hat{\mathcal{H}}|\alpha m\rangle \tag{VII.13}$$

$$\langle\Delta_0|\hat{\mathcal{H}}\hat{F}|\alpha m\rangle = \langle\Delta_0|\hat{\mathcal{H}}|\Delta_0\rangle\langle\Delta_0|\hat{F}|\alpha m\rangle + \sum_{\beta n} \langle\Delta_0|\hat{\mathcal{H}}|\alpha m, \beta n\rangle\langle\alpha m, \beta n|\hat{F}|\alpha m\rangle \tag{VII.14}$$

with analogous expressions for the parts of the matrix element

$\langle \alpha m \| [\hat{F}, \hat{\mathcal{H}}] \| \Delta_0 \rangle$. Using the notation

$$A_{\alpha m, \beta n} = \langle \alpha m | \hat{\mathcal{H}} | \beta n \rangle - \langle \Delta_0 | \hat{\mathcal{H}} | \Delta_0 \rangle \delta_{\alpha\beta} \delta_{mn} \tag{VII.15}$$

$$B_{\alpha m, \beta n} = \langle \alpha m, \beta n | \hat{\mathcal{H}} | \Delta_0 \rangle \tag{VII.16}$$

(VII.11) takes the form

$$\sum_{\alpha m} \left\{ X^b_{\alpha m} \sum_{\beta n} \left[\langle \Delta_0 | \hat{F} | \beta n \rangle A_{\beta n, \alpha m} - \langle \beta n | \hat{F} | \Delta_0 \rangle B^*_{\alpha m, \beta n} \right] \right.$$
$$\left. + Y^b_{\alpha m} \sum_{\beta n} \left[- \langle \beta n | \hat{F} | \Delta_0 \rangle A_{\alpha m, \beta n} + \langle \Delta_0 | \hat{F} | \beta n \rangle B_{\beta n, \alpha m} \right] \right\}$$
$$= \hbar \omega_{b0} \sum_{\beta n} \left\{ \langle \Delta_0 | \hat{F} | \beta n \rangle X^b_{\beta n} + \langle \beta n | \hat{F} | \Delta_0 \rangle Y^b_{\beta n} \right\} \tag{VII.17}$$

A general one-electron operator $\hat{F}$ is a sum of a Hermitian and an anti-Hermitian operator, and the matrix elements $\langle \Delta_0 | \hat{F} | \beta n \rangle$ and $\langle \beta n | \hat{F} | \Delta_0 \rangle$ are hence linearly independent in real as well as in complex orbital bases. Equation (VII.17) must then be satisfied term by term, and we obtain the two sets of equations[243]

$$\sum_{\alpha m} \left\{ A_{\beta n, \alpha m} X^b_{\alpha m} + B_{\beta n, \alpha m} Y^b_{\alpha m} \right\} = \hbar \omega_{b0} X^b_{\beta n} \tag{VII.18}$$

$$\sum_{\alpha m} \left\{ B^*_{\beta n, \alpha m} X^b_{\alpha m} + A^*_{\beta n, \alpha m} Y^b_{\alpha m} \right\} = -\hbar \omega_{b0} Y^b_{\beta n} \tag{VII.19}$$

where **A** (VII.15) is a Hermitian matrix, while **B** (VII.16) is symmetric. Equations (VII.18) and (VII.19) define a non-Hermitian eigenvalue problem whose dimensionality is twice the number of singly excited configurations. The coefficients fulfill the orthonormality and closure relations

$$\sum_{\alpha m} \left\{ X^{b*}_{\alpha m} X^d_{\alpha m} - Y^{b*}_{\alpha m} Y^d_{\alpha m} \right\} = \delta_{bd} \tag{VII.20}$$

$$\sum_{b} \left\{ Y^b_{\alpha m} X^{b*}_{\beta n} - X^{b*}_{\alpha m} Y^b_{\beta n} \right\} = 0 \tag{VII.21}$$

$$\sum_{b} \left\{ X^b_{\alpha m} X^{b*}_{\beta n} - Y^{b*}_{\alpha m} Y^b_{\beta n} \right\} = \delta_{\alpha\beta} \delta_{mn} \tag{VII.22}$$

of which the orthogonality relations follow from the structure of the non-Hermitian eigenvalue problem in (VII.18) and (VII.19),[247] whereas the normalization expressed in (VII.22) can be considered a consequence of requiring that intensity sum rules like (IV.12) be obeyed.[248]

Most intensity calculations are carried out in a basis of real Hartree–Fock orbitals (namely, for atoms and molecules in the absence of external magnetic fields), and with transition moment operators that are either purely Hermitian (such as $\mathbf{r}$) or purely anti-Hermitian (such as $\hat{\nabla}$). Under these conditions the matrix elements $\langle\Delta_0|\hat{F}|\alpha m\rangle$ and $\langle\alpha m|\hat{F}|\Delta_0\rangle$ are in fact linearly dependent, in contrast to the assumption made in the above derivation. However, for a Hermitian operator $\hat{F}$ and real orbitals, (VII.17) yields

$$\sum_{\alpha m}\left\{A_{\beta n,\alpha m}-B_{\beta n,\alpha m}\right\}\left\{X_{\alpha m}^{b}-Y_{\alpha m}^{b}\right\}=\hbar\omega_{b0}\left\{X_{\beta n}^{b}+Y_{\beta n}^{b}\right\} \qquad \text{(VII.23)}$$

where we observe that the commutator $[\hat{F},\hat{\mathcal{H}}]$ is anti-Hermitian for this case, while an anti-Hermitian operator $\hat{F}$ leads to

$$\sum_{\alpha m}\left\{A_{\beta n,\alpha m}+B_{\beta n,\alpha m}\right\}\left\{X_{\alpha m}^{b}+Y_{\alpha m}^{b}\right\}=\hbar\omega_{b0}\left\{X_{\beta n}^{b}-Y_{\beta n}^{b}\right\} \qquad \text{(VII.24)}$$

Equation (VII.23) is thus the condition that a hypervirial relation of the form given in (II.20) be fulfilled, whereas (VII.24) is the condition pertaining, for example, to a hypervirial relation connecting $\hat{\nabla}$ and $(\hat{\nabla}V)$ transition moment matrix elements.[243] If we now require that a given set of coefficients $X_{\alpha m}^{b}$, $Y_{\alpha m}^{b}$ and transition frequencies ω_{b0} provide transition moments that fulfill (II.25) for Hermitian as well as anti-Hermitian operators, then (VII.23) and (VII.24) must be satisfied simultaneously, and by taking sums and differences of these two equations the real versions of (VII.18) and (VII.19) are recovered.

The requirement that the transition moments $\langle 0|\hat{F}|b\rangle$ and $\langle 0|[\hat{F},\hat{\mathcal{H}}]|b\rangle$ be related by a hypervirial equation (II.25), in conjunction with (VII.9), is thus sufficient to determine the amplitudes $X_{\alpha m}^{b}$ and $Y_{\alpha m}^{b}$ as well as the transition frequency ω_{b0}, if (VII.18) and (VII.19) are solved in a complete Hartree–Fock basis.[249] These two equations are just the relations that in many-body theories define the so-called random phase approximation (RPA),[232,245,250–252] or equivalently the so-called time-dependent Hartree–Fock (TDHF) approximation.[244,247,253–254] The fact that a general one-electron hypervirial relation is fulfilled in the RPA method (in a complete basis) implies that the Thomas–Kuhn and Condon sum rules (IV.12) and (IV.14) are obeyed as well.[248,255] The applicability of other sum rules within the RPA and related methods is discussed in Ref. 254. In addition, the transition moment vectors $\langle 0|\mathbf{r}|b\rangle$ and $\langle 0|\hat{\nabla}|b\rangle$ are parallel, and the length form of the rotatory strength is origin-independent.

The coefficients $X_{\alpha m}^{b}$ and $Y_{\alpha m}^{b}$ obtained from (VII.18) and (VII.19) are not the same as those obtainable [via (VII.10a) and (VII.10b)] from a variational or perturbational determination of the coefficients $\boldsymbol{C}$ and $\boldsymbol{D}$ of (VII.6) and (VII.7). In fact, from (VII.19) we obtain

$$\mathbf{Y}^b = -(\mathbf{A}^* + \hbar\omega_{b0})^{-1}\mathbf{B}^*\mathbf{X}^b \equiv \mathbf{S}^b\mathbf{X}^b \tag{VII.25}$$

where $\mathbf{Y}^b$ and $\mathbf{X}^b$ are column vectors, and the square matrices $\mathbf{A}$ and $\mathbf{B}$ are defined in (VII.15) and (VII.16). Equations (VII.10b) and (VII.25) are of the same structure; however, they differ in two important respects. In the first place, the square matrix $\mathbf{S}^b$ in (VII.25) is specific for the bth excitation, which shows how each RPA excitation selects that part of the ground-state correlation necessary for fulfilling its own one-electron hypervirial requirement. Second, although the matrix elements $B_{\alpha m,\beta n}$ (VII.16) are antisymmetric with respect to interchange of α and β, or m and n, the elements of the matrix $(\mathbf{A}+\hbar\omega_{b0})^{-1}$ have no particular properties with respect to this interchange, and the elements of $\mathbf{S}^b$ therefore do not exhibit the antisymmetry required of correlation coefficients for fermion wave functions. This means that the elements of the square matrix $\mathbf{YX}^{-1}$ will not exhibit this antisymmetry either, and the formal solution, $\mathbf{D}^* = \mathbf{YX}^{-1}$, of (VII.10b) therefore does not provide a properly antisymmetric set of $D_{\alpha m,\beta n}$ coefficients. This characteristic feature of the RPA method is referred to as violation of the Pauli principle[251,252,256,257] or violation of N-representability.[258] In the present context it can be viewed as a consequence of the fact that the hypervirial requirement (II.25) is a condition on only the first-order transition density matrix,[256] and that this condition is not consistent with N-representability, as witnessed by (VII.25). On the other hand, the violation of N-representability of $\mathbf{S}^b$ is not very severe in practice (see, e.g., Refs. 256 and 257), essentially because the matrix $(\mathbf{A}+\hbar\omega_{b0})^{-1}$ is predominantly diagonal. It follows from the above that the wave functions in (VII.6) and (VII.7) in fact serve mainly a heuristic purpose in providing the form of the transition moment in (VII.9). However, if desired, the matrix $\mathbf{YX}^{-1}$ obtained from an RPA calculation can be antisymmetrized in the orbital indices, and used to construct a properly antisymmetric, but of course nonvariational, ground state (see Ref. 258 for a recent many-body discussion of the N-representability and nature of the RPA ground state).

The preceding discussion demonstrates how the RPA method is a balanced and consistent approximation scheme for excitation properties, at the expense of a fully consistent wave-function basis, in contrast to variational methods in which individual states are optimized, but the excitation properties are often not well balanced. Since the hypervirial requirement leads uniquely to the RPA equations, it follows that the RPA scheme is the only version of the single particle–hole approximation, based upon a single determinant Hartree–Fock reference state, for which a hypervirial relation *can* be fulfilled for a general one-electron operator. On the other hand, if the particle–hole states are defined with reference to a multiconfigurational ground state,[251,258,259] the hypervirial requirement may be fulfilled for schemes other than the normal RPA.[258]

We may use (VII.25) to rewrite the RPA equations (VII.18) and (VII.19) in the following form:

$$\left[\mathbf{A}-\mathbf{B}(\mathbf{A}^*+\hbar\omega_{b0})^{-1}\mathbf{B}^*\right]\mathbf{X}^b=\hbar\omega_{b0}\mathbf{X}^b \tag{VII.26}$$

or, equivalently,

$$\left[\mathbf{A}+\mathbf{V}(\omega_{b0})\right]\mathbf{X}^b=\hbar\omega_{b0}\mathbf{X}^b \tag{VII.27}$$

where the energy-dependent potential $V(\omega_{b0})$ is similar to the "optical potential" in the Dyson equation from electron scattering theory. Csanak[260] observed that the RPA method is a way of generating the proper V^{N-1} potential for the excited electron in the presence of the radiation field inducing the transition (cf. our discussion of the IVO and IGO methods above). In other words, one can view the RPA as an SCF calculation that is especially adapted for the one-electron perturbation generated by an external electromagnetic field. If all contributions due to the doubly excited configurations are neglected, so that $V(\omega_{b0})=0$, (VII.27) reduces to the diagonalization of the singly excited configuration interaction matrix **A**, which is the usual TDA eigenvalue equation (see Section VII.A). Alternatively, one may construct various higher-order RPA methods (HRPA),[252,254] which employ modified versions of the A and B matrices. A number of these HRPA methods can be considered steps in an iteration toward a self-consistent RPA (SCRPA).[261] In a wave-function language the SCRPA method is equivalent (at least for a two-electron system) to a variational determination of the coefficients $C^b_{\alpha m}$ and $D_{\alpha m,\beta n}$ of (VII.6) and (VII.7).[261] It follows from the discussion in the previous paragraph that the transition moments obtained in HRPA and SCRPA calculations cannot be expected to fulfill hypervirial relations, even in a complete basis.

If the hypervirial constraint is applied at the TDA level, where the ground state is approximated by the Hartree–Fock function $|\Delta_0\rangle$ while the excited state retains the singly excited CI form of (VII.7), no general solution is possible,[243] but there are two separate restricted solutions. For a Hermitian operator $\hat{F}$ in a basis set of real orbitals, one obtains

$$(\mathbf{A}-\mathbf{B})\mathbf{X}^b=\hbar\omega_{b0}\mathbf{X}^b \tag{VII.28}$$

whereas an anti-Hermitian operator leads to

$$(\mathbf{A}+\mathbf{B})\mathbf{X}^b = \hbar\omega_{b0}\mathbf{X}^b \tag{VII.29}$$

The former case corresponds to enforcing equivalence (in a complete MO basis) of dipole-length and dipole-velocity transition moments, while the latter equation corresponds to equal dipole-velocity and dipole-acceleration ($\hat{\nabla} V$) intensities. In the RPA, by contrast, both conditions may be met simultaneously.

For the special cases of spin singlet and triplet excitations, the **A** and **B** matrix elements reduce to the following general forms:[232]

$$A_{\alpha m,\beta n} = \mathfrak{F}_{mn}\delta_{\alpha\beta} - \mathfrak{F}_{\alpha\beta}\delta_{mn} + \left[1+(-1)^S\right](m\alpha|\beta n) - (mn|\beta\alpha) \tag{VII.30}$$

$$B_{\alpha m,\beta n} = (m\alpha|n\beta) + (-1)^S\left[(m\alpha|n\beta) - (n\alpha|\beta m)\right] \tag{VII.31}$$

where S is the spin quantum number (0 or 1), $\mathfrak{F}_{ij}$ is the matrix element $\langle\phi_i|\hat{\mathfrak{F}}|\phi_j\rangle$ of the Fock operator, and $(ij|kl)$ is an electron repulsion integral in Mulliken's notation:

$$(ij|kl) = \int\int \phi_i^*(1)\phi_j(1)r_{12}^{-1}\phi_k^*(2)\phi_l(2)\,d\tau_1\,d\tau_2$$

In a so-called canonical orbital basis, the Fock operator is diagonal (i.e., $\mathfrak{F}_{ij} = \varepsilon_i\delta_{ij}$), where ε_i is the orbital energy. We shall find it convenient later to use noncanonical (localized) MO's in describing excitation properties. Canonical and noncanonical sets of MO's are related by separate unitary transformations of the occupied and virtual sets:

$$\phi'_\alpha = \sum_\beta \phi_\beta T_{\beta\alpha'} \tag{VII.32}$$

$$\phi'_m = \sum_n \phi_n S_{nm}{}' \tag{VII.33}$$

where the primed quantities refer to the noncanonical set. The separate transformations in (VII.32) and (VII.33) imply that the Brillouin relation (VII.4) holds also in a noncanonical MO basis, and we shall refer to an RPA calculation in localized noncanonical MO basis as localized orbital RPA or LORPA.

In a spin adapted RPA calculation, that is, using (VII.30) and (VII.31) (in canonical or noncanonical MO bases), the actual electronic transition moments for a spin-singlet excitation are then calculated as

$$\langle 0|\mathbf{r}|b\rangle = \sqrt{2}\sum_{\alpha m}\langle\alpha|\mathbf{r}|m\rangle(X^b_{\alpha m}+Y^b_{\alpha m}) \tag{VII.34}$$

$$\langle 0|\hat{\nabla}|b\rangle = \sqrt{2}\sum_{\alpha m}\langle\alpha|\hat{\nabla}|m\rangle(X^b_{\alpha m}-Y^b_{\alpha m}) \tag{VII.35}$$

$$\langle 0|\mathbf{r}\times\hat{\nabla}|b\rangle = \sqrt{2}\sum_{\alpha m}\langle\alpha|\mathbf{r}\times\hat{\nabla}|m\rangle(X^b_{\alpha m}-Y^b_{\alpha m}), \tag{VII.36}$$

utilizing the Hermitian or anti-Hermitian character of the operators. McCurdy and Cusachs[331] have considered the possibility of computing the dipole length oscillator strengths (III.19) by transforming the RPA equations (VII.23) and (VII.24) into a form which yields the coefficient combination $(X^b_{\alpha m}+Y^b_{\alpha m})$ directly.

C. Random Phase Approximation for Molecular Excitons

The molecular exciton approach outlined in Section V.C is not a fully consistent first-order theory with regard to the interchromophoric coupling; the inconsistency results from the neglect of configurations in which two chromophores are excited simultaneously. Allowing for more than one excited state per chromophore, a consistent first-order exciton theory[21, 163, 165, 167, 170] requires the use of the ground-state wave function

$$\begin{aligned}|0\rangle &= \prod_\mu \psi^0_\mu + \tfrac{1}{2}\sum_{\mu,a}\sum_{\nu,b} D_{\mu a,\nu b}\left[\frac{\psi^a_\mu\psi^b_\nu}{\psi^0_\mu\psi^0_\nu}\right]\prod_\eta \psi^0_\eta \\ &\equiv |0) + \tfrac{1}{2}\sum_{\mu,a}\sum_{\nu,b} D_{\mu a,\nu b}|\mu a,\nu b)\end{aligned} \tag{VII.37}$$

and the low-lying excited-state wave functions

$$|q\rangle = \sum_{\mu,a} C^q_{\mu a}\left(\frac{\psi^a_\mu}{\psi^0_\mu}\right)\prod_\eta \psi^0_\eta \equiv \sum_{\mu,a} C^q_{\mu a}|\mu a) \tag{VII.38}$$

in place of (V.8a) and (V.8b). Again ψ^a_μ represents the ath state of the μth chromophore; interchromophoric overlap, charge transfer, and electron exchange are neglected,[112] and we have assumed a Brillouin-type relation [see (VII.50)]

$$(0|\hat{\mathcal{H}}|\mu a) = 0 \qquad \text{all } \mu \text{ and } a \tag{VII.39}$$

where $\hat{\mathcal{H}}$ is the total electronic Hamiltonian of the polymer (V.7). Equation

(VII.39) ensures the absence of a first-order mixing of $|0)$ and the singly excited configurations $|\mu a)$. Equations (VII.37) and (VII.38) are closely analogous to the molecular wave functions in (VII.6) and (VII.7), except that the doubly excited configurational functions $|\mu a,\nu b)$ in (VII.37) are symmetric in the interchange of μa and νb, whereas $|\alpha m,\beta n)$ in (VII.6) has an additional antisymmetry in the interchange of α and β or m and n (hence the difference between a factor of $\frac{1}{4}$ in one equation and $\frac{1}{2}$ in the other).

The first-order equations (VII.37) and (VII.38) have been used in variational and/or perturbational calculations of polymer chiroptical properties (see, e.g., Refs. 21, 163, 165, 170 and references therein). Such treatments are accordingly analogous to the state approach discussed in Section VII.A, and the treatment of molecular excitons given in Section V.C can be equated to the TDA method of ordinary molecular calculations. Here we shall use instead an excitation approach to derive an RPA for molecular excitons. To this end (VII.37) and (VII.38) are combined to yield the following expression for a transition moment generated by a general one-electron operator $\hat{F}=\Sigma_i\hat{f}_i$:

$$\begin{aligned}\langle 0|\hat{F}|q\rangle &= \sum_{\mu,a}\left\{(0|\hat{F}|\mu a)C^q_{\mu a}+(\mu a|\hat{F}|0)\sum_{\nu,b}D^*_{\mu a,\nu b}C^q_{\nu b}\right\}\\ &= \sum_{\mu,a}\{(0|\hat{F}|\mu a)X^q_{\mu a}+(\mu a|\hat{F}|0)Y^q_{\mu a}\}\end{aligned}\qquad\text{(VII.40)}$$

and we shall try to determine the coefficients $X^q_{\mu a}$ and $Y^q_{\mu a}$ such that the transition moments induced by $\hat{F}$ and by the commutator $[\hat{F},\hat{\mathcal{H}}]$ are related by a hypervirial equation, in complete analogy to the development in Section VII.B. Again the use of (VII.40) for the transition moments induced by $[\hat{F},\hat{\mathcal{H}}]$ requires that this commutator also be a one-electron operator (see below). Equations (II.25) and (VII.40) combine to yield

$$\begin{aligned}&\sum_{\mu,a}\{(0|[F,\mathcal{H}]|\mu a)X^q_{\mu a}+(\mu a|[\hat{F},\hat{\mathcal{H}}]|0)Y^q_{\mu a}\}\\ &\qquad=\hbar\omega_{q0}\sum_{\mu,a}\{(0|\hat{F}|\mu a)X^q_{\mu a}+(\mu a|\hat{F}|0)Y^q_{\mu a}\}\end{aligned}\qquad\text{(VII.41)}$$

and the commutator matrix elements can be evaluated using the resolution of the identity

$$1=|0)(0| + \sum_{\mu,a}|\mu a)(\mu a| + \tfrac{1}{2}\sum_{\mu,a}\sum_{\nu,b}|\mu a,\nu b)(\mu a,\nu b| + \cdots\qquad\text{(VII.42)}$$

in conjunction with (VII.39), which leads to

$$(0|[\hat{F},\hat{\mathcal{H}}]|\mu a)=\sum_{\nu,b}\{(0|\hat{F}|\nu b)A_{\nu b,\mu a}-(\nu b|\hat{F}|0)B^*_{\nu b,\mu a}\} \quad \text{(VII.43)}$$

$$(\mu a|[\hat{F},\hat{\mathcal{H}}]|0)=\sum_{\nu,b}\{(0|\hat{F}|\nu b)B_{\nu b,\mu a}-(\nu b|\hat{F}|0)A^*_{\nu b,\mu a}\} \quad \text{(VII.44)}$$

where

$$A_{\nu b,\mu a}=(\nu b|\hat{\mathcal{H}}|\mu a)-(0|\hat{\mathcal{H}}|0)\delta_{\nu\mu}\delta_{ba} \quad \text{(VII.45)}$$

$$B_{\nu b,\mu a}=(\nu b,\mu a|\hat{\mathcal{H}}|0) \quad \text{(VII.46)}$$

Inserting (VII.43) and (VII.44) into (VII.41), and utilizing that the matrix elements $(0|\hat{F}|\mu a)$ and $(\mu a|\hat{F}|0)$ are unrelated for a general one-electron operator $\hat{F}$, so that the resulting (VII.41) must be satisfied term by term, we find

$$\sum_{\nu b}\{A_{\mu a,\nu b}X^q_{\nu b}+B_{\mu a,\nu b}Y^q_{\nu b}\}=\hbar\omega_{q0}X^q_{\mu a} \quad \text{(VII.47)}$$

$$\sum_{\nu b}\{B^*_{\mu a,\nu b}X^q_{\nu b}+A^*_{\mu a,\nu b}Y^q_{\nu b}\}=-\hbar\omega_{q0}Y^q_{\mu a} \quad \text{(VII.48)}$$

in complete analogy to (VII.18) and (VII.19). This non-Hermitian eigenvalue problem defines a random phase approximation for exciton systems, and the eigenvectors fulfill orthonormality conditions similar to (VII.20) to (VII.22), *mutatis mutandis*.

The special form of the exciton Hamiltonian (V.7) allows some further reduction of (VII.47) and (VII.48). Introducing the notation

$$[\mu a,\mu c|\nu b,\nu d]=\int d\tau_\mu\int d\tau_\nu \psi^a_\mu\psi^b_\nu V_{\mu\nu}\psi^c_\mu\psi^d_\nu \quad \text{(VII.49)}$$

for real wave functions, the interchromophoric coupling is expressed in a form that reflects the multipolar interaction between a charge density (for $a=c$) or transition density[112] (for $a\neq c$) on chromophore μ with a charge or transition density on chromophore ν. If it is assumed that the chromophores are electrically neutral and without permanent electric dipole moments, we can write to a very good approximation

$$[\mu a,\mu a|\nu b,\nu d]=0 \qquad \text{all } a,b,d \quad \text{(VII.50)}$$

For $a=0$ this relation ensures the validity of (VII.39). For real wave functions (V.7), (VII.45), (VII.46), (VII.49) and (VII.50) then yield

$$A_{\nu b,\mu a}=\hbar\omega_{\mu}^{a}\delta_{\mu\nu}\delta_{ab}+[\nu 0,\nu b|\mu 0,\mu a] \tag{VII.51}$$

$$B_{\nu b,\mu a}=[\nu 0,\nu b|\mu 0,\mu a] \tag{VII.52}$$

where $\hbar\omega_{\mu}^{a}$ is the unperturbed excitation energy of chromophore μ (i.e., an eigenvalue of $\hat{\mathcal{H}}_{\mu}$), and the off-diagonal elements of the **A** and **B** matrices reduce to the multipolar interactions between the transition densities on the chromophores. Equations (VII.47) and (VII.48) reduce to

$$\hbar(\omega_{q0}-\omega_{\mu}^{a})X_{\mu a}^{q}=\sum_{\nu b}{}'[\mu 0,\mu a|\nu 0,\nu b]\{X_{\nu b}^{q}+Y_{\nu b}^{q}\} \tag{VII.53}$$

$$-\hbar(\omega_{q0}+\omega_{\mu}^{a})Y_{\mu a}^{q}=\sum_{\nu b}{}'[\mu 0,\mu a|\nu 0,\nu b]\{X_{\nu b}^{q}+Y_{\nu b}^{q}\} \tag{VII.54}$$

which combine into the equation

$$\sum_{\nu,b}\Big\{\hbar^{2}\big[(\omega_{\mu}^{a})^{2}-\omega_{q0}^{2}\big]\delta_{\mu\nu}\delta_{ab}$$

$$+2\hbar\omega_{\mu}^{a}[\mu 0,\mu a|\nu 0,\nu b]\Big\}\{X_{\nu b}^{q}+Y_{\nu b}^{q}\}=0 \tag{VII.55}$$

The summations in (VII.53) and (VII.54) exclude the diagonal contributions. This represents a non-Hermitian (in general) eigenvalue equation for the determination of the square of the excitation energy and the coefficients $\{X_{\mu a}^{q}+Y_{\mu a}^{q}\}$. The normalization of this combination of the coefficients, or the individual coefficients, can be obtained by subtracting (VII.53) and (VII.54) to yield

$$Y_{\mu a}^{q}=\frac{\omega_{\mu}^{a}-\omega_{q0}}{\omega_{\mu}^{a}+\omega_{q0}}X_{\mu a}^{q} \tag{VII.56}$$

which shows that the column vectors $\mathbf{X}^{q}$ and $\mathbf{Y}^{q}$ are proportional for a given exciton transition. Equations (VII.56) and (VII.20) then provide the orthonormalization relation

$$\sum_{\mu,a}(\omega_{\mu}^{a})^{-1}(X_{\mu a}^{q}+Y_{\mu a}^{q})\cdot(X_{\mu a}^{p}+Y_{\mu a}^{p})=\omega_{q0}^{-1}\delta_{pq} \tag{VII.57a}$$

or

$$\sum_{\mu,a}\left[\frac{(\omega_\mu^a\omega_{q0})^{1/2}}{(\omega_\mu^a+\omega_{q0})/2}X_{\mu a}^q\right]\left[\frac{(\omega_\mu^a\omega_{p0})^{1/2}}{(\omega_\mu^a+\omega_{p0})/2}X_{\mu a}^p\right]=\delta_{pq} \qquad \text{(VII.57b)}$$

and similarly from equations (VII.22) and (VII.56)

$$\sum_q \omega_{q0}(X_{\mu a}^q+Y_{\mu a}^q)(X_{\nu b}^q+Y_{\nu b}^q)=\omega_\mu^a\delta_{\mu\nu}\delta_{ab} \qquad \text{(VII.57c)}$$

or

$$\sum_q\left[\frac{(\omega_\mu^a\omega_{q0})^{1/2}}{(\omega_\mu^a+\omega_{q0})/2}X_{\mu a}^q\right]\left[\frac{(\omega_\nu^b\omega_{q0})^{1/2}}{(\omega_\nu^b+\omega_{q0})/2}X_{\nu b}^q\right]=\delta_{\mu\nu}\delta_{ab} \qquad \text{(VII.57d)}$$

In the present context a complete basis implies that all the individual chromophoric states are to be included, and this development therefore shows that the solutions of (VII.55) to (VII.57) in conjunction with (VII.40), in a complete basis, provide exciton transitions fulfilling a one-electron hypervirial relation [in particular (II.20)]. The completeness relation (VII.22) then ensures that sum rules such as the Thomas–Kuhn and Condon relations (IV.12) and (IV.14) are obeyed also. The non-Hermitian character of (VII.53) to (VII.55) shows that the results are generally not the same as a variational calculation based upon (VII.37) and (VII.38), and such variational calculations will therefore provide exciton transition moments that, in general, violate the hypervirial relations.

The exciton RPA relations (VII.53) to (VII.55) and (VII.57), obtained here via the hypervirial requirement, are the same as those obtained in the second-quantized exciton approach associated with the names Agranovich, Bogoliubov, and Davydov[109,110] (our relations are identical to equations 290, 291, 294, 295 of Ref. 110), and a calculation of polymer chiroptical properties within this approach was presented by Loxsom, Tterlikkis, and Rhodes.[263] In different guise the same approximation appears in the formalism of linear response theory,[264–267] Green's function approaches,[121,268] and time-dependent Hartree (TDH) theory;[40,269–273] the TDH theory has been applied recently to the chiroptical properties of nonpolymeric molecules.[87a,274] Also closely related is DeVoe's classical oscillator theory for polymeric optical properties[275] (see also Refs. 276 and 277).

In order to establish a closer connection to the treatment in Section V.C, consider again the restricted situation where we include only one excited state, ψ_μ^a say, per chromophore. Equation (VII.38) is then of the same form as (V.8b), and (VII.55) becomes

$$\sum_{\nu}\left\{\hbar^2\left[\omega_a^2-\omega_{q0}^2\right]\delta_{\mu\nu}+2\hbar\omega_a\left[\,\mu 0,\mu a|\nu 0,\nu a\right]\right\}\left\{X_{\nu a}^q+Y_{\nu a}^q\right\}=0 \quad \text{(VII.58)}$$

Notice now that the approximation $(\omega_a+\omega_{q0})=2\omega_a$ makes this relation identical to (V.9). In this approximation the excitation energies obtained from (V.9) and (VII.58) are therefore the same, and (VII.56) and (VII.40) combine to yield transition moments that are essentially the same as those obtained by a first-order perturbation treatment[21,166,167] based upon (VII.37) and (VII.38) (still restricted to one excited state per chromophore). In higher order, however, the two approaches tend to differ. To illustrate this, consider the oscillator strengths of equations (III.18) to (III.20), all of which produce the same result even in the restricted RPA corresponding to (VII.58), namely

$$f_q(\text{RPA})=\frac{2}{3}\,\frac{\omega_a\omega_{q0}}{(\omega_a+\omega_{q0})^2/4}$$

$$\times\sum_{\mu,\nu}\langle\psi_\mu^0|\hat{\nabla}_\mu|\psi_\mu^a\rangle\cdot\langle\psi_\nu^a|\mathbf{r}_\mu|\psi_\nu^0\rangle X_{\mu a}^q X_{\nu a}^q \quad \text{(VII.59a)}$$

from (VII.40), (VII.56) and (V.16) (see also (III.52b) and (V.11b) for notation). From this expression the completeness relation (VII.57d) leads to the sum rule

$$\sum_q f_q(\text{RPA})=N_0 f_a \quad \text{(II.59b)}$$

where N_0 is the number of chromophores, and f_a is the oscillator strength of the local $0\rightarrow a$ excitation. Strictly analogous results obtain for the rotatory strengths. Therefore, even the simple one-state-per-chromophore RPA is intensity conserving to all orders, whereas it has been demonstrated explicitly[332] that, for a model system, a perturbational state approach leads to sum rule violations in second order.

It should be added that the present derivation of the exciton RPA hinges upon the assumption that $\hat{F}$ and $[\hat{F},\hat{\mathcal{H}}]$ can both be considered general one-electron operators. There are indications[125] that a multipolar Hamiltonian of the form (V.7) may lead to problems in the evaluation of commutators such as $[\hat{F},\hat{\mathcal{H}}]$, even for $\hat{F}=\sum_i\hat{\nabla}_i$. It is therefore important to stress that it is fully legal within the present approach to use the correct many-electron Hamiltonian, rather than (V.7), for the evaluation of the commutators $[\hat{F},\hat{\mathcal{H}}]$. Equation (V.7) is only a convenient way of exhibiting the terms in the Hamiltonian that contribute in the matrix elements of (VII.45), (VII.46), (VII.51), and (VII.52). Once the molecular exciton approximation (i.e., neglect of interchromophoric charge transfer, overlap, and electron exchange) is introduced through the use of (VII.37) and (VII.38), then the matrix elements in (VII.45), (VII.46), (VII.51), and (VII.52) come out in the multipolar form, independent of whether the correct many-electron Hamiltonian or (V.7) is employed.[112] The requirement that $\hat{F}$ and $[\hat{F}, \hat{\mathcal{H}}]$ both be general one-electron operators is therefore not any more restrictive in the exciton RPA formalism than in the ordinary molecular RPA formalism (Section VII.B).

D. Remarks on Computation

In the foregoing we have assumed that the basis set of Hartree–Fock MO's is complete. In practice, of course, one deals with a finite—and often severely truncated—atomic and molecular orbital basis. A consequence of this truncation is that the hypervirial relation is not expected to be fulfilled exactly, even in the RPA methods. The degree to which the computed intensity forms agree may thus be used as an indicator of the completeness of the basis set for the particular excitation under consideration.

As we pointed out in Section IV.C, the failure of hypervirial equivalence also leads to violations of the intensity sum rules and to origin dependence of R^r.[54] Nonetheless, one can obtain sum-rule-like expressions for the energy-free intensities $f^{r\nabla}$ and R^r.

Let F^A denote an anti-Hermitian operator, such as $\hat{\nabla}$ or $\mathbf{r}\times\hat{\nabla}$, and F^H a Hermitian one, such as $\mathbf{r}$. Then the sum rule for $f_{0b}^{r\nabla}$ or R_{0b}^{r} can be written in the general form

$$S^{(0)} = k\sum_b \{\langle 0|F^A|b\rangle\cdot\langle b|F^H|0\rangle - \langle 0|F^H|b\rangle\cdot\langle b|F^A|0\rangle\} \tag{VII.60}$$

In this notation, the RPA transition moments (VII.9) become

$$\langle 0|F^A|b\rangle = \sum_{\alpha m}\{X_{\alpha m}^b - Y_{\alpha m}^b\}\langle\alpha|\hat{f}^A|m\rangle \tag{VII.61a}$$

$$\langle 0|F^H|b\rangle = \sum_{\alpha m}\{X_{\alpha m}^b + Y_{\alpha m}^b\}\langle\alpha|\hat{f}^H|m\rangle \tag{VII.61b}$$

Since the orthonormality and closure relations (VII.20) to (VII.22) are fulfilled *in finite as well as infinite bases*, we can combine (VII.20) to

(VII.22) with (VII.60) to (VII.61) to yield

$$S^{(0)} = k \sum_{\alpha m} \{\langle\alpha|\hat{f}^A|m\rangle\langle m|\hat{f}^H|\alpha\rangle - \langle\alpha|\hat{f}^H|m\rangle\langle m|\hat{f}^A|\alpha\rangle\} \equiv S^{(0)}\,(\mathrm{RPA}) \qquad \text{(VII.62)}$$

Since this equation is already in the form of $S^{(0)}$ (STA), and since the TDA coefficients $C^b_{\alpha m}$ form a unitary matrix, we find[248] that

$$S^{(0)}\,(\mathrm{STA}) = S^{(0)}\,(\mathrm{TDA}) = S^{(0)}\,(\mathrm{RPA}) \qquad \text{(VII.63)}$$

Equation (VII.63) holds for any size basis, and therefore Harris' proof[255] of the STA sum rules for $f^{r\nabla}$ and R^r holds also in the "exact" TDA. Jørgensen, Oddershede, and Beebe[278] have shown further that *any* approximation to the particle–hole propagator [i.e., the second-quantized formulation of the particle–hole terms in the transition moment expression, (VII.5)] exhibits the same property, so that $S^{(0)}$ is a property of the basis set alone. This suggests, in turn, that intensity sum rules may provide an additional criterion for selecting and optimizing basis sets for calculations of electronic excitation properties. Computed sum rules exhibiting these features are given in Refs. 248 and 278.

The problem of choosing a basis set is one aspect of a more general task of choosing among a number of methods for evaluating the necessary integrals and MO coefficients. In *ab initio* methods, all integrals are evaluated explicitly from a chosen atomic orbital basis set, while in semiempirical methods, a number of integrals are neglected and the remaining ones are either estimated from empirical data or evaluated over a simple AO basis. The *ab initio* methods afford the choice between Slater-type (STO) or Gaussian-type (GTO) atomic orbital basis sets, which in turn can be enlarged at will to achieve the desired degree of accuracy. [See, for example, the recent reviews in Ref. 279.] Formulas for the AO integrals of the various operators occurring in the molecular Hamiltonian are well known; the additional integrals needed for computation of oscillator and rotatory strengths according to (III.17) to (III.23) are available in both STO[280–284] and GTO[285,286] versions.

Semiempirical methods offer an equally rich, if not bewildering, variety of approximation and parameter fitting schemes, most of which are based upon some variant of the neglect of differential overlap (NDO) (e.g., CNDO, INDO, NDDO, CNDO/S). (For a recent review of these methods, see Ref. 287.) In the majority of applications of these semiempirical methods to chiroptical properties (see Section VIII), the Hamiltonian matrix elements are parametrized, but the "optical" integrals (involving $\mathbf{r}$, $\hat{\nabla}$, and $\mathbf{r}\times\hat{\nabla}$) are evaluated directly over an STO basis. In the case of the NDO models, this means that one in effect is retaining some integrals that are of the same order in overlap as others that are neglected. This mismatch of integrals has been discussed by several authors;[134,288–290] one can restore consistency by subjecting the NDO basis set to a "deorthogonalization" step[289] before evaluating the optical integrals. However, the effects of alleviating the imbalance may or may not be numerically significant. Semiempirical schemes in which some of the optical integrals are parametrized as well have been proposed,[291,292] and Linderberg and Seamans[293] have addressed the question of a consistent parametrization scheme for both the optical and the Hamiltonian matrix elements by requiring the fulfillment of hypervirial and other operator relations.

The *ab initio* methods clearly provide a systematic hierarchy of improvements to the molecular orbitals, and allow as well a realistic discussion of higher-lying and non-valence-shell excited states. However, exploiting this advantage for chiral systems of interest is expensive and requires access to large computer systems. Moreover, choosing a basis set that combines economy with the flexibility required for a balanced description of low-lying excitations can be difficult in its own right. The semiempirical methods are hence far from obsolete, in our opinion. Despite possible imbalances in parametrization and integral evaluation, one can still employ these methods fruitfully in chiroptical studies, both in gaining qualitative insights into larger systems at moderate cost and in providing guidance for *ab initio* studies.

Finally, we note that some caution must be exercised in using the RPA equations if the atomic orbital basis set is too poorly constituted. In certain cases, the excitation frequencies computed for triplet excitations turn out negative, and the RPA method fails. These occurrences are connected with the stability of the approximate restricted Hartree–Fock ground state for the basis set used; the problem has been discussed by Ostlund.[294] Moreover, Szabo and Ostlund[295] have pointed out that minimal basis sets may lead to RPA excited states that are too compact, at least for molecules with two or three atoms. In the next section we shall present RPA calculations of chiroptical properties that appear not to be affected by the caveats just mentioned.

VIII. COMPUTED CHIROPTICAL PROPERTIES

Even the simplest chiral molecule must contain at least four atoms, and resolvable chiral systems on which CD measurements can be taken are typically much larger. Hence computations of chiroptical properties have not, in general, achieved the level of sophistication currently possible for studies of the electronic structure of atoms and small molecules.[279] In addition, the literature on computed chiroptical properties, particularly when confined to all-electron or all-valence-electron calculations, is relatively small as compared with studies of ordinary electronic intensities. We do not, however, attempt complete coverage of such calculations. Instead, we shall discuss in some detail a number of calculations that illustrate the relative merits of the computational schemes outlined in Section VII, and/or which establish a connection to the models discussed in Section V. The outline of this section therefore parallels that of Section V, and we shall take advantage of the phrase "individual points of view" in the Editorial Introduction to draw heavily on our own recent results.

A. Chiral Chromophores

Leaving aside for the moment cases in which chirality results from the coupling of two or more achiral systems (i.e., dynamic coupling or exciton cases), essentially all the inherently chiral chromophores that have been studied may be envisioned as erstwhile achiral groups on which chirality has been conferred by distorting the chromophore. A simple model system that has been studied in some detail is a twisted conformation of H_2O_2,[231,296] where the small size of the system made feasible *ab initio* computations with extended basis sets. Rauk and Barriel[231] used a "first-order CI," which is essentially a truncated variational–perturbational calculation based upon (VII.6) to (VII.8), while Barbagli and Maestro[296] used a Coupled Hartree–Fock method that is equivalent to the RPA for the properties computed. The absence of resolvable enantiomers of H_2O_2 makes assessment of the results difficult, although Rauk and Barriel report reasonable agreement between R^r and R^∇ after some adjustments of the excitation energy. Semiempirical studies have been reported on chiral disulfides,[292,297] and on twisted amides.[298,299]

A simple chiral chromophore for which experimental comparisons are possible is the widely studied ethylenic chromophore in *trans*-cyclooctene (TCO), a highly strained ring system whose CD spectrum is dominated by two intense, oppositely signed bands.[300] The relatively small number of electrons in ethylene itself makes it amenable to *ab initio* MO approaches, and a number of studies have been reported[132,231,240,301,302] where a twisted ethylene group (as shown in Fig. 6) is used to mimic the chromophore in TCO. Two recent *ab initio* calculations[243,303] have employed an ethylene conformation more closely resembling the actual geometry of TCO,[304] and the effect of the saturated framework in TCO has been estimated by calculations on a chiral *trans*-2-butene.[132,303] In the *ab initio* calculations reported in Ref. 132 we computed the transition properties for the lower excitations in ethylene twisted 10° (in the sense indicated in Fig. 6) using the STA, TDA, RPA, and one version of the HRPA schemes discussed in Section VII. These calculations employed a Slater-type minimal (i.e., single-zeta, SZ) as well as a valence-shell double-zeta (DZ) atomic orbital basis (the actual orbital bases were the STO-4G and STO-5G bases provided by the GAUSSIAN 70 program[305]), and all the valence electron configurations (i.e., 36 in SZ and 108 in DZ) arising from the chosen orbital bases were included. A comparison among the results obtained in the various computational schemes for the two strong CD bands is given in Fig. 7 and in Table I. It is clear from these results that TDA is an unbalanced scheme as regards intensities, while the energy-free intensity results (III.20) and (III.23) are comparable in STA and RPA,[54] and that the weak intensities are somewhat more dependent upon the quality of the atomic basis set, as is perhaps to be expected. The RPA results most nearly satisfy the hypervirial relation, as anticipated from the derivation of the RPA method, whereas we find that the very poor agreement among equivalent intensities in the TDA scheme is a general feature in our calculations. Table I also shows how important the mixing of the $\sigma \rightarrow \sigma^*$ and the $\pi \rightarrow \pi^*$ configurations is for the correct assessment of the chiroptical properties of the lowest transitions. This observation points up the danger of relying only on an energy-based selection of configurations when a truncation of the set $|\alpha m\rangle$ is necessary. The $\sigma \rightarrow \sigma^*$ excitation is one of the highest-lying configurations in energy, and would likely be excluded on energy grounds in a truncation; the calculated oscillator and rotatory strengths would suffer accordingly. See Ref. 306 for further discussion.

Table II presents a comparison between our RPA results for the two characteristic bands in two conformations of ethylene, namely the 10° twisted form[132] (labeled 10°) and the conformation corresponding to the actual TCO geometry[243] (labeled TCO), and in a *trans*-2-butene where the double bond is twisted 10° (labeled T2B). The table also contains the recent results obtained in a large-scale perturbative CI calculation by Liskow and Segal,[303] and the actual experimental values.[300] The calculations all

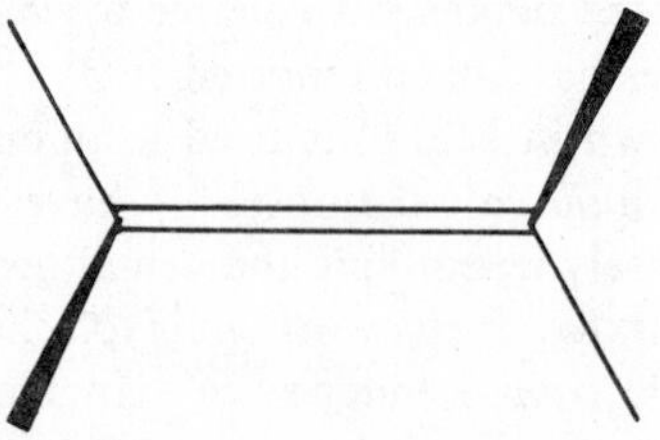

Fig. 6. Geometry and sense of helicity for twisted ethylene, as used in computations.

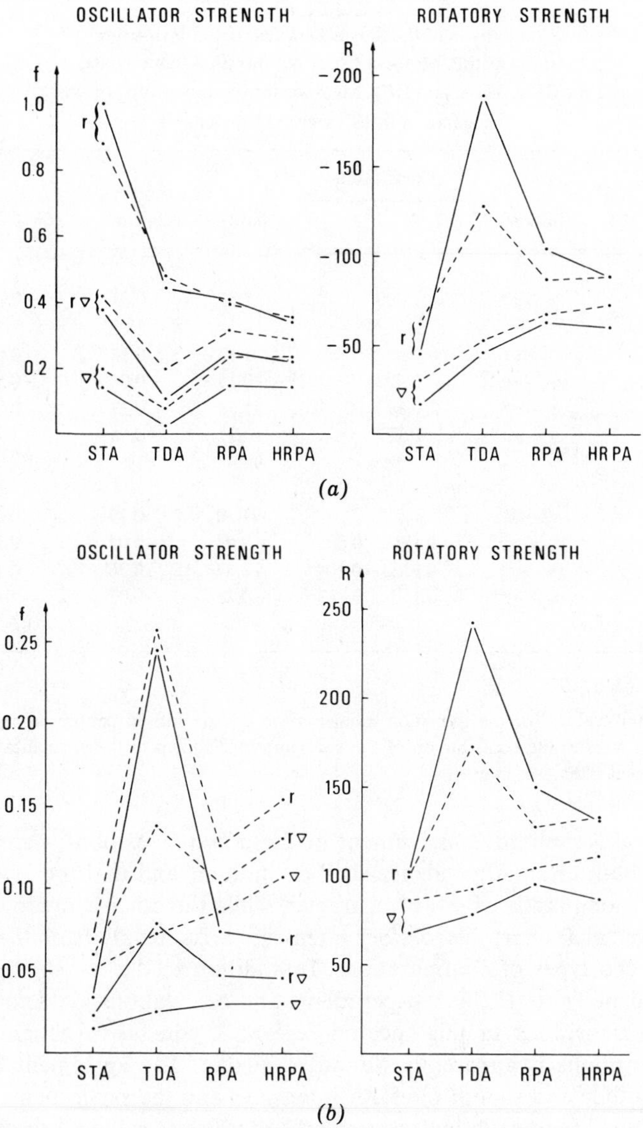

Fig. 7. (*a*) Comparison of STA, TDA, RPA, and HRPA results for the lowest $^1B_3$ transition in 10°-twisted ethylene. Solid lines refer to single-zeta and dashed lines to double-zeta Slater orbital basis sets. (*b*) Comparisons, as in (*a*), for the next-lowest $^1B_3$ transition.

TABLE I.
Comparison of the Single-Zeta Excitation Properties and Transition Moment Contributions (in Atomic Units) in the STA, TDA, and RPA Methods for the Lowest $^1B_3$ ($\pi \rightarrow \pi^*$) Excitation in 10°-Twisted Ethylene[a,b]

Method energy (eV)	$\|\alpha \rightarrow m\rangle$	Coefficients X	Y	$\langle 0\|\mathbf{r}\|b\rangle$	$\langle 0\|\hat{\nabla}\|b\rangle$	$\langle 0\|\mathbf{r}\times\hat{\nabla}\|b\rangle$
STA						
12.55	$\pi \rightarrow \pi^*$	1.000	0	1.81	0.31	−0.11
TDA	Total			1.32	0.12	−0.62
10.55	$\sigma_{CH} \rightarrow \pi^*$	0.433	0	−0.15	−0.04	−0.53
	$\pi \rightarrow \pi^*$	0.872	0	1.58	0.27	−0.09
	$\sigma_{CC} \rightarrow \sigma^*$	0.177	0	−0.21	−0.16	0.0
		98%		92%	61%	101%
RPA	Total			1.30	0.29	−0.34
9.86	$\sigma_{CH} \rightarrow \pi^*$	0.196	0.005	−0.07	−0.02	−0.23
	$\pi \rightarrow \pi^*$	0.972	−0.104	1.57	0.36	−0.12
	$\sigma_{CC} \rightarrow \sigma^*$	0.158	0.083	−0.28	−0.07	0.0
		99%		93%	87%	103%

[a]From Ref. 132.

[b]"Total" values for the transition moments are for the full 36 configurations. Percentages refer to the contribution of the configurations shown to the normalization or to the total transition moments.

agree on the qualitative assignment of these two transitions, namely that they are both essentially valence-shell excitations, and that the low-energy one is predominantly of $\pi \rightarrow \pi^*$ character, while the other is predominantly of $\sigma_{CH} \rightarrow \pi^*$ character; the rotatory intensities arise mainly from the mixing of these two types of configurations. In addition to the two strong bands contained in Table II, both experiments and calculations yield some low-intensity transitions in this spectral region; a conclusive assignment of these transitions is apparently not yet available. The agreement between the magnitude and sign of the RPA intensities and the experimental values is rather good (and is, in fact, somewhat better than for the CI results), and although the computed absolute transition energies are high, the splitting of the two bands is quite well reproduced. Finally, the similarity between the twisted ethylene and the twisted *trans*-2-butene results indicates that it is valid to correlate the chiroptical intensities of this system with its molecular structure on the basis of the chirality of the chromophore; the saturated molecular framework apparently does not significantly affect the sign and magnitude of the low-energy rotatory intensities. (Note added in

TABLE II
Summary of RPA Results for the Two Strong CD Bands in Distorted Ethylene and *Trans*-2-butene Conformers[a]

	ΔE (eV)	f^r	$f^{r\nabla}$	f^{∇}	$R^{[r]}$	R^{∇}
$^1B(\pi\to\pi^*)$						
10°, SZ[b]	9.86	0.41	0.25	0.15	−104	−63
10°, DZ[b]	8.32	0.41	0.32	0.25	−87	−68
T2B, SZ[b]	9.36	0.54	0.35	0.22	−78	−49
TCO[c]	7.15	0.25		0.18	−141	−119
Ref. 303	7.54	0.27		0.15	−215	−153
Exptl. (Ref. 300)	6.3		0.15		−70	
$^1B(\sigma_{CH}\to\pi^*)$						
10°, SZ[b]	11.34	0.07	0.05	0.03	151	95
10°, DZ[b]	9.85	0.13	0.10	0.09	128	105
T2B, SZ[b]	11.28	0.05	0.03	0.02	119	80
TCO[c]	9.05	0.06		0.04	152	135
Ref. 303	9.71	0.35		0.17	263	180
Exptl. (Ref. 300)	7.9		Weak		110	

[a]See text for notation. Units of R values are (cgs $\times 10^{40}$).

[b] From Ref. 132.

[c]From Ref. 243.

proof: after completing this manuscript we were able to perform a full valence electron RPA calculation for TCO, that is, 529 particle-hole excitations.[333] The results are in qualitative agreement with the results in Table II; however, the nonchromophoric parts of the molecule do lead to some redistribution of ordinary and rotatory intensities relative to the isolated chiral chromophore.)

Other inherently chiral chromophores include skewed α-diketones, which have been studied in semiempirical STA and TDA models,[307,308] and nonplanar *cisoid* butadiene, which is a well-studied model system[135,138,309–311] for a diene helicity rule. Recent experimental observations[139–141] indicate that the perturbation of the diene chromophore due to the saturated molecular framework may be just as important for the sign and magnitude of the chiroptical intensities as the effect of the inherent chirality due to nonplanarity, in contrast to the observation made above for *trans*-cyclooctene. Semiempirical calculations supporting the importance of the molecular framework for the chiroptical intensities of nonplanar diene systems have been presented by Rosenfield and Charney,[310] and in Ref. 311 we reported a set of *ab initio* minimal basis set calculations on planar *trans*- and 45° twisted *cis*-butadiene; Figure 8 shows the geometry used for the twisted conformer. Our STA, TDA, and RPA results are reproduced in Table III; again the results for the equivalent intensities are

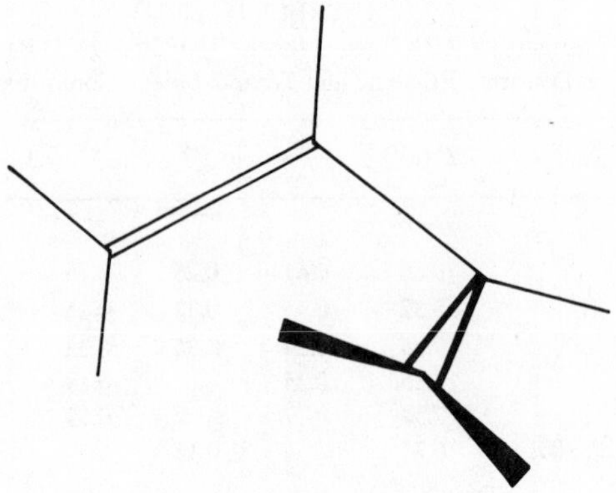

Fig. 8. Geometry and sense of helicity of *cis*-butadiene twisted 45° out of a planar conformation.

in much better agreement in RPA than in STA or TDA. Notice also that the RPA value of R^r for the lowest transition is in fact independent of origin, even though it is not so constrained by symmetry. For this lowest $\pi \rightarrow \pi^*$ transition, which has been the basis for the diene helicity rules, the computed rotatory strengths are all numerically small, and the TDA results are of opposite sign to the STA and RPA ones. The reason for this is indi-

TABLE III
STA, TDA, and RPA Results for the Properties of the Lowest-Energy Excitations in Twisted Butadiene (Fig. 8)[a,b]

Excitation	Method	E (eV)	f^r	$f^{r\nabla}$	f^{∇}	R^{∇}	$R^{[r]}$	ΔR^r
$^1B(\pi_2 \rightarrow \pi_3)$	STA	9.83	0.79	0.34	0.14	−13	−32	6.0
	TDA	8.84	0.62	0.23	0.08	15	43	5.6
	RPA	8.33	0.49	0.31	0.19	−7	−11	0.0
$^1A(\pi_1 \rightarrow \pi_3)$	STA	11.90	0.58	0.24	0.10	24	57	0
	TDA	10.10	0.05	0.01	0.00	10	39	0
	RPA	9.86	0.11	0.07	0.04	49	79	0
$^1B(\sigma \rightarrow \pi_3)$	STA	11.11	0.01	0.00	0.00	−15	−19	0.8
	TDA	10.42	0.00	0.00	0.00	−2	−1	0.2
	RPA	10.35	0.00	0.00	0.00	−5	−6	0.2
$^1A(\pi_2 \rightarrow \pi_4)$	STA	12.63	0.61	0.24	0.09	20	51	0
	TDA	11.61	0.68	0.13	0.03	−22	−118	0
	RPA	10.78	0.40	0.24	0.14	−24	−42	0

[a] From Ref. 311.

[b] Units of R are (cgs $\times 10^{40}$) and of ΔR^r, (cgs $\times 10^{40}$) Å^{-1}.

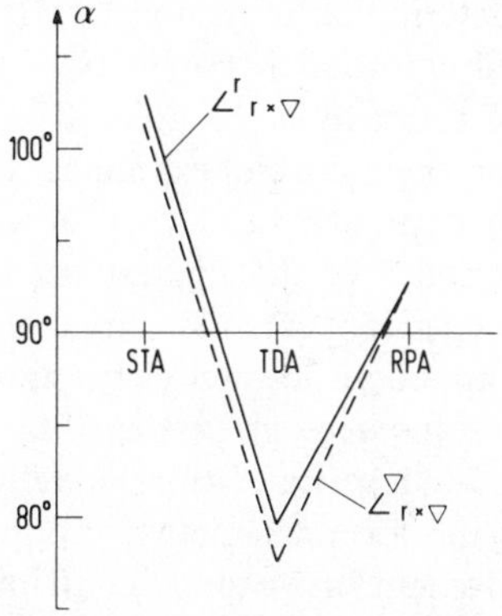

Fig. 9. Comparison of minimal-basis STA, TDA, and RPA magnetic dipole transition moments for the lowest transition in 45°-twisted *cis*-butadiene.

cated in Fig. 9. Bearing in mind the origin dependence of the magnetic dipole transition moment, we note that *for the origin chosen*, the angle between the electric dipole and magnetic dipole transition moments hovers around 90° and changes from a value less than 90° in TDA to values greater than 90° in STA and RPA. The projection of $\boldsymbol{\mu}_m$ on $\boldsymbol{\mu}_e$ is accordingly sensitive to the details of the model, and therefore presumably also sensitive to the effects of substituents, in agreement with the experimental violations of the diene chirality rule,[139–141] and with the results of Ref. 310. In other words, the experiments and calculations agree that for a molecule containing a twisted diene chromophore, a reliable configurational assignment cannot be expected from rules relating the chromophore helicity to the experimental rotatory strengths of the lowest transition.

B. Achiral Chromophores

A major rationale behind calculations on achiral chromophores in chiral environments is to assess the nature of the mechanisms by which rotatory intensity is induced, as well as to treat exceptions to the sector rules. Semiempirical calculations have been performed on a number of chirally perturbed chromophores by Richardson and co-workers, including derivatives of peptides,[312] carboxylic acids,[313] lactones,[314–316] benzene,[317] allenes,[318] and nitrosamines and nitramines.[319] However, by far the bulk of theoretical and experimental attention has been given to the long-wavelength absorption band ($n \rightarrow \pi^*$) of the carbonyl chromophore, the CD of which is governed by the so-called ketone octant rule of Moffitt, Woodward, Moscowitz, Klyne, and Djerassl.[145] This rule, which we described briefly in Section V.B, has prompted the development of a number of theoretical methods and calculations, which have attempted to account not only for the mechanism for the induced electric dipole transition moment, but also for the non-symmetry-determined third nodal surface. A study of the various papers on computed chiroptical properties of ketones mirrors in large part the development of the field as a whole.

In their 1966 paper on methylcyclohexanones, Pao and Santry[24] used a semiempirical (CNDO) approach, and they correlated the third octant nodal surface with the nodes of the highest occupied (n) orbital. Gould and Hoffmann[309] developed an extended Hückel method for evaluating rotatory strengths, and applied it to methylcyclohexanones, remarking that the computed (STA) results were at best qualitatively correct. Howell[320] applied the same method to a series of chiral aldehydes, where the chirality stems from rotation of the terminal methyl group; Howell also dissected the electric transition moment into atomic contributions, a procedure that has similarities to a localized MO treatment. Richardson, Shillady, and Bloor[289] developed another semiempirical approach (an INDO TDA-type model) for chiroptical properties and applied it to alkyl-substituted cyclopentanones. This time the analysis centered on substituent contributions vs. ring conformation as the dominant source of chirality. The work of Imamura, Hirano, Nagata, and Tsuruta,[290] again on methyl derivatives of cyclohexanone, is of interest from a computational standpoint because it showed the importance of including two-center optical matrix elements, even in zero-differential-overlap schemes. In a later paper,[321] Imamura, Hirano, Nagata, Tsuruta, and Kuriyama formally partitioned a chiral ketone into two fragments: the carbonyl group and the chiral remainder, and analyzed the $n \rightarrow \pi^*$ rotatory strength into inter- and intrafragment terms. Such an analysis is meaningful for the electric moment, but not for the magnetic moment due to its origin dependence. Nakamura and I'Haya[322] treated conformational equilibria and the temperature dependence of the rotatory strength.

The nature and shape of the third nodal surface in the octant rule has also been studied using a CNDO/S-STA computational model.[134,323] The study confirmed empirical correlations reported by Kirk and Klyne,[153,154] according to which substituents located along a zigzag path extending alternately equatorially and axially from the carbonyl carbon play a particularly important role in the enhancement of the rotatory strength of the $n \rightarrow \pi^*$ excitation (see also Refs. 324 and 325); this characteristic zigzag behavior of the contributions to the electric dipole transition moment was exhibited by analyzing the transition moment into one- and two-center contributions. The shape of the third nodal surface was probed with a series of conformers of some flexible ketones, in which the symmetry plane perpendicular to the plane of the carbonyl group was maintained save for the unbalanced terminal methyl group. The resulting surface, together with additional regions of sign change, is indicated schematically in Fig. 10. The

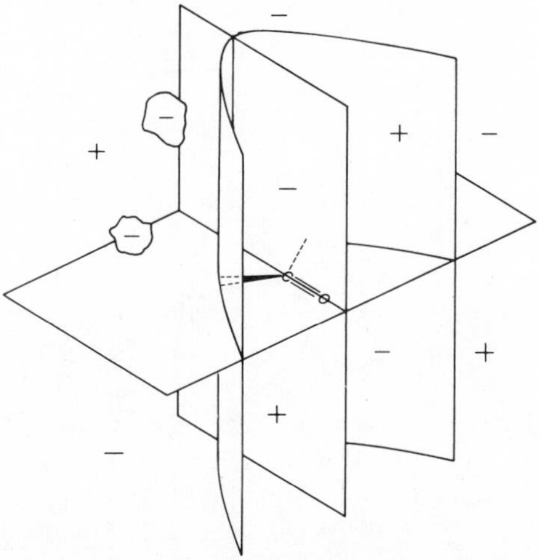

Fig. 10. Revised sign-determining regions of the ketone octant rule, as determined by CNDO/S-STA calculations. (From Ref. 134.)

additional "pockets" of sign change occur at just those positions known to exhibit "antioctant" behavior.

Ab initio (STO-3G) and semiempirical computations on methylcyclohexanones, using STA, IVO, and IGO models,[302] show some incorrect sign predictions, especially for the TDA-like schemes, and the magnitudes are too high, as is also the case with all the other studies reported above. A rather more detailed study of the $n \rightarrow \pi^*$ band in (+)-D-camphor was recently reported by Texter and Stevens.[87a] These authors used a time-dependent Hartree (TDH) theory (equivalent to the RPA without electron exchange terms; see Section VII.C) to compute $\kappa(\omega)$, $\Delta\kappa(\omega)$, and $g(\omega)$ across the band, and gave a detailed vibronic analysis of the results.

The study of the mechanisms generating molecular rotatory intensities in these systems can be facilitated significantly by use of localized molecular orbitals. In Ref. 130 we have presented *ab initio* minimal basis set computations on a chiral conformation of diethyl ketone (see Fig. 11), in which we first perform STA, TDA, and RPA calculations using all the valence-shell electronic configurations in a canonical SCF-MO basis, and subsequently transform the results into a localized MO basis according to (VII.32) and (VII.33). The localization procedure we used is based on the Foster–Boys criterion[326] for localization of occupied orbitals, supplemented with Coffey's suggestion[327] for the localization of the virtual orbitals. The chosen conformation mimics that of α-axial methyl groups in an idealized chair cyclohexanone geometry, and a C_2 rotation axis is maintained about the C=O group to allow some symmetry blocking in the

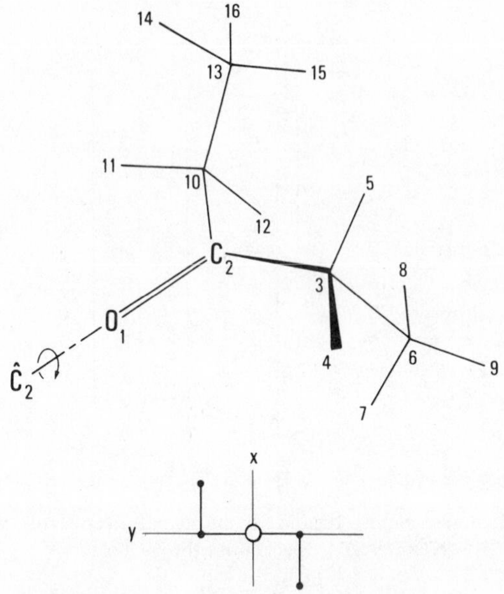

Fig. 11. Geometry, atom-numbering system, and octant projection used in the computations on a chiral conformation of diethyl ketone.

TABLE IV
Calculated Chiroptical Properties for the $^1A(n \rightarrow \pi^*)$ Excitation in Diethyl Ketone (Fig. 11)[a]

Method	ΔE (eV)	$f^{r\nabla}$	$R^{[r]}$	R^{∇}
STA[b]	5.36	0.0015	+29	+33
TDA[b]	4.44	0.0001	+7	+7
RPA[b]	4.34	0.0001	+7	+6
LORPA[c] (zigzag)	4.37	0.0001	+7	+6
LORPA[c] (C=O only)	5.09	0.0002	−1	−9
Exptl.[c]	4.3	$\sim 10^{-4}$	+8±2	

[a]Units of R are (cgs $\times 10^{40}$).
[b]From Ref. 130.
[c]Unpublished.
[d]From Ref. 153.

calculations. The STA, TDA, and RPA canonical orbital results for the $n \rightarrow \pi^*$ carbonyl excitation are shown in Table IV. The STA intensities are too high, as usual, whereas for once the TDA and the RPA are essentially equally good. The results of the analysis of the RPA electric dipole transition moment into localized orbital contributions are shown in Tables V and VI. The importance of the zigzag defined by atoms 1-2-10-13-16 and

TABLE V
Localized Analysis of the Electric Moment
of the $n \rightarrow \pi^*$ Excitation in Diethyl Ketone (atomic units)

Configuration description[a]	$\langle 0\|z\|b\rangle$	$\langle 0\|\nabla_z\|b\rangle$
LE in C═O (4 conf.)	0.00578	0.00307
C_α—C↔C═O CT (6 conf.)	0.00094	0.00246
LE in C_2H_5 (14 conf.)	−0.00921	−0.00310
C_α bonds↔C═O (18 conf.)	−0.00271	−0.00008
Nearest-neighbor CT in C_2H_5 (48 conf.)	−0.00354	−0.00161
Non-nearest-neighbor CT (54 conf.)	−0.01759	−0.00444
Total	−0.02821	−0.00348
Zigzag orbitals only	−0.02695	−0.00284
LE in zigzag	−0.00038	0.00061
CT in zigzag	−0.02657	−0.00345
Nearest-neighbor CT	−0.00447	0.00052
Non-nearest-neighbor CT	−0.02210	−0.00397

[a]LE means locally excited; CT means charge transfer.

TABLE VI
Analysis of $\langle 0|\nabla_z|b\rangle$ for the $^1A(n \rightarrow \pi^*)$ Excitation
in Diethyl Ketone in Terms of Bond Orbital Contributions[a]

	C═O	C_2—C_3	C_3—H_4	C_3—H_5	C_3—C_6	C_6—H_9	C_6—H_7	C_6—H_8
C═O	22							
C_2—C_3	18	−5						
C_3—H_4								
C_3—H_5	65	37	−5	29				
C_3—C_6	−64	−45	4	4	−30			
C_6—H_9	−21	−6		−2	−10	−11		
C_6—H_7					2		1	
C_6—H_8					3	−2		−5

[a]Units are $(au) \times 2^{-1/2} \times 10^4$. Terms less than 1 in these units have been omitted for clarity. Atom numbering is for only one ethyl group, but table entry includes both groups.

1-2-3-6-9 (Fig. 11) is clearly evident. Table IV also includes the results[328] of a localized orbital RPA (LORPA) calculation (see Section VII.B), including only configurations arising from occupied and virtual orbitals localized in the carbonyl chromophore and in the zigzag bonds (i.e., a total of 60 configurations, as compared to a total of 144 configurations of this symmetry in the full canonical RPA). Finally, the table contains the results of a LORPA calculation,[328] including only the local excitations (LE) out of the six occupied into the four virtual localized MO's of the $\rangle$C═O chromophore. The agreement among the RPA, the LORPA including zigzag, and the "experimental" chiroptical properties confirms the importance of these zigzag contributions, especially in view of the fact that the purely local LE-LORPA yields the wrong sign for the rotatory intensity. Since the electric dipole transition moment arises largely from $\sigma \rightarrow \sigma^*$ bond excitations outside the carbonyl group, as well as "charge-transfer" excitations from bond to bond, these results for the rotatory intensity of the carbonyl $n \rightarrow \pi^*$ transition do not confirm the static perturbation model envisaged in early calculations.[8,133]

C. Coupled Chromophore Systems

In previous sections we have discussed two models for the rotatory strengths of a molecule containing an extended chromophore that can be divided into (presumably) weakly interacting parts (i.e., a coupled chromophore system). The dynamical coupling model, Section V.A, is used for an extended chromophore containing nonidentical (and hence nondegenerate) fragments (e.g., a β,γ-unsaturated ketone),[143] whereas the exciton model (Sections II.B.2, V.C., and VII.C) is used for identical chromophores (e.g., an α-helical polypeptide). Underlying both models is the assumption[112] that electronic overlap, exchange, and charge transfer between the chromophores can be neglected. In addition, one commonly assumes that the inherent chiroptical properties of the individual chromophores are essentially unaffected by the nonchromophoric parts of the molecules; thus a monoolefinic fragment (such as one of the carbon-carbon double bonds in the molecules in Figs. 12 and 13), or the peptide fragment in a polypeptide, are treated as inherently optically inactive because they both possess a local plane of symmetry. The use of localized MO's allows an analysis of these

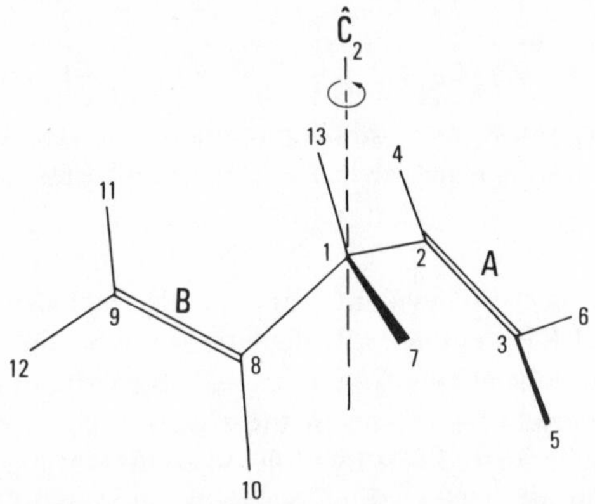

Fig. 12. Geometry and atom numbering system used in the computations of a chiral conformation of 1, 4-pentadiene. The planes of the two ethylenic chromophores form equal angles with the $C_1—H_7$ and $C_1—H_{13}$ bonds.

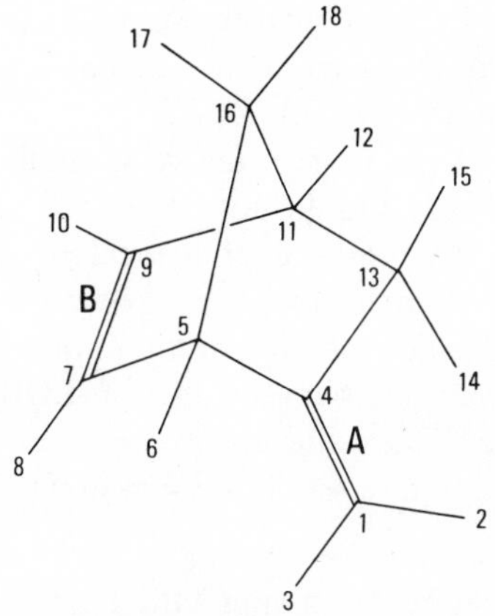

Fig. 13. Geometry and atom numbering system used in the computations on 5-methylenebicyclo [2.2.1] hept-2-ene.

assumptions within the context of all-(valence-)electron *ab initio* calculations; in this sense localized MO calculations can be viewed as an *ab initio* version of a coupled chromophore model, including charge transfer, that was proposed by Longuet-Higgins and Murrell[329] within a semiempirical molecular orbital scheme (see also Ref. 143).

Consider the molecules 1,4-pentadiene (PDE, in the chiral conformation shown in Fig. 12) and 5-methylenebicyclo[2.2.1]hept-2-ene (MBHE, Fig. 13). In both of these the two ethylenic chromophores are separated by two saturated bonds; in PDE the two chromophores are related by a C_2 operation and are hence identical, whereas the two chromophores in MBHE are not completely identical. Using a minimal atomic basis set, and our Foster-Boys-Coffey[130] procedure for the construction of the localized occupied and virtual MO's, we obtain[330] the chiroptical properties given in Table VII for the lowest two transitions of PDE and MBHE. The table also includes the chiroptical properties of the lowest transition in the molecule 1-pentene (PEN) in a conformation chosen to mimic the conformation of PDE (Fig. 12), where one of the two chromophores in PDE is now saturated. All the transitions displayed in Table VII are essentially of $\pi \rightarrow \pi^*$ character. LE-LORPA means a truncated RPA calculation, including only the configurations arising from 36 local excitations (LE) out of the six occupied into the six virtual localized orbitals of each $\rangle C{=}C\langle$ chromophore; for PDE and MBHE this represents a molecular exciton model in a highly restricted form (see below). Full RPA means that all the valence electron configurations arising from the minimal atomic basis set are included in the RPA calculations. Inspection of the table shows that the LE-LORPA oscillator strengths agree quite well with the full RPA results. On the other hand, the LE-LORPA rotatory strengths tend to be too small despite the fact that the local chromophoric excitations contribute more than 80% of the normalization in the full RPA.

In the present context the large rotatory strength of 1-pentene suggests that the ethylenic fragment should not be treated as an optically inactive chromophore, despite the presence of a local plane of symmetry, or in other words, that the extrachromophoric parts of the molecule are important for correct definition of the chromophoric parameters. From this perspective, the discrepancies between the LE-LORPA and full RPA results for PDE and MBHE are due to the neglect of a sizable part of the magnetic dipole transition moment in the LE-LORPA calculations. In fact,

TABLE VII
RPA and LORPA Results for Chiroptical Properties of 1-Pentene (PEN), 1,4-Pentadiene (PDE), and 5-Methylenebicyclo[2.2.1]hept-2-ene (MBHE)

System and method	ΔE (eV)	$f^{r\nabla}$	$R^{[r]}$ (cgs × 10^{40})	R^{∇} (cgs × 10^{40})	N_{LE}^{a} (%) A	N_{LE}^{a} (%) B
PEN[b]						
LE-LORPA[c]	10.23	0.16	−7	−1	100	—
Full RPA	9.59	0.22	−53	−37	93	—
PDE[b]						
LE-LORPA[c]	10.06	0.56	−138	−78	50	50
	10.37	0.03	113	57	50	50
Full RPA	9.15	0.57	−291	−183	43	43
	9.87	0.06	112	67	47	47
2×PEN(RPA)[d]	9.26	0.66	−244	−163		
	9.91	0.05	139	80		
MBHE[b]						
LE-LORPA[c]	10.41	0.44	−42	−22	81	19
	10.66	0.11	46	24	20	80
Full RPA	8.84	0.22	−178	−106	7	73
	9.38	0.35	120	74	80	5
Exptl. estimate[e]	6.0		−150			
	~6.7		150			

[a]Normalization contributions due to excitations localized in the chromophore(s). A and B refer to the chromophore labels in Figs. 12 and 13.

[b]PEN: See test and Fig. 12. PDE: Fig. 12. MBHE: Fig. 13.

[c]LORPA using only local excitations within the chromophore(s) (see the text).

[d]Molecular exciton results using the full RPA results for PEN for the chromophore data (see the text).

[e]From Ref. 143.

the LE-LORPA calculation in a sense neglects a significant part of the contributions from the first four terms in (V.22). This point of view is further confirmed by the results labeled 2×PEN (RPA) in Table VII. These results are obtained from a molecular exciton model, including all terms in (V.22) and in its dipole-length equivalent, using the full RPA results for PEN to define the electric and magnetic dipole transition moments of one of the chromophores in PDE. The transition moments of the other chromophore are obtained using the C_2 operation of the molecule, and the position vectors $\mathbf{R}_A$ and $\mathbf{R}_B$ locating the centers of the chromophores are taken as the centers of the C=C bonds. Finally, the energies are computed from the dipole coupling given in (V.25) based on the magnitude and direction of the electric dipole transition moments and the centers $\mathbf{R}_A$ and $\mathbf{R}_B$, and split around the RPA energy of PEN. The agreement between these molecular exciton results and the full RPA is quite satisfactory.

For MBHE again the LE-LORPA underestimates the magnetic dipole transition moments in the two chromophores. The local excitations account for a very large part of the total normalization; however, the very uneven distribution of the normalization contributions from the two chromophores shows that this is really more like a coupling case intermediate between the nondegenerate (i.e., dynamical) case and the true molecular exciton case. Notice further that the relative contributions from the two chromophores reverse from the LE-LORPA to the full RPA calculations. This feature again emphasizes the importance of the extrachromophoric parts for the definition of the chromophore properties. The agreement between the RPA and the "experimental" results is fair. The details of these coupled chromophore calculations will be published more fully elsewhere.[330]

In summary, these results do show a number of features characteristic of the dynamical coupling or the exciton model, namely that the localized excitations in the chromophores dominate the normalization of the two coupled states considered, and the success of the 2×PEN results of Table VII, which mimic an empirical exciton model where the "experimental data" for 1-pentene are used to define the chromophores in 1,4-pentadiene, also supports a model for the chiroptical properties based on the coupling of chromophores rather than on a completely delocalized description. On the other hand, the localized excitations do not suffice for the calculation of the appropriate chromophoric transition moments; for these, one needs the extrachromophoric parts as well. In particular, the difference between the LE-LORPA and 2×PEN results for 1,4-pentadiene shows that the ethylenic chromophore cannot be treated as an optically inactive chromophore, despite the presence of a local plane of symmetry.

D. Comments

The results discussed in this section show that it is possible to calculate chiroptical properties whose magnitude and sign compare well with the experimental results for the lower transitions in molecules that are large enough to be optically active in themselves (such as MBHE, Fig. 13), or that can serve as realistic model systems (such as shown in Figs. 6, 8, 11, and 12). The transitions considered here are essentially of valence-shell nature, and we found that minimal atomic orbital basis sets yield quite satisfactory results for the stronger ordinary and rotatory intensities (except for the transition energies, which tend to be too high, although the energy separations are reasonable). However, it is imperative that first-order electron correlation effects be included in a balanced scheme (e.g., through the use of the RPA or any of its equivalents), and we find also that it is very important to include either all the valence-shell configurations arising from the chosen atomic orbital basis set or at least a large, carefully selected subset of configurations. At the time of writing this article we must admit that our own criterion for the selection of configurations for truncated RPA (or LORPA) calculations is an after-the-fact rationalization. For example, the zigzag line of bonds contributing to the chiroptical properties of the lowest transition in diethyl ketone (Fig. 11, Table VI) was in fact extracted from the full calculation and then subsequently used in a truncated LORPA calculation. We hope nonetheless that the use of localized orbitals will allow the formulation of reliable criteria for the selection of the configurations that are important for chiroptical molecular properties.

ACKNOWLEDGMENT

Portions of this work were supported by grants from the Danish Natural Science Research Council, the Scientific Affairs Division of NATO, the U.S. Council for International Exchange of Scholars, and the Danish George C. Marshall Memorial Fund, as well as by the authors' respective universities. Fig. 10 is reprinted with permission from Ref. 134 (copyright by the American Chemical Society).

Finally, we wish to pay tribute here to the memory of our late colleague, Bjørn Voigt, who died suddenly on May 25, 1979. Bjørn's contributions to the development of the LORPA method and to criteria for selecting configurations were of crucial importance, and he will be sorely missed.

REFERENCES

1. A chiral object is one that can be distinguished from its mirror image.
2. F. Hund, *Z. Phys.*, **43**, 805 (1927).
3. L. Rosenfeld, *Z. Phys.*, **52**, 161 (1928).
4. E. U. Condon, *Rev. Mod. Phys.*, **9**, 432 (1937).
5. E. U. Condon, W. Altar, and H. Eyring, *J. Chem. Phys.*, **5**, 753 (1937).
6. W. Moffitt and A. Moscowitz, *J. Chem. Phys.*, **30**, 648 (1959).

7. A. Moscowitz, *Tetrahedron*, **13**, 48 (1961).
8. A. Moscowitz, *Adv. Chem. Phys.*, **4**, 67 (1962).
9. W. P. Healy and E. A. Power, *Am. J. Phys.*, **42**, 1070 (1974).
10. W. P. Healy, *J. Phys. B*, **7**, 1633 (1974).
11. W. J. A. Maaskant and L. J. Oosterhoff, *Mol. Phys.*, **8**, 319 (1964).
12. L. J. Oosterhoff, in O. Sinanoglu, Ed., *Modern Quantum Chemistry*, Vol. 3, Academic, New York, 1965.
13. S. I. Weissmann, *J. Org. Chem.*, **41**, 4040 (1978).
14. a. M. J. Harris, C. E. Loving, and P. G. H. Sandars, *J. Phys. B*, **11**, L749 (1978).
b. R. Conti, P. Bucksbaum, S. Chu, E. Commins, and L. Hunter, *Phys. Rev. Lett.* **42**, 343 (1979) and references therein.
15. R. A. Harris and L. Stodolski, *Phys. Lett.*, **78B**, 313 (1978).
16. C. Djerassi, *Optical Rotatory Dispersion*, McGraw-Hill, New York, 1960.
17. F. Woldbye, in H. B. Jonassen and A. Weissberger, Eds., *Technique of Inorganic Chemistry*, Vol. 4, Interscience, New York, 1965.
18. J. G. Kirkwood, *J. Chem. Phys.*, **5**, 479 (1937); **7**, 139 (1939).
19. W. J. Kauzmann, J. E. Walter, and H. Eyring, *Chem. Rev.*, **26**, 339 (1940).
20. W. Kuhn, *Annu. Rev. Phys. Chem.*, **9**, 417 (1958).
21. I. Tinoco, *Adv. Chem. Phys.*, **4**, 113 (1962).
22. A. D. Buckingham and P. J. Stiles, *Acc. Chem. Res.*, **7**, 258 (1974).
23. W. C. Johnson, *Annu. Rev. Phys. Chem.*, **29**, 93 (1978).
24. Y. H. Pao and D. P. Santry, *J. Am. Chem. Soc.*, **88**, 4157 (1966).
25. C. W. Deutsche, D. A. Lightner, R. W. Woody, and A. Moscowitz, *Annu. Rev. Phys. Chem.*, **20**, 407 (1969).
26. J. Schellman, *Chem. Rev.*, **75**, 323 (1975).
27. S. F. Mason, in F. Ciardelli and P. Salvadori, Eds., *Fundamental Aspects and Recent Developments in Optical Rotatory Dispersion and Circular Dichroism*, Heyden, London, 1973.
28. B. Bosnich, in Ref. 27.
29. M. J. Stephen, *Proc. Camb. Phil. Soc.*, **54**, 81 (1958).
30. P. W. Atkins and L. D. Barron, *Proc. R. Soc. Lond., Ser. A*, **304**, 303 (1968).
31. Y. -N. Chiu, *J. Chem. Phys.*, **50**, 5336 (1969).
32. P. W. Atkins and R. G. Woolley, *Proc. R. Soc. Lond., Ser. A*, **314**, 251 (1970).
33. E. A. Power and T. Thirunamachandran, *J. Chem. Phys.*, **55**, 5322 (1971).
34. I. Tobias, T. R. Brocki, and N. L. Balazs, *J. Chem. Phys.*, **62**, 4181 (1975).
35. A. Traspaderne and F. Castaño, *Chem. Phys. Lett.*, **56**, 318 (1978).
36. V. Lawetz and D. A. Hutchinson, *Can. J. Chem.*, **47**, 577 (1969).
37. E. A. Power and T. Thirunamachandran, *J. Chem. Phys.*, **60**, 3695 (1974).
38. S. H. Lin and R. Bersohn, *J. Chem. Phys.*, **44**, 3768 (1966).
39. W. R. Salzman, *Chem. Phys. Lett.* **25**, 302 (1974); *Chem Phys.*, **15**, 421 (1976); *J. Chem. Phys.*, **67**, 283 (1977).
40. R. A. Harris, *J. Chem. Phys.*, **43**, 959 (1965).
41. Aa. E. Hansen and J. Avery, *Chem. Phys. Lett.*, **13**, 396 (1972).
42. A. J. Duben, *Int. J. Quantum Chem.*, **6**, 789 (1972).
43. P. W. Atkins and R. G. Woolley, *J. Chem. Soc. A*, 515 (1969).
44. G. E. Desorbry and P. K. Kabir, *Am. J. Phys.*, **41**, 1350 (1973); **42**, 790 (1974).
45. I. Tinoco, *J. Chem. Phys.*, **62**, 1006 (1975).
46. E. A. Power, *J. Chem. Phys.*, **63**, 1348 (1975).
47. A. Messiah, *Quantum Mechanics*, North-Holland, Amsterdam, 1962.
48. The absorption cross-section σ and the molar extinction coefficient ε are related by $\varepsilon = L\sigma/(2.303\times 10^3)$, where L is Avogadro's number.

49. L. I. Schiff, *Quantum Mechanics*, McGraw-Hill, New York, 1968.
50. L. D. Barron, *Mol. Phys.*, **21**, 241 (1971).
51. E. N. Svendsen, *Mol. Phys.*, **23**, 213 (1972).
52. J. Snir and J. Schellman, *J. Phys. Chem.*, **77**, 1653 (1973).
53. A. D. Buckingham and M. B. Dunn, *J. Chem. Soc. A*, 1988 (1971).
54. Aa. E. Hansen, *Mol. Phys.*, **13**, 425 (1967).
55. Aa. E. Hansen and E. N. Svendsen, *Chem. Phys. Lett.*, **5**, 483 (1970).
56. Aa. E. Hansen and E. N. Svendsen, *Theor. Chim. Acta*, **20**, 303 (1971).
57. D. P. Chong, *J. Chem. Phys.*, **48**, 1413 (1968).
58. M. Goeppert-Mayer, *Ann. Phys.*, **9**, 273 (1931).
59. P. I. Richards, *Phys. Rev.*, **73**, 254 (1948).
60. J. Fiutak, *Can. J. Phys.*, **41**, 12 (1963).
61. Aa. E. Hansen, *Int. J. Quantum Chem.*, **4**, 473 (1971).
62. Aa. E. Hansen and E. N. Svendsen, *Mol. Phys.*, **28**, 1061 (1974).
63. Aa. E. Hansen, *Mol. Phys.*, **34**, 1473 (1977).
64. C. Møller and L. Rosenfeld, *K. Dan. Vidensk. Selsk. Mat.-Fys. Medd.*, **20**, (12) (1943).
65. E. A. Power and E. Zienau, *Phil. Trans. R. Soc. Lond. Ser. A*, **251**, 427 (1959).
66. E. A. Power, *Introductory Quantum Electrodynamics*, Longmans, London, 1964.
67. P. W. Atkins and R. G. Woolley, *Proc. R. Soc. Lond. Ser. A*, **319**, 549 (1970).
68. R. G. Woolley, *Proc. R. Soc. Lond. Ser. A*, **321**, 557 (1971).
69. R. G. Woolley, *Adv. Chem. Phys.*, **33**, 153 (1975).
70. Aa. E. Hansen, *Am. J. Phys.*, **39**, 653 (1971).
71. M. Born and R. Oppenheimer, *Ann. Phys.*, **84**, 457 (1927).
72. C. J. Ballhausen and Aa. E. Hansen, *Annu. Rev. Phys. Chem.*, **23**, 15 (1972).
73. I. Özkan and L. Goodman, *Chem. Rev.*, **79**, 275 (1979).
74. F. Duschinsky, *Acta Physico chim. URSS*, **7**, 551 (1937).
75. H. Frei and H. H. Günthard [*Chem. Phys.*, **15**, 155 (1976)], and J. Philippot and I. Sengers [*Int. J. Quantum Chem.*, **15**, 713 (1979)] discuss the extension of the chirality concept to nonrigid molecules, where vibrational amplitudes may be so large as to allow rapid interconversions among a set of nuclear configurations on the time scale of the experiment.
76. D. J. Caldwell, *J. Chem. Phys.*, **51**, 984 (1969).
77. H. P. J. M. Dekkers and L. E. Closs, *J. Am. Chem. Soc.*, **98**, 2210 (1976).
78. R. W. Nicholls, in D. Henderson, Ed., *Electronic Structure of Atoms and Molecules*, Vol. 3 of *Physical Chemistry*, Academic, New York, 1969.
79. A. C. Albrecht, *J. Chem. Phys.*, **33**, 156 (1960).
80. O. E. Weigang, J. A. Turner and P. A. Trouard, *J. Chem. Phys.*, **45**, 1126 (1966).
81. J. M. Hollas, E. Gregorek, and L. Goodman, *J. Chem. Phys.*, **49**, 1745 (1968).
82. a. O. E. Weigang, *J. Chem. Phys.*, **43**, 3609 (1965).
b. S. E. Harnung, E. C. Ong, and O. E. Weigang, *J. Chem. Phys.*, **55**, 5711 (1971).
c. O. E. Weigang and E. C. Ong, *Tetrahedron*, **30**, 1783 (1974).
83. W. Klyne and D. N. Kirk, in *Fundamental Aspects and Recent Developments in Optical Rotatory Dispersion and Circular Dichroism*, F. Ciardelli and P. Salvadori, Eds., Heyden, London, 1973.
84. A. Witkowski annd W. Moffitt, *J. Chem. Phys.*, **33**, 872 (1960).
85. P. J. Stephens, *Adv. Chem. Phys.*, **35**, 197 (1976).
86. K. M. Wellman, P. H. A. Laur, W. S. Briggs, A. Moscowitz, and C. Djerassi, *J. Am. Chem. Soc.*, **87**, 66 (1965).
87. a. J. Texter and E. S. Stevens, *J. Chem. Phys.*, **69**, 1680 (1978).
b. J. M. F. van Dijk, M. J. H. Kemper, J. H. M. Kerp, and H. M. Buck, *J. Chem. Phys.*, **69**, 2453 (1978).

88. R. S. Mulliken and C. A. Rieke, *Rept. Prog. Phys.*, **8**, 231 (1941).
89. W. C. M. C. Kokke and L. J. Oosterhoff, *J. Am. Chem. Soc.*, **94**, 7583 (1972); **95**, 7159 (1973). Isotopes ^{18}O and ^{2}D.
90. See S. -F. Lee, G. Barth, K. Kieslich, and C. Djerassi, *J. Am. Chem. Soc.*, **100**, 3965 (1978) for leading references on deuterium CD.
91. C. S. Pak and C. Djerassi, *Tetrahedron Lett.*, 4377 (1978). Isotope ^{13}C.
92. a. J. W. Simek, D. L. Mattern, and C. Djerassi, *Tetrahedron Lett.*, 3671 (1975).
b. D. A. Lightner, J. K. Gawroński, and T. D. Bouman, *J. Am. Chem. Soc.*, **102**, 1983 (1980).
93. R. F. R. Dezentje and H. P. J. M. Dekkers, *Chem. Phys.*, **18**, 189 (1976).
94. M. Born and K. Huang, *Dynamical Theory of Crystal Lattices*, Oxford University Press, London, 1954.
95. H. C. Longuet-Higgins, *Adv. Spectrosc.*, **2**, 429 (1961).
96. M. D. Sturge, *Solid State Phys.*, **20**, 91 (1967).
97. M. Farina and C. Morandi, *Tetrahedron*, **30**, 1819 (1974).
98. D. Caliga and F. S. Richardson, *Mol. Phys.*, **28**, 1145 (1974).
99. U. Fano, *Phys. Rev.*, **124**, 1866 (1961).
100. M. Bixon and J. Jortner, *J. Chem. Phys.*, **48**, 715 (1968).
101. J. Jortner and G. C. Morris, *J. Chem. Phys.*, **51**, 3689 (1969).
102. B. Sharf, *Chem. Phys. Lett.*, **5**, 456 (1970).
103. M. D. Sturge, H. J. Guggenheim, and M. M. L. Pryce, *Phys. Rev. B*, **2**, 2459 (1970).
104. D. Florida, R. Scheps, and S. A. Rice, *Chem. Phys. Lett.*, **15**, 490 (1972).
105. A. Shibatani and Y. Toyozawa, *J. Phys. Soc. Jap.*, **25**, 335 (1968).
106. Aa. E. Hansen, *Chem. Phys. Lett.*, **57**, 588 (1978).
107. R. A. Harris, *J. Chem. Phys.*, **39**, 978 (1963).
108. A. S. Davydov, *Theory of Molecular Excitons*, McGraw-Hill, New York, 1962.
109. A. S. Davydov, *Theory of Molecular Excitons*, Plenum, New York, 1971.
110. M. R. Philpott, *Adv. Chem. Phys.*, **23**, 277 (1973).
111. W. Moffitt, *J. Chem. Phys.*, **25**, 467 (1956).
112. H. C. Longuet-Higgins, *Proc. Roy. Soc. (London), Ser. A*, **235**, 537 (1956).
113. R. L. Fulton and M. Gouterman, *J. Chem. Phys.*, **35**, 1059 (1961); **41**, 2280 (1964).
114. E. G. McRae, *Australian J. Chem.*, **14**, 229, 344, 354 (1961).
115. W. T. Simpson and D. L. Peterson, *J. Chem. Phys.*, **26**, 588 (1957).
116. O. E. Weigang, *J. Chem. Phys.*, **43**, 71 (1965).
117. R. Lefebvre and M. GarciaSucre, *Int. J. Quantum Chem.*, **1S**, 339 (1967).
118. A. Witkowsky and M. Zgierski, *Int. J. Quantum Chem.*, **4**, 427 (1970).
119. J. S. Briggs and A. Herzenberg, *Mol. Phys.*, **23**, 203 (1972).
120. M. R. Philpott, *J. Chem. Phys.*, **47**, 2534, 4437 (1967); **55**, 2039 (1971).
121. J. S. Briggs and A. Herzenberg, *Proc. Phys. Soc.*, **92**, 159 (1967); *J. Phys.*, **B3**, 1663 (1970); *Mol. Phys.*, **21**, 111 (1971).
122. G. E. McCasland and S. Proskow, *J. Am. Chem. Soc.*, **77**, 4688 (1955).
123. H. A. Bethe and E. E. Salpeter, *Quantum Mechanics of One- and Two-Electron Atoms*, Springer-Verlag, Berlin, 1957.
124. J. C. Slater, *Quantum Theory of Molecules and Solids*, Vol. 1, McGraw-Hill, New York, 1963.
125. Aa. E. Hansen, *Mol. Phys.*, **33**, 483 (1977).
126. D. Caldwell, *Mol. Phys.*, **33**, 495 (1977).
127. R. A. Harris, *Chem. Phys. Lett.*, **45**, 477 (1977).
128. W. Kuhn, *Z. Phys.*, **54**, 14 (1929).
129. A. Moscowitz, in O. Sinanoglu, Ed., *Modern Quantum Chemistry*, Vol. 3, Academic, New York, 1965.

130. T. D. Bouman, B. Voigt, and Aa. E. Hansen, *J. Am. Chem. Soc.*, **101**, 550 (1979).
131. J. A. Schellman, *J. Chem. Phys.*, **44**, 55 (1966).
132. T. D. Bouman and Aa. E. Hansen, *J. Chem. Phys.*, **66**, 3460 (1977).
133. T. D. Bouman and A. Moscowitz, *J. Chem. Phys.*, **48**, 3115 (1968).
134. T. D. Bouman and D. A. Lightner, *J. Am. Chem. Soc.*, **98**, 3145 (1976).
135. A. Moscowitz, E. Charney, U. Weiss, and H. Ziffer, *J. Am. Chem. Soc.*, **83**, 4661 (1961).
136. E. Charney, H. Ziffer, and U. Weiss, *Tetrahedron*, **21**, 3121 (1965).
137. E. Charney, *Tetrahedron*, **21**, 3127 (1965).
138. W. Hug and G. Wagnière, *Helv. Chim. Acta*, **54**, 2920 (1971).
139. A. W. Burgstahler and R. C. Barkhurst, *J. Am. Chem. Soc.*, **92**, 7601 (1970).
140. A. W. Burgstahler, D. L. Boger, and N. C. Naik, *Tetrahedron*, **32**, 309 (1970).
141. A. W. Burgstahler, L. O. Weigel, and J. K. Gawroński, *J. Am. Chem. Soc.*, **98**, 3015 (1976).
142. E. G. Höhn and O. E. Weigang, *J. Chem. Phys.*, **48**, 1127 (1968).
143. A. Moscowitz, Aa. E. Hansen, L. S. Forster, and K. Rosenheck, *Biopolym. Symp.*, **1**, 75 (1964).
144. R. G. Parr, *Quantum Theory of Molecular Electronic Structure*, W. A. Benjamin, New York, 1963.
145. W. Moffitt, R. B. Woodward, A. Moscowitz, W. Klyne, and C. Djerassi, *J. Am. Chem. Soc.*, **83**, 4013 (1961).
146. D. A. Lightner and T. C. Chang, *J. Am. Chem. Soc.*, **96**, 3015 (1974).
147. J. A. Schellman, *Acc. Chem. Res.*, **1**, 144 (1968).
148. E. C. Ong, L. C. Cusachs, and O. E. Weigang, *J. Chem. Phys.*, **67**, 3289 (1977).
149. O. E. Weigang, Minisymposium on Optical Activity, Edwardsville, Ill., Jan. 26, 1978; *J. Am. Chem. Soc.*, **101**, 1965 (1979).
150. P. J. Stiles, *Nature Phys. Sci.*, **232**, 107 (1971).
151. C. Y. Yeh and F. S. Richardson, *Theor. Chim. Acta*, **43**, 253 (1977).
152. P. E. Schipper, *Chem. Phys.*, **23**, 159 (1977).
153. D. N. Kirk and W. Klyne, *J. Chem. Soc. Perkin Trans. 1*, 1076 (1974); 762, 2171 (1976).
154. D. N. Kirk, *J. Chem. Soc. Perkin Trans 1*, 2122 (1977).
155. E. Ruch and A. Schönhofer, *Theor. Chim. Acta*, **19**, 225 (1970).
156. E. Ruch, *Acc. Chem. Res.*, **5**, 49 (1972).
157. E. Ruch, *Angew. Chem. Int. Ed. Engl.* **16**, 65 (1977).
158. C. A. Mead, *Top. Curr. Chem.*, **49**, 1 (1974).
159. R. S. Cahn, C. K. Ingold, and V. Prelog, *Experientia*, **12**, 81 (1956).
160. D. Haase and E. Ruch, *Theor. Chim. Acta*, **29**, 189, 247 (1973).
161. W. Runge and G. Kresze, *J. Am. Chem. Soc.*, **99**, 5597 (1977).
162. R. S. Knox, *Theory of Excitons*, Academic, New York, 1963.
163. I. Tinoco, in B. Pullman and M. Weissbluth, Eds., *Molecular Biophysics*, Academic, New York, 1965; I. Tinoco, *J. Chim. Phys.*, **65**, 91 (1968).
164. Aa. E. Hansen and J. S. Avery, *Chem. Phys. Lett.*, **17**, 561 (1972).
165. J. S. Avery and Aa. E. Hansen, *Ann. Soc. Sci. Bruxelles*, **T89**, 253 (1975).
166. Aa. E. Hansen and J. S. Avery, *Ann. Soc. Sci. Bruxelles*, **T89**, 274 (1975).
167. W. Moffitt, D. D. Fitts, and J. G. Kirkwood, *Proc. Natl. Acad. Sci. USA*, **43**, 723 (1957).
168. I. Tinoco, *J. Am. Chem. Soc.*, **86**, 297 (1964).
169. W. Rhodes, *J. Am. Chem. Soc.*, **83**, 3609 (1961).
170. F. M. Loxsom, *J. Chem. Phys.*, **51**, 4899 (1969); *Int. J. Quantum Chem.*, **S3**, 147 (1969).
171. C. W. Deutsche, *J. Chem. Phys.*, **52**, 3703 (1970).
172. C. A. Busch and J. Brahms, *J. Chem. Phys.*, **46**, 79 (1967).

173. P. M. Bayley, E. B. Nielsen, and J. A. Schellman, *J. Phys. Chem.*, **73**, 228 (1969).
174. R. Mandel and G. Holzwarth, *J. Chem. Phys.*, **57**, 3469 (1972).
175. P. J. Stiles, *Mol. Phys.*, **22**, 731 (1971).
176. J. Tanaka, C. Katayama, F. Ogura, H. Tatemitsu, and M. Nagakawa, *J. Chem. Soc. Chem. Commun.*, 21 (1973).
177. H. Lipson and W. Cochran, *The Determination of Crystal Structures*, Bell, London, 1966.
178. A. M. F. Hezemans and M. P. Groenewege, *Tetrahedron*, **29**, 1223 (1973).
179. L. D. Barron and A. D. Buckingham, *Annu. Rev. Phys. Chem.*, **26**, 381 (1975).
180. B. N. Samoilov, *Zh. Eksp. Teor. Fiz.*, **18**, 1030 (1948).
181. C. A. Emeis and L. J. Oosterhoff, *J. Chem. Phys.*, **54**, 4809 (1971).
182. J. Snir and J. A. Schellman, *J. Phys. Chem.*, **78**, 387 (1974).
183. I. Steinberg and B. Ehrenberg, *J. Chem. Phys.*, **61**, 3382 (1974).
184. J. P. Riehl and F. S. Richardson, *J. Chem. Phys.*, **65**, 1011 (1976).
185. J. P. Riehl and F. S. Richardson, *J. Chem. Phys.*, **66**, 1988 (1977).
186. I. Steinberg, in R. F. Chen and H. Edelhoch, Eds., *Concepts in Biochemical Fluorescence*, Vol. 1, Dekker, New York, 1975.
187. F. S. Richardson and J. P. Riehl, *Chem. Rev.*, **77**, 773 (1977).
188. H. P. J. M. Dekkers, C. A. Emeis, and L. J. Oosterhoff, *J. Am. Chem. Soc.*, **91**, 4589 (1969).
189. C. K. Luk and F. S. Richardson, *J. Am. Chem. Soc.*, **96**, 2006 (1974).
190. H. P. J. M. Dekkers and L. E. Closs, *J. Am. Chem. Soc.*, **98**, 2210 (1976).
191. A. Gafni and I. Steinberg, *Photochem. Photobiol.*, **15**, 93 (1972).
192. J. Schlessinger, A. Gafni, and I. Z. Steinberg, *J. Am. Chem. Soc.*, **96**, 7396 (1974).
193. C. W. Deutsche and A. Moscowitz, *J. Chem. Phys.*, **49**, 3257 (1968); **53**, 2630 (1970).
194. G. Holzwarth and I. Chabay, *J. Chem. Phys.*, **57**, 1632 (1972).
195. J. A. Schellman, *J. Chem. Phys.*, **58**, 2882 (1973).
196. G. Holzwarth, E. C. Hsu, H. S. Mosher, T. R. Faulkner, and A. Moscowitz, *J. Am. Chem. Soc.*, **96**, 251 (1974).
197. L. A. Nafie, J. C. Cheng, and P. J. Stephens, *J. Am. Chem. Soc.*, **97**, 3842 (1975).
198. L. A. Nafie, T. A. Keiderling, and P. J. Stephens, *J. Am. Chem. Soc.*, **98**, 2715 (1976).
199. H. Sugeta, C. Marcott, T. R. Faulkner, J. Overend, and A. Moscowitz, *Chem. Phys. Lett.*, **40**, 397 (1976).
200. T. R. Faulkner, C. Marcott, A. Moscowitz, and J. Overend, *J. Am. Chem. Soc.*, **99**, 8160 (1977).
201. T. A. Keiderling and P. J. Stephens, *Chem. Phys. Lett.*, **41**, 46 (1976); see also *J. Am. Chem. Soc.*, **101**, 1396 (1979).
202. C. Marcott, T. R. Faulkner, A. Moscowitz, and J. Overend, *J. Am. Chem. Soc.*, **99**, 8169 (1977).
203. C. A. Mead and A. Moscowitz, *Int. J. Quantum Chem.*, **1**, 243 (1967).
204. a. T. H. Walnut and L. A. Nafie, *J. Chem. Phys.*, **67**, 1491 (1977).
b. T. H. Walnut and L. A. Nafie, *J. Chem. Phys.*, **67**, 1501 (1977). For a recent review, see L. A. Nafie and M. Diem, *Acct. Chem. Res.*, **12**, 296 (1979).
205. D. P. Craig and T. Thirunaramachandran, *Mol. Phys.*, **35**, 825 (1978).
206. P. J. Stephens, *J. Chem. Phys.*, **52**, 3489 (1970).
207. R. A. Harris, *J. Chem. Phys.*, **46**, 3398 (1967); **47**, 4481 (1967).
208. L. Seamans and A. Moscowitz, *J. Chem. Phys.*, **56**, 1099 (1972).
209. P. J. Stephens, *Annu. Rev. Phys. Chem.*, **25**, 201 (1974).
210. L. Seamans, A. Moscowitz, G. Barth, E. Bunnenberg, and C. Djerassi, *J. Am. Chem. Soc.*, **94**, 6464 (1972).

211. L. Seamans, A. Moscowitz, R. E. Linder, K. Morrill, J. S. Dixon, G. Barth, E. Bunnenberg, and C. Djerassi, *J. Am. Chem. Soc.*, **99**, 724 (1977).
212. R. E. Linder, K. Morrill, J. S. Dixon, G. Barth, E. Bunnenberg, C. Djerassi, L. Seamans, and A. Moscowitz, *J. Am. Chem. Soc.*, **99**, 727 (1977).
213. J. S. Rosenfield, A. Moscowitz, and R. E. Linder, *J. Chem. Phys.*, **61**, 2427 (1974).
214. J. S. Rosenfield, *Chem. Phys. Lett.*, **39**, 391 (1976); *J. Chem. Phys.*, **66**, 921 (1977).
215. S. T. Lee, Y. H. Yoon, H. Eyring, and S. H. Lin, *J. Chem. Phys.*, **66**, 4349 (1977).
216. S. M. Warnick and J. Michl, *J. Am. Chem. Soc.*, **96**, 6280 (1974).
217. J. Michl, *J. Am. Chem. Soc.*, **100**, 6801 (1978) and immediately following papers in that journal.
218. L. Seamans and J. Linderberg, *Mol. Phys.*, **24**, 1393 (1972).
219. A. D. Buckingham and P. J. Stephens, *Annu. Rev. Phys. Chem.*, **17**, 399 (1966).
220. F. London, *J. Phys. Radium*, **8**, 397 (1937).
221. J. C. Cheng, G. A. Osborne, P. J. Stephens, and W. A. Eaton, *Nature*, **241**, 193 (1973).
222. G. A. Osborne, *J. Mol. Spectrosc.*, **49**, 48 (1974).
223. L. Seamans, A. Moscowitz, R. E. Linder, E. Bunnenberg, G. Barth, and C. Djerassi, *J. Mol. Spectrosc.*, **54**, 412 (1975); **56**, 441 (1975).
224. L. Seamans, Minisymposium on Optical Activity, Edwardsville, Ill., Jan. 26, 1978.
225. See also S. H. Brown and Y. N. Chiu, *J. Chem. Phys.*, **69**, 3579 (1978).
226. C. Møller and M. S. Plesset, *Phys. Rev.*, **46**, 618 (1933).
227. F. Weinhold, *Adv. Quantum Chem.*, **6**, 299 (1972).
228. E. N. Svendsen, *Chem. Phys. Lett.*, **13**, 425 (1972).
229. G. J. Hoytink, in B. Pullman and P. O. Löwdin, Eds., *Molecular Orbitals in Chemistry, Physics, and Biology*, Academic, New York, 1964.
230. S. R. LaPaglia and O. Sinanoglu, *J. Chem. Phys.*, **44**, 1888 (1966).
231. A. Rauk and J. M. Barriel, *Chem. Phys.*, **25**, 409 (1977).
232. T. H. Dunning and V. McKoy, *J. Chem. Phys.*, **47**, 1735 (1967).
233. I. Shavitt, in H. F. Schaefer III, Ed., *Methods of Electronic Structure Theory*, Plenum, New York, 1977.
234. B. O. Roos and P. E. M. Siegbahn, in Ref. 233.
235. G. A. Segal, R. W. Wetmore, and K. Wolf, *Chem. Phys.*, **30**, 269 (1978).
236. S. Huzinaga and C. Arnau, *J. Chem. Phys.*, **54**, 1948 (1971); *Phys. Rev. A*, **1**, 1285 (1970).
237. K. Morokuma and S. Iwata, *Chem. Phys. Lett.*, **16**, 192 (1972).
238. E. J. Baerends, D. E. Ellis, and P. Ros, *Chem. Phys.*, **2**, 41 (1973).
239. M. Trsic, T. Ziegler, and W. G. Laidlaw, *Chem. Phys.*, **15**, 383 (1976).
240. A. Rauk, J. M. Barriel, and T. Ziegler, in I. G. Csizmadia, Ed., *Applications of MO Theory in Organic Chemistry*, Elsevier, New York, 1977.
241. V. Fock, *Z. Phys.*, **89**, 744 (1934).
242. M. Trsic, W. G. Laidlaw, and T. Ziegler, *Bull. Soc. Chim. Belg.*, **85**, 1027 (1976).
243. Aa. E. Hansen and T. D. Bouman, *Mol. Phys.*, **37**, 1713 (1979).
244. J. Linderberg and Y. Öhrn, *Propagators in Quantum Chemistry*, Academic, New York, 1973.
245. J. Avery, *Creation and Annihilation Operators*, McGraw-Hill, New York, 1976.
246. Caldwell, Ref. 126, has found that some of the higher-order hypervirial relations for electric and magnetic multipoles apparently lead to two-electron operators.
247. A. D. McLachlan and M. A. Ball, *Rev. Mod. Phys.*, **36**, 844 (1964).
248. Aa. E. Hansen and T. D. Bouman, *Chem. Phys. Lett.*, **45**, 326 (1977).
249. Notice that (7.18) and (7.19) will yield positive as well as negative eigenvalues ω_{b0}. Only the positive eigenvalues are physically meaningful.

250. P. L. Altick and A. E. Glassgold, *Phys. Rev.*, **A133**, 632 (1964).
251. D. J. Rowe, *Rev. Mod. Phys.*, **40**, 153 (1968).
252. C. W. McCurdy Jr., T. N. Rescigno, D. L. Yeager, and V. McKoy, in H. F. Schaefer III, Ed., *Methods of Electronic Structure Theory*, Plenum, New York, 1977.
253. P. Jørgensen, *Annu. Rev. Phys. Chem.*, **26**, 359 (1975).
254. J. Oddershede, *Adv. Quantum Chem.*, **11**, 275 (1978).
255. R. A. Harris, *J. Chem. Phys.*, **50**, 3947 (1969).
256. M. A. Ball and A. D. McLachlan, *Mol. Phys.*, **7**, 501 (1964).
257. T. I. Shibuya and V. McKoy, *J. Chem. Phys.*, **53**, 3308 (1970); **54**, 1738 (1971).
258. J. Linderberg and Y. Öhrn, *Int. J. Quantum Chem.*, **12**, 161 (1977).
259. M. Bouten, P. van Leuven, M. V. Mihailovic, and M. Rosina, *Nucl. Phys. A*, **202**, 127 (1973).
260. Gy. Csanak, *J. Phys. B*, **7**, 1289 (1974).
261. N. Ostlund and M. Karplus, *Chem. Phys. Lett.*, **11**, 450 (1971).
262. I. Ohmine, M. Karplus, and K. Schulten, *J. Chem. Phys.*, **68**, 2298 (1978).
263. F. M. Loxsom, L. Tterlikkis, and W. Rhodes, *Biopolymers*, **10**, 2405 (1971).
264. W. Rhodes and M. Chase, *Rev. Mod. Phys.*, **39**, 348 (1967).
265. D. G. Barnes and W. Rhodes, *J. Chem. Phys.*, **48**, 817 (1968).
266. W. Rhodes, *J. Chem. Phys.*, **53**, 3650 (1970).
267. D. A. Rabenold, *J. Chem. Phys.*, **62**, 376 (1975).
268. A. Herzenberg and A. Modinos, *Proc. Phys. Soc.*, **87**, 597 (1966).
269. A. D. McLachlan, R. D. Gregory, and M. A. Ball, *Mol. Phys.*, **7**, 119 (1964).
270. A. D. McLachlan and M. A. Ball, *Mol. Phys.*, **8**, 581 (1965).
271. A. S. Schneider and R. A. Harris, *J. Chem. Phys.*, **50**, 5204 (1969).
272. E. S. Pysh, *Biopolymers*, **13**, 1563 (1974).
273. N. P. Johnson and E. Switkes, *Biopolymers*, **17**, 857 (1978).
274. J. Texter and E. S. Stevens, *J. Chem. Phys.*, **70**, 1440 (1979).
275. H. DeVoe, *Biopolymers*, **S1**, 251 (1964); *J. Chem. Phys.*, **41**, 393 (1964); **43**, 3199 (1965).
276. A. I. Levin and I. Tinoco, *J. Chem. Phys.*, **66**, 3491 (1977).
277. J. Applequist, K. R. Sundberg, M. L. Olson, and L. C. Weiss, *J. Chem. Phys.*, **70**, 1240 (1979).
278. P. Jørgensen, J. Oddershede, and N. H. F. Beebe, *J. Chem. Phys.*, **68**, 2527 (1978).
279. H. F. Schaefer III, Ed., *Methods of Electronic Structure Theory*, Plenum, New York, 1977.
280. W. C. Hamilton, *J. Chem. Phys.*, **26**, 1018 (1957).
281. S. Fraga, *Can. J. Chem.*, **42**, 2509 (1964).
282. H. Ichimura and A. Rauk, *J. Chem. Phys.* **59**, 5720 (1973).
283. J. Avery and M. Cook, *Theor. Chim. Acta*, **35**, 99 (1974).
284. G. F. Tantardini and M. Raimondi, *Gazz. Chim. Ital.*, **105**, 361 (1975).
285. O. Matsuoka, *Int. J. Quantum Chem.*, **5**, 1 (1971); **7**, 365 (1973).
286. L. E. McMurchie and E. R. Davidson, *J. Comput. Phys.*, **26**, 218 (1978).
287. G. A. Segal, Ed., *Semiempirical Methods of Electronic Structure Calculations, Part A*, Plenum, New York, 1977.
288. Aa. E. Hansen, *Theor. Chim. Acta*, **6**, 341 (1966).
289. F. S. Richardson, D. D. Shillady, and J. E. Bloor, *J. Phys. Chem.*, **75**, 2466 (1971).
290. A. Imamura, T. Hirano, C. Nagata, and T. Tsuruta, *Bull. Chem. Soc. Jap.*, **45**, 396 (1972).
291. R. M. Lynden-Bell and V. R. Saunders, *J. Chem. Soc. A*, 2061 (1967).
292. J. Linderberg and J. Michl, *J. Am. Chem. Soc.*, **92**, 2619 (1970).
293. J. Linderberg and L. Seamans, *Int. J. Quantum Chem.*, **8**, 925 (1974).

294. N. S. Ostlund, *J. Chem. Phys.*, **57**, 2994 (1972).
295. A. Szabo and N. S. Ostlund, *Chem. Phys. Lett.*, **17**, 163 (1972).
296. E. Barbagli and M. Maestro, *Chem. Phys. Lett.*, **24**, 567 (1974).
297. J. Webb, R. W. Strickland, and F. S. Richardson, *J. Am. Chem. Soc.*, **95**, 4775 (1973).
298. P. Maloň and K. Bláha, *Coll. Czech. Chem. Commun.*, **42**, 687 (1977).
299. P. Maloň, S. Bystrický, and K. Bláha, *Coll. Czech. Chem. Commun.*, **43**, 781 (1978).
300. M. G. Mason and O. Schnepp, *J. Chem. Phys.*, **59**, 1092 (1973).
301. M. B. Robin, H. Basch, N. A. Kuebler, B. E. Kaplan, and J. Meinwald, *J. Chem. Phys.*, **48**, 5037 (1968).
302. A. Rauk, J. O. Jarvie, H. Ichimura, and J. M. Barriel, *J. Am. Chem. Soc.*, **97**, 5656 (1975).
303. D. H. Liskow and G. A. Segal, *J. Am. Chem. Soc.*, **100**, 2945 (1978).
304. M. Traetteberg, *Acta Chem. Scand.*, **B29**, 29 (1975).
305. W. J. Hehre, R. Ditchfield, M. D. Newton, and J. A. Pople, *QCPE*, **11**, 236 (1973).
306. J. W. Downing, J. Michl, P. Jørgensen, and E. W. Thulstrup, *Theor. Chim. Acta*, **32**, 203 (1974).
307. W. Hug and G. Wagnière, *Theor. Chim. Acta*, **18**, 57 (1970).
308. F. S. Richardson and D. Caliga, *Theor. Chim. Acta*, **36**, 49 (1974).
309. R. R. Gould and R. Hoffmann, *J. Am. Chem. Soc.*, **92**, 1813 (1969).
310. J. S. Rosenfield and E. Charney, *J. Am. Chem. Soc.*, **99**, 3209 (1977).
311. T. D. Bouman and Aa. E. Hansen, *Chem. Phys. Lett.*, **53**, 160 (1978).
312. J. Webb, R. W. Strickland, and F. S. Richardson, *Tetrahedron*, **29**, 2499 (1973).
313. F. S. Richardson and R. W. Strickland, *Tetrahedron*, **31**, 2309 (1975).
314. F. S. Richardson and N. Cox, *J. Chem. Soc. Perkin Trans. 2*, 1240 (1975).
315. F. S. Richardson and W. Pitts, *J. Chem. Soc. Perkin Trans. 2*, 1276 (1975).
316. H. Kreigh and F. S. Richardson, *J. Chem. Soc. Perkin Trans. 2*, 1674 (1976).
317. H. Dickerson and F. S. Richardson, *J. Phys. Chem.*, **80**, 2686 (1976).
318. H. Dickerson, S. Ferber, and F. S. Richardson, *Theor. Chim. Acta*, **42**, 333 (1976).
319. S. Ferber and F. S. Richardson, *Tetrahedron*, **33**, 1037 (1977).
320. J. M. Howell, *J. Chem. Phys.*, **53**, 4152 (1970).
321. A. Imamura, T. Hirano, C. Nagata, T. Tsuruta, and K. Kuriyama, *J. Am. Chem. Soc.*, **95**, 8621 (1973).
322. T. Nakamura and Y. J. I'Haya, *Bull. Chem. Soc. Jap.*, **49**, 3461 (1976).
323. T. D. Bouman, *J. Chem. Soc. Chem. Commun.*, 665 (1976).
324. E. E. Ernstbrunner, M. R. Giddings, and J. Hudec, *J. Chem. Soc. Chem. Commun.*, 953 (1976); M. R. Giddings, E. E. Ernstbrunner, and J. Hudec, *J. Chem. Soc. Chem. Commun.*, 954, 956 (1976).
325. E. E. Ernstbrunner and M. R. Giddings, *J. Chem. Soc. Perkin Trans. 2*, 989 (1978).
326. J. M. Foster and S. F. Boys, *Rev. Mod. Phys.*, **32**, 300 (1960).
327. P. Coffey, *Int. J. Quantum Chem.*, **8**, 777 (1974).
328. T. D. Bouman, Aa. E. Hansen, and B. Voigt, unpublished work.
329. H. C. Longuet-Higgins and J. N. Murrell, *Proc. Phys. Soc. Lond. Ser. A*, **68**, 601 (1955).
330. T. D. Bouman, Aa. E. Hansen, and B. Voigt, to be published.
331. C. W. McCurdy and L. C. Cusachs, *J. Chem. Phys.*, **55**, 1994 (1971).
332. R. K. Nesbet, *Mol. Phys.*, **5**, 63 (1962).
333. T. D. Bouman Aa. E. Hansen, to be published.

AUTHOR INDEX

Numbers in parentheses are reference numbers and show that an author's work is referred to although his name is not mentioned in the text. Numbers in *italics* indicate pages on which the full references appear.

SUBJECT INDEX